Algebraic Structures of Neutrosophic Triplets, Neutrosophic Duplets, or Neutrosophic Multisets

Algebraic Structures of Neutrosophic Triplets, Neutrosophic Duplets, or Neutrosophic Multisets

Volume 1

Special Issue Editors

Florentin Smarandache
Xiaohong Zhang
Mumtaz Ali

MDPI • Basel • Beijing • Wuhan • Barcelona • Belgrade

Special Issue Editors
Florentin Smarandache
University of New Mexico
USA

Xiaohong Zhang
Shaanxi University of Science and Technology
China

Mumtaz Ali
University of Southern Queensland
Australia

Editorial Office
MDPI
St. Alban-Anlage 66
4052 Basel, Switzerland

This is a reprint of articles from the Special Issue published online in the open access journal *Symmetry* (ISSN 2073-8994) in 2018 (available at: http://www.mdpi.com/journal/symmetry/special_issues/Algebraic_Structure_Neutrosophic_Triplet_Neutrosophic_Duplet_Neutrosophic_Multiset)

For citation purposes, cite each article independently as indicated on the article page online and as indicated below:

LastName, A.A.; LastName, B.B.; LastName, C.C. Article Title. *Journal Name* **Year**, *Article Number*, Page Range.

Volume 1
ISBN 978-3-03897-384-3 (Pbk)
ISBN 978-3-03897-385-0 (PDF)

Volume 1-2
ISBN 978-3-03897-477-2 (Pbk)
ISBN 978-3-03897-478-9 (PDF)

Contents

About the Special Issue Editors

Florentin Smarandache is a professor of mathematics at the University of New Mexico, USA. He got his M.Sc. in Mathematics and Computer Science from the University of Craiova, Romania, Ph.D. in Mathematics from the State University of Kishinev, and Post-Doctoral in Applied Mathematics from Okayama University of Sciences, Japan. He is the founder of neutrosophic set, logic, probability and statistics since 1995 and has published hundreds of papers on neutrosophic physics, superluminal and instantaneous physics, unmatter, absolute theory of relativity, redshift and blueshift due to the medium gradient and refraction index besides the Doppler effect, paradoxism, outerart, neutrosophy as a new branch of philosophy, Law of Included Multiple-Middle, degree of dependence and independence between the neutrosophic components, refined neutrosophic over-under-off-set, neutrosophic overset, neutrosophic triplet and duplet structures, DSmT and so on to many peer-reviewed international journals and many books and he presented papers and plenary lectures to many international conferences around the world.

Xiaohong Zhang is a professor of mathematics at Shaanxi University of Science and Technology, P. R. China. He got his bachelor's degree in Mathematics from Shaanxi University of Technology, P. R. China, and Ph.D. in Computer Science & Technology from the Northwestern Polytechnical University, P. R. China. He is a member of a council of Chinese Association for Artificial Intelligence (CAAI). He has published more than 100 international journals papers. His current research interests include non-classical logic algebras, fuzzy sets, rough sets, neutrosophic sets, data intelligence and decision-making theory.

Mumtaz Ali is a Ph.D. research scholar under Principal Supervision of Dr. Ravinesh Deo and also guided by Dr. Nathan Downs. He is originally from Pakistan where he completed his double masters (M.Sc. and M.Phil. in Mathematics) from Quaid-i-Azam University, Islamabad. Mumtaz has been an active researcher in Neutrosophic Set and Logic; proposed the Neutrosophic Triplets. Mumtaz is the author of three books on neutrosophic algebraic structures. Published more than 30 research papers in prestigious journals. He also published two chapters in the edited books. Research Interests: Currently, Mumtaz pursuing his doctoral studies in drought characteristic and atmospheric simulation models using artificial intelligence. He intends to apply probabilistic (copula-based) and machine learning modelling; fuzzy set and logic; neutrosophic set and logic; soft computing; decision support systems; data mining; clustering and medical diagnosis problems.

symmetry

MDPI

Editorial

Algebraic Structures of Neutrosophic Triplets, Neutrosophic Duplets, or Neutrosophic Multisets

Florentin Smarandache [1,*], Xiaohong Zhang [2,3] and Mumtaz Ali [4]

1 Department of Mathematics and Sciences, University of New Mexico, 705 Gurley Ave., Gallup, NM 87301, USA
2 Department of Mathematics, Shaanxi University of Science & Technology, Xi'an 710021, China; zxhonghz@263.net
3 Department of Mathematics, Shanghai Maritime University, Shanghai 201306, China
4 University of Southern Queensland, Springfield Campus, QLD 4300, Australia; Mumtaz.Ali@usq.edu.au
* Correspondence: smarand@unm.edu

Received: 29 January 2019; Accepted: 29 January 2019; Published: 1 February 2019

Neutrosophy (1995) is a new branch of philosophy that studies triads of the form (<A>, <neutA>, <antiA>), where <A> is an entity (i.e., element, concept, idea, theory, logical proposition, etc.), <antiA> is the opposite of <A>, while <neutA> is the neutral (or indeterminate) between them, i.e., neither <A> nor <antiA> [1].

Based on neutrosophy, the neutrosophic triplets were founded; they have a similar form: (x, neut(x), anti(x), that satisfy some axioms, for each element x in a given set [2–4].

This book contains the successful invited submissions [5–56] to a special issue of Symmetry, reporting on state-of-the-art and recent advancements of neutrosophic triplets, neutrosophic duplets, neutrosophic multisets, and their algebraic structures—that have been defined recently in 2016, but have gained interest from world researchers, and several papers have been published in first rank international journals.

The topics approached in the 52 papers included in this book are: neutrosophic sets; neutrosophic logic; generalized neutrosophic set; neutrosophic rough set; multigranulation neutrosophic rough set (MNRS); neutrosophic cubic sets; triangular fuzzy neutrosophic sets (TFNSs); probabilistic single-valued (interval) neutrosophic hesitant fuzzy set; neutro-homomorphism; neutrosophic computation; quantum computation; neutrosophic association rule; data mining; big data; oracle Turing machines; recursive enumerability; oracle computation; interval number; dependent degree; possibility degree; power aggregation operators; multi-criteria group decision-making (MCGDM); expert set; soft sets; LA-semihypergroups; single valued trapezoidal neutrosophic number; inclusion relation; Q-linguistic neutrosophic variable set; vector similarity measure; cosine measure; Dice measure; Jaccard measure; VIKOR model; potential evaluation; emerging technology commercialization; 2-tuple linguistic neutrosophic sets (2TLNSs); TODIM model; Bonferroni mean; aggregation operator; NC power dual MM (NCPDMM) operator; fault diagnosis; defuzzification; simplified neutrosophic weighted averaging operator; linear and non-linear neutrosophic number; de-neutrosophication methods; neutro-monomorphism; neutro-epimorphism; neutro-automorphism; fundamental neutro-homomorphism theorem; neutro-isomorphism theorem; quasi neutrosophic triplet loop; quasi neutrosophic triplet group; BE-algebra; cloud model; Maclaurin symmetric mean; pseudo-BCI algebra; hesitant fuzzy set; photovoltaic plan; decision-making trial and evaluation laboratory (DEMATEL); Choquet integral; fuzzy measure; clustering algorithm; and many more.

In the opening paper [5] of this book, the authors introduce refined concepts for neutrosophic quantum computing such as neutrosophic quantum states and transformation gates, neutrosophic Hadamard matrix, coherent and decoherent superposition states, entanglement and measurement notions based on neutrosophic quantum states. They also give some observations using these principles, and present a number of quantum computational matrix transformations based on neutrosophic logic, clarifying quantum mechanical notions relying on neutrosophic states. The paper is intended to extend the work of Smarandache [57–59] by introducing a mathematical framework for neutrosophic quantum computing and presenting some results.

The second paper [6] introduces oracle Turing machines with neutrosophic values allowed in the oracle information and then give some results when one is permitted to use neutrosophic sets and logic in relative computation. The authors also introduce a method to enumerate the elements of a neutrosophic subset of natural numbers.

In the third paper [7], a new approach and framework based on the interval dependent degree for MCGDM problems with SNSs is proposed. Firstly, the simplified dependent function and distribution function are defined. Then, they are integrated into the interval dependent function which contains interval computing and distribution information of the intervals. Subsequently, the interval transformation operator is defined to convert SNNs into intervals, and then the interval dependent function for SNNs is deduced. Finally, an example is provided to verify the feasibility and effectiveness of the proposed method, together with its comparative analysis. In addition, uncertainty analysis, which can reflect the dynamic change of the final result caused by changes in the decision makers' preferences, is performed in different distribution function situations. That increases the reliability and accuracy of the result.

Neutrosophic triplet structure yields a symmetric property of truth membership on the left, indeterminacy membership in the center and false membership on the right, as do points of object, center and image of reflection. As an extension of a neutrosophic set, the Q-neutrosophic set is introduced in the subsequent paper [8] to handle two-dimensional uncertain and inconsistent situations. The authors extend the soft expert set to the generalized Q-neutrosophic soft expert set by incorporating the idea of a soft expert set to the concept of a Q-neutrosophic set and attaching the parameter of fuzzy set while defining a Q-neutrosophic soft expert set. This pattern carries the benefits of Q-neutrosophic sets and soft sets, enabling decision makers to recognize the views of specialists with no requirement for extra lumbering tasks, thus making it exceedingly reasonable for use in decision-making issues that include imprecise, indeterminate and inconsistent two-dimensional data. Some essential operations, namely subset, equal, complement, union, intersection, AND and OR operations, and additionally several properties relating to the notion of a generalized Q-neutrosophic soft expert set are characterized. Finally, an algorithm on a generalized Q-neutrosophic soft expert set is proposed and applied to a real-life example to show the efficiency of this notion in handling such problems.

In the following paper [9], the authors extend the idea of a neutrosophic triplet set to non-associative semihypergroups and define neutrosophic triplet LA-semihypergroup. They discuss some basic results and properties, and provide an application of the proposed structure in football.

Single valued trapezoidal neutrosophic numbers (SVTNNs) are very useful tools for describing complex information, because of their advantage in describing the information completely, accurately and comprehensively for decision-making problems [60]. In the next paper [10], a method based on SVTNNs is proposed for dealing with MCGDM problems. Firstly, the new operation SVTNNs are developed for avoiding evaluation information aggregation loss and distortion. Then the possibility degrees and comparison of SVTNNs are proposed from the probability viewpoint for ranking and comparing the single valued trapezoidal neutrosophic information reasonably and accurately. Based on the new operations and possibility degrees of SVTNNs, the single valued trapezoidal neutrosophic power average (SVTNPA) and single valued trapezoidal neutrosophic power geometric (SVTNPG) operators are proposed to aggregate the single valued trapezoidal neutrosophic information. Furthermore, based on the developed aggregation operators, a single valued trapezoidal neutrosophic MCGDM method is developed. Finally, the proposed method is applied to solve the practical problem of the most appropriate green supplier selection and the rank results compared with the previous approach demonstrate the proposed method's effectiveness.

After the neutrosophic set (NS) was proposed [58], NS was used in many uncertainty problems. The single-valued neutrosophic set (SVNS) is a special case of NS that can be used to solve real-word problems. The next paper [11] mainly studies multigranulation neutrosophic rough sets (MNRSs) and their applications in multi-attribute group decision-making. Firstly, the existing definition of neutrosophic rough set (the authors call it type-I neutrosophic rough set (NRSI) in this paper) is analyzed, and then the definition of type-II neutrosophic rough set (NRSII), which is similar to NRSI, is given and its properties are studied. Secondly, a type-III neutrosophic rough set (NRSIII) is proposed and its differences from NRSI and NRSII are provided. Thirdly, single granulation NRSs are extended to multigranulation NRSs, and the type-I multigranulation neutrosophic rough set (MNRSI) is studied. The type-II multigranulation neutrosophic rough set (MNRSII) and type-III multigranulation neutrosophic rough set (MNRSIII) are proposed and their different properties are outlined. Finally, MNRSIII in two universes is proposed and an algorithm for decision-making based on MNRSIII is provided. A car ranking example is studied to explain the application of the proposed model.

Since language is used for thinking and expressing habits of humans in real life, the linguistic evaluation for an objective thing is expressed easily in linguistic terms/values. However, existing linguistic concepts cannot describe linguistic arguments regarding an evaluated object in two-dimensional universal sets (TDUSs). To describe linguistic neutrosophic arguments in decision making problems regarding TDUSs, the next article [12] proposes a Q-linguistic neutrosophic variable set (Q-LNVS) for the first time, which depicts its truth, indeterminacy, and falsity linguistic values independently corresponding to TDUSs, and vector similarity measures of Q-LNVSs. Thereafter, a linguistic neutrosophic MADM approach by using the presented similarity measures, including the cosine, Dice, and Jaccard measures, is developed under Q-linguistic neutrosophic setting. Lastly, the applicability and effectiveness of the presented MADM approach is presented by an illustrative example under Q-linguistic neutrosophic setting.

In the following article [13], the authors combine the original VIKOR model with a triangular fuzzy neutrosophic set [61] to propose the triangular fuzzy neutrosophic VIKOR method. In the extended method, they use the triangular fuzzy neutrosophic numbers (TFNNs) to present the criteria values in MCGDM problems. Firstly, they summarily introduce the fundamental concepts, operation formulas and distance calculating method of TFNNs. Then they review some aggregation operators of TFNNs. Thereafter, they extend the original VIKOR model to the triangular fuzzy neutrosophic environment and introduce the calculating steps of the TFNNs VIKOR method, the proposed method which is more reasonable and scientific for considering the conflicting criteria. Furthermore, a numerical example for potential evaluation of emerging technology commercialization is presented to illustrate the new method, and some comparisons are also conducted to further illustrate advantages of the new method.

Another paper [14] in this book aims to extend the original TODIM (Portuguese acronym for interactive multi-criteria decision making) method to the 2-tuple linguistic neutrosophic fuzzy environment [62] to propose the 2TLNNs TODIM method. In the extended method, the authors use 2-tuple linguistic neutrosophic numbers (2TLNNs) to present the criteria values in multiple attribute group decision making (MAGDM) problems. Firstly, they briefly introduce the definition, operational laws, some aggregation operators, and the distance calculating method of 2TLNNs. Then, the calculation steps of the original TODIM model are presented in simplified form. Thereafter, they extend the original TODIM model to the 2TLNNs environment to build the 2TLNNs TODIM model, the proposed method, which is more reasonable and scientific in considering the subjectivity of the decision makers' (DMs') behaviors and the dominance of each alternative over others. Finally, a numerical example for the safety assessment of a construction project is proposed to illustrate the new method, and some comparisons are also conducted to further illustrate the advantages of the new method.

The power Bonferroni mean (PBM) operator is a hybrid structure and can take the advantage of a power average (PA) operator, which can reduce the impact of inappropriate data given by the prejudiced decision makers (DMs) and Bonferroni mean (BM) operator, which can take into account the correlation between two attributes. In recent years, many researchers have extended the PBM operator to handle fuzzy information. The Dombi operations of T-conorm (TCN) and T-norm (TN), proposed by Dombi, have the supremacy of outstanding flexibility with general parameters. However, in the existing literature, PBM and the Dombi operations have not been combined for the above advantages for interval-neutrosophic sets (INSs) [63]. In the following paper [15], the authors define some operational laws for interval neutrosophic numbers (INNs) based on Dombi TN and TCN and discuss several desirable properties of these operational rules. Secondly, they extend the PBM operator based on Dombi operations to develop an interval-neutrosophic Dombi PBM (INDPBM) operator, an interval-neutrosophic weighted Dombi PBM (INWDPBM) operator, an interval-neutrosophic Dombi power geometric Bonferroni mean (INDPGBM) operator and an interval-neutrosophic weighted Dombi power geometric Bonferroni mean (INWDPGBM) operator, and discuss several properties of these aggregation operators. Then they develop a MADM method, based on these proposed aggregation operators, to deal with interval neutrosophic (IN) information. An illustrative example is provided to show the usefulness and realism of the proposed MADM method.

The neutrosophic cubic set (NCS) is a hybrid structure [64], which consists of INS [63] (associated with the undetermined part of information associated with entropy) and SVNS [60] (associated with the determined part of information). NCS is a better tool to handle complex DM problems with INS and SVNS. The main purpose of the next article [16] is to develop some new aggregation operators for cubic neutrosophic numbers (NCNs), which is a basic member of NCS. Taking the advantages of Muirhead mean (MM) operator and PA operator, the power Muirhead mean (PMM) operator is developed and is scrutinized under NC information. To manage the problems upstretched, some new NC aggregation operators, such as the NC power Muirhead mean (NCPMM) operator, weighted NC power Muirhead mean (WNCPMM) operator, NC power dual Muirhead mean (NCPMM) operator and weighted NC power dual Muirhead mean (WNCPDMM) operator are proposed and related properties of these proposed aggregation operators are conferred. The important advantage of the developed aggregation operator is that it can remove the effect of awkward data and it considers the interrelationship among aggregated values at the same time. Finally, a numerical example is given to show the effectiveness of the developed approach.

Smarandache defined a neutrosophic set [57] to handle problems involving incompleteness, indeterminacy, and awareness of inconsistency knowledge, and have further developed neutrosophic soft expert sets. In the next paper [17] of this book, this concept is further expanded to generalized neutrosophic soft expert set (GNSES). The authors then define its basic operations of complement, union, intersection, AND, OR, and study some related properties, with supporting proofs. Subsequently, they define a GNSES-aggregation operator to construct an algorithm for a GNSES decision-making method, which allows for a more efficient decision process. Finally, they apply the algorithm to a decision-making problem, to illustrate the effectiveness and practicality of the proposed concept. A comparative analysis with existing methods is done and the result affirms the flexibility and precision of the proposed method.

In the next paper [18], the authors define the neutrosophic valued (and generalized or G) metric spaces for the first time. Besides, they determine a mathematical model for clustering the neutrosophic big data sets using G-metric. Furthermore, relative weighted neutrosophic-valued distance and weighted cohesion measure are defined for neutrosophic big data set [65]. A very practical method for data analysis of neutrosophic big data is offered, although neutrosophic data type (neutrosophic big data) are in massive and detailed form when compared with other data types.

Bol-Moufang types of a particular quasi neutrosophic triplet loop (BCI-algebra), christened Fenyves BCI-algebras, are introduced and studied in another paper [19] of this book. 60 Fenyves BCI-algebras are introduced and classified. Amongst these 60 classes of algebras, 46 are found to be associative and 14 are found to be non-associative. The 46 associative algebras are shown to be Boolean groups. Moreover, necessary and sufficient conditions for 13 non-associative algebras to be associative are also obtained: p-semisimplicity is found to be necessary and sufficient for a F3, F5, F42, and F55 algebras to be associative while quasi-associativity is found to be necessary and sufficient for F19, F52, F56, and F59 algebras to be associative. Two pairs of the 14 non-associative algebras are found to be equivalent to associativity (F52 and F55, and F55 and F59). Every BCI-algebra is naturally a F54 BCI-algebra. The work is concluded with recommendations based on comparison between the behavior of identities of Bol-Moufang (Fenyves' identities) in quasigroups and loops and their behavior in BCI-algebra. It is concluded that results of this work are an initiation into the study of the classification of finite Fenyves' quasi neutrosophic triplet loops (FQNTLs) just like various types of finite loops have been classified. This research work has opened a new area of research finding in BCI-algebras, vis-a-vis the emergence of 540 varieties of Bol-Moufang type quasi neutrosophic triplet loops. A 'cycle of algebraic structures' which portrays this fact is provided.

The uncertainty and concurrence of randomness are considered when many practical problems are dealt with. To describe the aleatory uncertainty and imprecision in a neutrosophic environment and prevent the obliteration of more data, the concept of the probabilistic single-valued (interval) neutrosophic hesitant fuzzy set is introduced in the next paper [20]. By definition, the probabilistic single-valued neutrosophic hesitant fuzzy set (PSVNHFS) is a special case of the probabilistic interval neutrosophic hesitant fuzzy set (PINHFS). PSVNHFSs can satisfy all the properties of PINHFSs. An example is given to illustrate that PINHFS compared to PSVNHFS is more general. Then, PINHFS is the main research object. The basic operational relations of PINHFS are studied, and the comparison method of probabilistic interval neutrosophic hesitant fuzzy numbers (PINHFNs) is proposed. Then, the probabilistic interval neutrosophic hesitant fuzzy weighted averaging (PINHFWA) and the probability interval neutrosophic hesitant fuzzy weighted geometric (PINHFWG) operators are presented. Some basic properties are investigated. Next, based on the PINHFWA and PINHFWG operators, a decision-making method under a probabilistic interval neutrosophic hesitant fuzzy circumstance is established. Finally, the authors apply this method to the issue of investment options. The validity and application of the new approach is demonstrated.

Competition among different universities depends largely on the competition for talent. Talent evaluation and selection is one of the main activities in human resource management (HRM) which is critical for university development [21]. Firstly, linguistic neutrosophic sets (LNSs) are introduced to better express multiple uncertain information during the evaluation procedure. The authors further merge the power averaging operator with LNSs for information aggregation and propose a LN-power weighted averaging (LNPWA) operator and a LN-power weighted geometric (LNPWG) operator. Then, an extended technique for order preference by similarity to ideal solution (TOPSIS) method is developed to solve a case of university HRM evaluation problem. The main contribution and novelty of the proposed method rely on that it allows the information provided by different DMs to support and reinforce each other which is more consistent with the actual situation of university HRM evaluation. In addition, its effectiveness and advantages over existing methods are verified through sensitivity and comparative analysis. The results show that the proposal is capable in the domain of university HRM evaluation and may contribute to the talent introduction in universities.

The concept of a commutative generalized neutrosophic ideal in a BCK-algebra is proposed, and related properties are proved in another paper [22] of this book. Characterizations of a commutative generalized neutrosophic ideal are considered. Also, some equivalence relations on the family of all commutative generalized neutrosophic ideals in BCK-algebras are introduced, and some properties are investigated.

Fault diagnosis is an important issue in various fields and aims to detect and identify the faults of systems, products, and processes. The cause of a fault is complicated due to the uncertainty of the actual environment. Nevertheless, it is difficult to consider uncertain factors adequately with many traditional methods. In addition, the same fault may show multiple features and the same feature might be caused by different faults. In the next paper [23], a neutrosophic set based fault diagnosis method based on multi-stage fault template data is proposed to solve this problem. For an unknown fault sample whose fault type is unknown and needs to be diagnosed, the neutrosophic set based on multi-stage fault template data is generated, and then the generated neutrosophic set is fused via the simplified neutrosophic weighted averaging (SNWA) operator. Afterwards, the fault diagnosis results can be determined by the application of defuzzification method for a defuzzying neutrosophic set. Most kinds of uncertain problems in the process of fault diagnosis, including uncertain information and inconsistent information, could be handled well with the integration of multi-stage fault template data and the neutrosophic set. Finally, the practicality and effectiveness of the proposed method are demonstrated via an illustrative example.

The notions of neutrosophy, neutrosophic algebraic structures, neutrosophic duplet and neutrosophic triplet were introduced by Florentin Smarandache [57]. In another paper [24] of this book, some neutrosophic duplets are studied. A particular case is considered, and the complete characterization of neutrosophic duplets are given. Some open problems related to neutrosophic duplets are proposed.

In the next paper [25], the authors provide an application of neutrosophic bipolar fuzzy sets applied to daily life's problem related with the HOPE foundation, which is planning to build a children's hospital. They develop the theory of neutrosophic bipolar fuzzy sets, which is a generalization of bipolar fuzzy sets. After giving the definition they introduce some basic operation of neutrosophic bipolar fuzzy sets and focus on weighted aggregation operators in terms of neutrosophic bipolar fuzzy sets. They define neutrosophic bipolar fuzzy weighted averaging (NBFWA) and neutrosophic bipolar fuzzy ordered weighted averaging (NBFOWA) operators. Next they introduce different kinds of similarity measures of neutrosophic bipolar fuzzy sets. Finally, as an application, the authors give an algorithm for the multiple attribute decision making problems under the neutrosophic bipolar fuzzy environment by using the different kinds of neutrosophic bipolar fuzzy weighted/fuzzy ordered weighted aggregation operators with a numerical example related with HOPE foundation.

In the following paper [26], the authors introduce the concept of neutrosophic numbers from different viewpoints [57–65]. They define different types of linear and non-linear generalized triangular neutrosophic numbers which are very important for uncertainty theory. They introduce the de-neutrosophication concept for neutrosophic number for triangular neutrosophic numbers. This concept helps to convert a neutrosophic number into a crisp number. The concepts are followed by two applications, namely in an imprecise project evaluation review technique and a route selection problem.

In classical group theory, homomorphism and isomorphism are significant to study the relation between two algebraic systems. Through the next article [27], the authors propose neutro-homomorphism and neutro-isomorphism for the neutrosophic extended triplet group (NETG) which plays a significant role in the theory of neutrosophic triplet algebraic structures. Then, they define neutro-monomorphism, neutro-epimorphism, and neutro-automorphism. They give and prove some theorems related to these structures. Furthermore, the Fundamental homomorphism theorem for the NETG is given and some special cases are discussed. First and second neutro-isomorphism theorems are stated. Finally, by applying homomorphism theorems to neutrosophic extended triplet algebraic structures, the authors have examined how closely different systems are related.

It is an interesting direction to study rough sets from a multi-granularity perspective. In rough set theory, the multi-particle structure was represented by a binary relation. The next paper [28] considers a new neutrosophic rough set model, multi-granulation neutrosophic rough set (MGNRS). First, the concept of MGNRS on a single domain and dual domains was proposed. Then, their properties and operators were considered. The authors obtained that MGNRS on dual domains will degenerate into MGNRS on a single domain when the two domains are the same. Finally, a kind of special multi-criteria group decision making (MCGDM) problem was solved based on MGNRS on dual domains, and an example was given to show its feasibility.

As a new generalization of the notion of the standard group, the notion of the NTG is derived from the basic idea of the neutrosophic set and can be regarded as a mathematical structure describing generalized symmetry. In the next paper [29], the properties and structural features of NTG are studied in depth by using theoretical analysis and software calculations (in fact, some important examples in the paper are calculated and verified by mathematics software, but the related programs are omitted). The main results are obtained as follows: (1) by constructing counterexamples, some mistakes in the some literatures are pointed out; (2) some new properties of NTGs are obtained, and it is proved that every element has a unique neutral element in any neutrosophic triplet group; (3) the notions of NT-subgroups, strong NT-subgroups, and weak commutative neutrosophic triplet groups (WCNTGs) are introduced, the quotient structures are constructed by strong NT-subgroups, and a homomorphism theorem is proved in weak commutative neutrosophic triplet groups.

The aim of the following paper [30] is to introduce some new operators for aggregating single-valued neutrosophic (SVN) information and to apply them to solve the multi-criteria decision-making (MCDM) problems. The single-valued neutrosophic set, as an extension and generalization of an intuitionistic fuzzy set, is a powerful tool to describe the fuzziness and uncertainty [60], and MM is a well-known aggregation operator which can consider interrelationships among any number of arguments assigned by a variable vector. In order to make full use of the advantages of both, the authors introduce two new prioritized MM aggregation operators, such as the SVN prioritized MM (SVNPMM) and SVN prioritized dual MM (SVNPDMM) under an SVN set environment. In addition, some properties of these new aggregation operators are investigated and some special cases are discussed. Furthermore, the authors propose a new method based on these operators for solving the MCDM problems. Finally, an illustrative example is presented to testify the efficiency and superiority of the proposed method by comparing it with the existing method.

Making predictions according to historical values has long been regarded as common practice by many researchers. However, forecasting solely based on historical values could lead to inevitable over-complexity and uncertainty due to the uncertainties inside, and the random influence outside, of the data. Consequently, finding the inherent rules and patterns of a time series by eliminating disturbances without losing important details has long been a research hotspot. In the following paper [31], the authors propose a novel forecasting model based on multi-valued neutrosophic sets to find fluctuation rules and patterns of a time series. The contributions of the proposed model are: (1) using a multi-valued neutrosophic set (MVNS) to describe the fluctuation patterns of a time series, the model could represent the fluctuation trend of up, equal, and down with degrees of truth, indeterminacy, and falsity which significantly preserve details of the historical values; (2) measuring the similarities of different fluctuation patterns by the Hamming distance could avoid the confusion caused by incomplete information from limited samples; and (3) introducing another related time series as a secondary factor to avoid warp and deviation in inferring inherent rules of historical values, which could lead to more comprehensive rules for further forecasting. To evaluate the performance of the model, the authors explore the Taiwan Stock Exchange Capitalization Weighted Stock Index (TAIEX) as the major factor, and the Dow Jones Index as the secondary factor to facilitate the predicting of the TAIEX. To show the universality of the model, they apply the proposed model to forecast the Shanghai Stock Exchange Composite Index (SHSECI) as well.

The new notion of a neutrosophic triplet group (NTG) proposed by Smarandache is a new algebraic structure different from the classical group. The aim of the next paper [32] is to further expand this new concept and to study its application in related logic algebra systems. Some new notions of left (right)-quasi neutrosophic triplet loops and left (right)-quasi neutrosophic triplet groups are introduced, and some properties are presented. As a corollary of these properties, the following important result are proved: for any commutative neutrosophic triplet group, its every element has a unique neutral element. Moreover, some left (right)-quasi neutrosophic triplet structures in BE-algebras and generalized BE-algebras (including CI-algebras and pseudo CI-algebras) are established, and the adjoint semigroups of the BE-algebras and generalized BE-algebras are investigated for the first time.

In a neutrosophic triplet set, there is a neutral element and antielement for each element. In the following study [33], the concept of neutrosophic triplet partial metric space (NTPMS) is given and the properties of NTPMS are studied. The authors show that both classical metric and neutrosophic triplet metric (NTM) are different from NTPM. Also, they show that NTPMS can be defined with each

NTMS. Furthermore, the authors define a contraction for NTPMS and give a fixed point theory (FPT) for NTPMS. The FPT has been revealed as a very powerful tool in the study of nonlinear phenomena.

Another paper [34] of this book presents a modified Technique for Order Preference by Similarity to an Ideal Solution (TOPSIS) with maximizing deviation method based on the SVNS model [60]. A SVNS is a special case of a neutrosophic set which is characterized by a truth, indeterminacy, and falsity membership function, each of which lies in the standard interval of [0,1]. An integrated weight measure approach that takes into consideration both the objective and subjective weights of the attributes is used. The maximizing deviation method is used to compute the objective weight of the attributes, and the non-linear weighted comprehensive method is used to determine the combined weights for each attributes. The use of the maximizing deviation method allows our proposed method to handle situations in which information pertaining to the weight coefficients of the attributes are completely unknown or only partially known. The proposed method is then applied to a multi-attribute decision-making (MADM) problem. Lastly, a comprehensive comparative studies is presented, in which the performance of our proposed algorithm is compared and contrasted with other recent approaches involving SVNSs in literature.

One of the most significant competitive strategies for organizations is sustainable supply chain management (SSCM). The vital part in the administration of a sustainable supply chain is the sustainable supplier selection, which is a multi-criteria decision-making issue, including many conflicting criteria. The valuation and selection of sustainable suppliers are difficult problems due to vague, inconsistent, and imprecise knowledge of decision makers. In the literature on supply chain management for measuring green performance, the requirement for methodological analysis of how sustainable variables affect each other, and how to consider vague, imprecise and inconsistent knowledge, is still unresolved. The next research [35] provides an incorporated multi-criteria decision-making procedure for sustainable supplier selection problems (SSSPs). An integrated framework is presented via interval-valued neutrosophic sets to deal with vague, imprecise and inconsistent information that exists usually in real world. The analytic network process (ANP) is employed to calculate weights of selected criteria by considering their interdependencies. For ranking alternatives and avoiding additional comparisons of analytic network processes, the TOPSIS is used. The proposed framework is turned to account for analyzing and selecting the optimal supplier. An actual case study of a dairy company in Egypt is examined within the proposed framework. Comparison with other existing methods is implemented to confirm the effectiveness and efficiency of the proposed approach.

The concept of interval neutrosophic sets has been studied [63] and the introduction of a new kind of set in topological spaces called the interval valued neutrosophic support soft set is suggested in the next paper [36]. The authors also study some of its basic properties. The main purpose of the paper is to give the optimum solution to decision-making in real life problems the using interval valued neutrosophic support soft set.

In inconsistent and indeterminate settings, as a usual tool, the NCS containing single-valued neutrosophic numbers [60] and interval neutrosophic numbers [64] can be applied in decision-making to present its partial indeterminate and partial determinate information. However, a few researchers have studied neutrosophic cubic decision-making problems, where the similarity measure of NCSs is one of the useful measure methods. For the following work [37] in this book, the authors propose the Dice, cotangent, and Jaccard measures between NCSs, and indicate their properties. Then, under an NCS environment, the similarity measures-based decision-making method of multiple attributes is developed. In the decision-making process, all the alternatives are ranked by the similarity measure of each alternative and the ideal solution to obtain the best one. Finally, two practical examples are applied to indicate the feasibility and effectiveness of the developed method.

In real-world diagnostic procedures, due to the limitation of human cognitive competence, a medical expert may not conveniently use some crisp numbers to express the diagnostic information, and plenty of research has indicated that generalized fuzzy numbers play a significant role in describing complex diagnostic information. To deal with medical diagnosis problems based on generalized fuzzy sets (FSs), the notion of single-valued neutrosophic multisets (SVNMs) [60] is firstly used to express the diagnostic information [38]. Then the model of probabilistic rough sets (PRSs) over two universes is applied to analyze SVNMs, and the concepts of single-valued neutrosophic rough multisets (SVNRMs) over two universes and probabilistic rough single-valued neutrosophic multisets (PRSVNMs) over two universes are introduced. Based on SVNRMs over two universes and PRSVNMs over two universes, single-valued neutrosophic probabilistic rough multisets (SVNPRMs) over two universes are further established. Next, a three-way decision model by virtue of SVNPRMs over two universes in the context of medical diagnosis is constructed. Finally, a practical case study along with a comparative study are carried out to reveal the accuracy and reliability of the constructed three-way decisions model.

The next article [39] is based on new developments on a NTG and applications earlier introduced in 2016 by Smarandache and Ali. NTG sprang up from neutrosophic triplet set X: a collection of triplets (b,neut(b),anti(b)) for an b∈X that obeys certain axioms (existence of neutral(s) and opposite(s)). Some results that are true in classical groups are investigated in NTG and shown to be either universally true in NTG or true in some peculiar types of NTG. Distinguishing features between an NTG and some other algebraic structures such as: generalized group (GG), quasigroup, loop, and group are investigated. Some neutrosophic triplet subgroups (NTSGs) of a neutrosophic triplet group are studied. Applications of the neutrosophic triplet set, and our results on NTG in relation to management and sports, are highlighted and discussed.

Neutrosophic cubic sets [64] are the more generalized tool by which one can handle imprecise information in a more effective way as compared to fuzzy sets and all other versions of fuzzy sets. Neutrosophic cubic sets have the more flexibility, precision and compatibility to the system as compared to previous existing fuzzy models. On the other hand, the graphs represent a problem physically in the form of diagrams and matrices, etc., which is very easy to understand and handle. Therefore, the authors of the subsequent paper [40] apply the neutrosophic cubic sets to graph theory in order to develop a more general approach where they can model imprecise information through graphs. One of very important futures of two neutrosophic cubic sets is the R-union that R-union of two neutrosophic cubic sets is again a neutrosophic cubic set. Since the purpose of this new model is to capture the uncertainty, the authors provide applications in industries to test the applicability of the defined model based on present time and future prediction which is the main advantage of neutrosophic cubic sets.

Thereafter, another paper [41] presents a deciding technique for robotic dexterous hand configurations. This algorithm can be used to decide on how to configure a robotic hand so it can grasp objects in different scenarios. Receiving as input from several sensor signals that provide information on the object's shape, the DSmT decision-making algorithm passes the information through several steps before deciding what hand configuration should be used for a certain object and task. The proposed decision-making method for real time control will decrease the feedback time between the command and grasped object, and can be successfully applied on robot dexterous hands. For this, the authors have used the Dezert–Smarandache theory which can provide information even on contradictory or uncertain systems.

The study [42] that follows introduces simplified neutrosophic linguistic numbers (SNLNs) to describe online consumer reviews in an appropriate manner. Considering the defects of studies on SNLNs in handling linguistic information, the cloud model is used to convert linguistic terms in SNLNs to three numerical characteristics. Then, a novel simplified neutrosophic cloud (SNC) concept is presented, and its operations and distance are defined. Next, a series of simplified neutrosophic cloud aggregation operators are investigated, including the simplified neutrosophic clouds Maclaurin symmetric mean (SNCMSM) operator, weighted SNCMSM operator, and generalized weighted SNCMSM operator. Subsequently, a MCDM model is constructed based on the proposed aggregation operators. Finally, a hotel selection problem is presented to verify the effectiveness and validity of our developed approach.

In recent years, typhoon disasters have occurred frequently and the economic losses caused by them have received increasing attention. The next study [43] focuses on the evaluation of typhoon disasters based on the interval neutrosophic set theory. An interval neutrosophic set (INS) [63] is a subclass of a NS [57]. However, the existing exponential operations and their aggregation methods are primarily for the intuitionistic fuzzy set. So, this paper mainly focus on the research of the exponential operational laws of INNs in which the bases are positive real numbers and the exponents are interval neutrosophic numbers. Several properties based on the exponential operational law are discussed in detail. Then, the interval neutrosophic weighted exponential aggregation (INWEA) operator is used to aggregate assessment information to obtain the comprehensive risk assessment. Finally, a multiple attribute decision making (MADM) approach based on the INWEA operator is introduced and applied to the evaluation of typhoon disasters in Fujian Province, China. Results show that the proposed new approach is feasible and effective in practical applications.

In the coming paper [44] of this book, the authors study the neutrosophic triplet groups for a∈Z2p and prove this collection of triplets (a,neut(a),anti(a)) if trivial forms a semigroup under product, and semi-neutrosophic triplets are included in that collection. Otherwise, they form a group under product, and it is of order (p−1), with (p+1,p+1,p+1) as the multiplicative identity. The new notion of pseudo primitive element is introduced in Z2p analogous to primitive elements in Zp, where p is a prime. Open problems based on the pseudo primitive elements are proposed. The study is restricted to Z2p and take only the usual product modulo 2p.

Fuzzy graph theory plays an important role in the study of the symmetry and asymmetry properties of fuzzy graphs. With this in mind, in the next paper [45], the authors introduce new neutrosophic graphs called complex neutrosophic graphs of type 1 (abbr. CNG1). They then present a matrix representation for it and study some properties of this new concept. The concept of CNG1 is an extension of the generalized fuzzy graphs of type 1 (GFG1) and generalized single-valued neutrosophic graphs of type 1 (GSVNG1). The utility of the CNG1 introduced here is applied to a multi-attribute decision making problem related to Internet server selection.

The purpose of the subsequent paper [46] is to study new algebraic operations and fundamental properties of totally dependent-neutrosophic sets and totally dependent-neutrosophic soft sets. Firstly, the in-coordination relationships among the original inclusion relations of totally dependent-neutrosophic sets (called type-1 and typ-2 inclusion relations in this paper) and union (intersection) operations are analyzed, and then type-3 inclusion relation of totally dependent-neutrosophic sets and corresponding type-3 union, type-3 intersection, and complement operations are introduced. Secondly, the following theorem is proved: all totally dependent-neutrosophic sets (based on a certain universe) determined a generalized De Morgan algebra with respect to type-3 union, type-3 intersection, and complement operations. Thirdly, the relationships among the type-3 order relation, score function, and accuracy function of totally dependent-neutrosophic sets are discussed. Finally, some new operations and properties of totally dependent-neutrosophic soft sets are investigated, and another generalized De Morgan algebra induced by totally dependent-neutrosophic soft sets is obtained.

In the recent years, school administrators often come across various problems while teaching, counseling, and promoting and providing other services which engender disagreements and interpersonal conflicts between students, the administrative staff, and others. Action learning is an effective way to train school administrators in order to improve their conflict-handling styles. In the next paper [47], a novel approach is used to determine the effectiveness of training in school administrators who attended an action learning course based on their conflict-handling styles. To this end, a Rahim Organization Conflict Inventory II (ROCI-II) instrument is used that consists of both the demographic information and the conflict-handling styles of the school administrators. The proposed method uses the neutrosophic set (NS) and support vector machines (SVMs) to construct an efficient classification scheme neutrosophic support vector machine (NS-SVM). The neutrosophic c-means (NCM) clustering algorithm is used to determine the neutrosophic memberships and then a weighting parameter is calculated from the neutrosophic memberships. The calculated weight value is then used in SVM as handled in the fuzzy SVM (FSVM) approach. Various experimental works are carried in a computer environment out to validate the proposed idea. All experimental works are simulated in a MATLAB environment with a five-fold cross-validation technique. The classification performance is measured by accuracy criteria. The prediction experiments are conducted based on two scenarios. In the first one, all statements are used to predict if a school administrator is trained or not after attending an action learning program. In the second scenario, five independent dimensions are used individually to predict if a school administrator is trained or not after attending an action learning program. According to the obtained results, the proposed NS-SVM outperforms for all experimental works.

The notions of the neutrosophic hesitant fuzzy subalgebra and neutrosophic hesitant fuzzy filter in pseudo-BCI algebras are introduced, and some properties and equivalent conditions are investigated in the next paper [48]. The relationships between neutrosophic hesitant fuzzy subalgebras (filters) and hesitant fuzzy subalgebras (filters) are discussed. Five kinds of special sets are constructed by a neutrosophic hesitant fuzzy set, and the conditions for the two kinds of sets to be filters are given. Moreover, the conditions for two kinds of special neutrosophic hesitant fuzzy sets to be neutrosophic hesitant fuzzy filters are proved.

To solve the problems related to inhomogeneous connections among the attributes, the authors of the following paper [49] introduce a novel multiple attribute group decision-making (MAGDM) method based on the introduced linguistic neutrosophic generalized weighted partitioned Bonferroni mean operator (LNGWPBM) for linguistic neutrosophic numbers (LNNs). First of all, inspired by the merits of the generalized partitioned Bonferroni mean (GPBM) operator and LNNs, they combine the GPBM operator and LNNs to propose the linguistic neutrosophic GPBM (LNGPBM) operator, which supposes that the relationships are heterogeneous among the attributes in MAGDM. In addition, aimed at the different importance of each attribute, the weighted form of the LNGPBM operator is investigated. Then, the authors discuss some of its desirable properties and special examples accordingly. Finally, they propose a novel MAGDM method on the basis of the introduced LNGWPBM operator, and illustrate its validity and merit by comparing it with the existing methods.

Based on the multiplicity evaluation in some real situations, the next paper [50] firstly introduces a single-valued neutrosophic multiset (SVNM) as a subclass of neutrosophic multiset (NM) to express the multiplicity information and the operational relations of SVNMs. Then, a cosine measure between SVNMs and weighted cosine measure between SVNMs are presented to measure the cosine degree between SVNMs, and their properties are investigated. Based on the weighted cosine measure of SVNMs, a multiple attribute decision-making method under a SVNM environment is proposed, in which the evaluated values of alternatives are taken in the form of SVNMs. The ranking order of all alternatives and the best one can be determined by the weighted cosine measure between every alternative and the ideal alternative. Finally, an actual application on the selecting problem illustrates the effectiveness and application of the proposed method.

Rooftop distributed photovoltaic projects have been quickly proposed in China because of policy promotion. Before, the rooftops of the shopping mall had not been occupied, and it was urged to have a decision-making framework to select suitable shopping mall photovoltaic plans. However, a traditional MCDM method failed to solve this issue at the same time, due to the following three defects: the interactions problems between the criteria, the loss of evaluation information in the conversion process, and the compensation problems between diverse criteria. In the subsequent paper [51], an integrated MCDM framework is proposed to address these problems. First of all, the compositive evaluation index is constructed, and the application of DEMATEL method helped analyze the internal influence and connection behind each criterion. Then, the interval-valued neutrosophic set is utilized to express the imperfect knowledge of experts group and avoid the information loss. Next, an extended elimination et choice translation reality (ELECTRE) III method is applied, and it succeed in avoiding the compensation problem and obtaining the scientific result. The integrated method used maintained symmetry in the solar photovoltaic (PV) investment. Last but not least, a comparative analysis using Technique for Order Preference by Similarity to an Ideal Solution (TOPSIS) method and VIKOR method is carried out, and alternative plan X1 ranks first at the same. The outcome certified the correctness and rationality of the results obtained in this study.

In the next paper [52], by utilizing the concept of a neutrosophic extended triplet (NET), the authors define the neutrosophic image, neutrosophic inverse-image, neutrosophic kernel, and the NET subgroup. The notion of the neutrosophic triplet coset and its relation with the classical coset are defined and the properties of the neutrosophic triplet cosets are given. Furthermore, the neutrosophic triplet normal subgroups, and neutrosophic triplet quotient groups are studied.

The following paper [53] in the book proposes novel skin lesion detection based on neutrosophic clustering and adaptive region growing algorithms applied to dermoscopic images, called NCARG. First, the dermoscopic images are mapped into a neutrosophic set domain using the shearlet transform results for the images. The images are described via three memberships: true, indeterminate, and false memberships. An indeterminate filter is then defined in the neutrosophic set for reducing the indeterminacy of the images. A neutrosophic c-means clustering algorithm is applied to segment the dermoscopic images. With the clustering results, skin lesions are identified precisely using an adaptive region growing method. To evaluate the performance of this algorithm, a public data set (ISIC 2017) is employed to train and test the proposed method. Fifty images are randomly selected for training and 500 images for testing. Several metrics are measured for quantitatively evaluating the performance of NCARG. The results establish that the proposed approach has the ability to detect a lesion with high accuracy, 95.3% average value, compared to the obtained average accuracy, 80.6%, found when employing the neutrosophic similarity score and level set (NSSLS) segmentation approach.

Every organization seeks to set strategies for its development and growth and to do this, it must take into account the factors that affect its success or failure. The most widely used technique in strategic planning is SWOT analysis. SWOT examines strengths (S), weaknesses (W), opportunities (O), and threats (T), to select and implement the best strategy to achieve organizational goals. The chosen strategy should harness the advantages of strengths and opportunities, handle weaknesses, and avoid or mitigate threats. SWOT analysis does not quantify factors (i.e., strengths, weaknesses, opportunities, and threats) and it fails to rank available alternatives. To overcome this drawback, the authors of the next paper [54] integrate it with the analytic hierarchy process (AHP). The AHP is able to determine both quantitative and the qualitative elements by weighting and ranking them via comparison matrices. Due to the vague and inconsistent information that exists in the real world, they apply the proposed model in a neutrosophic environment. A real case study of Starbucks Company is presented to validate the model.

Big Data is a large-sized and complex dataset, which cannot be managed using traditional data processing tools. The mining process of big data is the ability to extract valuable information from these large datasets. Association rule mining is a type of data mining process, which is intended to determine interesting associations between items and to establish a set of association rules whose support is greater than a specific threshold. The classical association rules can only be extracted from binary data where an item exists in a transaction, but it fails to deal effectively with quantitative attributes, through decreasing the quality of generated association rules due to sharp boundary problems. In order to overcome the drawbacks of classical association rule mining, the authors of the following research [55] propose a new neutrosophic association rule algorithm. The algorithm uses a new approach for generating association rules by dealing with membership, indeterminacy, and non-membership functions of items, conducting to an efficient decision-making system by considering all vague association rules. To prove the validity of the method, they compare the fuzzy mining and the neutrosophic mining [65]. The results show that the proposed approach increases the number of generated association rules.

The INS is a subclass of the NS and a generalization of the interval-valued intuitionistic fuzzy set (IVIFS), which can be used in real engineering and scientific applications. The last paper [56] in the book aims at developing new generalized Choquet aggregation operators for INSs, including the generalized interval neutrosophic Choquet ordered averaging (G-INCOA) operator and generalized interval neutrosophic Choquet ordered geometric (G-INCOG) operator. The main advantages of the proposed operators can be described as follows: (i) during decision-making or analyzing process, the positive interaction, negative interaction or non-interaction among attributes can be considered by the G-INCOA and G-INCOG operators; (ii) each generalized Choquet aggregation operator presents a unique comprehensive framework for INSs, which comprises a bunch of existing interval neutrosophic aggregation operators; (iii) new multi-attribute decision making (MADM) approaches for INSs are established based on these operators, and decision makers may determine the value of λ by different MADM problems or their preferences, which makes the decision-making process more flexible; (iv) a new clustering algorithm for INSs are introduced based on the G-INCOA and G-INCOG operators, which proves that they have the potential to be applied to many new fields in the future.

The individual articles of this book can be downloaded from here:

https://www.mdpi.com/journal/symmetry/special_issues/Algebraic_Structure_Neutrosophic_Triplet_Neutrosophic_Duplet_Neutrosophic_Multiset.

Our authors' geographical distribution (published papers) is:

China (51)
Turkey (15)
India (11)
Pakistan (8)
Malaysia (6)
USA (3)
Romania (3)
Egypt (3)
Morocco (3)
Nigeria (3)
Iran (2)
Korea (2)
Denmark (1)
Saudi Arabia (1)

We found the edition and selections of papers for this book very inspiring and rewarding. We also thank the editorial staff and reviewers for their efforts and help during the process.

Conflicts of Interest: The authors declare no conflict of interest.

References

1. Neutrosophy. Available online: http://fs.gallup.unm.edu/neutrosophy.htm (accessed on 30 January 2019).
2. Neutrosophic Triplet Structures. Available online: http://fs.gallup.unm.edu/NeutrosophicTriplets.htm (accessed on 30 January 2019).
3. Neutrosophic Duplet Structures. Available online: http://fs.gallup.unm.edu/NeutrosophicDuplets.htm (accessed on 30 January 2019).
4. Neutrosophic Multiset Structures. Available online: http://fs.gallup.unm.edu/NeutrosophicMultisets.htm (accessed on 30 January 2019).
5. Çevik, A.; Topal, S.; Smarandache, F. Neutrosophic Logic Based Quantum Computing. *Symmetry* **2018**, *10*, 656. [CrossRef]
6. Çevik, A.; Topal, S.; Smarandache, F. Neutrosophic Computability and Enumeration. *Symmetry* **2018**, *10*, 643. [CrossRef]
7. Xu, L.; Li, X.; Pang, C.; Guo, Y. Simplified Neutrosophic Sets Based on Interval Dependent Degree for Multi-Criteria Group Decision-Making Problems. *Symmetry* **2018**, *10*, 640.

8. Abu Qamar, M.; Hassan, N. Generalized Q-Neutrosophic Soft Expert Set for Decision under Uncertainty. *Symmetry* **2018**, *10*, 621. [CrossRef]
9. Gulistan, M.; Nawaz, S.; Hassan, N. Neutrosophic Triplet Non-Associative Semihypergroups with Application. *Symmetry* **2018**, *10*, 613. [CrossRef]
10. Wu, X.; Qian, J.; Peng, J.; Xue, C. A Multi-Criteria Group Decision-Making Method with Possibility Degree and Power Aggregation Operators of Single Trapezoidal Neutrosophic Numbers. *Symmetry* **2018**, *10*, 590. [CrossRef]
11. Bo, C.; Zhang, X.; Shao, S.; Smarandache, F. New Multigranulation Neutrosophic Rough Set with Applications. *Symmetry* **2018**, *10*, 578. [CrossRef]
12. Ye, J.; Fang, Z.; Cui, W. Vector Similarity Measures of Q-Linguistic Neutrosophic Variable Sets and Their Multi-Attribute Decision Making Method. *Symmetry* **2018**, *10*, 531. [CrossRef]
13. Wang, J.; Wei, G.; Lu, M. An Extended VIKOR Method for Multiple Criteria Group Decision Making with Triangular Fuzzy Neutrosophic Numbers. *Symmetry* **2018**, *10*, 497. [CrossRef]
14. Wang, J.; Wei, G.; Lu, M. TODIM Method for Multiple Attribute Group Decision Making under 2-Tuple Linguistic Neutrosophic Environment. *Symmetry* **2018**, *10*, 486. [CrossRef]
15. Khan, Q.; Liu, P.; Mahmood, T.; Smarandache, F.; Ullah, K. Some Interval Neutrosophic Dombi Power Bonferroni Mean Operators and Their Application in Multi–Attribute Decision–Making. *Symmetry* **2018**, *10*, 459. [CrossRef]
16. Khan, Q.; Hassan, N.; Mahmood, T. Neutrosophic Cubic Power Muirhead Mean Operators with Uncertain Data for Multi-Attribute Decision-Making. *Symmetry* **2018**, *10*, 444. [CrossRef]
17. Uluçay, V.; Şahin, M.; Hassan, N. Generalized Neutrosophic Soft Expert Set for Multiple-Criteria Decision-Making. *Symmetry* **2018**, *10*, 437. [CrossRef]
18. Taş, F.; Topal, S.; Smarandache, F. Clustering Neutrosophic Data Sets and Neutrosophic Valued Metric Spaces. *Symmetry* **2018**, *10*, 430. [CrossRef]
19. Jaíyéọlá, T.G.; Ilojide, E.; Olatinwo, M.O.; Smarandache, F. On the Classification of Bol-Moufang Type of Some Varieties of Quasi Neutrosophic Triplet Loop (Fenyves BCI-Algebras). *Symmetry* **2018**, *10*, 427. [CrossRef]
20. Shao, S.; Zhang, X.; Li, Y.; Bo, C. Probabilistic Single-Valued (Interval) Neutrosophic Hesitant Fuzzy Set and Its Application in Multi-Attribute Decision Making. *Symmetry* **2018**, *10*, 419. [CrossRef]
21. Liang, R.-X.; Jiang, Z.-B.; Wang, J.-Q. A Linguistic Neutrosophic Multi-Criteria Group Decision-Making Method to University Human Resource Management. *Symmetry* **2018**, *10*, 364. [CrossRef]
22. Borzooei, R.A.; Zhang, X.; Smarandache, F.; Jun, Y.B. Commutative Generalized Neutrosophic Ideals in BCK-Algebras. *Symmetry* **2018**, *10*, 350. [CrossRef]
23. Jiang, W.; Zhong, Y.; Deng, X. A Neutrosophic Set Based Fault Diagnosis Method Based on Multi-Stage Fault Template Data. *Symmetry* **2018**, *10*, 346. [CrossRef]
24. Kandasamy, W.B.V.; Kandasamy, I.; Smarandache, F. Neutrosophic Duplets of $\{Zpn,\times\}$ and $\{Zpq,\times\}$ and Their Properties. *Symmetry* **2018**, *10*, 345. [CrossRef]
25. Hashim, R.M.; Gulistan, M.; Smarandache, F. Applications of Neutrosophic Bipolar Fuzzy Sets in HOPE Foundation for Planning to Build a Children Hospital with Different Types of Similarity Measures. *Symmetry* **2018**, *10*, 331. [CrossRef]
26. Chakraborty, A.; Mondal, S.P.; Ahmadian, A.; Senu, N.; Alam, S.; Salahshour, S. Different Forms of Triangular Neutrosophic Numbers, De-Neutrosophication Techniques, and their Applications. *Symmetry* **2018**, *10*, 327. [CrossRef]
27. Çelik, M.; Shalla, M.M.; Olgun, N. Fundamental Homomorphism Theorems for Neutrosophic Extended Triplet Groups. *Symmetry* **2018**, *10*, 321. [CrossRef]
28. Bo, C.; Zhang, X.; Shao, S.; Smarandache, F. Multi-Granulation Neutrosophic Rough Sets on a Single Domain and Dual Domains with Applications. *Symmetry* **2018**, *10*, 296. [CrossRef]
29. Zhang, X.; Hu, Q.; Smarandache, F.; An, X. On Neutrosophic Triplet Groups: Basic Properties, NT-Subgroups, and Some Notes. *Symmetry* **2018**, *10*, 289. [CrossRef]
30. Garg, H.; Nancy. Multi-Criteria Decision-Making Method Based on Prioritized Muirhead Mean Aggregation Operator under Neutrosophic Set Environment. *Symmetry* **2018**, *10*, 280. [CrossRef]
31. Guan, H.; He, J.; Zhao, A.; Dai, Z.; Guan, S. A Forecasting Model Based on Multi-Valued Neutrosophic Sets and Two-Factor, Third-Order Fuzzy Fluctuation Logical Relationships. *Symmetry* **2018**, *10*, 245. [CrossRef]

32. Zhang, X.; Wu, X.; Smarandache, F.; Hu, M. Left (Right)-Quasi Neutrosophic Triplet Loops (Groups) and Generalized BE-Algebras. *Symmetry* **2018**, *10*, 241. [CrossRef]
33. Şahin, M.; Kargın, A.; Çoban, M.A. Fixed Point Theorem for Neutrosophic Triplet Partial Metric Space. *Symmetry* **2018**, *10*, 240. [CrossRef]
34. Selvachandran, G.; Quek, S.G.; Smarandache, F.; Broumi, S. An Extended Technique for Order Preference by Similarity to an Ideal Solution (TOPSIS) with Maximizing Deviation Method Based on Integrated Weight Measure for Single-Valued Neutrosophic Sets. *Symmetry* **2018**, *10*, 236. [CrossRef]
35. Abdel-Basset, M.; Mohamed, M.; Smarandache, F. A Hybrid Neutrosophic Group ANP-TOPSIS Framework for Supplier Selection Problems. *Symmetry* **2018**, *10*, 226. [CrossRef]
36. Mani, P.; Muthusamy, K.; Jafari, S.; Smarandache, F.; Ramalingam, U. Decision-Making via Neutrosophic Support Soft Topological Spaces. *Symmetry* **2018**, *10*, 217. [CrossRef]
37. Tu, A.; Ye, J.; Wang, B. Multiple Attribute Decision-Making Method Using Similarity Measures of Neutrosophic Cubic Sets. *Symmetry* **2018**, *10*, 215. [CrossRef]
38. Zhang, C.; Li, D.; Broumi, S.; Sangaiah, A.K. Medical Diagnosis Based on Single-Valued Neutrosophic Probabilistic Rough Multisets over Two Universes. *Symmetry* **2018**, *10*, 213. [CrossRef]
39. Jaíyéolá, T.G.; Smarandache, F. Some Results on Neutrosophic Triplet Group and Their Applications. *Symmetry* **2018**, *10*, 202. [CrossRef]
40. Gulistan, M.; Yaqoob, N.; Rashid, Z.; Smarandache, F.; Wahab, H.A. A Study on Neutrosophic Cubic Graphs with Real Life Applications in Industries. *Symmetry* **2018**, *10*, 203. [CrossRef]
41. Gal, I.-A.; Bucur, D.; Vladareanu, L. DSmT Decision-Making Algorithms for Finding Grasping Configurations of Robot Dexterous Hands. *Symmetry* **2018**, *10*, 198. [CrossRef]
42. Wang, J.-Q.; Tian, C.-Q.; Zhang, X.; Zhang, H.-Y.; Wang, T.-L. Multi-Criteria Decision-Making Method Based on Simplified Neutrosophic Linguistic Informatiosn with Cloud Model. *Symmetry* **2018**, *10*, 197. [CrossRef]
43. Tan, R.; Zhang, W.; Chen, S. Exponential Aggregation Operator of Interval Neutrosophic Numbers and Its Application in Typhoon Disaster Evaluation. *Symmetry* **2018**, *10*, 196. [CrossRef]
44. WB, V.K.; Kandasamy, I.; Smarandache, F. A Classical Group of Neutrosophic Triplet Groups Using {Z2p, ×}. *Symmetry* **2018**, *10*, 194.
45. Quek, S.G.; Broumi, S.; Selvachandran, G.; Bakali, A.; Talea, M.; Smarandache, F. Some Results on the Graph Theory for Complex Neutrosophic Sets. *Symmetry* **2018**, *10*, 190. [CrossRef]
46. Zhang, X.; Bo, C.; Smarandache, F.; Park, C. New Operations of Totally Dependent-Neutrosophic Sets and Totally Dependent-Neutrosophic Soft Sets. *Symmetry* **2018**, *10*, 187. [CrossRef]
47. Turhan, M.; Şengür, D.; Karabatak, S.; Guo, Y.; Smarandache, F. Neutrosophic Weighted Support Vector Machines for the Determination of School Administrators Who Attended an Action Learning Course Based on Their Conflict-Handling Styles. *Symmetry* **2018**, *10*, 176. [CrossRef]
48. Shao, S.; Zhang, X.; Bo, C.; Smarandache, F. Neutrosophic Hesitant Fuzzy Subalgebras and Filters in Pseudo-BCI Algebras. *Symmetry* **2018**, *10*, 174. [CrossRef]
49. Wang, Y.; Liu, P. Linguistic Neutrosophic Generalized Partitioned Bonferroni Mean Operators and Their Application to Multi-Attribute Group Decision Making. *Symmetry* **2018**, *10*, 160. [CrossRef]
50. Fan, C.; Fan, E.; Ye, J. The Cosine Measure of Single-Valued Neutrosophic Multisets for Multiple Attribute Decision-Making. *Symmetry* **2018**, *10*, 154. [CrossRef]
51. Feng, J.; Li, M.; Li, Y. Study of Decision Framework of Shopping Mall Photovoltaic Plan Selection Based on DEMATEL and ELECTRE III with Symmetry under Neutrosophic Set Environment. *Symmetry* **2018**, *10*, 150. [CrossRef]
52. Bal, M.; Shalla, M.M.; Olgun, N. Neutrosophic Triplet Cosets and Quotient Groups. *Symmetry* **2018**, *10*, 126. [CrossRef]
53. Guo, Y.; Ashour, A.S.; Smarandache, F. A Novel Skin Lesion Detection Approach Using Neutrosophic Clustering and Adaptive Region Growing in Dermoscopy Images. *Symmetry* **2018**, *10*, 119. [CrossRef]
54. Abdel-Basset, M.; Mohamed, M.; Smarandache, F. An Extension of Neutrosophic AHP–SWOT Analysis for Strategic Planning and Decision-Making. *Symmetry* **2018**, *10*, 116. [CrossRef]
55. Abdel-Basset, M.; Mohamed, M.; Smarandache, F.; Chang, V. Neutrosophic Association Rule Mining Algorithm for Big Data Analysis. *Symmetry* **2018**, *10*, 106. [CrossRef]
56. Li, X.; Zhang, X.; Park, C. Generalized Interval Neutrosophic Choquet Aggregation Operators and Their Applications. *Symmetry* **2018**, *10*, 85. [CrossRef]

57. Smarandache, F. *Neutrosophy. Neutrosophic Probability, Set, and Logic*; American Research Press: Rehoboth, DE, USA, 1998.
58. Smarandache, F. A generalization of the intuitionistic fuzzy set. *Int. J. Pure Appl. Math.* **2005**, *24*, 287–297.
59. Smarandache, F. Neutrosophic Quantum Computer. Intern. *J. Fuzzy Math. Arch.* **2016**, *10*, 139–145.
60. Wang, H.B.; Smarandache, F.; Zhang, Y.Q.; Sunderraman, R. Single Valued Neutrosophic Sets. Available online: http://citeseerx.ist.psu.edu/viewdoc/download;jsessionid=65C7521427055BA55C102843C01F668C?doi=10.1.1.640.7072&rep=rep1&type=pdf (accessed on 30 January 2019).
61. Biswas, P.; Pramanik, S.; Giri, B.C. Value and ambiguity index based ranking method of single-valued trapezoidal neutrosophic numbers and its application to multi-attribute decision making. *Neutrosophic Sets Syst.* **2016**, *12*, 127–138.
62. Wu, Q.; Wu, P.; Zhou, L.; Chen, H.; Guan, X. Some new Hamacher aggregation operators under single-valued neutrosophic 2-tuple linguistic environment and their applications to multi-attribute group decision making. *Comput. Ind. Eng.* **2018**, *116*, 144–162. [CrossRef]
63. Wang, H.; Madiraju, P. Interval-neutrosophic Sets. *J. Mech.* **2004**, *1*, 274–277.
64. Ali, M.; Deli, I.; Smarandache, F. The theory of neutrosophic cubic sets and their applications in pattern recognition. *J. Intell. Fuzzy Syst.* **2018**, *30*, 1957–1963. [CrossRef]
65. Mondal, K.; Pramanik, S.; Giri, B.C. Role of Neutrosophic Logic in Data Mining. *New Trends Neutrosophic Theory Appl.* **2016**, *1*, 15.

symmetry

MDPI

Article

Generalized Interval Neutrosophic Choquet Aggregation Operators and Their Applications

Xin Li [1], **Xiaohong Zhang** [1,2,*] and **Choonkil Park** [3]

[1] College of Arts and Sciences, Shanghai Maritime University, Shanghai 201306, China; Smu201631010013@163.com
[2] School of Arts and Sciences, Shaanxi University of Science & Technology, Xi'an 710021, China
[3] Department of Mathematics, Hanyang University, Seoul 04763, Korea; baak@hanyang.ac.kr
* Correspondence: zhangxiaohong@sust.edu.cn or zhangxh@shmtu.edu.cn

Received: 14 March 2018; Accepted: 26 March 2018; Published: 28 March 2018

Abstract: The interval neutrosophic set (INS) is a subclass of the neutrosophic set (NS) and a generalization of the interval-valued intuitionistic fuzzy set (IVIFS), which can be used in real engineering and scientific applications. This paper aims at developing new generalized Choquet aggregation operators for INSs, including the generalized interval neutrosophic Choquet ordered averaging (G-INCOA) operator and generalized interval neutrosophic Choquet ordered geometric (G-INCOG) operator. The main advantages of the proposed operators can be described as follows: (i) during decision-making or analyzing process, the positive interaction, negative interaction or non-interaction among attributes can be considered by the G-INCOA and G-INCOG operators; (ii) each generalized Choquet aggregation operator presents a unique comprehensive framework for INSs, which comprises a bunch of existing interval neutrosophic aggregation operators; (iii) new multi-attribute decision making (MADM) approaches for INSs are established based on these operators, and decision makers may determine the value of λ by different MADM problems or their preferences, which makes the decision-making process more flexible; (iv) a new clustering algorithm for INSs are introduced based on the G-INCOA and G-INCOG operators, which proves that they have the potential to be applied to many new fields in the future.

Keywords: generalized aggregation operators; interval neutrosophic set (INS); multi-attribute decision making (MADM); Choquet integral; fuzzy measure; clustering algorithm

1. Introduction

The neutrosophic set (NS) is a powerful comprehensive framework that comprises the concepts of the classic set, fuzzy set (FS), intuitionistic fuzzy set (IFS), hesitant fuzzy set (HFS), paraconsistent set, paradoxist set, and interval-valued fuzzy set (IVFS) [1–4]. It was introduced by Smarandache to deal with incomplete, indeterminate, and inconsistent decision information, which includes the truth membership, falsity membership, and indeterminacy membership, and their functions are non-standard subsets of $]^{-}0, 1^{+}[$ [5]. However, without a specific description, it is difficult to apply the NS in practical application. Therefore, scholars proposed the interval neutrosophic set (INS), single-valued neutrosophic set (SVNS), rough neutrosophic set (RNS), multi-valued neutrosophic set (MVNS) as some special cases of the NS, and studied their related properties in [6–9]. Recently, numbers of new neutrosophic theories have been proposed and applied to image segmentation, image processing, rock mechanics, stock market, computational intelligence, multi-attribute decision making (MADM), medical diagnosis, fault diagnosis, and optimization design as described in [10–13].

The INS is a subclass of the NS and generalization of the IFS and IVIFS, which was proposed by Wang [6]. Motivated by some aggregation operators and decision-making methods for IFSs,

IVIFSs, and NSs [14–19], a lot of theories about INSs have been put forward successively, and their basic concepts and aggregation tools play important roles in practical applications. For instance, Wang et al. [6] defined the basic operational relations for INSs and Zhang et al. [20] pointed out some drawbacks of these operational laws and improved them. Then they also put forward some basic aggregation operators to deal with MADM problems with interval neutrosophic information. Besides, Broumi [21] introduced the definition of correlation coefficient between INSs. Then Zhang et al. [22] pointed out some shortcomings of the existing correlation coefficient and they also proposed the definition of improved weighted correlation coefficient. Ye [23] defined some distance measures and similarity measures for INSs and applied these measures in practical MADM problems, and he also [24] proposed the interval neutrosophic ordered weighted arithmetic and geometric averaging operators, and further constructed a possibility degree ranking method under the interval neutrosophic environment. Moreover, Liu et al. [25–27] proposed the power generalized aggregation operators, the prioritized ordered weighted aggregation operators and induced generalized interval neutrosophic Shapley hybrid geometric averaging/mean operators for INSs under an interval neutrosophic environment.

For some practical problems, there exists mutual influence and interaction among attributes, which should be considered in decision-making or other analyzing process. The interaction between attributes can be classified into three types, which are positive interaction, negative interaction, and non-interaction [28,29]. Failure to consider the interactions among attributes may directly lead to errors of decision results. To solve this problem under the interval neutrosophic environment, we first intend to define some aggregation operators in this paper by combining the definition of Choquet integral to process the mutual influence and interaction among attributes with respect to fuzzy measure [30,31].

Besides, cluster analysis, or clustering, is defined as the unsupervised process of group (a set of data objects) in such a way that objects in the same group (called a cluster) are somehow more similar to each other than those in other groups (clusters) [32]. There are many algorithms for clustering which differ significantly in their notion of what constitutes a cluster and how to efficiently find them. Under a hesitant fuzzy environment, Chen et al. [33] proposed an algorithm to cluster hesitant fuzzy data into different clusters. Using the algorithm as a reference, we also intend to propose an effective new clustering algorithm under the interval neutrosophic environment.

Moreover, the generalized aggregation operators are a new class of operators, which have been widely applied in fuzzy areas, since they can be used to synthesize multi-dimensional evaluation values represented by kinds of hesitant fuzzy values or intuitionstic fuzzy values into collective values. Overall, this paper aims at proposing new generalized Choquet aggregation operators for INSs—namely, the G-INCOA operator and G-INCOG operator—which can be applied in MADM and clustering using interval neutrosophic information. In some special cases, each generalized aggregation operator reduces to various existing non-generalized interval neutrosophic aggregation operators.

To do so, the rest of this paper is organized as follows: Section 2 introduces some basic definitions about the Choquet integral and INS. In Section 3, the G-INCOA operator and G-INCOG operator are put forward and some desirable properties of them are discussed and proved. We also consider special cases of these operators and distinguish them in two main classes, the first class focuses on the parameter λ, and the second class on the fuzzy measure $\mu(x_j)$. In Section 4, we put forward some novel MADM methods based on the proposed operators to deal with interval neutrosophic information and utilize an illustrative example to validate the proposed MADM approaches by taking different values of parameter λ of the proposed operators. In Section 5, a new clustering algorithm for INSs is introduced based on the G-INCOA operator and the G-INCOG operator. Then, a numerical example concerning clustering is utilized as the demonstration of the application and effectiveness of the proposed clustering algorithm. Finally, conclusions and future research directions are drawn in Section 6.

2. Preliminaries

To facilitate the following discussion, some basic definitions about the Choquet integral and INS are briefly introduced in this section.

2.1. Interval Neutrosophic Sets (INS)

The NS was firstly introduced by Smarandache [5], which is a comprehensive framework for expressing and processing incomplete and indeterminate information.

Definition 1. ([5]) Let X be a non-empty fixed set, a NS on X is defined as:

$$T_A(x), I_A(x), F_A(x) : \ X \rightarrow]^-0, 1^+[, \tag{1}$$

where $T_A(x), I_A(x), F_A(x)$ representing the truth membership function, indeterminacy membership function and falsity membership function, respectively, and satisfying the limit: $0^- \leq sup\ T_A(x) + supI_A(x) + supF_A(x) \leq 3^+$.

It is not difficult to find that the NS is difficult to apply in the real applications. Therefore, Wang et al. [6] proposed the interval neutrosophic set (INS) as an instance of the NS, which is defined as:

Definition 2. ([6]) Let X be a non-empty finite set, an INS in X is expressed by:

$$N = \{\langle x, [\tilde{t}^L(x), \tilde{t}^U(x)], [\tilde{i}^L(x), \tilde{i}^U(x)], [\tilde{f}^L(x), \tilde{f}^U(x)\rangle] |\ x \in X\}, \tag{2}$$

where $\tilde{t}(x) = [\tilde{t}^L(x), \tilde{t}^U(x)] \subseteq [0,1]$, $\tilde{i}(x) = [\tilde{i}^L(x), \tilde{i}^U(x)] \subseteq [0,1]$, $\tilde{f}(x) = [\tilde{f}^L(x), \tilde{f}^U(x)] \subseteq [0,1]$ representing truth, indeterminacy, and falsity membership functions of the element $x \in X$, and satisfying limits: $0 \leq \tilde{t}^U(x) + \tilde{i}^U(x) + \tilde{f}^U(x) \leq 3$.

For convenience, we call $\tilde{n} = \langle [\tilde{t}^L, \tilde{t}^U], [\tilde{i}^L, \tilde{i}^U], [\tilde{f}^L, \tilde{f}^U] \rangle$ an interval neutrosophic element (INN). The basic operational relations of INNs are defined as:

Definition 3. ([6]) Let $\tilde{n_1} = \langle [\tilde{t}_1^L, \tilde{t}_1^U], [\tilde{i}_1^L, \tilde{i}_1^U], [\tilde{f}_1^U, \tilde{f}_1^U] \rangle$ and $\tilde{n_2} = \langle [\tilde{t}_2^L, \tilde{t}_2^U], [\tilde{i}_2^L, \tilde{i}_2^U], [\tilde{f}_2^U, \tilde{f}_2^U] \rangle$ be two INNs, then:

1. $\tilde{n_1} \oplus \tilde{n_2} = \langle [\tilde{t}_1^L + \tilde{t}_2^L - \tilde{t}_1^L \tilde{t}_2^L, \tilde{t}_1^U + \tilde{t}_2^U - \tilde{t}_1^U \tilde{t}_2^U], [\tilde{i}_1^L \tilde{i}_2^L, \tilde{i}_1^U \tilde{i}_2^U], [\tilde{f}_1^L \tilde{f}_2^L, \tilde{f}_1^U \tilde{f}_2^U] \rangle$;
2. $\tilde{n_1} \oplus \tilde{n_2} = \langle [\tilde{t}_1^L \tilde{t}_2^L, \tilde{t}_1^U \tilde{t}_2^U], [\tilde{i}_1^L + \tilde{i}_2^L - \tilde{i}_1^L \tilde{i}_2^L, \tilde{i}_1^U + \tilde{i}_2^U - \tilde{i}_1^U \tilde{i}_2^U], [\tilde{f}_1^L + \tilde{f}_2^L - \tilde{f}_1^L \tilde{i}_2^L, \tilde{f}_1^U + \tilde{f}_2^U - \tilde{f}_1^U \tilde{f}_2^U] \rangle$;
3. $r\tilde{n_1} = \langle [1 - (1 - \tilde{t}_1^L)^r, 1 - (1 - \tilde{t}_1^U)^r], [(\tilde{i}_1^L)^r, (\tilde{i}_1^U)^r], [(\tilde{f}_1^L)^r, (\tilde{f}_1^U)^r] \rangle$;
4. $\tilde{n_1}^r = \langle [(\tilde{t}_1^L)^r, (\tilde{t}_1^U)^r], [1 - (1 - \tilde{i}_1^L)^r, 1 - (1 - \tilde{i}_1^U)^r], [1 - (1 - \tilde{f}_1^L)^r, 1 - (1 - \tilde{f}_1^U)^r] \rangle$.

2.2. Some Concepts of INSs

On the basis of the distance measures of INSs [23], Ye defined some similarity measures between INSs $\tilde{n_1}$ and $\tilde{n_2}$, which can be given as:

Definition 4. Let $\tilde{n_1} = \langle [\tilde{t}_1^L, \tilde{t}_1^U], [\tilde{i}_1^L, \tilde{i}_1^U], [\tilde{f}_1^U, \tilde{f}_1^U] \rangle$ and $\tilde{n_2} = \langle [\tilde{t}_2^L, \tilde{t}_2^U], [\tilde{i}_2^L, \tilde{i}_2^U], [\tilde{f}_2^U, \tilde{f}_2^U] \rangle$ be two INNs, thus, the similarity function between $\tilde{n_1}$ and $\tilde{n_2}$ is defined by:

$$C(\tilde{n_1}, \tilde{n_2}) = 1 - \frac{\left((\tilde{t}_1^L - \tilde{t}_2^L)^2 + (\tilde{i}_1^L - \tilde{i}_2^L)^2 + (\tilde{f}_1^L - \tilde{f}_2^L)^2 + (\tilde{t}_1^U - \tilde{t}_2^U)^2 + (\tilde{i}_1^U - \tilde{i}_2^U)^2 + (\tilde{f}_1^U - \tilde{f}_2^U)^2 \right)}{6}. \tag{3}$$

According to the value range of the similarity measures, we can obtain the value range of the cosine function, we can obtain the following property $0 \leq C(\tilde{n_1}, \tilde{n_2}) \leq 1$. Suppose the best ideal alternative $\tilde{n}^+ = \langle [\tilde{t}_1^{L+}, \tilde{t}_1^{U+}], [\tilde{i}_1^{L+}, \tilde{i}_1^{U+}], [\tilde{f}_1^{L+}, \tilde{f}_1^{U+}] \rangle = \langle [1,1], [0,0], [0,0] \rangle$, then, the similarity measures between $\tilde{n}_1$ and $\tilde{n}^+$ can be described as:

$$C(\tilde{n}_1, \tilde{n}_2) = 1 - \frac{\left((1-\tilde{t}_1^L)^2 + (\tilde{i}_1^L)^2 + (\tilde{f}_1^L)^2 + (1-\tilde{t}_1^U)^2 + (\tilde{i}_1^U)^2 + (\tilde{f}_1^U)^2\right)}{6}. \quad (4)$$

The score function are effective tools to rank INNs, and here we give its definition:

Definition 5. ([25]) For $\tilde{n}$, the score function $s(\tilde{n})$ is defined as:

$$s(\tilde{n}) = \left(\frac{(\tilde{t}_1^L + \tilde{t}_1^U)}{2} + \left(1 - \frac{(\tilde{i}_1^L + \tilde{i}_1^U)}{2}\right) + \left(1 - \frac{(\tilde{f}_1^L + \tilde{f}_1^U)}{2}\right)\right)/3 \quad (5)$$

obviously, $s(\tilde{n}) \in [0,\ 1]$. *If* $s(\tilde{n}_1) > s(\tilde{n}_2)$, then $\tilde{n}_1 > \tilde{n}_2$.

2.3. The Fuzzy Measure and Choquet Integral

The Choquet integral is a powerful operator to aggregate kinds of fuzzy information in MADM with respect to fuzzy measure.

Definition 6. ([30]) Let $(X,\ \mathcal{A},\ \mu)$ be a measurable space and $\mu :\ \mathcal{A} \to [0,\ 1]$, if it satisfies the conditions:

1. $\mu(\varnothing) = 0$;
2. $\mu(A) \leq \mu(B)$ whenever $A \subset B, A, B \in \mathcal{A}$;
3. $If\ A_1 \subset A_2 \subset \ldots \subset A_n \subset \ldots,\quad A_n \in \mathcal{A}$, then $\mu\left(\cup_{n=1}^{\infty} A_n\right) = lim_{n\to\infty}\mu(A_n)\,;$
4. $If\ A_1 \supset A_2 \supset \ldots \supset A_n \supset \ldots,\quad A_n \in \mathcal{A}$, then $\mu\left(\cup_{n=1}^{\infty} A_n\right) = lim_{n\to\infty}\mu(A_n)\,;$

then, we call μ be a fuzzy measure defined by Sugeno M.

To avoid the problems with computational complexity in paractical applications, g_λ fuzzy measure also called λ-fuzzy measure, was proposed by Sugeno M [30], which satisfies an additional properties: $\mu(X \cup Y) = \mu(X) + \mu(Y) + g_\lambda\, \mu(X)\mu(Y)$, $g_\lambda \in (-1,\ \infty)$ for all $X, Y \in \mathcal{A}$ and $X \cap Y = \varnothing$. Specially, the expression of g_λ fuzzy measure defined on a finite set $X = \{x_1, x_2, \ldots, x_m\}$ can be simplified as:

Theorem 1. ([30]) Let X be a set ($X = \{x_1, x_2, \ldots, x_m\}$), λ-fuzzy measure defined on X is expressed as:

$$\mu(X) = \begin{cases} \frac{1}{\lambda_g}\left(\prod\limits_{i\in X}(1+\lambda_g \mu(x_i)) - 1\right),\ if\ \lambda_g \neq 0, \\ \sum\limits_{i\in X} \mu(x_i),\ if\ \lambda_g\ = 0, \end{cases} \quad (6)$$

where $x_i \cap x_j = \varnothing$ for all $i,\ j = 1,\ 2, 3,\ \ldots,\ m$ and $i \neq j$.

Then, the Choquet integral with respect to fuzzy measures, is defined as:

Definition 7. ([31]) When μ is a fuzzy measure, $X = \{x_1, x_2, \ldots, x_m\}$ is a finite set. The Choquet integral of a function $f :\ X \to [0,\ 1]$ with respect to fuzzy measure μ can be expressed as:

$$\int f d\mu = \sum_{i=1}^{m}\left(\mu\left(F_{\phi(i)}\right) - \mu\left(F_{\phi(i-1)}\right)\right) \oplus f\left(x_{\phi(i)}\right), \quad (7)$$

where $(\phi(1), \phi(2), \ldots \phi(i), \ldots,\ \phi(m))$ is a permutation of $(1,\ 2,\ \ldots i, \ldots,\ m)$ such that $f\left(x_{\phi(1)}\right) \leq f\left(x_{\phi(2)}\right) \leq, \ldots, \leq\ f\left(x_{\phi(i)}\right) \leq, \ldots, \leq\ f\left(x_{\phi(m)}\right)$, $F_{\phi(i)} = \{x_{\phi(1)}, x_{\phi(2)}, \ldots, x_{\phi(i)}\}$ and $F_{\phi(0)} = \varnothing$.

3. Generalized Interval Neutrosphic Choquet Aggregation Operators

In what follows, based on the operational relations of INNs and Choquet aggregation operator, we shall develop new generalized Choquet aggregation operators under the interval neutrosophic environment, such as the generalized interval neutrosophic Choquet ordered averaging (G-INCOA) operator and generalized interval neutrosophic Choquet ordered geometric (G-INCOG) operator.

3.1. The G-INCOA and G-INCOG Operators

Definition 8. When $\widetilde{n}_j(j = 1,\ 2,\ 3,\ \ldots,\ m)$ is a collection of INNs, X= $\{x_1, x_2, x_3, \ldots, x_m\}$ is the set of attributes and μ measure on X, the G-INCOA and G-INCOG operators are defined as:

$$G - INCOA_{\mu,\lambda}\{\widetilde{n}_1, \widetilde{n}_2, \ldots, \widetilde{n}_m\} = \left(\oplus_{j=1}^{m}\left(\mu\left(F_{\phi(j)}\right) - \mu\left(F_{\phi(j-1)}\right)\right)\widetilde{n}_{\phi(j)}{}^{\lambda}\right)^{\frac{1}{\lambda}}, \tag{8}$$

$$G - INCOG_{\mu,\lambda}\{\widetilde{n}_1, \widetilde{n}_2, \ldots, \widetilde{n}_m\} = \left(\otimes_{j=1}^{m}\left(\mu\left(F_{\phi(j)}\right) - \mu\left(F_{\phi(j-1)}\right)\right)\widetilde{n}_{\phi(j)}{}^{\lambda}\right)^{\frac{1}{\lambda}}, \tag{9}$$

where $\lambda > 0$, $\mu_{\phi(i)} = \mu\left(F_{\phi(i)}\right) - \mu\left(F_{\phi(i-1)}\right)$. where $(\phi(1), \phi(2), \ldots \phi(i), \ldots,\ \phi(m))$ is a permutation of $(1,\ 2,\ \ldots i, \ldots,\ m)$ such that $f\left(x_{\phi(1)}\right) \leq f\left(x_{\phi(2)}\right) \leq, \ldots, \leq f\left(x_{\phi(i)}\right) \leq, \ldots, \leq f\left(x_{\phi(m)}\right)$, $F_{\phi(0)} = \varnothing$ and $F_{\phi(i)} = \{x_{\phi(1)}, x_{\phi(2)}, \ldots, x_{\phi(i)}\}$.

Theorem 2. When $\widetilde{n}_j(j = 1,\ 2,\ 3,\ \ldots,\ m)$ is a collection of INNs, then the aggregated value obtained by the G-INCOA operator is also a INN, and:

$$\begin{aligned} &G - INCOA_{\mu,\lambda}\{\widetilde{n}_1, \widetilde{n}_2, \ldots, \widetilde{n}_m\} = \left(\oplus_{j=1}^{m}\left(\mu\left(F_{\phi(j)}\right) - \mu\left(F_{\phi(j-1)}\right)\right)\widetilde{n}_{\phi(j)}{}^{\lambda}\right)^{\frac{1}{\lambda}} \\ &= \left\{\left[\left(1 - \prod_{j=1}^{m}\left(1 - \left(t_{\phi(j)}^{L}\right)^{\lambda}\right)^{\mu_{\phi(j)}}\right)^{\frac{1}{\lambda}}, \left(1 - \prod_{j=1}^{m}\left(1 - \left(t_{\phi(j)}^{U}\right)^{\lambda}\right)^{\mu_{\phi(j)}}\right)^{\frac{1}{\lambda}}\right],\right. \\ &\left[1 - \left(1 - \prod_{j=1}^{m}\left(1 - \left(1 - i_{\phi(j)}^{L}\right)^{\lambda}\right)^{\mu_{\phi(j)}}\right)^{\frac{1}{\lambda}}, 1 - \left(1 - \prod_{j=1}^{m}\left(1 - \left(1 - i_{\phi(j)}^{U}\right)^{\lambda}\right)^{\mu_{\phi(j)}}\right)^{\frac{1}{\lambda}}\right], \\ &\left.\left[1 - \left(1 - \prod_{j=1}^{m}\left(1 - \left(1 - f_{\phi(j)}^{L}\right)^{\lambda}\right)^{\mu_{\phi(j)}}\right)^{\frac{1}{\lambda}}, 1 - \left(1 - \prod_{j=1}^{m}\left(1 - \left(1 - f_{\phi(j)}^{U}\right)^{\lambda}\right)^{\mu_{\phi(j)}}\right)^{\frac{1}{\lambda}}\right]\right\}. \end{aligned} \tag{10}$$

Similarly, the aggregated value obtained by the G-INCOG operator is also a INN,

$$\begin{aligned} &G - INCOG_{\mu,\lambda}\{\widetilde{n}_1, \widetilde{n}_2, \ldots, \widetilde{n}_m\} = \left(\oplus_{j=1}^{m}\left(\mu\left(F_{\phi(j)}\right) - \mu\left(F_{\phi(j-1)}\right)\right)\widetilde{n}_{\phi(j)}{}^{\lambda}\right)^{\frac{1}{\lambda}}, \\ &= \left\{\left[1 - \left(1 - \prod_{j=1}^{m}\left(1 - \left(1 - t_{\phi(j)}^{L}\right)^{\lambda}\right)^{\mu_{\phi(j)}}\right)^{\frac{1}{\lambda}}, 1 - \left(1 - \prod_{j=1}^{m}\left(1 - \left(1 - t_{\phi(j)}^{U}\right)^{\lambda}\right)^{\mu_{\phi(j)}}\right)^{\frac{1}{\lambda}}\right],\right. \\ &\left[\left(1 - \prod_{j=1}^{m}\left(1 - \left(i_{\phi(j)}^{L}\right)^{\lambda}\right)^{\mu_{\phi(j)}}\right)^{\frac{1}{\lambda}}, \left(1 - \prod_{j=1}^{m}\left(1 - \left(i_{\phi(j)}^{U}\right)^{\lambda}\right)^{\mu_{\phi(j)}}\right)^{\frac{1}{\lambda}}\right], \\ &\left.\left[\left(1 - \prod_{j=1}^{m}\left(1 - \left(f_{\phi(j)}^{L}\right)^{\lambda}\right)^{\mu_{\phi(j)}}\right)^{\frac{1}{\lambda}}, \left(1 - \prod_{j=1}^{m}\left(1 - \left(f_{\phi(j)}^{U}\right)^{\lambda}\right)^{\mu_{\phi(j)}}\right)^{\frac{1}{\lambda}}\right]\right\}. \end{aligned} \tag{11}$$

Proof. The result of $m = 1$ follows quickly from Definition 8, below we prove Equations (10) and (11) by means of mathematical induction on m, here, take Equation (11) as an example.

(a) For $m = 2$, based on the operation relations of INNs defined in Definition 3, we have:

$$\left(\mu_{\phi(1)}\widetilde{n}_{\phi(1)}{}^{\lambda}\right)^{\frac{1}{\lambda}}=\left\{\left[1-\left(1-\left(1-\left(1-t^{L}_{\phi(1)}\right)^{\lambda}\right)^{\mu_{\phi(1)}}\right)^{\frac{1}{\lambda}},1-\left(1-\left(1-\left(1-t^{U}_{\phi(1)}\right)^{\lambda}\right)^{\mu_{\phi(1)}}\right)^{\frac{1}{\lambda}}\right],\right.$$
$$\left.\left[\left(1-\left(1-\left(i^{L}_{\phi(1)}\right)^{\lambda}\right)^{\mu_{\phi(1)}}\right)^{\frac{1}{\lambda}},\left(1-\left(1-\left(i^{U}_{\phi(1)}\right)^{\lambda}\right)^{\mu_{\phi(1)}}\right)^{\frac{1}{\lambda}}\right],\left[\left(1-\left(1-\left(f^{L}_{\phi(1)}\right)^{\lambda}\right)^{\mu_{\phi(1)}}\right)^{\frac{1}{\lambda}},\left(1-\left(1-\left(f^{U}_{\phi(1)}\right)^{\lambda}\right)^{\mu_{\phi(1)}}\right)^{\frac{1}{\lambda}}\right]\right\};$$

$$\left(\mu_{\phi(2)}\widetilde{n}_{\phi(2)}{}^{\lambda}\right)^{\frac{1}{\lambda}}=\left\{\left[1-\left(1-\left(1-\left(1-t^{L}_{\phi(2)}\right)^{\lambda}\right)^{\mu_{\phi(2)}}\right)^{\frac{1}{\lambda}},1-\left(1-\left(1-\left(1-t^{U}_{\phi(2)}\right)^{\lambda}\right)^{\mu_{\phi(2)}}\right)^{\frac{1}{\lambda}}\right],\right.$$
$$\left.\left[\left(1-\left(1-\left(i^{L}_{\phi(2)}\right)^{\lambda}\right)^{\mu_{\phi(2)}}\right)^{\frac{1}{\lambda}},\left(1-\left(1-\left(i^{U}_{\phi(2)}\right)^{\lambda}\right)^{\mu_{\phi(2)}}\right)^{\frac{1}{\lambda}}\right],\left[\left(1-\left(1-\left(f^{L}_{\phi(2)}\right)^{\lambda}\right)^{\mu_{\phi(2)}}\right)^{\frac{1}{\lambda}},\left(1-\left(1-\left(f^{U}_{\phi(2)}\right)^{\lambda}\right)^{\mu_{\phi(2)}}\right)^{\frac{1}{\lambda}}\right]\right\};$$

thus, for $m=2$, the $G-INCOG_{\mu,\ \lambda}\{\widetilde{n}_1,\widetilde{n}_2\}$ can be obtained as:

$$G-INCOG_{\mu,\lambda}\{\widetilde{n}_1,\widetilde{n}_2\}=\left(\mu_{\phi(1)}\widetilde{n}_{\phi(1)}{}^{\lambda}\right)^{\frac{1}{\lambda}}\oplus\left(\mu_{\phi(2)}\widetilde{n}_{\phi(2)}{}^{\lambda}\right)^{\frac{1}{\lambda}}=$$
$$\left\{\left[1-\left(1-\left(1-\left(1-t^{L}_{\phi(1)}\right)^{\lambda}\right)^{\mu_{\phi(1)}}\left(1-\left(1-t^{L}_{\phi(2)}\right)^{\lambda}\right)^{\mu_{\phi(2)}}\right)^{\frac{1}{\lambda}},1-\left(1-\left(1-\left(1-t^{U}_{\phi(1)}\right)^{\lambda}\right)^{\mu_{\phi(1)}}\left(1-\left(1-t^{U}_{\phi(2)}\right)^{\lambda}\right)^{\mu_{\phi(2)}}\right)^{\frac{1}{\lambda}}\right],\right.$$
$$\left[\left(1-\left(1-\left(i^{L}_{\phi(1)}\right)^{\lambda}\right)^{\mu_{\phi(1)}}\left(1-\left(i^{L}_{\phi(2)}\right)^{\lambda}\right)^{\mu_{\phi(2)}}\right)^{\frac{1}{\lambda}},\left(1-\left(1-\left(i^{U}_{\phi(1)}\right)^{\lambda}\right)^{\mu_{\phi(1)}}\left(1-\left(i^{U}_{\phi(2)}\right)^{\lambda}\right)^{\mu_{\phi(2)}}\right)^{\frac{1}{\lambda}}\right],$$
$$\left.\left[\left(1-\left(1-\left(f^{L}_{\phi(1)}\right)^{\lambda}\right)^{\mu_{\phi(1)}}\left(1-\left(f^{L}_{\phi(2)}\right)^{\lambda}\right)^{\mu_{\phi(2)}}\right)^{\frac{1}{\lambda}},\left(1-\left(1-\left(f^{U}_{\phi(1)}\right)^{\lambda}\right)^{\mu_{\phi(1)}}\left(1-\left(f^{U}_{\phi(2)}\right)^{\lambda}\right)^{\mu_{\phi(2)}}\right)^{\frac{1}{\lambda}}\right]\right\},$$

thus, Equation (11) holds for $m=2$.

(b) If Equation (11) holds for $m=k$, then:

$$G-INCOG_{\mu,\lambda}\{\widetilde{n}_1,\widetilde{n}_2,\ldots,\widetilde{n}_k\}=\left\{\left[1-\left(1-\prod_{j=1}^{k}\left(1-\left(1-t^{L}_{\phi(j)}\right)^{\lambda}\right)^{\mu_{\phi(j)}}\right)^{\frac{1}{\lambda}},1-\left(1-\prod_{j=1}^{k}\left(1-\left(1-t^{U}_{\phi(j)}\right)^{\lambda}\right)^{\mu_{\phi(j)}}\right)^{\frac{1}{\lambda}}\right],\right.$$
$$\left[\left(1-\prod_{j=1}^{k}\left(1-\left(i^{L}_{\phi(j)}\right)^{\lambda}\right)^{\mu_{\phi(j)}}\right)^{\frac{1}{\lambda}},\left(1-\prod_{j=1}^{k}\left(1-\left(i^{U}_{\phi(j)}\right)^{\lambda}\right)^{\mu_{\phi(j)}}\right)^{\frac{1}{\lambda}}\right],$$
$$\left.\left[\left(1-\prod_{j=1}^{k}\left(1-\left(f^{L}_{\phi(j)}\right)^{\lambda}\right)^{\mu_{\phi(j)}}\right)^{\frac{1}{\lambda}},\left(1-\prod_{j=1}^{k}\left(1-\left(f^{U}_{\phi(j)}\right)^{\lambda}\right)^{\mu_{\phi(j)}}\right)^{\frac{1}{\lambda}}\right]\right\}.$$

For $m=k+1$,

$$G-INCOG_{\mu,\lambda}\{\widetilde{n}_1,\widetilde{n}_2,\ldots,\widetilde{n}_k,\widetilde{n}_{k+1}\}=\left(\oplus_{j=1}^{k}\left(\mu\left(F_{\phi(j)}\right)-\mu\left(F_{\phi(j-1)}\right)\right)\widetilde{n}_{\phi(j)}{}^{\lambda}\right)^{\frac{1}{\lambda}}\oplus\left(\mu_{\phi(k+1)}\widetilde{n}_{\phi(k+1)}{}^{\lambda}\right)^{\frac{1}{\lambda}}=$$
$$\left\{\left[1-\left(1-\prod_{j=1}^{k}\left(1-\left(1-t^{L}_{\phi(j)}\right)^{\lambda}\right)^{\mu_{\phi(j)}}\right)^{\frac{1}{\lambda}},1-\left(1-\prod_{j=1}^{k}\left(1-\left(1-t^{U}_{\phi(j)}\right)^{\lambda}\right)^{\mu_{\phi(j)}}\right)^{\frac{1}{\lambda}}\right],\right.$$
$$\left[\left(1-\prod_{j=1}^{k}\left(1-\left(i^{L}_{\phi(j)}\right)^{\lambda}\right)^{\mu_{\phi(j)}}\right)^{\frac{1}{\lambda}},\left(1-\prod_{j=1}^{k}\left(1-\left(i^{U}_{\phi(j)}\right)^{\lambda}\right)^{\mu_{\phi(j)}}\right)^{\frac{1}{\lambda}}\right],$$
$$\left.\left[\left(1-\prod_{j=1}^{k}\left(1-\left(f^{L}_{\phi(j)}\right)^{\lambda}\right)^{\mu_{\phi(j)}}\right)^{\frac{1}{\lambda}},\left(1-\prod_{j=1}^{k}\left(1-\left(f^{U}_{\phi(j)}\right)^{\lambda}\right)^{\mu_{\phi(j)}}\right)^{\frac{1}{\lambda}}\right]\right\}\otimes$$
$$\left\{\left[1-\left(1-\left(1-\left(1-t^{L}_{\phi(k+1)}\right)^{\lambda}\right)^{\mu_{\phi(k+1)}}\right)^{\frac{1}{\lambda}},1-\left(1-\left(1-\left(1-t^{U}_{\phi(k+1)}\right)^{\lambda}\right)^{\mu_{\phi(k+1)}}\right)^{\frac{1}{\lambda}}\right],\right.$$
$$\left[\left(1-\left(1-\left(i^{L}_{\phi(k+1)}\right)^{\lambda}\right)^{\mu_{\phi(k+1)}}\right)^{\frac{1}{\lambda}},\left(1-\left(1-\left(i^{U}_{\phi(k+1)}\right)^{\lambda}\right)^{\mu_{\phi(k+1)}}\right)^{\frac{1}{\lambda}}\right]$$
$$\left.\left[\left(1-\left(1-\left(f^{L}_{\phi(k+1)}\right)^{\lambda}\right)^{\mu_{\phi(k+1)}}\right)^{\frac{1}{\lambda}},\left(1-\left(1-\left(f^{U}_{\phi(k+1)}\right)^{\lambda}\right)^{\mu_{\phi(k+1)}}\right)^{\frac{1}{\lambda}}\right]\right\}$$
$$=\left\{\left[1-\left(1-\prod_{j=1}^{k+1}\left(1-\left(1-t^{L}_{\phi(j)}\right)^{\lambda}\right)^{\mu_{\phi(j)}}\right)^{\frac{1}{\lambda}},1-\left(1-\prod_{j=1}^{k+1}\left(1-\left(1-t^{U}_{\phi(j)}\right)^{\lambda}\right)^{\mu_{\phi(j)}}\right)^{\frac{1}{\lambda}}\right],\right.$$
$$\left[\left(1-\prod_{j=1}^{k+1}\left(1-\left(i^{L}_{\phi(j)}\right)^{\lambda}\right)^{\mu_{\phi(j)}}\right)^{\frac{1}{\lambda}},\left(1-\prod_{j=1}^{k+1}\left(1-\left(i^{U}_{\phi(j)}\right)^{\lambda}\right)^{\mu_{\phi(j)}}\right)^{\frac{1}{\lambda}}\right],$$
$$\left.\left[\left(1-\prod_{j=1}^{k+1}\left(1-\left(f^{L}_{\phi(j)}\right)^{\lambda}\right)^{\mu_{\phi(j)}}\right)^{\frac{1}{\lambda}},\left(1-\prod_{j=1}^{k+1}\left(1-\left(f^{U}_{\phi(j)}\right)^{\lambda}\right)^{\mu_{\phi(j)}}\right)^{\frac{1}{\lambda}}\right]\right\}.$$

That is, for $m=k+1$, the Equation (11) still holds, by the proof Equation (11), it is not difficult to get Equation (10).

This completes the proof of Theorem 2. □

Theorem 4. The G-INCOA and G-INCOG operators have the following desirable properties, taking the G-INCOA operator as:

1. (Idempotency) Let $\widetilde{n}_j = \widetilde{n}$ *for all* $j = 1, 2, 3, \ldots, m$, and $\widetilde{n} = \{[\widetilde{t}^L, \widetilde{t}^U], [\widetilde{i}^L, \widetilde{i}^U], [\widetilde{f}^L, \widetilde{f}^U]\}$, then:

$$G - INCOA_{\mu,\lambda}\{\widetilde{n}_1, \widetilde{n}_2, \ldots, \widetilde{n}_m\} = \{[\widetilde{t}^L, \widetilde{t}^U], [\widetilde{i}^L, \widetilde{i}^U], [\widetilde{f}^L, \widetilde{f}^U]\}.$$

2. (Boundedness) Let $\widetilde{n}^- = \{[\widetilde{t}^{L^-}, \widetilde{t}^{U^-}], [\widetilde{i}^{L^+}, \widetilde{i}^{U^+}], [\widetilde{f}^{L^+}, \widetilde{f}^{U^+}]\}, \widetilde{n}^+ = \{[\widetilde{t}^{L^+}, \widetilde{t}^{U^+}], [\widetilde{i}^{L^-}, \widetilde{i}^{U^-}], [\widetilde{f}^{L^-}, \widetilde{f}^{U^-}]\}$,

$$\widetilde{n}^- \leq G - INCOA_{\mu,\lambda}\{\widetilde{n}_1, \widetilde{n}_2, \ldots, \widetilde{n}_m\} \leq \widetilde{n}^+.$$

3. (Commutativity) If $\{\widetilde{n}'_1, \widetilde{n}'_2, \ldots, \widetilde{n}'_m\}$ is a permutation of $\{\widetilde{n}_1, \widetilde{n}_2, \ldots, \widetilde{n}_m\}$, then,

$$G - INCOA_{\mu,\lambda}\{\widetilde{n}_1, \widetilde{n}_2, \ldots, \widetilde{n}_m\} = G - INCOA_{\mu,\lambda}\{\widetilde{n}'_1, \widetilde{n}'_2, \ldots, \widetilde{n}'_m\}.$$

4. (Monotonity) If $\widetilde{n}_j \leq \widetilde{n}'_j$ for $\forall j \in \{1, 2, \ldots, n\}$, then,

$$G - INCOA_{\mu,\lambda}\{\widetilde{n}_1, \widetilde{n}_2, \ldots, \widetilde{n}_m\} \leq G - INCOA_{\mu,\lambda}\{\widetilde{n}'_1, \widetilde{n}'_2, \ldots, \widetilde{n}'_m\}.$$

Proof. Suppose $(1, 2, 3, \ldots, m)$ is a permutation such that $\widetilde{n}_1 \leq \widetilde{n}_2 \leq \widetilde{n}_3 \ldots, \leq \widetilde{n}_m$.

1. For $\widetilde{n} = \{[\widetilde{t}^L, \widetilde{t}^U], [\widetilde{i}^L, \widetilde{i}^U], [\widetilde{f}^L, \widetilde{f}^U]\}$, according to Theorem 1, it follows that:

$$G - INCOA_{\mu,\lambda}\{\widetilde{n}_1, \widetilde{n}_2, \ldots, \widetilde{n}_m\} = \left\{\left[\left(1-\left(1-\left(t^L_{\phi(j)}\right)^{\lambda}\right)^{\sum_{j=1}^{m}(\mu(F_j)-\mu(F_{j-1}))}\right)^{\frac{1}{\lambda}}, \left(1-\left(1-\left(t^U_{\phi(j)}\right)^{\lambda}\right)^{\sum_{j=1}^{m}(\mu(F_j)-\mu(F_{j-1}))}\right)^{\frac{1}{\lambda}}\right],\right.$$
$$\left[1-\left(1-\left(1-\left(1-i^L_{\phi(j)}\right)^{\lambda}\right)^{\sum_{j=1}^{m}(\mu(F_j)-\mu(F_{j-1}))}\right)^{\frac{1}{\lambda}}, 1-\left(1-\left(1-\left(1-i^U_{\phi(j)}\right)^{\lambda}\right)^{\sum_{j=1}^{m}(\mu(F_j)-\mu(F_{j-1}))}\right)^{\frac{1}{\lambda}}\right],$$
$$\left.\left[1-\left(1-\left(1-\left(1-f^L_{\phi(j)}\right)^{\lambda}\right)^{\sum_{j=1}^{m}(\mu(F_j)-\mu(F_{j-1}))}\right)^{\frac{1}{\lambda}}, 1-\left(1-\left(1-\left(1-f^L_{\phi(j)}\right)^{\lambda}\right)^{\sum_{j=1}^{m}(\mu(F_j)-\mu(F_{j-1}))}\right)^{\frac{1}{\lambda}}\right]\right\}.$$

Since $\sum_{j=1}^{m}(\mu(F_j) - \mu(F_{j-1})) = 1$, thus, $G - INCOA_{\mu,\lambda}\{\widetilde{n}_1, \widetilde{n}_2, \ldots, \widetilde{n}_m\} = \{[\widetilde{t}^L, \widetilde{t}^U], [\widetilde{i}^L, \widetilde{i}^U], [\widetilde{f}^L, \widetilde{f}^U]\}$.

2 For any $\widetilde{t}_j = [\widetilde{t}^L_j, \widetilde{t}^U_j], \widetilde{i}_j = [\widetilde{i}^L_j, \widetilde{i}^U_j]$ and $\widetilde{f}_j = [\widetilde{f}^U_j, \widetilde{f}^U_j]$, $j = 1, 2, \ldots, m$, we have,

$$\widetilde{t}^{L^-} \leq \widetilde{t}^L_j \leq \widetilde{t}^{L^+};\ \widetilde{t}^{U^-} \leq \widetilde{t}^U_j \leq \widetilde{t}^{U^+}; \widetilde{i}^{L^-} \leq \widetilde{i}^L_j \leq \widetilde{i}^{L^+};$$

$$\widetilde{i}^{U^-} \leq \widetilde{i}^U_j \leq \widetilde{i}^{U^+}; \widetilde{f}^{L^-} \leq \widetilde{f}^L_j \leq \widetilde{f}^{L^+};\ \widetilde{f}^{U^-} \leq \widetilde{f}^U_j \leq \widetilde{f}^{U^+}.$$

Since $y = x^a (0 < a < 1)$ is a monotone increasing function when $x > 0$ and values in the *G-INCOA* operator are all valued in [0, 1], therefore,

$$\left(1-\left(1-\left(\widetilde{t}^{L^-}\right)^{\lambda}\right)^{\sum_{j=1}^{m}(\mu(F_j)-\mu(F_{j-1}))}\right)^{\frac{1}{\lambda}} + \left(1-\left(1-\left(\widetilde{t}^{U^-}\right)^{\lambda}\right)^{\sum_{j=1}^{m}(\mu(F_j)-\mu(F_{j-1}))}\right)^{\frac{1}{\lambda}}$$
$$\leq \left(1-\prod_{j=1}^{m}\left(1-\left(t^L_{\phi(j)}\right)^{\lambda}\right)^{(\mu(F_j)-\mu(F_{j-1}))}\right)^{\frac{1}{\lambda}} + \left(1-\prod_{j=1}^{m}\left(1-\left(t^U_{\phi(j)}\right)^{\lambda}\right)^{(\mu(F_j)-\mu(F_{j-1}))}\right)^{\frac{1}{\lambda}}$$
$$\leq \left(1-\left(1-\left(\widetilde{t}^{L^+}\right)^{\lambda}\right)^{\sum_{j=1}^{m}(\mu(F_j)-\mu(F_{j-1}))}\right)^{\frac{1}{\lambda}} + \left(1-\left(1-\left(\widetilde{t}^{U^+}\right)^{\lambda}\right)^{\sum_{j=1}^{m}(\mu(F_j)-\mu(F_{j-1}))}\right)^{\frac{1}{\lambda}}.$$

Since $\sum_{j=1}^{m}(\mu(F_j) - \mu(F_{j-1})) = 1$, the above equation is equivalent to:

$$\tilde{t}^{L^-}+\tilde{t}^{U^-}\le\left(1-\prod_{j=1}^{m}\left(1-\left(t_{\phi(j)}^{L}\right)^{\lambda}\right)^{(\mu(F_j)-\mu(F_{j-1}))}\right)^{\frac{1}{\lambda}}+\left(1-\prod_{j=1}^{m}\left(1-\left(t_{\phi(j)}^{U}\right)^{\lambda}\right)^{(\mu(F_j)-\mu(F_{j-1}))}\right)^{\frac{1}{\lambda}}.$$
$$\le\tilde{t}^{L^+}+\tilde{t}^{U^+}$$

Analogously, we have:

$$\tilde{i}^{L^-}+\tilde{i}^{U^-}\ge1-\left(1-\prod_{j=1}^{m}\left(1-\left(1-i_{\phi(j)}^{L}\right)^{\lambda}\right)^{\mu_{\phi(j)}}\right)^{\frac{1}{\lambda}}+1-\left(1-\prod_{j=1}^{m}\left(1-\left(1-i_{\phi(j)}^{U}\right)^{\lambda}\right)^{\mu_{\phi(j)}}\right)^{\frac{1}{\lambda}}\ge\tilde{i}^{L^+}+\tilde{i}^{U^+};$$

and

$$\tilde{f}^{L^-}+\tilde{f}^{U^-}\ge1-\left(1-\prod_{j=1}^{m}\left(1-\left(1-f_{\phi(j)}^{L}\right)^{\lambda}\right)^{\mu_{\phi(j)}}\right)^{\frac{1}{\lambda}}+1-\left(1-\prod_{j=1}^{m}\left(1-\left(1-f_{\phi(j)}^{L}\right)^{\lambda}\right)^{\mu_{\phi(j)}}\right)^{\frac{1}{\lambda}}.$$
$$\ge\tilde{f}^{L^+}+\tilde{f}^{U^+};$$

Since $s(\tilde{n}^-)\le s(\tilde{n})\le s(\tilde{n}^+)$, namely, $\tilde{n}^-\le G-INCOA_{\mu,\lambda}\{\tilde{n}_1,\tilde{n}_2,\ldots,\tilde{n}_m\}\le\tilde{n}^+$.

3 Suppose $(\phi(1),\phi(2),\ldots,\phi(m))$ is a permutation of both $\{\tilde{n}'_1,\tilde{n}'_2,\ldots,\tilde{n}'_m\}$ and $\{\tilde{n}_1,\tilde{n}_2,\ldots,\tilde{n}_m\}$, such that $\tilde{n}_{\phi(1)}\le\tilde{n}_{\phi(2)},\ldots,\le\tilde{n}_{\phi(m)}$, $F_{\phi(i)}=\{x_{\phi(1)},x_{\phi(2)},\ldots,x_{\phi(i)}\}$, then,

$$G-INCOA_{\mu,\lambda}\{\tilde{n}_1,\tilde{n}_2,\ldots,\tilde{n}_m\}=G-INCOA_{\mu,\lambda}\{\tilde{n}'_1,\tilde{n}'_2,\ldots,\tilde{n}'_m\}=\oplus_{j=1}^{m}\left(\left(\mu\left(F_{\phi(j)}\right)-\mu\left(F_{\phi(j-1)}\right)\right)\tilde{n}_{\phi(j)}\right).$$

4 In general, it can be derived from the second theorem.

This completes the proof of Theorem 4. □

3.2. *Families of G-INCOA and G-INCOG Operators*

In this section, we consider special cases of the G-INCOA and G-INCOG operators and distinguish them in two main classes, the first class focuses on the parameter λ, and the second class on the fuzzy measure $\mu(x_j)$.

3.2.1. Analyzing the Parameter λ

Like other generalized operators, both the G-INCOA and G-INCOG can reduce to some general circumstances when the parameter λ takes different values, which are described as:

(1) When $\lambda=1$, the G-INCOA operator reduces to the interval neutrosophic Choquet ordered averaging (INCOA) operator,

$$INCOA_{\mu}\{\tilde{n}_1,\tilde{n}_2,\ldots,\tilde{n}_m\}=\oplus_{j=1}^{m}\left(\left(\mu\left(F_{\phi(j)}\right)-\mu\left(F_{\phi(j-1)}\right)\right)\tilde{n}_{\phi(j)}\right)$$
$$=\left\{\left[1-\prod_{j=1}^{m}\left(1-t_{\phi(j)}^{L}\right)^{\mu_{\phi(j)}},1-\prod_{j=1}^{m}\left(1-t_{\phi(j)}^{U}\right)^{\mu_{\phi(j)}}\right],\left[\prod_{j=1}^{m}\left(i_{\phi(j)}^{L}\right)^{\mu_{\phi(j)}},\prod_{j=1}^{m}\left(i_{\phi(j)}^{U}\right)^{\mu_{\phi(j)}}\right],\left[\prod_{j=1}^{m}\left(f_{\phi(j)}^{L}\right)^{\mu_{\phi(j)}},\prod_{j=1}^{m}\left(f_{\phi(j)}^{U}\right)^{\mu_{\phi(j)}}\right].$$

Similarly, the G-INCOG operator reduces to the interval neutrosophic Choquet ordered geometric (INCOG) operator when $\lambda=1$.

(2) If $\lambda\to0$, the G-INCOA operator reduces to the INCOG operator,

$$INCOG_{\mu}\{\tilde{n}_1,\tilde{n}_2,\ldots,\tilde{n}_m\}=\otimes_{j=1}^{m}\left(\tilde{n}_{\phi(j)}{}^{(\mu(F_{\phi(j)})-\mu(F_{\phi(j-1)}))}\right)=\left\{\left[\prod_{j=1}^{m}\left(t_{\phi(j)}^{L}\right)^{\mu_{\phi(j)}},\prod_{j=1}^{m}\left(t_{\phi(j)}^{U}\right)^{\mu_{\phi(j)}}\right],\right.$$
$$\left.\left[1-\prod_{j=1}^{m}\left(1-i_{\phi(j)}^{L}\right)^{\mu_{\phi(j)}},1-\prod_{j=1}^{m}\left(1-i_{\phi(j)}^{U}\right)^{\mu_{\phi(j)}}\right],\left[1-\prod_{j=1}^{m}\left(1-f_{\phi(j)}^{L}\right)^{\mu_{\phi(j)}},1-\prod_{j=1}^{m}\left(1-f_{\phi(j)}^{U}\right)^{\mu_{\phi(j)}}\right]\right\}.$$

Similarly, the G-INCOG operator reduces to the INCOA operator.

(3) When $\lambda = 2$, the G-INCOA operator can reduce to the interval neutrosophic Choquet ordered quadratic averaging (INCOQA) operator,

$$INCOQA_{\mu}\{\tilde{n}_1, \tilde{n}_2, \ldots, \tilde{n}_m\} = \left(\mu_{\phi(1)}\tilde{n}^2_{\phi(1)} \oplus \mu_{\phi(2)}\tilde{n}^2_{\phi(2)} \oplus \ldots \oplus \mu_{\phi(m)}\tilde{n}^2_{\phi(m)}\right)^{1/2}.$$

Similarly, then the G-INCOG operator can reduce to the interval neutrosophic Choquet ordered quadratic geometric (INCOQG) operator.

(4) If $\lambda = 3$, then the G-INCOA operator can reduce to the interval neutrosophic Choquet ordered cubic averaging (INCOCA) operator,

$$INCOCA_{\mu}\{\tilde{n}_1, \tilde{n}_2, \ldots, \tilde{n}_m\} = \left(\mu_{\phi(1)}\tilde{n}^3_{\phi(1)} \oplus \mu_{\phi(2)}\tilde{n}^3_{\phi(2)} \oplus \ldots \oplus \mu_{\phi(m)}\tilde{n}^3_{\phi(m)}\right)^{1/3}.$$

Similarly, then the G-INCOG operator can reduce to the interval neutrosophic Choquet ordered cubic geometric (INCOCG) operator.

3.2.2. Analyzing the Fuzzy Measure $\mu(x_j)$

When considering different circumstances of the fuzzy measure $\mu(x_j)$, some special cases of the G-INCOA and G-INCOG operators are given as:

(1) When $\mu(F) \equiv 1$, then $G - INCOA_{\mu,\lambda}\{\tilde{n}_1, \tilde{n}_2, \ldots, \tilde{n}_m\} = max\{\tilde{n}_1, \tilde{n}_2, \ldots, \tilde{n}_m\}$;
(2) When $\mu(F) \equiv 0$, then $G - INCOA_{\mu,\lambda}\{\tilde{n}_1, \tilde{n}_2, \ldots, \tilde{n}_m\} = min\{\tilde{n}_1, \tilde{n}_2, \ldots, \tilde{n}_m\}$;
(3) The G-INCOA operator reduces to the generalized interval neutrosophic weighted averaging (G-INWA) operator, if the independent condition $\mu\left(x_{\phi(j)}\right) = \mu\left(F_{\phi(j)}\right) - \mu\left(F_{\phi(j-1)}\right)$ holds.

$$\begin{gathered} G - INWA\{\tilde{n}_1, \tilde{n}_2, \ldots, \tilde{n}_m\} = \left(\oplus_{j=1}^{m}\mu(x_j) \oplus \tilde{n}_{\phi(j)}{}^{\lambda}\right)^{\frac{1}{\lambda}} = \\ \left\{\left[\left(1 - \prod_{j=1}^{m}\left(1 - \left(t^L_{\phi(j)}\right)^{\lambda}\right)^{\mu(x_j)}\right)^{\frac{1}{\lambda}}, \left(1 - \prod_{j=1}^{m}\left(1 - \left(t^U_{\phi(j)}\right)^{\lambda}\right)^{\mu(x_j)}\right)^{\frac{1}{\lambda}}\right],\right. \\ \left[1 - \left(1 - \prod_{j=1}^{m}\left(1 - \left(1 - i^L_{\phi(j)}\right)^{\lambda}\right)^{\mu(x_j)}\right)^{\frac{1}{\lambda}}, 1 - \left(1 - \prod_{j=1}^{m}\left(1 - \left(1 - i^U_{\phi(j)}\right)^{\lambda}\right)^{\mu(x_j)}\right)^{\frac{1}{\lambda}}\right], \\ \left.\left[1 - \left(1 - \prod_{j=1}^{m}\left(1 - \left(1 - f^L_{\phi(j)}\right)^{\lambda}\right)^{\mu(x_j)}\right)^{\frac{1}{\lambda}}, 1 - \left(1 - \prod_{j=1}^{m}\left(1 - \left(1 - f^U_{\phi(j)}\right)^{\lambda}\right)^{\mu(x_j)}\right)^{\frac{1}{\lambda}}\right]\right\}. \end{gathered}$$

(4) When $\mu(x_j) = 1/m$, for $j = 1, 2, 3, \ldots, m$, both the G-INCOA and G-INWA operators reduce to the generalized interval neutrosophic averaging (G-INA) operator, which is defined as:

$$\begin{gathered} G - INWA\{\tilde{n}_1, \tilde{n}_2, \ldots, \tilde{n}_m\} = \left(\oplus_{j=1}^{m}\tfrac{1}{m} \oplus \tilde{n}_{\phi(j)}{}^{\lambda}\right)^{\frac{1}{\lambda}} = \\ \left\{\left[\left(1 - \prod_{j=1}^{m}\left(1 - \left(t^L_{\phi(j)}\right)^{\lambda}\right)^{\frac{1}{m}}\right)^{\frac{1}{\lambda}}, \left(1 - \prod_{j=1}^{m}\left(1 - \left(t^U_{\phi(j)}\right)^{\lambda}\right)^{\frac{1}{m}}\right)^{\frac{1}{\lambda}}\right],\right. \\ \left[1 - \left(1 - \prod_{j=1}^{m}\left(1 - \left(1 - i^L_{\phi(j)}\right)^{\lambda}\right)^{\frac{1}{m}}\right)^{\frac{1}{\lambda}}, 1 - \left(1 - \prod_{j=1}^{m}\left(1 - \left(1 - i^U_{\phi(j)}\right)^{\lambda}\right)^{\frac{1}{m}}\right)^{\frac{1}{\lambda}}\right], \\ \left.\left[1 - \left(1 - \prod_{j=1}^{m}\left(1 - \left(1 - f^L_{\phi(j)}\right)^{\lambda}\right)^{\frac{1}{m}}\right)^{\frac{1}{\lambda}}, 1 - \left(1 - \prod_{j=1}^{m}\left(1 - \left(1 - f^U_{\phi(j)}\right)^{\lambda}\right)^{\frac{1}{m}}\right)^{\frac{1}{\lambda}}\right]\right\}. \end{gathered}$$

(5) When $\mu(F) = \sum_{j=1}^{|F|} \omega_j$ for all $F \subseteq X$, where $|F|$ is the number of elements in F, then $\omega_j = \mu\left(F_{\phi(j)}\right) - \mu\left(F_{\phi(j-1)}\right)$, $j = 1,2,\ldots,m$, where $\omega = (\omega_1, \omega_2, \ldots, \omega_m)^T$ such that $\omega_j \geq 0$ and $\sum_{j=1}^{m} \omega_j = 1$. In such a situation, the G-INCOA operator reduces to the generalized interval neutrosophic ordered weighted averaging (G-INOWA) operator as:

$$G-INOWA\{\tilde{n}_1,\tilde{n}_2,\ldots,\tilde{n}_m\} = \left(\oplus_{j=1}^{m} \omega_j \oplus \tilde{n}_{\phi(j)}{}^{\lambda}\right)^{\frac{1}{\lambda}}$$

$$= \left\{\left[\left(1-\prod_{j=1}^{m}\left(1-\left(t^L_{\phi(j)}\right)^{\lambda}\right)^{\omega_j}\right)^{\frac{1}{\lambda}}, \left(1-\prod_{j=1}^{m}\left(1-\left(t^U_{\phi(j)}\right)^{\lambda}\right)^{\omega_j}\right)^{\frac{1}{\lambda}}\right],\right.$$

$$\left[1-\left(1-\prod_{j=1}^{m}\left(1-\left(1-i^L_{\phi(j)}\right)^{\lambda}\right)^{\omega_j}\right)^{\frac{1}{\lambda}}, 1-\left(1-\prod_{j=1}^{m}\left(1-\left(1-i^U_{\phi(j)}\right)^{\lambda}\right)^{\omega_j}\right)^{\frac{1}{\lambda}}\right],$$

$$\left.\left[1-\left(1-\prod_{j=1}^{m}\left(1-\left(1-f^L_{\phi(j)}\right)^{\lambda}\right)^{\omega_j}\right)^{\frac{1}{\lambda}}, 1-\left(1-\prod_{j=1}^{m}\left(1-\left(1-f^U_{\phi(j)}\right)^{\lambda}\right)^{\omega_j}\right)^{\frac{1}{\lambda}}\right]\right\}.$$

Particularly, when $\mu(F) = |F|/m$, for all $F \subseteq X$, then the G-INCOA operator and G-INOWA operator can reduce to the G-INA operator. Similarly, the G-INCOG can reduce to G-INOWG operator, the G-ING operator, the G-INOWG operator and others.

4. Application in MADM under Interval Neutrosophic Environment

This section puts forward new approaches based on the G-INCOA and G-INCOG operators for MADM problems with interval neutrosophic information, where the characteristics of the alternatives are represented by INSs and the interaction relationship among attributes can be considered. Thus, the remaining issue is to use these aggregation operators in practical MADM problems to verify the correctness and practicality of them.

4.1. Approaches Based on the G-INCOA and G-INCOG Operators for MADM

Let $X = \{X_1, X_2, \ldots, X_m\}$ be a finite set of m inter-related attributes and $C = \{C_1, C_2, \ldots, C_n\}$ be a set of n choices. Suppose that with respect to the attributes, the alternatives $C = \{C_1, C_2, \ldots, C_n\}$ denoted by an interval neutrosophic matrix $N = \left(\tilde{n}_{ij} = \{\tilde{t}_{ij}, \tilde{i}_{ij}, \tilde{f}_{ij}\}\right)_{n\times m}$, in detail, $\tilde{t}_{ij}, \tilde{i}_{ij}, \tilde{f}_{ij}$ indicate the truth, indeterminacy and falsity membership function of C_i satisfying x_j given by decision-makers, respectively. Next, to get the best choice, the G-INCOA and G-INCOG operators are utilized to establish MADM methods with interval neutrosophic information, which involves the following steps:

Step 1. Reorder the decision matrix

With respect to attributes $X = \{X_1, X_2, \ldots, X_m\}$, reorder m INNs $\tilde{n}_{ij}$ of C_i $(i = 1, 2, \ldots, m)$ from smallest to largest, according to their score function values $s(\tilde{n}_{ij})$ calculated by Equation (5), the reorder sequence for $i = 1, 2, \ldots, m$ is $(\phi(1), \phi(2), \ldots, \phi(m))$;

Step 2. Confirm fuzzy measures of m attributes

Use g_λ fuzzy measure defined in Equation (6) to determine fuzzy measures μ of X, in which the interaction relationship among attributes is considered;

Step 3. Aggregate decision information by the G-INCOA or G-INCOG operators

Aggregate m INNs $\tilde{n}_{i\phi(j)}$ of C_i based on the G-INCOA and G-INCOG operator defined in Equation (8) or (9), with respect to attributes $X = \{X_1, X_2, \ldots, X_m\}$, as proved by Theorem 2, the aggregated values obtained by the G-INCOA and G-INCOG operators are also INNs;

Step 4. Rank all alternatives

Rank all alternatives to select the most desirable one by their score function values between $\tilde{n}_i$, described in Equation (5).

4.2. Numerical Example

An illustrative example concerning selecting is utilized to verify feasibility of the proposed MADM approaches. Suppose that a fund manager in a wealth management firm is assessing four potential investment opportunities, there is a panel with four possible alternatives denoted by C_1, C_2, C_3, C_4. During MADM process, some attributes should be taken into account: (1) X_1 is risk; (2) X_2 is growth; (3) X_3 is socio-political issues and environmental impacts. Experts are required to evaluate the four possible enterprises $C_i(i = 1,2,3,4)$ under these attributes, and interval neutrosophic decision matrix $N = (\tilde{n}_{ij})_{4\times 3}$ is constructed as:

$$N = \begin{pmatrix} ([0.4,0.5],[0.1,0.2],[0.2,0.4]) & ([0.3,0.5],[0.2,0.3],[0.3,0.5]) & ([0.5,0.6],[0.2,0.3],[0.2,0.3]) \\ ([0.3,0.5],[0.2,0.3],[0.2,0.4]) & ([0.2,0.4],[0.2,0.3],[0.3,0.3]) & ([0.3,0.4],[0.3,0.4],[0.1,0.4]) \\ ([0.5,0.8],[0.1,0.2],[0.1,0.2]) & ([0.5,0.6],[0.1,0.3],[0.2,0.4]) & ([0.5,0.7],[0.1,0.2],[0.1,0.2]) \\ ([0.3,0.5],[0.2,0.4],[0.3,0.4]) & ([0.3,0.5],[0.3,0.4],[0.2,0.5]) & ([0.2,0.5],[0.3,0.4],[0.3,0.4]) \end{pmatrix}.$$

Step 1. Get score function values of $\tilde{n}_{ij}$ calculated by Equation (5), shown as Table 1,

Table 1. Score values of $\tilde{n}_{ij}$.

C_i \ X_j	X_1	X_2	X_3
C_1	0.667	0.583	0.683
C_2	0.617	0.65	0.583
C_3	0.817	0.683	0.767
C_4	0.538	0.567	0.55

To facilitate the following calculation and accord to their score function values, the reordered decision matrix N' can be constructed as:

$$N' = \begin{pmatrix} ([0.3,0.5],[0.2,0.3],[0.3,0.5]) & ([0.4,0.5],[0.1,0.2],[0.2,0.4]) & ([0.5,0.6],[0.2,0.3],[0.2,0.3]) \\ ([0.3,0.4],[0.3,0.4],[0.1,0.4]) & ([0.3,0.5],[0.2,0.3],[0.2,0.4]) & ([0.2,0.4],[0.2,0.3],[0.3,0.3]) \\ ([0.5,0.6],[0.1,0.3],[0.2,0.4]) & ([0.5,0.7],[0.1,0.2],[0.1,0.2]) & ([0.5,0.8],[0.1,0.2],[0.1,0.2]) \\ ([0.3,0.5],[0.2,0.4],[0.3,0.4]) & ([0.2,0.5],[0.3,0.4],[0.3,0.4]) & ([0.3,0.5],[0.3,0.4],[0.2,0.5]) \end{pmatrix}.$$

Step 2. First, if the fuzzy measures of all inter-related attributes are given as follows: $\mu(x_1) = 0.25$, $\mu(x_2) = 0.38, \mu(x_3) = 0.46$. According to Equation (6), the value of λ_g is obtained: $\lambda_g = -0.24$. Thus, we have $\mu(x_1,\ x_2) = 0.6072$, $\mu(x_2,\ x_3) = 0.798$, $\mu(x_1,\ x_3) = 0.6824$, $\mu(X) = 1$.

Step 3. Aggregate $\tilde{n}_{ij}(j = 1,2,3; i = 1,2,3,4)$ by utilizing the G-INCOA operator (in which $\lambda = 1$) to derive the comprehensive score values $\tilde{n}_i$ for $a_i(i = 1,2,3,4)$.

$$\tilde{n}_1 = \{[0.431,0.549],[0.172,0.277],[0.158,0.367]\};$$

$$\tilde{n}_2 = \{[0.327,0.427],[0.267,0.338],[0.195,0.354]\};$$

$$\tilde{n}_3 = \{[0.500,0.669],[0.100,0.228],[0.125,0.249]\};$$

$$\tilde{n}_4 = \{[0.260,0.500],[0.348,0.400],[0.264,0.430]\}.$$

Step 4. Ranking the comprehensive score values $\tilde{n}_i$ for $a_i(i = 1,2,3,4)$, we get:

$$s(\tilde{n}_1) = 0.618, s(\tilde{n}_2) = 0.6, s(\tilde{n}_3) = 0.745, s(\tilde{n}_4) = 0.553.$$

Therefore, we have $a_3 > a_1 > a_2 > a_4$ and a_3 is the best choice.

If we utilize the G-INCOG operator for this MADM problem, aggregate $\widetilde{n}_{ij}(j = 1,2,3; i = 1,2,3,4)$ to derive the comprehensive score value $\widetilde{n}_i$ for $a_i(i = 1,2,3,4)$.

$$\widetilde{n}'_1 = \{[0.418, 0.544], [0.182, 0.290], [0.178, 0.379]\};$$
$$\widetilde{n}'_2 = \{[0.322, 0.454], [0.270, 0.370], [0.247, 0.360]\};$$
$$\widetilde{n}'_3 = \{[0.500, 0.689], [0.100, 0.232], [0.133, 0.267]\};$$
$$\widetilde{n}'_4 = \{[0.253, 0.500], [0.364, 0.400], [0.270, 0.434]\}.$$

Then, ranking the score function values of INNs, we get:

$$s(\widetilde{n}'_1) = 0.656, s(\widetilde{n}'_2) = 0.588, s(\widetilde{n}'_3) = 0.743, s(\widetilde{n}'_4) = 0.548.$$

Rank a_i according to the score values $a_3 > a_1 > a_2 > a_4$. Therefore, we can see that a_3 is the best choice. Obviously, the above two kinds of ranking orders are the same, therefore, the above example clearly indicates that the proposed MADM methods are applicable and effective under an interval neutrosophic environment.

4.3. Rank Alternatives for Different Values of λ

In real life, decision makers may determine the value of λ by different MADM problems or their preferences, which makes the decision-making process more flexible. In this section, we use different values of parameter λ of the G-INCOA and G-INCOG operators, such as $\lambda \to 0$ or $\lambda = 1 - 10$, to rank alternatives of the numerical example in Section 4.2.

Combined with the proposed approaches for MADM with interval neutrosophic information, we can obtain their score function values of four alternatives, the ranking results for different values of λ determined by the G-INCOA and G-INCOG operator are shown in Figures 1 and 2, respectively. As shown in Figures 1 and 2, the best choice is always a_3 and the worst alternative is always a_4, which means they have higher accuracy and greater reference value. Besides, the changing trends of decision results with parameter λ calculated by the G-INCOA operator presents an increasing trend, meanwhile, the changing trends of decision results with λ calculated by the G-INCOG operator shows a declining trend, which further validates the duality of the proposed operators.

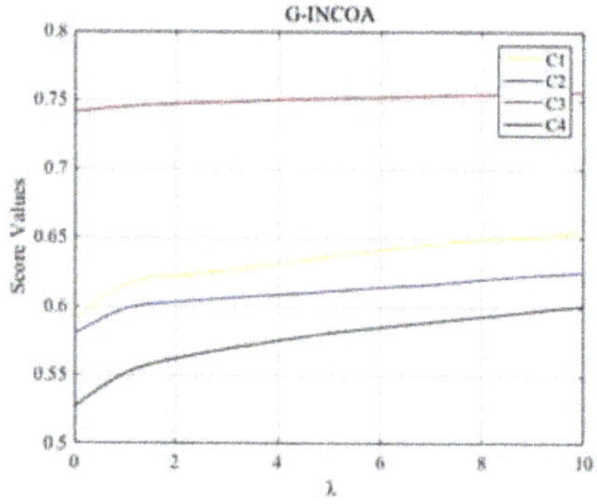

Figure 1. The changing trends of decision results with λ calculated by the G-INCOA operator.

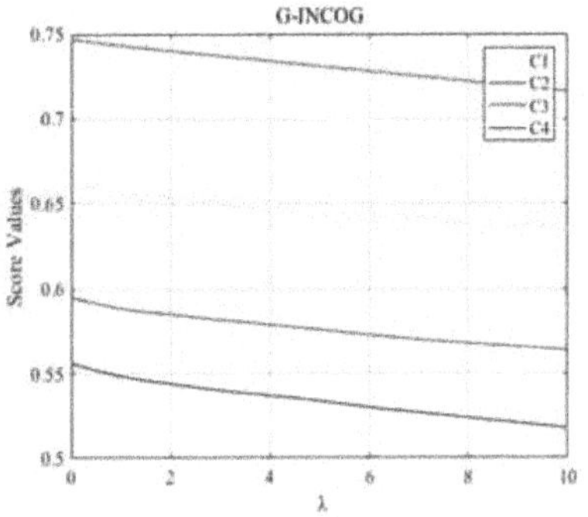

Figure 2. The changing trends of decision results with λ calculated by the G-INCOG operator.

5. Apply the Proposed Operators for INSs to Cluster Analysis

5.1. New Clustering Algorithm for INSs

In this section, we intend to propose a new clustering algorithm for INSs to illustrate the efficiency of the proposed operators. Let $N = (\widetilde{n}_{ij} = \{\widetilde{t}_{ij},\ \widetilde{i}_{ij},\ \widetilde{f}_{ij}\})_{n\times m}$ be a matrix of INNs on $X = \{X_1, X_2, \ldots, X_m\}$, the algorithm can be described as:

Step 1. Using the proposed operator, here, take the G-INCOA operator as an example, to aggregate m INNs of each alternative to an comprehensive INN $\widetilde{n}_i$; Using the similarity measures function defined in Equation (3) to calculate measures between $\widetilde{n}_j$ and $\widetilde{n}_k$ $(j,\ k = 1,\ 2,\ \ldots, m)$, the corresponding results are recorded in a matrix $S_{m\times m} = S_{jk}$;

Step 2. Check whether the measure matrix S satisfies $S^2 \subseteq S$, where $S^2 = S \circ S = \left(S'_{jk}\right)_{m\times m}$, and $\widetilde{n}'_{jk} = max_p\left\{min\left\{S_{jp}, S_{pk}\right\}\right\}$, $(j,\ k = 1,2,\ldots,m)$. If it does not hold, then construct the equivalent matrix: $S^{2^p} : S \to S^2 \to S^4 \to \ldots \to$ until $S^{2^p} = S^{2^{(p+1)}}$;

Step 3. For a given confident level $\alpha \in [0,1]$, construct a α-cutting matrix $S_\alpha = \left(S^\alpha_{jk}\right)_{m\times m}$, where S^α_{jk} is defined as:

$$S^\alpha_{jk} = \begin{cases} 0, \text{ if } S_{jk} < \alpha; \\ 1, \text{ if } S_{jk} \geq \alpha. \end{cases}$$

Step 4. Classify the INSs by the rule: if all elements of the jth line in S_α are the same as the corresponding elements of the kth line, thus, the INSs $\widetilde{n}_j$ and $\widetilde{n}_k$ are supposed as the same type.

5.2. Numerical Example

A numerical example concerning investing is utilized to demonstrate the application of these aggregation operators, as well as the effectiveness of them. Suppose there are five attributes to be considered: (1) X_1 : profitability; (2) X_2 : operating capacity; (3) X_3 : market competition. The fuzzy measures of attributes in X are given as follows: $\mu(x_1) = 0.362$, $\mu(x_2) = 0.2$, $\mu(x_3) = 0.438$. Firstly, according to Equation (7), the value of λ_g is obtained: $\lambda_g = 0.856$. Thus, $\mu(x_1,\ x_2) = 0.626$, $\mu(x_2,\ x_3) = 0.713$, $\mu(x_1,\ x_3) = 0.936$, $\mu(X) = 1$. Experts are required to evaluate 10 firms $C_i(i = 1,\ 2,\ \ldots, 10)$ under the three attributes, and interval neutrosophic decision matrix $N = (\widetilde{n}_{ij})_{10\times 3}$ is constructed as:

$$N = \begin{pmatrix} ([0.3,0.5],[0.5,0.6],[0.4,0.6]) & ([0.4,0.5],[0.4,0.6],[0.4,0.5]) & ([0.7,0.8],[0.2,0.3],[0.3,0.5]) \\ ([0.4,0.6],[0.3,0.5],[0.3,0.7]) & ([0.6,0.8],[0.3,0.4],[0.2,0.4]) & ([0.2,0.3],[0.8,0.9],[0.7,0.8]) \\ ([0.5,0.7],[0.2,0.3],[0.4,0.5]) & ([0.7,0.9],[0.2,0.4],[0.1,0.2]) & ([0.3,0.4],[0.5,0.7],[0.7,0.8]) \\ ([0.3,0.5],[0.4,0.6],[0.5,0.8]) & ([0.8,0.9],[0.1,0.2],[0.2,0.3]) & ([0.7,0.9],[0.2,0.3],[0.2,0.3]) \\ ([0.8,1.0],[0.2,0.3],[0.1,0.3]) & ([0.8,1.0],[0.1,0.2],[0.1,0.3]) & ([0.4,0.6],[0.3,0.4],[0.4,0.6]) \\ ([0.4,0.6],[0.4,0.6],[0.5,0.7]) & ([0.2,0.3],[0.6,0.8],[0.7,0.9]) & ([0.9,1.0],[0.1,0.2],[0.1,0.2]) \\ ([0.5,0.6],[0.4,0.5],[0.5,0.6]) & ([0.7,0.9],[0.2,0.3],[0.2,0.4]) & ([0.6,0.8],[0.3,0.4],[0.2,0.5]) \\ ([0.9,1.0],[0.1,0.2],[0.1,0.2]) & ([0.7,0.8],[0.2,0.3],[0.3,0.4]) & ([0.4,0.5],[0.4,0.6],[0.5,0.7]) \\ ([0.4,0.6],[0.6,0.7],[0.2,0.4]) & ([0.9,1.0],[0.1,0.2],[0.1,0.2]) & ([0.6,0.7],[0.3,0.4],[0.3,0.5]) \\ ([0.8,0.9],[0.2,0.4],[0.2,0.3]) & ([0.6,0.8],[0.3,0.5],[0.3,0.4]) & ([0.5,0.8],[0.3,0.6],[0.4,0.5]) \end{pmatrix}.$$

In the following, we use the proposed clustering algorithm to cluster these alternatives:

Step 1. Aggregated the G-INCOA operator defined in Equation (8) and calculated by the similarity measure function defined in Equation (5), the weighted measures S_{jk} between each pair of alternatives are recorded in a matrix $S_{10\times 10}$.

$$S = \begin{pmatrix} 1 & 0.5984 & 0.458 & 0.4635 & 0.3964 & 0.7100 & 0.5572 & 0.4761 & 0.4143 & 0.3984 \\ 0.5984 & 1 & 0.5 & 0.5136 & 0.4456 & 0.4667 & 0.5409 & 0.5456 & 0.5051 & 0.3851 \\ 0.4580 & 0.5 & 1 & 0.6811 & 0.5596 & 0.4080 & 0.6994 & 0.6875 & 0.6682 & 0.4753 \\ 0.4635 & 0.5136 & 0.6811 & 1 & 0.5421 & 0.4540 & 0.7236 & 0.6744 & 0.731 & 0.4747 \\ 0.3964 & 0.4456 & 0.5596 & 0.5421 & 1 & 0.3762 & 0.5734 & 0.6517 & 0.625 & 0.7511 \\ 0.7100 & 0.4667 & 0.4080 & 0.4540 & 0.3762 & 1 & 0.5431 & 0.4647 & 0.3813 & 0.4019 \\ 0.5572 & 0.5409 & 0.6994 & 0.7236 & 0.5734 & 0.5431 & 1 & 0.7023 & 0.6726 & 0.5211 \\ 0.4761 & 0.5456 & 0.6875 & 0.6744 & 0.6517 & 0.4647 & 0.7023 & 1 & 0.6615 & 0.6063 \\ 0.4143 & 0.5051 & 0.6682 & 0.7310 & 0.6250 & 0.3813 & 0.6726 & 0.6615 & 1 & 0.5372 \\ 0.3984 & 0.3851 & 0.4753 & 0.4747 & 0.7511 & 0.4019 & 0.5211 & 0.6063 & 0.5372 & 1 \end{pmatrix}.$$

Step 2. The equivalent measure matrix can be constructed as follows, as $S^8 = S^4$, therefore, S^4 is an equivalent measure matrix.

$$S^2 = \begin{pmatrix} 1 & 0.5984 & 0.5572 & 0.5572 & 0.5572 & 0.5431 & 0.5572 & 0.5572 & 0.5572 & 0.5211 \\ 0.5984 & 1 & 0.5456 & 0.5456 & 0.5456 & 0.5984 & 0.5572 & 0.5456 & 0.5456 & 0.5456 \\ 0.5572 & 0.5456 & 1 & 0.6994 & 0.6517 & 0.5431 & 0.6994 & 0.6994 & 0.6811 & 0.6063 \\ 0.5572 & 0.5456 & 0.6944 & 1 & 0.6517 & 0.5431 & 0.7236 & 0.7023 & 0.7310 & 0.6063 \\ 0.5572 & 0.5456 & 0.6517 & 0.6517 & 1 & 0.5431 & 0.6517 & 0.6517 & 0.6517 & 0.7511 \\ 0.5431 & 0.5984 & 0.5431 & 0.5431 & 0.5431 & 1 & 0.5572 & 0.5431 & 0.5431 & 0.5211 \\ 0.5572 & 0.5572 & 0.6994 & 0.7236 & 0.6517 & 0.5572 & 1 & 0.7023 & 0.7236 & 0.6063 \\ 0.5572 & 0.5456 & 0.6994 & 0.7023 & 0.6517 & 0.5431 & 0.7023 & 1 & 0.6744 & 0.6517 \\ 0.5572 & 0.5456 & 0.6811 & 0.7310 & 0.6517 & 0.5431 & 0.7236 & 0.6744 & 1 & 0.6520 \\ 0.5211 & 0.5456 & 0.6063 & 0.6063 & 0.7511 & 0.5211 & 0.6063 & 0.6517 & 0.6250 & 1 \end{pmatrix}.$$

$$S^4 = \begin{pmatrix} 1 & 0.5984 & 0.5572 & 0.5572 & 0.5572 & 0.5984 & 0.5572 & 0.5572 & 0.5572 & 0.5572 \\ 0.5984 & 1 & 0.5572 & 0.5572 & 0.5572 & 0.5984 & 0.5572 & 0.5572 & 0.5572 & 0.5572 \\ 0.5572 & 0.5572 & 1 & 0.6994 & 0.6517 & 0.5431 & 0.6994 & 0.6994 & 0.6944 & 0.6517 \\ 0.5572 & 0.5572 & 0.6944 & 1 & 0.6517 & 0.5572 & 0.7236 & 0.7023 & 0.7310 & 0.6517 \\ 0.5572 & 0.5572 & 0.6517 & 0.6517 & 1 & 0.5572 & 0.6517 & 0.6517 & 0.6517 & 0.7511 \\ 0.5984 & 0.5984 & 0.5572 & 0.5572 & 0.5572 & 1 & 0.5572 & 0.5572 & 0.5572 & 0.5572 \\ 0.5572 & 0.5572 & 0.6994 & 0.7236 & 0.6517 & 0.5572 & 1 & 0.7023 & 0.7236 & 0.6517 \\ 0.5572 & 0.5572 & 0.6994 & 0.7023 & 0.6517 & 0.5572 & 0.7023 & 1 & 0.7023 & 0.6517 \\ 0.5572 & 0.5572 & 0.6944 & 0.7310 & 0.6517 & 0.5572 & 0.7236 & 0.7023 & 1 & 0.6520 \\ 0.5572 & 0.5572 & 0.6517 & 0.6517 & 0.7511 & 0.5572 & 0.6517 & 0.6517 & 0.6517 & 1 \end{pmatrix}.$$

$$S^8 = \begin{pmatrix} 1 & 0.5984 & 0.5572 & 0.5572 & 0.5572 & 0.5984 & 0.5572 & 0.5572 & 0.5572 & 0.5572 \\ 0.5984 & 1 & 0.5572 & 0.5572 & 0.5572 & 0.5984 & 0.5572 & 0.5572 & 0.5572 & 0.5572 \\ 0.5572 & 0.5572 & 1 & 0.6994 & 0.6517 & 0.5431 & 0.6994 & 0.6994 & 0.6944 & 0.6517 \\ 0.5572 & 0.5572 & 0.6944 & 1 & 0.6517 & 0.5572 & 0.7236 & 0.7023 & 0.7310 & 0.6517 \\ 0.5572 & 0.5572 & 0.6517 & 0.6517 & 1 & 0.5572 & 0.6517 & 0.6517 & 0.6517 & 0.7511 \\ 0.5984 & 0.5984 & 0.5572 & 0.5572 & 0.5572 & 1 & 0.5572 & 0.5572 & 0.5572 & 0.5572 \\ 0.5572 & 0.5572 & 0.6994 & 0.7236 & 0.6517 & 0.5572 & 1 & 0.7023 & 0.7236 & 0.6517 \\ 0.5572 & 0.5572 & 0.6994 & 0.7023 & 0.6517 & 0.5572 & 0.7023 & 1 & 0.7023 & 0.6517 \\ 0.5572 & 0.5572 & 0.6944 & 0.7310 & 0.6517 & 0.5572 & 0.7236 & 0.7023 & 1 & 0.6520 \\ 0.5572 & 0.5572 & 0.6517 & 0.6517 & 0.7511 & 0.5572 & 0.6517 & 0.6517 & 0.6517 & 1 \end{pmatrix}.$$

Step 3. For a given confident level $\alpha \in [0,1]$, we can construct a α-cutting matrix $S_\alpha = \left(S^\alpha_{jk}\right)_{10\times 10}$ for $S^8 = \left(S_{jk}\right)_{10\times 10}$, different α produces different α-cutting matrix, for example, if $\alpha = 0$, the α-cutting matrix can be constructed as $S_\alpha = \left(S^\alpha_{jk} = 1\right)_{10\times 10}$, since $S^8 = \left(S_{jk} > 0\right)_{10\times 10}$.

Step 4. Based on the α-cutting matrix S_α, we can classify 10 alternatives into different clusters, the possible classification of these choices is shown in Table 2.

Table 2. Different clusters of 10 alternatives with respect to different α.

Class	Confidence Level	Clusters
8	$0.731 < \alpha \leq 1$	$\{\{C_1\},\{C_2\},\{C_3\},\{C_4\},\{C_5\},\{C_6\},\{C_7\},\{C_8\},\{C_9\},\{C_{10}\}\}$
7	$0.7236 < \alpha \leq 0.731$	$\{\{C_1\},\{C_2\},\{C_3\},\{C_4,C_9\},\{C_5\},\{C_6\},\{C_7\},\{C_8\},\{C_{10}\}\}$
6	$0.7023 < \alpha \leq 0.7236$	$\{\{C_1\},\{C_2\},\{C_3\},\{C_4,C_7,C_9\},\{C_5\},\{C_6\},\{C_8\},\{C_{10}\}\}$
5	$0.6994 < \alpha \leq 0.7023$	$\{\{C_1\},\{C_2\},\{C_3\},\{C_4,C_7,C_8,C_9\},\{C_5\},\{C_6\},\{C_{10}\}\}$
4	$0.6517 < \alpha \leq 0.6994$	$\{\{C_1\},\{C_2\},\{C_3,C_4,C_7,C_8,C_9\},\{C_5,C_{10}\},\{C_6\}\}$
3	$0.5984 < \alpha \leq 0.6517$	$\{\{C_1\},\{C_2\},\{C_3,C_4,C_5,C_7,C_8,C_9,C_{10}\},\{C_6\}\}$
2	$0.5572 < \alpha \leq 0.5984$	$\{\{C_1,C_2,C_6\},\{C_3,C_4,C_5,C_7,C_8,C_9,C_{10}\}\}$
1	$0 < \alpha \leq 0.5572$	$\{\{C_1,C_2,C_3,C_4,C_5,C_6,C_7,C_8,C_9,C_{10}\}\}$

With respect to different values of α, different clusters of 10 alternatives are shown in Table 2. When $0 < \alpha \leq 0.5572$, all alternatives belong to the same cluster, then $0.5572 < \alpha \leq 0.5984$, 10 alternatives are divided in to two clusters, namely, $\{C_3, C_4, C_5, C_7, C_8, C_9, C_{10}\}$ and $\{C_1, C_2, C_6\}$, until $0.731 < \alpha \leq 1$, each alternative is an independent cluster.

6. Conclusions

This paper studies new MADM methods and clustering algorithm under an interval neutrosophic environment, in which the attributes are inter-related. Motivated by the idea of the generalized operator, we proposed the G-INCOA and G-INCOG operators based on the related research of the NS and SVNS theories, which can reduce to the existing aggregation operators of INSs and have some desirable properties. By taking different values of the parameters and comparing them with existing methods for MADM problems, under interval neutrosophic environment, results obtained by the proposed operators are consistent and accurate, which illustrates their practicability in application. The new clustering algorithm are established on the G-INCOA and G-INCOG operators, a numerical example concerning investing is utilized as the demonstration of the application of the proposed aggregation operators, as well as the effectiveness of them. In the future, motivated by different MADM methods under linguistic environment [34,35], it is worth investigating the use granular computing techniques to develop new MADM methods with interval neutrosophic linguistic information.

Acknowledgments: This work was supported by the National Natural Science Foundation of China (Grant Numbers 61573240, 61473239).

Author Contributions: All authors have contributed equally to this paper. The individual responsibilities and contribution of all authors can be described as follows: the idea of this whole thesis was put forward by Xiaohong Zhang, he also completed the preparatory work of the paper. Xin Li analyzed the existing work of interval neutrosophic sets and wrote part of the paper. The revision and submission of this paper was completed by Choonkil Park.

Conflicts of Interest: The authors declare no conflicts of interest.

References

1. Ye, J. Multicriteria decision-making method using the correlation coefficient under single-valued neutrosophic environment. *Int. J. Gen. Syst.* **2013**, *42*, 386–394. [CrossRef]
2. Zadeh, L.A. Fuzzy sets. *Inf. Control* **1965**, *8*, 338–353. [CrossRef]
3. Torra, V.; Narukawa, Y. On hesitant fuzzy sets and decision. In Proceedings of the 18th IEEE International Conference on Fuzzy Systems, Jeju Island, Korea, 20–24 August 2009; pp. 1378–1382. [CrossRef]
4. Zhang, X.H. Fuzzy anti-grouped filters and fuzzy normal filters in pseudo-BCI algebras. *J. Intell. Fuzzy Syst.* **2017**, *33*, 1767–1774. [CrossRef]
5. Smarandache, F. *A Unifying Field in Logics: Neutrosophic Logic, Neutrosophy, Neutrosophic Set, Neutrosophic Probability*; American Research Press: Rehoboth, DE, USA, 1999.
6. Wang, H.; Madiraju, P. Interval-neutrosophic Sets. *J. Mech.* **2004**, *1*, 274–277.
7. Wang, H.; Smarandache, F.; Sunderraman, R. Single-valued neutrosophic sets. *Rev. Air Force Acad.* **2013**, *17*, 10–13.
8. Broumi, S.; Smarandache, F.; Dhar, M. Rough neutrosophic sets. *Ital. J. Pure. Appl. Math.* **2014**, *32*. [CrossRef]
9. Peng, J.; Wang, J.; Wu, X. Multi-valued neutrosophic sets and power aggregation operators with their applications in multi-criteria group decision-making problems. *Int. J. Comput. Int. Syst.* **2015**, *8*, 345–363. [CrossRef]
10. Guo, Y.H.; Ümit, B.; Şengür, A.; Smarandache, F. A Retinal Vessel Detection Approach Based on Shearlet Transform and Indeterminacy Filtering on Fundus Images. *Symmetry* **2017**, *9*, 235–244. [CrossRef]
11. Chen, J.Q.; Ye, J.; Du, S.G. Scale Effect and Anisotropy Analyzed for Neutrosophic Numbers of Rock Joint Roughness Coefficient Based on Neutrosophic Statistics. *Symmetry* **2017**, *9*, 14–27. [CrossRef]
12. Zhang, X.H.; Bo, C.X.; Smarandache, F.; Dai, J.H. New inclusion relation of neutrosophic sets with applications and related lattice structrue. *Int. J. Mach. Learn. Cybern.* **2018**, accepted.
13. Akbulut, Y.; Sengur, A.; Guo, Y.H.; Smarandache, F. A Novel Neutrosophic Weighted Extreme Learning Machine for Imbalanced Data Set. *Symmetry* **2017**, *9*, 142. [CrossRef]
14. Xu, Z.S.; Gou, X.J. An overview of interval-valued intuitionistic fuzzy information aggregations and applications. *Granul. Comput.* **2017**, *2*, 13–39. [CrossRef]
15. Wang, C.Q.; Fu, X.G.; Meng, S.; He, Y. Multi-attribute decision making based on the SPIFGIA operators. *Granul. Comput.* **2017**, *2*, 321–331. [CrossRef]
16. Jiang, Y.; Xu, Z.S.; Shu, Y.H. Interval-valued intuitionistic multiplicative aggregation in group decision making. *Granul. Comput.* **2017**, *2*, 387–407. [CrossRef]
17. Joshi, B.P. Moderator intuitionistic fuzzy sets with applications in multi-criteria decision-making. *Granul. Comput.* **2018**, *3*, 61–73. [CrossRef]
18. Chen, J.Q.; Ye, J.; Du, S.G. Vector Similarity Measures between Refined Simplified Neutrosophic Sets and Their Multiple Attribute Decision-Making Method. *Symmetry* **2017**, *9*, 153. [CrossRef]
19. Hu, K.L.; Fan, E.; Ye, J.; Fan, C.; Shen, S.; Gu, Y. Neutrosophic Similarity Score Based Weighted Histogram for Robust Mean-Shift Tracking. *Information* **2017**, *8*, 122. [CrossRef]
20. Zhang, H.Y.; Wang, J.Q.; Chen, X.H. Inverval neutrosophic sets and their application in multicriteria decision making problems. *Sci. World J.* **2014**, *2014*, 645953. [CrossRef]
21. Broumi, S.; Smarandache, F. Correlation Coefficient of Interval Neutrosophic Set. *Appl. Mech. Mater.* **2013**, *436*, 511–517. [CrossRef]
22. Zhang, H.Y.; Ji, P.; Wang, J.Q.; Chen, X.H. An Improved Weighted Correlation Coefficient Based on Integrated Weight for Interval Neutrosophic Sets and its Application in Multi-criteria Decision-making Problems. *Int. J. Comput. Int. Syst.* **2015**, *8*, 1027–1043. [CrossRef]
23. Ye, J. Similarity measures between interval neutrosophic sets and their multicriteria decision-making method. *J. Intell. Fuzzy Syst.* **2014**, *26*, 165–172.
24. Ye, J. Multiple attribute decision-making method based on the possibility degree ranking method and ordered weighted aggregation operators of interval neutrosophic numbers. *J. Intell. Syst.* **2015**, *28*, 1307–1317. [CrossRef]

25. Liu, P.D.; Tang, G.L. Some power generalized aggregation operators based on the interval neutrosophic numbers and their application to decision making. *J. Intell. Fuzzy. Syst.* **2015**, *30*, 2517–2528. [CrossRef]
26. Liu, P.D.; Teng, F. Multiple attribute decision making method based on normal neutrosophic generalized weighted power averaging operator. *Int. J. Mach. Learn. Cybern.* **2015**, 1–13. [CrossRef]
27. Liu, P.D.; Wang, Y.M. Interval neutrosophic prioritized OWA operator and its application to multiple attribute decision making. *J. Syst. Sci. Complex.* **2016**, *29*, 681–697. [CrossRef]
28. Zhang, X.H.; She, Y.H. *Fuzzy Quantifies with Integral Semantics*; Science Press: Beijing, China, 2017; ISBN 978-7-03053-480-4. Available online: http://product.dangdang.com/25113577.html (accessed on 1 July 2017).
29. Ju, Y.B.; Yang, S.H.; Liu, X.Y. Some new dual hesitant fuzzy aggregation operators based on Choquet integral and their applications to multiple attribute decision making. *J. Intell. Fuzzy Syst.* **2014**, *27*, 2857–2868. [CrossRef]
30. Choquet, G. Theory of capacities. *Ann. Inst. Fourier* **1953**, *5*, 131–295. [CrossRef]
31. Sugeno, M. Theory of Fuzzy Integral and its Application. Ph.D. Thesis, Tokyo Institute of Technology, Tokyo, Japan, 22 January 1975.
32. Liao, H.C.; Xu, Z.S.; Zeng, X.J. Novel correlation coefficients between hesitant fuzzy sets and their application in decision making. *Knowl. Based Syst.* **2015**, *82*, 115–127. [CrossRef]
33. Chen, N.; Xu, Z.S.; Xia, M.M. Correlation coefficients of hesitant fuzzy sets and their applications to clustering analysis. *Appl. Math. Model.* **2013**, *37*, 2197–2211. [CrossRef]
34. Liu, P.D.; You, X.L. Probabilistic linguistic TODIM approach for multiple attribute decision making. *Granul. Comput.* **2017**, *2*, 333–342. [CrossRef]
35. Xu, Z.S.; Wang, H. Managing multi-granularity linguistic information in qualitative group decision making: An overview. *Granul. Comput.* **2016**, *1*, 21–35. [CrossRef]

symmetry

MDPI

Article

Neutrosophic Association Rule Mining Algorithm for Big Data Analysis

Mohamed Abdel-Basset [1,*], Mai Mohamed [1], Florentin Smarandache [2,*] and Victor Chang [3]

1 Department of Operations Research, Faculty of Computers and Informatics, Zagazig University, Sharqiyah 44519, Egypt; analyst_mohamed@yahoo.com
2 Math & Science Department, University of New Mexico, Gallup, NM 87301, USA
3 International Business School Suzhou, Xi'an Jiaotong-Liverpool University, Wuzhong, Suzhou 215123, China; ic.victor.chang@gmail.com
* Correspondence: analyst_mohamed@zu.edu.eg (M.A.-B.); smarand@unm.edu (F.S.)

Received: 5 March 2018; Accepted: 9 April 2018; Published: 11 April 2018

Abstract: Big Data is a large-sized and complex dataset, which cannot be managed using traditional data processing tools. Mining process of big data is the ability to extract valuable information from these large datasets. Association rule mining is a type of data mining process, which is indented to determine interesting associations between items and to establish a set of association rules whose support is greater than a specific threshold. The classical association rules can only be extracted from binary data where an item exists in a transaction, but it fails to deal effectively with quantitative attributes, through decreasing the quality of generated association rules due to sharp boundary problems. In order to overcome the drawbacks of classical association rule mining, we propose in this research a new neutrosophic association rule algorithm. The algorithm uses a new approach for generating association rules by dealing with membership, indeterminacy, and non-membership functions of items, conducting to an efficient decision-making system by considering all vague association rules. To prove the validity of the method, we compare the fuzzy mining and the neutrosophic mining. The results show that the proposed approach increases the number of generated association rules.

Keywords: neutrosophic association rule; data mining; neutrosophic sets; big data

---

1. Introduction

The term 'Big Data' originated from the massive amount of data produced every day. Each day, Google receives cca. 1 billion queries, Facebook registers more than 800 million updates, and YouTube counts up to 4 billion views, and the produced data grows with 40% every year. Other sources of data are mobile devices and big companies. The produced data may be structured, semi-structured, or unstructured. Most of the big data types are unstructured; only 20% of data consists in structured data. There are four dimensions of big data:

(1) Volume: big data is measured by petabytes and zettabytes.
(2) Velocity: the accelerating speed of data flow.
(3) Variety: the various sources and types of data requiring analysis and management.
(4) Veracity: noise, abnormality, and biases of generated knowledge.

Consequently, Gartner [1] outlines that big data's large volume requires cost-effective, innovative forms for processing information, to enhance insights and decision-making processes.

Prominent domains among applications of big data are [2,3]:

(1) Business domain.

(2) Technology domain.
(3) Health domain.
(4) Smart cities designing.

These various applications help people to obtain better services, experiences, or be healthier, by detecting illness symptoms much earlier than before [2]. Some significant challenges of managing and analyzing big data are [4,5]:

(1) Analytics Architecture: The optimal architecture for dealing with historic and real-time data at the same time is not obvious yet.
(2) Statistical significance: Fulfill statistical results, which should not be random.
(3) Distributed mining: Various data mining methods are not fiddling to paralyze.
(4) Time evolving data: Data should be improved over time according to the field of interest.
(5) Compression: To deal with big data, the amount of space that is needed to store is highly relevant.
(6) Visualization: The main mission of big data analysis is the visualization of results.
(7) Hidden big data: Large amounts of beneficial data are lost since modern data is unstructured data.

Due to the increasing volume of data at a matchless rate and of various forms, we need to manage and analyze uncertainty of various types of data. Big data analytics is a significant function of big data, which discovers unobserved patterns and relationships among various items and people interest on a specific item from the huge data set. Various methods are applied to obtain valid, unknown, and useful models from large data. Association rule mining stands among big data analytics functionalities. The concept of association rule (AR) mining already returns to H'ajek et al. [6]. Each association rule in database is composed from two different sets of items, which are called antecedent and consequent. A simple example of association rule mining is "if the client buys a fruit, he/she is 80% likely to purchase milk also". The previous association rule can help in making a marketing strategy of a grocery store. Then, we can say that association rule-mining finds all of the frequent items in database with the least complexities. From all of the available rules, in order to determine the rules of interest, a set of constraints must be determined. These constraints are support, confidence, lift, and conviction. Support indicates the number of occurrences of an item in all transactions, while the confidence constraint indicates the truth of the existing rule in transactions. The factor "lift" explains the dependency relationship between the antecedent and consequent. On the other hand, the conviction of a rule indicates the frequency ratio of an occurring antecedent without a consequent occurrence. Association rules mining could be limited to the problem of finding large itemsets, where a large itemset is a collection of items existing in a database transactions equal to or greater than the support threshold [7–20]. In [8], the author provides a survey of the itemset methods for discovering association rules. The association rules are positive and negative rules. The positive association rules take the form $X \rightarrow Y$, $X \subseteq I$, $Y \subseteq I$ and $X \cap Y = \varphi$, where X, Y are antecedent and consequent and I is a set of items in database. Each positive association rule may lead to three negative association rules, $\rightarrow Y$, $X \rightarrow Y$, and $X \rightarrow Y$. Generating association rules in [9] consists of two problems. The first problem is to find frequent itemsets whose support satisfies a predefined minimum value. Then, the concern is to derive all of the rules exceeding a minimum confidence, based on each frequent itemset. Since the solution of the second problem is straightforward, most of the proposed work goes in for solving the first problem. An a priori algorithm has been proposed in [19], which was the basis for many of the forthcoming algorithms. A two-pass algorithm is presented in [11]. It consumes only two database scan passes, while a priori is a multi-pass algorithm and needs up to c+1 database scans, where c is the number of items (attributes). Association rules mining is applicable in numerous database communities. It has large applications in the retail industry to improve market basket analysis [7]. Streaming-Rules is an algorithm developed by [9] to report an association between pairs of elements in streams for predictive caching and detecting the previously undetectable hit inflation attacks in advertising networks. Running mining algorithms on numerical attributes may result in a large set of candidates. Each candidate has small support and many rules have been generated with useless information, e.g., the age attribute, salary attribute, and students' grades. Many partitioning algorithms have been developed to solve the numerical attributes problem. The proposed algorithms faced two problems. The first problem was the partitioning of attribute domain into meaningful partitions. The second problem was the loss of many useful rules due to the sharp boundary problem. Consequently, some rules may fail to achieve the minimum support threshold because of the separating of its domain into two partitions.

Fuzzy sets have been introduced to solve these two problems. Using fuzzy sets make the resulted association rules more meaningful. Many mining algorithms have been introduced to solve the quantitative attributes problem using fuzzy sets proposed algorithms in [13–27] that can be separated into two types related to the kind of minimum support threshold, fuzzy mining based on single-minimum support threshold, and fuzzy mining based on multi-minimum support threshold [21]. Neutrosophic theory was introduced in [28] to generalize fuzzy theory. In [29–32], the neutrosophic theory has been proposed to solve several applications and it has been used to generate a solution based on neutrosophic sets. Single-valued neutrosophic set was introduced in [33] to transfer the neutrosophic theory from the philosophic field into the mathematical theory, and to become applicable in engineering applications. In [33], a differentiation has been proposed between intuitionistic fuzzy sets and neutrosophic sets based on the independence of membership functions (truth-membership function, falsity-membership function, and indeterminacy-membership function). In neutrosophic sets, indeterminacy is explicitly independent, and truth-membership function and falsity-membership function are independent as well. In this paper, we introduce an approach that is based on neutrosophic sets for mining association rules, instead of fuzzy sets. Also, a comparison resulted association rules in both of the scenarios has been presented. In [34], an attempt to express how neutrosophic sets theory could be used in data mining has been proposed. They define SVNSF (single-valued neutrosophic score function) to aggregate attribute values. In [35], an algorithm has been introduced to mining vague association rules. Items properties have been added to enhance the quality of mining association rules. In addition, almost sold items (items has been selected by the customer, but not checked out) were added to enhance the generated association rules. AH-pair Database consisting of a traditional database and the hesitation information of items was generated. The hesitation information was collected, depending online shopping stores, which make it easier to collect that type of information, which does not exist in traditional stores. In this paper, we are the first to convert numerical attributes (items) into neutrosophic sets. While vague association rules add new items from the hesitating information, our framework adds new items by converting the numerical attributes into linguistic terms. Therefore, the vague association rule mining can be run on the converted database, which contains new linguistic terms.

Research Contribution

Detecting hidden and affinity patterns from various, complex, and big data represents a significant role in various domain areas, such as marketing, business, medical analysis, etc. These patterns are beneficial for strategic decision-making. Association rules mining plays an important role as well in detecting the relationships between patterns for determining frequent itemsets, since classical association rules cannot use all types of data for the mining process. Binary data can only be used to form classical rules, where items either exist in database or not. However, when classical association rules deal with quantitative database, no discovered rules will appear, and this is the reason for innovating quantitative association rules. The quantitative method also leads to the sharp boundary problem, where the item is below or above the estimation values. The fuzzy association rules are introduced to overcome the classical association rules drawbacks. The item in fuzzy association rules has a membership function and a fuzzy set. The fuzzy association rules can deal with vague rules, but not in the best manner, since it cannot consider the indeterminacy of rules. In order to overcome drawbacks of previous association rules, a new neutrosophic association rule algorithm has been introduced in this research. Our proposed algorithm deals effectively and efficiently with vague rules by considering not only the membership function of items, but also the indeterminacy and the falsity functions. Therefore, the proposed algorithm discovers all of the possible association rules and minimizes the losing processes of rules, which leads to building efficient and reliable decision-making system. By comparing our proposed algorithm with fuzzy approaches, we note that the number of association rules is increased, and negative rules are also discovered. The separation of negative association rules from positive ones is not a simple process, and it helps in various fields. As an example, in the medical domain, both positive and negative association rules help not only in the diagnosis of disease, but also in detecting prevention manners.

The rest of this research is organized as follows. The basic concepts and definitions of association rules mining are presented in Section 2. A quick overview of fuzzy association rules is described in Section 3. The neutrosophic association rules and the proposed model are presented in Section 4. A case study of Telecom Egypt Company is presented in Section 5. The experimental results and comparisons between fuzzy and proposed association rules are discussed in Section 6. The conclusions are drawn in Section 7.

2. Association Rules Mining

In this section, we formulate the $|D|$ transactions from the mining association rules for a database D. We used the following notations:

(i) $I = \{i_1, i_2, \ldots i_m\}$ represents all the possible data sets, called items.
(ii) Transaction set T is the set of domain data resulting from transactional processing such as $T \subseteq I$.
(iii) For a given itemset $X \subseteq I$ and a given transaction T, we say that T contains X if and only if $X \subseteq T$.
(iv) σ_X: the support frequency of X, which is defined as the number of transactions out of D that contain X.
(v) s: the support threshold.

X is considered a large itemset, if $\sigma_X \geq |D| \times s$. Further, an association rule is an implication of the form $X \Rightarrow Y$, where $X \subseteq I$, $Y \subseteq I$ and $X \cap Y = \varphi$.

An association rule $X \Rightarrow Y$ is addressed in D with confidence c if at least c transactions out of D contain both X and Y. The rule $X \Rightarrow Y$ is considered as a large itemset having a minimum support s if: $\sigma_{X \cup Y} \geq |D| \times s$.

For a specific confidence and specific support thresholds, the problem of mining association rules is to find out all of the association rules having confidence and support that is larger than the corresponding thresholds. This problem can simply be expressed as finding all of the large itemsets, where a large itemset L is:

$$L = \{X | X \subseteq I \wedge \sigma_X \geq |D| \times s\}.$$

3. Fuzzy Association Rules

Mining of association rules is considered as the main task in data mining. An association rule expresses an interesting relationship between different attributes. Fuzzy association rules can deal with both quantitative and categorical data and are described in linguistic terms, which are understandable terms [26].

Let $T = \{t_1, \ldots, t_n\}$ be a database transactions. Each transaction consists of a number of attributes (items). Let $I = \{i_1, \ldots, i_m\}$ be a set of categorical or quantitative attributes. For each attribute i_k, $(k = 1, \ldots, m)$, we consider $\{n_1, \ldots, n_k\}$ associated fuzzy sets. Typically, a domain expert determines the membership function for each attribute.

The tuple $< X, A >$ is called the fuzzy itemset, where $X \subseteq I$ (set of attributes) and A is a set of fuzzy sets that is associated with attributes from X.

Following is an example of fuzzy association rule:

IF salary is high and age is old THEN insurance is high

Before the mining process starts, we need to deal with numerical attributes and prepare them for the mining process. The main idea is to determine the linguistic terms for the numerical attribute and define the range for every linguistic term. For example, the temperature attribute is determined by the linguistic terms {very cold, cold, cool, warm, hot}. Figure 1 illustrates the membership function of the temperature attribute.

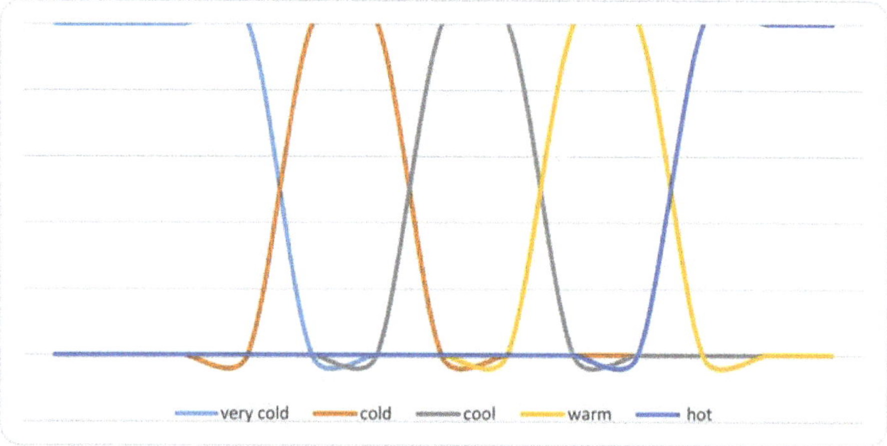

Figure 1. Linguistic terms of the temperature attribute.

The membership function has been calculated for the following database transactions illustrated in Table 1.

Table 1. Membership function for Database Transactions.

Transaction	Temp.	Membership Degree
T1	18	1 cool
T2	13	0.6 cool, 0.4 cold
T3	12	0.4 cool, 0.6 cold
T4	33	0.6 warm, 0.4 hot
T5	21	0.2 warm, 0.8 cool
T6	25	1 warm

We add the linguistic terms {very cold, cold, cool, warm, hot} to the candidate set and calculate the support for those itemsets. After determining the linguistic terms for each numerical attribute, the fuzzy candidate set have been generated.

Table 2 contains the support for each itemset individual one-itemsets. The count for every linguistic term has been calculated by summing its membership degree over the transactions. Table 3 shows the support for two-itemsets. The count for the fuzzy sets is the summation of degrees that resulted from the membership function of that itemset. The count for two-itemset has been calculated by summing the minimum membership degree of the 2 items. For example, {cold, cool} has count 0.8, which resulted from transactions T2 and T3. For transaction T2, membership degree of cool is 0.6 and membership degree for cold is 0.4, so the count for set {cold, cool} in T2 is 0.4. Also, T3 has the same count for {cold, cool}. So, the count of set {cold, cool} over all transactions is 0.8.

Table 2. 1-itemset support.

1-itemset	Count	Support
Very cold	0	0
Cold	1	0.17
Cool	2.8	0.47
Warm	1.6	0.27
Hot	0.6	0.1

Table 3. 2-itemset support.

2-itemset	Count	Support
{Cold, cool}	0.8	0.13
{Warm, hot}	0.4	0.07
{warm, cool}	0.2	0.03

In subsequent discussions, we denote an itemset that contains k items as k-itemset. The set of all k-itemsets in L is referred as L_k.

4. Neutrosophic Association Rules

In this section, we overview some basic concepts of the NSs and SVNSs over the universal set X, and the proposed model of discovering neutrosophic association rules.

4.1. Neutrosophic Set Definitions and Operations

Definition 1 ([33]). *Let X be a space of points and x∈X. A neutrosophic set (NS) A in X is definite by a truth-membership function $T_A(x)$, an indeterminacy-membership function $I_A(x)$ and a falsity-membership function $F_A(x)$. $T_A(x)$, $I_A(x)$ and $F_A(x)$ are real standard or real nonstandard subsets of $]^-0, 1^+[$. That is $T_A(x)$: $X \to]^-0, 1^+[$, $I_A(x)$: $X \to]^-0, 1^+[$ and $F_A(x)$: $X \to]^-0, 1^+[$. There is no restriction on the sum of $T_A(x)$, $I_A(x)$ and $F_A(x)$, so $0^- \leq \sup T_A(x) + \sup I_A(x) + \sup F_A(x) \leq 3^+$.*

Neutrosophic is built on a philosophical concept, which makes it difficult to process during engineering applications or to use it to real applications. To overcome that, Wang et al. [31], defined the SVNS, which is a particular case of NS.

Definition 2. *Let X be a universe of discourse. A single valued neutrosophic set (SVNS) A over X is an object taking the form $A = \{\langle x, T_A(x), I_A(x), F_A(x)\rangle : x \in X\}$, where $T_A(x)$: $X \to [0, 1]$, $I_A(x)$: $X \to [0, 1]$ and $F_A(x)$: $X \to [0, 1]$ with $0 \leq T_A(x) + I_A(x) + F_A(x) \leq 3$ for all x∈X. The intervals $T_A(x)$, $I_A(x)$ and $F_A(x)$ represent the truth-membership degree, the indeterminacy-membership degree and the falsity membership degree of x to A, respectively. For convenience, a SVN number is represented by A = (a, b, c), where a, b, c∈[0, 1] and $a + b + c \leq 3$.*

Definition 3 (Intersection) ([31]). *For two SVNSs $A = \langle T_A(x), I_A(x), F_A(x)\rangle$ and $B = \langle T_B(x), I_B(x), F_B(x)\rangle$, the intersection of these SVNSs is again an SVNSs which is defined as $C = A \cap B$ whose truth, indeterminacy and falsity membership functions are defined as $T_C(x) = min(T_A(x), T_B(x))$, $I_A(x) = min(I_A(x), I_B(x))$ and $F_C(x) = max(F_A(x), F_B(x))$.*

Definition 4 (Union) ([31]). *For two SVNSs $A = \langle T_A(x), I_A(x), F_A(x)\rangle$ and $B = \langle T_B(x), I_B(x), F_B(x)\rangle$, the union of these SVNSs is again an SVNSs which is defined as $C = A \cup B$ whose truth, indeterminacy and falsity membership functions are defined as $T_C(x) = max(T_A(x), T_B(x))$, $I_A(x) = max(I_A(x), I_B(x))$ and $F_C(x) = min(F_A(x), F_B(x))$.*

Definition 5 (*Containment*) ([31]). A single valued neutrosophic set A contained in the other SVNS B, denoted by $A \subseteq B$ if and only if $T_A(x) \leq T_B(x)$, $I_A(x) \leq I_B(x)$ and $F_A(x) \geq F_B(x)$ for all x in X.

Next, we propose a method for generating the association rule under the SVNS environment.

4.2. Proposed Model for Association Rule

In this paper, we introduce a model to generate association rules of form:
$X \rightarrow Y$ where $X \cap Y = \varphi$ and X, Y are neutrosophic sets.

Our aim is to find the frequent itemsets and their corresponding support. Generating an association rule from its frequent itemsets, which are dependent on the confidence threshold, are also discussed here. This has been done by adding the neutrosophic set into I, where I is all of the possible data sets, which are referred as items. So $I = N \cup M$ where N is neutrosophic set and M is classical set of items. The general form of an association rule is an implication of the form $X \rightarrow Y$, where $X \subseteq I$, $Y \subseteq I$, $X \cap Y = \varphi$.

Therefore, an association rule $X \rightarrow Y$ is addressed in Database D with confidence 'c' if at least c transactions out of D contains both X and Y. On the other hand, the rule $X \rightarrow Y$ is considered a large item set having a minimum support s if $\sigma_{X \cup Y} \geq |D| \times s$. Furthermore, the process of converting the quantitative values into the neutrosophic sets is proposed, as shown in Figure 2.

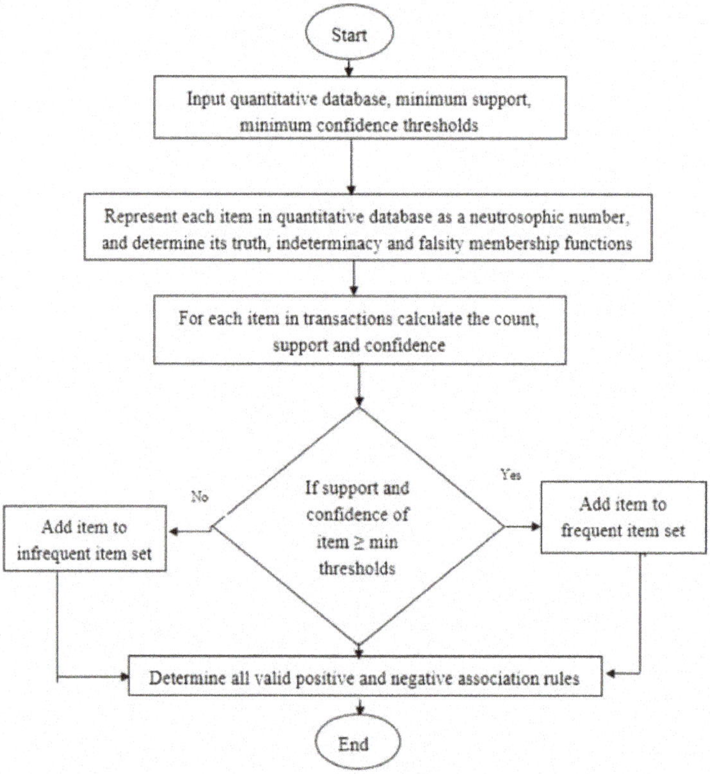

Figure 2. The proposed model.

The proposed model for the construction of the neutrosophic numbers is summarized in the following steps:

Step 1 Set linguistic terms of the variable, which will be used for quantitative attribute.

Step 2 Define the truth, indeterminacy, and the falsity membership functions for each constructed linguistic term.

Step 3 For each transaction t in T, compute the truth-membership, indeterminacy-membership and falsity-membership degrees.

Step 4 Extend each linguistic term l in set of linguistic terms L into T_L, I_L, and F_L to denote truth-membership, indeterminacy-membership, and falsity-membership functions, respectively.

Step 5 For each k-item set where $k = \{1, 2, \ldots, n\}$, and n number of iterations.

- calculate count of each linguistic term by summing degrees of membership for each transaction as $Count(A) = \sum_{i=1}^{i=t} \mu_A(x)$ where μ_A is T_A, I_A or F_A.
- calculate support for each linguistic term $s = \frac{Count(A)}{No.\ of\ trnsactions}$.

Step 6 The above procedure has been repeated for every quantitative attribute in the database.

In order to show the working procedure of the approach, we consider the temperature as an attribute and the terms "very cold", "cold", "cool", "warm", and "hot" as their linguistic terms to represent the temperature of an object. Then, following the steps of the proposed approach, construct their membership function as below:

Step 1 The attribute temperature' has set the linguistic terms "very cold", "cold", "cool", "warm", and "hot", and their ranges are defined in Table 4.

Table 4. Linguistic terms ranges.

Linguistic Term	Core Range	Left Boundary Range	Right Boundary Range
Very Cold	−∞–0	N/A	0–5
Cold	5–10	0–5	10–15
Cool	15–20	10–15	20–25
Warm	25–30	20–25	30–35
Hot	35–∞	30–35	N/A

Step 2 Based on these linguistic term ranges, the truth-membership functions of each linguistic variable are defined, as follows:

$$T_{very-cold}(x) = \begin{cases} 1 & ;for\ x \leq 0 \\ (5-x)/5 & ;for\ 0 < x < 5 \\ 0 & ;for\ x \geq 5 \end{cases}$$

$$T_{cold}(x) = \begin{cases} 1 & ;for\ 5 \leq x \leq 10 \\ (15-x)/5 & ;for\ 10 < x < 15 \\ x/5 & ;for\ 0 < x < 5 \\ 0 & ;for\ x \geq 15\ or\ x \leq 0 \end{cases}$$

$$T_{cool}(x) = \begin{cases} 1 & ;for\ 15 \leq x \leq 20 \\ (25-x)/5 & ;for\ 20 < x < 25 \\ (x-10)/5 & ;for\ 10 < x < 15 \\ 0 & ;\ otherwise \end{cases}$$

$$T_{warm}(x) = \begin{cases} 1 & ;for\ 25 \leq x \leq 30 \\ (35-x)/5 & ;for\ 30 < x < 35 \\ (x-20)/5 & ;for\ 20 < x < 25 \\ 0 & ;\ otherwise \end{cases}$$

$$T_{hot}(x) = \begin{cases} 1 & ;for\ x \geq 35 \\ (x-30)/5 & ;for\ 30 < x < 35 \\ 0 & ;\ otherwise \end{cases}$$

The falsity-membership functions of each linguistic variable are defined as follows:

$$F_{very-cold}(x) = \begin{cases} 0 & ;for\ x \leq 0 \\ x/5 & ;for\ 0 < x < 5\ ; \\ 1 & ;for\ x \geq 5 \end{cases}$$

$$F_{cold}(x) = \begin{cases} 0 & ;for\ 5 \leq x \leq 10 \\ (x-10)/5 & ;for\ 10 < x < 15 \\ (5-x)/5 & ;for\ 0 < x < 5 \\ 1 & ;for\ x \geq 15\ or\ x \leq 0 \end{cases}$$

$$F_{cool}(x) = \begin{cases} 0 & ;for\ 15 \leq x \leq 20 \\ (x-20)/5 & ;for\ 20 < x < 25 \\ (15-x)/5 & ;for\ 10 < x < 15 \\ 1 & ;\ otherwise \end{cases}$$

$$F_{warm}(x) = \begin{cases} 0 & ;for\ 25 \leq x \leq 30 \\ (x-30)/5 & ;for\ 30 < x < 35 \\ (25-x)/5 & ;for\ 20 < x < 25 \\ 1 & ;\ otherwise \end{cases}$$

$$F_{hot}(x) = \begin{cases} 0 & ;\ for\ x \geq 35 \\ (35-x)/5 & ;\ for\ 30 < x < 35 \\ 1 & ;\ otherwise \end{cases}$$

The indeterminacy membership functions of each linguistic variables are defined as follows:

$$I_{very-cold}(x) = \begin{cases} 0 & ;for\ x \leq -2.5 \\ (x+2.5)/5 & ;for -2.5 \leq x \leq 2.5 \\ (7.5-x)/5 & ;for\ 2.5 \leq x \leq 7.5 \\ 0 & ;for\ x \geq 7.5 \end{cases}$$

$$I_{cold}(x) = \begin{cases} (x+2.5)/5 & ;for\ 2.5 \leq x \leq 2.5 \\ (7.5-x)/5 & ;for\ 2.5 \leq x \leq 7.5 \\ (x-7.5)/5 & ;for\ 7.5 \leq x \leq 12.5 \\ (17.5-x)/5 & ;for\ 12.5 \leq x \leq 17.5 \\ 0 & ;otherwise \end{cases}$$

$$I_{cool}(x) = \begin{cases} (x-7.5)/5 & ;for\ 7.5 \leq x \leq 12.5 \\ (17.5-x)/5 & ;for\ 12.5 \leq x \leq 17.5 \\ (x-17.5)/5 & ;for\ 17.5 \leq x \leq 22.5 \\ (27.5-x)/5 & ;for\ 22.5 \leq x \leq 27.5 \\ 0 & ;otherwise \end{cases}$$

$$I_{warm}(x) = \begin{cases} (x-17.5)/5 & ;for\ 17.5 \leq x \leq 22.5 \\ (27.5-x)/5 & ;for\ 22.5 \leq x \leq 27.5 \\ (x-27.5)/5 & ;for\ 27.5 \leq x \leq 32.5 \\ (37.5-x)/5 & ;for\ 32.5 \leq x \leq 37.5 \\ 0 & ;otherwise \end{cases}$$

$$I_{hot}(x) = \begin{cases} (x-27.5)/5 & ; for\ 27.5 \le x \le 32.5 \\ (37.5-x)/5 & ; for\ 32.5 \le x \le 37.5 \\ 0 & ; otherwise \end{cases}$$

The graphical membership degrees of these variables are summarized in Figure 3. The graphical falsity degrees of these variables are summarized in Figure 4. Also, the graphical indeterminacy degrees of these variables are summarized in Figure 5. On the other hand, for a particular linguistic term, 'Cool' in the temperature attribute, their neutrosophic membership functions are represented in Figure 6.

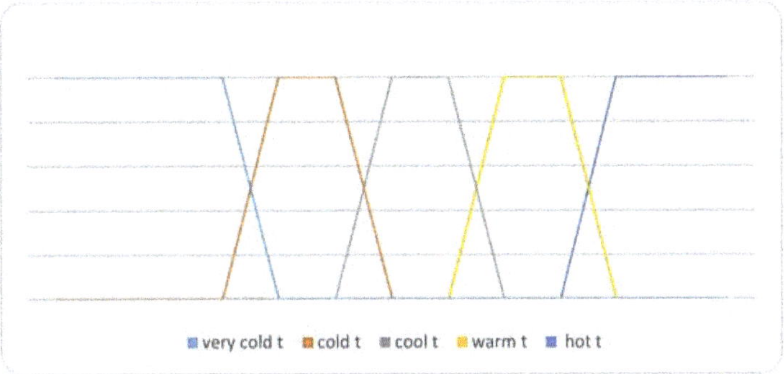

Figure 3. Truth-membership function of temperature attribute.

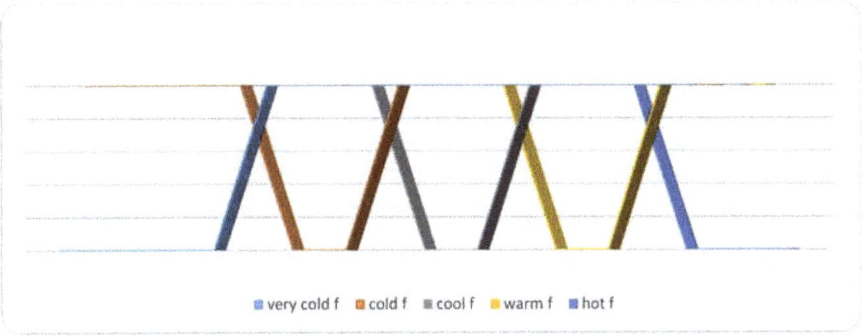

Figure 4. Falsity-membership function of temperature attribute.

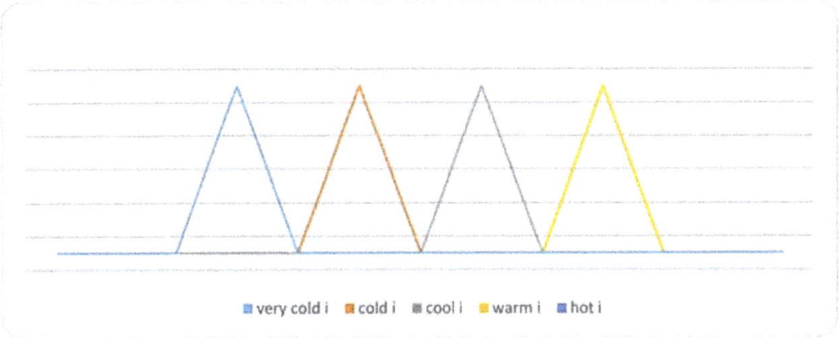

Figure 5. Indeterminacy-membership function of temperature attribute.

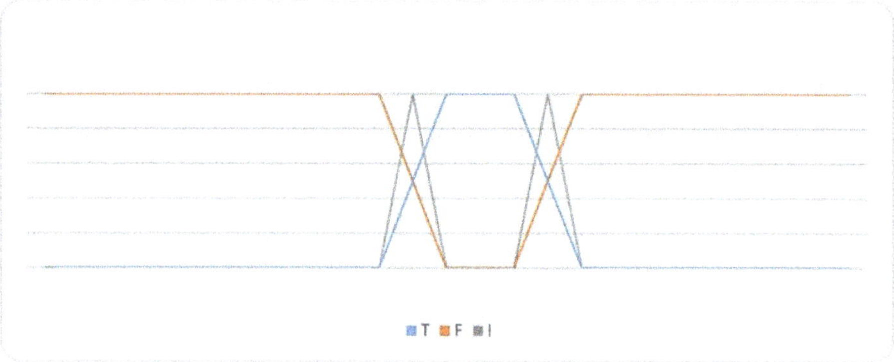

Figure 6. Cool (T, I, F) for temperature attribute.

Step 3 Based on the membership grades, different transaction has been set up by taking different sets of the temperatures. The membership grades in terms of the neutrosophic sets of these transactions are summarized in Table 5.

Table 5. Membership function for database Transactions.

Transaction	Temp.	Membership Degree
T1	18	Very-cold <0,0,1> cold <0,0,1> cool <1,0.1,0> warm <0,0.1,1> hot <0,0,1>
T2	13	Very cold <0,0,1> cold <0.4,0.9,0.6> cool <0.6,0.9,0.4> warm <0,0,1> hot <0,0,1>
T3	12	Very cold <0,0,1> cold <0.6,0.9,0.4> cool <0.4,0.9,0.6> warm <0,0,1> hot <0,0,1>
T4	33	Very cold <0,0,1> cold <0,0,1> cool <0,0,1> warm <0.4,0.9,0.6> hot <0.6,0.9,0.4>
T5	21	Very cold <0,0,1> cold <0,0,1> cool <0.8,0.7,0.2> warm <0.2,0.7,0.8> hot <0,0,1>
T6	25	Very cold <0,0,1> cold <0,0,1> cool <0,0,1> warm <1,0.5,0> hot <0,0,1>

Step 4 Now, we count the set of linguistic terms {very cold, cold, cool, warm, hot} for every element in transactions. Since the truth, falsity, and indeterminacy-memberships are independent functions, the set of linguistic terms can be extended to $\left\{ T_{very-cold}, T_{cold}, T_{cool}, T_{warm}, T_{hot} F_{very-cold}, F_{cold}, F_{cool}, F_{warm}, F_{hot} I_{very-cold}, I_{cold}, I_{cool}, I_{warm}, I_{hot} \right\}$ where F_{warm} means not worm and I_{warm} means not sure of warmness. This enhances dealing with negative association rules, which is handled as positive rules without extra calculations.

Step 5 By using the membership degrees that are given in Table 5 for candidate 1-itemset, the count and support has been calculated, respectively. The corresponding results are summarized in Table 6.

Table 6. Support for candidate 1-itemset neutrosophic set.

1-itemset	Count	Support
$T_{verycold}$	0	0
T_{Cold}	1	0.17
T_{Cool}	2.8	0.47
T_{Warm}	1.6	0.27
T_{Hot}	0.6	0.1
$I_{verycold}$	0	0
I_{Cold}	1.8	0.3
I_{Cool}	2.6	0.43
I_{Warm}	2.2	0.37
I_{Hot}	0.9	0.15
$F_{verycold}$	6	1
F_{Cold}	5	0.83
F_{Cool}	3.2	0.53
F_{Warm}	4.4	0.73
F_{Hot}	5.4	0.9

Similarly, the two-itemset support is illustrated in Table 7 and the rest of itemset generation (k-itemset for $k = 3, 4 \ldots 8$) are obtained similarly. The count for k-item set in database record is defined by minimum count of each one-itemset exists.

For example: $\{T_{Cold}, T_{Cool}\}$ count is 0.8

Because they exists in both T2 and T3.

In T2: $T_{Cold} = 0.4$ and $T_{Cool} = 0.6$ so, count for $\{T_{Cold}, T_{Cool}\}$ in T2 = 0.4

In T3: $T_{Cold} = 0.6$ and $T_{Cool} = 0.4$ so, count for $\{T_{Cold}, T_{Cool}\}$ in T2 = 0.4

Thus, count of $\{T_{Cold}, T_{Cool}\}$ in (Database) DB is 0.8.

Table 7. Support for candidate 2-itemset neutrosophic set.

2-itemset	Count	Support	2-itemset	Count	Support
$\{T_{Cold}, T_{Cool}\}$	0.8	0.13	$\{I_{Cold}, I_{Cool}\}$	1.8	0.30
$\{T_{Cold}, I_{Cold}\}$	1	0.17	$\{I_{Cold}, F_{verycold}\}$	1.8	0.30
$\{T_{Cold}, I_{Cool}\}$	1	0.17	$\{I_{Cold}, F_{Cold}\}$	1	0.17
$\{T_{Cold}, F_{verycold}\}$	1	0.17	$\{I_{Cold}, F_{Cool}\}$	1	0.17
$\{T_{Cold}, F_{Cold}\}$	0.8	0.13	$\{I_{Cold}, F_{Warm}\}$	1.8	0.30
$\{T_{Cold}, F_{Cool}\}$	1	0.17	$\{I_{Cold}, F_{Hot}\}$	1.8	0.30
$\{T_{Cold}, F_{Warm}\}$	1	0.17	$\{I_{Cool}, I_{Warm}\}$	0.8	0.13
$\{T_{Cold}, F_{Hot}\}$	1	0.17	$\{I_{Cool}, F_{verycold}\}$	2.6	0.43
$\{T_{Cool}, T_{Warm}\}$	0.2	0.03	$\{I_{Cool}, F_{Cold}\}$	1.8	0.30
$\{T_{Cool}, I_{Cold}\}$	1	0.17	$\{I_{Cool}, F_{Cool}\}$	1.2	0.20
$\{T_{Cool}, F_{Cool}\}$	1.8	0.30	$\{I_{Cool}, F_{Warm}\}$	2.6	0.43
$\{T_{Cool}, I_{Warm}\}$	0.8	0.13	$\{I_{Cool}, F_{Hot}\}$	2.6	0.43
$\{T_{Cool}, F_{verycold}\}$	2.8	0.47	$\{I_{Warm}, I_{Hot}\}$	0.9	0.15

Table 7. *Cont.*

$\{T_{Cool}, F_{Cold}\}$	2.8	0.47	$\{I_{Warm}, F_{verycold}\}$	2.2	0.37
$\{T_{Cool}, F_{Cool}\}$	1	0.17	$\{I_{Warm}, F_{Cold}\}$	2.2	0.37
$\{T_{Cool}, F_{Warm}\}$	2.8	0.47	$\{I_{Warm}, F_{Cool}\}$	1.6	0.27
$\{T_{Cool}, F_{Hot}\}$	2.8	0.47	$\{I_{Warm}, F_{Warm}\}$	1.4	0.23
$\{T_{Warm}, T_{Hot}\}$	0.4	0.07	$\{I_{Warm}, F_{Hot}\}$	1.7	0.28
$\{T_{Warm}, I_{Cool}\}$	0.2	0.03	$\{I_{Hot}, F_{verycold}\}$	0.9	0.15
$\{T_{Warm}, I_{Warm}\}$	1.1	0.18	$\{I_{Hot}, F_{Cold}\}$	0.9	0.15
$\{T_{Warm}, I_{Hot}\}$	0.4	0.07	$\{I_{Hot}, F_{Cool}\}$	0.9	0.15
$\{T_{Warm}, F_{verycold}\}$	1.6	0.27	$\{I_{Hot}, F_{Warm}\}$	0.6	0.10
$\{T_{Warm}, F_{Cold}\}$	1.6	0.27	$\{I_{Hot}, F_{Hot}\}$	0.4	0.07
$\{T_{Warm}, F_{Cool}\}$	1.6	0.27	$\{F_{verycold}, F_{Cold}\}$	5	0.83
$\{T_{Warm}, F_{Warm}\}$	0.6	0.10	$\{F_{verycold}, F_{Cool}\}$	3.2	0.53
$\{T_{Warm}, F_{Hot}\}$	1.6	0.27	$\{F_{verycold}, F_{Warm}\}$	4.4	0.73
$\{T_{Hot}, I_{Warm}\}$	0.6	0.10	$\{F_{verycold}, F_{Hot}\}$	5.4	0.90
$\{T_{Hot}, I_{Hot}\}$	0.6	0.10	$\{F_{Cold}, F_{Cool}\}$	3	0.50
$\{T_{Hot}, F_{verycold}\}$	0.6	0.10	$\{F_{Cold}, F_{Warm}\}$	3.4	0.57
$\{T_{Hot}, F_{Cold}\}$	0.6	0.10	$\{F_{Cold}, F_{Hot}\}$	4.4	0.73
$\{T_{Hot}, F_{Cool}\}$	0.6	0.10	$\{F_{Cool}, F_{Warm}\}$	1.8	0.30
$\{T_{Hot}, F_{Warm}\}$	0.6	0.10	$\{F_{Cool}, F_{Hot}\}$	2.6	0.43
$\{T_{Hot}, F_{Hot}\}$	0.4	0.07	$\{F_{Warm}, F_{Hot}\}$	4.2	0.70

5. Case Study

In this section, the case of Telecom Egypt Company stock records has been studied. Egyptian stock market has many companies. One of the major questions for stock market users is when to buy or to sell a specific stock. Egyptian stock market has three indicators, EGX30, EGX70, and EGX100. Each indicator gives a reflection of the stock market. Also, these indicators have an important impact on the stock market users, affecting their decisions of buying or selling stocks. We focus in our study on the relation between the stock and the three indicators. Also, we consider the month and quarter of the year to be another dimension in our study, while the sell/buy volume of the stock per day is considered to be the third dimension.

In this study, the historical data has been taken from the Egyptian stock market program (Mist) during the program September 2012 until September 2017. For every stock/indicator, Mist keeps a daily track of number of values (opening price, closing price, high price reached, low price reached, and volume). The collected data of Telecom Egypt Stock are summarized in Figure 7.

Telecom Egypt

Symbolid	Ts_date	Open	Close	High	Low	Volume
S48031C016	012-09-13	13.98	14.07	14.29	13.85	490,997
S48031C016	012-09-16	14.08	14.09	14.20	14.02	347,258
S48031C016	012-09-17	14.11	14.60	15.01	14.11	2,050,226
S48031C016	012-09-18	14.52	14.72	15.00	14.52	500,290
S48031C016	012-09-19	14.90	14.68	14.95	14.41	362,893
S48031C016	012-09-20	14.77	14.84	15.00	14.50	913,683
S48031C016	012-09-23	14.56	15.09	15.37	14.56	283,008
S48031C016	012-09-24	14.80	14.73	15.00	14.68	896,914
S48031C016	012-09-25	14.80	14.39	14.80	14.00	2,604,812
S48031C016	012-09-26	14.25	14.53	14.71	14.25	648,102

Figure 7. Telecom Egypt stock records.

In this study, we use the open price and close price values to get price change rate, which are defined as follows:

$$\text{price change rate} = \frac{\text{close price} - \text{open price}}{\text{open price}} \times 100$$

and change the volume to be a percentage of total volume of the stock with the following relation:

$$\text{percentage of volume} = \frac{\text{volume}}{\text{total volume}} \times 100$$

The same was performed for the stock market indicators. Now, we take the attributes as "quarter", "month", "stock change rate", "volume percentage", and "indicators change rate". Table 8 illustrates the segment of resulted data after preparation.

Table 8. Segment of data after preparation.

Ts_Date	Month	Quarter	Change	Volume	Change30	Change70	Change100
13 September 2012	September	3	0.64	0.03	−1.11	0.01	−0.43
16 September 2012	September	3	0.07	0.02	2.82	4.50	3.67
17 September 2012	September	3	3.47	0.12	1.27	0.76	0.81
18 September 2012	September	3	1.38	0.03	−0.08	−0.48	−0.43
19 September 2012	September	3	−1.48	0.02	0.35	−1.10	−0.64
20 September 2012	September	3	0.47	0.05	−1.41	−1.64	−1.55
23 September 2012	September	3	3.64	0.02	−0.21	1.00	0.41
24 September 2012	September	3	−0.47	0.05	0.27	−0.09	0.03
25 September 2012	September	3	−2.77	0.15	2.15	1.79	1.85
26 September 2012	September	3	1.96	0.04	0.22	0.96	0.57
27 September 2012	September	3	0.90	0.05	−1.38	−0.88	−0.92
30 September 2012	September	3	−0.14	0.00	−1.11	−0.79	−0.75
1 October 2012	October	4	−1.60	0.02	−2.95	−4.00	−3.51

Based on these linguistic terms, define the ranges under the SVNSs environment. For this, corresponding to the attribute in "change rate" and "volume", the truth-membership functions by defining their linguistic terms as {"high up", "high low", "no change", "low down", "high down"} corresponding to attribute "change rate", while for the attribute "volume", the linguistic terms are (low, medium, and high) and their ranges are summarized in Figures 8 and 9, respectively. The falsity-membership function and indeterminacy-membership function have been calculated and applied as well for change rate attribute.

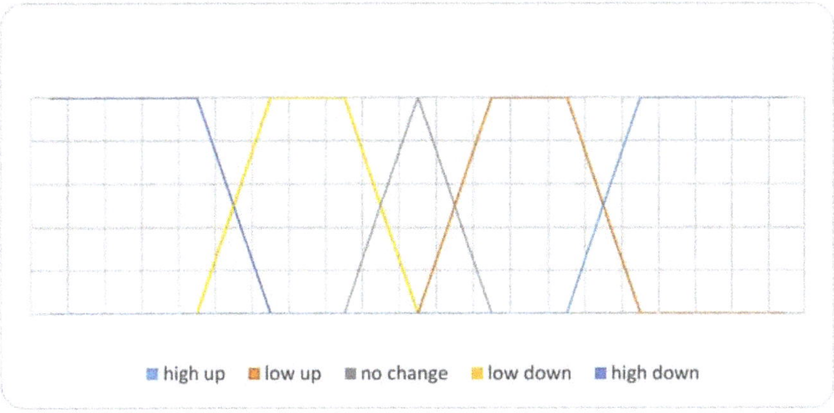

Figure 8. Change rate attribute truth-membership function.

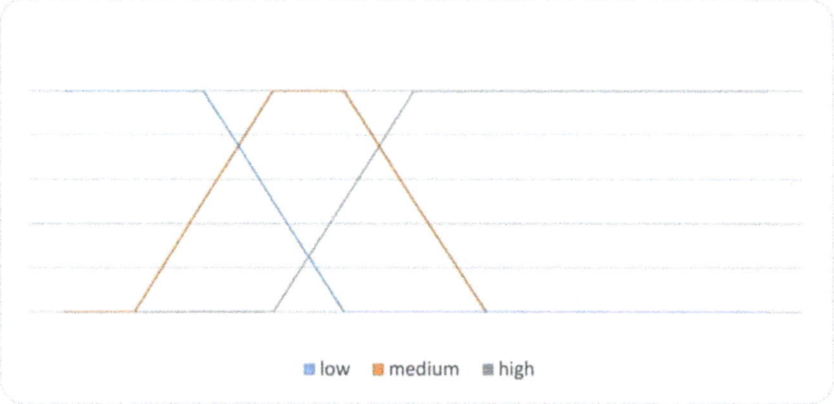

Figure 9. Volume attribute truth-membership function.

6. Experimental Results

We proceeded to a comparison between fuzzy mining and neutrosophic mining algorithms, and we found out that the number of generated association rules increased in neutrosophic mining.

A program has been developed to generate large itemsets for Telecom Egypt historical data. VB.net has been used in creating this program. The obtained data have been stored in an access database. The comparison depends on the number of generated association rules in a different min-support threshold. It should be noted that the performance cannot be part of the comparison because of the number of items (attributes) that are different in fuzzy vs. neutrosophic association rules mining. In fuzzy mining, the number of items was 14, while in neutrosophic mining it is 34. This happens because the number of attributes increased. Spreading each linguistic term into three (True, False, Indeterminacy) terms make the generated rules increase. The falsity-generated association rules can be considered a negative association rules. As pointed out in [36], the conviction of a rule $conv(X \to Y)$ is defined as the ratio of the expected frequency that X happened without Y falsity-association rules to be used to generate negative association rules if $T(x) + F(x) = 1$. In Table 9, the number of generated fuzzy rules in each $k-$itemset using different min-support threshold are reported, while the total generated fuzzy association rule is presented in Figure 10.

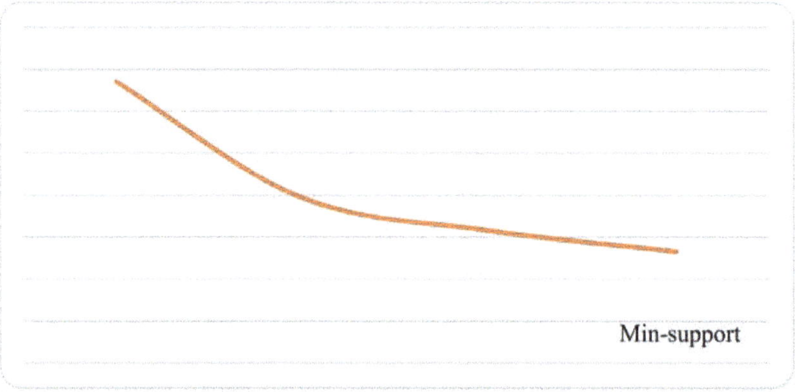

Figure 10. No. of fuzzy association rules with different min-support threshold.

Table 9. No. of resulted fuzzy rules with different min-support.

Min-Support	0.02	0.03	0.04	0.05
1-itemset	10	10	10	10
2-itemset	37	36	36	33
3-itemset	55	29	15	10
4-itemset	32	4	2	0

As compared to the fuzzy approach, by applying the same min-support threshold, we get a huge set of neutrosophic association rules. Table 10 illustrates the booming that happened to generated neutrosophic association rules. We stop generating itemsets at iteration 4 due to the noted expansion in the results shown in Figure 11, which shows the number of neutrosophic association rules.

Table 10. No. of neutrosophic rules with different min-support threshold.

Min-Support	0.02	0.03	0.04	0.05
1-itemset	26	26	26	26
2-itemset	313	311	309	300
3-itemset	2293	2164	2030	1907
4-itemset	11,233	9689	8523	7768

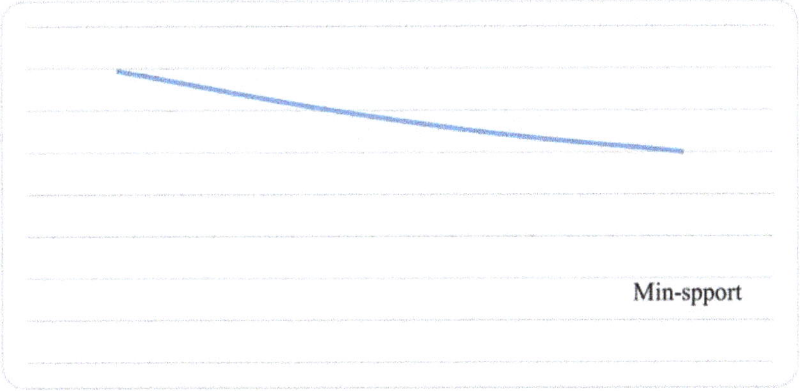

Figure 11. No. of neutrosophic association rules with different min-support threshold.

Experiment has been re-run using different min-support threshold values and the resulted neutrosophic association rules counts has been noted and listed in Table 11. Note the high values that are used for min-support threshold. Figure 12 illustrates the generated neutrosophic association rules for min-support threshold from 0.5 to 0.9.

Table 11. No. of neutrosophic rules with different min-support threshold.

Min-Support	0.5	0.6	0.7	0.8	0.9
1-itemset	11	9	9	6	5
2-itemset	50	33	30	11	10
3-itemset	122	64	50	10	10
4-itemset	175	71	45	5	5
5-itemset	151	45	21	1	1
6-itemset	88	38	8	0	0

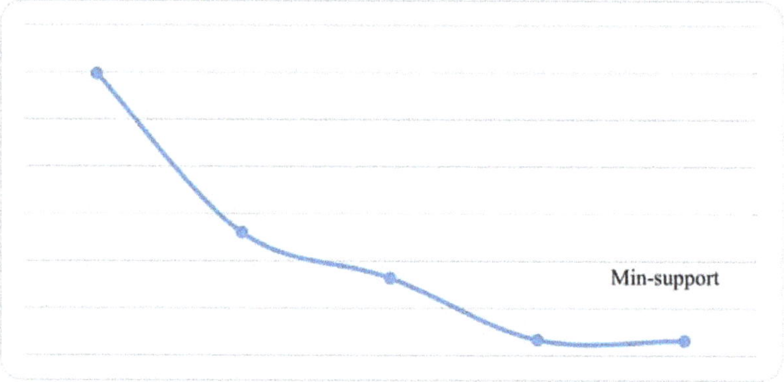

Figure 12. No. of neutrosophic rules for min-support threshold from 0.5 to 0.9.

Using the neutrosophic mining approach makes association rules exist for most of the min-support threshold domain, which may be sometimes misleading. We found that using the neutrosophic approach is useful in generating negative association rules beside positive association rules minings. Huge generated association rules provoke the need to re-mine generated rules (mining of mining association rules). Using suitable high min-support values may help in the neutrosophic mining process.

7. Conclusions and Future Work

Big data analysis will continue to grow in the next years. In order to efficiently and effectively deal with big data, we introduced in this research a new algorithm for mining big data using neutrosophic association rules. Converting quantitative attributes is the main key for generating such rules. Previously, it was performed by employing the fuzzy sets. However, due to fuzzy drawbacks, which we discussed in the introductory section, we preferred to use neutrosophic sets. Experimental results showed that the proposed approach generated an increase in the number of rules. In addition, the indeterminacy-membership function has been used to prevent losing rules from boundaries problems. The proposed model is more effective in processing negative association rules. By comparing it with the fuzzy association rules mining approaches, we conclude that the proposed model generates a larger number of positive and negative association rules, thus ensuring the construction of a real and efficient decision-making system. In the future, we plan to extend the comparison between the neutrosophic association rule mining and other interval fuzzy association rule minings. Furthermore, we seized the falsity-membership function capacity to generate negative association rules. Conjointly, we availed of the indeterminacy-membership function to prevent losing rules from boundaries problems. Many applications can emerge by adaptions of truth-membership function, indeterminacy-membership function, and falsity-membership function. Future work will benefit from the proposed model in generating negative association rules, or in increasing the quality of the generated association rules by using multiple support thresholds and multiple confidence thresholds for each membership function. The proposed model can be developed to mix positive association rules (represented in the truth-membership function) and negative association rules (represented in the falsity-membership function) in order to discover new association rules, and the indeterminacy-membership function can be put forth to help in the automatic adoption of support thresholds and confidence thresholds. Finally, yet importantly, we project to apply the proposed model in the medical field, due to its capability in effective diagnoses through discovering both positive and negative symptoms of a disease. All future big data challenges could be handled by combining neutrosophic sets with various techniques.

Author Contributions: All authors have contributed equally to this paper. The individual responsibilities and contribution of all authors can be described as follows: the idea of this whole paper was put forward by Mohamed Abdel-Basset and Mai Mohamed, Victor Chang completed the preparatory work of the paper. Florentin Smarandache analyzed the existing work. The revision and submission of this paper was completed by Mohamed Abdel-Basset.

Conflicts of Interest: The authors declare no conflict of interest.

References

1. Gartner. Available online: http://www.gartner.com/it-glossary/bigdata (accessed on 3 December 2017).
2. Intel. Big Thinkers on Big Data (2012). Available online: http://www.intel.com/content/www/us/en/bigdata/big-thinkers-on-big-data.html (accessed on 3 December 2017).
3. Aggarwal, C.C.; Ashish, N.; Sheth, A. The internet of things: A survey from the data-centric perspective. In *Managing and Mining Sensor Data*; Springer: Berlin, Germany, 2013; pp. 383–428.
4. Parker, C. Unexpected challenges in large scale machine learning. In Proceedings of the 1st International Workshop on Big Data, Streams and Heterogeneous Source Mining: Algorithms, Systems, Programming Models and Applications, Beijing, China, 12 August 2012; pp. 1–6.
5. Gopalkrishnan, V.; Steier, D.; Lewis, H.; Guszcza, J. Big data, big business: Bridging the gap. In Proceedings of the 1st International Workshop on Big Data, Streams and Heterogeneous Source Mining: Algorithms, Systems, Programming Models and Applications, Beijing, China, 12 August 2012; pp. 7–11.
6. Chytil, M.; Hajek, P.; Havel, I. The GUHA method of automated hypotheses generation. *Computing* **1966**, *1*, 293–308.
7. Park, J.S.; Chen, M.-S.; Yu, P.S. An effective hash-based algorithm for mining association rules. *SIGMOD Rec.* **1995**, *24*, 175–186. [CrossRef]
8. Aggarwal, C.C.; Yu, P.S. Mining large itemsets for association rules. *IEEE Data Eng. Bull.* **1998**, *21*, 23–31.
9. Savasere, A.; Omiecinski, E.R.; Navathe, S.B. An efficient algorithm for mining association rules in large databases. In Proceedings of the 21th International Conference on Very Large Data Bases, Georgia Institute of Technology, Zurich, Swizerland, 11–15 September 1995.
10. Agrawal, R.; Srikant, R. Fast algorithms for mining association rules. In Proceedings of the 20th International Conference of Very Large Data Bases (VLDB), Santiago, Chile, 12–15 September 1994; pp. 487–499.
11. Hidber, C. Online association rule mining. In Proceedings of the 1999 ACM SIGMOD International Conference on Management of Data, Philadelphia, PA, USA, 31 May–3 June 1999; Volume 28.
12. Valtchev, P.; Hacene, M.R.; Missaoui, R. A generic scheme for the design of efficient on-line algorithms for lattices. In *Conceptual Structures for Knowledge Creation and Communication*; Springer: Berlin/Heidelberg, Germany, 2003; pp. 282–295.
13. Verlinde, H.; de Cock, M.; Boute, R. Fuzzy versus quantitative association rules: A fair data-driven comparison. *IEEE Trans. Syst. Man Cybern. Part B* **2005**, *36*, 679–684. [CrossRef]
14. Huang, C.-H.; Lien, H.-L.; Wang, L.S.-L. An Empirical Case Study of Internet Usage on Student Performance Based on Fuzzy Association Rules. In Proceedings of the 3rd Multidisciplinary International Social Networks Conference on Social Informatics 2016 (Data Science 2016), Union, NJ, USA, 15–17 August 2016; p. 7.
15. Hui, Y.Y.; Choy, K.L.; Ho, G.T.; Lam, H. A fuzzy association Rule Mining framework for variables selection concerning the storage time of packaged food. In Proceedings of the 2016 IEEE International Conference on Fuzzy Systems (FUZZ-IEEE), Vancouver, BC, Canada, 24–29 July 2016; pp. 671–677.
16. Huang, T.C.-K. Discovery of fuzzy quantitative sequential patterns with multiple minimum supports and adjustable membership functions. *Inf. Sci.* **2013**, *222*, 126–146. [CrossRef]
17. Hong, T.-P.; Kuo, C.-S.; Chi, S.-C. Mining association rules from quantitative data. *Intell. Data Anal.* **1999**, *3*, 363–376. [CrossRef]
18. Pei, B.; Zhao, S.; Chen, H.; Zhou, X.; Chen, D. FARP: Mining fuzzy association rules from a probabilistic quantitative database. *Inf. Sci.* **2013**, *237*, 242–260. [CrossRef]
19. Siji, P.D.; Valarmathi, M.L. Enhanced Fuzzy Association Rule Mining Techniques for Prediction Analysis in Betathalesemia's Patients. *Int. J. Eng. Res. Technol.* **2014**, *4*, 1–9.
20. Lee, Y.-C.; Hong, T.-P.; Wang, T.-C. Multi-level fuzzy mining with multiple minimum supports. *Expert Syst. Appl.* **2008**, *34*, 459–468. [CrossRef]

21. Chen, C.-H.; Hong, T.-P.; Li, Y. Fuzzy association rule mining with type-2 membership functions. In Proceedings of the Asian Conference on Intelligent Information and Database Systems, Bali, Indonesia, 23–25 March 2015; pp. 128–134.
22. Sheibani, R.; Ebrahimzadeh, A. An algorithm for mining fuzzy association rules. In Proceedings of the International Multi Conference of Engineers and Computer Scientists, Hong Kong, China, 19–21 March 2008.
23. Lee, Y.-C.; Hong, T.-P.; Lin, W.-Y. Mining fuzzy association rules with multiple minimum supports using maximum constraints. In *Knowledge-Based Intelligent Information and Engineering Systems*; Springer: Berlin/Heidelberg, Germany, 2004; pp. 1283–1290.
24. Han, J.; Pei, J.; Kamber, M. *Data Mining: Concepts and Techniques*; Elsevier: New York, NY, USA, 2011.
25. Hong, T.-P.; Lin, K.-Y.; Wang, S.-L. Fuzzy data mining for interesting generalized association rules. *Fuzzy Sets Syst.* **2003**, *138*, 255–269. [CrossRef]
26. Au, W.-H.; Chan, K.C. Mining fuzzy association rules in a bank-account database. *IEEE Trans. Fuzzy Syst.* **2003**, *11*, 238–248.
27. Dubois, D.; Prade, H.; Sudkamp, T. On the representation, measurement, and discovery of fuzzy associations. *IEEE Trans. Fuzzy Syst.* **2005**, *13*, 250–262. [CrossRef]
28. Smarandache, F. Neutrosophic set-a generalization of the intuitionistic fuzzy set. *J. Def. Resour. Manag.* **2010**, *1*, 107.
29. Abdel-Basset, M.; Mohamed, M. The Role of Single Valued Neutrosophic Sets and Rough Sets in Smart City: Imperfect and Incomplete Information Systems. *Measurement* **2018**, *124*, 47–55. [CrossRef]
30. Ye, J. A multicriteria decision-making method using aggregation operators for simplified neutrosophic sets. *J. Intell. Fuzzy Syst.* **2014**, *26*, 2459–2466.
31. Ye, J. Vector Similarity Measures of Simplified Neutrosophic Sets and Their Application in Multicriteria Decision Making. *Int. J. Fuzzy Syst.* **2014**, *16*, 204–210.
32. Hwang, C.-M.; Yang, M.-S.; Hung, W.-L.; Lee, M.-G. A similarity measure of intuitionistic fuzzy sets based on the Sugeno integral with its application to pattern recognition. *Inf. Sci.* **2012**, *189*, 93–109. [CrossRef]
33. Wang, H.; Smarandache, F.; Zhang, Y.; Sunderraman, R. Single valued neutrosophic sets. *Rev. Air Force Acad.* **2010**, *10*, 11–20.
34. Mondal, K.; Pramanik, S.; Giri, B.C. Role of Neutrosophic Logic in Data Mining. *New Trends Neutrosophic Theory Appl.* **2016**, *1*, 15.
35. Lu, A.; Ke, Y.; Cheng, J.; Ng, W. Mining vague association rules. In *Advances in Databases: Concepts, Systems and Applications*; Kotagiri, R., Krishna, P.R., Mohania, M., Nantajeewarawat, E., Eds.; Springer: Berlin/Heidelberg, Germany, 2007; Volume 4443, pp. 891–897.
36. Srinivas, K.; Rao, G.R.; Govardhan, A. Analysis of coronary heart disease and prediction of heart attack in coal mining regions using data mining techniques. In Proceedings of the 5th International Conference on Computer Science and Education (ICCSE), Hefei, China, 24–27 August 2010; pp. 1344–1349.

symmetry

MDPI

Article

An Extension of Neutrosophic AHP–SWOT Analysis for Strategic Planning and Decision-Making

Mohamed Abdel-Basset [1,*], **Mai Mohamed** [1] **and Florentin Smarandache** [2]

[1] Department of Operations Research, Faculty of Computers and Informatics, Zagazig University, Markaz El-Zakazik 44519, Egypt; analyst_mohamed@zu.edu.eg

[2] Math & Science Department, University of New Mexico, Gallup, NM 87301, USA; smarand@unm.edu

* Correspondence: analyst_mohamed@yahoo.com

Received: 2 March 2018; Accepted: 12 April 2018; Published: 17 April 2018

Abstract: Every organization seeks to set strategies for its development and growth and to do this, it must take into account the factors that affect its success or failure. The most widely used technique in strategic planning is SWOT analysis. SWOT examines strengths (S), weaknesses (W), opportunities (O) and threats (T), to select and implement the best strategy to achieve organizational goals. The chosen strategy should harness the advantages of strengths and opportunities, handle weaknesses, and avoid or mitigate threats. SWOT analysis does not quantify factors (i.e., strengths, weaknesses, opportunities and threats) and it fails to rank available alternatives. To overcome this drawback, we integrated it with the analytic hierarchy process (AHP). The AHP is able to determine both quantitative and the qualitative elements by weighting and ranking them via comparison matrices. Due to the vague and inconsistent information that exists in the real world, we applied the proposed model in a neutrosophic environment. A real case study of Starbucks Company was presented to validate our model.

Keywords: analytic hierarchy process (AHP); SWOT analysis; multi-criteria decision-making (MCDM) techniques; neutrosophic set theory

1. Introduction

To achieve an organization's goals, the strategic factors affecting its performance should be considered. These strategic factors are classified as internal factors, that are under its control, and external factors, that are not under its control.

The most popular technique for analyzing strategic cases is SWOT analysis. SWOT is considered a decision-making tool. The SWOT acronym stands for Strengths, Weaknesses, Opportunities and Threats [1]. Strengths and weaknesses are internal factors, while opportunities and threats are external factors. The successful strategic plan of an organization should focus on strengths and opportunities, try to handle weaknesses, and avoid or mitigate threats.

By using SWOT analysis, an organization can choose one of four strategic plans as follows:

- SO: The good use of opportunities through existing strengths.
- ST: The good use of strengths to eliminate or reduce the impact of threats.
- WO: Taking into account weaknesses to obtain the benefits of opportunities.
- WT: Seeking to reduce the impact of threats by considering weaknesses.

SWOT analysis can be used to build successful company strategies, but it fails to provide evaluations and measures. Therefore, in the present research, we integrated it with the neutrosophic analytic hierarchy process (AHP).

The analytic hierarchy process (AHP) is a multi-criterion decision-making technique (MCDM) used for solving and analyzing complex problems. MCDM is an important branch in operations research, when seeking to construct mathematical and programming tools to select the superior alternative between various choices, according to particular criteria.

The AHP consists of several steps. The first step is structuring the hierarchy of the problem to understand it more clearly. The hierarchy of the AHP consists of a goal (objective), decision criteria, sub-criteria, and, finally, all available alternatives.

After structuring the AHP hierarchy, pair-wise comparison matrices are constructed by decision makers to weight criteria using Saaty's scale [2].

Finally, the final weight of alternatives are determined and ranked.

Then, the AHP is able to estimate both qualitative and the quantitative elements. For this reason, it is one of the most practical multi-criteria decision-making techniques [3].

In real life applications, decision criteria are often vague, complex and inconsistent in nature. In addition, using crisp values in a comparison matrix is not always accurate due to uncertainty and the indeterminate information available to decision makers. Many researchers have begun to use fuzzy set theory [4]. However, fuzzy set theory considers only a truth-membership degree. Atanassov introduced intuitionistic fuzzy set theory [5], which considers both truth and falsity degrees, but it fails to consider indeterminacy. To deal with the previous drawbacks of fuzzy and intuitionistic fuzzy sets, Smarandache introduced neutrosophic sets [6], which consider truth, indeterminacy and falsity degrees altogether to represent uncertain and inconsistent information. Therefore, neutrosophic sets are a better representation of reality. For this reason, in our research, we employed the AHP under a neutrosophic environment.

This research represents the first attempt at combining SWOT analysis with a neutrosophic analytic hierarchy process.

The structure of this paper is as follows: a literature review of SWOT analysis and the AHP is presented in Section 2; the basic definitions of neutrosophic sets are introduced in Section 3; the proposed model is discussed in Section 4; a real case study illustrates the applicability of the model proposed in Section 5; and, finally, Section 6 concludes the paper, envisaging future work.

2. Literature Review

In this section, we present an overview of the AHP technique and SWOT analysis, which are used across various domains.

SWOT analysis [7] is a practical methodology pursued by managers to construct successful strategies by analyzing strengths, weaknesses, opportunities and threats. SWOT analysis is a powerful methodology for making accurate decisions [8]. Organization's construct strategies to enhance their strengths, remove weaknesses, seize opportunities, and avoid threats.

Kotler et al. used SWOT analysis to attain an orderly approach to decision-making [9–11]. Many researchers in different fields [4] apply SWOT analysis. An overview of the applications of SWOT analysis is given by Helms and Nixon [8]. SWOT analysis has been applied in the education domain by Dyson [12]. It has also been applied to healthcare, government and not-for-profit organizations, to handle country-level issues [13] and for sustainable investment-related decisions [14]. It has been recommended for use when studying the relationships among countries [15]. SWOT analysis is mainly qualitative. This is the main disadvantage of SWOT, because it cannot assign strategic factor weights to alternatives. In order to overcome this drawback, many researchers have integrated it with the analytic hierarchy process (AHP).

Since the AHP is convenient and easy to understand, some managers find it a very useful decision-making technique. Vaidya and Kumar reviewed 150 publications, published in international journals between 1983 and 2003, and concluded that the AHP technique was useful for solving, selecting, evaluating and making decisions [16]. Achieving a consensus decision despite the large number of decision makers is another advantage of the AHP [17].

Several researchers have combined SWOT analysis methodology with the analytic hierarchy process (AHP). Leskinen et al. integrated SWOT with the AHP in an environmental domain [18–20], Kajanus used SWOT–AHP in tourism [21], and Setwart used SWOT–AHP in project management [22]. Competitive strength, environment and company strategy, were integrated by Chan and Heide [23]. Because the classical version of the AHP fails to handle uncertainty, many researchers have integrated SWOT analysis with the fuzzy AHP (FAHP). Demirtas et al. used SWOT with the fuzzy AHP for project management methodology selection [24]. Lumaksono used SWOT-FAHP to define the best strategy of expansion for a traditional shipyard [25]. Tavana et al. integrated SWOT analysis with intuitionistic fuzzy AHP to outsource reverse logistics [26].

Fuzzy sets focus only on the membership function (truth degree) and do not take into account the non-membership (falsity degree) and the indeterminacy degrees, so fail to represent uncertainty and indeterminacy. To overcome these drawbacks of the fuzzy set, we integrated SWOT analysis with the analytic hierarchy process in a neutrosophic environment.

A neutrosophic set is an extension of a classical set, fuzzy set, and intuitionistic fuzzy set, and it effectively represents real world problems by considering all facets of a decision situation, (i.e., truthiness, indeterminacy and falsity) [27–48]. This research attempted, for the first time, to present the mathematical representation of SWOT analysis with an AHP in a neutrosophic environment. The neutrosophic set acted as a symmetric tool in the proposed method, since membership was the symmetric equivalent of non-membership, with respect to indeterminacy.

3. Definition of a Neutrosophic Set

In this section, some important definitions of neutrosophic sets are introduced.

Definition 1. *[33,34] The neutrosophic set N is characterized by three membership functions, which are the truth-membership function $T_{Ne}(x)$, indeterminacy-membership function $I_{Ne}(x)$ and falsity-membership function $F_{Ne}(x)$, where $x \in X$ and X are a space of points. Also, $T_{Ne}(x)$:X→$[^-0, 1^+]$, $I_{Ne}(x)$:X→$[^-0, 1^+]$ and $F_{Ne}(x)$:X→$[^-0, 1^+]$. There is no restriction on the sum of $T_{Ne}(x)$, $I_{Ne}(x)$ and $F_{Ne}(x)$, so $0^- \leq \sup T_{Ne}(x) + \sup I_{Ne}(x) + \sup F_{Ne}(x) \leq 3^+$.*

Definition 2. *[33,35] A single valued neutrosophic set Ne over X takes the following form: $A = \{\langle x, T_{Ne}(x), I_{Ne}(x), F_{Ne}(x)\rangle : x \in X\}$, where $T_{Ne}(x)$:X→[0,1], $I_{Ne}(x)$:X→[0,1] and $F_{Ne}(x)$:X→[0,1], with $0 \leq T_{Ne}(x)$: $+ I_{Ne}(x) + F_{Ne}(x) \leq 3$ for all $x \in X$. The single valued neutrosophic (SVN) number is symbolized by $Ne = (d, e, f)$, where $d, e, f \in [0,1]$ and $d + e + f \leq 3$.*

Definition 3. *[36,37] The single valued triangular neutrosophic number, $\tilde{a} = \langle (a_1, a_2, a_3); \alpha_{\tilde{a}}, \theta_{\tilde{a}}, \beta_{\tilde{a}} \rangle$, is a neutrosophic set on the real line set R, whose truth, indeterminacy and falsity membership functions are as follows:*

$$T_{\tilde{a}}(x) = \begin{cases} \alpha_{\tilde{a}}\left(\frac{x-a_1}{a_2-a_1}\right) & (a_1 \leq x \leq a_2) \\ \alpha_{\tilde{a}} & (x = a_2) \\ \alpha_{\tilde{a}}\left(\frac{a_3-x}{a_3-a_2}\right) & (a_2 < x \leq a_3) \\ 0 & \text{otherwise} \end{cases} \tag{1}$$

$$I_{\tilde{a}}(x) = \begin{cases} \frac{(a_2-x+\theta_{\tilde{a}}(x-a_1))}{(a_2-a_1)} & (a_1 \leq x \leq a_2) \\ \theta_{\tilde{a}} & x = a_2 \\ \frac{(x-a_2+\theta_{\tilde{a}}(a_3-x))}{(a_3-a_2)} & (a_2 < x \leq a_3) \\ 1 & \text{otherwise} \end{cases} \tag{2}$$

$$F_{\tilde{a}}(x) = \begin{cases} \frac{(a_2 - x + \beta_{\tilde{a}}(x - a_1))}{(a_2 - a_1)} & (a_1 \le x \le a_2) \\ \beta_{\tilde{a}} & (x = a_2) \\ \frac{(x - a_2 + \beta_{\tilde{a}}\,(a_3 - x))}{(a_3 - a_2)} & (a_2 < x \le a_3) \\ 1 & \text{otherwise} \end{cases} \tag{3}$$

where $\alpha_{\tilde{a}}, \theta_{\tilde{a}}, \beta_{\tilde{a}} \in [0,1]$ and $a_1, a_2, a_3 \in R$, $a_1 \le a_2 \le a_3$.

Definition 4. *[34,36] Let* $\tilde{a} = \langle (a_1,\ a_2,\ a_3); \alpha_{\tilde{a}}, \theta_{\tilde{a}},\ \beta_{\tilde{a}} \rangle$ *and* $\tilde{b} = \langle (b_1,\ b_2,\ b_3); \alpha_{\tilde{a}}, \theta_{\tilde{a}},\ \beta_{\tilde{a}} \rangle$ *be two single-valued triangular neutrosophic numbers and* $\gamma \neq 0$ *be any real number. Then:*

1. Addition of two triangular neutrosophic numbers

$$\tilde{a} + \tilde{b} = \langle (a_1 + b_1, a_2 + b_2, a_3 + b_3); \alpha_{\tilde{a}} \wedge \alpha_{\tilde{b}},\ \theta_{\tilde{a}} \vee \theta_{\tilde{b}},\ \beta_{\tilde{a}} \vee \beta_{\tilde{b}} \rangle$$

2. Subtraction of two triangular neutrosophic numbers

$$\tilde{a} - \tilde{b} = \langle (a_1 - b_3, a_2 - b_2\ , a_3 - b_1); \alpha_{\tilde{a}} \wedge \alpha_{\tilde{b}},\ \theta_{\tilde{a}} \vee \theta_{\tilde{b}},\ \beta_{\tilde{a}} \vee \beta_{\tilde{b}} \rangle$$

3. Inverse of a triangular neutrosophic number

$$\tilde{a}^{-1} = \langle \left(\frac{1}{a_3}, \frac{1}{a_2}, \frac{1}{a_1} \right); \alpha_{\tilde{a}}, \theta_{\tilde{a}}, \beta_{\tilde{a}} \rangle, \text{ where } (\tilde{a} \neq 0)$$

4. Multiplication of a triangular neutrosophic number by a constant value

$$\gamma\tilde{a} = \begin{cases} \langle (\gamma a_1,\ \gamma a_2,\ \gamma a_3); \alpha_{\tilde{a}}, \theta_{\tilde{a}}, \beta_{\tilde{a}} \rangle\ if\ (\gamma > 0) \\ \langle (\gamma a_3,\ \gamma a_2, \gamma a_1); \alpha_{\tilde{a}}, \theta_{\tilde{a}}, \beta_{\tilde{a}} \rangle\ if\ (\gamma < 0) \end{cases}$$

5. Division of a triangular neutrosophic number by a constant value

$$\frac{\tilde{a}}{\gamma} = \begin{cases} \langle \left(\frac{a_1}{\gamma}, \frac{a_2}{\gamma}, \frac{a_3}{\gamma} \right); \alpha_{\tilde{a}}, \theta_{\tilde{a}}, \beta_{\tilde{a}} \rangle\ if\ (\gamma > 0) \\ \langle \left(\frac{a_3}{\gamma}, \frac{a_2}{\gamma}, \frac{a_1}{\gamma} \right); \alpha_{\tilde{a}}, \theta_{\tilde{a}}, \beta_{\tilde{a}} \rangle\ if\ (\gamma < 0) \end{cases}$$

6. Division of two triangular neutrosophic numbers

$$\frac{\tilde{a}}{\tilde{b}} = \begin{cases} \langle \left(\frac{a_1}{b_3}, \frac{a_2}{b_2}, \frac{a_3}{b_1} \right); \alpha_{\tilde{a}} \wedge \alpha_{\tilde{b}},\ \theta_{\tilde{a}} \vee \theta_{\tilde{b}}, \beta_{\tilde{a}} \vee \beta_{\tilde{b}} \rangle\ if\ (a_3 > 0, b_3 > 0) \\ \langle \left(\frac{a_3}{b_3}, \frac{a_2}{b_2}, \frac{a_1}{b_1} \right); \alpha_{\tilde{a}} \wedge \alpha_{\tilde{b}},\ \theta_{\tilde{a}} \vee \theta_{\tilde{b}}, \beta_{\tilde{a}} \vee \beta_{\tilde{b}} \rangle\ if\ (a_3 < 0, b_3 > 0) \\ \langle \left(\frac{a_3}{b_1}, \frac{a_2}{b_2}, \frac{a_1}{b_3} \right); \alpha_{\tilde{a}} \wedge \alpha_{\tilde{b}},\ \theta_{\tilde{a}} \vee \theta_{\tilde{b}}, \beta_{\tilde{a}} \vee \beta_{\tilde{b}} \rangle\ if\ (a_3 < 0, b_3 < 0) \end{cases}$$

7. Multiplication of two triangular neutrosophic numbers

$$\tilde{a}\tilde{b} = \begin{cases} \langle (a_1 b_1, a_2 b_2, a_3 b_3); \alpha_{\tilde{a}} \wedge \alpha_{\tilde{b}},\ \theta_{\tilde{a}} \vee \theta_{\tilde{b}}, \beta_{\tilde{a}} \vee \beta_{\tilde{b}} \rangle\ if\ (a_3 > 0, b_3 > 0) \\ \langle (a_1 b_3, a_2 b_2, a_3 b_1); \alpha_{\tilde{a}} \wedge \alpha_{\tilde{b}},\ \theta_{\tilde{a}} \vee \theta_{\tilde{b}}, \beta_{\tilde{a}} \vee \beta_{\tilde{b}} \rangle\ if\ (a_3 < 0, b_3 > 0) \\ \langle (a_3 b_3, a_2 b_2, a_1 b_1); \alpha_{\tilde{a}} \wedge \alpha_{\tilde{b}},\ \theta_{\tilde{a}} \vee \theta_{\tilde{b}}, \beta_{\tilde{a}} \vee \beta_{\tilde{b}} \rangle\ if\ (a_3 < 0, b_3 < 0) \end{cases}$$

4. Neutrosophic AHP (N-AHP) in SWOT Analysis

This section describes the proposed model of integrating SWOT analysis with the neutrosophic AHP. A step-by-step procedure for the model described is provided in this section.

Step 1 Select a group of experts at performing SWOT analysis.

In this step, experts identify the internal and the external factors of the SWOT analysis by employing questionnaires/interviews.

Figure 1 presents the SWOT analysis diagram:

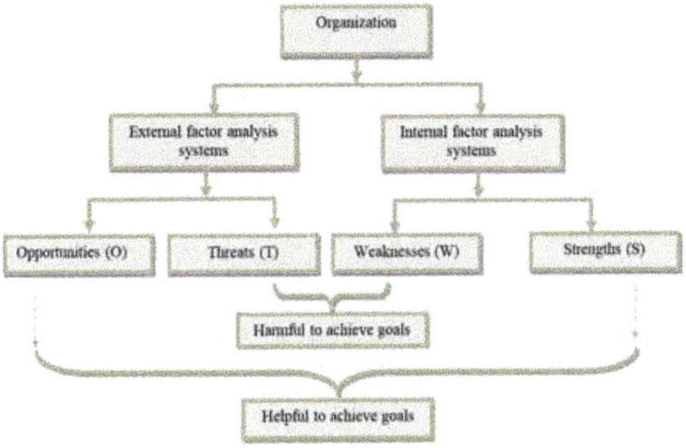

Figure 1. Strengths, Weaknesses, Opportunities and Threats (SWOT) analysis diagram.

To transform a complex problem to a simple and easy to understand problem, the following step is applied:

Step 2 Structure the hierarchy of the problem.

The hierarchy of the problem has four levels:

- The first level is the goal the organization wants to achieve.
- The second level consists of the four strategic criteria that are defined by the SWOT analysis (i.e., criteria).
- The third level are the factors that are included in each strategic factor of the previous level (i.e., sub-criteria).
- The final level includes the strategies that should be evaluated and compared.

The general hierarchy is presented in Figure 2.

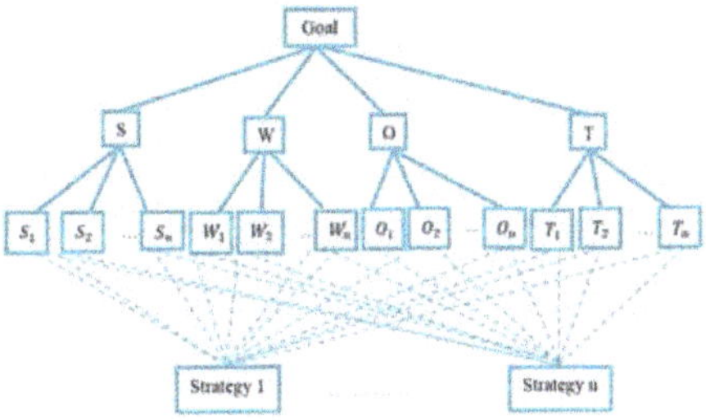

Figure 2. The hierarchy of a problem.

The next step is applied for weighting factors (criteria), sub-factors (sub-criteria) and strategies (alternatives), according to experts' opinions.

Step 3 Structure the neutrosophic pair-wise comparison matrix of factors, sub-factors and strategies, through the linguistic terms which are shown in Table 1.

Table 1. Linguistic terms and the identical triangular neutrosophic numbers.

Saaty Scale	Explanation	Neutrosophic Triangular Scale
1	Equally influential	$\tilde{1} = \langle(1, 1, 1); 0.50, 0.50, 0.50\rangle$
3	Slightly influential	$\tilde{3} = \langle(2, 3, 4); 0.30, 0.75, 0.70\rangle$
5	Strongly influential	$\tilde{5} = \langle(4, 5, 6); 0.80, 0.15, 0.20\rangle$
7	Very strongly influential	$\tilde{7} = \langle(6, 7, 8); 0.90, 0.10, 0.10\rangle$
9	Absolutely influential	$\tilde{9} = \langle(9, 9, 9); 1.00, 0.00, 0.00\rangle$
2 4 6 8	Sporadic values between two close scales	$\tilde{2} = \langle(1, 2, 3); 0.40, 0.65, 0.60\rangle$ $\tilde{4} = \langle(3, 4, 5); 0.60, 0.35, 0.40\rangle$ $\tilde{6} = \langle(5, 6, 7); 0.70, 0.25, 0.30\rangle$ $\tilde{8} = \langle(7, 8, 9); 0.85, 0.10, 0.15\rangle$

The neutrosophic scale is attained according to expert opinion.

The neutrosophic pair-wise comparison matrix of factors, sub-factors and strategies are as follows:

$$\tilde{A} = \begin{bmatrix} \tilde{1} & \tilde{a}_{12} \cdots & \tilde{a}_{1n} \\ \vdots & \ddots & \vdots \\ \tilde{a}_{n1} & \tilde{a}_{n2} \cdots & \tilde{1} \end{bmatrix} \tag{4}$$

where $\tilde{a}_{ji} = \tilde{a}_{ij}^{-1}$, and is the triangular neutrosophic number that measures the decision makers vagueness.

Step 4 Check the consistency of experts' judgments.

If the pair-wise comparison matrix has a transitive relation, i.e., $a_{ik} = a_{ij}a_{jk}$ for all i, j and k, then the comparison matrix is consistent [38], focusing only on the lower, median and upper values of the triangular neutrosophic number of the comparison matrix.

Step 5 Calculate the weight of the factors (S, W, O, T), sub-factors $\{(S_1, \ldots, S_n), (W_1, \ldots, W_n), (O_1, \ldots, O_n), (T_1, \ldots, T_n)\}$ and strategies/alternatives ($Alt_1, \ldots, Alt_n$) from the neutrosophic pair-wise comparison matrix, by transforming it to a deterministic matrix using the following equations.

Let $\tilde{a}_{ij} = \langle(a_1, b_1, c_1), \alpha_{\tilde{a}}, \theta_{\tilde{a}}, \beta_{\tilde{a}}\rangle$ be a single valued triangular neutrosophic number; then,

$$S(\tilde{a}_{ij}) = \frac{1}{8}[a_1 + b_1 + c_1] \times (2 + \alpha_{\tilde{a}} - \theta_{\tilde{a}} - \beta_{\tilde{a}}) \tag{5}$$

and

$$A(\tilde{a}_{ij}) = \frac{1}{8}[a_1 + b_1 + c_1] \times (2 + \alpha_{\tilde{a}} - \theta_{\tilde{a}} + \beta_{\tilde{a}}) \tag{6}$$

which are the score and accuracy degrees of $\tilde{a}_{ij}$ respectively.

To get the score and the accuracy degree of $\tilde{a}_{ji}$, we use the following equations:

$$S(\tilde{a}_{ji}) = 1/\ S(\tilde{a}_{ij}) \tag{7}$$

$$A(\tilde{a}_{ji}) = 1/A(\tilde{a}_{ij}) \tag{8}$$

With compensation by score value of each triangular neutrosophic number in the neutrosophic pair-wise comparison matrix, we derive the following deterministic matrix:

$$\mathrm{A} = \begin{bmatrix} 1 & a_{12}\cdots & a_{1n} \\ \vdots & \ddots & \vdots \\ a_{n1} & a_{n2}\cdots & 1 \end{bmatrix} \tag{9}$$

Determine the ranking of priorities, namely the Eigen Vector X, from the previous matrix as follows:

1. Normalize the column entries by dividing each entry by the sum of the column.
2. Take the total of the row averages.

Step 6 Calculate the total priority of each strategy (alternative) for the final ranking of all strategies using Equation (10).

The total weight value of the alternative j $(j = 1, \ldots, n)$ can be written as follows:

$$Tw_{\mathrm{Alt}_j} = w_S * \sum_{i=1}^{n} w_{S_i} * w_{\mathrm{Alt}_j} + w_W * \sum_{i=1}^{n} w_{W_i} * w_{\mathrm{Alt}_j} + w_O * \sum_{i=1}^{n} w_{O_i} * w_{\mathrm{Alt}_j} + w_T * \sum_{i=1}^{n} w_{T_i} * w_{\mathrm{Alt}_j} \tag{10}$$

where $(i = 1, \ldots, n)$ and (w_S, w_W, w_O, w_T) are the weights of Strengths, Weaknesses, Opportunities and Threats; $(w_{S_i}, w_{W_i}, w_{O_i}, w_{T_i})$ are the sub-factor weights; and w_{Alt_j} is the weight of the alternative j, corresponding to its sub-factor.

From previous steps, we obtain the phases of integrating SWOT analysis with neutrosophic analytic hierarchy processes, as shown in Figure 3.

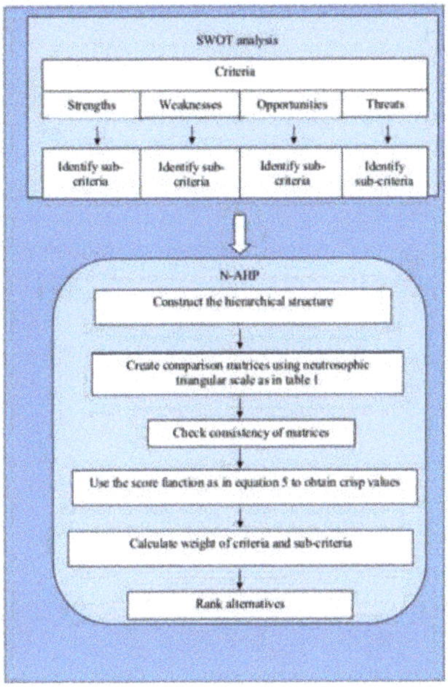

Figure 3. SWOT-neutrosophic analytic hierarchy process (N-AHP) diagram.

5. Illustrative Example

The model proposed in Section 3 is used to solve a real case study in this section.

Every company should analyze and dissect itself from time to time, in order to face competition. This is important especially when a company wants to launch a new product, or open a new market, in order to measure its presumptive success. A company can appraise itself honestly and effectively by performing SWOT analysis, which will help it examine its performance by analyzing internal and external factors. Once SWOT analysis is complete, a company will gain more information about its capabilities. For the evaluation process, a multi-criteria decision-making technique should be used. In this research, we used a neutrosophic AHP. A case study is offered in this section to illustrate this process in detail.

The phases for implementing a N-AHP in SWOT analysis are shown in Figure 4.

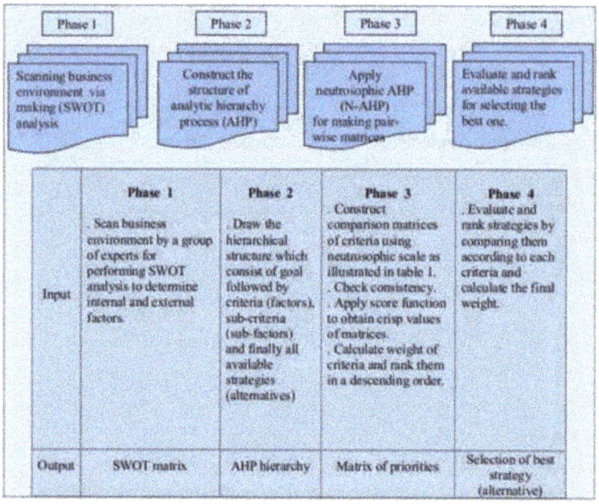

Figure 4. The phases for implementing a N-AHP in SWOT analysis.

Starbucks Company is the most widely prolific marketer and retailer of coffee in the world. The company has branches in 75 countries, with more than 254,000 employees. The company also sells different types of coffee and tea products and has a licensed trademark. The company offers food, in addition to coffee, and this makes it an attractive spot for snacks and breakfast. The company has different competitors, such as Caribou Coffee Company, Costa Coffee, Green Mountain Coffee Roasters and many others. To face competition, a group of experts perform Starbucks SWOT analysis, as shown in Figure 5. Depending on the SWOT factors and sub-factors, a set of alternatives strategies is developed. Our aim was to prioritize the strategies suggested by company indicators.

These strategies were:

❖ SO strategies

- Amplifying global stores
- Seeking higher growth markets

❖ WO strategies

- Adding different forms, new categories and diverse channels of products
- Trying to minimize the coffee price

❖ ST strategies

- Taking precautions to mitigate economic crises and maintain profitability

❖ WT strategies

- Competing with other companies by offering different coffee and creating brand loyalty
- Diversifying stores around the world and minimizing raw materials prices

By applying our proposed model to Starbucks Company, the evaluation process and the selection of different strategies was anticipated to become simpler and more valuable.

Step 1 Perform SWOT analysis.

Four experts were selected to perform Starbucks Company SWOT analysis, as they had experience in the coffee industry.

To implement the SWOT analysis, we prepared a questionnaire (see Appendix A) and sent it out online to experts. After obtaining the answers, the internal (Strengths and Weaknesses) and external (Opportunities and Threats) factors were identified, as shown in Figure 5.

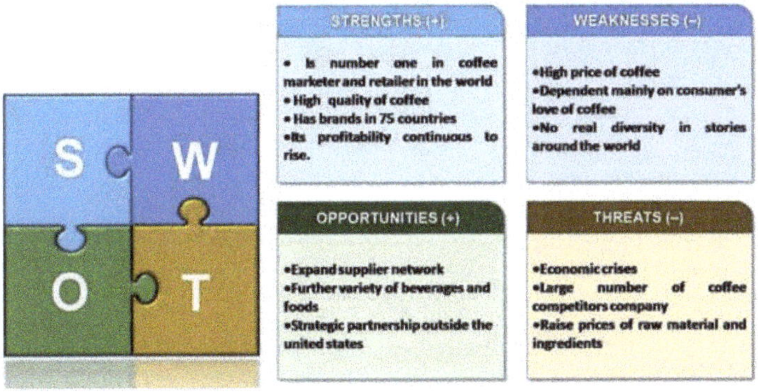

Figure 5. Starbucks SWOT analysis.

Step 2 Structure the hierarchy of the problem.

The hierarchical structure of Starbucks Company, according to the proposed methodology, is presented in Figure 6.

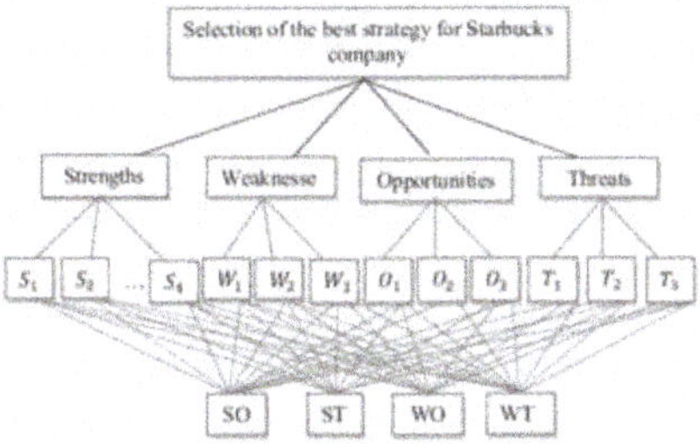

Figure 6. The hierarchical structure of the problem.

In Figure 6, $S_1, \ldots, S_4$ were the strengths sub-factors, as listed in the SWOT analysis. Also, $W_1, \ldots, W_3$, $O_1, \ldots, O_3$ and $T_1, \ldots, T_3$ were the weaknesses, opportunities and threats sub-factors of the SWOT analysis, as shown in Figure 5.

Step 3 Structure the neutrosophic pair-wise comparison matrix of factors, sub-factors and strategies, through the linguistic terms which are shown in Table 1. The values in Table 2 pertain to the experts' opinions.

The pair-wise comparison matrix of SWOT factors is presented in Table 2.

Table 2. The neutrosophic comparison matrix of factors.

Factors	Strengths	Weaknesses	Opportunities	Threats
Strengths	$\langle(1, 1, 1); 0.50, 0.50, 0.50\rangle$	$\langle(1, 1, 1); 0.50, 0.50, 0.50\rangle$	$\langle(4, 5, 6); 0.80, 0.15, 0.20\rangle$	$\langle(6, 7, 8); 0.90, 0.10, 0.10\rangle$
Weaknesses	$\langle(1, 1, 1); 0.50, 0.50, 0.50\rangle$	$\langle(1, 1, 1); 0.50, 0.50, 0.50\rangle$	$\langle(4, 5, 6); 0.80, 0.15, 0.20\rangle$	$\langle(6, 7, 8); 0.90, 0.10, 0.10\rangle$
Opportunities	$\langle(\frac{1}{6}, \frac{1}{5}, \frac{1}{4}); 0.80, 0.15, 0.20\rangle$	$\langle(\frac{1}{6}, \frac{1}{5}, \frac{1}{4}); 0.80, 0.15, 0.20\rangle$	$\langle(1, 1, 1); 0.50, 0.50, 0.50\rangle$	$\langle(\frac{1}{4}, \frac{1}{3}, \frac{1}{2}); 0.30, 0.75, 0.70\rangle$
Threats	$\langle(\frac{1}{8}, \frac{1}{7}, \frac{1}{6}); 0.90, 0.10, 0.10\rangle$	$\langle(\frac{1}{8}, \frac{1}{7}, \frac{1}{6}); 0.90, 0.10, 0.10\rangle$	$\langle(2, 3, 4); 0.30, 0.75, 0.70\rangle$	$\langle(1, 1, 1); 0.50, 0.50, 0.50\rangle$

Step 4 Check the consistency of experts' judgments.

The previous comparison matrix was consistent when applying the method proposed in [38].

Step 5 Calculate the weight of the factors, sub-factors and strategies.

To calculate weight, we first transformed the neutrosophic comparison matrix to its crisp form by using Equation (5). The crisp matrix is presented in Table 3.

Table 3. The crisp comparison matrix of factors.

Factors	Strengths	Weaknesses	Opportunities	Threats
Strengths	1	1	4	7
Weaknesses	1	1	4	7
Opportunities	$\frac{1}{4}$	$\frac{1}{4}$	1	1
Threats	$\frac{1}{7}$	$\frac{1}{7}$	1	1

Then, we determined the ranking of the factors, namely the Eigen Vector X, from the previous matrix, as illustrated previously in the detailed steps of the proposed model.

The normalized comparison matrix of factors is presented in Table 4.

Table 4. The normalized comparison matrix of factors.

Factors	Strengths	Weaknesses	Opportunities	Threats
Strengths	0.4	0.4	0.4	0.44
Weaknesses	0.4	0.4	0.4	0.44
Opportunities	0.1	0.1	0.1	0.06
Threats	0.06	0.06	0.1	0.06

By taking the total of the row averages:

$$X = \begin{bmatrix} 0.41 \\ 0.41 \\ 0.1 \\ 0.1 \end{bmatrix}$$

The neutrosophic comparison matrix of strengths is presented in Table 5.

Table 5. The neutrosophic comparison matrix of strengths.

Strengths	S_1	S_2	S_3	S_4
S_1	$\langle(1, 1, 1); 0.50, 0.50, 0.50\rangle$	$\langle(4, 5, 6); 0.80, 0.15, 0.20\rangle$	$\langle(2, 3, 4); 0.30, 0.75, 0.70\rangle$	$\langle(2, 3, 4); 0.30, 0.75, 0.70\rangle$
S_2	$\langle(\frac{1}{6}, \frac{1}{5}, \frac{1}{4}); 0.80, 0.15, 0.20\rangle$	$\langle(1, 1, 1); 0.50, 0.50, 0.50\rangle$	$\langle(\frac{1}{6}, \frac{1}{5}, \frac{1}{4}); 0.80, 0.15, 0.20\rangle$	$\langle(2, 3, 4); 0.30, 0.75, 0.70\rangle$
S_3	$\langle(\frac{1}{4}, \frac{1}{3}, \frac{1}{2}); 0.30, 0.75, 0.70\rangle$	$\langle(4, 5, 6); 0.80, 0.15, 0.20\rangle$	$\langle(1, 1, 1); 0.50, 0.50, 0.50\rangle$	$\langle(2, 3, 4); 0.30, 0.75, 0.70\rangle$
S_4	$\langle(\frac{1}{4}, \frac{1}{3}, \frac{1}{2}); 0.30, 0.75, 0.70\rangle$	$\langle(\frac{1}{4}, \frac{1}{3}, \frac{1}{2}); 0.30, 0.75, 0.70\rangle$	$\langle(\frac{1}{4}, \frac{1}{3}, \frac{1}{2}); 0.30, 0.75, 0.70\rangle$	$\langle(1, 1, 1); 0.50, 0.50, 0.50\rangle$

The crisp pair-wise comparison matrix of strengths is presented in Table 6 and the normalized comparison matrix of strengths is presented in Table 7.

Table 6. The crisp comparison matrix of strengths.

Strengths	S_1	S_2	S_3	S_4
S_1	1	3	1	1
S_2	$\frac{1}{3}$	1	$\frac{1}{4}$	1
S_3	1	4	1	1
S_4	1	1	1	1

Table 7. The normalized comparison matrix of strengths.

Strengths	S_1	S_2	S_3	S_4
S_1	0.3	0.3	0.3	0.25
S_2	0.1	0.1	0.1	0.25
S_3	0.3	0.4	0.3	0.25
S_4	0.3	0.1	0.3	0.25

By taking the total of the row averages:

$$X = \begin{bmatrix} 0.29 \\ 0.14 \\ 0.31 \\ 0.24 \end{bmatrix}$$

The neutrosophic comparison matrix of weaknesses is presented in Table 8.

Table 8. The neutrosophic comparison matrix of weaknesses.

Weaknesses	W_1	W_2	W_3
W_1	$\langle(1, 1, 1); 0.50, 0.50, 0.50\rangle$	$\langle(\frac{1}{6}, \frac{1}{5}, \frac{1}{4}); 0.80, 0.15, 0.20\rangle$	$\langle(\frac{1}{4}, \frac{1}{3}, \frac{1}{2}); 0.30, 0.75, 0.70\rangle$
W_2	$\langle(4, 5, 6); 0.80, 0.15, 0.20\rangle$	$\langle(1, 1, 1); 0.50, 0.50, 0.50\rangle$	$\langle(4, 5, 6); 0.80, 0.15, 0.20\rangle$
W_3	$\langle(2, 3, 4); 0.30, 0.75, 0.70\rangle$	$\langle(\frac{1}{6}, \frac{1}{5}, \frac{1}{4}); 0.80, 0.15, 0.20\rangle$	$\langle(1, 1, 1); 0.50, 0.50, 0.50\rangle$

The crisp comparison matrix of weaknesses is presented in Table 9.

Table 9. The crisp comparison matrix of weaknesses.

Weaknesses	W_1	W_2	W_3
W_1	1	$\frac{1}{4}$	1
W_2	4	1	4
W_3	1	$\frac{1}{4}$	1

The normalized comparison matrix of weaknesses is presented in Table 10.

Table 10. The normalized comparison matrix of weaknesses.

Weaknesses	W_1	W_2	W_3
W_1	0.2	0.2	0.2
W_2	0.7	0.7	0.7
W_3	0.2	0.2	0.2

By taking the total of the row averages:

$$X = \begin{bmatrix} 0.2 \\ 0.35 \\ 0.2 \end{bmatrix}$$

The neutrosophic comparison matrix of opportunities is presented in Table 11.

Table 11. The neutrosophic comparison matrix of opportunities.

Opportunities	O_1	O_2	O_3
O_1	$\langle(1, 1, 1); 0.50, 0.50, 0.50\rangle$	$\langle(\frac{1}{4}, \frac{1}{3}, \frac{1}{2}); 0.30, 0.75, 0.70\rangle$	$\langle(\frac{1}{6}, \frac{1}{5}, \frac{1}{4}); 0.80, 0.15, 0.20\rangle$
O_2	$\langle(2, 3, 4); 0.30, 0.75, 0.70\rangle$	$\langle(1, 1, 1); 0.50, 0.50, 0.50\rangle$	$\langle(\frac{1}{4}, \frac{1}{3}, \frac{1}{2}); 0.30, 0.75, 0.70\rangle$
O_3	$\langle(4, 5, 6); 0.80, 0.15, 0.20\rangle$	$\langle(2, 3, 4); 0.30, 0.75, 0.70\rangle$	$\langle(1, 1, 1); 0.50, 0.50, 0.50\rangle$

The crisp comparison matrix of opportunities is presented in Table 12.

Table 12. The crisp comparison matrix of opportunities.

Opportunities	O_1	O_2	O_3
O_1	1	1	$\frac{1}{4}$
O_2	1	1	1
O_3	4	1	1

The normalized comparison matrix of opportunities is presented in Table 13.

Table 13. The normalized comparison matrix of opportunities.

Opportunities	O_1	O_2	O_3
O_1	0.2	0.3	0.1
O_2	0.2	0.3	0.4
O_3	0.7	0.3	0.4

By taking the total of the row averages:

$$X = \begin{bmatrix} 0.2 \\ 0.3 \\ 0.5 \end{bmatrix}$$

The neutrosophic comparison matrix of threats is presented in Table 14.

Table 14. The neutrosophic comparison matrix of threats.

Threats	T_1	T_2	T_3
T_1	$\langle(1, 1, 1); 0.50, 0.50, 0.50\rangle$	$\langle(2, 3, 4); 0.30, 0.75, 0.70\rangle$	$\langle(4, 5, 6); 0.80, 0.15, 0.20\rangle$
T_2	$\langle(\frac{1}{4}, \frac{1}{3}, \frac{1}{2}); 0.30, 0.75, 0.70\rangle$	$\langle(1, 1, 1); 0.50, 0.50, 0.50\rangle$	$\langle(\frac{1}{4}, \frac{1}{3}, \frac{1}{2}); 0.30, 0.75, 0.70\rangle$
T_3	$\langle(\frac{1}{6}, \frac{1}{5}, \frac{1}{4}); 0.80, 0.15, 0.20\rangle$	$\langle(2, 3, 4); 0.30, 0.75, 0.70\rangle$	$\langle(1, 1, 1); 0.50, 0.50, 0.50\rangle$

The crisp comparison matrix of threats is presented in Table 15.

Table 15. The crisp comparison matrix of threats.

Threats	T_1	T_2	T_3
T_1	1	1	4
T_2	1	1	1
T_3	4	1	1

The normalized comparison matrix of threats is presented in Table 16.

Table 16. The normalized comparison matrix of threats.

Opportunities	T_1	T_2	T_3
T_1	0.2	0.3	0.7
T_2	0.2	0.3	0.2
T_3	0.7	0.3	0.2

By taking the total of the row averages:

$$X = \begin{bmatrix} 0.4 \\ 0.2 \\ 0.4 \end{bmatrix}$$

Similar to the factors and sub-factors calculation methodology, the weights of alternatives (strategies), with respect to sub-factors, were as follows:

$$\text{The Eigen Vector X of strategies with respect to } S_1 = \begin{bmatrix} 0.4 \\ 0.1 \\ 0.3 \\ 0.2 \end{bmatrix}$$

$$\text{The Eigen Vector X of strategies with respect to } S_2 = \begin{bmatrix} 0.4 \\ 0.3 \\ 0.2 \\ 0.1 \end{bmatrix}$$

$$\text{The Eigen Vector X of strategies with respect to } S_3 = \begin{bmatrix} 0.5 \\ 0.3 \\ 0.1 \\ 0.1 \end{bmatrix}$$

$$\text{The Eigen Vector X of strategies with respect to } S_4 = \begin{bmatrix} 0.3 \\ 0.2 \\ 0.4 \\ 0.1 \end{bmatrix}$$

$$\text{The Eigen Vector X of strategies with respect to } W_1 = \begin{bmatrix} 0.2 \\ 0.2 \\ 0.3 \\ 0.3 \end{bmatrix}$$

$$\text{The Eigen Vector X of strategies with respect to } W_2 = \begin{bmatrix} 0.4 \\ 0.1 \\ 0.3 \\ 0.2 \end{bmatrix}$$

$$\text{The Eigen Vector X of strategies with respect to } W_3 = \begin{bmatrix} 0.6 \\ 0.1 \\ 0.2 \\ 0.1 \end{bmatrix}$$

$$\text{The Eigen Vector X of strategies with respect to } O_1 = \begin{bmatrix} 0.1 \\ 0.4 \\ 0.2 \\ 0.3 \end{bmatrix}$$

$$\text{The Eigen Vector X of strategies with respect to } O_2 = \begin{bmatrix} 0.1 \\ 0.4 \\ 0.2 \\ 0.3 \end{bmatrix}$$

$$\text{The Eigen Vector X of strategies with respect to } O_3 = \begin{bmatrix} 0.3 \\ 0.2 \\ 0.3 \\ 0.2 \end{bmatrix}$$

$$\text{The Eigen Vector X of strategies with respect to } T_1 = \begin{bmatrix} 0.1 \\ 0.4 \\ 0.2 \\ 0.3 \end{bmatrix}$$

$$\text{The Eigen Vector X of strategies with respect to } T_2 = \begin{bmatrix} 0.6 \\ 0.2 \\ 0.1 \\ 0.1 \end{bmatrix}$$

$$\text{The Eigen Vector X of strategies with respect to } T_3 = \begin{bmatrix} 0.5 \\ 0.1 \\ 0.2 \\ 0.2 \end{bmatrix}$$

Step 6 Determine the total priority of each strategy (alternative) and define the final ranking of all strategies using Equation (10).

The weights of SWOT factors, sub-factors and alternative strategies are presented in Table 17.

According to our analysis of Starbucks Company using SWOT–N-AHP, the strategies were ranked as follows: SO, WO, ST and WT, as presented in detail in Table 17 and in Figure 7. In conclusion, SO was the best strategy for achieving Starbuck's goals since it had the greatest weight value.

Table 17. The weights of SWOT factors, sub-factors, alternatives strategies and their ranking.

Factors/Sub-Factors	Weight	Alternatives (Strategies)			
		SO	ST	WO	WT
Strengths	0.41				
S_1	0.29	0.4	0.1	0.3	0.2
S_2	0.14	0.4	0.3	0.2	0.1
S_3	0.31	0.5	0.3	0.1	0.1
S_4	0.24	0.3	0.2	0.4	0.1
Weaknesses	0.41				
W_1	0.2	0.2	0.2	0.3	0.3
W_2	0.35	0.4	0.1	0.3	0.2
W_3	0.2	0.6	0.1	0.2	0.1
Opportunities	0.1				
O_1	0.2	0.1	0.4	0.2	0.3
O_2	0.3	0.1	0.4	0.2	0.3
O_3	0.5	0.3	0.2	0.3	0.2
Threats	0.1				
T_1	0.4	0.1	0.4	0.2	0.3
T_2	0.2	0.6	0.2	0.1	0.1
T_3	0.4	0.5	0.1	0.2	0.2
Total		0.34	0.2	0.22	0.15
Rank of strategies		1	3	2	4

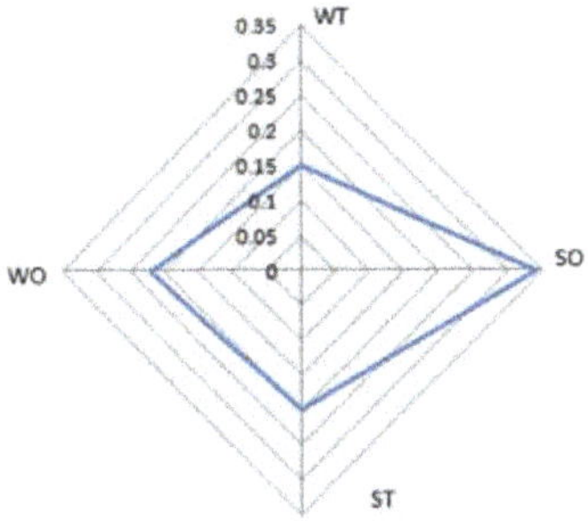

Figure 7. The final ranking of strategies.

To evaluate the quality of the proposed model, we compared it with other existing methods:

- The authors in [18–21] combined the AHP with SWOT analysis to solve the drawbacks of SWOT analysis, as illustrated in the introduction section, but in the comparison matrices of the AHP they used crisp values, which were not accurate due to the vague and uncertain information of decision makers.
- In order to solve the drawbacks of classical AHP, several researchers combined SWOT analysis with the fuzzy AHP [24–26]. Since fuzzy sets consider only the truth degree and fail to deal with the indeterminacy and falsity degrees, it also does not offer the best representation of vague and uncertain information.
- Since neutrosophic sets consider truth, indeterminacy and falsity degrees altogether, it is the best representation for the vague and uncertain information that exists in the real world. We were the first to integrate the neutrosophic AHP with SWOT analysis. In addition, our model considered all aspects of vague and uncertain information by creating a triangular neutrosophic scale for comparing factors and strategies. Due to its versatility, this method can be applied to various problems across different fields.

6. Conclusions and Future Works

SWOT analysis is an important tool for successful planning, but it has some drawbacks because it fails to provide measurements and evaluations of factors (criteria) and strategies (alternatives). In order to deal with SWOT analysis drawbacks, this research integrated the neutrosophic AHP (N-AHP) approach. Using the N-AHP in SWOT analysis produced both quantitative and qualitative measurements of factors. The reasons for applying an AHP in a neutrosophic environment are as follows: due to vague, uncertain and inconsistent information, which usually exists in real world applications, the crisp values in the classical AHP are not accurate; in the fuzzy AHP, only the truth degree is considered, which makes it incompatible with real world applications; and the intuitionistic AHP holds only truth and falsity degrees, therefore failing to deal with indeterminacy. The neutrosophic AHP is useful to interpret vague, inconsistent and incomplete information by deeming the truth, indeterminacy and falsity degrees altogether. Therefore, by integrating the N-AHP with SWOT analysis we were able to effectively and efficiently deal with vague information better than fuzzy and intuitionistic fuzzy set theories. The parameters of the N-AHP comparison matrices were triangular neutrosophic numbers and a score function was used to transform the neutrosophic AHP parameters to deterministic values. By applying our proposed model to Starbucks Company, the evaluation process of its performance was effective, and the selection between the different strategies became simpler and more valuable.

In the future, this research should be extended by employing different multi-criteria decision-making (MCDM) techniques and studying their effect on SWOT analysis. In particular, it would be useful to integrate SWOT analysis with the neutrosophic network process (ANP) to effectively deal with interdependencies between decision criteria and handle the vague, uncertain and inconsistent information that exists in real world applications.

Acknowledgments: The authors would like to thank the anonymous referees, Chief-Editor, and support Editors for their constructive suspensions and propositions that have helped to improve the quality of this research.

Author Contributions: All authors contributed equally to this paper. The individual responsibilities and contributions of each author is described as follows: the idea of the paper was put forward by Mohamed Abdel-Basset; Florentin Smarandache completed the preparatory work of the paper; Mai Mohamed analyzed the existing work; and the revision and submission of the paper was completed by Mohamed Abdel-Basset.

Conflicts of Interest: The authors declare no conflict of interest.

Appendix A

Four experts were selected to perform the SWOT analysis to determine the four strategic factors of Starbucks Company. The experts were specialized in manufacturing, sales and quality. To implement the SWOT analysis, we prepared the following questionnaire and sent it out online to the experts:

1. What is your specialty?
2. How many years of experience in coffee industry you have?
3. What are in your opinion the strengths of the Starbucks Company?
4. What are in your opinion the weaknesses of the Starbucks Company?
5. What are in your opinion the opportunities of the Starbucks Company?
6. What are in your opinion the threats of the Starbucks Company?
7. Please use the triangular neutrosophic scale introduced in Table 1 to compare all factors and present your answers in a table format.
8. Please use the triangular neutrosophic scale introduced in Table 1 to compare all strategies and present your answers in a table format.
9. In your opinion, which strategy from below will achieve the Starbucks goals:
    - SO, A strategic plan involving a good use of opportunities through existing strengths.
    - ST, A good use of strengths to remove or reduce the impact of threats.

- WO, Taking into accounts weaknesses to gain benefit from opportunities.
- WT, Reducing threats by becoming aware of weaknesses.

References

1. Killen, C.P.; Walker, M.; Hunt, R.A. Strategic planning using QFD. *Int. J. Qual. Reliab. Manag.* **2005**, *22*, 17–29. [CrossRef]
2. Saaty, T.L. What is the analytic hierarchy process? In *Mathematical Models for Decision Supported*; Springer: Berlin, Germany, 1988; pp. 109–121.
3. Saaty, T.L. *Theory and Applications of the Analytic Network Process: Decision Making with Benefits, Opportunities, Costs, and Risks*; RWS: Chalfont St Peter, UK, 2005.
4. Taghavifard, M.T.; Amoozad Mahdiraji, H.; Alibakhshi, A.M.; Zavadskas, E.K.; Bausys, R. An Extension of Fuzzy SWOT Analysis: An Application to Information Technology. *Information* **2018**, *9*, 46. [CrossRef]
5. Atanassov, K.T. Intuitionistic fuzzy sets. *Fuzzy Sets Syst.* **1986**, *20*, 87–96. [CrossRef]
6. Abdel-Basset, M.; Mohamed, M.; Chang, V. NMCDA: A framework for evaluating cloud computing services. *Future Gener. Comput. Syst.* **2018**, *86*, 12–29. [CrossRef]
7. Arslan, O.; Er, I.D. SWOT analysis for safer carriage of bulk liquid chemicals in tankers. *J. Hazard. Mater.* **2008**, *154*, 901–913. [CrossRef] [PubMed]
8. Helms, M.M.; Nixon, J. Exploring SWOT analysis—Where are we now? A review of academic research from the last decade. *J. Strategy Manag.* **2010**, *3*, 215–251. [CrossRef]
9. Kotler, P. *Marketing Management, Analysis, Planning, Implementation, and Control, Philip Kotler*; Prentice-Hall International: London, UK, 1994.
10. Wheelen, T.L.; Hunger, J.D. *Strategic Management and Business Policy*; Addison-Wesley: Boston, MA, USA, 1986.
11. Kajanus, M.; Leskinen, P.; Kurttila, M.; Kangas, J. Making use of MCDS methods in SWOT analysis—Lessons learnt in strategic natural resources management. *For. Policy Econ.* **2012**, *20*, 1–9. [CrossRef]
12. Dyson, R.G. Strategic development and SWOT analysis at the University of Warwick. *Eur. J. Oper. Res.* **2004**, *152*, 631–640. [CrossRef]
13. Chang, L.; Lin, C. The exploratory study of competitive advantage of Hsin-Chu city government by using diamond theory. *Bus. Rev.* **2005**, *3*, 180–185.
14. Khatri, J.K.; Metri, B. SWOT-AHP approach for sustainable manufacturing strategy selection: A case of Indian SME. *Glob. Bus. Rev.* **2016**, *17*, 1211–1226. [CrossRef]
15. Mehta, A.; Armenakis, A.; Mehta, N.; Irani, F. Challenges and opportunities of business process outsourcing in India. *J. Lab. Res.* **2006**, *27*, 323–338. [CrossRef]
16. Vaidya, O.S.; Kumar, S. Analytic hierarchy process: An overview of applications. *Eur. J. Oper. Res.* **2006**, *169*, 1–29. [CrossRef]
17. Akyuz, E.; Celik, M. A hybrid decision-making approach to measure effectiveness of safety management system implementations on-board ships. *Saf. Sci.* **2014**, *68*, 169–179. [CrossRef]
18. Leskinen, L.A.; Leskinen, P.; Kurttila, M.; Kangas, J.; Kajanus, M. Adapting modern strategic decision support tools in the participatory strategy process—A case study of a forest research station. *For. Policy Econ.* **2006**, *8*, 267–278. [CrossRef]
19. Pesonen, M.; Kurttila, M.; Kangas, J.; Kajanus, M.; Heinonen, P. Assessing the priorities using A'WOT among resource management strategies at the Finnish Forest and Park Service. *For. Sci.* **2001**, *47*, 534–541.
20. Masozera, M.K.; Alavalapati, J.R.; Jacobson, S.K.; Shrestha, R.K. Assessing the suitability of community-based management for the Nyungwe Forest Reserve, Rwanda. *For. Policy Econ.* **2006**, *8*, 206–216. [CrossRef]
21. Shrestha, R.K.; Alavalapati, J.R.; Kalmbacher, R.S. Exploring the potential for silvopasture adoption in south-central Florida: An application of SWOT–AHP method. *Agric. Syst.* **2004**, *81*, 185–199. [CrossRef]
22. Stewart, R.A.; Mohamed, S.; Daet, R. Strategic implementation of IT/IS projects in construction: A case study. *Autom. Constr.* **2002**, *11*, 681–694. [CrossRef]
23. Chan, P.S.; Heide, D. Information technology and the new environment: Developing and sustaining competitive advantage. *SAM Adv. Manag. J.* **1992**, *57*, 4.
24. Demirtas, N.; Tuzkaya, U.R.; Seker, S. Project Management Methodology Selection Using SWOT-Fuzzy AHP. In Proceedings of the World Congress on Engineering, London, UK, 2–4 July 2014.

25. Lumaksono, H. Implementation of SWOT-FAHP method to determine the best strategy on development of traditional shipyard in Sumenep. *Acad. Res. Int.* **2014**, *5*, 56.
26. Tavana, M.; Zareinejad, M.; Di Caprio, D.; Kaviani, M.A. An integrated intuitionistic fuzzy AHP and SWOT method for outsourcing reverse logistics. *Appl. Soft Comput.* **2016**, *40*, 544–557. [CrossRef]
27. Zhang, C.; Li, D.; Sangaiah, A.K.; Broumi, S. Merger and Acquisition Target Selection Based on Interval NeutrosophicMulti-granulation Rough Sets over Two Universes. *Symmetry* **2017**, *9*, 126. [CrossRef]
28. Goyal, R.K.; Kaushal, S.; Sangaiah, A.K. The utility based non-linear fuzzy AHP optimization model for network selection in heterogeneous wireless networks. *Appl. Soft Comput.* **2017**. [CrossRef]
29. Abdel-Basset, M.; Mohamed, M.; Hussien, A.N.; Sangaiah, A.K. A novel group decision-making model based on triangular neutrosophic numbers. *Soft Comput.* **2017**, 1–15. [CrossRef]
30. Wang, L.; Wang, Y.; Sangaiah, A.K.; Liao, B. Intuitionistic linguistic group decision-making methods based on generalized compensative weighted averaging aggregation operators. *Soft Comput.* **2017**, 1–13. [CrossRef]
31. Abdel-Basset, M.; Mohamed, M.; Sangaiah, A.K. Neutrosophic AHP-Delphi Group decision-making model based on trapezoidal neutrosophic numbers. *J. Ambient Intell. Humaniz. Comput.* **2017**, 1–17. [CrossRef]
32. Chan, F.T.; Chan, H.K.; Lau, H.C.; Ip, R.W. An AHP approach in benchmarking logistics performance of the postal industry. *Benchmarking* **2006**, *13*, 636–661. [CrossRef]
33. Gallego Lupiáñez, F. Interval neutrosophic sets and topology. *Kybernetes* **2009**, *38*, 621–624. [CrossRef]
34. Abdel-Basset, M.; Mohamed, M. The role of single valued neutrosophic sets and rough sets in smart city: imperfect and incomplete information systems. *Measurement* **2018**, *124*, 47–55. [CrossRef]
35. El-Hefenawy, N.; Metwally, M.A.; Ahmed, Z.M.; El-Henawy, I.M. A review on the applications of neutrosophic sets. *J. Comput. Theor. Nanosci.* **2016**, *13*, 936–944. [CrossRef]
36. Deli, I.; Subas, Y. Single valued neutrosophic numbers and their applications to multicriteria decision making problem. *Neutrosophic Sets Syst.* **2014**, *2*, 1–3.
37. Abdel-Baset, M.; Hezam, I.M.; Smarandache, F. Neutrosophic goal programming. *Neutrosophic Sets Syst.* **2016**, *11*, 112–118.
38. Buckley, J.J. Fuzzy hierarchical analysis. *Fuzzy Sets Syst.* **1985**, *17*, 233–247. [CrossRef]
39. Li, X.; Zhang, X.; Choonkil, P. Generalized Interval Neutrosophic Choquet Aggregation Operators and Their Applications. *Symmetry* **2018**, *10*, 85. [CrossRef]
40. Li, X.; Zhang, X. Single-Valued Neutrosophic Hesitant Fuzzy Choquet Aggregation Operators for Multi-Attribute Decision Making. *Symmetry* **2018**, *10*, 50. [CrossRef]
41. Zhang, X.; Florentin, S.; Liang, X. Neutrosophic Duplet Semi-Group and Cancellable Neutrosophic Triplet Groups. *Symmetry* **2017**, *9*, 275. [CrossRef]
42. Guan, H.; Guan, S.; Zhao, A. Forecasting model based on neutrosophic logical relationship and Jaccard similarity. *Symmetry* **2017**, *9*, 191. [CrossRef]
43. Chen, J.; Ye, J.; Du, S. Vector similarity measures between refined simplified neutrosophic sets and their multiple attribute decision-making method. *Symmetry* **2017**, *9*, 153. [CrossRef]
44. Liang, W.; Zhao, G.; Wu, H. Evaluating investment risks of metallic mines using an extended TOPSIS method with linguistic neutrosophic numbers. *Symmetry* **2017**, *9*, 149. [CrossRef]
45. Jiang, W.; Shou, Y. A Novel Single-Valued Neutrosophic Set Similarity Measure and Its Application in Multicriteria Decision-Making. *Symmetry* **2017**, *9*, 127. [CrossRef]
46. Fang, Z.; Ye, J. Multiple attribute group decision-making method based on linguistic neutrosophic numbers. *Symmetry* **2017**, *9*, 111. [CrossRef]
47. Ye, J. Multiple attribute decision-making method using correlation coefficients of normal neutrosophic sets. *Symmetry* **2017**, *9*, 80. [CrossRef]
48. Guo, Z.-L.; Liu, Y.-L.; Yang, H.-L. A Novel Rough Set Model in Generalized Single Valued Neutrosophic Approximation Spaces and Its Application. *Symmetry* **2017**, *9*, 119. [CrossRef]

symmetry

MDPI

Article

A Novel Skin Lesion Detection Approach Using Neutrosophic Clustering and Adaptive Region Growing in Dermoscopy Images

Yanhui Guo [1,*], Amira S. Ashour [2] and Florentin Smarandache [3]

1 Department of Computer Science, University of Illinois at Springfield, Springfield, IL 62703, USA
2 Department of Electronics and Electrical Communications Engineering, Faculty of Engineering, Tanta University, Tanta 31527, Egypt; amirasashour@yahoo.com
3 Department of Mathematics, University of New Mexico, 705 Gurley Ave., Gallup, NM 87301, USA; fsmarandache@gmail.com
* Correspondence: yguo56@uis.edu; Tel.: +1-435-227-5882

Received: 26 March 2018; Accepted: 14 April 2018; Published: 18 April 2018

Abstract: This paper proposes novel skin lesion detection based on neutrosophic clustering and adaptive region growing algorithms applied to dermoscopic images, called NCARG. First, the dermoscopic images are mapped into a neutrosophic set domain using the shearlet transform results for the images. The images are described via three memberships: true, indeterminate, and false memberships. An indeterminate filter is then defined in the neutrosophic set for reducing the indeterminacy of the images. A neutrosophic c-means clustering algorithm is applied to segment the dermoscopic images. With the clustering results, skin lesions are identified precisely using an adaptive region growing method. To evaluate the performance of this algorithm, a public data set (ISIC 2017) is employed to train and test the proposed method. Fifty images are randomly selected for training and 500 images for testing. Several metrics are measured for quantitatively evaluating the performance of NCARG. The results establish that the proposed approach has the ability to detect a lesion with high accuracy, 95.3% average value, compared to the obtained average accuracy, 80.6%, found when employing the neutrosophic similarity score and level set (NSSLS) segmentation approach.

Keywords: neutrosophic clustering; image segmentation; neutrosophic c-means clustering; region growing; dermoscopy; skin cancer

1. Introduction

Dermoscopy is an in-vivo and noninvasive technique to assist clinicians in examining pigmented skin lesions and investigating amelanotic lesions. It visualizes structures of the subsurface skin in the superficial dermis, the dermoepidermal junction, and the epidermis [1]. Dermoscopic images are complex and inhomogeneous, but they have a significant role in early identification of skin cancer. Recognizing skin subsurface structures is performed by visually searching for individual features and salient details [2]. However, visual assessment of dermoscopic images is subjective, time-consuming, and prone to errors [3]. Consequently, researchers are interested in developing automated clinical assessment systems for lesion detection to assist dermatologists [4,5]. These systems require efficient image segmentation and detection techniques for further feature extraction and skin cancer lesion classification. However, skin cancer segmentation and detection processes are complex due to dissimilar lesion color, texture, size, shape, and type; as well as the irregular boundaries of various lesions and the low contrast between skin and the lesion. Moreover, the existence of dark hair that covers skin and lesions leads to specular reflections.

Traditional skin cancer detection techniques implicate image feature analysis to outline the cancerous areas of the normal skin. Thresholding techniques use low-level features, including intensity and color to separate the normal skin and cancerous regions. Garnavi et al. [6] applied Otsu's method to identify the core-lesion; nevertheless, such process is disposed to skin tone variations and lighting. Moreover, dermoscopic images include some artifacts due to water bubble, dense hairs, and gel that are a great challenge for accurate detection. Silveira et al. [7] evaluated six skin lesions segmentation techniques in dermoscopic images, including the gradient vector flow (GVF), level set, adaptive snake, adaptive thresholding, fuzzy-based split and merge (FSM), and the expectation–maximization level set (EMLV) methods. The results established that adaptive snake and EMLV were considered the superior semi-supervised techniques, and that FSM achieved the best fully computerized results.

In dermoscopic skin lesion images, Celebi et al. [8] applied an unsupervised method using a modified JSEG algorithm for border detection, where the original JSEG algorithm is an adjusted version of the generalized Lloyd algorithm (GLA) for color quantization. The main idea of this method is to perform the segmentation process using two independent stages, namely color quantization and spatial segmentation. However, one of the main limitations occurs when the bounding box does not entirely include the lesion. This method was evaluated on 100 dermoscopic images, and border detection error was calculated. Dermoscopic images for the initial consultation were analyzed by Argenziano et al. [9] and were compared with images from the last follow-up consultation and the symmetrical/asymmetrical structural changes. Xie and Bovik [10] implemented a dermoscopic image segmentation approach by integrating the genetic algorithm (GA) and self-generating neural network (SGNN). The GA was used to select the optimal samples as initial neuron trees, and then the SGNF was used to train the remaining samples. Accordingly, the number of clusters was determined by adjusting the SD of cluster validity. Thus, the clustering is accomplished by handling each neuron tree as a cluster. A comparative study between this method and other segmentation approaches—namely *k*-means, statistical region merging, Otsu's thresholding, and the fuzzy *c*-means methods—has been conducted revealing that the optimized method provided improved segmentation and more accurate results.

Barata et al. [11] proposed a machine learning based, computer-aided diagnosis system for melanoma using features having medical importance. This system used text labels to detect several significant dermoscopic criteria, where, an image annotation scheme was applied to associate the image regions with the criteria (texture, color, and color structures). Features fusion was then used to combine the lesions' diagnosis and the medical information. The proposed approach achieved 84.6% sensitivity and 74.2% specificity on 804 images of a multi-source data set.

Set theory, such as the fuzzy set method, has been successfully employed into image segmentation. Fuzzy sets have been introduced into image segmentation applications to handle uncertainty. Several researchers have been developing efficient clustering techniques for skin cancer segmentation and other applications based on fuzzy sets. Fuzzy c-means (FCM) uses the membership function to segment the images into one or several regions. Lee and Chen [12] proposed a segmentation technique on different skin cancer types using classical FCM clustering. An optimum threshold-based segmentation technique using type-2 fuzzy sets was applied to outline the skin cancerous areas. The results established the superiority of this method compared to Otsu's algorithm, due its robustness to skin tone variations and shadow effects. Jaisakthi et al. [13] proposed an automated skin lesion segmentation technique in dermoscopic images using a semi-supervised learning algorithm. A *k*-means clustering procedure was employed to cluster the pre-processed skin images, where the skin lesions were identified from these clusters according to the color feature. However, the fuzzy set technique cannot assess the indeterminacy of each element in the set. Zhou et al. [14] introduced the fuzzy c-means (FCM) procedure based on mean shift for detecting regions within the dermoscopic images.

Recently, neutrosophy has provided a prevailing technique, namely the neutrosophic set (NS), to handle indeterminacy during the image processing. Guo and Sengur [15] integrated the NS and FCM frameworks to resolve the inability of FCM for handling uncertain data. A clustering approach called neutrosophic c-means (NCM) clustering was proposed to cluster typical data points. The results proved the efficiency of the NCM for image segmentation and data clustering. Mohan et al. [16] proposed automated brain tumor segmentation based on a neutrosophic and *k*-means clustering technique. A non-local neutrosophic Wiener filter was used to improve the quality of magnetic resonance images (MRI) before applying the *k*-means clustering approach. The results found detection rates of 100% with 98.37% accuracy and 99.52% specificity. Sengur and Guo [17] carried out an automated technique using a multiresolution wavelet transform and NS. The color/texture features have been mapped on the NS and wavelet domain. Afterwards, the *c*-*k*-means clustering approach was employed for segmentation. Nevertheless, wavelets [18] are sensitive to poor directionality during the analysis of supplementary functions in multi-dimensional applications. Hence, wavelets are relatively ineffectual to represent edges and anisotropic features in the dermoscopic images. Subsequently, enhanced multi-scale procedures have been established, including the curvelets and shearlets to resolve the limitations of wavelet analysis. These methods have the ability to encode directional information for multi-scale analysis. Shearlets provides a sparse representation of the two-dimensional information with edge discontinuities [19]. Shearlet-based techniques were established to be superior to wavelet-based methods [20].

Dermoscopic images include several artifacts such as hair, air bubbles, and other noise factors that are considered indeterminate information. The above-mentioned skin lesion segmentation methods either need a preprocessing to deal with the indeterminate information, or their detection results must be affected by them. To overcome this disadvantage, we introduce the neutrosophic set to deal with indeterminate information in dermoscopic images; we use a shearlet transform and the neutrosophic *c*-means (NCM) method along with an indeterminacy filter (IF) to eliminate the indeterminacy for accurate skin cancer segmentation. An adaptive region growing method is also employed to identify the lesions accurately.

The rest of the paper is organized as follows. In the second section, the proposed method is presented. Then the experimental results are discussed in the third section. The conclusions are drawn in the final section.

2. Methodology

The current work proposes a skin lesion detection algorithm using neutrosophic clustering and adaptive region growing in dermoscopic images. In this study, the red channel is used to detect the lesion, where healthy skin regions tend to be reddish, while darker pixels often occur in skin lesion regions [21]. First, the shearlet transform is employed on the red channel of dermoscopic image to extract the shearlet features. Then, the red channel of the image is mapped into the neutrosophic set domain, where the map functions are defined using the shearlet features. In the neutrosophic set, an indeterminacy filtering operation is performed to remove indeterminate information, such as noise and hair without using any de-noising or hair removal approaches. Then, the segmentation is performed through the neutrosophic *c*-means (NCM) clustering algorithm. Finally, the lesions are identified precisely using adaptive region growing on the segmentation results.

2.1. Shearlet Transform

Shearlets are based on a rigorous and simple mathematical framework for the geometric representation of multidimensional data and for multiresolution analysis [22]. The shearlet transform (ST) resolves the limitations of wavelet analysis; where wavelets fail to represent the geometric regularities and yield surface singularities due to their isotropic support. Shearlets include nearly parallel elongated functions to achieve surface anisotropy along the edges. The ST is an innovative two-dimensional wavelet transformation extension using directional and multiscale filter banks to capture smooth contours corresponding to the prevailing features in an image. Typically, the ST is a function with three parameters a, s, and t denoting the scale, shear, and translation parameters, respectively. The shearlet can fix both the locations of singularities and the singularities' curve tracking automatically. For $a > 0, s \in R, t \in R^2$, the ST can be defined using the following expression [23]:

$$ST_{\varsigma}p(a,s,t) =< \langle p, \varsigma_{a,s,t} \rangle, \tag{1}$$

where $\varsigma_{a,s,t}(f) = |\det N_{a,s}|^{-1/2} \varsigma(N_{a,s}^{-1}(f-t))$ and $N_{a,s} = \begin{bmatrix} a & s \\ 0 & \sqrt{a} \end{bmatrix}$. Each matrix $N_{a,s}$ can be defined as:

$$N_{a,s} = V_s D_a, \tag{2}$$

where the shear matrix is expressed by:

$$V_s = \begin{bmatrix} 1 & s \\ 0 & 1 \end{bmatrix} \tag{3}$$

and the anisotropic dilation matrix is given by:

$$D_a = \begin{bmatrix} a & s \\ 0 & \sqrt{a} \end{bmatrix}. \tag{4}$$

During the selection of a proper decomposition function for any $\tau = (\tau_1, \tau_2) \in R^2$, and $\tau_2 \neq 0$, ς can be expressed by:

$$\widehat{\varsigma}(\tau) = \widehat{\varsigma}(\tau_1, \tau_2) = \widehat{\varsigma}_1(\tau_1)\widehat{\varsigma}_2\left(\frac{\tau_1}{\tau_2}\right), \tag{5}$$

where $\widehat{\varsigma}_1 \in L^2(R)$ and $\|\varsigma_2\|_{L_2} = 1$.

From the preceding equations, the discrete shearlet transform (DST) is formed by translation, shearing, and scaling to provide the precise orientations and locations of edges in an image. The DST is acquired by sampling the continuous ST. It offers a decent anisotropic feature extraction. Thus, the DST system is properly definite by sampling the continuous ST on a discrete subset of the shearlet group as follows, where $j, k, m \in Z \times Z \times Z^2$ [24]:

$$ST(\varsigma) = \left\{ \varsigma_{j,k,m} = a^{-\frac{3}{4}} \varsigma\left(D_a{}^{-1} V_s{}^{-1}(. - t) \right) : (j,k,m) \in \wedge \right\}. \tag{6}$$

The DST can be divided into two steps: multi-scale subdivision and direction localization [25], where the Laplacian pyramid algorithm is first applied to an image in order to obtain the low-and-high-frequency components at any scale j, and then direction localization is achieved with a shear filter on a pseudo polar grid.

2.2. Neutrosophic Images

Neutrosophy has been successfully used for many applications to describe uncertain or indeterminate information. Every event in the neutrosophy set (NS) has a certain degree of truth (T), indeterminacy (I), and falsity (F), which are independent from each other. Previously reported studies have demonstrated the role of NS in image processing [26,27].

A pixel $P(i,j)$ in an image is denoted as $P_{NS}(i,j) = \{T(i,j), I(i,j), F(i,j)\}$ in the NS domain, where $T(i,j)$, $I(i,j)$, and $F(i,j)$ are the membership values belonging to the brightest pixel set, indeterminate set, and non-white set, respectively.

In the proposed method, the red channel of the dermoscopic image is transformed into the NS domain using shearlet feature values as follows:

$$T(x,y) = \frac{ST_L(x,y) - ST_{L\min}}{ST_{L\max} - ST_{L\min}}$$
$$I(x,y) = \frac{ST_H(x,y) - ST_{H\min}}{ST_{H\max} - ST_{H\min}} \tag{7}$$

where T and I are the true and indeterminate membership values in the NS. $ST_L(x,y)$ is the low-frequency component of the shearlet feature at the current pixel $P(x, y)$. In addition, $ST_{L\max}$ and $ST_{L\min}$ are the maximum and minimum of the low-frequency component of the shearlet feature in the whole image, respectively. $ST_H(x,y)$ is the high-frequency component of the shearlet feature at the current pixel $P(x, y)$. Moreover, $ST_{H\max}$ and $ST_{H\min}$ are the maximum and minimum of the high-frequency component of the shearlet feature in the whole image, respectively. In the proposed method, we only use T and I for segmentation because we are only interested in the degree to which a pixel belongs to the high intensity set of the red channel.

2.3. Neutrosophic Indeterminacy Filtering

In an image, noise can be considered as indeterminate information, which can be handled efficiently using NS. Such noise and artifacts include the existence of hair, air bubbles, and blurred boundaries. In addition, NS can be integrated with different clustering approaches for image segmentation [16,28], where the boundary information, as well as the details, may be blurred due to the principal low-pass filter leading to inaccurate segmentation of the boundary pixels. A novel NS based clustering procedure, namely the NCM has been carried out for data clustering [15], which defined the neutrosophic membership subsets using attributes of the data. Nevertheless, when it is applied to the image processing area, it does not account for local spatial information. Several side effects can affect the image when using classical filters in the NS domain, leading to blurred edge information, incorrect boundary segmentation, and an inability to combine the local spatial information with the global intensity distribution.

After the red channel of the dermoscopic image is mapped into the NS domain, an indeterminacy filter (IF) is defined based on the neutrosophic indeterminacy value, and the spatial information is utilized to eliminate the indeterminacy. The IF is defined by using the indeterminate value $I_s(x,y)$, which has the following kernel function [28]:

$$O_I(u,v) = \frac{1}{2\pi\sigma_I^2} e^{-\frac{u^2+v^2}{2\sigma_I^2}} \tag{8}$$

$$\sigma_I(x,y) = f(I(x,y)) = rI(x,y) + q, \tag{9}$$

where σ_I represents the Gaussian distribution's standard deviation, which is defined as a linear function $f(.)$ associated with the indeterminacy degree. Since σ_I becomes large with a high indeterminacy degree, the IF can create a smooth current pixel by using its neighbors. On the other hand, with a low indeterminacy degree, the value of σ_I is small and the IF performs less smoothing on the current pixel with its neighbors.

2.4. Neutrosophic C-Means (NCM)

In the NCM algorithm, an objective function and membership are considered as follows [29]:

$$J(T,I,F,A) = \sum_{i=1}^{N}\sum_{j=1}^{A} (\varpi_1 T_{ij})^m ||x_i - a_j||^2 + \sum_{i=1}^{N} (\varpi_2 I_i)^m ||x_i - \bar{a}_{i\max}||^2 + \sum_{i=1}^{N} \delta^2 (\varpi_3 F_i)^m \tag{10}$$

$$\begin{aligned} \bar{a}_{i\max} &= \tfrac{a_{p_i}+a_{q_i}}{2} \\ p_i &= \underset{j=1,2,\cdots,A}{\operatorname{argmax}}\,(T_{ij}) \\ q_i &= \underset{j\neq p_i \cap j=1,2,\cdots,A}{\operatorname{argmax}}\,(T_{ij}) \end{aligned} \tag{11}$$

where m is a constant and usually equal to 2. The value of $\bar{a}_{i\max}$ is calculated, since p_i and q_i are identified as the cluster numbers with the largest and second largest values of T, respectively. The parameter δ is used for controlling the number of objects considered as outliers, and ϖ_i is a weight factor.

In our NS domain, we only defined the membership values of T and I. Therefore, the objective function reduces to:

$$J(T,I,F,A) = \sum_{i=1}^{N}\sum_{j=1}^{A} (\varpi_1 T_{ij})^m ||x_i - a_j||^2 + \sum_{i=1}^{N} (\varpi_2 I_i)^m ||x_i - \bar{a}_{i\max}||^2. \tag{12}$$

To minimize the objective function, three membership values are updated on each iteration as:

$$\begin{aligned} T_{ij} &= \tfrac{K}{\varpi_1}(x_i - a_j)^{-\frac{2}{m-1}} \\ I_i &= \tfrac{K}{\varpi_2}(x_i - \bar{a}_{i\max})^{-\frac{2}{m-1}} \\ K &= \left[\tfrac{1}{\varpi_1}\sum_{j=1}^{A}(x_i - a_j)^{-\frac{2}{m-1}} + \tfrac{1}{\varpi_2}(x_i - \bar{a}_{i\max})^{-\frac{2}{m-1}}\right]^{-1} \end{aligned} \tag{13}$$

where $\bar{a}_{i\max}$ is calculated based on the indexes of the largest and the second largest value of T_{ij}. The iteration does not stop until $\left|T_{ij}^{(k+1)} - T_{ij}^{(k)}\right| < \varepsilon$, where ε is a termination criterion between 0 and 1, and k is the iteration step. In the proposed method, the neutrosophic image after indeterminacy filtering is used as the input for NCM algorithm, and the segmentation procedure is performed using the final clustering results. Since the pixels whose indeterminacy membership values are higher than their true membership values, it is hard to determine which group they belong to. To solve this problem, the indeterminacy filter is employed again on all pixels, and the group is determined according to their biggest true membership values for each cluster after the IF operation.

2.5. Lesion Detection

After segmentation, the pixels in an image are grouped into several groups according to their true membership values. Due to the fact that the lesions have low intensities, especially for the core part inside a lesion, the cluster with lowest true membership value is initially considered as the lesion candidate pixels. Then an adaptive region growing algorithm is employed to precisely detect the lesion boundary parts having higher intensity and lower contrast than the core ones. A contrast ratio is defined adaptively to control the growing speed:

$$DR(t) = \frac{\text{mean}(R_a - R_b)}{\text{mean}(R_b)}, \tag{14}$$

where *DR(t)* is the contrast ratio at the t-th iteration of growing, and R_b and R_a are the regions before and after the t-th iteration of growing, respectively.

A connected component analysis is taken to extract the components' morphological features. Due to the fact that there is only one lesion in a dermoscopic image, the region with the biggest area is identified as the final lesion region. The block diagram of the proposed neutrosophic clustering and adaptive region growing (NCARG) method is illustrated in Figure 1.

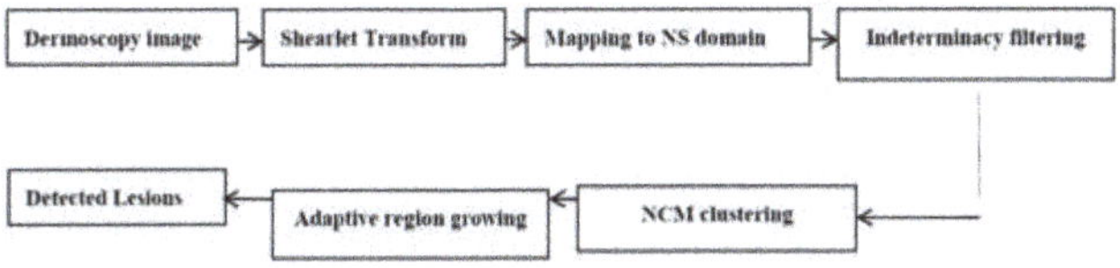

Figure 1. Flowchart of the proposed neutrosophic clustering and adaptive region growing (NCARG) skin lesion detection algorithm.

Figure 1 illustrates the steps of the proposed skin lesion segmentation method (NCARG) using neutrosophic *c*-means and region growing algorithms. Initially, the red channel of the dermoscopic image is transformed using a shearlet transform, and the shearlet features of the image are used to map the image into the NS domain. In the NS domain, an indeterminacy filtering operation is taken to remove the indeterminate information. Afterward, the segmentation is performed through NCM clustering on the filtered image. Finally, the lesion is accurately identified using an adaptive region growing algorithm where the growing speed is controlled by a newly defined contrast ratio.

To illustrate the steps in the proposed method, we use an example to demonstrate the intermediate results in Figure 2. Figure 2a,b are the original image and its ground truth image of segmentation. Figure 2c is its red channel. Figure 2d,e are the results after indeterminacy filtering and the NCM. In Figure 2f, the final detection result is outlined in blue and ground truth in red where the detection result is very close to its ground truth result.

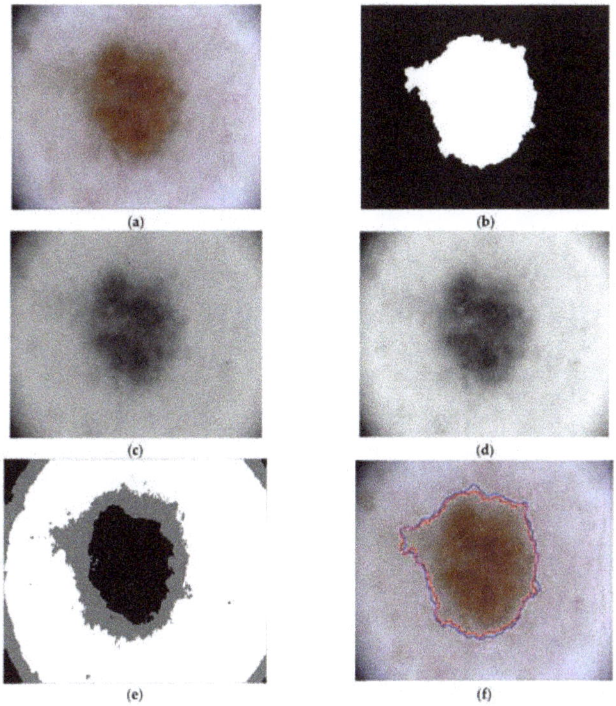

Figure 2. Intermediate results of an example image: ISIC_0000015: (**a**) Original skin lesion image; (**b**) Ground truth image; (**c**) Red channel of the original image; (**d**) Result after indeterminate filtering; (**e**) Result after NCM; (**f**) Detected lesion region after adaptive region growing, where the blue line is for the boundary of the detection result and the red line is the boundary of the ground truth result.

2.6. Evaluation Metrics

Several performance metrics are measured to evaluate the proposed skin cancer segmentation approach, namely the Jaccard index (JAC), Dice coefficient, sensitivity, specificity, and accuracy [30]. Each of these metric is defined in the remainder of this section. JAC is a statistical metric to compare diversity between the sample sets based upon the union and intersection operators as follows:

$$JAC(Y,Q) = \frac{Ar_Y \cap Ar_Q}{Ar_Y \cup Ar_Q}, \tag{15}$$

where $\cap$ and $\cup$ are the intersection and union of two sets, respectively. In addition, Ar_Y and Ar_Q are the automated segmented skin lesion area and the reference golden standard skin lesion area enclosed by the boundaries Y and Q; respectively. Typically, a value of 1 specifies complete similarity, while a JAC value of 0 specifies no similarity.

The Dice index compares the similarity of two sets, which is given as following for two sets X and Y:

$$DSC = \frac{2|X \cap Y|}{|X| + |Y|} \tag{16}$$

Furthermore, the sensitivity, specificity, and accuracy are related to the detection of the lesion region. The sensitivity indicates the true positive rate, showing how well the algorithm successfully predicts the skin lesion region, which is expressed as follows:

$$\text{Sensitivity} = \frac{\text{Number of true positives}}{\text{Number of true positives} + \text{Number of false negatives}}. \tag{17}$$

The specificity indicates the true negative rate, showing how well the algorithm predicts the non-lesion regions, which is expressed as follows:

$$\text{Specificity} = \frac{\text{Number of true negative}}{\text{Number of conditionnegative}}. \tag{18}$$

The accuracy is the proportion of true results (either positive or negative), which measures the reliability degree of a diagnostic test:

$$\text{Accuracy} = \frac{\text{Number of true positive} + \text{Number of true negative}}{\text{Number of total population}}. \tag{19}$$

These metrics are measured to evaluate the proposed NCARG method compared to another efficient segmentation algorithm that is based on the neutrosophic similarity score (NSS) and level set (LS), called NSSLS [31]. In the NSSLS segmentation method, the three membership subsets are used to transfer the input image to the NS domain, and then the NSS is applied to measure the fitting degree to the true tumor region. Finally, the LS method is employed to segment the tumor in the NSS image. In the current work, when the NSSLS is applied to the skin images, the images are interpreted using NSS, and the skin lesion boundary is extracted using the level set algorithm. Moreover, the statistical significance between the evaluated metrics using both segmentation methods is measured by calculating the significant difference value (p-value) to estimate the difference between the two methods. The p-value refers to the probability of error, where the two methods are considered statistically significant when $p \leq 0.05$.

3. Experimental Results and Discussion

3.1. Dataset

The International Skin Imaging Collaboration (ISIC) Archive [32] contains over 13,000 dermoscopic images of skin lesions. Using the images in the ISIC Archive, the 2017 ISBI Challenge on Skin Lesion Analysis

Towards Melanoma Detection was proposed to help participants develop image analysis tools to enable the automated diagnosis of melanoma from dermoscopic images. Image analysis of skin lesions includes lesion segmentation, detection and localization of visual dermoscopic features/patterns, and disease classification. All cases contain training, and binary mask images as ground truth files.

In our experiment, 50 images were selected to tune the parameters in the proposed NCARG algorithm and 500 images were used as the testing dataset. In the experiment, the parameters are set to $r = 1$, $q = 0.05$, $w1 = 0.75$, $w2 = 0.25$, and $\varepsilon = 0.001$.

3.2. Detection Results

Skin lesions are visible by the naked eye; however, early-stage detection of melanomas is complex and difficult to distinguish from benign skin lesions with similar appearances. Detecting and recognizing melanoma at its earliest stages reduces melanoma mortality. Skin lesion digital dermoscopic images are employed in the present study to detect skin lesions for accurate automated diagnosis and clinical decision support. The ISIC images are used to test and to validate the proposed approach of skin imaging. Figure 3 demonstrates the detection results using the proposed NCARG approach compared to the ground truth images. In the Figure 3d, the boundary detection results are marked in blue and the ground truth results are in red. The detection results match the ground truth results, and their boundaries are very close. Figure 3 establishes that the proposed approach accurately detects skin lesion regions, even with lesions of different shapes and sizes.

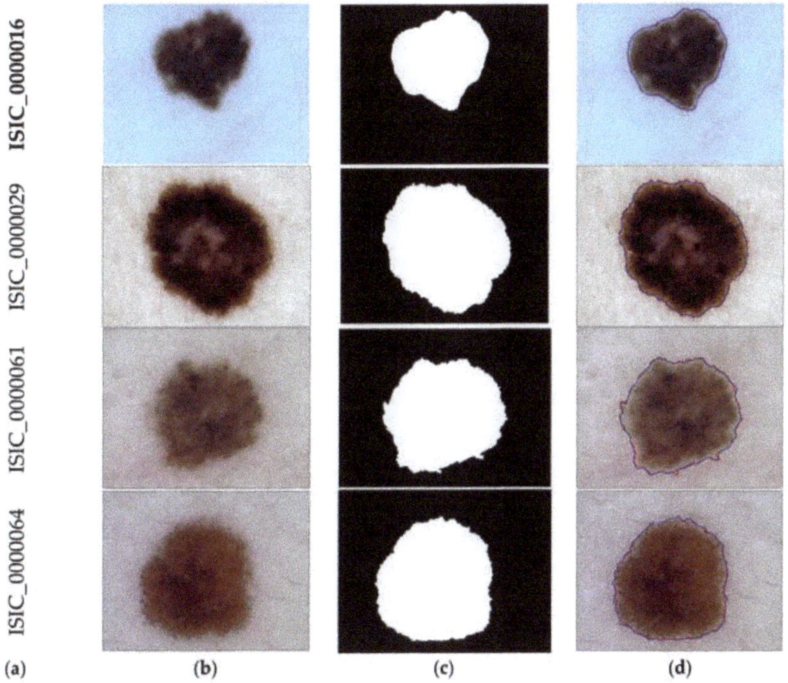

Figure 3. Detection results: (**a**) Skin cancer image number; (**b**) Original skin lesion image; (**c**) Ground truth image; and (**d**) Detected lesion region using the proposed approach.

3.3. Evaluation

Table 1 reports the average values as well as the standard deviations (SD) of the evaluation metrics on the proposed approach's performance over 500 images.

Table 1. The performance of computer segmentation using the proposed NCARG method with reference to ground truth boundaries (Average ± SD).

Metric Value	Accuracy (%)	Dice (%)	JAC (%)	Sensitivity (%)	Specificity (%)
Average	95.3	90.38	83.2	97.5	88.8
Standard deviation	6	7.6	10.5	3.5	11.4

Table 1 establishes that the proposed approach achieved a detection accuracy for the skin lesion regions of 95.3% with a 6% standard deviation, compared to the ground truth images. In addition, the mean values of the Dice index, Jaccard index, sensitivity, and specificity are 90.38%, 83.2%, 97.5%, and 88.8%; respectively, with standard deviations (SD) of 7.6%, 10.5%, 3.5%, and 11.4%; respectively. These reported experimental test results proved that the proposed NCARG approach correctly detects skin lesions of different shapes and sizes with high accuracy. Ten dermoscopic images were randomly selected; their segmentation results are shown in Figure 4, and the evaluation metrics are reported in Table 2.

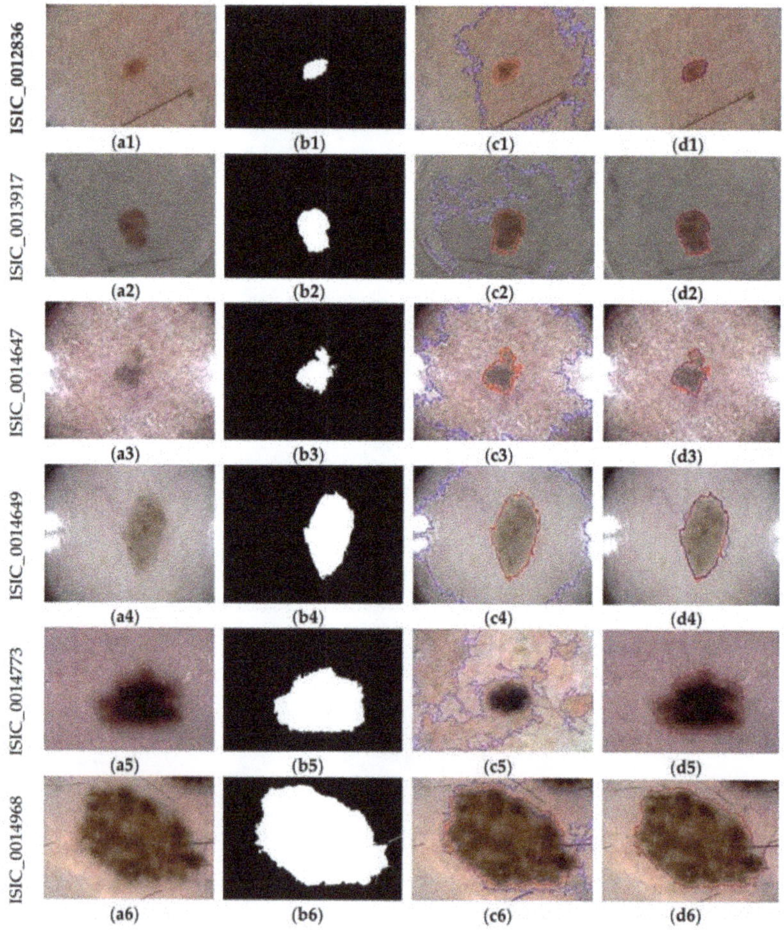

Figure 4. *Cont.*

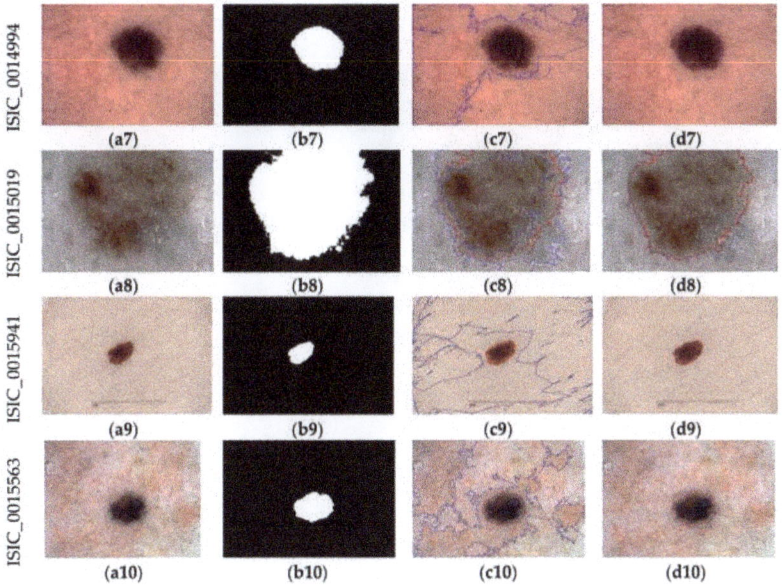

Figure 4. Comparative segmentation results, where (**a1–a10**): original dermoscopic test images; (**b1–b10**): ground truth images; (**c1–c10**): segmented images using the neutrosophic similarity score and level set (NSSLS) algorithm, and (**d1–d10**): NCARG proposed approach.

Table 2. The performance of computer segmentation using the proposed method with reference to the ground truth boundaries (Average ± SD) of ten images during the test phase.

Image ID	Accuracy (%)	Dice (%)	JAC (%)	Sensitivity (%)	Specificity (%)
ISIC_0012836	99.7819	93.2747	87.397	99.9909	87.851
ISIC_0013917	99.1485	90.4852	82.6237	1	82.6237
ISIC_0014647	99.4684	92.8643	86.6791	99.7929	91.2339
ISIC_0014649	98.8823	95.2268	90.8886	98.8313	99.2854
ISIC_0014773	98.9017	97.3678	94.8707	98.6294	99.9692
ISIC_0014968	89.5888	89.2267	80.5489	81.7035	99.9913
ISIC_0014994	98.9242	93.0613	87.023	1	87.023
ISIC_0015019	93.8788	93.9689	88.6239	88.6218	99.602
ISIC_0015941	99.7687	94.3589	89.3203	1	89.3203
ISIC_0015563	98.0344	83.939	72.3232	97.928	1
Average (%)	97.63777	92.3774	86.0298	96.54978	93.68998
SD (%)	3.31069	3.7373	6.2549	6.26068	6.76053

3.4. Comparative Study with NSSLS Method

The proposed NCARG approach is compared with the NSSLS algorithm [31] for detecting skin lesions. Figure 4(a1–a10), Figure 4(b1–b10), Figure 4(c1–c10) and Figure 4(d1–d10) include the original dermoscopic images, the ground truth images, the segmented images using the NSSLS algorithm, and the NCARG proposed approach; respectively.

Figure 4 illustrates different samples from the test images with different size, shape, light illumination, skin surface roughness/smoothness, and the existence of hair and/or air bubbles. For these different samples, the segmented image using the proposed NCARG algorithm is matched with the ground truth; while, the NSSLS failed to accurately match the ground truth. Thus, Figure 4 demonstrates that the proposed approach accurately detects the skin lesion under the different cases compared with the NSSLS method. The superiority of the proposed approach is due to the ability of the NCM along with the IF to handle indeterminate information. In addition, shearlet transform achieved the surface anisotropic regularity along the edges leading the algorithm to capture the smooth contours corresponding to the dominant features in the image. For the same images in Figure 4, the comparative results of the previously mentioned evaluation metrics are plotted for the NCARG and NSSLS in Figures 5 and 6; respectively. In both figures, the X-axis denotes the image name under study, and the Y-axis denotes the value of the corresponding metric in the bar graph.

Figure 5 along with Table 2 illustrate the accuracy of the proposed algorithm, which achieves an average accuracy of 97.638% for the segmentation of the different ten skin lesion samples, while Figure 6 illustrates about 44% average accuracy of the NSSLS method. Thus, Figures 5 and 6 establish the superiority of the proposed approach compared with the NSSLS method, owing to the removal the indeterminate information and the efficiency of the shearlet transform. The same results are confirmed by measuring the same metrics using 500 images, as reported in Figure 7.

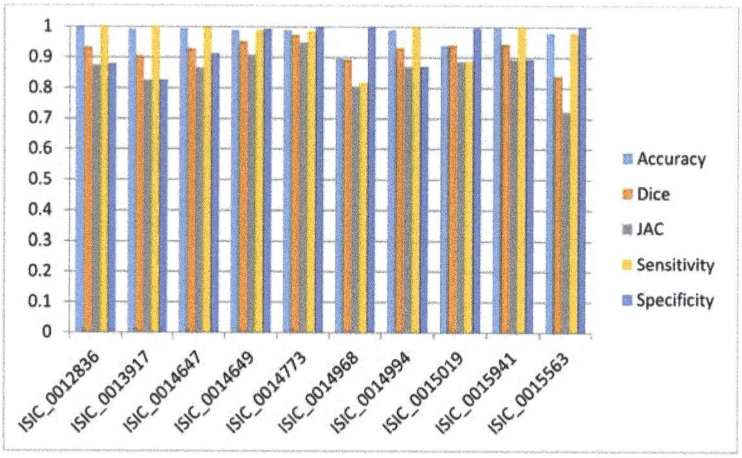

Figure 5. Evaluation metrics of the ten test images using the proposed segmentation NCARG approach.

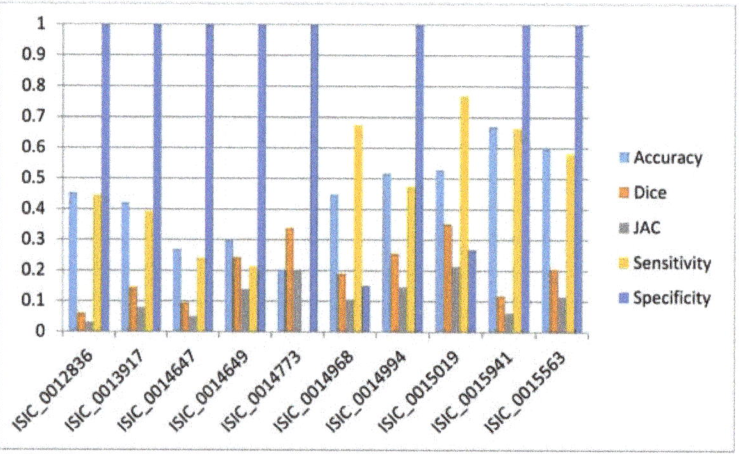

Figure 6. Evaluation metrics of the ten test images using the NSSLS segmentation approach for comparison.

Figure 7 reports that the proposed method achieves about 15% improvement on the accuracy and about 25% improvement in the JAC over the NSSLS method. Generally, Figure 7 proves the superiority of the proposed method compared with the NSSLS method. In addition, Table 3 reports the statistical results on the testing images; it compares the detection performance with reference to the ground truth segmented images for the NSSLS and the proposed NCARG method. The *p*-values are used to estimate the differences between the metric results of the two methods. The statistical significance was set at a level of 0.05; a *p*-value of <0.05 refers to the statistically significant relation.

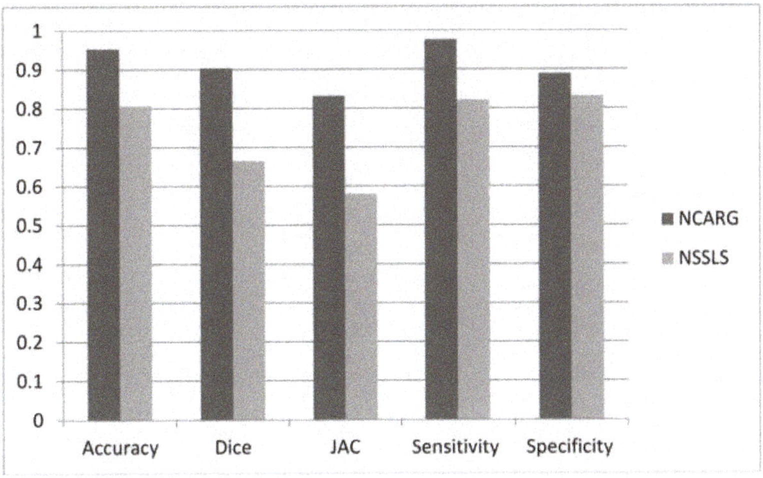

Figure 7. Comparative results of the performance evaluation metrics of the proposed NCARG and NSSLS methods.

The *p*-values reported in Table 3 establish a significant difference in the performance metric values when using the proposed NCARG and NSSLS methods. The mean and standard deviation of the accuracy, Dice, JAC, sensitivity, and specificity for the NSSLS and NCARG methods, along with the *p*-values, establish that the proposed NCARG method improved skin lesion segmentation compared with the NSSLS method. Figure 7 along with Table 3 depicts that the NCARG achieved 95.3% average accuracy, which is superior to the 80.6% average accuracy of the NSSLS approach. Furthermore, the proposed algorithm achieved a 90.4% average Dice coefficient value, 83.2% average JAC value, 97.5% average sensitivity value, and 88.8% average specificity value. The segmentation accuracy improved from 80.6 $\pm$ 22.1 using the NSSLS to 95.3 $\pm$ 6 using the proposed method, which is a significant difference. The skin lesion segmentation improvement is statistically significant ($p < 0.05$) for all measured performances metrics by SPSS software.

Table 3. The average values (mean $\pm$ SD) of the evaluation metrics using the NCARG approach compared to the NSSLS approach.

Method	Accuracy (%)	Dice (%)	JAC (%)	Sensitivity (%)	Specificity (%)
NSSLS method	80.6 $\pm$ 22.1	66.4 $\pm$ 32.6	57.9 $\pm$ 33.7	82.1 $\pm$ 24	83.1 $\pm$ 30.4
Proposed NCARG method	95.3 $\pm$ 6	90.4 $\pm$ 7.6	83.2 $\pm$ 10.5	97.5 $\pm$ 6.3	88.8 $\pm$ 11.4
p-value	<0.0001	<0.0001	<0.0001	<0.0001	<0.0001

The cumulative percentage is used to measure the percentage of images, which have a metric value less than a threshold value. The cumulative percentage (CP) curves of the measured metrics are plotted for comparing the performance of the NSSLS and NCARG algorithms. Figures 8–12 show the cumulative percentage of images having five measurements less than a certain value; the X-axis represents the different threshold values on the metric and the Y-axis is the percentage of the number of images whose metric values are greater than this threshold value. These figures demonstrate the comparison of performances in terms of the cumulative percentage of the different metrics, namely the accuracy, Dice value, JAC, sensitivity, and specificity; respectively.

Figure 8 illustrates a comparison of performances in terms of the cumulative percentage of the NCARG and NSSLS segmentation accuracy. About 80% of the images have a 95% accuracy for the segmentation using the proposed NCARG, while the achieved cumulative accuracy percentage using the NSSLS is about 65% for 80% of the images.

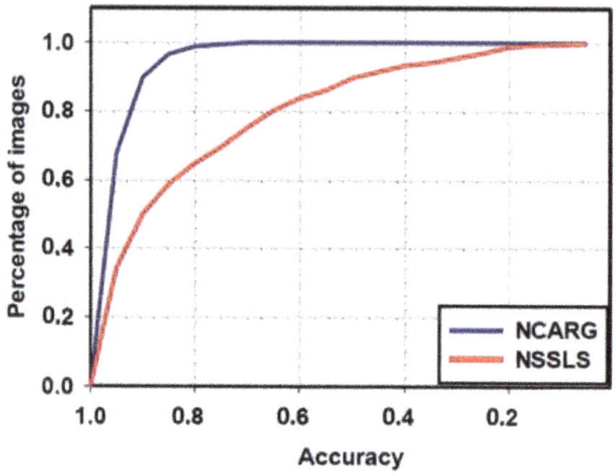

Figure 8. Comparison of performances in terms of the cumulative percentage of the accuracy using the NCARG and NSSLS segmentation methods.

Figure 9 compares the performances, in terms of the cumulative percentage of the Dice index values, of the NCARG and NSSLS segmentations. Figure 9 depicts that 100% of the images have about 82% Dice CP values using the NCARG method, while 58% of the images achieved the same 82% Dice CP values when using the NSSLS method.

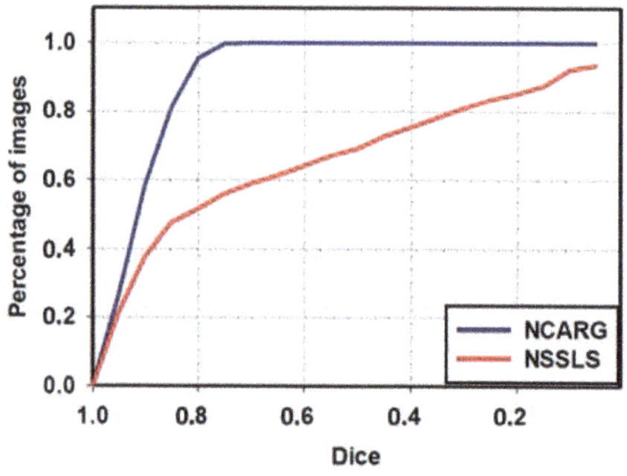

Figure 9. Comparison of performances in terms of the cumulative percentage of the Dice values using the NCARG and NSSLS segmentation methods.

Figure 10 compares the performances, in terms of the cumulative percentage of the JAC values, of the NCARG and NSSLS segmentation. About 50% of the images have 83% CP JAC values using the NCARG method, while the obtained CP JAC using the NSSLS for the same number of images is about 72%.

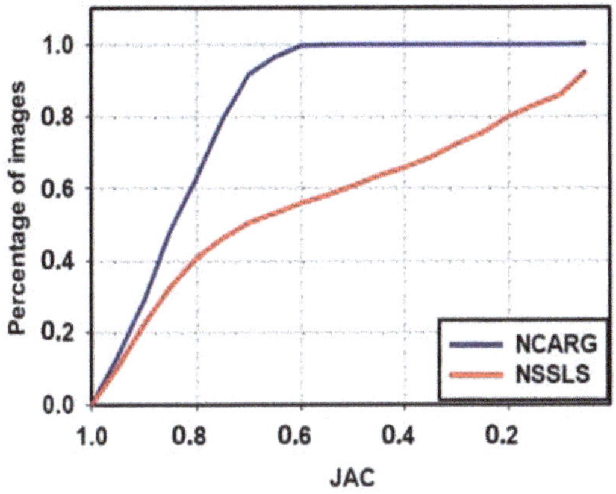

Figure 10. Comparison of performances in terms of the cumulative percentage of the JAC values using NCARG and NSSLS segmentation methods.

Figure 11 compares the performances, in terms of the cumulative percentage of the sensitivity, using the NCARG and NSSLS segmentation methods. About 50% of the images have 97% sensitivity value using the NCARG method, while the NSSLS achieves about 92% sensitivity value.

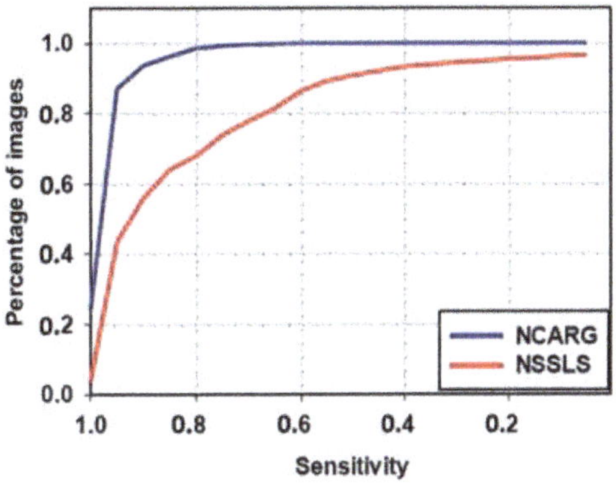

Figure 11. Comparison of performances in terms of the cumulative percentage of the sensitivity using the NCARG and NSSLS segmentation methods.

Figure 12 demonstrates the comparison of performances, in terms of the cumulative percentage of the specificity, using the NCARG and NSSLS segmentation methods. A larger number of images have accuracies in the range of 100% to 85% when using the NSSLS compared to the proposed method. However, about 100% of the images have 63% CP specificity values using the NCARG method, while the NSSLS achieved about 20% cumulative specificity values with 90% of the images. Generally, the cumulative percentage of each metric establishes the superiority of the proposed NCARG method compared with the NSSLS method.

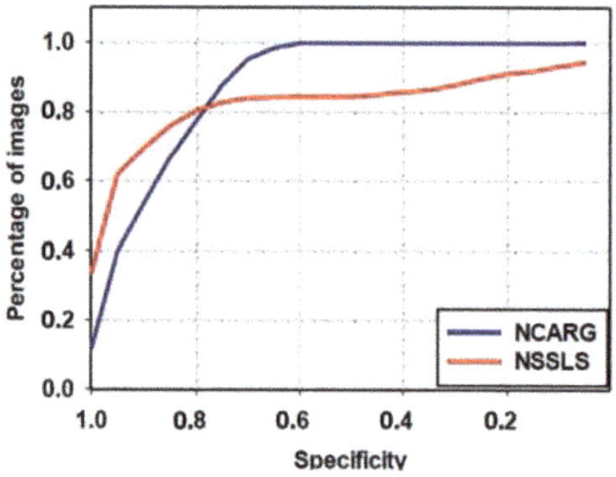

Figure 12. Comparison of performances in terms of the cumulative percentage of the specificity using the NCARG and NSSLS segmentation methods.

3.5. Comparison with Other Segmentation Methods Using the ISIC Archive

In case of lesion segmentation, variability in the images is very high; therefore, performance results highly depend on the data set that is used in the experiments. Several studies and challenges have been conducted to resolve such trials [33]. In order to validate the performance of the proposed NCARG method, a comparison is conducted on the results of previously published studies on the same ISIC dermoscopic image data set. Yu et al. [34] leveraged very deep convolutional neural networks (CNN) for melanoma image recognition using the ISIC data set. The results proved that deeper networks, of more than 50 layers, provided more discriminating features with more accurate recognition. For accurate skin lesion segmentation, fully convolutional residual networks (FCRN) with a multi-scale contextual information integration structure were applied to the further classification stage. The network depth increase achieved enhanced discrimination capability of CNN. The FCRNs of 38 layers achieved 0.929 accuracy, 0.856 Dice, 0.785 JAC, and 0.882 sensitivity. Thus, our proposed NCARG provides superior performance in terms of these metrics. However, with an increased FCRN layer depth of 50, the performance improvement increased compared to our proposed method. However, the complexity also increases. In addition, Yu et al. have compared their study with other studies, namely the fully convolutional VGG-16 network [34,35] and the fully convolutional GoogleNet [34,36] establishing the superiority of our work compared to both of those studies. Table 4 reports a comparative study between the preceding studies, which have used the same ISIC data set, and the proposed NCARG method.

Table 4. Performance metrics comparison of different studies using the ISIC dataset for segmentation.

Method	Accuracy (%)	Dice (%)	JAC (%)	Sensitivity (%)	Specificity (%)
FCRNs of 38 layers [34]	92.9	85.6	78.5	88.2	93.2
FCRNs of 101 layers [34]	93.7	87.2	80.3	90.3	93.5
VGG-16 [34,35]	90.3	79.4	70.7	79.6	94.5
GoogleNet [34,36]	91.6	84.8	77.6	90.1	91.6
Proposed NCARG method	95.3	90.4	83.2	97.5	88.8

The preceding results and the comparative study establish the superiority of the proposed NCARG method compared with other methods. This superiority arises due to the effectiveness of the shearlet transform, the indeterminacy filtering, and the adaptive region growing, yielding an overall accuracy of 95.3%. Moreover, in comparison with previously conducted studies on the same ISIC dermoscopic image data set, the proposed method can be considered an effective method. In addition, the studies in References [37,38] can be improved and compared with the proposed method on the same dataset.

4. Conclusions

In this study, a novel skin lesion detection algorithm is proposed based on neutrosophic *c*-means and adaptive region growing algorithms applied to dermoscopic images. The dermoscopic images are mapped into the neutrosophic domain using the shearlet transform results of the image. An indeterminate filter is used for reducing the indeterminacy on the image, and the image is segmented via a neutrosophic *c*-means clustering algorithm. Finally, the skin lesion is accurately identified using a newly defined adaptive region growing algorithm. A public data set was employed to test the proposed method. Fifty images were selected randomly for tuning, and five hundred images were used to test the process. Several metrics were measured for evaluating the proposed method performance. The evaluation results demonstrate the proposed method achieves better performance to detect the skin lesions when compared to the neutrosophic similarity score and level set (NSSLS) segmentation approach.

The proposed NCARG approach achieved average 95.3% accuracy of 500 dermoscopic images including, ones with different shape, size, color, uniformity, skin surface roughness, light illumination during the image capturing process, and existence of air bubbles. The significant difference in the *p*-values of the measured metrics using the NSSLS and the proposed NCARG proved the superiority of the proposed method. This proposed method determines possible skin lesions in dermoscopic images which can be employed for further accurate automated diagnosis and clinical decision support.

Author Contributions: Yanhui Guo, Amira S. Ashour and Florentin Smarandache conceived and worked together to achieve this work.

Conflicts of Interest: The authors declare no conflict of interest.

References

1. Marghoob, A.A.; Swindle, L.D.; Moricz, C.Z.; Sanchez, F.A.; Slue, B.; Halpern, A.C.; Kopf, A.W. Instruments and new technologies for the in vivo diagnosis of melanoma. *J. Am. Acad. Dermatol.* **2003**, *49*, 777–797. [CrossRef]
2. Wolfe, J.M.; Butcher, S.J.; Lee, C.; Hyle, M. Changing your mind: On the contributions of top-down and bottom-up guidance in visual search for feature singletons. *J. Exp. Psychol. Hum. Percept. Perform.* **2003**, *29*, 483–502. [CrossRef] [PubMed]
3. Binder, M.; Schwarz, M.; Winkler, A.; Steiner, A.; Kaider, A.; Wolff, K.; Pehamberger, M. Epiluminescence microscopy. A useful tool for the diagnosis of pigmented skin lesions for formally trained dermatologists. *Arch. Dermatol.* **1995**, *131*, 286–291. [CrossRef] [PubMed]
4. Celebi, M.E.; Wen, Q.; Iyatomi, H.; Shimizu, K.; Zhou, H.; Schaefer, G. A State-of-the-Art Survey on Lesion Border Detection in Dermoscopy Images. In *Dermoscopy Image Analysis*; Celebi, M.E., Mendonca, T., Marques, J.S., Eds.; CRC Press: Boca Raton, FL, USA, 2015; pp. 97–129.
5. Celebi, M.E.; Iyatomi, H.; Schaefer, G.; Stoecker, W.V. Lesion Border Detection in Dermoscopy Images. *Comput. Med. Imaging Graph.* **2009**, *33*, 148–153. [CrossRef] [PubMed]
6. Garnavi, R.; Aldeen, M.; Celebi, M.E.; Varigos, G.; Finch, S. Border detection in dermoscopy images using hybrid thresholding on optimized color channels. *Comput. Med. Imaging Graph.* **2011**, *35*, 105–115. [CrossRef] [PubMed]
7. Silveira, M.; Nascimento, J.C.; Marques, J.S.; Marçal, A.R.; Mendonça, T.; Yamauchi, S.; Maeda, J.; Rozeira, J. Comparison of segmentation methods for melanoma diagnosis in dermoscopy images. *IEEE J. Sel. Top. Signal Process.* **2009**, *3*, 35–45. [CrossRef]
8. Celebi, M.E.; Aslandogan, Y.A.; Stoecker, W.V.; Iyatomi, H.; Oka, H.; Chen, X. Unsupervised border detection in dermoscopy images. *Skin Res. Technol.* **2007**, *13*, 454–462. [CrossRef] [PubMed]
9. Argenziano, G.; Kittler, H.; Ferrara, G.; Rubegni, P.; Malvehy, J.; Puig, S.; Cowell, L.; Stanganelli, I.; de Giorgi, V.; Thomas, L.; et al. Slow-growing melanoma: A dermoscopy follow-up study. *Br. J. Dermatol.* **2010**, *162*, 267–273. [CrossRef] [PubMed]
10. Xie, F.; Bovik, A.C. Automatic segmentation of dermoscopy images using self-generating neural networks seeded by genetic algorithm. *Pattern Recognit.* **2013**, *46*, 1012–1019. [CrossRef]

11. Barata, C.; Celebi, M.E.; Marques, J.S. Development of a clinically oriented system for melanoma diagnosis. *Pattern Recognit.* **2017**, *69*, 270–285. [CrossRef]
12. Lee, H.; Chen, Y.P.P. Skin cancer extraction with optimum fuzzy thresholding technique. *Appl. Intell.* **2014**, *40*, 415–426. [CrossRef]
13. Jaisakthi, S.M.; Chandrabose, A.; Mirunalini, P. Automatic Skin Lesion Segmentation using Semi-supervised Learning Technique. *arXiv*, **2017**.
14. Zhou, H.; Schaefer, G.; Sadka, A.H.; Celebi, M.E. Anisotropic mean shift based fuzzy c-means segmentation of dermoscopy images. *IEEE J. Sel. Top. Signal Process.* **2009**, *3*, 26–34. [CrossRef]
15. Guo, Y.; Sengur, A. NCM: Neutrosophic c-means clustering algorithm. *Pattern Recognit.* **2015**, *48*, 2710–2724. [CrossRef]
16. Mohan, J.; Krishnaveni, V.; Guo, Y. Automated Brain Tumor Segmentation on MR Images Based on Neutrosophic Set Approach. In Proceedings of the 2015 2nd International Conference on Electronics and Communication Systems (ICECS), Coimbatore, India, 26–27 February 2015; pp. 1078–1083.
17. Sengur, A.; Guo, Y. Color texture image segmentation based on neutrosophic set and wavelet transformation. *Comput. Vis. Image Underst.* **2011**, *115*, 1134–1144. [CrossRef]
18. Khalid, S.; Jamil, U.; Saleem, K.; Akram, M.U.; Manzoor, W.; Ahmed, W.; Sohail, A. Segmentation of skin lesion using Cohen–Daubechies–Feauveau biorthogonal wavelet. *SpringerPlus* **2016**, *5*, 1603. [CrossRef] [PubMed]
19. Guo, K.; Labate, D. Optimally Sparse Multidimensional Representation using Shearlets. *SIAM J. Math. Anal.* **2007**, *39*, 298–318. [CrossRef]
20. Guo, K.; Labate, D. Characterization and analysis of edges using the continuous shearlet transform. *SIAM J. Imaging Sci.* **2009**, *2*, 959–986. [CrossRef]
21. Cavalcanti, P.G.; Scharcanski, J. Automated prescreening of pigmented skin lesions using standard cameras. *Comput. Med. Imaging Graph.* **2011**, *35*, 481–491. [CrossRef] [PubMed]
22. Labate, D.; Lim, W.; Kutyniok, G.; Weiss, G. Sparse Multidimensional Representation Using Shearlets. In Proceedings of the Wavelets XI, San Diego, CA, USA, 31 July–4 August 2005; SPIE: Bellingham, WA, USA, 2005; Volume 5914, pp. 254–262.
23. Zhou, H.; Niu, X.; Qin, H.; Zhou, J.; Lai, R.; Wang, B. Shearlet Transform Based Anomaly Detection for Hyperspectral Image. In Proceedings of the 6th International Symposium on Advanced Optical Manufacturing and Testing Technologies: Optoelectronic Materials and Devices for Sensing, Imaging, and Solar Energy, Xiamen, China, 26–29 April 2012; Volume 8419.
24. Theresa, M.M. Computer aided diagnostic (CAD) for feature extraction of lungs in chest radiograph using different transform features. *Biomed. Res.* **2017**, S208–S213. Available online: http://www.biomedres.info/biomedical-research/computer-aided-diagnostic-cad-for-feature-extraction-of-lungs-in-chest-radiograph-using-different-transform-features.html (accessed on 26 March 2018).
25. Liu, X.; Zhou, Y.; Wang, Y.J. Image fusion based on shearlet transform and regional features. *AEU Int. J. Electron. Commun.* **2014**, *68*, 471–477. [CrossRef]
26. Mohan, J.; Krishnaveni, V.; Guo, Y. A new neutrosophic approach of Wiener filtering for MRI denoising. *Meas. Sci. Rev.* **2013**, *13*, 177–186. [CrossRef]
27. Mohan, J.; Guo, Y.; Krishnaveni, V.; Jeganathan, K. MRI Denoising Based on Neutrosophic Wiener Filtering. In Proceedings of the 2012 IEEE International Conference on Imaging Systems and Techniques, Manchester, UK, 16–17 July 2012; pp. 327–331.
28. Cheng, H.; Guo, Y.; Zhang, Y. A novel image segmentation approach based on neutrosophic set and improved fuzzy c-means algorithm. *New Math. Nat. Comput.* **2011**, *7*, 155–171. [CrossRef]
29. Guo, Y.; Xia, R.; Şengür, A.; Polat, K. A novel image segmentation approach based on neutrosophic c-means clustering and indeterminacy filtering. *Neural Comput. Appl.* **2017**, *28*, 3009–3019. [CrossRef]
30. Guo, Y.; Zhou, C.; Chan, H.P.; Chughtai, A.; Wei, J.; Hadjiiski, L.M.; Kazerooni, E.A. Automated iterative neutrosophic lung segmentation for image analysis in thoracic computed tomography. *Med. Phys.* **2013**, *40*. [CrossRef] [PubMed]
31. Guo, Y.; Şengür, A.; Tian, J.W. A novel breast ultrasound image segmentation algorithm based on neutrosophic similarity score and level set. *Comput. Methods Programs Biomed.* **2016**, *123*, 43–53.
32. ISIC. Available online: http://www.isdis.net/index.php/isic-project (accessed on 26 March 2018).

33. Gutman, D.; Codella, N.C.; Celebi, E.; Helba, B.; Marchetti, M.; Mishra, N.; Halpern, A. Skin lesion analysis toward melanoma detection: A challenge at the international symposium on biomedical imaging (ISBI) 2016, hosted by the international skin imaging collaboration (ISIC). *arXiv*, 2016.
34. Yu, L.; Chen, H.; Dou, Q.; Qin, J.; Heng, P.A. Automated melanoma recognition in dermoscopy images via very deep residual networks. *IEEE Trans. Med. Imaging* **2017**, *36*, 994–1004. [CrossRef] [PubMed]
35. Simonyan, K.; Zisserman, A. Very deep convolutional networks for large-scale image recognition. *arXiv*, **2014**.
36. Szegedy, C.; Liu, W.; Jia, Y.; Sermanet, P.; Reed, S.; Anguelov, D.; Erhan, D.; Vanhoucke, V.; Rabinovich, A. Going Deeper with Convolutions. In Proceedings of the IEEE Conference on Computer Vision and Pattern Recognition, Boston, MA, USA, 7–12 June 2015; pp. 1–9.
37. Ma, Z.; Tavares, J.M.R. A novel approach to segment skin lesions in dermoscopic images based on a deformable model. *IEEE J. Biomed. Health Inform.* **2016**, *20*, 615–623.
38. Codella, N.C.; Gutman, D.; Celebi, M.E.; Helba, B.; Marchetti, M.A.; Dusza, S.W.; Kalloo, A.; Liopyris, K.; Mishra, N.; Kittler, H.; et al. Skin lesion analysis toward melanoma detection: A challenge at the 2017 international symposium on biomedical imaging (ISBI), 2017, hosted by the international skin imaging collaboration (ISIC). *arXiv*, **2017**.

symmetry

MDPI

Article

Neutrosophic Triplet Cosets and Quotient Groups

Mikail Bal, Moges Mekonnen Shalla and Necati Olgun *

Faculty of Arts and Sciences, Department of Mathematics, Gaziantep University, 27310 Gaziantep, Turkey; mikailbal46@hotmail.com (M.B.); moges6710@gmail.com (M.M.S.)
* Correspondence: olgun@gantep.edu.tr; Tel.: +90-536-321-4006

Received: 29 March 2018; Accepted: 17 April 2018; Published: 20 April 2018

Abstract: In this paper, by utilizing the concept of a neutrosophic extended triplet (NET), we define the neutrosophic image, neutrosophic inverse-image, neutrosophic kernel, and the NET subgroup. The notion of the neutrosophic triplet coset and its relation with the classical coset are defined and the properties of the neutrosophic triplet cosets are given. Furthermore, the neutrosophic triplet normal subgroups, and neutrosophic triplet quotient groups are studied.

Keywords: neutosophic extended triplet subgroups; neutrosophic triplet cosets; neutrosophic triplet normal subgroups; neutrosophic triplet quotient groups

1. Introduction

Neutrosophy was first introduced by Smarandache (Smarandache, 1999, 2003) as a branch of philosophy, which studied the origin, nature, and scope of neutralities, as well as their interactions with different ideational spectra: (A) is an idea, proposition, theory, event, concept, or entity; anti(A) is the opposite of (A); and (neut-A) means neither (A) nor anti(A), that is, the neutrality in between the two extremes. A notion of neutrosophic set theory was introduced by Smarandache in [1]. By using the idea of the neutrosophic theory, Kandasamy and Smarandache introduced neutrosophic algebraic structures in [2,3]. The neutrosophic triplets were first introduced by Florentin Smarandache and Mumtaz Ali [4–10], in 2014–2016. Florentin Smarandache and Mumtaz Ali introduced neutrosophic triplet groups in [6,11]. A lot of researchers have been dealing with neutrosophic triplet metric space, neutrosophic triplet vector space, neutrosophic triplet inner product, and neutrosophic triplet normed space in [12–22].

A neutrosophic extended triplet, introduced by Smarandache [7,20] in 2016, is defined as the neutral of x (denoted by $e^{neut(x)}$ and called "extended neutral"), which is equal to the classical algebraic unitary element (if any). As a result, the "extended opposite" of x (denoted by $e^{anti(x)}$) is equal to the classical inverse element from a classical group. Thus, the neutrosophic extended triplet (NET) has a form $\left(x, e^{neut(x)}, e^{anti(x)}\right)$ for $x \in$ N, where $e^{neut(x)} \in$ N is the extended neutral of x. Here, the neutral element can be equal to or different from the classical algebraic unitary element, if any, such that: $x * e^{neut(x)} = e^{neut(x)} * x = x$, and $e^{anti(x)} \in$ N is the extended opposite of x, where $x * e^{anti(x)} = e^{anti(x)} * x = e^{neut(x)}$. Therefore, we used NET to define these new structures.

In this paper, we deal with neutosophic extended triplet subgroups, neutrosophic triplet cosets, neutrosophic triplet normal subgroups, and neutrosophic triplet quotient groups for the purpose to develop new algebraic structures on NET groups. Additionally, we define the neutrosophic triplet image, neutrosophic triplet kernel, and neutrosophic triplet inverse image. We give preliminaries and results with examples in Section 2, and we introduce neutrosophic extended triplet subgroups in Section 3. Section 4 is dedicated to introduing neutrosophic triplet cosets, with some of their properties, and we show that neutrosophic triplet cosets are different from classical cosets. In Section 5, we introduce neutrosophic triplet normal subgroups and the neutrosophic triplet normal subgroup test. In Section 6, we define the neutrosophic triplet quotient groups and we examine the relationships of these structures with each other. In Section 7, we provide some conclusions.

2. Preliminaries

In this section, the definition of neutrosophic triplets, NET's, and the concepts of NET groups have been outlined.

2.1. Neutrosophic Triplet

Let U be a universe of discourse, and (N, $*$) a set included in it, endowed with a well-defined binary law $*$.

Definition 1 ([1–3]). *A neutrosophic triplet has a form (x, neut(x), anti(x)), for x in N, where neut(x) and anti(x) ∈ N are neutral and opposite to x, which are different from the classical algebraic unitary element, if any, such that:* $x * neut(x) = neut(x) * x = x$ *and* $x * anti(x) = anti(x) * x = neut(x)$*, respectively. In general, x may have more than one neut's and anti's.*

2.2. NET

Definition 2 ([4,7]). *A neutrosophic extended triplet is a neutrosophic triplet, as defined in Definition 1, where the neutral of x (denoted by* $e^{neut(x)}$ *and called extended neutral) is equal to the classical algebraic unitary element, if any. As a consequence, the extended opposite of x (denoted by* $e^{anti(x)}$*) is also equal to the classical inverse element from a classical group. Thus, an NET has a form* $\left(x, e^{neut(x)}, e^{anti(x)}\right)$*, for x ∈ N, where* $e^{neut(x)}$ *and* $e^{anti(x)}$ *in N are the extended neutral and opposite of x, respectively, such that:* $x * e^{neut(x)} = e^{neut(x)} * x = x$*, which can be equal to or different from the classical algebraic unitary element, if any, and* $x * e^{anti(x)} = e^{anti(x)} * x = e^{neut(x)}$*. In general, for each x ∈ N there are many* $e^{neut(x)}$*'s and* $e^{anti(x)}$*'s.*

Definition 3 ([1–3]). *The element y in* $(N, *)$ *is the second coordinate of a neutrosophic extended triplet (denoted as neut(y) of a neutrosophic triplet), if there are other elements exist, x and z ∈ N such that:* $x * y = y * x = x$ *and* $x * z = z * x = y$*. The formed neutrosophic triplet is (x, y, z). The element* $z \in (N, *)$*, as the third coordinate, can be defined in the same way.*

Example 1. *Let X = (0, 1, 2, 3, 4, 5, 6, 7, 8, 9, 10, 11), enclosed with the classical multiplication law, (x) modulo 12, which is well defined on X, with the classical unitary element 1. X iss an NET "weak commutative set" see "Table 1".*

Table 1. Neutrosophic triplets of (*x*) modulo 12.

$*$	0	1	2	3	4	5	6	7	8	9	10	11
0	0	0	0	0	0	0	0	0	0	0	0	0
1	0	1	2	3	4	5	6	7	8	9	10	11
2	0	2	4	6	8	10	0	2	4	6	8	10
3	0	3	6	9	0	3	6	9	0	3	6	9
4	0	4	8	0	4	8	0	4	8	0	4	8
5	0	5	10	3	8	1	6	11	4	9	2	7
6	0	6	0	6	0	6	0	6	0	6	0	6
7	0	7	2	9	4	11	6	1	8	3	10	5
8	0	8	4	0	8	4	0	8	4	0	8	4
9	0	9	6	3	0	9	6	3	0	9	6	3
10	0	10	8	6	4	2	0	10	8	6	4	2
11	0	11	10	9	8	7	6	5	4	3	2	1

The formed NETs of X are: (0, 0, 0), (0, 0, 1), (0, 0, 2), ... , (0, 0, 11), (1, 1, 1), (3, 9, 3), (3, 9, 7), (3, 9, 11), (4, 4, 4), (4, 4, 7), (4, 4, 10), (5, 1, 5), (7, 1, 7), (8, 4, 2), (8, 4, 5), (8, 4, 8), (8, 4, 11), (9, 9, 5), (9, 9, 9), (11, 1, 11).

Here, 2, 6, and 10 did not give rise to a neutrosophic triplet, as neut(2) = 1 and 7, however anti(2) did not exist in Z_{12}. In addition, neut(6) = 1, 3, 5, 7, 9, and 11, however anti(6) did not exist in Z_{12}. The neut(10) = 1, however anti(10) did not exist in Z_{12}.

Definition 4 ([4,7]). *The set N is called a strong neutrosophic extended triplet set if, for any x in N, $e^{neut(x)} \in$ N and $e^{anti(x)} \in$ N exists.*

Example 2. *The NET's of (x) modulo 12 were as follows:*

(0, 0, 0), (0, 0, 1), (0, 0, 2), ... , (0, 0, 11), (1, 1, 1), (3, 9, 3), (3, 9, 7), (3, 9, 11), (4, 4, 4), (4, 4, 7), (4, 4, 10), (5, 1, 5), (7, 1, 7), (8, 4, 2), (8, 4, 5), (8, 4, 8), (8, 4, 11), (9, 9, 5), (9, 9, 9), (11, 1, 11).

Definition 5 ([4,7]). *The set N is called an NET weak set if, for any $x \in$ N, an NET $\left(y,\ e^{neut(y)},\ e^{anti(y)}\right)$ included in N exists, such that:*

$$x = y$$

or

$$x = e^{neut(y)}$$

or

$$x = e^{anti(y)}.$$

Definition 6. *A neutrosophic extended triplet (x, y, z) for x, y, z $\in$ N, is called a neutrosophic perfect triplet if both (z, y, x) and (y, y, y) are also neutrosophic triplets.*

Example 3. *The neutrosophic perfect triplets of (x) modulo 12 are described in "Table 1" as follows:*

Here, (0, 0, 0), (1, 1, 1), (3, 9, 3), (4, 4, 4), (5, 1, 5), (7, 1, 7), (8, 4, 8), (9, 9, 9), (11, 1, 11) are neutrosophic perfect triplets of (x) modulo 12.

Definition 7. *An NET (x, y, z) for x, y, z $\in$ N, is called a neutrosophic imperfect triplet if at least one of (z, y, x) or (y, y, y) is not a neutrosophic triplet(s).*

Example 4. *The neutrosophic imperfect triplets of (x) modulo 12, from the above table, were as follows:*

(0, 0, 1), (0, 0, 2),..., (0, 0, 11), (3, 9, 7), (3, 9, 11), (4, 4, 7), (4, 4, 10), (8, 4, 2), (8, 4, 5), (8, 4, 11), (9, 9, 5).

2.3. Neutrosophic Triplet Group (NTG)

Definition 8 ([1–3]). *Let (N, ∗) be a neutrosophic strong triplet set. Then, (N, ∗) is called a neutrosophic strong triplet group, if the following classical axioms are satisfied:*

(1) *(N, ∗) is well-defined, that is, for any x, y $\in$ N, one has $x * y \in$ N.*
(2) *(N, ∗) is associative, that is, for any x, y, z $\in$ N, one has $x * (y * z) = (x * y) * z$.*

Example 5. *We let Y = (Z_{12}, ×) be a semi-group under product 12. The neutral elements of Z_{12} were 4 and 9. The elements (8, 4, 8), (4, 4, 4), (3, 9, 3), and (9, 9, 9) were NETs.*

NTG, in general, was not a group in the classical sense, because it might not have had a classical unitary element, nor the classical inverse elements. We considered that the neutrosophic neutrals replaced the classical unitary element, and the neutrosophic opposites replaced the classical inverse elements.

Proposition 1 ([3]). *Let (N, $*$) be an NTG with respect to $*$ and a, b, c $\in$ N:*

(1) $a * b = a * c \Leftrightarrow neut(a) * b = neut(a) * c.$
(2) $b * a = c * a \Leftrightarrow b * neut(a) = c * neut(a).$
(3) *if* $anti(a) * b = anti(a) * c$, *then* $neut(a) * b = neut(a) * c.$
(4) *if* $b * anti(a) = c * anti(a)$, *then* $b * neut(a) = c * neut(a).$

Theorem 1 ([3]). *Let (N, $*$) be a commutative NET, with respect to $*$ and a, b $\in$ N:*

(i) $neut(a) * neut(b) = neut(a * b);$
(ii) $anti(a) * anti(b) = anti(a * b);$

Theorem 2 ([3]). *Let (N, $*$) be a commutative NET, with respect to $*$ and a $\in$ N:*

(i) $neut(a) * neut(a) = neut(a);$
(ii) $anti(a) * neut(a) = neut(a) * anti(a) = anti(a);$

Definition 9 ([3]). *An NET (N, $*$) is called to be cancellable, if it satisfies the following conditions:*

(a) $\forall\, x, y, z \in N,\ x * y = y * z \Rightarrow y = z.$
(b) $\forall\, x, y, z \in N,\ y * x = z * x \Rightarrow y = z.$

Definition 10 ([3]). *Let N be an NTG and x $\in$ N. N is then called a neutro-cyclic triplet group if N $= \langle a \rangle$. We can say that a is the neutrosophic triplet generator of N.*

Example 6. *We let N = (2, 4, 6) be an NTG with respect to (Z_8, .). Then, N was clearly a neutro-cyclic triplet group as N = $\langle a \rangle$. Therefore, 2 was the neutrosophic triplet generator of N.*

2.4. Neutrosophic Extended Triplet Group (NETG)

Definition 11 ([4,7]). *Let (N, $*$) be an NET strong set. Then, (N, $*$) is called an NETG, if the following classical axioms are satisfied:*

(1) *(N, $*$) is well-defined, that is, for any x, y $\in$ N, one has $x * y \in N$.*
(2) *(N, $*$) is associative, that is, for any x, y, z $\in$ N, one has*

$$x * (y * z) = (x * y) * z.$$

For NETG, the neutrosophic extended neutrals replaced the classical unitary element, and the neutrosophic extended opposites replaced the classical inverse elements. In the case where NETG included a classical group, then NETG enriched the structure of a classical group, since there might have been elements with more extended neutrals and more extended opposites.

Definition 12. *A permutation of a set X is a function σ: $x \to x$ that is one to one and onto, that is, a bijective map. Permutation maps, being bijective, have anti neutrals and the maps combine neutrally under composition of maps, which are associative. There is natural neutral permutation σ: $x \to x$, X = (1, 2, 3, ... , n), which is $\sigma(k) = k$. Therefore, all of the permutations of a set X = (1, 2, 3, ... , n) form an NETG under composition. This group is called the symmetric NETG (e^{Sn}) of degree n.*

Example 7. *We let A = (1, 2, 3). The elements of symmetric group of S_3 were as follows:*

$$\sigma_0 = \begin{pmatrix} 1\,2\,3 \\ 1\,2\,3 \end{pmatrix}, \sigma_1 = \begin{pmatrix} 1\,2\,3 \\ 2\,3\,1 \end{pmatrix}, \sigma_2 = \begin{pmatrix} 1\,2\,3 \\ 3\,1\,2 \end{pmatrix}$$

$$\mu_1 = \begin{pmatrix} 1\,2\,3 \\ 1\,2\,3 \end{pmatrix}, \mu_2 = \begin{pmatrix} 1\,2\,3 \\ 1\,2\,3 \end{pmatrix}, \mu_3 = \begin{pmatrix} 1\,2\,3 \\ 1\,2\,3 \end{pmatrix}$$

The operartion of S_3 is defined in Table 2 as follows:

1. *$(S_3, \bigcirc)$ is well-defined, that is, for any σ_i, $\mu_i \in S_3$, i = 1,2,3 one has $\sigma_i \bigcirc \mu_i \in S_3$.*
2. *$(S_3, \bigcirc)$ is associative, that is, for any σ_1, μ_1, $\mu_3 \in S_3$, one has the following:*

$$(\sigma_1 \bigcirc \mu_1) \bigcirc \mu_3 = \sigma_1 \bigcirc (\mu_1 \bigcirc \mu_3)$$

$$(\mu_1 \bigcirc \mu_3) = (\sigma_1 \bigcirc \sigma_1) = \sigma_2.$$

Table 2. Neutrosophic triplets of X.

$\bigcirc$	σ_0	σ_1	σ_2	μ_1	μ_2	μ_3
σ_0	σ_0	σ_1	σ_2	μ_1	μ_2	μ_3
σ_1	σ_1	σ_2	σ_0	μ_2	μ_3	μ_1
σ_2	σ_2	σ_0	σ_1	μ_3	μ_1	μ_2
μ_1	μ_1	μ_2	μ_3	σ_0	σ_2	σ_1
μ_2	μ_2	μ_1	μ_3	σ_1	σ_0	σ_2
μ_3	μ_3	μ_2	μ_1	σ_2	σ_1	σ_0

The NET's of S_3 (e^{S3}) are as follows:

$$(\sigma_0, \sigma_0, \sigma_0), (\sigma_1, \sigma_0, \sigma_2), (\sigma_2, \sigma_0, \sigma_1), (\mu_1, \sigma_0, \mu_1), (\mu_2, \sigma_0, \mu_2), (\mu_3, \sigma_0, \mu_3).$$

Hence, (S3, $\bigcirc$) is an NET strong group.

Definition 13 ([9–11]). *Let (N_1 *, N_2 $\bigcirc$) be two NETGs. A mapping f: $N_1 \to N_2$ is called a neutro-homomorphism if:*

(1) *For any $x, y \in N_1$, we have $f(x * y) = f(x)\ f(y)$*
(2) *If (x, neut[x], anti[x]) is an NET from N_1, then,*

$$f(neut[x]) = neut(f[x]) \text{ and } f(anti[x]) = anti(f[x]).$$

Example 8. *We let N_1 be an NETG with respect multiplication modulo 6 in (Z_6, ×), where N_1 = (0, 2, 4), and we let N_2 be another NETG in (Z_{10}, ×), where N_2 = (0, 2, 4, 6, 8). We let f: $N_1 \to N_2$ be a mapping defined as f(0) = 0, f (2) = 4, f (4) = 6. Then, f was clearly a neutro-homomorphism, because condition (1) and (2) were satisfied easily.*

Definition 14. *Let f: $N_1 \to N_2$ be a neutro-homomorphism from an NETG (N_1, *) to an NETG (N_2, *). The neutrosophic image of f is a subset, as follows:*

$$Im(f) = (f(g){:}g \in N_1, *) \text{ of } N_2.$$

Definition 15. *Let f: $N_1 \rightarrow N_2$ be a neutro-homomorphism from an NETG (N_1, $*$) to an NETG ($N*$, $\bigcirc$) and B $\subseteq N_2$. Then*

$$f^{-1}(B) = (x \in N_1\text{: } f(x) \in B)$$

is the neutrosophic inverse image of B under f.

Definition 16. *Let f: $N_1 \rightarrow N_2$ be a neutro-homomorphism from am NETG (N_1, $*$) to an NETG (N_2, $\bigcirc$). The neutrosophic kernel off is a subset*

$$ker(f) = \{x \in N_1 : f(x) = neut(x)\}$$

of N_1, where $neut(x)$ denotes the neutral element of N_2.

Example 9. *We took D_4, the symmetry NETG of the square, which consisted of four rotations and four reflections. We took a set of the four lines through the origin at angles 0, $\pi/4$, $\pi/2$, and $3\pi/4$, numbered 1, 2, 3, 4, respectively. We let S_4 be the permutation NETG of the set of four lines. Each symmetry s, of the square in particular, gave a permutation $\varphi(s)$ of the four lines. Then we defined a mapping, as follows:*

$$\Phi\text{: } D_4 \rightarrow S_4$$

whose value at the symmetry s $\in D_4$ was the permutation $\varphi(s)$ of the four lines. Such a process would always define a neutro-homomorphism. We found the kernel and image of φ. The neutral permutation of the square gave the neutral of the four lines. The rotation (1234) of the square gave the permutation (13)(24) of the four lines; the rotation (13)(24) by 180 degrees gave the neutral permutation e^{neut} of the four lines; the rotation (4321) of the square gave the permutation (13)(24) of the four lines again. Thus, the neutrosophic image of the rotation NET subgroup R_4 of D_4 was the NET subgroup (neut, [13][24]) of S_4. The reflections of the square were given by the compositions of the rotations of the square with a reflection, for example, the reflection (13). The reflection (13) of the square (in the vertical axis) gave the permutation (24) of the lines. Thus, the homomorphism φ took the set of reflections $R_4 \bigcirc$ (13) to the following:

$$\varphi(R4) \bigcirc \phi(13) = (neut, [13][24] \bigcirc [24]) = ([24], [13]).$$

The neutrosophic image of φ was the union of the neutrosophic image of the rotations and the reflections, which was Im(φ) = (neut, [13][24], [13], [24]) $\in S_4$. In the work above, we saw that the neutrosophic kernel of φ was as follows:

$$ker(\varphi) = (neut, [13][24]) \text{ of } D_4$$

3. Neutrosophic Extended Triplet Subgroup

In this section, a definition of the neutrosophic extended triplet subgroup and its example have been given.

Definition 17. *Given an NETG (N, $*$), a subset H is called an NET subgroup of N, if it forms an NETG itself under $*$. Explicitly, this means the following:*

(1) *The extended neutral element $e^{neut(x)}$ lies $\in$ H.*
(2) *For any x, y $\in$ H, $x * y \in$ H (H is closed under $*$).*
(3) *If x $\in$ H, then $e^{anti(x)} \in$ H (H has extended opposites).*

We wrote H $\leq$ N whenever H was an NET subgroup of N. $\emptyset \neq$ H $\subseteq$ N, satisfying (2) and (3) of Definition 17, would be an NET subgroup, as we took $x \in$ H and then (2) gave $e^{anti(x)} \in$ H, after which (3) gave $x * e^{anti(x)} = e^{neut(x)} \in$ H.

Example 10. *We let* $S_4 = (neut, \sigma_1, \sigma_2, \ldots, \sigma_9, \tau_1, \tau_2, \ldots, \tau_8, \delta_1, \delta_2, \ldots, \delta_6)$ *with* $\sigma_1 = (1234)$, $\sigma_2 = (13)(24)$, $\sigma_3 = (1432)$, $\sigma_4 = (1243)$, $\sigma_5 = (14)(23)$, $\sigma_6 = (1342)$, $\sigma_7 = (1324)$, $\sigma_8 = (12)(34)$,$\sigma_9 = (1432)$,$\tau_1 = (234)$,$\tau_2 = (243)$, $\tau_3 = (134)$, $\tau_4 = (143)$, $\tau_5 = (124)$, $\tau_6 = (142)$, $\tau_7 = (123)$, $\tau_8 = (132)$, $\delta_1 = (12)$, $\delta_2 = (13)$, $\delta_3 = (14)$, $\delta_4 = (23)$, $\delta_5 = (24)$, $\delta_6 = (34)$. *The trivial neutrosophic extended subgroups of* S_4 *were the neutral elements, and the non-trivial neutrosophic extended subgroups* S_4 *of order 2 were as follows:* $(neut, \sigma_2)$, $(neut, \sigma_5)$, $(neut, \sigma_8)$, $(neut, \delta_1)$, $(neut, \delta_2)$, $(neut, \delta_3)$, $(neut, \delta_4)$, $(neut, \delta_5)$, $(neut, \delta_6)$, *and the neutrosophic extended subgroups,* S_4, *of order 3 were as follows:*

$$L_{11} = \langle \tau_1 \rangle = \langle \tau_2 \rangle = (neut, \tau_1, \tau_2)$$

$$L_{12} = \langle \tau_3 \rangle = \langle \tau_{14} \rangle = (neut, \tau_3, \tau_4)$$

$$L_{13} = \langle \tau_5 \rangle = \langle \tau_6 \rangle = (neut, \tau_5, \tau_6)$$

$$L_{14} = \langle \tau_7 \rangle = \langle \tau_8 \rangle = (neut, \tau_7, \tau_8)$$

it was straightforward to find the neutrosophic extended subgroups of order 4, 6, 8, and 12 of S_4.

4. Neutrosophic Triplet Cosets

In this section, the neutrosophic triplet coset and its properties have been outlined. Furthermore, the difference between the neutrosophic triplet coset and the classical one have been given.

Definition 18. *Let N be an NETG and* $H \subseteq N$. $\forall\, x \in N$, *the set* $xh/\, h \in H$, *is denoted by xH, analogously, as follows:*

$$Hx = hx/h \in H$$

and

$$(xH)anti(x) = (xh)anti(x)/h \in H.$$

When $h \leq N$, *xH is called the left neutrosophic triplet coset of* $H \in N$ *containing x, and Hx is called the right neutrosophic triplet coset of* $H \in N$ *containing x. In this case, the element x is called the neutrosophic triplet coset representative of xH or Hx.* $|xH|$ *and* $|Hx|$ *are used to denote the number of elements in xH or Hx, respectively.*

Example 11. *When* $N = S_3$ *and* $H = ([1], [12])$, *the "Table 3" lists the left and right neutrosophic triplet H-cosets of every element of the NETG.*

Table 3. Neutrosophic triplet left and right cosets of S_3.

g	*gH*	*Hg*
(1)	([1], [12])	([1], [12])
(12)	([1], [12])	([1], [12])
(13)	([13], [123])	([13], [132])
(23)	([23], [132])	([23], [123])
(123)	([13], [123])	([23], [123])
(132)	([23], [132])	([23], [123])

First of all, cosets were not usually neutrosophic extended triplet subgroups (some did not even contain the extended neutral). In addition, since (13) ≠ H(13), a particular element could have different left and right neutrosophic triplet H-cosets. Since (13)H = H(13), different elements could have the same left neutrosophic triplet H-cosets.

Example 12. *We calculated the neutrosophic triplet cosets of N = (Z_4, +) under addition and let H = (0, 2). The elements (0, 0, 0), (0, 0, 1), (0, 0, 2), (0, 0, 3), (1, 1, 1), and (3, 3, 3) were NET's of Z_4 and the classical cosets of N were as follows:*

$$H = H + 0 = H + 2 = (0, 2).$$

and

$$H + 1 = H + 3 = (1, 3).$$

Here, 2 did not give rise to NET, because the neut's of 2 were 1 and 3, however there were no anti's. Therefore, we could not obtain the neutrosophic triplet coset of N. In general, classical cosets were not neutrosophic triplet cosets, because they might not have satisfied the NET conditions.

Similarly to Definition 16, we could define neutrosophic triplet cosets as follows:

Definition 19. *Let N be a neutrosophic triplet group and $H \leq N$. We defined a relation $\equiv \ell(modH)$ on N as follows:*

if $x_1, x_2 \in N$ and $anti(x_1)x_2 \in N$, Then

$$x_1 = l\, x_2(modH)$$

Or, equivalently, if there exists an $h \in H$, such that:

$$anti(x_1) * x_2 = h$$

That is, if $x_2 = x_1 h$ for some $h \in H$.

Proposition 2. *The relation $\equiv \ell(modH)$ is a neutrosophic triplet equivalence relation. The neutrosophic triplet equivalence class containing x is the set $xH = xh/h \in H$.*

Proof.

(1) $\forall x \in N_1$, anti(x) $*$ x = neut(x) $\in$ H. Hence, x = ℓx_1(modH)} and $\equiv \ell$(modH) is reflexive.

(2) İf x = ℓx_2(modH), then anti(x_1) $*$ $x_2 \in$ H. However, since an anti of an element of H is also in H, anti(anti[x_1] $*$ x_2) = anti(x_2) $*$ anti(anti[x_1]) = anti(x_2) $*$ $x_1 \in$ H. Thus, $x_2 = \ell x_1$(modH), hence $\equiv \ell$(modH) is symmetric.

(3) Finally, if $x_1 = \ell x_2$(modH) and $x_2 = \ell x_3$(modH), then anti(x_1) $*$ $x_2 \in$ H and anti(x_2) $*$ $x_3 \in$ H. Since H is closed under taking products, anti(x_1)$x_2$anti(x_2)x_3 = anti(x_1)$x_3 \in$ H. Hence, $x_1 = \ell x_3$(modH) so that $\equiv \ell$(modH) is transitive. Thus, $\equiv \ell$(modH) is a neutrosophic triplet equivalence relation. □

4.1. Properties of Neutrosophic Triplet Cosets

Lemma 1. *Let $H \leq N$ and let $x, y \in N$. Then,*

(1) $x \in xH$.

(2) $xH = H \Leftrightarrow x \in H$.

(3) $xH = yH \Leftrightarrow x \in yH$.

(4) $xH = yH$ *or* $xH \cap yH = \emptyset$.

(5) $xH = yH \Leftrightarrow anti(x)y \in H$.

(6) $xH = Hx \Leftrightarrow H = (xH)anti(x)$.

(7) $xH \subseteq N \Leftrightarrow x \in H$.

(8) $(xy)H = x(yH)$ *and* $H(xy) = (Hx)y$.

(9) $|xH| = |YH|$.

Proof.

(1) $x = x(neut(x)) \in xH$

(2) $\Rightarrow$ Suppose $xH = H$. Then $x = x(neut(x)) \in xH = H$.

$\Leftarrow$ Now assume x in H. Since H is closed, $xH \subseteq H$.

Next, also assume $h \in H$, so $anti(x)h \in H$, since $H \leq N$. Then,

$$h = neut(x)h = x * anti(x)h = x(anti[x])h \in xH,$$

So $H \subseteq xH$. By mutual inclusion, $xH = h$.

(3) $xH = Yh$

$\Rightarrow x = x(neut(x)) \in xH = yH$.

$\Leftarrow x \in yH \Rightarrow x = yh$, where $h \in H \Rightarrow h \in H$, $xH = (yh)H = y(hH) = yH$.

(4) Suppose that $xH \cap yH \neq \emptyset$. Then, $\exists a \in xH \cap yH \Rightarrow \exists h_1 h_2 \in H \ni a = xh_1$

and

$a = yh_2$. Thus, $x = a(anti(h_1)) = yh_2(antih_1)$ and $xH = yh_2(anti(h_1))H$

$= yh_2(anti(h_1)H) = yH$ by (2) of Lemma 1.

(5) $xH = yH \Leftrightarrow H = anti(x)yH \Leftrightarrow$ (2) of Lemma 1, $anti(x)y \in H$.

(6) $xH = Hx \Leftarrow (xH)anti(x) = (Hx)anti(x) = H(x * anti(x) = H \Leftarrow xH(anti(x)) = H$.

(7) (That is, $xH = H$)

Suppose thay xH is a neutrosophic extended triplet subgroup of N. Then

xH contains the identity, so $xH = H$ by (3) of Lemma 1, which holds $\Leftrightarrow x \in H$ by (2) of Lemma 1.

Conversely, if $x \in H$, then $xH = H \leq N$ by (2) of Lemma 1.

(8) $(xy)H = x(yH)$ and $H(xy) = (Hx)y$ follows from the associative

property of group multiplication.

(9) (Find a map α: $xH \longrightarrow xH$ that is one to one and onto)

Consider α: $xH \longrightarrow xH$ defined by $\alpha(xh) = yh$. This is clearly onto yH. Suppose $\alpha(xh_1)$

$= \alpha(xh_2)$. Then $yh_1 = yh_2 \Rightarrow h_1 = h_2$ by left cancellation $\Rightarrow xh_1 = xh_2$, therefore α is one to one. Since α provides a one to one correspondence between xH and yH, $|xH| = |yH|$. □

In classical group theory, cosets were used in the construction of vitali sets (a type of non-measurable set), and in computational group theory cosets were used to decode received data in linear error-correcting codes, to prove Lagrange's theorem. The neutrosophic triplet coset plays a similar role in the theory of neutrosophic extended triplet group, as in the classical group theory. Neutrosophic triplet cosets could be used in areas, such as neutrosophic computational modelling, to prove Lagrange's theorem in the neutrosophic extended triplet, etc.

4.2. The Index and Lagrange's Theorem: $|H|$ divides $|N|$

Theorem 3 *If N is a finite neutrosophic extended triplet group and $H \leq N$, then $|H|/|N|$. Moreover, the number of the distinct left neutrosophic triplet cosets of H in N is $|N|/|H|$.*

Proof. Let $x_1H, x_2H, \ldots, x_rH$ denote the distinct left neutrosophic triplet cosets of H in N. Then, $\forall$ x $\in$ N. xH = x_iH for some i = 1, 2, . . . , r. Considering (1) of Lemma 1, x $\in$ xH. Thus, N = $x_1H \cup x_2H \cup, \ldots, \cup x_rH$. Considering (4) of Lemma 1, this union is disjointed:

$$|N| = |x_1H| + |x_2H| + \ldots + |x_rH = r|H|.$$

Therefore: $|x_iH| = |xH|$ for i = 1, 2, . . . , r. □

Example 13. *We let H = ([1], [12]), it had three left neutrosophic triplet cosets in S3, see example 11, [S3:H] = 3 = (H,* [13]*H,* [23]*H) = (H,* [13]*H,* [23]*H).*

5. Neutrosophic Triplet Normal Subgroups

In this section, the neutrosophic triplet normal subgroup and neutrosophic triplet normal subgroup test have been outlined.

Definition 20. *A neutrosophic extended triplet subgroup H of a neutrosophic extended triplet group N is called a neutrosophic triplet normal subgroup of N, if xH = Hx,* $\forall$ *x* $\in$ *N and we denote it as* $H \trianglelefteq N$.

Example 14. *The set* $A_n = \sigma \in S_n/\sigma$ *was even a normal subgroup of* S_n*. It was called the alternating neutrosophic extended triplet group on n letters. It was enough to notice that* A_n *= ker(sgn). Since* $|S_n| = n!$*, thus,*

$$|A_n| = n!/2.$$

$$S_n/A_n = n!/n!/2 = 2.$$

Neutrosophic Triplet Normal Subgroup Test

Theorem 4 *A neutrosophic extended triplet subgroup H of N is normal in N if, and only if, anti(x)Hx* $\subseteq$ *H,*

$$\forall x \in N.$$

Proof. *Let H be a neutrosphic extended triplet subgroup of N. Suppose H is neutrosophic extended triplet subgroup of N. Then* $\forall x \in N, y \in H : \exists z \in H : xy = zx$*. Thus (xy)anti(x) = z* $\in$ *H implying (xH)anti(x)* $\subseteq$ *H.*
□

Conversly, suppose $\forall$ x $\in$ N :(xH)anti(x) $\subseteq$ H. Then for n $\in$ N, we have (nH)anti(n) $\subseteq$ H, which implies nH $\subseteq$ Hn. Also, for anti(n) $\in$ N, we have anti(n)H(anti[anti{n}]) = anti(n)Hn $\subseteq$ H, which implies Hn $\subseteq$ nH. Therefore, nH = Hn, meaning that $H \trianglelefteq N$.

Example 15. *We let f: N* $\rightarrow$ *H be a neutro-homomorphism from a neutrosophic extended triplet group N to a neutrosophic extended triplet group H, Kerf* $\trianglelefteq$ *N.*

(1) *If* $\forall$ *a, b* $\in$ *kerf, we had to show that a(anti[b])* $\in$ *kerf. This meant that kerf was a neutrosophic extended triplet subgroup of N. If a* $\in$ *kerf, then*

$$f(a) = neut_H$$

and

b $\in$ *kerf, then*

$$f(b) = neut_H$$

Then, we showed that f(a(anti[b]) = $neut_H$*. (f is neutro-homomorphism)*

$f(a(anti(b)) = f(a) . f(anti(b))$

$= f(a) . f(anti(b))$

$= neut_H . anti(neut_H)$

$= neut_H . neut_H$

$= neut_H$

$\Rightarrow a(anti(b)) \in kerf.$

(2) *We let $n \in N$ and $a \in kerf$. We had to show that $n . a . (anti(n)) \in kerf$. (f is neutro-homomorphism)*

$f(n . a . (anti(n) = f(n) . f(a) . f(anti(n))$

$= f(n) f(a) anti(f(n))$

$= h\ neut_H\ (anti(h))$

$= neut_H$

$\Rightarrow n . a . (anti(n)) \in kerf$

$\Rightarrow kerf \triangleleft N.$

Theorem 5 . *A neutrosophic triplet subgroup H of N is a neutrosophic triplet normal subgroup of N if, and only if, each left neutrosophic triplet coset of H in N is a right neutrosophic triplet coset of $H \in N$.*

Proof. *Let H be a neutrosophic triplet normal subgroup of N, then $xH(anti[x])=H$, $\forall\ x \in N \Rightarrow xH(anti[x])x = Hx$, $\forall x \in N \Rightarrow xH = Hx$, $\forall x \in N$, since each left neutrosophic triplet coset xH is the right neutrosophic triplet coset Hx.* □

Conversely, let each left neutrosophic triplet coset of H in N be a right neutrosophic triplet coset of H in N. This means that if x is any element of N, then the left neutrosophic triplet coset xH is also a right neutrosophic triplet coset. Now neut(x) $\in$ H, therefore x $*$ neut(x) = x $\in$ xH. Consequently x must also belong to that right neutrosophic triplet coset, which is equal to left neutrosophic triplet coset xH. However, x is a left neutrosophic triplet coset and needs to contain one common element before they are identical. Therefore, Hx is the unique right neutrosophic triplet coset which is equal to the left neutrosophic triplet coset xH. Therefore, we have xH = xH, $\forall x \in N \Rightarrow$ xH(anti(x)) = Hx(anti(x), $\forall x \in N \Rightarrow$ xH(anti(x)) = H, $\forall x \in$ N, since H is a neutrosophic triplet normal subgroup of N.

6. Neutrosophic Triplet Quotient (Factor) Groups

The notion of quotient (factor) groups was one of the central concepts of classical group theory and played an important role in the study of the general structure of groups. Just as in a classical group theory, quotient groups played a similar role in the theory of neutrosophic extended triplet group. In this section, we have introduced the notion of neutrosophic triplet quotient group and its relation to the neutrosophic extended triplet group.

Definition 21. *If N is a neutrosophic extended triplet group and $H \trianglelefteq N$ is a neutrosophic triplet normal subgroup, then the neutrosophic triplet quotient group N/H has elements xH: $x \in N$, the neutrosophic triplet cosets of H in N, and an operation of (xH)(yH) = (xy)H.*

Example 16. *Let's find all of the possible neutrosophic triplet quotient groups for the dihedral group D_3.*

$D_3 = (1, r, r^2, s, sr, sr^2)$, where $r^3 = s^2 = rsrs = 1$. A quotient set D_3/N is a neutrosophic triplet group if, and only if, $N \trianglelefteq D_3$. Then, all of neutrosophic triplet normal subgroups are $D_{3\ itself}$. We always have the trivial ones $D_3/D_3 = 1 \cong 1$ and $D_3/1 \cong D_3$. The subgroup $\langle r\rangle = \langle r^2\rangle = (1, r, r^2)$ is that of index 2 and thus is

normal. Therefore, $D_3/\langle r\rangle$ *is also a neutrosophic triplet quotient group. If* $N \trianglelefteq D_3$ *is a different neutrosophic triplet normal subgroup, then* $\langle N\rangle. = 2$*, so either* $N = \langle s\rangle, N = \langle sr\rangle.$ *or* $N = \langle sr^2\rangle$*. However, none of them are normal, since* $(sr)s(anti(sr)) = sr^2$ *not in* $\langle s\rangle$*. Hence, the only non-triavial neutrosophic triplet quotient group is* $D_{3/}\langle r\rangle$*.*

Theorem 6 *Let N be a neutrosophic extended triplet group and H be a neutrosophic triplet normal subgroup of N. In the set N/H = xH, x ∈ N is a neutrosophic extended triplet group under the operation of (xH)(yH) = xyH.*

Proof. $N/H \times N/H \rightarrow N/H$

1. $xH = x'H$ and $yH = y'H$

   $Xh_1 = x'$ and $yh_2 = y'$, $h_1, h_2 \in H$

   $x'y'H = xh_1yh_2H = xh_1yH = x\, h_1Hy = xHy = xyH.$
2. The neutral, for any $x \in H$, is $neut(x)H = H$. That is, $xH * H = xH * neut(x)H = x * neut(x)H = xH$.
3. An anti of a neutrosophic triplet coset xH is anti(x)H, since $xH* anti(x)H = (x * anti(x)H) = neut(x)H = H$.
4. Associativity, $(xHyH)zH = (xy)HzH = (xy)zH = xH(yz)H = xH(yHzH)$, $\forall x, y, z \in N$. □

7. Conclusions

The main theme of this paper was to introduce the neutrosophic extended triplets and then to utilize these neutrosophic extended triplets in order to introduce the neutrosophic triplet cosets, neutrosophic triplet normal subgroup, and finally, the neutrosophic triplet quotient group. We also studied some interesting properties of these newly created structures and their application to neutrosophic extended triplet group. We further defined the neutrosophic kernel, neutrosophic-image, and inverse image for neutrosophic extended triplets. As a further generalization, we created a new field of research, called Neutrosophic Triplet Structures (namely, the neutrosophic triplet cosets, neutrosophic triplet normal subgroup, and neutrosophic triplet quotient group).

Author Contributions: All authors have contributed equally to this paper. The individual responsibilities and contribution of all authors can be described as follows: the idea of this whole paper was put forward by Mikail Bal, he also completed the preparatory work of the paper. Moges Mekonnen Shalla analyzed the existing work of symmetry 292516 neutrosophic triplet coset and quotient group and wrote part of the paper. The revision and submission of this paper was completed by Necati Olgun.

Conflicts of Interest: The authors declare no conflict of interest.

References and Note

1. Smarandache, F. *A Unifying Field in Logics: Neutrosophic Logic*; Philosophy; American Research Press: Santa Fe, NM, USA, 1999; pp. 1–141.
2. Kandasamy Vasantha, W.B.; Smarandache, F. *Basic Neutrosophic Algebraic Structures and Their Application to Fuzzy and Neutrosophic Models*; ProQuest Information & Learning: Ann Harbor, MI, USA, 2004.
3. Kandasamy Vasantha, W.B.; Smarandache, F. *Some Neutrosophic Algebraic Structures and Neutrosophic N-Algebraic Structures*; ProQuest Information & Learning: Ann Harbor, MI, USA, 2006.
4. Smarandache, F.; Mumtaz, A. Neutrosophic triplet group. *Neural Comput. Appl.* **2018**, *29*, 595–601. [CrossRef]
5. Smarandache, F.; Mumtaz, A. Neutrosophic triplet as extension of matter plasma, unmatter plasma, and antimatter plasma. In Proceedings of the 69th Annual Gaseous Electronics Conference on APS Meeting Abstracts, Bochum, Germany, 10–14 October 2016.
6. Smarandache, F.; Mumtaz, A. The Neutrosophic Triplet Group and Its Application to Physics. Available online: https://scholar.google.com.hk/scholar?hl=en&as_sdt=0%2C5&q=The+Neutrosophic+Triplet+Group+and+Its+Application+to+Physics&btnG= (accessed on 20 March 2018).
7. Smarandache, F. Neutrosophic Extended Triplets; Special Collections. Arizona State University: Tempe, AZ, USA, 2016.

8. Smarandache, F.; Mumtaz, A. Neutrosophic Triplet Field used in Physical Applications. In Proceedings of the 18th Annual Meeting of the APS Northwest Section, Pacific University, Forest Grove, OR, USA, 1–3 June 2017.
9. Smarandache, F.; Mumtaz, A. Neutrosophic Triplet Ring and its Applications. In Proceedings of the 18th Annual Meeting of the APS Northwest Section, Pacific University, Forest Grove, OR, USA, 1–3 June 2017.
10. Smarandache, F. (Ed.) A Unifying Field in Logics: Neutrosophic Logic. Neutrosophy, Neutrosophic Set, Neutrosophic Probability: Neutrosophic Logic: Neutrosophy, Neutrosophic Set, Neutrosophic Probability; Infinite Study; 2003. Available online: https://www.google.com.hk/url?sa=t&rct=j&q=&esrc=s&source=web&cd=1&cad=rja&uact=8&ved=0ahUKEwjatuT1w8jaAhVIkpQKHeezDpIQFgglMAA&url=https%3A%2F%2Farxiv.org%2Fpdf%2Fmath%2F0101228&usg=AOvVaw0eRjk3emNhoohTA4tbq5cc (accessed on 20 March 2018).
11. Smarandache, F.; Mumtaz, A. Neutrosophic triplet group. *Neural Comput. Appl.* **2016**, *29*, 1–7. [CrossRef]
12. Şahin, M.; Abdullah, K. Neutrosophic triplet normed space. *Open Phys.* **2017**, *15*, 697–704. [CrossRef]
13. Sahin, M.; Kargin, A. Neutrosophic operational research. *Pons Publ. House/Pons Asbl* **2017**, *I*, 1–13.
14. Uluçay, V.; Irfan, D.; Mehmet, Ş. Similarity measures of bipolar neutrosophic sets and their application to multiple criteria decision making. *Neural Comput. Appl.* **2018**, *29*, 739–748. [CrossRef]
15. Sahin, M.; Ecemis, O.; Ulucay, V.; Deniz, H. Refined NeutrosophicHierchical Clustering Methods. *Asian J. Math. Comput. Res.* **2017**, *15*, 283–295.
16. Sahin, M.; Necati, O.; Vakkas, U.; Abdullah, K.; Smarandache, F. A New Similarity Measure Based on Falsity Value between Single Valued Neutrosophic Sets Based on the Centroid Points of Transformed Single Valued Neutrosophic Numbers with Applications to Pattern Recognition; Infinite Study. 2017. Available online: https://www.google.com/books?hl=en&lr=&id=x707DwAAQBAJ&oi=fnd&pg=PA43&dq=A+New+Similarity+Measure+Based+on+Falsity+Value+between+Single+Valued+Neutrosophic+Sets+Based+on+the+Centroid+Points+of+Transformed+Single+Valued+Neutrosophic+Numbers+with+Applications+to+Pattern+Recognition&ots=M_GVzltdc6&sig=uW5uUsVDAOUJPQex6D8JBj-ZZjw (accessed on 20 March 2018).
17. Uluçay, V.; Mehmet, Ş.; Necati, O.; Adem, K. On neutrosophic soft lattices. *Afr. Matematika* **2017**, *28*, 379–388. [CrossRef]
18. Sahin, M.; Shawkat, A.; Vakkas, U. Neutrosophic soft expert sets. *Appl. Math.* **2015**, *6*, 116. [CrossRef]
19. Sahin, M.; Ecemis, O.; Ulucay, V.; Kargin, A. Some New Generalized Aggregation Operators Based on Centroid Single Valued Triangular Neutrosophic Numbers and Their Applications in Multi-Attribute Decision Making. *Asian J. Math. Comput. Res.* **2017**, *16*, 63–84.
20. Smarandache, F. Seminar on Physics (unmatter, absolute theory of relativity, general theory—Distinction between clock and time, superluminal and instantaneous physics, neutrosophic and paradoxist physics), Neutrosophic Theory of Evolution, Breaking Neutrosophic Dynamic Systems, and Neutrosophic Triplet Algebraic Structures, Federal University of Agriculture, Communication Technology Resource Centre, Abeokuta, Ogun State, Nigeria, 19 May 2017.
21. Smarandache, F. Hybrid Neutrosophic Triplet Ring in Physical Structures. *Bull. Am. Phys. Soc.* **2017**, *62*, 17.
22. Smarandache, F. Neutrosophic Perspectives: Triplets, Duplets, Multisets, Hybrid Operators, Modal Logic, Hedge Algebras. And Applications; Infinite Study. 2017. Available online: https://www.google.com/books?hl=en&lr=&id=4803DwAAQBAJ&oi=fnd&pg=PA15&dq=Neutrosophic+Perspectives:+Triplets,+Duplets,+Multisets,+Hybrid+Operators,+Modal+Logic,+Hedge+Algebras.+And+Applications&ots=SzcTAGL10B&sig=kJMEHWsQ_tYjTJZbOk5TaNahnt0 (accessed on 20 March 2018).

symmetry

MDPI

Article

Study of Decision Framework of Shopping Mall Photovoltaic Plan Selection Based on DEMATEL and ELECTRE III with Symmetry under Neutrosophic Set Environment

Jiangbo Feng [1,*], Min Li [2] and Yansong Li [1]

[1] School of Electrical and Electronic Engineering, North China Electric Power University, Beijing 102206, China; fyh@ncepu.edu.cn

[2] Sinopec Nanjing Eegineering and Construction Inc., Nanjing 210000, China; yxglc@sina.com

* Correspondence: fengjiangbo1995@163.com; Tel.: +86-18810727102

Received: 23 April 2018; Accepted: 3 May 2018; Published: 9 May 2018

Abstract: Rooftop distributed photovoltaic projects have been quickly proposed in China because of policy promotion. Before, the rooftops of the shopping mall had not been occupied, and it was urged to have a decision-making framework to select suitable shopping mall photovoltaic plans. However, a traditional multi-criteria decision-making (MCDM) method failed to solve this issue at the same time, due to the following three defects: the interactions problems between the criteria, the loss of evaluation information in the conversion process, and the compensation problems between diverse criteria. In this paper, an integrated MCDM framework was proposed to address these problems. First of all, the compositive evaluation index was constructed, and the application of decision-making trial and evaluation laboratory (DEMATEL) method helped analyze the internal influence and connection behind each criterion. Then, the interval-valued neutrosophic set was utilized to express the imperfect knowledge of experts group and avoid the information loss. Next, an extended elimination et choice translation reality (ELECTRE) III method was applied, and it succeed in avoiding the compensation problem and obtaining the scientific result. The integrated method used maintained symmetry in the solar photovoltaic (PV) investment. Last but not least, a comparative analysis using Technique for Order Preference by Similarity to an Ideal Solution (TOPSIS) method and VIKOR method was carried out, and alternative plan X1 ranks first at the same. The outcome certified the correctness and rationality of the results obtained in this study.

Keywords: shopping mall; photovoltaic plan; decision-making trial and evaluation laboratory (DEMATEL); interval-valued neutrosophic set; extended ELECTRE III; symmetry

1. Introduction

The frequent occurrence of fog or haze and other negative types of climate change in recent decades is the grave reality that the whole world is experiencing. The pivotal reason behind these environmental problems is atmospheric pollutants and greenhouse gas emissions, mainly produced by fossil fuel consumption. Fossil fuel supplies approximately 80% of the world's energy, and it is drying up with the rapid increase of the world energy demand [1].

To face this situation, many countries have endorsed policies to submit fossil fuel utilization with renewable energy generation. Among diverse types of alternative energy, solar photovoltaic (PV hereinafter) energy is recognized as promising, since sunlight is unlimited and widespread and the converting efficiencies of photovoltaic are getting higher and higher while the manufacturing costs are becoming lower and lower [2].

The past five years has witnessed the astonishing increase in installed cumulative globe solar PV capacity, which nearly quintupled from 70 GW in 2011, to 275 GW in 2016. China is following the worldwide trend with solar PV rapid development and gained the number one cumulative PV capacity of 65.57 GW in 2016. It is worth noting that the large-scale ground PV power station, newly installed, had 28.45 GW capacity this year, and grew by 75% compared to the last year, accounting for 89% of all new PV power plants in 2016, while the distributed PV, newly installed, with 3.66 GW capacity this year, increased by 45% in comparison to last year, and accounted for 11% of capacity of new installations in 2016. The scale of distributed PV development is significantly lower than that of the large-scale ground PV.

Under these circumstances, the China authorities have launched the feed in tariff adjustment. The feed in tariff of ground solar PV generation will decrease to some extent, but in contrast to that of distributed PV, will not decrease at all. As shown in Figure 1, according to the 13th five-year (2016–2020) solar energy planning objectives of China, the goal is to build a total installed capacity of 150 GW solar PV, in which more than 40% of the new installed capacity will be from distributed PV, and to build 100 distributed PV demonstration areas. In fact, China is a country with high potential of solar radiation, and of generous policy subsidies to promote achieving the ambitious plan. Obviously, in China, distributed solar PV generation is government encouraged, well-resourced, environmentally friendly, and closely following the world trend. It is worth considering investment in such a promising project.

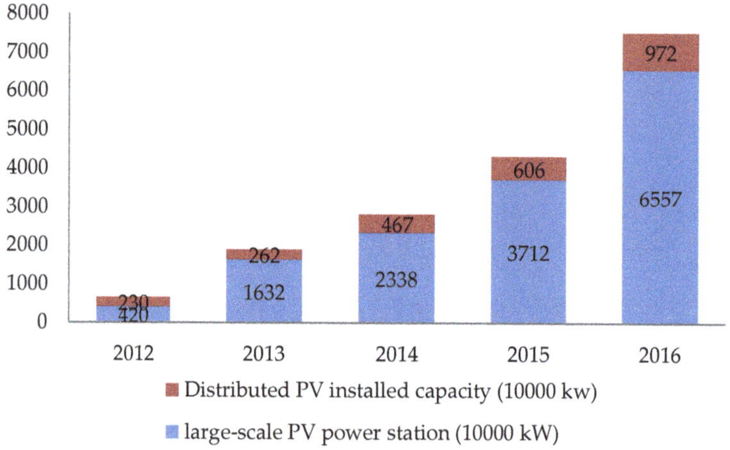

Figure 1. Chinese photovoltaic power structure, 2012–2016.

Numerous vacant roofs of building or structures in the cities provide the best-fitting location for distributed solar PV, and farsighted investors have been preempting outstanding roof resources. There are various types of buildings, such as government building, hospitals, schools, coliseums, or residential and industrial buildings. Among all these types, the shopping centers possesses plenty of advantages, superior to the other types of buildings, and are one of the most promising places worth preempting for PV installation.

First of all, previous construction characters of shopping centers facilitate the rooftop PV installation. According to literature [3], the availability rate of roof space is between 60% and 65% in shopping malls, but just 22% and 50% in residential buildings. That is to say, double or triple PV capacity can be installed in the former roofs, compared to the latter one. Secondly, the shopping malls need a large amount of electricity consumption daily, the most generation can feed the themselves-consumption. Besides, it is usually cited in downtown areas and populated areas. On the one hand, there have a so complete transmission and distribution network that the surplus generation can efficiently feed local electricity consumption. On the other hand, brand advertisement and promotion of the PV manufacturer and the rising level of awareness of the citizen towards sustainable efforts can be greatly obtained for the large visitors there. Last but not least, facing the non-manageable feature through electricity generation in PV facilities, technical management measures need to be taken for protecting distribution system [4]. Fortunately, the shopping malls equipped professional staff and equipment for energy supply, security, air conditioning, and so on. So, the previous staff in shopping malls can condemn the whole new challenge to reduce the extra expenditure in PV management.

It is desirable to adopt a proper methodology for evaluating the shopping mall PV plan in order to demonstrate the optimal possible selections for an investment. Since the following three problems existed in this issue, which are the interaction problems between the criteria, the loss of evaluation information in the conversion process, and the compensation problems between diverse criteria. There is no doubt that traditional multi-criteria decision-making (MCDM) method failed to solve the three problems at the same time. Nevertheless, the decision-making trial and evaluation laboratory (DEMATEL) method can help analyze the internal influence and connection behind each criterion, and the interval-valued neutrosophic set is accomplished in expressing the imperfect knowledge of experts group. Besides, an extended ELECTRE III method as an outstanding outranking method can succeed in avoiding the compensation problem, and the integrated method used maintained symmetry in the solar PV investment. Therefore, in this context, the integrated DEMATEL method and extended ELECTRE III method under interval-valued neutrosophic set environment for searching the optimal shopping mall solar PV plan has been devised.

2. Literature Review

MCDM (multi-criteria decision-making) has been successfully applied in energy planning problems. For example, a review of MCDM methods towards renewable energy development identified MCDM methods as one of the most suitable tools to finding optimal results concerned with energy planning progress in complex scenarios, including various indicators, and conflicting objectives and criteria [5]. For example, Fausto Cavallaro et al. use an intuitionistic fuzzy multi-criteria approach combined with fuzzy entropy to rank different solar-hybrid power plants successfully [6].

In a real case, it is different for decision makers to express preferences when facing inaccurate, uncertain, or incomplete information. Although the fuzz set, intuitionistic fuzzy sets, interval-valued intuitionistic fuzzy sets, and hesitant fuzzy sets can address the situation. However, when being asked the evaluation on a certain statement, the experts can use the interval-valued neutrosophic set (IVNNS) expressing the probability that the statement is true, false, and the degree of uncertainty can be accurately described, respectively [7]. The IVNNS, combined with outrank methods, has addressed many MCDM problems successfully [8]. For example, the IVNNS combined with VIKOR was applied to solve selection of location for a logistic terminal problem [9]. Hong-yu Zhang et al. developed two interval neutrosophic number aggregation operators and applied them to explore multi-criteria decision-making problems [10].

The independence of criteria remains in most of the MCDM methodologies. In recent years, lots of methods appeared to solve the problem, and the DEMATEL method is popularly used. According to the statistic censused in article [11], of the use of MCDM methods in hybrid MCDM methods, the top five methods are Analytical Network Process (ANP), DEMANTEL, Analytic Hierarchy Process(AHP), TOPSIS, and VIKOR.

The DEMATEL methodology has been acknowledged as a proper tool for drawing the relationships concerning interdependencies and the intensity of interdependence between complex criteria in an evaluation index system [12,13]. As a powerful tool to describe the effect relationship, it help evaluate the enablers in solar power developments [11] and evaluate factors which influencing industries' electric consumption [14]. The application of DEMATEL contributed to determining the weight coefficient of the evaluation criteria, and successfully helped identify the suitable locations for installation of wind farms. It is the DEMATEL method that helped the investors improve their decisions when there are many interrelated criteria. Thus, the connection relationship of the climate and economy criteria that exists in this study is of great need of the application of DEMATEL method.

The commonly used outrank models are TOPSIS, AHP, ANP, preference ranking organization method for enrichment evaluations(PROMETHE), and ELECTRE, of which ELECTRE methods are preferred by decision makers in energy planning progress. Among the ELECTRE methods, ELECTRE III method conveys much more information than the ELECTRE I and ELECTRE II methods [15]. In the literature [16], economics of investment in the field of PV, an inclusive decision-making structure using ELECTRE III that would help photo voltaic (PV) system owners, bureaucrats, and the business communities to decide on PV technologies, financial support systems and business strategies were featured. ELECTRE III was used to structure a multi-criteria framework to evaluate the impact of different financial support policies on their attractiveness for domestic PV system deployment on a multinational level [17]. Due to the compensation problem in information processing, incomplete utilization of decision information, and information loss, ELECTRE III was chosen to build a framework for offshore wind farm site selection decision in the intuitionistic fuzzy environment. These literature studies improve the application of ELECTRE III method in the energy planning process, and terrify the effectiveness of evaluation in decision making progress [18].

In conclusion, based on the mentioned evolvement, the shopping mall PV plan evaluation result will be more scientific and reasonable than before.

3. Decision Framework of SMPV Plan Selection

The evaluation criteria are basic to the entire evaluation, so that they are of great importance to the shopping mall photovoltaic plan selection. In view of the special characteristics of photovoltaic plan and the shopping malls, six factors were taken into consideration, namely architectural elements, climate, photovoltaic array, economy, risk, contribution. Table 1 shows six criteria and twenty-one subcriteria.

Table 1. Analysis of evaluation attributes of shopping centers photovoltaic plan selection.

Criteria		Subcriteria	Resources
(a) Architectural elements	a1	Roof pitch and orientation	[2]
	a2	Covering ratio	[2]
	a3	PV roof space	[2]
(b) Economy	b1	Total investment	[19]
	b2	Total profit	[19]
	b3	Annual rate of return	[20]
	b4	Payback year	[20]
(c) Climate	c1	Annual average solar radiation (kwh/m^2/year)	[19,21]
	c2	Land surface temperature (°C)	[19,21]
	c3	Annual sunshine utilization hours (h)	[19,21]
(d) Photovoltaic array	d1	Suitability of the local solar regime	[22]
	d2	PV area	[2]
	d3	PV generation (yearly electricity generation) MWh/year	[2,19]
	d4	Repair and clean rate	[20]
(e) Contribution	e1	Increase in local economy and employment	[22]
	e2	Publicity effects	Own
	e3	Environment protection	[23]
(f) Risk	f1	Grid connection risk	[19]
	f2	Rooftop ownership and occupancy disputes	Own
	f3	Bad climate	[22]
	f4	Government subsidies reduction	Own

3.1. Architectural Elements

Not all the shopping malls are suitable to allocate the photovoltaics, and architectural elements are the primary intrinsic limitations. Steep roof pitch and wrong orientation will increase the difficulty of allocation and maintenance. In addition, building obstructions, and vegetation shading part of the space are not able to allow for the allocation of PV equipment, which can be measured by the covering ratio estimated by Equation (2). Last but not least, the photovoltaic roof space needed to be calculated by the Equation (1). Colmenar-Santos, Antonio et al. [15] assessed the photovoltaic potential in shopping malls by calculating the photovoltaic roof space. We decided to refer to this research method.

$$RS_{PV} = RS \times \alpha \tag{1}$$

RS_{PV} is PV roof space, RS stand for roof space, α is availability ratio

$$CR = L/(a+b) = \frac{\tan\varphi \tan(\alpha_s)^{\alpha}}{\sin\varphi \tan(\alpha_s)^{\alpha} + \sin\varphi \tan\varphi \sin(\gamma_s)^{a}} \tag{2}$$

CR is covering ratio. $(\alpha_s)^{\alpha}$ solar altitude angle $(\gamma_s)^{a}$ φ solar azimuth angle. The value is at nine o'clock on the winter solstice in each location.

3.2. Climate

Not all the locations of the shopping malls have the optimal climate for solar power generation. It is undoubted that the solar resource depends on local climate. Thus, the annual average solar radiation, land surface temperature, and annual sunshine utilization hours, are the four typical criteria to judge whether the local solar energy resource is in abundance.

3.3. Photovoltaic Array

It is well known that the performance of photovoltaic arrays will impact the electricity generation reliability and stability. The dust in the photovoltaic cell panel will reduce the solar energy conversion efficiency, meanwhile, since the photovoltaic cell panel damage and faults are directly related to the electricity supply reliability, the repair and clean rate should be pondered. In addition, it is worth concerning whether the specified type of the photovoltaic panel is absolutely suitable to the local solar regime. The total PV area affects the electricity generated which is calculated by Equation (3). PV generation (yearly electricity generation) is estimated by Equation (4).

$$A_{\text{panel}} = RS_{PV} \times CR \tag{3}$$

$$E = H \times A_{\text{panel}} \times \varepsilon \times \kappa \tag{4}$$

H is the total yearly solar irradiation. A_{panel} is the total area of PV panel. ε is the efficiency of the panel. κ is the comprehensive facility performance efficiency, which is 0.8 [19], and 0.28 is the empirical conversion coefficient of the PV module area to the horizontal area.

3.4. Economy

It is beyond doubt that the economy of the shopping mall photovoltaic plans ought to be taken into account by the decision makers. There are plenty of studies to assess the financial aspects of the photovoltaic projects. Indrajit Das et al. [23] presented an investor-oriented planning model for optimum selection of solar PV investment decisions. Rodrigues Sandy et al. [24] conducted economic analysis of photovoltaic systems under China's new regulation. The economic assessment methods there are so suitable and scientific that they are worth referring to in this paper. The significant economic attributes we considered are pay pack period, total investment, total profit, and annual rate of return, which are calculated by Equations (5)–(7).

$$I = C \times A_{\text{panel}} \times \frac{p_{\text{panel}}}{a_{\text{panel}}} \quad (5)$$

I is the total investment, C is the average cost of building per W roof PV projects. p_{panel} is the max power pin (W) under STC situation of the solar panel, a_{panel} is the area of per photovoltaic panels.

$$B = P_E \times t_{LC} + P_S \times t_S - I \times t_{LC} - O \times t_{LC} \quad (6)$$

B is the total profit, P_E is the electricity price buying from the power supply company, P_S is the electricity price subsidy, t_{LC} is the time of PV projects life cycle, t_S is the time subsidy lasting, O is the cost for operation and maintenance.

$$ROI = \frac{B}{I \times t_{LC}} \quad (7)$$

ROI is the annual rate of return.

$$T_{PB} = \frac{I}{P_E + P_S - I - O} \quad (8)$$

T_{PB} is the pay pack year.

3.5. Contribution

Although the environmental and social contributions the SMPV projects made may not be calculated explicitly as economic profit, there is no denying that these benefits result in increase in local economy and employment, and environment protection and publicity effects are worth the focus of attention. Particularly, the shopping mall holds a great number of visitors. On the one hand, it obtains, easily, the brand advertisement and promotion when PV equipment of a particular company occupies the rooftop of a large commercial building. On the other hand, it is effective to raise the level of awareness of the citizen towards renewable energy and sustainable efforts.

3.6. Risk

Expect that for the above attributes, the risk faced cannot be neglected. First of all, the government subsides policy is likely to change, and the impartiality, sufficiency, stability, and constancy of the subsidy is unable to be ensured. Secondly, the generating capacity is influenced by the climate heavily, so the profits will reduce when facing consecutive rainy days. Thirdly, the rooftop usage needs the allowance from all the owners, however, the rooftop ownership and occupancy disputes are a very common risk. Last but not least, the connected photovoltaic grid is unable to bring any benefits to the grid enterprise because of the intermittent power output. Thus, how long the support to photovoltaic grid connected from the grid enterprise can exist is uncertain.

All in all, the SCPV plan alternatives ought to be appraised from architectural elements, climate, photovoltaic array performance, economy, risk, contribution attributes. The unique custom-made framework of criteria and subcriteria is set up in view of the actual SMPV plans and national conditions.

4. Research Methodology

A decision framework of SMPV selection has been proposed in this section, and there are four phases in this framework, as shown in Figure 2. The research framework is described in the following subsections.

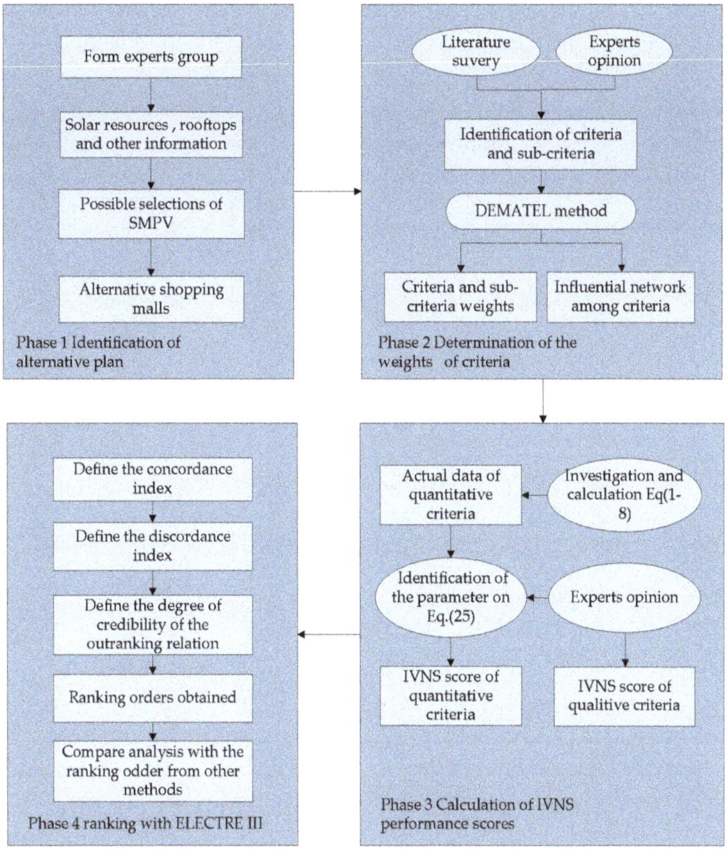

Figure 2. The flowchart of the research methodology. DEMATEL: decision-making trial and evaluation laboratory.

4.1. Preliminary Knowledge in the Neutrosophic Set Environment

Due to the existence of many uncertainties in real decision-making problems, such as indeterminate and inconsistent, the neutrosophic set (NS) is used in the MCDM method, and definition of NS is introduced in this section.

Definition 1 [25]. *Let X be a space of objects with a generic element in X denoted by x. A NS A in X is defined using three functions: truth-membership function* $T_A(x)$, *indeterminacy-membership function* $I_A(x)$ *and falsity-membership function* $F_A(x)$. *These functions are real standard or nonstandard subsets of* $]0^-,1^+[$, *that is,* $T_A(x): X \to]0^-,1^+[$, $I_A(x): X \to]0^-,1^+[$ *and* $F_A(x): X \to]0^-,1^+[$. *And that the sum of* $T_A(x)$, $I_A(x)$ *and* $F_A(x)$ *satisfies the condition* $0^- \leq supT_A(x) + supI_A(x) + supF_A(x) \leq 3^+$.

Since the non-standard unit interval $]0^-,1^+[$ is hard to apply in practice, and the degree of truth, falsity, and indeterminacy about a certain statement could not be described precisely in the practical evaluation, the interval-valued neutrosophic set (IVNNS) of standard intervals has been proposed by Wang [26], and a few definitions and operations of IVNNS are introduced in the GPP technology selection MCDM problem.

Definition 2 [26]. *Let X be a space of objects with a generic element in X denoted by x. An IVNNS A can be defined as*

$$A = \{\langle x, T_A(x), I_A(x), F_A(x)\rangle : x \in X\} \tag{9}$$

where $T_A(x) : X \to [0,1]$, $I_A(x) : X \to [0,1]$, $F_A(x) : X \to [0,1]$. *For each element x in X, these functions can be expressed as* $T_A(x) = [infT_A(x), supT_A(x)] \subseteq [0,1]$, $I_A(x) = [infI_A(x), supI_A(x)] \subseteq [0,1]$, $F_A(x) = [infF_A(x), supF_A(x)] \subseteq [0,1]$ *and* $0 \le supT_A(x) + supI_A(x) + supF_A(x) \le 3, x \in X$. *For convenience, the interval-valued neutrosophic number (IVNN) can be expressed as* $\tilde{a} = \langle [T^L_{\tilde{a}}, T^U_{\tilde{a}}], [I^L_{\tilde{a}}, I^U_{\tilde{a}}], [F^L_{\tilde{a}}, F^U_{\tilde{a}}]\rangle$, *and L, U represent the inferiors and superiors of IVNN respectively.*

Definition 3 [10]. *Let* $\tilde{a} = \langle [T^L_{\tilde{a}}, T^U_{\tilde{a}}], [I^L_{\tilde{a}}, I^U_{\tilde{a}}], [F^L_{\tilde{a}}, F^U_{\tilde{a}}]\rangle$ *and* $\tilde{b} = \left\langle \left[T^L_{\tilde{b}}, T^U_{\tilde{b}}\right], \left[I^L_{\tilde{b}}, I^U_{\tilde{b}}\right], \left[F^L_{\tilde{b}}, F^U_{\tilde{b}}\right]\right\rangle$ *be two IVNNs, and* λ *is a real number for not less than 0. Which operational rules can be expressed as follows*

$$\tilde{a} \oplus \tilde{b} = \left\langle \left[T^L_{\tilde{a}} + T^L_{\tilde{b}} - T^L_{\tilde{a}} \cdot T^L_{\tilde{b}}, T^U_{\tilde{a}} + T^U_{\tilde{b}} - T^U_{\tilde{a}} \cdot T^U_{\tilde{b}}\right], \left[I^L_{\tilde{a}} \cdot I^L_{\tilde{b}}, I^U_{\tilde{a}} \cdot I^U_{\tilde{b}}\right], \left[F^L_{\tilde{a}} \cdot F^L_{\tilde{b}}, F^U_{\tilde{a}} \cdot F^U_{\tilde{b}}\right]\right\rangle \tag{10}$$

$$\tilde{a} \otimes \tilde{b} = \left\langle \left[T^L_{\tilde{a}} \cdot T^L_{\tilde{b}}, T^U_{\tilde{a}} \cdot T^U_{\tilde{b}}\right], \left[I^L_{\tilde{a}} + I^L_{\tilde{b}} - I^L_{\tilde{a}} \cdot I^L_{\tilde{b}}, I^U_{\tilde{a}} + I^U_{\tilde{b}} - I^U_{\tilde{a}} \cdot I^U_{\tilde{b}}\right], \left[F^L_{\tilde{a}} + F^L_{\tilde{b}} - F^L_{\tilde{a}} \cdot F^L_{\tilde{b}}, F^U_{\tilde{a}} + F^U_{\tilde{b}} - F^U_{\tilde{a}} \cdot F^U_{\tilde{b}}\right]\right\rangle \tag{11}$$

$$\lambda\tilde{a} = \left\langle \left[1 - (1 - T^L_{\tilde{a}})^\lambda, 1 - (1 - T^U_{\tilde{a}})^\lambda\right], \left[(I^L_{\tilde{a}})^\lambda, (I^U_{\tilde{a}})^\lambda\right], \left[(F^L_{\tilde{a}})^\lambda, (F^U_{\tilde{a}})^\lambda\right]\right\rangle \tag{12}$$

$$\tilde{a}^\lambda = \left\langle \left[(T^L_{\tilde{a}})^\lambda, (T^U_{\tilde{a}})^\lambda\right], \left[1 - (1 - I^L_{\tilde{a}})^\lambda, 1 - (1 - I^U_{\tilde{a}})^\lambda\right], \left[1 - (1 - F^L_{\tilde{a}})^\lambda, 1 - (1 - F^U_{\tilde{a}})^\lambda\right]\right\rangle \tag{13}$$

The original data of selection of SPPV are collected and processed in this phase. The alternative plans of the GPP project are evaluated by the experts, firstly according to the local technical condition data and practical experience. Then, the decision matrices are expressed in the form of IVNNs, which can handle incomplete and indeterminate information. Finally, a comprehensive decision matrix is formed based on interval-valued neutrosophic number weighted geometric operator (IVNNWG) operator. Let A_i denote the technology alternatives ($i = 1, 2, \cdots m$), and C_j denote the criteria ($j = 1, 2, \cdots n$). It is assumed that $\tilde{a}^k_{ij}$ can be used to represent the evaluation value of attribute of alternative from every expert $E_k (k = 1, 2, \cdots h)$.

Definition 4 [27]. *Let* $\tilde{A}^k_{ij} = \left(\tilde{a}^k_{ij}\right)_{m \times n}$ *be the IVNN-decision matrix of the k-th DM,* $k = 1, 2, \cdots h$, *and* $\tilde{a}^k_{ij} = \left\langle \left[T^L_{\tilde{a}^k_{ij}}, T^U_{\tilde{a}^k_{ij}}\right], \left[I^L_{\tilde{a}^k_{ij}}, I^U_{\tilde{a}^k_{ij}}\right], \left[F^L_{\tilde{a}^k_{ij}}, F^U_{\tilde{a}^k_{ij}}\right]\right\rangle$. *An IVNNWG operator is a mapping:* $IVNN^n \to IVNN$, *such that*

$$\begin{aligned} IVNNWG_\omega\left(\tilde{a}^1_{ij}, \tilde{a}^2_{ij}, \cdots, \tilde{a}^h_{ij}\right) &= \prod_{k=1}^{k=h}\left(\tilde{a}^k_{ij}\right)^{\omega_k} \\ &= \left\langle \left[\prod_{k=1}^{k=h}\left(T^L_{\tilde{a}^k_{ij}}\right)^{\omega_k}, \prod_{k=1}^{k=h}\left(T^U_{\tilde{a}^k_{ij}}\right)^{\omega_k}\right], \left[1 - \prod_{k=1}^{k=h}\left(1 - I^L_{\tilde{a}^k_{ij}}\right)^{\omega_k}, 1 - \prod_{k=1}^{k=h}\left(1 - I^U_{\tilde{a}^k_{ij}}\right)^{\omega_k}\right], \left[1 - \prod_{k=1}^{k=h}\left(1 - F^L_{\tilde{a}^k_{ij}}\right)^{\omega_k}, 1 - \prod_{k=1}^{k=h}\left(1 - F^U_{\tilde{a}^k_{ij}}\right)^{\omega_k}\right]\right\rangle \end{aligned} \tag{14}$$

where $\omega = (\omega_1, \omega_2, \cdots, \omega_h)^T$ *represents the weight vector of DMs, satisfying* $\Sigma^h_{k=1}\omega_k = 1$, $\omega_k \in [0,1]$.

4.2. Phase I Identification of Alternative SMPV Plans

At this stage, a group of experts consisting of several doctorate engineers will be constituted by the investor. All the experts possess abundant working experience in solar energy investment field, and are specialized in solar photovoltaic and power grid technologies.

More than twenty famous influential large-scale shopping malls located in those cities with both abundant sunshine and general policy subsidies need to be collected, and based on that information, less than ten alternative plans roughly screened out, based on the plentitude of documents and investigation. After that, the investigation will be carried out by the experts group, and involve meeting the Development and Reform Commission, the Meteorological Bureau, and the local power supply companies, in order to gather information about solar resources, city planning and construction, local solar subsidy policies, distributed solar power planning, economic assessment, and approved shopping mall rooftops for construction. Lastly, there will be less than five of the most potential alternative places presented for the next evaluation.

4.3. Phase II Determination of the Weights of Criteria Based on DEMATEL Method

The importance of the criteria on GPP technology selection is different, so the DEMATEL method is used to decide the weight of criteria in this phase. The direct and indirect causal relations among criteria are considered in the DEMATEL method, and the subjective judgment of DMs is also considered. The steps of determining the weights based on the DEMATEL method are shown as follows [28]:

Step 1. Determine the influence factors in the system

The influence factors of GPP technology selection system are determined based on expert opinions and literature reviews, which is called the criteria, as shown in Table 1.

Step 2. Construct the direct-relation matrix among the criteria

The direct-relation matrix $X = (x_{pq})_{n \times n}$ is constructed in stages, where x_{pq} is used to represent the degree of direct influence of pth criterion on qth criterion which is evaluated by the experts, and n is the number of criteria.

Step 3. Normalize the direct-relation matrix

The direct-relation matrix $X = (x_{pq})_{n \times n}$ is normalized into $Y = (y_{pq})_{n \times n}$. A normalization factor [29] s is applied in the normalized calculation, and the normalized direct-relation matrix Y is calculated by using Equations (1) and (2).

$$Y = s \cdot X \quad (15)$$

$$s = \text{Min}\left(\frac{1}{\text{Max}_{1 \le p \le n}\left(\sum_{q=1}^{n} x_{pq}\right)}, \frac{1}{\text{Max}_{1 \le q \le n}\left(\sum_{p=1}^{n} x_{pq}\right)}\right) \quad (16)$$

Step 4. Calculate the comprehensive-relation matrix.

The comprehensive-relation matrix T is obtained by using Equation (3).

$$T = \sum_{\lambda=1}^{\infty} Y^{\lambda} = Y(I-Y)^{-1} \quad (17)$$

where $T = (t_{pq})_{n \times n}$, $p, q = 1, 2, \cdots n$, and t_{pq} is used to represent the degree of total influence of pth criterion on qth criterion. I represents for the identity matrix.

Step 5. Determine the influence relation among criteria

The influence degree and influenced degree of the criteria is determined after obtaining the comprehensive-relation matrix T. The sum of the row and column values of matrix T can be obtained by the Equations (4) and (5). The sum of row values of T, denoted by D, which represents the overall influence of a given criterion on other criteria. The sum of column values of T, denoted by R, which implies the overall influence of other criteria on a given criterion.

$$D = (d_p)_{n \times 1} = \sum_{q=1}^{n} t_{pq} \quad (18)$$

$$R = (r_q)_{1 \times n} = \sum_{p=1}^{n} t_{pq} \quad (19)$$

The causal diagram is obtained based on the $D + R$ and $D - R$ values. The $D + R$ value indicates the importance of indicator in the SMPV plan selection system, the greater the $D + R$ value, the more important the corresponding indicator is. On the other hand, the $D - R$ value indicates the influence between a certain indicator and the other indicators, which can be separated into cause and effect groups. The indicator, which has positive values of $D-R$, belongs to the cause group, and dispatches effects to the other indicators. Otherwise, the indicator, which has negative values of $D-R$, belongs to the effect group, and receives effects to the other indicators.

Step 6. Determine the weight of criteria

The criteria are represented by j, and satisfying $j = 1, \cdots p, \cdots q, \cdots n$, then the weights of criteria are determined based on the following Equations (6) and (7) [30].

$$w_j{}' = \left[(d_j + r_j)^2 + (d_j - r_j)^2 \right]^{1/2} \tag{20}$$

$$w_j = \frac{w_j{}'}{\sum_{j=1}^{n} w_j{}'} \tag{21}$$

where $w_j{}'$ denotes the relative importance of the indicators, and w_j denotes the weights of the indicators in SMPV technology selection.

4.4. Phase III Calculation IVNNs Performance Score

ELECTRE is a family of methods used for choosing, sorting and ranking, multi-criteria problems. ELECTRE III was developed by Roy in 1978m, which is valued outranking relation.

$A = \{a_1, a_2, \ldots, a_n\}$ is the finite set of alternatives, $Y = \{y_1, y_2, \ldots, y_m\}$ is the finite set of criteria, $y_j(a)$ represents the performance of alternative a on criterion $y_j \in Y$. Assume that all the criteria are of the gain type, which means the greater the value, the better. q_j is the indifference threshold, which represents two alternatives in terms of their evaluations on criterion y_j. In general, q_j is a function of attribute value $q_j(a_i)$, which can be denoted as $q_j(y_j(a_i))$; $p_j(y_j(a_i))$ is preference threshold, which indicates that there is a clear strict preference of one alternative over the other in terms of their evaluations on criterion y_j. In addition, $v_j(y_j(a_i))$ is a veto threshold that indicates that the attribute value $y_j(a_i)$ of scheme a_i is lower than the attribute value $y_j(a_k)$ of scheme a_k, and when it reaches or exceeds $v_j(y_j(a_i))$, it is not recognized that the a_i is preferred to the a_k. $y_j(a_i) - y_j(a_k)$ which indicates the situation of preference of a_i over a_k for criterion C_j. A weight w_j expresses the relative importance of criterion y_j, as it can be interpreted as the voting power of each criterion to the outranking relation.

Where $y_j(a_i)$ and $y_j(a_k)$ are expressed in the form of IVNNs in the paper, that is, $y_j(a_i) = \langle [T_i^L, T_i^U], [I_i^L, I_i^U], [F_i^L, F_i^U] \rangle$, $y_j(a_k) = \langle [T_k^L, T_k^U], [I_k^L, I_k^U], [F_k^L, F_k^U] \rangle$. In the calculation of Equation (8), let

$$y_j(a_i) = (T_i^L + T_i^U) - (I_i^L + I_i^U) - (F_i^L + F_i^U) \tag{22}$$

$$y_j(a_k) = (T_k^L + T_k^U) - (I_k^L + I_k^U) - (F_k^L + F_k^U) \tag{23}$$

and so

$$y_j(a_i) - y_j(a_k) = (T_i^L + T_i^U - T_k^L - T_k^U) + (I_i^L + I_i^U - I_k^L - I_k^U) + (F_i^L + F_i^U - F_k^L - F_k^U) \tag{24}$$

$$T_{ij} = \begin{cases} \delta_i \frac{a_{ij}}{a_i^{\max}} (i \in \theta_B) \\ \lambda_i \frac{a_i^{\min}}{a_{ij}} (i \in \theta_C, a_i^{\min} \neq 0) \\ \lambda_i (1 - \frac{a_{ij}}{a_i^{\max}}) (i \in \theta_C, a_i^{\min} \neq 0) \end{cases}$$

$$I_{ij} = \begin{cases} \varepsilon_i \frac{a_{ij}}{a_i^{\max}} (i \in \theta_B) \\ \mu_i \frac{a_i^{\min}}{a_{ij}} (i \in \theta_C, a_i^{\min} \neq 0) and \\ \mu_i (1 - \frac{a_{ij}}{a_i^{\max}}) (i \in \theta_C, a_i^{\min} \neq 0) \end{cases} \tag{25}$$

$$F_{ij} \begin{cases} F_{ij}{}^U = 1 - T_{ij}^L - I_{ij}^L \\ F_{ij}{}^L = 1 - T_{ij}^U - I_{ij}^U \end{cases}$$

The δ, λ refer to the certainty parameters of the benefit criteria and cost criteria, respectively [23], while the ε, μ stand for the uncertainty parameters of the benefit criteria and cost criteria respectively, and they obey rule $0 < \delta_i + \varepsilon_i \leq 1, 0 < \lambda_i + \mu_i \leq 1$.

4.5. Phase IV Calculation of Outranking Relation of IVNNs Based on Extended ELECTRE-III

Step 1. Define the concordance index

The concordance index $c(a_i, a_k)$ that measures the strength of the coalition of criteria the support the hypothesis "is at least as good as", $c(a_i, a_k)$ is computed for each ordered pair $a_i, a_k \in A$ as follows:

$$c(a_i, a_k) = \sum_{j=1}^{n} w_j c_j(a_i, a_k) / \sum_{i=1}^{n} w_j \tag{26}$$

and the partial concordance index $c(a_i, a_k)$ is defined as

$$c_j(a_i, a_k) = \begin{cases} 0 & (y_j(a_i) - y_j(a_k) \leq q_j[y_j(a_i)]) \\ 1 & (y_j(a_i) - y_j(a_k) \leq q_j[y_j(a_i)]) \\ \frac{y_j(a_i) - y_j(a_k) - q_j[y_j(a_i)]}{p_j[y_j(a_i)] - q_j[y_j(a_i)]} & (others) \end{cases} \tag{27}$$

Step 2. Define the discordance index

The discordance index $d_j(a_i, a_k)$ is defined as follows:

$$d_j(a_i, a_k) = \begin{cases} 0 & (y_j(a_k) - y_j(a_i) \leq -q_j[y_j(a_i)]) \\ 1 & (y_j(a_k) - y_j(a_i) \geq v_j[y_j(a_i)]) \\ \frac{y_j(a_k) - y_j(a_i) + q_j[y_j(a_i)]}{v_j[y_j(a_i)] + q_j[y_j(a_i)]} & (others) \end{cases} \tag{28}$$

Step 3. Define the degree of credibility of the outranking relation

The overall concordance and partial discordance indices are combined to obtain a valued outranking relation with credibility $s(a_i, a_k) \in [0, 1]$ defined by:

$$s(a_i, a_k) = \begin{cases} c(a_i, a_k) & (\forall j, d_j(a_i, a_k) \leq c(a_i, a_k)) \\ c(a_i, a_k) \prod\limits_{j \in J(a_i, a_j)} \frac{1 - d_j(a_i, a_k)}{1 - c(a_i, a_k)} & (others) \end{cases} \tag{29}$$

where $j(a_i, a_k)$ is the set of criteria for which $d_j(a_i, a_k) > c(a_i, a_k)$.

Step 4. Define the ranking of the alternatives

$\sum s(a_i \succ a_k)$ means the sum degree of credibility that alternative a_i outranks all the other alternatives, and $\sum s(a_k \succ a_i)$ means the sum degree of credibility that all the other alternatives outrank alternative a_i. Thus, $\Delta S(a_i)$ represents the ranking of the alternative a_i, and the higher the $\Delta S(a_i)$ value is, the more superior the outranking order is.

$$\Delta S(a_i) = \sum s(a_i \succ a_k) - \sum s(a_k \succ a_i) \ (k = 1, 2 \ldots n) \tag{30}$$

5. A Real Case Study

A Chinese renewable energy investment company wants to build a shopping center rooftop photovoltaic power project. In order to seek the optimal shopping mall for rooftop photovoltaic power plants, furthermore, one must judge the weight and the influence network of the criteria. A group of experts consisting of three doctorate engineers (referred to as E1, E2, E3) was constituted by the company. All three experts possess more than 15 years' working experience in solar energy investment field, and are specialized in solar photovoltaic and power grid technologies. The collaboration of the all the experts was needed, thus, a pseudo-delphi method was applied in which each expert has no interaction.

Considering the development target and the investment capacity of the company, the famous influential large-scale shopping malls located in those cities, with both abundant sunshine and general policy subsidies, have been roughly screened out based on the plentitude of documents and investigation. The investigation involved several potential shopping malls located in Beijing, Shanghai, Guangzhou, Chengdu, Hangzhou, and Nanjing. The experts group met the Development and Reform Commission, the Meteorological Bureau, and the local power supply companies, in order to gather information about solar resources, city planning and construction, local solar subsidy policies, distributed solar power planning, economic assessment, and approved shopping mall rooftops for construction. There are four potential shopping malls picked out after the first filter, and they are the Golden Resources shopping mall in Beijing, Super Brand Mall in Shanghai, Deji Plaza in Nanjing, Jiangsu province, The Mixc shopping mall in Shenzhen, Guangdong province (hereafter referred to as X_1, X_2, X_3, X_4), as shown in Figure 3.

Figure 3. The alternative shopping malls geography distribution.

Firstly, based on the evaluation criteria, the influence of each criteria to the other one criteria was accessed by the experts. Then, the three experts discussed with each other and obtained a consensus about the influence of each criteria, as shown in Table 2. According to the DEMANTEL method, the weight of criteria and subcriteria was calculated based on Equations (15)–(21), and shown in Table 2. For the intuitive and simple understanding and analysis of the criteria and subcriteria, Figures 4 and 5 were drawn. As shown in Figure 4, the horizontal axis represents the importance of a criteria, while the vertical axis indicates the influence between the criteria, and the arrow is from the sender of this influence to the receiver. As we can see, the (b) economy (0.286) obtained the most importance, but was vulnerable to other criteria. The (a) architectural element and (c) climate (0.88) seemed not particularly important, however, they had significant direct impacts to the other four criteria. The (f) risk (0.188), (d) photovoltaic array (0.162), and (e) contribution (0.151) were considered of medium importance, and among them, (f) and (d) had more of an impact, while (e) received more impact. Horizontal histogram clearly and intuitively shows the weight of each subcriteria in Figure 5. It is obvious that the (b1) Total investment, (b2) Total profit, and (b3). Annual rate of return acquired the highest weight, in addition to the (f4) Government subsidies reduction, (e2) Publicity effects and (d2). PV area was considered to be less but also very important. Therefore, it can be imagined that the SMPV plan alternatives which obtained high scores in these criteria are more likely to win the competition.

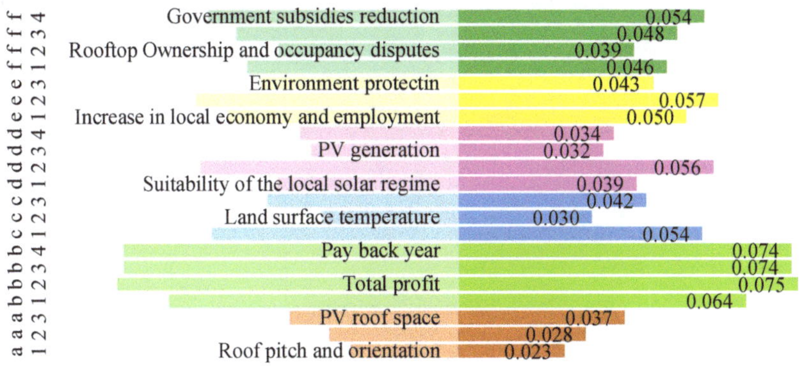

Figure 4. The weights of subcriteria.

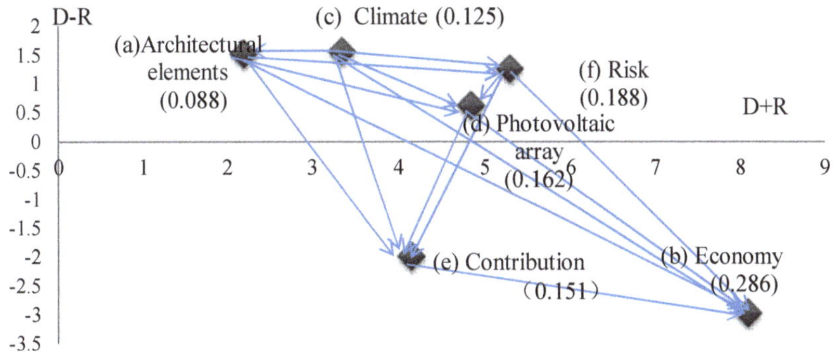

Figure 5. Influential network relationship map within systems.

Secondly, there are two types of criteria, one is quantitative, and the other is qualitative. On one hand, for the quantitative subcriteria, searching from the NASA atmospheric science data center, the data for c1, c2, c3 were obtained. Through using the Google earth map, the roof space data were obtained. Then, according to Equations (1)–(8), the data for a3, b1–b4, d2, d3 were estimated. Because the Equations (1)–(8) are just for rough estimate, these data were not highly accurate. Considering the uncertainty and fuzziness of the data, Equation (25) was used to turn the numerical value into an IVNN value. The performance scores of the quantitative subcriteria are shown in Table 3. On the other hand, for the qualitative subcriteria, the experts group devoted their efforts to investigate the alternative plans and evaluate the performance score for the subcriteria a1, a2, d1, d4, e1–e3, f1–f4. The performance scores of the qualitative subcriteria were shown in Table 4. In addition, the subcriteria were divided into positive and negative. The score of positive criteria higher and negative criteria lower means the alternative better. In this paper, subcriteria b1, d4, f1, f3, f4 are negative and the others are positive.

Table 2. The score of influence among each subcriteria.

	a1	a2	a3	b1	b2	b3	b4	c1	c2	c3	d1	d2	d3	d4	e1	e2	e3	f1	f2	f3	f4	D	R	W
a1	0	3	4	1	1	1	1	0	0	4	0	0	0	0	0	4	1	0	0	0	0	0.50	0.03	0.023
a2	2	0	5	3	2	2	2	0	2	4	0	1	0	0	0	0	0	0	0	0	0	0.60	0.05	0.028
a3	0	0	0	5	5	4	4	0	0	0	0	3	1	1	2	3	2	0	0	0	0	0.75	0.25	0.037
b1	0	0	0	0	3	5	5	0	0	0	3	3	2	2	4	4	4	1	1	0	0	0.89	1.05	0.064
b2	0	0	0	0	0	5	5	0	0	0	0	1	1	1	5	4	2	0	2	0	3	0.68	1.47	0.075
b3	0	0	0	0	0	0	2	0	0	0	0	1	1	1	5	4	2	0	2	0	3	0.49	1.51	0.074
b4	0	0	0	0	0	2	0	0	0	0	0	1	2	2	4	3	3	1	1	0	2	0.50	1.51	0.074
c1	0	0	0	5	5	5	5	0	5	5	3	3	0	2	1	0	2	1	0	1	2	1.16	0.08	0.054
c2	0	0	0	4	4	4	4	0	0	0	2	2	0	1	0	0	0	0	0	1	0	0.58	0.27	0.030
c3	0	0	0	4	4	4	4	0	3	0	0	3	0	3	0	0	2	0	0	1	0	0.72	0.54	0.042
d1	0	0	0	3	3	3	3	0	0	4	0	3	1	1	0	0	1	3	0	0	2	0.71	0.47	0.039
d2	0	0	0	2	5	3	3	0	0	0	3	0	0	1	2	2	4	5	0	0	3	0.82	0.90	0.056
d3	0	0	0	1	5	3	3	0	0	4	3	2	0	3	1	0	0	0	0	0	0	0.64	0.26	0.032
d4	0	0	0	1	5	3	3	0	0	3	3	3	0	0	1	0	0	0	0	0	0	0.57	0.48	0.034
e1	0	0	0	1	1	0	0	0	0	0	0	0	0	0	0	4	1	2	0	0	3	0.30	1.05	0.050
e2	0	0	0	2	1	0	0	0	0	0	0	0	0	0	2	0	1	3	3	0	4	0.42	1.17	0.057
e3	0	0	0	0	0	0	0	0	0	0	0	0	0	0	0	5	0	3	2	0	4	0.36	0.86	0.043
f1	0	0	0	3	5	4	4	0	0	0	0	5	0	0	1	2	2	0	0	0	3	0.74	0.66	0.046
f2	0	0	5	5	5	4	4	0	0	0	0	2	0	0	2	2	0	0	0	0	0	0.74	0.39	0.039
f3	0	0	0	1	5	4	4	5	5	5	4	4	0	0	1	0	0	1	0	0	0	1.04	0.06	0.048
f4	0	0	0	5	5	5	5	0	0	0	0	1	0	0	2	2	2	3	0	0	0	0.75	0.91	0.054

Table 3. IVNN performance scores of alternative SMPV plans on the quantitative subcriteria.

Subcriteria	S1	S2	S3	S4
(a3)	([0.90,0.90],[0.10,0.10],[0.00,0.00])	([0.32,0.32],[0.04,0.04],[0.65,0.65])	([0.26,0.26],[0.03,0.03],[0.71,0.71])	([0.42,0.42],[0.05,0.05],[0.54,0.54])
(b1)	([0.41,0.41],[0.10,0.10],[0.49,0.49])	([0.70,0.70],[0.08,0.06],[0.23,0.25])	([0.26,0.26],[0.03,0.03],[0.71,0.71])	([0.50,0.50],[0.06,0.06],[0.44,0.44])
(b2)	([0.90,0.90],[0.03,0.03],[0.07,0.07])	([0.40,0.40],[0.04,0.06],[0.56,0.54])	([0.90,0.90],[0.10,0.10],[0.00,0.00])	([0.53,0.53],[0.05,0.05],[0.42,0.42])
(b3)	([0.90,0.90],[0.03,0.03],[0.07,0.07])	([0.65,0.65],[0.07,0.07],[0.28,0.28])	([0.53,0.53],[0.06,0.06],[0.41,0.41])	([0.63,0.63],[0.07,0.07],[0.30,0.30])
(b4)	([0.90,0.90],[0.03,0.03],[0.07,0.07])	([0.72,0.72],[0.08,0.08],[0.20,0.20])	([0.60,0.60],[0.07,0.07],[0.33,0.33])	([0.60,0.60],[0.07,0.07],[0.33,0.33])
(c1)	([0.90,0.90],[0.03,0.03],[0.07,0.07])	([0.79,0.79],[0.10,0.10],[0.11,0.11])	([0.81,0.81],[0.09,0.09],[0.10,0.10])	([0.81,0.81],[0.09,0.09],[0.09,0.09])
(c2)	([0.53,0.53],[0.06,0.06],[0.41,0.41])	([0.70,0.70],[0.08,0.08],[0.22,0.22])	([0.62,0.62],[0.07,0.07],[0.31,0.31])	([0.90,0.90],[0.10,0.10],[0.00,0.00])
(c3)	([0.90,0.90],[0.03,0.03],[0.07,0.07])	([0.71,0.71],[0.08,0.08],[0.21,0.21])	([0.69,0.69],[0.08,0.08],[0.23,0.23])	([0.73,0.73],[0.08,0.08],[0.19,0.19])
(d2)	([0.90,0.90],[0.03,0.03],[0.07,0.07])	([0.53,0.53],[0.06,0.06],[0.41,0.41])	([0.41,0.41],[0.05,0.05],[0.54,0.54])	([0.74,0.74],[0.08,0.08],[0.17,0.17])
(d3)	([0.90,0.90],[0.03,0.03],[0.07,0.07])	([0.47,0.47],[0.05,0.05],[0.48,0.48])	([0.37,0.37],[0.04,0.04],[0.59,0.59])	([0.67,0.67],[0.07,0.07],[0.25,0.25])

Table 4. IVNN performance scores of alternative SMPV plans on the qualitative subcriteria.

Subcriteria	X1	X2	X3	X4
(a1)	([0.45,0.56],[0.18,0.30],[0.13,0.50])	([0.68,0.79],[0.18,0.30],[0.13,0.50])	([0.56,0.68],[0.18,0.24],[0.38,0.67])	([0.79,0.90],[0.12,0.18],[0.25,0.50])
(a2)	([0.50,0.50],[0.06,0.06],[0.44,0.44])	([0.85,0.85],[0.09,0.09],[0.06,0.06])	([0.80,0.80],[0.09,0.09],[0.12,0.12])	([0.90,0.90],[0.10,0.10],[0.00,0.00])
(d1)	([0.80,0.90],[0.15,0.23],[0.13,0.33])	([0.50,0.60],[0.23,0.30],[0.38,0.67])	([0.40,0.60],[0.15,0.23],[0.25,0.50])	([0.60,0.80],[0.23,0.30],[0.25,0.50])
(d4)	([0.20,0.23],[0.12,0.20],[0.25,0.33])	([0.45,0.60],[0.20,0.30],[0.13,0.17])	([0.60,0.90],[0.15,0.20],[0.13,0.14])	([0.30,0.45],[0.10,0.12],[0.20,0.50])
(e1)	([0.70,0.90],[0.08,0.15],[0.21,0.33])	([0.50,0.60],[0.23,0.30],[0.36,0.58])	([0.30,0.50],[0.08,0.15],[0.43,0.58])	([0.60,0.80],[0.15,0.30],[0.14,0.25])
(e2)	([0.50,0.60],[0.23,0.30],[0.33,0.75])	([0.70,0.80],[0.15,0.23],[0.33,0.75])	([0.70,0.90],[0.15,0.23],[0.17,0.50])	([0.80,0.90],[0.15,0.23],[0.17,0.50])
(e3)	([0.79,0.90],[0.12,0.18],[0.17,0.50])	([0.68,0.90],[0.06,0.12],[0.17,0.75])	([0.68,0.79],[0.18,0.24],[0.33,0.75])	([0.68,0.79],[0.24,0.30],[0.33,0.75])
(f1)	([0.30,0.45],[0.20,0.26],[0.33,0.43])	([0.45,0.90],[0.26,0.30],[0.38,0.43])	([0.23,0.45],[0.20,0.23],[0.38,0.43])	([0.30,0.90],[0.23,0.26],[0.38,0.50])
(f2)	([0.45,0.68],[0.08,0.15],[0.44,0.56])	([0.45,0.68],[0.15,0.23],[0.39,0.56])	([0.23,0.45],[0.15,0.23],[0.44,0.56])	([0.68,0.90],[0.23,0.30],[0.33,0.44])
(f3)	([0.23,0.26],[0.12,0.20],[0.25,0.33])	([0.60,0.90],[0.20,0.30],[0.13,0.17])	([0.60,0.90],[0.15,0.20],[0.13,0.14])	([0.30,0.45],[0.10,0.12],[0.20,0.50])
(f4)	([0.30,0.34],[0.15,0.30],[0.25,0.50])	([0.30,0.39],[0.15,0.30],[0.17,0.25])	([0.68,0.90],[0.03,0.04],[0.13,0.25])	([0.68,0.90],[0.04,0.05],[0.17,0.25])

Thirdly, based on the weight of the sub-criteria and the IVNN scores of each alternative on each subcriteria, the final composite scores were calculated by improved ELECTRIC III method. After being told that q_j is the indifference threshold, p_j is the preference threshold and v_j is the veto threshold, the experts group suggested that $\sum s(a_i, a_k) - \sum s(a_k, a_i)$ $(k = 1, 2 \ldots n)$, respectively. The concordance index $c(X_i, X_k)$ and the partial concordance index $c_j(X_i, X_k)$ were calculated by Equations (26) and (27), as shown in Table 5. The discordance index $d_j(X_i, X_k)$ was achieved by Equation (28), as shown in Table 6. Then, overall concordance and partial discordance indices was obtained by Equation (29) as shown in Table 7. Finally, the degree of credibility of the outranking relation was calculated by Equation (30), and the rankings of alternative X1, X2, X3, X4 was shown in Table 8.

From Table 8, the SMPV plan X1 of the Golden Resources shopping mall in Beijing is the optimal selection. The alternative X1 is particularly superior to other alternative plans in terms of the economy, photovoltaic array, and contribution criteria, while these three criteria weighed more than a half of the entire criteria weights, so there is no doubt that plan X1 obtained the best position. However, plan X1 performs badly in the risk and architectural elements criteria. Respectively, Plan X2 have strength on the economy, but are weak on photovoltaic criteria. Yet, plan X2 is much better than plan X3 and X4, so that it can be the stand-by choice.

Table 5. The concordance index and the partial concordance index for each pair of SMPV plans.

	a1	a2	a3	b1	b2	b3	b4	c1	c2	c3	d1
c(X1 ≥ X2)	0.00	0.00	1.00	0.00	1.00	0.62	0.43	0.24	0.00	0.46	0.83
c(X1 ≥ X3)	0.05	0.00	1.00	0.36	0.00	0.92	0.74	0.20	0.00	0.50	0.61
c(X1 ≥ X4)	0.00	0.00	1.00	0.00	0.93	0.68	0.74	0.19	0.00	0.41	0.45
c(X2 ≥ X1)	0.26	0.85	0.00	0.70	0.00	0.00	0.00	0.00	0.41	0.00	0.00
c(X2 ≥ X2)	0.34	0.10	0.11	1.00	0.00	0.26	0.28	0.00	0.19	0.01	0.00
c(X2 ≥ X3)	0.00	0.00	0.00	0.47	0.00	0.02	0.28	0.00	0.00	0.00	0.00
c(X3 ≥ X1)	0.00	0.72	0.00	0.00	0.00	0.00	0.00	0.00	0.20	0.00	0.00
c(X3 ≥ X2)	0.00	0.00	0.00	0.00	1.00	0.00	0.00	0.01	0.00	0.00	0.19
c(X3 ≥ X4)	0.00	0.00	0.00	0.00	0.93	0.00	0.00	0.00	0.00	0.00	0.00
c(X4 ≥ X1)	0.44	0.99	0.00	0.19	0.00	0.00	0.00	0.00	0.92	0.00	0.00
c(X4 ≥ X2)	0.15	0.11	0.22	0.00	0.30	0.00	0.00	0.02	0.48	0.02	0.35
c(X4 ≥ X3)	0.52	0.24	0.37	0.59	0.00	0.21	0.00	0.00	0.69	0.05	0.13

	d2	d3	d4	e1	e2	e3	f1	f2	f3	f4	C
c(X1 ≥ X2)	0.91	1.00	0.00	0.74	0.00	0.12	0.00	0.03	0.00	0.00	0.38
c(X1 ≥ X3)	1.00	1.00	0.00	0.78	0.00	0.46	0.02	0.35	0.00	0.00	0.40
c(X1 ≥ X4)	0.37	0.55	0.00	0.14	0.00	0.54	0.00	0.00	0.00	0.00	0.32
c(X2 ≥ X1)	0.00	0.00	0.44	0.00	0.32	0.00	0.26	0.00	0.70	0.21	0.18
c(X2 ≥ X2)	0.28	0.23	0.00	0.01	0.00	0.30	0.32	0.29	0.00	0.00	0.19
c(X2 ≥ X3)	0.00	0.00	0.24	0.00	0.00	0.38	0.06	0.00	0.53	0.00	0.11
c(X3 ≥ X1)	0.00	0.00	0.85	0.00	0.66	0.00	0.00	0.00	0.81	1.00	0.19
c(X3 ≥ X2)	0.00	0.00	0.37	0.00	0.30	0.00	0.00	0.00	0.08	0.81	0.16
c(X3 ≥ X4)	0.00	0.00	0.65	0.00	0.00	0.05	0.00	0.00	0.65	0.00	0.13
c(X4 ≥ X1)	0.00	0.00	0.17	0.00	0.72	0.00	0.17	0.22	0.13	1.00	0.20
c(X4 ≥ X2)	0.51	0.49	0.00	0.56	0.37	0.00	0.00	0.28	0.00	0.78	0.21
c(X4 ≥ X3)	0.82	0.75	0.00	0.61	0.03	0.00	0.22	0.60	0.00	0.00	0.25

Table 6. The discordance index for each pair of SMPV plans.

d(Xi ≥ Xj)	a1	a2	a3	b1	b2	b3	b4	c1	c2	c3	d1
d(X1 ≥ X2)	0.00	0.00	1.00	0.00	0.91	0.47	0.34	0.21	0.00	0.36	0.61
d(X1 ≥ X3)	0.08	0.00	1.00	0.30	0.02	0.68	0.56	0.19	0.00	0.39	0.46
d(X1 ≥ X4)	0.00	0.00	0.88	0.00	0.68	0.51	0.56	0.17	0.00	0.33	0.35
d(X2 ≥ X1)	0.22	0.63	0.00	0.52	0.00	0.00	0.00	0.00	0.33	0.00	0.00
d(X2 ≥ X2)	0.28	0.11	0.12	0.80	0.00	0.23	0.24	0.00	0.17	0.05	0.00
d(X2 ≥ X3)	0.00	0.00	0.00	0.37	0.00	0.06	0.24	0.00	0.00	0.00	0.00
d(X3 ≥ X1)	0.00	0.54	0.00	0.00	0.02	0.00	0.00	0.00	0.18	0.00	0.00
d(X3 ≥ X2)	0.00	0.00	0.00	0.00	0.91	0.00	0.00	0.05	0.00	0.00	0.17
d(X3 ≥ X4)	0.00	0.00	0.00	0.00	0.68	0.00	0.02	0.01	0.00	0.00	0.00
d(X4 ≥ X1)	0.35	0.72	0.00	0.18	0.00	0.00	0.00	0.00	0.68	0.00	0.00
d(X4 ≥ X2)	0.15	0.12	0.20	0.00	0.25	0.00	0.00	0.06	0.37	0.06	0.29
d(X4 ≥ X3)	0.41	0.21	0.30	0.45	0.00	0.19	0.02	0.03	0.52	0.08	0.13

d(Xi ≥ Xj)	d2	d3	d4	e1	e2	e3	f1	f2	f3	f4
d(X1 ≥ X2)	0.67	0.78	0.00	0.55	0.00	0.13	0.00	0.06	0.00	0.00
d(X1 ≥ X3)	0.89	0.96	0.00	0.58	0.00	0.36	0.06	0.29	0.00	0.00
d(X1 ≥ X4)	0.30	0.43	0.00	0.14	0.00	0.41	0.00	0.00	0.00	0.00
d(X2 ≥ X1)	0.00	0.00	0.35	0.00	0.27	0.00	0.23	0.00	0.52	0.19
d(X2 ≥ X2)	0.24	0.20	0.00	0.05	0.00	0.25	0.26	0.25	0.00	0.00
d(X2 ≥ X3)	0.00	0.00	0.21	0.00	0.00	0.31	0.09	0.00	0.41	0.00
d(X3 ≥ X1)	0.00	0.00	0.63	0.00	0.50	0.00	0.00	0.00	0.60	0.77
d(X3 ≥ X2)	0.00	0.00	0.30	0.00	0.25	0.00	0.00	0.00	0.10	0.60
d(X3 ≥ X4)	0.00	0.00	0.49	0.00	0.00	0.08	0.00	0.00	0.49	0.05
d(X4 ≥ X1)	0.00	0.00	0.16	0.00	0.54	0.00	0.16	0.19	0.13	0.75
d(X4 ≥ X2)	0.39	0.38	0.00	0.43	0.30	0.00	0.00	0.24	0.00	0.58
d(X4 ≥ X3)	0.61	0.56	0.00	0.46	0.07	0.00	0.20	0.46	0.00	0.00

Table 7. The overall concordance and partial discordance indices for each pair of SMPV plans.

s(Xi ≥ Xj)	a1	a2	a3	b1	b2	b3	b4	c1	c2	c3	d1
s(X1 ≥ X2)	1.62	1.67	0.00	1.67	0.15	0.88	1.10	1.32	1.67	1.07	0.64
s(X1 ≥ X3)	1.54	1.67	0.00	1.18	1.63	0.54	0.74	1.36	1.67	1.02	0.90
s(X1 ≥ X4)	1.48	1.48	0.18	1.48	0.47	0.73	0.66	1.22	1.48	1.00	0.96
s(X2 ≥ X1)	0.95	0.45	1.22	0.58	1.22	1.22	1.22	1.22	0.82	1.22	1.22
s(X2 ≥ X2)	0.88	1.09	1.08	0.25	1.23	0.95	0.94	1.23	1.02	1.17	1.23
s(X2 ≥ X3)	1.12	1.12	1.12	0.70	1.12	1.05	0.85	1.12	1.12	1.12	1.12
s(X3 ≥ X1)	1.23	0.57	1.23	1.23	1.20	1.23	1.23	1.23	1.01	1.23	1.23
s(X3 ≥ X2)	1.19	1.19	1.19	1.19	0.10	1.19	1.19	1.13	1.19	1.19	0.98
s(X3 ≥ X4)	1.14	1.14	1.14	1.14	0.36	1.14	1.12	1.13	1.14	1.14	1.14
s(X4 ≥ X1)	0.82	0.34	1.25	1.03	1.25	1.25	1.25	1.25	0.40	1.25	1.25
s(X4 ≥ X2)	1.09	1.12	1.02	1.27	0.95	1.27	1.27	1.20	0.80	1.20	0.91
s(X4 ≥ X3)	0.79	1.06	0.94	0.73	1.33	1.08	1.30	1.29	0.64	1.22	1.16

s(Xi ≥ Xj)	d2	d3	d4	e1	e2	e3	f1	f2	f3	f4	s
s(X1 ≥ X2)	0.55	0.36	1.67	0.75	1.67	1.45	1.67	1.56	1.67	1.67	0.38
s(X1 ≥ X3)	0.19	0.06	1.67	0.69	1.67	1.07	1.57	1.19	1.67	1.67	0.40
s(X1 ≥ X4)	1.04	0.85	1.48	1.27	1.48	0.87	1.48	1.48	1.48	1.48	0.32
s(X2 ≥ X1)	1.22	1.22	0.79	1.22	0.89	1.22	0.94	1.22	0.58	0.98	0.18
s(X2 ≥ X2)	0.94	0.98	1.23	1.16	1.23	0.92	0.91	0.93	1.23	1.23	0.19
s(X2 ≥ X3)	1.12	1.12	0.88	1.12	1.12	0.78	1.02	1.12	0.66	1.12	0.11
s(X3 ≥ X1)	1.23	1.23	0.46	1.23	0.62	1.23	1.23	1.23	0.49	0.28	0.19
s(X3 ≥ X2)	1.19	1.19	0.83	1.19	0.89	1.19	1.19	1.19	1.07	0.47	0.16
s(X3 ≥ X4)	1.14	1.14	0.58	1.14	1.14	1.06	1.14	1.14	0.58	1.09	0.13
s(X4 ≥ X1)	1.25	1.25	1.05	1.25	0.58	1.25	1.05	1.01	1.08	0.32	0.20
s(X4 ≥ X2)	0.77	0.79	1.27	0.72	0.89	1.27	1.27	0.97	1.27	0.54	0.21
s(X4 ≥ X3)	0.52	0.59	1.33	0.71	1.24	1.33	1.07	0.72	1.33	1.33	0.25

Table 8. Final composite scores and rankings of alternative SMPV plans.

Xi	$\sum s(a_i \succ a_k)$	$\sum s(a_k \succ a_i)$	$\Delta S(a_i)$
X1	0.73	0.29	0.44
X2	0.47	0.37	0.1
X3	032	0.39	−0.07
X4	0.56	1.02	−0.47

The ranking order using ELECTRI III is X1 > X2 > X3 > X4. In order to check the validity of the results, TOPSIS and VIKOR methods were used to reorder the alternative SMPV plans as shown in Table 9. The result obtained by TOPSIS method is X1 > X4 > X3 > X2, while the result achieved by VIKOR method is X1 > X4 > X3 > X2. from these three rankings, alternative plan X1 is the optimal selection, no matter what method was used. That is to say, the alternative X1 is much better than the remaining alternatives, and there is no doubt in choosing X1 first. However, the rankings for X2, X3, and X4 are different between these three methods. There is the veto threshold, which indicates when the value of alternative Xi is lower than the value of alternative Xj, and the lower value exceeds the veto threshold, and it is not recognized that Xi is preferred to the Xj in general. However, it is the other term in TOPSIS method and VIKOR method, and some really bad performance in a certain criterion can be tolerated and remedied by other good performances in other criteria. In that case, an alternative with some fatal defect in a certain criterion of an alternative may be neglected, which leads to an unsatisfactory selection. When an alternative is vetoed better than the other alternative in ELECTRE III, it can still come out in front in the TOPSIS and VIKOR methods. That why the X2, X3, X4 ranked differently in TOPSIS and VIKOR methods.

Table 9. The rankings of SMPV plans using TOPSIS and VIKOR.

Method	TOPSIS				VIKOR			
	y+	y−	C	Rankings	s	r	Q	Rankings
X1	0.62	1.43	0.70	1	−2.57	0.12	0.00	1
X2	1.17	0.47	0.29	4	1.88	1.18	0.922	4
X3	1.35	0.71	0.35	3	−1.74	1.06	0.526	2
X4	0.76	0.74	0.49	2	2.74	1.01	0.919	3

6. Conclusions

The selection of SMPV plan is crucial to the entire life of SMPV project. Although there has been some research on this issue, several questions still need addressing. Firstly, the interaction of the criteria lay in the evaluation criteria. Secondly, the loss of evaluation information exited in the information conversion process. Thirdly, the compensation problem between best and worst performance in diverse criteria was not easily to avoided.

In this paper, an integrated MCDM framework was proposed to address the SMPV plan selection problem. First of all, the compositive evaluation index was constructed, and the application of DEMATEL method helped analyze the internal influence and connection behind each criterion. From the influential network-relationship map, we discovered that the criteria (b) economy obtained the most importance but was vulnerable to other criteria as well as the (a) architectural element and (c) climate had significant direct impacts to the other four criteria. These three criteria should be the first for the decision maker to consider when selecting the SMPV plan. Then, the interval-valued neutrosophic set is utilized to express the imperfect knowledge of experts group. Since the application of IVNNS, the experts can clearly express their evaluation information, including their certainty, uncertainty, as well as hesitation attitude. Following this, an extended ELECTRE III method as an outstanding outranking method was applied, and it succeed in avoiding the compensation problem and obtaining the scientific result. In the case of China, the integrated method has been successfully applied to select the SMPV plan X1 as the optimal selection which is particularly superior to other alternative plans in terms of the economy, photovoltaic array, and contribution criteria. Also, the integrated method used maintained symmetry in the solar PV investment. Last but not least, a comparative analysis using TOPSIS method and VIKOR method was carried out, and alternative plan X1 ranks first at the same. The outcome certified the correctness and rationality of the results obtained from this paper.

Therefore, this study has not only served to evaluate the SMPV plans, it has also demonstrated how it is possible to combine IVNNS, DEMATEL method, and ELECTRE III method for application in handling MCDM problems in the field of solar energy.

Author Contributions: M.L. designed the decision framework of shopping mall photovoltaic plan selection, and studied the proposed method. Then, J.F. collected the relative data, calculated the result, and drafted the paper. Next, Y.L. adjusted the format of the paper and formatted the manuscript for submission.

Conflicts of Interest: The authors declare no conflict of interest.

References

1. Zhang, L.; Wang, J.; Wen, H.; Fu, Z.; Li, X. Operating Performance, Industry Agglomeration and Its Spatial Characteristics of Chinese Photovoltaic Industry. *Renew. Sustain. Energy Rev.* **2016**, *65*, 373–386. [CrossRef]
2. Sánchez-Lozano, J.M.; Teruel-Solano, J.; Soto-Elvira, P.L.; García-Cascales, M.S. Geographical Information Systems (GIS) and Multi-Criteria Decision Making (MCDM) Methods for the Evaluation of Solar Farms Locations: Case Study in South-Eastern Spain. *Renew. Sustain. Energy Rev.* **2013**, *24*, 544–556. [CrossRef]
3. Zhang, L.; Zhou, P.; Newton, S.; Fang, J.; Zhou, D.; Zhang, L. Evaluating Clean Energy Alternatives for Jiangsu, China: An Improved Multi-Criteria Decision Making Method. *Energy* **2015**, *90*, 953–964. [CrossRef]
4. Liu, P.; Yu, X. 2-Dimension Uncertain Linguistic Power Generalized Weighted Aggregation Operator and Its Application in Multiple Attribute Group Decision Making. *Knowl. Based Syst.* **2014**, *57*, 69–80. [CrossRef]
5. Kumar, A.; Sah, B.; Singh, A.R.; Deng, Y.; He, X.; Kumar, P.; Bansal, R.C. A Review of Multi Criteria Decision Making (MCDM) towards Sustainable Renewable Energy Development. *Renew. Sustain. Energy Rev.* **2017**, *69*, 596–609. [CrossRef]
6. Cavallaro, F.; Zavadskas, E.K.; Streimikiene, D. Concentrated Solar Power (CSP) Hybridized Systems. Ranking Based on An Intuitionistic Fuzzy Multi-Criteria Algorithm. *J. Clean. Prod.* **2018**, *179*, 40416. [CrossRef]
7. Peng, J.J.; Wang, J.Q.; Zhang, H.Y.; Chen, X.H. An Outranking Approach for Multi-Criteria Decision-Making Problems with Simplified Neutrosophic Sets. *Appl. Soft Comput.* **2014**, *25*, 336–346. [CrossRef]
8. Zhang, H.; Wang, J.; Chen, X. An Outranking Approach for Multi-Criteria Decision-Making Problems with Interval-Valued Neutrosophic Sets. *Neural Comput. Appl.* **2016**, *27*, 615–627. [CrossRef]
9. Bausys, R.; Zavadskas, E.K. Multicriteria Decision Making Approach by Vikor under Interval Neutrosophic Set Environment. *Econ. Comput. Econ. Cybern. Stud. Res.* **2017**, *49*, 33–48.
10. Zhang, H.Y.; Wang, J.Q.; Chen, X.H. Interval neutrosophic sets and their application in multicriteria decision making problems. *Sci. World J.* **2014**, *2014*, 645953. [CrossRef] [PubMed]
11. Zavadskas, E.K.; Govindan, K.; Antucheviciene, J.; Turskis, Z. Hybrid Multiple Criteria Decision-Making Methods: A Review of Applications for Sustainability Issues. *Econ. Res.-Ekon. Istraž.* **2016**, *29*, 857–887. [CrossRef]
12. Tsai, W.H.; Chou, W.C. Selecting Management Systems for Sustainable Development in Smes: A Novel Hybrid Model Based on DEMATEL, ANP, and ZOGP. *Expert Syst. Appl.* **2009**, *36*, 1444–1458. [CrossRef]
13. Xia, X.; Govindan, K.; Zhu, Q. Analyzing Internal Barriers for Automotive Parts Remanufacturers in China Using Grey-DEMATEL Approach. *J. Clean. Prod.* **2015**, *87*, 811–825. [CrossRef]
14. George-Ufot, G.; Qu, Y.; Orji, I.J. Sustainable Lifestyle Factors Influencing Industries' Electric Consumption Patterns Using Fuzzy Logic and DEMATEL: The Nigerian Perspective. *J. Clean. Prod.* **2017**, *162*, 624–634. [CrossRef]
15. Azzopardi, B.; Martinez-Cesena, E.A.; Mutale, J. Decision Support System for Ranking Photovoltaic Technologies. *IET Renew. Power Gener.* **2013**, *7*, 669–679. [CrossRef]
16. Matulaitis, V.; Straukaitė, G.; Azzopardi, B.; Martinez-Cesena, E.A. Multi-Criteria Decision Making for PV Deployment on a Multinational Level. *Sol. Energy Mater. Sol. Cells* **2016**, *156*, 122–127. [CrossRef]

17. Wu, Y.; Zhang, J.; Yuan, J.; Shuai, G.; Haobo, Z. Study of Decision Framework of Offshore Wind Power Station Site Selection Based on ELECTRE-III under Intuitionistic Fuzzy Environment: A Case of China. *Energy Convers. Manag.* **2016**, *113*, 66–81. [CrossRef]
18. Colmenar-Santos, A.; Campinez-Romero, S.; Perez-Molina, C.; Mur-Pérez, F. An Assessment of Photovoltaic Potential in Shopping Centres. *Sol. Energy* **2016**, *135*, 662–673. [CrossRef]
19. Wu, Y.; Geng, S. Multi-Criteria Decision Making on Selection of Solar–Wind Hybrid Power Station Location: A Case of China. *Energy Convers. Manag.* **2014**, *81*, 527–533.
20. Tahri, M.; Hakdaoui, M.; Maanan, M. The Evaluation of Solar Farm Locations Applying Geographic Information System and Multi-Criteria Decision-Making Methods: Case Study in Southern Morocco. *Renew. Sustain. Energy Rev.* **2015**, *51*, 1354–1362. [CrossRef]
21. Long, S.; Geng, S. Decision Framework of Photovoltaic Module Selection under Interval-Valued Intuitionistic Fuzzy Environment. *Energy Convers. Manag.* **2015**, *106*, 1242–1250. [CrossRef]
22. Patlitzianas, K.D.; Skylogiannis, G.K.; Papastefanakis, D. Assessing The PV Business Opportunities in Greece. *Energy Convers. Manag.* **2013**, *75*, 651–657. [CrossRef]
23. Das, I.; Bhattacharya, K.; Canizares, C.; Muneer, W. Sensitivity-Indices-Based Risk Assessment of Large-Scale Solar PV Investment Projects. *IEEE Trans. Sustain. Energy* **2013**, *4*, 370–378. [CrossRef]
24. Rodrigues, S.; Chen, X.J.; Morgado-Dias, F. Economic Analysis of Photovoltaic Systems for the Residential Market under China's New Regulation. *Energy Policy* **2017**, *101*, 467–472. [CrossRef]
25. Smarandache, F. *A Unifying Field in Logics. Neutrosophy: Neutrosophic Probability, Set and Logic*; American Research Press: Rehoboth, NM, USA, 1999.
26. Wang, H.; Smarandache, F.; Zhang, Y.-Q.; Sunderraman, R. *Interval Neutrosophic Sets and Logic: Theory and Applications in Computing*; Hexis: Phoenix, AZ, USA, 2005.
27. Lee, Y.C.; Li, M.L.; Yen, T.M.; Huang, T.H. Analysis of adopting an integrated decision making trial and evaluation laboratory on a technology acceptance model. *Expert Syst. Appl.* **2010**, *37*, 1745–1754. [CrossRef]
28. Tzeng, G.H.; Chiang, C.H.; Li, C.W. Evaluating intertwined effects in e-learning programs: A novel hybrid MCDM model based on factor analysis and DEMATEL. *Expert Syst. Appl.* **2007**, *32*, 1028–1044. [CrossRef]
29. Quader, M.A.; Ahmed, S.; Ghazilla, R.A.R.; Ahmed, S.; Dahari, M. Evaluation of criteria for CO_2 capture and storage in the iron and steel industry using the 2-tuple DEMATEL technique. *J. Clean. Prod.* **2015**, *120*, 207–220. [CrossRef]
30. Bouyssou, D.; Marchant, T. An axiomatic approach to noncompensatory sorting methods in MCDM, I: The case of two categories. *Eur. J. Oper. Res.* **2007**, *178*, 217–245. [CrossRef]

symmetry

MDPI

Article

The Cosine Measure of Single-Valued Neutrosophic Multisets for Multiple Attribute Decision-Making

Changxing Fan [1,*], En Fan [1] and Jun Ye [2]

[1] Department of Computer Science, Shaoxing University, 508 Huancheng West Road, Shaoxing 312000, China; efan@usx.edu.cn

[2] Department of Electrical and Information Engineering, Shaoxing University, 508 Huancheng West Road, Shaoxing 312000, China; yehjun@aliyun.com

* Correspondence: fcxjszj@usx.edu.cn; Tel.: +86-575-8820-2669

Received: 29 April 2018; Accepted: 10 May 2018; Published: 11 May 2018

Abstract: Based on the multiplicity evaluation in some real situations, this paper firstly introduces a single-valued neutrosophic multiset (SVNM) as a subclass of neutrosophic multiset (NM) to express the multiplicity information and the operational relations of SVNMs. Then, a cosine measure between SVNMs and weighted cosine measure between SVNMs are presented to measure the cosine degree between SVNMs, and their properties are investigated. Based on the weighted cosine measure of SVNMs, a multiple attribute decision-making method under a SVNM environment is proposed, in which the evaluated values of alternatives are taken in the form of SVNMs. The ranking order of all alternatives and the best one can be determined by the weighted cosine measure between every alternative and the ideal alternative. Finally, an actual application on the selecting problem illustrates the effectiveness and application of the proposed method.

Keywords: single valued neutrosophic set (SVNS); neutrosophic multiset (NM); single valued neutrosophic multiset (SVNM); cosine measure; multiple attribute decision-making

1. Introduction

In 1965, Zadeh [1] proposed the theory of fuzzy sets (FS), in which every fuzzy element is expressed by the membership degree $T(x)$ belonging to the scope of [0, 1]. While the fuzzy membership degree of $T(x)$ is difficult to be determined, or cannot be expressed by an exact real number, the practicability of FS is limited. In order to avoid the above situation, Turksen [2] extended a single-value membership to an interval-valued membership. Generally, when the membership degree $T(x)$ is determined, the non-membership degree can be calculated by $1 - T(x)$. Considering the role of the non-membership degree, Atanassov [3] put forward the intuitionistic fuzzy sets (IFS) and introduced the related theory of IFS. Since then, IFS has been widely used for solving the decision-making problems. Although the FS theory and IFS theory have been constantly extended and completed, they are not applicable to all the fuzzy problems. In 1998, Smarandache [4] added the uncertain degree to the IFS and put forward the theory of the neutrosophic set (NS), which is a general form of the FS and IFS. NS is composed of the neutrosophic components of truth, indeterminacy, and falsity denoted by T, I, F, respectively. Since then, many forms of the neutrosophic set were proposed as extensions of the neutrosophic set. Wang and Smarandache [5,6] introduced a single-valued neutrosophic set (SVNS) and an interval neutrosophic set (INS). Smarandache [7] and Smarandache and Ye [8] presented n-value and refined-single valued neutrosophic sets (R-SVNSs). Fan and Ye [9] presented a refined-interval neutrosophic set (R-INS). Ye [10] presented a dynamic single-valued neutrosophic multiset (DSVM), and so on.

Now, more researches have been done on the NS theory by experts and scholars. Ye [11,12] proposed the correlation coefficient and the weighted coefficient correlation of SVNS and proved that cosine similarity is a special case of the SVNS correlation coefficient. Broumi and Smarandache [13] proposed three vector similarity methods to simplify the similarity of SVNS, including Jaccard similarity, Dice similarity, and cosine similarity. Majumdar and Samanta [14] gave the similarity formula of SVNSs. Broumi and Smarandache [15] gave the correlation coefficient of INSs. Based on the Hamming and Euclidean distances, Ye [16] defined the similarity of INSs. For the operation rules of NSs, Smarandache, Ye, and Chi [4,16,17] gave different operation rules, respectively, where they all have certain rationality and applicability.

Recently, Smarandache [18] introduced the neutrosophic multiset and the neutrosophic multiset algebraic structures, in which one or more elements are repeated for some times, keeping the same or different neutrosophic components. Its concept is different from the concept of single-valued neutrosophic multiset in [10,19]. Until now, there are few studies and applications of neutrosophic multisets (NM) in science and engineering fields, so we introduce a single valued neutrosophic multiset (SVNM) as a subclass of the neutrosophic multiset (NM) to express the multiplicity information and propose a decision-making method based on the weighted cosine measures of SVNMs, and then provide a decision-making example to show its application under SVNM environments.

The remaining sections of this article are organized as follows. Section 2 describes some basic concepts of SVNS, NM, and the cosine measure of SVNSs. Section 3 presents a SVNM and its basic operational relations. Section 4 proposes a cosine measure between SVNMs and a weighted cosine measure between SVNMs and investigates their properties. Section 5 establishes a multiple attribute decision-making method using the weighted cosine measure of SVNMs under SVNM environment. Section 6 presents an actual example to demonstrate the application of the proposed methods under SVNM environment. Section 7 gives a conclusion and further research.

2. Some Concepts of SVNS and NM

Definition 1 [5]. *Let X be a space of points (objects), with a generic element x in X. A SVNS R in X can be characterized by a truth-membership function $T_R(x)$, an indeterminacy-membership function $I_R(x)$, and a falsity-membership function $F_R(x)$, where $T_R(x), I_R(x), F_R(x) \in [0,1]$ for each point x in X. Then, a SVNS R can be expressed by the following form:*

$$R = \{\langle x, T_R(x), I_R(x), F_R(x)\rangle | x \in X\}.$$

Thus, the SVNS R satisfies the condition $0 \le T_R(x) + I_R(x) + F_R(x) \le 3$.

For two SVNSs M and N, the relations of them are defined as follows [5]:

(1) $M \subseteq N$ if and only if $T_M(x) \le T_N(x)$, $I_M(x) \ge I_N(x)$, $F_M(x) \ge F_N(x)$ for any x in X;
(2) $M = N$ if and only if $M \subseteq N$ and $N \subseteq M$;
(3) $M^c = \{\langle x, F_M(x), 1 - I_M(x), T_M(x)\rangle | x \in X\}$.

For writing convenience, an element called single-valued neutrosophic number (SVNN) in the SVNS R can be denoted by $R = \langle T_R(x), I_R(x), F_R(x)\rangle$ for any x in X. For two SVNNs M and N, the operational relations of them can be defined as follows [5]:

(1) $M \cup N = <\max(T_M(x), T_N(x)), \min(I_M(x), I_N(x)), \min(F_M(x), F_N(x)) >$ for any x in X;
(2) $M \cap N = <\min(T_M(x), T_N(x)), \max(I_M(x), I_N(x)), \max(F_M(x), F_N(x)) >$ for any x in X.

For two SVNNs M and N, the operational rules of them can be defined as follows [5]:

$$M + N = \langle T_M(x) + T_N(x) - T_M(x)T_N(x), I_M(x)I_N(x), F_M(x)F_N(x)\rangle \text{ for any } x \text{ in } X; \quad (1)$$

$$M \times N = < T_M(x)T_N(x), I_M(x) + I_N(x) - I_M(x)I_N(x), F_M(x) + F_N(x) - F_M(x)F_N(x) > \text{ for any } x \text{ in } X; \quad (2)$$

$$\varphi M = < 1 - (1 - T_M(x))^{\varphi}, (I_M(x))^{\varphi}, (F_M(x))^{\varphi} > \text{ for } \varphi > 0 \text{ and any } x \text{ in } X; \quad (3)$$

$$M^{\varphi} =< (T_M(x))^{\varphi}, (1 - I_M(x))^{\varphi}, (1 - F_M(x))^{\varphi} >, \text{ for } \varphi > 0 \text{ and any } x \text{ in } X. \quad (4)$$

Definition 2 [20]. *Let* $X = \{x_1, x_2, \ldots, x_n\}$ *be a space of points (objects), L and M be two SVNSs. The cosine measure between L and M is defined as follows:*

$$\rho(L, M) = \frac{1}{n}\sum_{i=1}^{n} cos\{\frac{\pi}{6}(|T_L(x_i) - T_M(x_i)| + |I_L(x_i) - I_M(x_i)| + |F_L(x_i) - F_M(x_i)|)\}. \quad (5)$$

Obviously, the cosine measure between L and M satisfies the following properties [20]:

① $0 \leq \rho(L, M) \leq 1$;
② $\rho(L, M) = 1$ if and only if $L = M$;
③ $\rho(L, M) = \rho(M, L)$.

Definition 3 [18]. *Let X be a space of points (objects), and a neutrosophic multiset is repeated by one or more elements with the same or different neutrosophic components.*

For example, $M = \{(m_1, \langle 0.7, 0.2, 0.1\rangle), (m_2, \langle 0.6, 0.4, 0.1\rangle), (m_3, \langle 0.8, 0.3, 0.2\rangle)\}$ is a neutrosophic set rather than a neutrosophic multiset; while $K = \{(k_1, \langle 0.7, 0.2, 0.1\rangle), (k_1, \langle 0.7, 0.2, 0.1\rangle), (k_1\langle, 0.7, 0.2, 0.1\rangle), (k_2, \langle 0.6, 0.4, 0.1\rangle)\}$ is a neutrosophic multiset, where the element k_1 is repeated. Then, we can say that the element k_1 has neutrosophic multiplicity 3 with the same neutrosophic components.

Meanwhile, $L = \{(l_1, \langle 0.7, 0.2, 0.1\rangle), (l_1, \langle 0.6, 0.3, 0.1\rangle), (l_1, \langle 0.8, 0.1, 0.1\rangle), (l_2, \langle 0.6, 0.4, 0.1\rangle)\}$ is also a neutrosophic multiset since the element l_1 is repeated, and then we can say that the element l_1 has neutrosophic multiplicity 3 with different neutrosophic components.

If the element l_1 is repeated times with the same neutrosophic comonents, we say l_1 has multiplicity. If the element l_1 is repeated times with different neutrosophic comonents, we say l_1 has the neutrosophic multiplicity (*nm*). The *nm* function can be defined as follows:

nm: $X \rightarrow N = \{1, 2, 3, \ldots, \infty\}$ for any $r \in R$
$nm(r) = \{(p_1, \langle T_1, I_1, F_1\rangle), (p_2, \langle T_2, I_2, F_2\rangle), \ldots, (p_i, \langle T_i, I_i, F_i\rangle), \ldots\}$,

which means that r is repeated by p_1 times with the neutrosophic components $\langle T_1, I_1, F_1\rangle$; r is repeated by p_2 times with the neutrosophic components $\langle T_2, I_2, F_2\rangle$; … ; r is repeated by p_i times with the neutrosophic components $\langle T_i, I_i, F_i\rangle$; and so on. $p_1, p_2, \ldots, p_i, \ldots \in N$, and $\langle T_j, I_j, F_j\rangle \neq \langle T_k, I_k, F_k\rangle$, for $j \neq k$ and j, $k \in N$. Then a neutrosophic multiset R can be written as:

$$(R, nm(r)) \text{ or } \{(r,\ nm(r),\ for\, r \in R)\}. \quad (6)$$

Now, with respect to the previous neutrosophic multisets K, L, we compute the neutrosophic multiplicity function:

$nm_K : K \rightarrow N$;
$nm_K(k_1) = \{(3, \langle 0.7, 0.2, 0.1\rangle)\}$;
$nm_K(k_2) = \{(1, \langle 0.6, 0.4, 0.1\rangle)\}$;
$nm_L : L \rightarrow N$;
$nm_L(l_1) = \{(1, \langle 0.7, 0.2, 0.1\rangle), (1, \langle 0.6, 0.3, 0.1\rangle), (1, \langle 0.8, 0.1, 0.1\rangle)\}$;
$nm_L(l_2) = \{(1, \langle 0.6, 0.4, 0.1\rangle)\}$.

3. Single Valued Neutrosophic Multiset

Definition 4. *Let X be a space of points (objects) with a generic element x in X and N = {1, 2, 3, … , ∞}. A SVNM R in X can be defined as follows:*

$R = \{x, ((p_{R1}, \langle T_{R1}(x), I_{R1}(x), F_{R1}(x)\rangle), (p_{R2}, \langle T_{R2}(x), I_{R2}(x), F_{R2}(x)\rangle), \ldots, (p_{Rj}, \langle T_{Rj}(x), I_{Rj}(x), F_{Rj}(x)\rangle))|x \in X\}$, *where* $T_{Rk}(x), I_{Rk}(x), F_{Rk}(x)$ *express the truth-membership function, the indeterminacy-membership function, and the falsity-membership function, respectively.* $T_{R1}(x), T_{R2}(x), \ldots, T_{Rk}(x) \in [0,1]$, $I_{R1}(x), I_{R2}(x), \ldots, I_{Rk}(x) \in [0,1]$, $F_{R1}(x), F_{R2}(x), \ldots, F_{Rk}(x) \in [0,1]$ *and* $0 \leq T_{Rk}(x) + I_{Rk}(x) + F_{Rk}(x) \leq 3$, *for* $k = 1, 2, \ldots j$, $j \in N$, $p_{R1}, p_{R2}, \ldots, p_{Rj} \in N$ *and* $p_{R1} + p_{R2} + \ldots + p_{Rj} \geq 2$.

For convenience, a SVNM R can be denoted by the following simplified form:

$$R = \{x, (p_{Rk}, \langle T_{Rk}(x), I_{Rk}(x), F_{Rk}(x)\rangle)|x \in X\}, for\ k = 1,\ 2,\ \ldots,\ j.$$

For example, with a universal set $X = \{x_1, x_2\}$, a SVNM R is given as:

$$R = \{(x_1, (2, \langle 0.6, 0.2, 0.1\rangle), (1, \langle 0.8, 0.2, 0.2\rangle)), (x_2, (1, \langle 0.7, 0.3, 0.1\rangle), (2, \langle 0.7, 0.2, 0.3\rangle))\}.$$

Then

$$nm_R(x_1) = \{(2, \langle 0.6, 0.2, 0.1\rangle), (1, \langle 0.8, 0.2, 0.2\rangle)\};$$
$$nm_R(x_2) = \{(1, \langle 0.7, 0.3, 0.1\rangle), (2, \langle 0.7, 0.2, 0.3\rangle)\}.$$

Definition 5. *Let X be a space of points (objects) with a generic element x in X, M and L be two SVNMs,*

$$M = \{x, (p_{Mk}, \langle T_{Mk}(x), I_{Mk}(x), F_{Mk}(x)\rangle)|x \in X\},\ for\ k = 1,\ 2,\ \ldots j,$$

$$L = \{x, (p_{Lk}, \langle T_{Lk}(x), I_{Lk}(x), F_{Lk}(x)\rangle)|x \in X\}, for\ k = 1,\ 2,\ \ldots j,$$

Then the relations of them are given as follows:

① $M = L$, if and only if $p_{Mk} = p_{Lk}, T_{Mk}(x) = T_{Lk}(x)$, $I_{Mk}(x) = I_{Lk}(x), F_{Mk}(x) = F_{Lk}(x)$, $for\ k = 1,\ 2,\ \ldots,\ j$;

② $M \cup L = \{x, ((p_{Mk} \vee p_{Lk}), \langle T_{Mk}(x) \vee T_{Lk}(x), I_{Mk}(x) \wedge I_{Lk}(x), F_{Mk}(x) \wedge F_{Lk}(x)\rangle)|x \in X\}$, $for\ k = 1,\ 2, \ldots,\ j$;

③ $M \cap L = \{x, ((p_{Mk} \wedge p_{Lk}), \langle T_{Mk}(x) \wedge T_{Lk}(x), I_{Mk}(x) \vee I_{Lk}(x), F_{Mk}(x) \vee F_{Lk}(x)\rangle)|x \in X\}$, $for\ k = 1,\ 2, \ldots,\ j$.

For convenience, we can use $r = ((p_{r1}, < T_{r1}(x), I_{r1}(x), F_{r1}(x) >), (p_{r2}, < T_{r2}(x), I_{r2}(x), F_{r2}(x) >), \ldots, (p_{rj}, < T_{rj}(x), I_{rj}(x), F_{rj}(x) >))$ to express a basic element in a SVNM R and call r a single valued neutrosophic multiset element (SVNME).

For example, with a universal set $X = \{x_1, x_2\}$, then two SVNMs M and L are given as:

$$M = \{(x_1, (2, \langle 0.6, 0.2, 0.1\rangle), (1, \langle 0.4, 0.1, 0.2\rangle)), (x_2, (1, \langle 0.7, 0.3, 0.1\rangle))\};$$

$$L = \{(x_1, (1, \langle 0.6, 0.2, 0.1\rangle), (1, \langle 0.8, 0.2, 0.1\rangle)), (x_2, (1, \langle 0.9, 0.3, 0.1\rangle))\};$$

$$M \cup L = \{(x_1, (2, \langle 0.6, 0.2, 0.1\rangle), (1, \langle 0.4, 0.1, 0.2\rangle), (1, \langle 0.8, 0.2, 0.1\rangle)), \ (x_2, (1, \langle 0.7, 0.3, 0.1\rangle), (1, \langle 0.9, 0.3, 0.1\rangle))\}$$

$$M \cap L = \{x_1, (1, \langle 0.6, 0.2, 0.1\rangle)\}.$$

Definition 6. *Let X be a space of points (objects) with a generic element x in X and M be a SVNM, we can change a SVNM M into a SVNS* $\widetilde{M}$ *by using the operational rules of SVNS.*

$$M = \{x, (p_{Mk}, \langle T_{Mk}(x), I_{Mk}(x), F_{Mk}(x)\rangle)|x \in X\}, for\ k = 1, 2, \ldots, j.$$

Then

$$\widetilde{M} = \{\langle x, 1 - \prod_{k=1}^{j}(1 - T_{Mk}(x))^{p_{Mk}}, \prod_{k=1}^{j}(I_{Mk}(x))^{p_{Mk}}, \prod_{k=1}^{j}(F_{Mk}(x))^{p_{Mk}}\rangle | x \in X\}. \quad (7)$$

Proof. Set $m_1, m_2, \ldots m_j$ are basic elements in M.

When k = 1, we can get

$m_1 = (\langle T_{m1}(x), I_{m1}(x), F_{m1}(x)\rangle, \langle T_{m1}(x), I_{m1}(x), F_{m1}(x)\rangle, \ldots, \langle T_{m1}(x), I_{m1}(x), F_{m1}(x)\rangle)$, which has neutrosophic multiplicity p_{m1}.

According to the operational rules of SVNSs, we can get:

$$+m_1 = 1 - (1 - T_{m1}(x))^{p_{m1}}, (I_{m1}(x))^{p_{m1}}, (F_{m1}(x))^{p_{m1}}.$$

As the same reason, when k = 2, we can get

$$+m_2 = 1 - (1 - T_{m2}(x))^{p_{m2}}, (I_{m1}(x))^{p_{m2}}, (F_{m2}(x))^{p_{m2}}.$$

Then

$$\begin{aligned} m_1 + m_2 &= 1 - (1 - T_{m1}(x))^{p_{m1}}(1 - T_{m2}(x))^{p_{m2}}, (I_{m1}(x))^{p_{m1}}(I_{m1}(x))^{p_{m2}}, (F_{m2}(x))^{p_{m1}}(F_{m2}(x))^{p_{m2}} \\ &= 1 - \prod_{k=1}^{2}(1 - T_{mk}(x))^{p_{mk}}, \prod_{k=1}^{2}(I_{mk}(x))^{p_{mk}}, \prod_{k=1}^{2}(F_{mk}(x))^{p_{mk}}; \end{aligned}$$

Suppose when $k = i$, the Equation (7) is established, then we can get:

$$m_1 + m_2 + \ldots + m_i = 1 - \prod_{k=1}^{i}(-T_{mk}(x))^{p_{mk}}, \prod_{k=1}^{i}(I_{mk}(x))^{p_{mk}}, \prod_{k=1}^{i}(F_{mk}(x))^{p_{mk}};$$

Then

$$\begin{aligned} m_1 + m_2 + \ldots + m_i \quad &+ m_{i+1} \\ &= 1 - \prod_{k=1}^{i}(1 - T_{mk}(x))^{p_{mk}} + 1 - \left(1 - T_{m(i+1)}(x)\right)^{p_{m(i+1)}} \\ &- \left(1 - \prod_{k=1}^{i}(1 - T_{mk}(x))^{p_{mk}}\right)\left(1 - \left(1 - T_{m(i+1)}(x)\right)^{p_{m(i+1)}}\right), \\ &\left(\prod_{k=1}^{i}(I_{mk}(x))^{p_{mk}}\right)\left(I_{m(i+1)}(x)\right)^{p_{m(i+1)}}, \\ &\left(\prod_{k=1}^{i}(F_{mk}(x))^{p_{mk}}\right)\left(F_{m(i+1)}(x)\right)^{p_{m(i+1)}} \\ &= 1 - \prod_{k=1}^{i+1}(1 - T_{mk}(x))^{p_{mk}}, \prod_{k=1}^{i+1}(I_{mk}(x))^{p_{mk}}, \prod_{k=1}^{i+1}(F_{mk}(x))^{p_{mk}}. \end{aligned}$$

To sum up, when $k = i + 1$, Equation (7) is true, and then according to the mathematical induction, we can get that the aggregation result is also true. □

Definition 7. *Let* $X = \{x_1, x_2, \ldots, x_n\}$ *be a universe of discourse, and M and N be two SVNMs, and then the operational rules of SVNMs are defined as follows:*

$$M = \{x, (p_{Mk}, \langle T_{Mk}(x), I_{Mk}(x), F_{Mk}(x)\rangle) | x \in X\},\ for\ k = 1,\ 2, \ldots j;$$

$$N = \{x, (p_{Nk}, \langle T_{Nk}(x), I_{Nk}(x), F_{Nk}(x)\rangle) | x \in X\}, for\ k = 1,\ 2, \ldots j;$$

$$M \oplus N = \{\langle x, 1 - \prod_{k=1}^{j}(1 - T_{Mk}(x))^{p_{Mk}} \prod_{k=1}^{j}(1 - T_{Nk}(x))^{p_{Nk}}, \prod_{k=1}^{j}(I_{Mk}(x))^{p_{Mk}} \prod_{k=1}^{j}(I_{Nk}(x))^{p_{Nk}}, \prod_{k=1}^{j}(F_{Mk}(x))^{p_{Mk}} \prod_{k=1}^{j}(F_{Nk}(x))^{p_{Nk}} \rangle | x \in X\}$$

$$\begin{aligned} M \otimes N \; = \; & \{\langle x, \left(1 - \prod_{k=1}^{j}(1 - T_{Mk}(x))^{p_{Mk}}\right)\left(1 - \prod_{k=1}^{j}(1 - T_{Nk}(x))^{p_{Nk}}\right), \prod_{k=1}^{j}(I_{Mk}(x))^{p_{Mk}} \\ & + \prod_{k=1}^{j}(I_{Nk}(x))^{p_{Nk}} - \prod_{k=1}^{j}(I_{Mk}(x))^{p_{Mk}} \prod_{k=1}^{j}(I_{Nk}(x))^{p_{Nk}}, \prod_{k=1}^{j}(F_{Mk}(x))^{p_{Mk}} \\ & + \prod_{k=1}^{j}(F_{Nk}(x))^{p_{Nk}} - \prod_{k=1}^{j}(F_{Mk}(x))^{p_{Mk}} \prod_{k=1}^{j}(F_{Nk}(x))^{p_{Nk}} \rangle | x \in X\} \end{aligned}$$

$$\varphi M = \{\langle x, \left(1 - \left(\prod_{k=1}^{j}(1 - T_{Mk}(x))^{p_{Mk}}\right)^{\varphi}\right), \left(\prod_{k=1}^{j}(I_{Mk}(x))^{p_{Mk}}\right)^{\varphi}, \left(\prod_{k=1}^{j}(F_{Nk}(x))^{p_{Nk}}\right)^{\varphi} \rangle | x \in X\}$$

$$\begin{aligned} M^{\varphi} = \{\langle x, & \left(1 - \prod_{k=1}^{j}(1 - T_{Mk}(x))^{p_{Mk}}\right)^{\varphi}, 1 - \left(1 - \prod_{k=1}^{j}(I_{Mk}(x))^{p_{Mk}}\right)^{\varphi}, 1 \\ & - \left(1 - \prod_{k=1}^{j}(F_{Nk}(x))^{p_{Nk}}\right)^{\varphi} \rangle | x \in X\} \end{aligned}$$

4. Cosine Measures of Single-Value Neutrosophic Multisets

Cosine measures are usually used in science and engineering applications. In this section, we propose a cosine measure of SVNMs and a weighted cosine measure of SVNMs.

Definition 8. *Let $X = \{x_1, x_2, \ldots, x_n\}$ be a universe of discourse, M and N be two SVNMs,*

$$M = \{x_i, (p_{M1}, \langle T_{M1}(x_i), I_{M1}(x_i), F_{M1}(x_i)\rangle), (p_{M2}, \langle T_{M2}(x_i), I_{M2}(x_i), F_{M2}(x_i)\rangle), \ldots, (p_{Mj}, \langle T_{Mj}(x_i), I_{Mj}(x_i), F_{Mj}(x_i)\rangle) | x_i \in X\},$$

$$N = \left\{ x_i, (p_{N1}, \langle T_{N1}(x_i), I_{N1}(x_i), F_{N1}(x_i)\rangle), (p_{N2}, \langle T_{N2}(x_i), I_{N2}(x_i), F_{N2}(x_i)\rangle), \ldots, (p_{Nj}, \langle T_{Nj}(x_i), I_{Nj}(x_i), F_{Nj}(x_i)\rangle) \; \middle| x_i \in X \right\}$$

Then, a cosine measure between two SVNMs M and N is defined as follows:

$$\begin{aligned} \rho(M, N) \; = \; & \frac{1}{n}\sum_{i=1}^{n} cos\left\{ \frac{\pi}{6} \left| \prod_{k=1}^{j}(1 - T_{Mk}(x_i))^{p_{Mk}} - \prod_{k=1}^{j}(1 - T_{Nk}(x_i))^{p_{Nk}} \right| \right. \\ & + \left| \prod_{k=1}^{j}(I_{Mk}(x_i))^{p_{Mk}} - \prod_{k=1}^{j}(I_{Nk}(x_i))^{p_{Nk}} \right| \\ & \left. + \left| \prod_{k=1}^{j}(F_{Mk}(x_i))^{p_{Mk}} - \prod_{k=1}^{j}(F_{Nk}(x_i))^{p_{Nk}} \right| \right\} \end{aligned} \tag{8}$$

Theorem 1. *The cosine measure $\rho(M, N)$ between two SVNMs M and N satisfies the following properties:*

① $\rho(M, N) = \rho(N, M)$;
② $0 \leq \rho(M, N) \leq 1$;
③ $\rho(M, N) = 1, if\ and\ only\ if\ M = N$.

Proof. ①: For $\left|\prod_{k=1}^{j}(1 - T_{Mk}(x_i))^{p_{Mk}} - \prod_{k=1}^{j}(1 - T_{Nk}(x_i))^{p_{Nk}}\right| + \left|\prod_{k=1}^{j}(I_{Mk}(x_i))^{p_{Mk}} - \prod_{k=1}^{j}(I_{Nk}(x_i))^{p_{Nk}}\right| + \left|\prod_{k=1}^{j}(F_{Mk}(x_i))^{p_{Mk}} - \prod_{k=1}^{j}(F_{Nk}(x_i))^{p_{Nk}}\right| = \left|\prod_{k=1}^{j}(1 - T_{Nk}(x_i))^{p_{Nk}} - \prod_{k=1}^{j}(1 - T_{Mk}(x_i))^{p_{Mk}}\right| + \left|\prod_{k=1}^{j}(I_{Nk}(x_i))^{p_{Nk}} - \prod_{k=1}^{j}(I_{Mk}(x_i))^{p_{Mk}}\right| + \left|\prod_{k=1}^{j}(F_{Nk}(x_i))^{p_{Nk}} - \prod_{k=1}^{j}(F_{Mk}(x_i))^{p_{Mk}}\right|$, so we can get $\rho(M, N) = \rho(N, M)$.

②: For $0 \le T_{Mk}(x_i) \le 1$, $0 \le I_{Mk}(x_i) \le 1$, $0 \le F_{Mk}(x_i) \le 1$, $0 \le T_{Nk}(x_i) \le 1$, $0 \le I_{Nk}(x_i) \le 1$, $0 \le F_{Nk}(x_i) \le 1$;
Then, we can get

$$0 \le 1 - T_{Mk}(x_i) \le 1 \; and \; 0 \le 1 - T_{Nk}(x_i) \le 1;$$

$$0 \le \prod_{k=1}^{j}(1 - T_{Mk}(x_i))^{p_{Mk}} \le 1 \; and \; 0 \le \prod_{k=1}^{j}(1 - T_{Nk}(x_i))^{p_{Nk}} \le 1;$$

So,

$$0 \le \left|\prod_{k=1}^{j}(1 - T_{Mk}(x_i))^{p_{Mk}} - \prod_{k=1}^{j}(1 - T_{Nk}(x_i))^{p_{Nk}}\right| \le 1.$$

For the same reason, we can get

$$0 \le \left|\prod_{k=1}^{j}(I_{Mk}(x_i))^{p_{Mk}} - \prod_{k=1}^{j}(I_{Nk}(x_i))^{p_{Nk}}\right| \le 1 and 0 \le \left|\prod_{k=1}^{j}(F_{Mk}(x_i))^{p_{Mk}} - \prod_{k=1}^{j}(F_{Nk}(x_i))^{p_{Nk}}\right| \le 1$$

Above all, we can get $0 \le \left|\prod_{k=1}^{j}(1 - T_{Mk}(x_i))^{p_{Mk}} - \prod_{k=1}^{j}(1 - T_{Nk}(x_i))^{p_{Nk}}\right| + \left|\prod_{k=1}^{j}(I_{Mk}(x_i))^{p_{Mk}} - \prod_{k=1}^{j}(I_{Nk}(x_i))^{p_{Nk}}\right| + \left|\prod_{k=1}^{j}(F_{Mk}(x_i))^{p_{Mk}} - \prod_{k=1}^{j}(F_{Nk}(x_i))^{p_{Nk}}\right| \le 3$ and $0 \le \sum_{i=1}^{n} cos\left\{\frac{\pi}{6}\left|\prod_{k=1}^{j}(1 - T_{Mk}(x_i))^{p_{Mk}} - \prod_{k=1}^{j}(1 - T_{Nk}(x_i))^{p_{Nk}}\right| + \left|\prod_{k=1}^{j}(I_{Mk}(x_i))^{p_{Mk}} - \prod_{k=1}^{j}(I_{Nk}(x_i))^{p_{Nk}}\right| + \left|\prod_{k=1}^{j}(F_{Mk}(x_i))^{p_{Mk}} - \prod_{k=1}^{j}(F_{Nk}(x_i))^{p_{Nk}}\right|\right\} \le 1$;

$$\begin{aligned}
\rho(M,N) = & \frac{1}{n}\sum_{i=1}^{n} cos\left\{\frac{\pi}{6}\left|\prod_{k=1}^{j}(1 - T_{Mk}(x_i))^{p_{Mk}} - \prod_{k=1}^{j}(1 - T_{Nk}(x_i))^{p_{Nk}}\right|\right. \\
& + \left|\prod_{k=1}^{j}(I_{Mk}(x_i))^{p_{Mk}} - \prod_{k=1}^{j}(I_{Nk}(x_i))^{p_{Nk}}\right| \\
& \left. + \left|\prod_{k=1}^{j}(F_{Mk}(x_i))^{p_{Mk}} - \prod_{k=1}^{j}(F_{Nk}(x_i))^{p_{Nk}}\right|\right\} \\
= & \frac{1}{n}\left(cos\left\{\frac{\pi}{6}\left(\left|\prod_{k=1}^{j}(1 - T_{Mk}(x_1))^{p_{Mk}} - \prod_{k=1}^{j}(1 - T_{Nk}(x_1))^{p_{Nk}}\right|\right.\right.\right. \\
& + \left|\prod_{k=1}^{j}(I_{Mk}(x_1))^{p_{Mk}} - \prod_{k=1}^{j}(I_{Nk}(x_1))^{p_{Nk}}\right| \\
& \left.\left. + \left|\prod_{k=1}^{j}(F_{Mk}(x_1))^{p_{Mk}} - \prod_{k=1}^{j}(F_{Nk}(x_1))^{p_{Nk}}\right|\right)\right\} \\
& + cos\left\{\frac{\pi}{6}\left(\left|\prod_{k=1}^{j}(1 - T_{Mk}(x_2))^{p_{Mk}} - \prod_{k=1}^{j}(1 - T_{Nk}(x_2))^{p_{Nk}}\right|\right.\right. \\
& + \left|\prod_{k=1}^{j}(I_{Mk}(x_2))^{p_{Mk}} - \prod_{k=1}^{j}(I_{Nk}(x_2))^{p_{Nk}}\right| \\
& \left.\left. + \left|\prod_{k=1}^{j}(F_{Mk}(x_2))^{p_{Mk}} - \prod_{k=1}^{j}(F_{Nk}(x_2))^{p_{Nk}}\right|\right)\right\} + \cdots \\
& + cos\left\{\frac{\pi}{6}\left(\left|\prod_{k=1}^{j}(1 - T_{Mk}(x_n))^{p_{Mk}} - \prod_{k=1}^{j}(1 - T_{Nk}(x_n))^{p_{Nk}}\right|\right.\right. \\
& + \left|\prod_{k=1}^{j}(I_{Mk}(x_n))^{p_{Mk}} - \prod_{k=1}^{j}(I_{Nk}(x_n))^{p_{Nk}}\right| \\
& \left.\left.\left. + \left|\prod_{k=1}^{j}(F_{Mk}(x_n))^{p_{Mk}} - \prod_{k=1}^{j}(F_{Nk}(x_n))^{p_{Nk}}\right|\right)\right\}\right)
\end{aligned}$$

Let $\cos\left\{\frac{\pi}{6}\left(\left|\prod_{k=1}^{j}(1-T_{Mk}(x_i))^{p_{Mk}}-\prod_{k=1}^{j}(1-T_{Nk}(x_i))^{p_{Nk}}\right|+\left|\prod_{k=1}^{j}(I_{Mk}(x_i))^{p_{Mk}}-\prod_{k=1}^{j}(I_{Nk}(x_i))^{p_{Nk}}\right|+\left|\prod_{k=1}^{j}(F_{Mk}(x_i))^{p_{Mk}}-\prod_{k=1}^{j}(F_{Nk}(x_i))^{p_{Nk}}\right|\right)\right\}=a_i\ (i=1,2,\ldots n)$, then $\rho(M,N)=\frac{1}{n}\{a_1+a_2+\cdots+a_n\}$.

According to $0\le\cos\left\{\frac{\pi}{6}\left(\left|\prod_{k=1}^{j}(1-T_{Mk}(x_i))^{p_{Mk}}-\prod_{k=1}^{j}(1-T_{Nk}(x_i))^{p_{Nk}}\right|+\left|\prod_{k=1}^{j}(I_{Mk}(x_i))^{p_{Mk}}-\prod_{k=1}^{j}(I_{Nk}(x_i))^{p_{Nk}}\right|+\left|\prod_{k=1}^{j}(F_{Mk}(x_i))^{p_{Mk}}-\prod_{k=1}^{j}(F_{Nk}(x_i))^{p_{Nk}}\right|\right)\right\}=a_i\le1$, we can obtain $0\le a_1+a_2+\cdots+a_n\le n$ and $0\le\frac{1}{n}(a_1+a_2+\cdots+a_n)\le1$, so we can get $0\le\rho(M,N)\le1$.

③: If M = N then $T_{Mk}(x_i)=T_{Nk}(x_i), I_{Mk}(x_i)=I_{Nk}(x_i)$, and $F_{Mk}(x_i)=F_{Nk}(x_i)$ for any $x_i\in X$ and $i=1,2,\ldots n$, so we can get $\rho(M,N)=1$, if and only if $M=N$.

Now, we consider different weights for each element $x_i(i=1,2,\ldots,n)$ in X. Then, let $w=(w_1,w_2,\ldots,w_n)^T$ be the weight vector of each element $x_i(i=1,2,\ldots,n)$ with $w_i\in[0,1]$, and $\sum_{i=1}^{n}w_i=1$. Hence, we further extend the cosine measure of Equation (8) to the following weighted cosine measure of SVNM:

$$\rho_w(M,N)=\sum_{i=1}^{n}w_i\cos\left\{\frac{\pi}{6}\left(\left|\prod_{k=1}^{j}(1-T_{Mk}(x_i))^{p_{Mk}}-\prod_{k=1}^{j}(1-T_{Nk}(x_i))^{p_{Nk}}\right|+\left|\prod_{k=1}^{j}(I_{Mk}(x_i))^{p_{Mk}}-\prod_{k=1}^{j}(I_{Nk}(x_i))^{p_{Nk}}\right|+\left|\prod_{k=1}^{j}(F_{Mk}(x_i))^{p_{Mk}}-\prod_{k=1}^{j}(F_{Nk}(x_i))^{p_{Nk}}\right|\right)\right\}\quad(9)$$

□

Theorem 2. *The cosine measure $\rho_w(M,N)$ between two SVNMs M and N satisfies the following properties:*

① $\rho_w(M,N)=\rho_w(N,M)$;
② $0\le\rho_w(M,N)\le1$;
③ $\rho_w(M,N)=1$, *if and only if* $M=N$.

The proof of Theorem 2 is similar to that of the Theorem 1, so we omitted it here.

5. Cosine Measure of SVNM for Multiple Attribute Decision-Making

In this section, we use the weighted cosine measure of SVNM to deal with the multiple attribute decision-making problems with SVNM information.

Let $G=\{g_1,g_2,\ldots,g_m\}$ as a set of alternatives and $X=\{x_1,x_2,\ldots,x_n\}$ as a set of attributes, then they can be established in a decision-making problem. However, sometimes $x_i(i=1,2,\ldots,n)$ may have multiplicity, and then we can use the form of a SVNM to represent the evaluation value.

Let $g_r=\{x_i,(p_{g_r1},\langle T_{g_r1}(x_i),I_{g_r1}(x_i),F_{g_r1}(x_i)\rangle),(p_{g_r2},\langle T_{g_r2}(x_i),I_{g_r2}(x_i),F_{g_r2}(x_i)\rangle),\ldots,(p_{g_rj},\langle T_{g_rj}(x_i),I_{g_rj}(x_i),F_{g_rj}(x_i)\rangle)|x_i\in X\}$, for $r=1,2,\ldots,m$ and $i=1,2,\ldots,n$. Then we can establish the SVNM decision matrix D, which is shown in Table 1.

Table 1. The single-valued neutrosophic multiset (SVNM) decision matrix D.

	x_1	...
g_1	$x_1, (p_{g_11}, \langle T_{g_11}(x_1), I_{g_11}(x_1), F_{g_11}(x_1)\rangle), \ldots, \left(p_{g_1j}, \left\langle T_{g_1j}(x_1), I_{g_1j}(x_1), F_{g_1j}(x_1)\right\rangle\right)$	...
g_2	$x_1, (p_{g_21}, \langle T_{g_21}(x_1), I_{g_21}(x_1), F_{g_21}(x_1)\rangle), \ldots, \left(p_{g_2j}, \left\langle T_{g_2j}(x_1), I_{g_2j}(x_1), F_{g_2j}(x_1)\right\rangle\right)$	...
...	...	...
g_m	$x_1, (p_{g_m1}, \langle T_{g_m1}(x_1), I_{g_m1}(x_1), F_{g_m1}(x_1)\rangle), \ldots, \left(p_{g_mj}, \left\langle T_{g_mj}(x_1), I_{g_mj}(x_1), F_{g_mj}(x_1)\right\rangle\right)$	...
	x_n	
	$x_n, (p_{g_11}, \langle T_{g_11}(x_n), I_{g_11}(x_n), F_{g_11}(x_n)\rangle), \ldots, \left(p_{g_1j}, \left\langle T_{g_1j}(x_n), I_{g_1j}(x_n), F_{g_1j}(x_n)\right\rangle\right)$	
	$x_n, (p_{g_21}, \langle T_{g_21}(x_n), I_{g_21}(x_n), F_{g_21}(x_n)\rangle), \ldots, \left(p_{g_2j}, \left\langle T_{g_2j}(x_n), I_{g_2j}(x_n), F_{g_2j}(x_n)\right\rangle\right)$	
	...	
	$x_n, (p_{g_m1}, \langle T_{g_m1}(x_n), I_{g_m1}(x_n), F_{g_m1}(x_n)\rangle), \ldots, \left(p_{g_mj}, \left\langle T_{g_mj}(x_n), I_{g_mj}(x_n), F_{g_mj}(x_n)\right\rangle\right)$	

Step 1: By using Equation (7), we change the SVNM decision matrix D into SVNS decision matrix $\widetilde{D}$, which is shown in Table 2.

Table 2. The single-valued neutrosophic set (SVNS) decision matrix $\widetilde{D}$.

	x_1	...
$\widetilde{g}_1$	$\langle x_1, 1-\prod_{k=1}^{j}\left(1-T_{g_1k}(x_1)\right)^{p_{g_1k}}, \prod_{k=1}^{j}\left(I_{g_1k}(x_1)\right)^{p_{g_1k}}, \prod_{k=1}^{j}\left(F_{g_1k}(x_1)\right)^{p_{g_1k}}\rangle$	...
$\widetilde{g}_2$	$\langle x_1, 1-\prod_{k=1}^{j}\left(1-T_{g_2k}(x_1)\right)^{p_{g_2k}}, \prod_{k=1}^{j}\left(I_{g_2k}(x_1)\right)^{p_{g_2k}}, \prod_{k=1}^{j}\left(F_{g_2k}(x_1)\right)^{p_{g_2k}}\rangle$	...
...	...	...
$\widetilde{g}_m$	$\langle x_1, 1-\prod_{k=1}^{j}\left(1-T_{g_mk}(x_1)\right)^{p_{g_mk}}, \prod_{k=1}^{j}\left(I_{g_mk}(x_1)\right)^{p_{g_mk}}, \prod_{k=1}^{j}\left(F_{g_mk}(x_1)\right)^{p_{g_mk}}\rangle$	...
	x_n	
	$\langle x_n, 1-\prod_{k=1}^{j}\left(1-T_{g_1k}(x_n)\right)^{p_{g_1k}}, \prod_{k=1}^{j}\left(I_{g_1k}(x_n)\right)^{p_{g_1k}}, \prod_{k=1}^{j}\left(F_{g_1k}(x_n)\right)^{p_{g_1k}}\rangle$	
	$\langle x_n, 1-\prod_{k=1}^{j}\left(1-T_{g_2k}(x_n)\right)^{p_{g_2k}}, \prod_{k=1}^{j}\left(I_{g_2k}(x_n)\right)^{p_{g_2k}}, \prod_{k=1}^{j}\left(F_{g_2k}(x_n)\right)^{p_{g_2k}}\rangle$	
	...	
	$\langle x_n, 1-\prod_{k=1}^{j}\left(1-T_{g_mk}(x_n)\right)^{p_{g_mk}}, \prod_{k=1}^{j}\left(I_{g_mk}(x_n)\right)^{p_{g_mk}}, \prod_{k=1}^{j}\left(F_{g_mk}(x_n)\right)^{p_{g_mk}}\rangle$	

Step 2: Setting $T_{g^*}(x_i)$ is the maximum truth value in each column x_i of the decision matrix $\widetilde{D}$, $I_{g^*}(x_i)$ and $F_{g^*}(x_i)$ are the minimum indeterminate and falsity values in each column x_i of the decision matrix $\widetilde{D}$, respectively, the ideal solution can be determined as x_i^*.

$x_i^* = \langle T_{g^*}(x_i),\ I_{g^*}(x_i), F_{g^*}(x_i)\rangle, for\ i = 1, 2, \ldots, n.$

So, we can get the ideal alternative $g^* = \{x_1^*, x_2^*, \ldots, x_n^*\}$.

Step 3: When the weight vector of attributes for the different importance of each attribute $x_i(i = 1, 2, \ldots, n)$ is given by $w = (w_1, w_2, \ldots, w_n)^T$ with $w_i \geq 0$ and $\sum_{i=1}^{n} w_i = 1$, then we utilize the weighted cosine measure to deal with multiple attribute decision-making problems with SVNM information. The weighted cosine measure between an alternative $\widetilde{g}_r(r = 1, 2, \ldots, m)$ and the ideal alternative g^* can be calculated by using the following formula:

$$\begin{aligned} \rho_w(g_r, g^*) &= \rho_w(\widetilde{g}_r, g^*) \\ &= \sum_{i=1}^{n} w_i \cos\{\tfrac{\pi}{6}(\left|T_{\widetilde{g}_r}(x_i) - T_{g^*}(x_i)\right| + \left|I_{\widetilde{g}_r}(x_i) - I_{g^*}(x_i)\right| \\ &\quad + \left|F_{\widetilde{g}_r}(x_i) - F_{g^*}(x_i)\right|)\}. \end{aligned} \tag{10}$$

Step 4: According to the values of $\rho_w(\tilde{g}_r, g^*)$ for $r = 1, 2, \ldots, m$, we rank the alternatives and select the best one.

Step 5: End.

The formalization of the steps is illustrated in Figure 1.

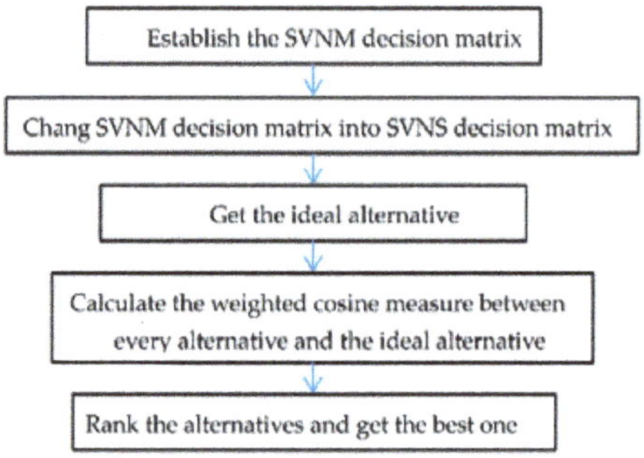

Figure 1. Flowchart of the decision steps.

6. Numerical Example and Comparative Analysis

6.1. Numerical Example

Now, we utilize a practical example for the decision-making problem adapted from the literature [21] to demonstrate the applications of the proposed method under a SVNM environment. Now, one customer wants to buy a car, he selects four types of cars and evaluates them according to four attributes. Then, we build a decision model. There are four possible alternatives (g_1, g_2, g_3, g_4) to be considered. The decision should be taken according to four attributes: fuel economy (x_1), price (x_2), comfort (x_3), and safety (x_4). The weight vector of these four attributes is given by $w = (0.5, 0.25, 0.125, 0.125)^T$. Then, the customer tests the four cars on the road with less obstacles and on the road with more obstacles, respectively, and after testing, some attributes may have two different evaluated values or the same value. So, the customer evaluates the four cars (alternatives) under the four attributes by the form of SVNMs.

Step 1: Establish the SVNM decision matrix D provided by the customer, which is given as the following SVNM decision matrix D in Table 3.

Table 3. The SVNM decision matrix D.

	x_1	x_2	x_3	x_4
g_1	$(1, \langle 0.5, 0.7, 0.2\rangle), (1, \langle 0.7, 0.3, 0.6\rangle)$	$1, \langle 0.4, 0.4, 0.5\rangle$	$(1, \langle 0.7, 0.7, 0.5\rangle), (1, \langle 0.8, 0.7, 0.6\rangle)$	$(1, \langle 0.1, 0.5, 0.7\rangle), (1, \langle 0.5, 0.2, 0.8\rangle)$
g_2	$(1, \langle 0.9, 0.7, 0.5\rangle), (1, \langle 0.7, 0.7, 0.1\rangle)$	$1, \langle 0.7, 0.6, 0.8\rangle$	$2, \langle 0.9, 0.4, 0.6\rangle$	$(1, \langle 0.5, 0.2, 0.7\rangle, (1, \langle 0.5, 0.1, 0.9\rangle)$
g_3	$(1, \langle 0.3, 0.4, 0.2\rangle, (1, \langle 0.6, 0.3, 0.7\rangle)$	$1, \langle 0.2, 0.2, 0.2\rangle$	$(1, \langle 0.9, 0.5, 0.5\rangle, (1, \langle 0.6, 0.5, 0.2\rangle)$	$(1, \langle 0.7, 0.5, 0.3\rangle, (1, \langle 0.4, 0.2, 0.2\rangle)$
g_4	$(1, \langle 0.9, 0.7, 0.2\rangle, (1, \langle 0.8, 0.6, 0.1\rangle)$	$1, \langle 0.3, 0.5, 0.2\rangle$	$(1, \langle 0.5, 0.4, 0.5\rangle, (1, \langle 0.1, 0.7, 0.2\rangle)$	$2, \langle 0.4, 0.2, 0.8\rangle$

Step 2: By using Equation (7), we change the SVNM decision matrix D into SVNS decision matrix $\tilde{D}$, which is shown in Table 4.

Table 4. The SVNS decision matrix $\tilde{D}$.

	x_1	x_2	x_3	x_4
$\tilde{g}_1$	$\langle 0.85, 0.21, 0.12\rangle$	$\langle 0.4, 0.4, 0.5\rangle$	$\langle 0.94, 0.49, 0.3\rangle$	$\langle 0.55, 0.1, 0.56\rangle$
$\tilde{g}_2$	$\langle 0.97, 0.49, 0.05\rangle$	$\langle 0.7, 0.6, 0.8\rangle$	$\langle 0.99, 0.16, 0.36\rangle$	$\langle 0.75, 0.02, 0.63\rangle$
$\tilde{g}_3$	$\langle 0.72, 0.12, 0.14\rangle$	$\langle 0.2, 0.2, 0.2\rangle$	$\langle 0.96, 0.25, 0.1\rangle$	$\langle 0.82, 0.1, 0.06\rangle$
$\tilde{g}_4$	$\langle 0.98, 0.42, 0.02\rangle$	$\langle 0.3, 0.5, 0.2\rangle$	$\langle 0.55, 0.28, 0.1\rangle$	$\langle 0.64, 0.04, 0.64\rangle$

Step 3: According to the decision matrix $\tilde{D}$, we can get the ideal alternative g^*:

$$g^* = \{\langle 0.98, 0.12, 0.02\rangle\langle 0.7, 0.2, 0.2\rangle\langle 0.99, 0.16, 0.1\rangle\langle 0.82, 0.02, 0.06\rangle\}. \tag{11}$$

Step 4: By applying the Equation (10), we can obtain the values of the weighted cosine measure between each alternative and the ideal alternative g^* as follows:

$$\begin{aligned} &\rho_w(g_1, g^*) = 0.9535, \rho_w(g_2, g^*) = 0.9511, \\ &\rho_w(g_3, g^*) = 0.9813 \text{ and } \rho_w(g_4, g^*) = 9616. \end{aligned} \tag{12}$$

Step 5: According to the above values of weighted cosine measure, we can rank the four alternatives: $g_3 \succ g_4 \succ g_1 \succ g_2$. Therefore, the alternative g_3 is the best choice.

This example clearly indicates that the proposed decision-making method based on the weighted cosine measure of SVNMs is relatively simple and easy for dealing with multiple attribute decision-making problems under SVNM environment.

6.2. Comparative Analysis

In what follows, we compare the proposed method for SVNM with other existing related methods for SVNM; all the results are shown in Table 5.

Table 5. The ranking orders by utilizing four different methods.

Method	Result	Ranking Order	The Best Alternative
Method 1 based on correlation coefficient in [11]	$\rho_w(g_1, g^*) = 0.9053$, $\rho_w(g_2, g^*) = 0.9017$, $\rho_w(g_3, g^*) = 0.9516$, $\rho_w(g_4, g^*) = 0.8816$.	$g_3 \succ g_1 \succ g_2 \succ g_4$	g_3
Method 2 based on similarity in [16]	$\rho_w(g_1, g^*) = 0.8204$, $\rho_w(g_2, g^*) = 0.8108$, $\rho_w(g_3, g^*) = 0.8867$, $\rho_w(g_4, g^*) = 0.8358$.	$g_3 \succ g_4 \succ g_2 \succ g_1$	g_3
Method 3 based on similarity in [16]	$\rho_w(g_1, g^*) = 0.7898$, $\rho_w(g_2, g^*) = 0.7121$, $\rho_w(g_3, g^*) = 0.8125$, $\rho_w(g_4, g^*) = 0.7553$.	$g_3 \succ g_1 \succ g_4 \succ g_2$	g_3
The proposed method	$\rho_w(g_1, g^*) = 0.9535$, $\rho_w(g_2, g^*) = 0.9511$, $\rho_w(g_3, g^*) = 0.9813$, $\rho_w(g_4, g^*) = 9616$.	$g_3 \succ g_4 \succ g_1 \succ g_2$	g_3

From Table 5, these four methods have the same best alternative g_3. Many methods such as similarity measure, correlation coefficient, and cosine measure can all be used in SVNM to handle the multiple attribute decision-making problems and can get the similar results.

The proposed decision-making method can express and handle the multiplicity evaluated data given by decision makers or experts, while various existing neutrosophic decision-making methods cannot deal with these problems.

7. Conclusions

Based on the multiplicity evaluation in some real situations, this paper introduced a SVNM as a subclass of NM to express the multiplicity information and the operational relations of SVNMs. The SVNM is expressed by its one or more elements, which may have multiplicity. Therefore, SVNM has the desirable advantages and characteristics of expressing and handling the multiplicity problems, while existing neutrosophic sets cannot deal with them.

Then, we proposed the cosine measure of SVNMs and weighted cosine measure of SVNMs and investigated their properties. Based on the weighted cosine measure of SVNMs, the multiple attribute decision-making methods under SVNM environments was proposed, in which the evaluated values were taken the form of SVNMEs. Through the weighed cosine measure between each alternative and the ideal alternative, one can determine the ranking order of all alternatives and can select the best one. Finally, a practical example adapted from the literature [21] about buying cars was presented to demonstrate the effectiveness and practicality of the proposed method in this paper. According to the ranking orders, we can find that the ranking result with weighted cosine measures is agreement with the ranking results in literature [21]. Then, the proposed method is suitable for actual applications in multiple attribute decision-making problems with single-value neutrosophic multiplicity information.

In the future, we shall extend SVNMs to interval neutrosophic multisets and develop the application of interval neutrosophic multisets for handling the decision-making methods or other domains.

Author Contributions: C.F. originally proposed the LNNNWBM and LNNNWGBM operators and investigated their properties; J.Y. and E.F. provided the calculation and comparative analysis; all authors wrote the paper together.

Funding: This research was funded by the National Natural Science Foundation of China grant number [61703280] and Science and Technology Planning Project of Shaoxing City of China grant number [2017B70056].

Acknowledgments: This work was supported by the National Natural Science Foundation of China under grant Nos. 61703280, and Science and Technology Planning Project of Shaoxing City of China (No. 2017B70056).

Conflicts of Interest: The author declares no conflict of interest.

References

1. Zadeh, L.A. Fuzzy sets. *Inf. Control* **1965**, *8*, 338–353. [CrossRef]
2. Turksen, I.B. Interval valued fuzzy sets based on normal forms. *Fuzzy Sets Syst.* **1986**, *20*, 191–210. [CrossRef]
3. Atanassov, K.T. Intuitionistic fuzzy sets. *Fuzzy Sets Syst.* **1986**, *20*, 87–96. [CrossRef]
4. Smarandache, F. *Neutrosophy: Neutrosophic Probability, Set, and Logic*; American Research Press: Rehoboth, DE, USA, 1998.
5. Wang, H.; Smarandache, F.; Zhang, Y.Q.; Sunderraman, R. Single Valued Neutrosophic Sets. *Multispace MultiStruct.* **2010**, *4*, 410–413.
6. Wang, H.; Smarandache, F.; Zhang, Y.Q.; Sunderraman, R. *Interval Neutrosophic Sets and Logic: Theory and Applications in Computing*; Hexis: Phoenix, AZ, USA, 2005.
7. Smarandache, F. n-Valued Refined Neutrosophic Logic and Its Applications in Physics. *Prog. Phys.* **2013**, *4*, 143–146.
8. Ye, J.; Smarandache, F. Similarity Measure of Refined Single-Valued Neutrosophic Sets and Its Multicriteria Decision Making Method. *Neutrosophic Sets Syst.* **2016**, *12*, 41–44.
9. Fan, C.X.; Ye, J. The Cosine Measure of Refined-Single Valued Neutrosophic Sets and Refined-Interval Neutrosophic Sets for Multiple Attribute Decision-Making. *J. Intell. Fuzzy Syst.* **2017**, *33*, 2281–2289. [CrossRef]
10. Ye, J. Correlation Coefficient between Dynamic Single Valued Neutrosophic Multisets and Its Multiple Attribute Decision-Making Method. *Information* **2017**, *8*, 41. [CrossRef]
11. Ye, J. Multicriteria decision-making method using the correlation coefficient under single-valued neutrosophic environment. *Int. J. Gen. Syst.* **2013**, *42*, 386–394. [CrossRef]
12. Ye, J. A multicriteria decision-making method using aggregation operators for simplified neutrosophic sets. *J. Intell. Fuzzy Syst.* **2014**, *26*, 2459–2466.
13. Broumi, S.; Smarandache, F. Several similarity measures of neutrosophic sets. *Neutrosophic Sets Syst.* **2013**, *1*, 54–62.
14. Majumdar, P.; Samanta, S.K. On similarity and entropy of neutrosophic sets. *J. Intell. Fuzzy Syst.* **2014**, *26*, 1245–1252.
15. Broumi, S.; Smarandache, F. Correlation coefficient of interval neutrosophic set. *Appl. Mech. Mater.* **2013**, *436*, 511–517. [CrossRef]

16. Ye, J. Similarity measures between interval neutrosophic sets and their applications in multicriteria decision-making. *J. Intell. Fuzzy Syst.* **2014**, *26*, 165–172.
17. Chi, P.P.; Liu, P.D. An extended TOPSIS method for the multiple attribute decision making problems based on interval neutrosophic set. *Neutrosophic Sets Syst.* **2013**, *1*, 63–70.
18. Smarandache, F. *Neutrosophic Perspectives: Triplets, Duplets, Multisets, Hybrid Operators, Modal Logic, Hedge Algebras and Applications*; PonsEditions: Bruxelles, Brussels, 2017; pp. 115–123.
19. Ye, S.; Ye, J. Dice Similarity Measure between Single Valued Neutrosophic Multisets and Its Application in Medical Diagnosis. *Neutrosophic Sets Syst.* **2014**, *6*, 48–53.
20. Ye, J. Improved cosine similarity measures of simplified neutrosophic sets for medical diagnoses. *Artif. Intell. Med.* **2015**, *63*, 171–179. [CrossRef] [PubMed]
21. Deli, I.; Ali, M.; Smarandache, F. Bipolar neutrosophic sets and their application based on multi-criteria decision making problems. In Proceedings of the International Conference on Advanced Mechatronic Systems, Beijing, China, 5 October 2015; pp. 249–254.

symmetry

MDPI

Article

Linguistic Neutrosophic Generalized Partitioned Bonferroni Mean Operators and Their Application to Multi-Attribute Group Decision Making

Yumei Wang and Peide Liu *

School of Management Science and Engineering, Shandong University of Finance and Economics, Jinan 250014, China; wangyumei@mail.sdufe.edu.cn

* Corresponding Author: peide.liu@gmail.com; Tel.: +86-531-82222188

Received: 10 April 2018; Accepted: 26 April 2018; Published: 14 May 2018

Abstract: To solve the problems related to inhomogeneous connections among the attributes, we introduce a novel multiple attribute group decision-making (MAGDM) method based on the introduced linguistic neutrosophic generalized weighted partitioned Bonferroni mean operator (*LNGWPBM*) for linguistic neutrosophic numbers (LNNs). First of all, inspired by the merits of the generalized partitioned Bonferroni mean (GPBM) operator and LNNs, we combine the GPBM operator and LNNs to propose the linguistic neutrosophic GPBM (*LNGPBM*) operator, which supposes that the relationships are heterogeneous among the attributes in MAGDM. Then, we discuss its desirable properties and some special cases. In addition, aimed at the different importance of each attribute, the weighted form of the *LNGPBM* operator is investigated, which we call the *LNGWPBM* operator. Then, we discuss some of its desirable properties and special examples accordingly. In the end, we propose a novel MAGDM method on the basis of the introduced *LNGWPBM* operator, and illustrate its validity and merit by comparing it with the existing methods.

Keywords: *LNGPBM* operator; *LNGWPBM* operator; Linguistic neutrosophic sets; generalized partitioned Bonferroni mean operator; multiple attribute group decision-making (MAGDM)

---

1. Introduction

The goal of the multiple attribute group decision-making (MAGDM) method is to select the optimal scheme from finite alternatives. First of all, decision makers (DMs) evaluate each alternative under the different attributes. Then, based on the DMs' evaluation information, the alternatives are ranked in a certain way. As a research hotspot in recent decades, the MAGDM theory and methods have widely been used in all walks of life, such as supplier selection [1–3], medical diagnosis, clustering analysis, pattern recognition, and so on [4–11]. When evaluating alternatives, DMs used to evaluate alternatives by crisp numbers, but sometimes it is hard to use precise numbers because the surrounding environment has too much redundant data or interfering information. As a result, DMs have difficulty fully understanding the object of evaluation and exploiting exact information. As an example, when we evaluate people's morality or vehicle performance, we can easily use linguistic term such as good, fair, or poor, or fuzzy concepts such as slightly, obviously, or mightily, to give evaluation results. For this reason, Zadeh [12] put forward the concept of linguistic variables (LVs) in 1975. Later, Herrera and Herrera-Viedma [5,6] proposed a linguistic assessments consensus model and further developed the steps of linguistic decision analysis. Subsequently, it has become an area of wide concern, and resulted in several in-depth studies, especially in MAGDM [8,11,13–15]. In addition, for the reason of fuzziness, Atanassov [16] introduced the intuitionistic fuzzy set (IFS) on the basis of the fuzzy set developed by Zadeh [17]. IFS can embody the degrees of satisfaction and dissatisfaction to judge alternatives, synchronously, and has been studied by large numbers of scholars in many fields [1,2,9,10,18–23]. However, intuitionistic fuzzy numbers (IFNs) use the two real numbers of the interval [0,1] to represent membership degree and non-membership degree, which is not adequate or sufficient to quantify DMs' opinions. Hence, Chen et al. [24] used LVs to express the degrees of satisfaction and dissatisfaction instead of the real numbers of the interval [0,1], and proposed the linguistic intuitionistic fuzzy number (LIFN). LIFNs contain the advantages of both linguistic term sets and IFNs, so that it can address vague or imprecise information more accurately than LVs and IFNs. Since the birth of LIFNs, some scholars have proposed some improved aggregation operators and have applied them to MAGDM problems [10,25–28].

With the further development of fuzzy theory, Fang and Ye [29] noted while LIFNs can deal with vague or imprecise information more accurately than LVs and IFNs, it can only express incomplete information rather than indeterminate or inconsistent information. Since the indeterminacy of LIFN $I_A(x)$ is reckoned by $1 - T_A(x) - F_A(x)$ in default, evaluating the indeterminate or inconsistent information, i.e., $I_A(x) < 1 - T_A(x) - F_A(x)$ or $I_A(x) > 1 - T_A(x) - F_A(x)$, is beyond the scope of the LIFN. Hence, a new form of information expression needs to be found. Fortunately, the neutrosophic sets (NSs) developed by Smarandache [30] are able to quantify the indeterminacy clearly, which is independent of truth-membership and false-membership, but NSs are not easy to apply to the MAGDM. So, some stretched form of NS was proposed for solving MAGDM, such as single-valued neutrosophic sets (SVNSs) [31], interval neutrosophic sets (INSs) [32], simplified neutrosophic sets (SNSs) [33], and so on. Meanwhile, they have attracted a lot of research, especially related to MAGDM [34–41]. Due to the characteristic of SNSs that use three crisp numbers of the interval [0,1] to depict truth-membership, indeterminacy-membership, and false-membership, motivated by the narrow scope of the LIFN, Fang and Ye [29] put forward the concept of linguistic neutrosophic numbers (LNNs) by combining linguistic terms and a simplified neutrosophic number (SNN). LNNs use LVs in the predefined linguistic term set to express the truth-membership, indeterminacy-membership, and falsity-membership of SNNs. So, LNNs are more appropriate to depict qualitative information than SNNs, and are also an extension of the LIFNs, obviously. Therefore, in this paper, we tend to study the MAGDM problems with LNNs.

In MAGDM, the key step is how to select the optimal alternative according to the existing information. Usually, we adopt the traditional evaluation methods or the information aggregation operators. The common traditional evaluation methods include TOPSIS [7,9], VIKOR [19], ELECTRE [42], TODIM [20,43], PROMETHE [18], etc., and they can only give the priorities in order regarding alternatives. However, the information aggregation operators first integrate DMs' evaluation information into a comprehensive value, and then rank the alternatives. In other words, they not only give the prioritization orders of alternatives, they also give each alternative an integrated assessment value, so that the information aggregation operators are more workable than the traditional evaluation approaches in solving MAGDM problems. Hence, our study is concentrated on how to use information aggregation operators to solve the MAGDM problems with LNNs. In addition, in real MAGDM problems, there are often homogeneous connections among the attributes. Using a common example, quality is related to customer satisfaction when picking goods on the Internet. In order to solve this MAGDM problems where the attributes are interrelated, many related results have been achieved as a result, especially information aggregation operators such as the Bonferroni mean (BM) operator [23,44], the Maclaurin symmetric mean (MSM) operator [45], the Hamy mean operator [46], the generalized MSM operator [47], and so forth. However, the heterogeneous connections among the attributes may also exist in real MAGDM problems. For instance, in order to choose a car, we may consider the following attributes: the basic requirements (G_1), the physical property (G_2), the brand influence (G_3), and the user appraisal (G_4), where the attribute G_1 is associated with the attribute G_2, and the attribute G_3 is associated with the attribute G_4, but the attributes G_1 and G_2 are independent of the attributes G_3 and G_4. So, the four attributes can be sorted into two clusters, P_1 and P_2, namely $P_1 = \{G_1, G_2\}$ and $P_2 = \{G_3, G_4\}$ meeting the condition where P_1 and P_2 have no relationship. To solve this issue, Dutta and Guha [48] proposed the partition Bonferroni mean (*PBM*) operator, where all attributes are sorted into several clusters, and the members have an inherent connection in the same clusters, but independence in different clusters. Subsequently, Banerjee et al. [4] extended the *PBM* operator to the general form that was called the generalized partitioned Bonferroni mean (GPBM) operator, which further clarified the heterogeneous relationship and individually processed the elements that did not belong to any cluster of correlated elements, so the GPBM operator can model the average of the respective satisfaction of the independent and dependent input arguments. Besides, the GPBM operator can be translated into the BM operator, arithmetic mean operator, and *PBM* operator, so the GPBM operator is a wider range of applications for solving MAGDM problems with related attributes. Therefore, in this paper, we are further focused on how to combine the GPBM operator with LNNs to address the MAGDM problems with heterogeneous relationships among attributes. Inspired by the aforementioned ideas, we aim at:

(1) establishing a linguistic neutrosophic GPBM (*LNGPBM*) operator and the weighted form of the *LNGPBM* operator (the form of shorthand is *LNGWPBM*).
(2) discussing their properties and particular cases.
(3) proposing a novel MAGDM method in light of the proposed *LNGWPBM* operator to address the MAGDM problems with LNNs and the heterogeneous relationships among its attributes.
(4) showing the validity and merit of the developed method.

The arrangement of this paper is as follows. In Section 2, we briefly retrospect some elementary knowledge, including the definitions, operational rules, and comparison method of the LNNs. We also review some definitions and characteristics of the *PBM* operator and GPBM operator. In Section 3, we construct the *LNGPBM* operator and *LNGWPBM* operator for LNNs, including their characteristics and some special cases. In Section 4, we propose a novel MAGDM method based on the proposed *LNGWPBM* operator to address the MAGDM problems where heterogeneous connections exist among the attributes. In Section 5, we give a practical application related to the selection of green suppliers to show the validity and the generality of the MAGDM method, and compare the experimental results of the proposed MAGDM method with the ones of Fang and Ye's MAGDM method [29] and Liang et al.'s MAGDM method [7]. Section 6 presents the conclusions.

2. Preliminaries

To understand this article much better, this section intends to retrospect some elementary knowledge, including the definitions, operational rules, and comparison method of the LNNs, *PBM* operator, and generalized *PBM* operator.

2.1. Linguistic Neutrosophic Set (LNS)

Definition 1 [29]. *Let Z be the universe of discourse, and z be a generic element in Z, and let $L = (l_0, l_1, \cdots, l_s)$ be a linguistic term set. A LNS X in Z is represented by:*

$$X = \left\{ \left(z, l_{T_X(z)}, l_{I_X(z)}, l_{F_X(z)}\right) \middle| z \in Z \right\} \tag{1}$$

where T_X, I_X, and F_X denote the truth-membership function, indeterminacy-membership function, and falsity-membership function of z in the set X, respectively, and $T_X, I_X, F_X : Z \to [0, s]$ with s is an even number.

In [29], Fang and Ye called the pair $(l_\alpha, l_\beta, l_\gamma)$ an LNN, which meets $\alpha, \beta, \gamma : Z \to [0, s]$, and s is an even number.

Definition 2 [29]. *Let $z = (l_\alpha, l_\beta, l_\gamma)$ be an optional LNN in L, where the score function $C(z)$ of the LNN z is defined as shown:*

$$C(z) = \frac{2s + \alpha - \beta - \gamma}{3s} \tag{2}$$

where $\alpha, \beta, \gamma \in [0, s]$ and $C(z) \in [0, 1]$.

Definition 3 [29]. *Let* $z = (l_\alpha, l_\beta, l_\gamma)$ *be an optional LNN in L, where the accuracy function* $A(z)$ *of the LNN z is defined as shown:*

$$A(z) = \frac{\alpha - \gamma}{s} \tag{3}$$

where $\alpha, \beta, \gamma \in [0, s]$ *and* $A(z) \in [-1, 1]$.

Definition 4 [29]. *Let* $z_1 = (l_{\alpha_1}, l_{\beta_1}, l_{\gamma_1})$ *and* $z_2 = (l_{\alpha_2}, l_{\beta_2}, l_{\gamma_2})$ *be two optional LNNs in L. Then, the order between* z_1 *and* z_2 *is given by the following rules:*

(1) *If* $C(z_1) > C(z_2)$, *then* $z_1 > z_2$;
(2) *If* $C(z_1) = C(z_2)$, *then*

If $A(z_1) > A(z_2)$, *then* $z_1 > z_2$;

If $A(z_1) = A(z_2)$, *then* $z_1 = z_2$.

Example 1. *Suppose* $L = (l_0, l_1, \cdots, l_6)$ *is a linguistic term set, and* $z_1 = (l_6, l_2, l_3)$ *and* $z_2 = (l_4, l_1, l_1)$ *are two LNNs in L. Then, we can calculate the values of their score functions and accuracy functions as* $C(z_1) = 0.7222$, $C(z_2) = 0.7778$, $A(z_1) = 0.5$, and $A(z_2) = 0.5$. *According to Definition 4, it is easy to find that* $z_1 < z_2$.

Definition 5 [29]. *Let* $L = (l_0, l_1, \cdots, l_s)$ *be a linguistic term set, and* $z_1 = (l_{\alpha_1}, l_{\beta_1}, l_{\gamma_1})$ *and* $z_2 = (l_{\alpha_2}, l_{\beta_2}, l_{\gamma_2})$ *be two haphazard LNNs in L. The basic operational laws between the two LNNs are shown as below:*

$$z_1 \oplus z_2 = \left(l_{\alpha_1+\alpha_2-\alpha_1\alpha_2/s}, l_{\beta_1\beta_2/s}, l_{\gamma_1\gamma_2/s}\right), \tag{4}$$

$$z_1 \otimes z_2 = \left(l_{\alpha_1\alpha_2/s}, l_{\beta_1+\beta_2-\beta_1\beta_2/s}, l_{\gamma_1+\gamma_2-\gamma_1\gamma_2/s}\right), \tag{5}$$

$$\theta z_1 = \left(l_{s-s(1-\alpha_1/s)^\theta}, l_{s(\beta_1/s)^\theta}, l_{s(\gamma_1/s)^\theta}\right), \text{ where } \theta > 0, \tag{6}$$

$$z_1{}^\theta = \left(l_{s(\alpha_1/s)^\theta}, l_{s-s(1-\beta_1/s)^\theta}, l_{s-s(1-\gamma_1/s)^\theta}\right), \text{ where } \theta > 0 \tag{7}$$

It is easy to prove the following operational properties of the LNNs, according to Definition 5. Let $z_1 = (l_{\alpha_1}, l_{\beta_1}, l_{\gamma_1})$ and $z_2 = (l_{\alpha_2}, l_{\beta_2}, l_{\gamma_2})$ be any two LNNs in L. Then:

$$z_1 \oplus z_2 = z_2 \oplus z_1, \tag{8}$$

$$z_1 \otimes z_2 = z_2 \otimes z_1, \tag{9}$$

$$\theta(z_1 \oplus z_2) = \theta z_1 \oplus \theta z_2, \text{ where } \theta > 0, \tag{10}$$

$$\theta_1 z_1 \oplus \theta_2 z_1 = (\theta_1 + \theta_2) z_1, \text{ where } \theta_1, \theta_2 > 0, \tag{11}$$

$$z_1{}^{\theta_1} \otimes z_1{}^{\theta_2} = z_1{}^{\theta_1+\theta_2}, \text{ where } \theta_1, \theta_2 > 0, \tag{12}$$

$$z_1^\theta \otimes z_2^\theta = (z_1 \otimes z_2)^\theta, \text{ where } \theta > 0 \tag{13}$$

2.2. Generalized Partitioned Bonferroni Mean Operators

Definition 6 [48]. *Suppose the non-negative real set* $A = \{a_1, a_2, \cdots, a_n\}$ *is divided into t clusters* $P_1, P_2, \cdots, P_t$, *which satisfies* $P_x \cap P_y = \varnothing$, $x \neq y$ *and* $\bigcup_{r=1}^{t} P_r = A$. *Then, the partitioned Bonferroni mean (PBM) operator is defined as follows:*

$$PBM^{p,q}(a_1,a_2,\cdots,a_n)=\frac{1}{t}\left(\sum_{r=1}^{t}\left(\frac{1}{h_r}\sum_{i=1}^{h_r}a_i^p\left(\frac{1}{h_r-1}\sum_{\substack{j=1\\ j\neq i}}^{h_r}a_j^q\right)\right)\right)^{\frac{1}{p+q}} \tag{14}$$

where $p,q\geq 0$ *and* $p+q>0$, h_r *indicates the number of elements in partition* P_r *and* $\sum\limits_{r=1}^{t}h_r=n$.

The *PBM* operator is used to integrate the input arguments of the different clusters, which satisfies that the data has inherent connections in the same clusters, but independence in different clusters. However, sometimes, some of the input arguments have nothing to do with any other argument, that is, it does not exist in any cluster. We can part these arguments and deal with them individually. Hence, we sort the input arguments into two groups: F_1 contains the relevant arguments, and F_2 contains the input arguments that are irrelevant to any argument. These easily derive $F_1\cap F_2=\varnothing$ and $|F_1|+|F_2|=n$ where $|F_1|$ and $|F_2|$ denote the numbers of arguments in F_1 and F_2, respectively. According to the upper description, we suppose that the arguments of F_1 are divided into t partitions $P_1,P_2,\cdots,P_t$ on the basis of the interrelationship pattern [4]. To address this issue, the *PBM* operator is modified, and the GPBM operator is proposed, as shown in the following.

Definition 7 [4]. *Suppose that the non-negative real set* $A=\{a_1,a_2,\cdots,a_n\}$ *is sorted into two groups:* F_1 *and* F_2. *In* F_1, *the elements are divided into t clusters* $P_1,P_2,\cdots,P_t$, *which satisfies* $P_x\cap P_y=\varnothing, x\neq y$ *and* $\bigcup\limits_{r=1}^{t}P_r=F_1$; *in* F_2, *the elements are irrelevant to any element. Then, the GPBM operator is defined as follows:*

$$GPBM^{p,q}(a_1,a_2,\cdots,a_n)=\left(\frac{n-|F_2|}{n}\left(\frac{1}{t}\sum_{r=1}^{t}\left(\frac{1}{h_r}\sum_{i=1}^{h_r}a_i^p\left(\frac{1}{h_r-1}\sum_{\substack{j=1\\ j\neq i}}^{h_r}a_j^q\right)\right)\right)^{\frac{p}{p+q}}+\frac{|F_2|}{n}\left(\frac{1}{|F_2|}\sum_{i=1}^{|F_2|}a_i^p\right)\right)^{\frac{1}{p}} \tag{15}$$

where $p,q\geq 0$ *and* $p+q>0$, $|F_2|$ *denotes the number of elements in* F_2, h_r *indicates the number of elements in cluster* P_r *and* $\sum\limits_{r=1}^{t}h_r=n-|F_2|$.

Remark 1. *If* $|F_2|=0$, *we consider the first sum, and if* $|F_2|=n$, *we consider the last sum. At the same time, we have made the convention* $\frac{0}{0}=0$ *(we only need to define* $\frac{0}{0}$; *its conventional real value is not important here).*

The interpretation of the GPBM operator is detailed by Banerjee et al. in [4], and the GPBM operator has the following characteristics: idempotency, monotonicity, and boundedness [4].

Based on the characteristics of F_2, there are some special cases of GPBM operator, which are described as follows [4]:

(1) When $|F_2|=0$, all elements belong to the group F_1 and are divided into t clusters.

$$GPBM^{p,q}(a_1,a_2,\cdots,a_n)=\frac{1}{t}\sum_{r=1}^{t}\left(\frac{1}{h_r}\sum_{i=1}^{h_r}a_i^p\left(\frac{1}{h_r-1}\sum_{\substack{j=1\\ j\neq i}}^{h_r}a_j^q\right)\right)^{\frac{1}{p+q}}=PBM^{p,q}(a_1,a_2,\cdots,a_n)$$

It is simplified as the *PBM* operator described in Formula (15).

(2) When $|F_2|=0$ and $t=1$, all elements belong to the same cluster.

$$GPBM^{p,q}(a_1,a_2,\cdots,a_n)=\left(\frac{1}{h_r}\sum_{i=1}^{h_r}a_i^p\left(\frac{1}{h_r-1}\sum_{\substack{j=1\\j\neq i}}^{h_r}a_j^q\right)\right)^{\frac{1}{p+q}}=\left(\frac{1}{n}\sum_{i=1}^{n}a_i^p\left(\frac{1}{n-1}\sum_{\substack{j=1\\j\neq i}}^{n}a_j^q\right)\right)^{\frac{1}{p+q}}=BM^{p,q}(a_1,a_2,\cdots,a_n)$$

It becomes the *BM* operator [44].

(3) When $|F_2|=n$, all elements are independent.

$$GPBM^{p,q}(a_1,a_2,\cdots,a_n)=\left(\frac{1}{|F_2|}\sum_{i=1}^{|F_2|}a_i^p\right)^{\frac{1}{p}}=\left(\frac{1}{n}\sum_{i=1}^{n}a_i^p\right)^{\frac{1}{p}}$$

It is simplified as the power root arithmetic mean operator [4].

3. The Linguistic Neutrosophic GPBM Operators

In this section, we will construct the *LNGPBM* operator from the GPBM operator and LNNs. Moreover, with respect to the different weights of different attributes in real life, we will propose the corresponding weighted operators, and call it the *LNGWPBM* operator. They are defined as follows.

3.1. The LNGPBM Operator

Definition 8. *Let $z_1,z_2,\cdots$ and z_n be LNNs, which are sorted into two groups: F_1 and F_2. In F_1, the elements are divided into t clusters $P_1,P_2,\cdots,P_t$, which satisfies $P_x\cap P_y=\varnothing, x\neq y$ and $\bigcup_{r=1}^{t}P_r=F_1$; in F_2, the elements are irrelevant to any element. The LNGPBM operator of the LNNs $z_1,z_2,\cdots$ and z_n is defined as follows:*

$$LNGPBM^{p,q}(z_1,z_2,\cdots,z_n)=\left(\frac{n-|F_2|}{n}\left(\frac{1}{t}\bigoplus_{r=1}^{t}\left(\frac{1}{h_r}\bigoplus_{i=1}^{h_r}z_i^p\otimes\left(\frac{1}{h_r-1}\bigoplus_{\substack{j=1\\j\neq i}}^{h_r}z_j^q\right)\right)^{\frac{p}{p+q}}\right)\oplus\frac{|F_2|}{n}\left(\frac{1}{|F_2|}\bigoplus_{i=1}^{|F_2|}z_i^p\right)\right)^{\frac{1}{p}} \quad (16)$$

where $z_i=(l_{\alpha_i},l_{\beta_i},l_{\gamma_i})$ and $\alpha_i,\beta_i,\gamma_i\in[0,s]$ $(i=1,2,\cdots,n)$; $p,q\geq 0$ and $p+q>0$; $|F_2|$ denotes the number of elements in F_2, h_r indicates the number of elements in cluster P_r and $\sum_{r=1}^{t}h_r=n-|F_2|$.

Theorem 1. *Let $z_1,z_2,\cdots$ and z_n be LNNs, where $z_i=(l_{\alpha_i},l_{\beta_i},l_{\gamma_i})$ and $\alpha_i,\beta_i,\gamma_i\in[0,s]$ $(i=1,2,\cdots,n)$. The synthesized result of the LNGPBM operator of the LNNs $z_1,z_2,\cdots$ and z_n is still a LNN, which is shown as follows:*

$$LNGPBM^{p,q}(z_1, z_2, \cdots, z_n) = \left(l_{s\left(1-\left(\left(\left(\prod_{r=1}^{t}\left(1-(1-H_\alpha)^{\frac{p}{p+q}}\right)\right)^{\frac{1}{t}}\right)^{\frac{n-|F_2|}{n}}\right)\left(\left(\left(\prod_{i=1}^{|F_2|}\left(1-(\alpha_i/s)^p\right)\right)^{\frac{1}{|F_2|}}\right)^{\frac{|F_2|}{n}}\right)\right)^{\frac{1}{p}}}, \right.$$

$$\left. l_{s-s\left(1-\left(\left(\prod_{r=1}^{t}\left(1-(1-H_\beta)^{\frac{p}{p+q}}\right)\right)^{\frac{1}{t}}\right)^{\frac{n-|F_2|}{n}}\left(\left(\prod_{i=1}^{|F_2|}\left(1-(1-\beta_i/s)^p\right)\right)^{\frac{1}{|F_2|}}\right)^{\frac{|F_2|}{n}}\right)^{\frac{1}{p}}}, l_{s-s\left(1-\left(\left(\prod_{r=1}^{t}\left(1-(1-H_\gamma)^{\frac{p}{p+q}}\right)\right)^{\frac{1}{t}}\right)^{\frac{n-|F_2|}{n}}\left(\left(\prod_{i=1}^{|F_2|}\left(1-(1-\gamma_i/s)^p\right)\right)^{\frac{1}{|F_2|}}\right)^{\frac{|F_2|}{n}}\right)^{\frac{1}{p}}} \right) \tag{17}$$

where $H_\alpha = \left(\prod_{i=1}^{h_r}\left(1-(\alpha_i/s)^p\left(1-\left(\prod_{\substack{j=1\\ j\neq i}}^{h_r}\left(1-(\alpha_j/s)^q\right)\right)^{\frac{1}{h_r-1}}\right)\right)\right)^{\frac{1}{h_r}}$, $H_\beta = \left(\prod_{i=1}^{h_r}\left(1-(1-\beta_i/s)^p\left(1-\left(\prod_{\substack{j=1\\ j\neq i}}^{h_r}\left(1-(1-\beta_j/s)^q\right)\right)^{\frac{1}{h_r-1}}\right)\right)\right)^{\frac{1}{h_r}}$ and $H_\gamma = \left(\prod_{i=1}^{h_r}\left(1-(1-\gamma_i/s)^p\left(1-\left(\prod_{\substack{j=1\\ j\neq i}}^{h_r}\left(1-(1-\gamma_j/s)^q\right)\right)^{\frac{1}{h_r-1}}\right)\right)\right)^{\frac{1}{h_r}}$.

Proof. According to Formula (16), first of all, we can part two steps: the processing of F_1 and F_2, and then combine them to prove.

(i) The processing of F_1:

Based on the operational rules of LNNs, we can get $z_j^q = \left(l_{s(\alpha_j/s)^q}, l_{s-s(1-\beta_j/s)^q}, l_{s-s(1-\gamma_j/s)^q}\right)$ and

$$\mathop{\oplus}_{\substack{j=1\\ j\neq i}}^{h_r} z_j^q = \left(l_{s-s\prod_{\substack{j=1\\ j\neq i}}^{h_r}\left(1-(\alpha_j/s)^q\right)}, l_{s\prod_{\substack{j=1\\ j\neq i}}^{h_r}\left(1-(1-\beta_j/s)^q\right)}, l_{s\prod_{\substack{j=1\\ j\neq i}}^{h_r}\left(1-(1-\gamma_j/s)^q\right)} \right).$$

Then, we can calculate the average satisfaction of the elements in P_r except z_i:

$$\frac{1}{h_r-1}\mathop{\oplus}_{\substack{j=1\\ j\neq i}}^{h_r} z_j^q = \left(l_{s-s\left(\prod_{\substack{j=1\\ j\neq i}}^{h_r}\left(1-(\alpha_j/s)^q\right)\right)^{\frac{1}{h_r-1}}}, l_{s\left(\prod_{\substack{j=1\\ j\neq i}}^{h_r}\left(1-(1-\beta_j/s)^q\right)\right)^{\frac{1}{h_r-1}}}, l_{s\left(\prod_{\substack{j=1\\ j\neq i}}^{h_r}\left(1-(1-\gamma_j/s)^q\right)\right)^{\frac{1}{h_r-1}}} \right)$$

and the conjunction of the satisfaction of element z_i with the average satisfaction of the rest of elements in P_r:

$$z_i^p \otimes \left(\frac{1}{h_r - 1} \mathop{\oplus}_{\substack{j=1 \\ j \neq i}}^{h_r} z_j^q \right) = \left(l_{s(\alpha_i/s)^p \left(1 - \left(\prod_{\substack{j=1 \\ j \neq i}}^{h_r} \left(1-(\alpha_j/s)^q\right) \right)^{\frac{1}{h_r-1}} \right)}, l_{s - s\left((1-\beta_i/s)^p\right) \left(1 - \left(\prod_{\substack{j=1 \\ j \neq i}}^{h_r} \left(1-(1-\beta_j/s)^q\right) \right)^{\frac{1}{h_r-1}} \right)}, l_{s - s\left((1-\gamma_i/s)^p\right) \left(1 - \left(\prod_{\substack{j=1 \\ j \neq i}}^{h_r} \left(1-(1-\gamma_j/s)^q\right) \right)^{\frac{1}{h_r-1}} \right)} \right)$$

Then, the satisfaction of the interrelated elements of P_r is:

$$\frac{1}{h_r} \mathop{\oplus}_{i=1}^{h_r} z_i^p \otimes \left(\frac{1}{h_r - 1} \mathop{\oplus}_{\substack{j=1 \\ j \neq i}}^{h_r} z_j^q \right) = \left(l_{s - s\left(\prod_{i=1}^{h_r} \left(1-(\alpha_i/s)^p \left(1 - \left(\prod_{\substack{j=1 \\ j \neq i}}^{h_r} \left(1-(\alpha_j/s)^q\right) \right)^{\frac{1}{h_r-1}} \right) \right) \right)^{\frac{1}{h_r}}}, l_{s\left(\prod_{i=1}^{h_r} \left(1-(1-\beta_i/s)^p \left(1 - \left(\prod_{\substack{j=1 \\ j \neq i}}^{h_r} \left(1-(1-\beta_j/s)^q\right) \right)^{\frac{1}{h_r-1}} \right) \right) \right)^{\frac{1}{h_r}}}, l_{s\left(\prod_{i=1}^{h_r} \left(1-(1-\gamma_i/s)^p \left(1 - \left(\prod_{\substack{j=1 \\ j \neq i}}^{h_r} \left(1-(1-\gamma_j/s)^q\right) \right)^{\frac{1}{h_r-1}} \right) \right) \right)^{\frac{1}{h_r}}} \right)$$

So, the average satisfaction of all of the elements of the t clusters is:

$$A = \frac{1}{t} \mathop{\oplus}_{r=1}^{t} \left(\frac{1}{h_r} \mathop{\oplus}_{i=1}^{h_r} z_i^p \otimes \left(\frac{1}{h_r - 1} \mathop{\oplus}_{\substack{j=1 \\ j \neq i}}^{h_r} z_j^q \right) \right)^{\frac{p}{p+q}} = \left(l_{s - s\left(\prod_{r=1}^{t} \left(1 - \left(1 - \left(\prod_{i=1}^{h_r} \left(1-(\alpha_i/s)^p \left(1 - \left(\prod_{\substack{j=1 \\ j \neq i}}^{h_r} \left(1-(\alpha_j/s)^q\right) \right)^{\frac{1}{h_r-1}} \right) \right) \right)^{\frac{1}{h_r}} \right)^{\frac{p}{p+q}} \right) \right)^{\frac{1}{t}}}, l_{s\left(\prod_{r=1}^{t} \left(1 - \left(1 - \left(\prod_{i=1}^{h_r} \left(1-(1-\beta_i/s)^p \left(1 - \left(\prod_{\substack{j=1 \\ j \neq i}}^{h_r} \left(1-(1-\beta_j/s)^q\right) \right)^{\frac{1}{h_r-1}} \right) \right) \right)^{\frac{1}{h_r}} \right)^{\frac{p}{p+q}} \right) \right)^{\frac{1}{t}}}, l_{s\left(\prod_{r=1}^{t} \left(1 - \left(1 - \left(\prod_{i=1}^{h_r} \left(1-(1-\gamma_i/s)^p \left(1 - \left(\prod_{\substack{j=1 \\ j \neq i}}^{h_r} \left(1-(1-\gamma_j/s)^q\right) \right)^{\frac{1}{h_r-1}} \right) \right) \right)^{\frac{1}{h_r}} \right)^{\frac{p}{p+q}} \right) \right)^{\frac{1}{t}}} \right)$$

We suppose $H_\alpha = \left(\prod_{i=1}^{h_r} \left(1 - (\alpha_i/s)^p \left(1 - \left(\prod_{\substack{j=1 \\ j \neq i}}^{h_r} \left(1 - (\alpha_j/s)^q \right) \right)^{\frac{1}{h_r - 1}} \right) \right) \right)^{\frac{1}{h_r}}$,

$H_\beta = \left(\prod_{i=1}^{h_r} \left(1 - (1 - \beta_i/s)^p \left(1 - \left(\prod_{\substack{j=1 \\ j \neq i}}^{h_r} \left(1 - (1 - \beta_j/s)^q \right) \right)^{\frac{1}{h_r - 1}} \right) \right) \right)^{\frac{1}{h_r}}$ and $H_\gamma = \left(\prod_{i=1}^{h_r} \left(1 - (1 - \gamma_i/s)^p \left(1 - \left(\prod_{\substack{j=1 \\ j \neq i}}^{h_r} \left(1 - (1 - \gamma_j/s)^q \right) \right)^{\frac{1}{h_r - 1}} \right) \right) \right)^{\frac{1}{h_r}}$, then the upper formula can be rewritten as:

$$A = \frac{1}{t} \bigoplus_{r=1}^{t} \left(\frac{1}{h_r} \bigoplus_{i=1}^{h_r} z_i^p \otimes \left(\frac{1}{h_r - 1} \bigoplus_{\substack{j=1 \\ j \neq i}}^{h_r} z_j^q \right) \right)^{\frac{p}{p+q}} = \left(l_{s - s\left(\prod_{r=1}^{t} \left(1 - (1 - H_\alpha)^{\frac{p}{p+q}} \right) \right)^{\frac{1}{t}}}, l_{s\left(\prod_{r=1}^{t} \left(1 - (1 - H_\beta)^{\frac{p}{p+q}} \right) \right)^{\frac{1}{t}}}, l_{s\left(\prod_{r=1}^{t} \left(1 - (1 - H_\gamma)^{\frac{p}{p+q}} \right) \right)^{\frac{1}{t}}} \right)$$

(ii) The processing of F_2:

The average satisfaction of all the elements that are irrelevant to any element is:

$$B = \frac{1}{|F_2|} \bigoplus_{i=1}^{|F_2|} z_i^p = \left(l_{s - s\left(\prod_{i=1}^{|F_2|} \left(1 - (\alpha_i/s)^p \right) \right)^{\frac{1}{|F_2|}}}, l_{s\left(\prod_{i=1}^{|F_2|} \left(1 - (1 - \beta_i/s)^p \right) \right)^{\frac{1}{|F_2|}}}, l_{s\left(\prod_{i=1}^{|F_2|} \left(1 - (1 - \gamma_i/s)^p \right) \right)^{\frac{1}{|F_2|}}} \right)$$

Finally, we can compute the average satisfaction of the elements $z_1, z_2, \cdots$ and z_n:

$$\left(\frac{n - |F_2|}{n} A \oplus \frac{|F_2|}{n} B \right)^{\frac{1}{p}} = \left(\begin{array}{l} l_{s\left(1 - \left(\left(\left(\prod_{r=1}^{t} \left(1 - (1 - H_\alpha)^{\frac{p}{p+q}} \right) \right)^{\frac{1}{t}} \right)^{\frac{n - |F_2|}{n}} \right) \left(\left(\left(\prod_{i=1}^{|F_2|} \left(1 - (\alpha_i/s)^p \right) \right)^{\frac{1}{|F_2|}} \right)^{\frac{|F_2|}{n}} \right) \right)^{\frac{1}{p}}}, l_{s - s\left(1 - \left(\left(\prod_{r=1}^{t} \left(1 - (1 - H_\beta)^{\frac{p}{p+q}} \right) \right)^{\frac{1}{t}} \right)^{\frac{n - |F_2|}{n}} \left(\left(\prod_{i=1}^{|F_2|} \left(1 - (1 - \beta_i/s)^p \right) \right)^{\frac{1}{|F_2|}} \right)^{\frac{|F_2|}{n}} \right)^{\frac{1}{p}}}, \\ l_{s - s\left(1 - \left(\left(\prod_{r=1}^{t} \left(1 - (1 - H_\gamma)^{\frac{p}{p+q}} \right) \right)^{\frac{1}{t}} \right)^{\frac{n - |F_2|}{n}} \left(\left(\prod_{i=1}^{|F_2|} \left(1 - (1 - \gamma_i/s)^p \right) \right)^{\frac{1}{|F_2|}} \right)^{\frac{|F_2|}{n}} \right)^{\frac{1}{p}}} \end{array} \right)$$

That proves that Formula (17) is kept. Then, we prove that the aggregated result of Formula (17) is a LNN. It is easy to prove the following inequalities:

$$0 \leq s \left(1 - \left(\left(\left(\prod_{r=1}^{t} \left(1 - (1 - H_\alpha)^{\frac{p}{p+q}} \right) \right)^{\frac{1}{t}} \right)^{\frac{n - |F_2|}{n}} \right) \left(\left(\left(\prod_{i=1}^{|F_2|} \left(1 - (\alpha_i/s)^p \right) \right)^{\frac{1}{|F_2|}} \right)^{\frac{|F_2|}{n}} \right) \right)^{\frac{1}{p}} \leq s,$$

$$0 \leq s - s\left(1 - \left(\left(\prod_{r=1}^{t}\left(1-(1-H_\beta)^{\frac{p}{p+q}}\right)\right)^{\frac{1}{t}}\right)^{\frac{n-|F_2|}{n}}\left(\left(\prod_{i=1}^{|F_2|}(1-(1-\beta_i/s)^p)\right)^{\frac{1}{|F_2|}}\right)^{\frac{|F_2|}{n}}\right)^{\frac{1}{p}} \leq s,$$

and:

$$0 \leq s - s\left(1 - \left(\left(\prod_{r=1}^{t}\left(1-(1-H_\gamma)^{\frac{p}{p+q}}\right)\right)^{\frac{1}{t}}\right)^{\frac{n-|F_2|}{n}}\left(\left(\prod_{i=1}^{|F_2|}(1-(1-\gamma_i/s)^p)\right)^{\frac{1}{|F_2|}}\right)^{\frac{|F_2|}{n}}\right)^{\frac{1}{p}} \leq s.$$

Firstly, we prove $0 \leq H_\alpha \leq 1, 0 \leq H_\beta \leq 1$ and $0 \leq H_\gamma \leq 1$.

Since $\alpha_j, \beta_j, \gamma_j \in [0, s]$ and $q \geq 0$, we can get $0 \leq 1-(\alpha_j/s)^q \leq 1, 0 \leq 1-(1-\beta_j/s)^q \leq 1$ and $0 \leq 1-(1-\gamma_j/s)^q \leq 1$. Owing to $h_r > 0$, the following inequalities are established:

$$0 \leq 1 - \left(\prod_{\substack{j=1 \\ j\neq i}}^{h_r}\left(1-(\alpha_j/s)^q\right)\right)^{\frac{1}{h_r-1}} \leq 1, 0 \leq 1 - \left(\prod_{\substack{j=1 \\ j\neq i}}^{h_r}\left(1-(1-\beta_j/s)^q\right)\right)^{\frac{1}{h_r-1}} \leq 1, \text{ and } 0 \leq 1 - \left(\prod_{\substack{j=1 \\ j\neq i}}^{h_r}\left(1-(1-\gamma_j/s)^q\right)\right)^{\frac{1}{h_r-1}} \leq 1.$$

According to $p \geq 0$, it is easy to obtain the below inequality: $0 \leq H_\alpha \leq 1, 0 \leq H_\beta \leq 1$ and $0 \leq H_\gamma \leq 1$.

In addition, because $p+q > 0, t > 0$, and $|F_2| > 0$, we can get the following inequalities:

$$0 \leq \left(\left(\prod_{r=1}^{t}\left(1-(1-H_\alpha)^{\frac{p}{p+q}}\right)\right)^{\frac{1}{t}}\right)^{\frac{n-|F_2|}{n}} \leq 1, 0 \leq \left(\left(\prod_{r=1}^{t}\left(1-(1-H_\beta)^{\frac{p}{p+q}}\right)\right)^{\frac{1}{t}}\right)^{\frac{n-|F_2|}{n}} \leq 1, \text{and} 0 \leq \left(\left(\prod_{r=1}^{t}\left(1-(1-H_\gamma)^{\frac{p}{p+q}}\right)\right)^{\frac{1}{t}}\right)^{\frac{n-|F_2|}{n}} \leq 1.$$

$$0 \leq \left(\left(\prod_{i=1}^{|F_2|}(1-(\alpha_i/s)^p)\right)^{\frac{1}{|F_2|}}\right)^{\frac{|F_2|}{n}} \leq 1, 0 \leq \left(\left(\prod_{i=1}^{|F_2|}(1-(1-\beta_i/s)^p)\right)^{\frac{1}{|F_2|}}\right)^{\frac{|F_2|}{n}} \leq 1, \text{and} 0 \leq \left(\left(\prod_{i=1}^{|F_2|}(1-(1-\gamma_i/s)^p)\right)^{\frac{1}{|F_2|}}\right)^{\frac{|F_2|}{n}} \leq 1.$$

Besides, on the basis of the upper inequalities, we can get:

$$0 \leq \left(1 - \left(\left(\left(\prod_{r=1}^{t}\left(1-(1-H_\alpha)^{\frac{p}{p+q}}\right)\right)^{\frac{1}{t}}\right)^{\frac{n-|F_2|}{n}}\right)\left(\left(\left(\prod_{i=1}^{|F_2|}(1-(\alpha_i/s)^p)\right)^{\frac{1}{|F_2|}}\right)^{\frac{|F_2|}{n}}\right)\right)^{\frac{1}{p}} \leq 1,$$

$$0 \leq \left(1 - \left(\left(\prod_{r=1}^{t}\left(1-(1-H_\beta)^{\frac{p}{p+q}}\right)\right)^{\frac{1}{t}}\right)^{\frac{n-|F_2|}{n}}\left(\left(\prod_{i=1}^{|F_2|}(1-(1-\beta_i/s)^p)\right)^{\frac{1}{|F_2|}}\right)^{\frac{|F_2|}{n}}\right)^{\frac{1}{p}} \leq 1, \text{ and}$$

$$0 \leq \left(1 - \left(\left(\prod_{r=1}^{t}\left(1-(1-H_\gamma)^{\frac{p}{p+q}}\right)\right)^{\frac{1}{t}}\right)^{\frac{n-|F_2|}{n}}\left(\left(\prod_{i=1}^{|F_2|}(1-(1-\gamma_i/s)^p)\right)^{\frac{1}{|F_2|}}\right)^{\frac{|F_2|}{n}}\right)^{\frac{1}{p}} \leq 1.$$

which can derive directly:

$$0 \le s\left(1-\left(\left(\left(\prod_{r=1}^{t}\left(1-(1-H_\alpha)^{\frac{p}{p+q}}\right)\right)^{\frac{1}{t}}\right)^{\frac{n-|F_2|}{n}}\right)\left(\left(\left(\prod_{i=1}^{|F_2|}\left(1-(\alpha_i/s)^p\right)\right)^{\frac{1}{|F_2|}}\right)^{\frac{|F_2|}{n}}\right)\right)^{\frac{1}{p}} \le s,$$

$$0 \le s-s\left(1-\left(\left(\prod_{r=1}^{t}\left(1-(1-H_\beta)^{\frac{p}{p+q}}\right)\right)^{\frac{1}{t}}\right)^{\frac{n-|F_2|}{n}}\left(\left(\prod_{i=1}^{|F_2|}\left(1-(1-\beta_i/s)^p\right)\right)^{\frac{1}{|F_2|}}\right)^{\frac{|F_2|}{n}}\right)^{\frac{1}{p}} \le s,$$

$$\text{and } 0 \le s-s\left(1-\left(\left(\prod_{r=1}^{t}\left(1-(1-H_\gamma)^{\frac{p}{p+q}}\right)\right)^{\frac{1}{t}}\right)^{\frac{n-|F_2|}{n}}\left(\left(\prod_{i=1}^{|F_2|}\left(1-(1-\gamma_i/s)^p\right)\right)^{\frac{1}{|F_2|}}\right)^{\frac{|F_2|}{n}}\right)^{\frac{1}{p}} \le s.$$

Therefore, Theorem 1 is kept if some of the partitions only contain one element. □

In the following, we will demonstrate the desired properties of the proposed *LNGPBM* operator:

(1) **Idempotency**: If $z_1, z_2, \cdots$ and z_n are LNNs meeting the condition $z_i = \left(l_{\alpha_i}, l_{\beta_i}, l_{\gamma_i}\right) = z = \left(l_\alpha, l_\beta, l_\gamma\right)$ $(i = 1, 2, \cdots, n)$; then, $LNGPBM^{p,q}(z_1, z_2, \cdots, z_n) = z$.

Proof. Since $z_i = \left(l_{\alpha_i}, l_{\beta_i}, l_{\gamma_i}\right) = z = \left(l_\alpha, l_\beta, l_\gamma\right)$, we can get:

$$H_\alpha = \left(\prod_{i=1}^{h_r}\left(1-(\alpha_i/s)^p\left(1-\left(\prod_{\substack{j=1\\ j\ne i}}^{h_r}\left(1-(\alpha_j/s)^q\right)\right)^{\frac{1}{h_r-1}}\right)\right)\right)^{\frac{1}{h_r}} = \left(\prod_{i=1}^{h_r}\left(1-(\alpha/s)^p\left(1-\left(\prod_{\substack{j=1\\ j\ne i}}^{h_r}\left(1-(\alpha/s)^q\right)\right)^{\frac{1}{h_r-1}}\right)\right)\right)^{\frac{1}{h_r}}.$$

$$= \left(\prod_{i=1}^{h_r}\left(1-(\alpha/s)^p\left(1-\left(\left(1-(\alpha/s)^q\right)^{h_r-1}\right)^{\frac{1}{h_r-1}}\right)\right)\right)^{\frac{1}{h_r}} = \left(\prod_{i=1}^{h_r}\left(1-(\alpha/s)^p(\alpha/s)^q\right)\right)^{\frac{1}{h_r}} = 1-(\alpha/s)^{p+q}$$

In the same way, we can obtain $H_\beta = 1-(1-\beta/s)^{p+q}$ and $H_\gamma = 1-(1-\gamma/s)^{p+q}$. According to Theorem 1, we can obtain:

$$LNGPBM^{p,q}(z_1, z_2, \cdots, z_n) = \left(l_{s\left(1-\left(\left(\left(\prod_{r=1}^{t}\left(1-(1-H_\alpha)^{\frac{p}{p+q}}\right)\right)^{\frac{1}{t}}\right)^{\frac{n-|F_2|}{n}}\right)\left(\left(\left(\prod_{i=1}^{|F_2|}\left(1-(\alpha_i/s)^p\right)\right)^{\frac{1}{|F_2|}}\right)^{\frac{|F_2|}{n}}\right)\right)^{\frac{1}{p}}},\right.$$

$$\left. l_{s-s\left(1-\left(\left(\prod_{r=1}^{t}\left(1-(1-H_\beta)^{\frac{p}{p+q}}\right)\right)^{\frac{1}{t}}\right)^{\frac{n-|F_2|}{n}}\left(\left(\prod_{i=1}^{|F_2|}\left(1-(1-\beta_i/s)^p\right)\right)^{\frac{1}{|F_2|}}\right)^{\frac{|F_2|}{n}}\right)^{\frac{1}{p}}}, l_{s-s\left(1-\left(\left(\prod_{r=1}^{t}\left(1-(1-H_\gamma)^{\frac{p}{p+q}}\right)\right)^{\frac{1}{t}}\right)^{\frac{n-|F_2|}{n}}\left(\left(\prod_{i=1}^{|F_2|}\left(1-(1-\gamma_i/s)^p\right)\right)^{\frac{1}{|F_2|}}\right)^{\frac{|F_2|}{n}}\right)^{\frac{1}{p}}} \right)$$

$$= \left(l_{s\left(1-\left(\left(\left(\prod_{r=1}^{t}\left(1-\left(1-\left(1-(\alpha/s)^{p+q}\right)\right)^{\frac{p}{p+q}}\right)\right)^{\frac{1}{t}}\right)^{\frac{n-|F_2|}{n}}\right)\left(\left(\left(\prod_{i=1}^{|F_2|}\left(1-(\alpha/s)^{p}\right)\right)^{\frac{1}{|F_2|}}\right)^{\frac{|F_2|}{n}}\right)\right)^{\frac{1}{p}}}, \right.$$

$$\left. l_{s-s\left(1-\left(\left(\prod_{r=1}^{t}\left(1-\left(1-\left(1-(1-\beta/s)^{p+q}\right)\right)^{\frac{p}{p+q}}\right)\right)^{\frac{1}{t}}\right)^{\frac{n-|F_2|}{n}}\left(\left(\prod_{i=1}^{|F_2|}\left(1-(1-\beta/s)^{p}\right)\right)^{\frac{1}{|F_2|}}\right)^{\frac{|F_2|}{n}}\right)^{\frac{1}{p}}}, l_{s-s\left(1-\left(\left(\prod_{r=1}^{t}\left(1-\left(1-\left(1-(1-\gamma/s)^{p+q}\right)\right)^{\frac{p}{p+q}}\right)\right)^{\frac{1}{t}}\right)^{\frac{n-|F_2|}{n}}\left(\left(\prod_{i=1}^{|F_2|}\left(1-(1-\gamma/s)^{p}\right)\right)^{\frac{1}{|F_2|}}\right)^{\frac{|F_2|}{n}}\right)^{\frac{1}{p}}} \right)$$

$$= \left(l_{s\left(1-\left(\left(\left(\prod_{r=1}^{t}\left(1-(\alpha/s)^{p}\right)\right)^{\frac{1}{t}}\right)^{\frac{n-|F_2|}{n}}\right)\left(\left(1-(\alpha/s)^{p}\right)^{\frac{|F_2|}{n}}\right)\right)^{\frac{1}{p}}}, l_{s-s\left(1-\left(\left(\prod_{r=1}^{t}\left(1-(1-\beta/s)^{p}\right)\right)^{\frac{1}{t}}\right)^{\frac{n-|F_2|}{n}}\left(\left(1-(1-\beta/s)^{p}\right)^{\frac{|F_2|}{n}}\right)\right)^{\frac{1}{p}}}, l_{s-s\left(1-\left(\left(\prod_{r=1}^{t}\left(1-(1-\gamma/s)^{p}\right)\right)^{\frac{1}{t}}\right)^{\frac{n-|F_2|}{n}}\left(\left(1-(1-\gamma/s)^{p}\right)^{\frac{|F_2|}{n}}\right)\right)^{\frac{1}{p}}} \right)$$

$$= \left(l_{s\left(1-\left(\left(1-(\alpha/s)^{p}\right)^{\frac{n-|F_2|}{n}}\right)\left(\left(1-(\alpha/s)^{p}\right)^{\frac{|F_2|}{n}}\right)\right)^{\frac{1}{p}}}, l_{s-s\left(1-\left(\left(1-(1-\beta/s)^{p}\right)^{\frac{n-|F_2|}{n}}\right)\left(\left(1-(1-\beta/s)^{p}\right)^{\frac{|F_2|}{n}}\right)\right)^{\frac{1}{p}}}, l_{s-s\left(1-\left(\left(1-(1-\gamma/s)^{p}\right)^{\frac{n-|F_2|}{n}}\right)\left(\left(1-(1-\gamma/s)^{p}\right)^{\frac{|F_2|}{n}}\right)\right)^{\frac{1}{p}}} \right)$$

$$= \left(l_{s\left(1-\left(1-(\alpha/s)^{p}\right)\right)^{\frac{1}{p}}}, l_{s-s\left(1-\left(1-(1-\beta/s)^{p}\right)\right)^{\frac{1}{p}}}, l_{s-s\left(1-\left(1-(1-\gamma/s)^{p}\right)\right)^{\frac{1}{p}}} \right) = \left(l_{s\left((\alpha/s)^{p}\right)^{\frac{1}{p}}}, l_{s-s\left((1-\beta/s)^{p}\right)^{\frac{1}{p}}}, l_{s-s\left((1-\gamma/s)^{p}\right)^{\frac{1}{p}}} \right).$$
$$= (l_\alpha, l_\beta, l_\gamma)$$

□

(2) **Monotonicity**: If $z_i = (l_{\alpha_i}, l_{\beta_i}, l_{\gamma_i})$ $(i = 1, 2, \cdots, n)$ and $y_i = (l_{\delta_i}, l_{\eta_i}, l_{\sigma_i})$ $(i = 1, 2, \cdots, n)$ are any two sets of LNNs; they satisfy the condition $\alpha_i \geq \delta_i$, $\beta_i \leq \eta_i$ and $\gamma_i \leq \sigma_i$, then $LNGPBM^{p,q}(z_1, z_2, \cdots, z_n) \geq LNGPBM^{p,q}(y_1, y_2, \cdots, y_n)$.

Proof. Suppose that $LNGPBM^{p,q}(z_1, z_2, \cdots, z_n) = z = (l_\alpha, l_\beta, l_\gamma)$ and $LNGPBM^{p,q}(y_1, y_2, \cdots, y_n) = y = (l_\delta, l_\eta, l_\sigma)$, then:

$$\alpha = s\left(1-\left(\left(\left(\prod_{r=1}^{t}\left(1-(1-H_\alpha)^{\frac{p}{p+q}}\right)\right)^{\frac{1}{t}}\right)^{\frac{n-|F_2|}{n}}\right)\left(\left(\left(\prod_{i=1}^{|F_2|}\left(1-(\alpha_i/s)^{p}\right)\right)^{\frac{1}{|F_2|}}\right)^{\frac{|F_2|}{n}}\right)\right)^{\frac{1}{p}},$$

$$\delta = s\left(1-\left(\left(\left(\prod_{r=1}^{t}\left(1-(1-H_\delta)^{\frac{p}{p+q}}\right)\right)^{\frac{1}{t}}\right)^{\frac{n-|F_2|}{n}}\right)\left(\left(\left(\prod_{i=1}^{|F_2|}\left(1-(\delta_i/s)^{p}\right)\right)^{\frac{1}{|F_2|}}\right)^{\frac{|F_2|}{n}}\right)\right)^{\frac{1}{p}},$$

$$\beta = s - s\left(1-\left(\left(\prod_{r=1}^{t}\left(1-(1-H_\beta)^{\frac{p}{p+q}}\right)\right)^{\frac{1}{t}}\right)^{\frac{n-|F_2|}{n}}\left(\left(\prod_{i=1}^{|F_2|}\left(1-(1-\beta_i/s)^{p}\right)\right)^{\frac{1}{|F_2|}}\right)^{\frac{|F_2|}{n}}\right)^{\frac{1}{p}},$$

$$\eta = s - s\left(1-\left(\left(\prod_{r=1}^{t}\left(1-(1-H_\eta)^{\frac{p}{p+q}}\right)\right)^{\frac{1}{t}}\right)^{\frac{n-|F_2|}{n}}\left(\left(\prod_{i=1}^{|F_2|}\left(1-(1-\eta_i/s)^{p}\right)\right)^{\frac{1}{|F_2|}}\right)^{\frac{|F_2|}{n}}\right)^{\frac{1}{p}},$$

$$\gamma = s - s\left(1-\left(\left(\prod_{r=1}^{t}\left(1-(1-H_\gamma)^{\frac{p}{p+q}}\right)\right)^{\frac{1}{t}}\right)^{\frac{n-|F_2|}{n}}\left(\left(\prod_{i=1}^{|F_2|}\left(1-(1-\gamma_i/s)^{p}\right)\right)^{\frac{1}{|F_2|}}\right)^{\frac{|F_2|}{n}}\right)^{\frac{1}{p}},$$

$$\sigma = s - s\left(1 - \left(\left(\prod_{r=1}^{t}\left(1-(1-H_{\sigma})^{\frac{p}{p+q}}\right)\right)^{\frac{1}{t}}\right)^{\frac{n-|F_2|}{n}}\left(\left(\prod_{i=1}^{|F_2|}(1-(1-\sigma_i/s)^p)\right)^{\frac{1}{|F_2|}}\right)^{\frac{|F_2|}{n}}\right)^{\frac{1}{p}}.$$

In order to prove this property, we need to compute their score function values $C(z)$ and $C(y)$, and their accuracy values $A(z)$ and $A(y)$ to compare their synthesized result, i.e., $z \geq y$. Firstly, on the basis of the condition $\alpha_i \geq \delta_i$, $\beta_i \leq \eta_i$, and $\gamma_i \leq \sigma_i$, we can get the compared result of their truth-membership degrees, indeterminacy-membership degrees, and falsity-membership degrees, respectively.

(i) The comparison of the truth-membership degrees:

Based on $\alpha_i \geq \delta_i$, we can get:

$$H_\alpha = \left(\prod_{i=1}^{h_r}\left(1-(\alpha_i/s)^p\left(1-\left(\prod_{\substack{j=1\\ j\neq i}}^{h_r}\left(1-(\alpha_j/s)^q\right)\right)^{\frac{1}{h_r-1}}\right)\right)\right)^{\frac{1}{h_r}} \leq \left(\prod_{i=1}^{h_r}\left(1-(\delta_i/s)^p\left(1-\left(\prod_{\substack{j=1\\ j\neq i}}^{h_r}\left(1-(\delta_j/s)^q\right)\right)^{\frac{1}{h_r-1}}\right)\right)\right)^{\frac{1}{h_r}} = H_\delta$$

$$\Rightarrow \left(\left(\prod_{r=1}^{t}\left(1-(1-H_\alpha)^{\frac{p}{p+q}}\right)\right)^{\frac{1}{t}}\right)^{\frac{n-|F_2|}{n}} \leq \left(\left(\prod_{r=1}^{t}\left(1-(1-H_\delta)^{\frac{p}{p+q}}\right)\right)^{\frac{1}{t}}\right)^{\frac{n-|F_2|}{n}}$$

$$\text{and } \left(\left(\prod_{i=1}^{|F_2|}(1-(\alpha_i/s)^p)\right)^{\frac{1}{|F_2|}}\right)^{\frac{|F_2|}{n}} \leq \left(\left(\prod_{i=1}^{|F_2|}(1-(\delta_i/s)^p)\right)^{\frac{1}{|F_2|}}\right)^{\frac{|F_2|}{n}}$$

In accordance with the upper two inequalities, we have:

$$s\left(1-\left(\left(\left(\prod_{r=1}^{t}\left(1-(1-H_\alpha)^{\frac{p}{p+q}}\right)\right)^{\frac{1}{t}}\right)^{\frac{n-|F_2|}{n}}\right)\left(\left(\left(\prod_{i=1}^{|F_2|}(1-(\alpha_i/s)^p)\right)^{\frac{1}{|F_2|}}\right)^{\frac{|F_2|}{n}}\right)\right)^{\frac{1}{p}} \geq s\left(1-\left(\left(\left(\prod_{r=1}^{t}\left(1-(1-H_\delta)^{\frac{p}{p+q}}\right)\right)^{\frac{1}{t}}\right)^{\frac{n-|F_2|}{n}}\right)\left(\left(\left(\prod_{i=1}^{|F_2|}(1-(\delta_i/s)^p)\right)^{\frac{1}{|F_2|}}\right)^{\frac{|F_2|}{n}}\right)\right)^{\frac{1}{p}}$$

That is, $\alpha \geq \delta$.

(ii) The comparision of indeterminacy-membership degrees and falsity-membership degrees, respectively:

Based on $\beta_i \leq \eta_i$ and $\gamma_i \leq \sigma_i$, we can also obtain $\beta \leq \eta$ and $\gamma \leq \sigma$; this process is similar to the process of the truth-membership degrees.

Thus, it can be obtained that $C(z) = \frac{2s+\alpha-\beta-\gamma}{3s} \geq \frac{2s+\delta-\eta-\sigma}{3s} = C(y)$. In the following, we discuss two cases.

(i) If $C(z) > C(y)$, then $z > y$, according to Definition 2.
(ii) If $C(z) = C(y)$, then $(\alpha-\gamma)-\beta = (\delta-\sigma)-\eta$. Since $\alpha-\gamma \geq \delta-\sigma$ in the light of $\alpha \geq \delta$ and $\gamma \leq \sigma$, now we assume $\alpha-\gamma > \delta-\sigma$, then $\beta > \eta$, which is in contradiction with the previous proof $\beta \leq \eta$. So, we can conclude that $\alpha-\gamma = \delta-\sigma$. That is, $A(z) = \frac{\alpha-\gamma}{s} = \frac{\delta-\sigma}{s} = A(y)$, which testifies $z = y$.

In conclusion, the synthesized result $z \geq y$, which explains:

$$LNGPBM^{p,q}(z_1, z_2, \cdots, z_n) \geq LNGPBM^{p,q}(y_1, y_2, \cdots, y_n)$$

□

(3) **Boundedness**: Let $z_i = (l_{\alpha_i}, l_{\beta_i}, l_{\gamma_i})$ $(i = 1, 2, \cdots, n)$ be an arbitrary set of LNNs, then:

$$\min_i z_i \leq LNGPBM^{p,q}(z_1, z_2, \cdots, z_n) \leq \max_i z_i$$

Proof. Since $z_i \geq \underset{i}{min} z_i$, according to the monotonicity and idempotency of the proposed *LNGPBM* operator, we can obtain the following result:

$$LNGPBM^{p,q}(z_1, z_2, \cdots, z_n) \geq LNGPBM^{p,q}\left(\underset{i}{min} z_i, \underset{i}{min} z_i, \cdots, \underset{i}{min} z_i\right) = \underset{i}{min} z_i$$

Similarly, we can obtain the corresponding result for $\underset{i}{max} z_i$:

$$LNGPBM^{p,q}(z_1, z_2, \cdots, z_n) \leq LNGPBM^{p,q}\left(\underset{i}{max} z_i, \underset{i}{max} z_i, \cdots, \underset{i}{max} z_i\right) = \underset{i}{max} z_i$$

Therefore, $\underset{i}{min} z_i \leq LNGPBM^{p,q}(z_1, z_2, \cdots, z_n) \leq \underset{i}{max} z_i$. □

Based on the character of F_2, some special cases are discussed about the *LNGPBM* operator, and shown in the following.

(1) When $|F_2| = 0$, all arguments belong to the group F_1, and are divided into t clusters; then, the proposed *LNGPBM* operator is simplified as the following form:

$$LNGPBM^{p,q}(z_1, z_2, \cdots, z_n) = \frac{1}{t}\overset{t}{\underset{r=1}{\oplus}}\left(\frac{1}{h_r}\overset{h_r}{\underset{i=1}{\oplus}} z_i^p \otimes \left(\frac{1}{h_r - 1}\overset{h_r}{\underset{\substack{j=1\\ j\neq i}}{\oplus}} z_j^q\right)\right)^{\frac{1}{p+q}} = LNPBM^{p,q}(z_1, z_2, \cdots, z_n)$$

The *LNPBM* is called the linguistic neutrosophic *PBM* operator.

(2) When $|F_2| = 0$ and $t = 1$, all arguments belong to the same cluster, i.e., $h_r = n$; then, the proposed *LNGPBM* operator becomes the following form:

$$LNGPBM^{p,q}(z_1, z_2, \cdots, z_n) = \left(\frac{1}{h_r}\overset{h_r}{\underset{i=1}{\oplus}} z_i^p \otimes \left(\frac{1}{h_r - 1}\overset{h_r}{\underset{\substack{j=1\\ j\neq i}}{\oplus}} z_j^q\right)\right)^{\frac{1}{p+q}} = \left(\frac{1}{n}\overset{n}{\underset{i=1}{\oplus}} z_i^p \otimes \left(\frac{1}{n - 1}\overset{n}{\underset{\substack{j=1\\ j\neq i}}{\oplus}} z_j^q\right)\right)^{\frac{1}{p+q}} = LNBM^{p,q}(z_1, z_2, \cdots, z_n)$$

The *LNBM* is called the linguistic neutrosophic *BM* operator.

(3) When $|F_2| = n$, there is no element in group F_1 and all elements are independent; then, the proposed *LNGPBM* operator reduces to the following form:

$$LNGPBM^{p,q}(z_1, z_2, \cdots, z_n) = \left(\frac{1}{|F_2|}\overset{|F_2|}{\underset{i=1}{\oplus}} z_i^p\right)^{\frac{1}{p}} = \left(\frac{1}{n}\overset{n}{\underset{i=1}{\oplus}} z_i^p\right)^{\frac{1}{p}} = LNPRAM^{p}(z_1, z_2, \cdots, z_n)$$

The *LNPRAM* is called the linguistic neutrosophic power root arithmetic mean operator.

Moreover, we can also get some special cases by distributing different values to the parameters p and q.

(1) When $q \to 0$, the proposed *LNGPBM* operator becomes the *LNPRAM* operator, which was described in the previous discussion. Since there is no inner connection in group F_1, all of the elements are independent.

(2) When $p = 1$ and $q \to 0$, the proposed *LNGPBM* operator reduces to the linguistic neutrosophic arithmetic mean (*LNAM*) operator, which is shown as follows:

$$LNGPBM^{p=1,q\to 0}(z_1, z_2, \cdots, z_n) = \left(\frac{1}{n}\overset{n}{\underset{i=1}{\oplus}} z_i^p\right)^{\frac{1}{p}} = \frac{1}{n}\overset{n}{\underset{i=1}{\oplus}} z_i = LNAM(z_1, z_2, \cdots, z_n)$$

(3) When $p = 2$ and $q \to 0$, the proposed *LNGPBM* operator is transformed into the linguistic neutrosophic square root arithmetic mean (*LNSRAM*) operator, which is shown as follows:

$$LNGPBM^{p=2,q\to 0}(z_1, z_2, \cdots, z_n) = \left(\frac{1}{n}\overset{n}{\underset{i=1}{\oplus}} z_i^2\right)^{\frac{1}{2}} = LNSRAM(z_1, z_2, \cdots, z_n).$$

(4) When $p = q = 1$, the proposed *LNGPBM* operator is simplified as the simplest form of the *LNGPBM* operator, which is shown as follows:

$$LNGPBM^{p=1,q=1}(z_1, z_2, \cdots, z_n) = \frac{n-|F_2|}{n}\left(\frac{1}{t}\overset{t}{\underset{r=1}{\oplus}}\left(\frac{1}{h_r}\overset{h_r}{\underset{i=1}{\oplus}} z_i \otimes\left(\frac{1}{h_r-1}\overset{h_r}{\underset{\substack{j=1\\ j\neq i}}{\oplus}} z_j\right)\right)\right)^{\frac{1}{2}} \oplus \frac{|F_2|}{n}\left(\frac{1}{|F_2|}\overset{|F_2|}{\underset{i=1}{\oplus}} z_i\right)$$

It is often used to simplify the calculation in a problem.

3.2. The LNGWPBM Operator

In Definitions 8, we assume that all the input arguments have the same position. However, in many realistic decision-makings, every input argument may have different importance. Accordingly, we give different values to the weights of input arguments, and propose the weighted form of the *LNGPBM* operator. Let the weight of input argument $z_i = (l_{\alpha_i}, l_{\beta_i}, l_{\gamma_i})$ $(i = 1, 2, \cdots, n)$ be ω_i, where $\omega_i \in [0, 1]$ and $\sum_{i=1}^{n} \omega_i = 1$. The weighted form of the *LNGPBM* operator is shown in the following.

Definition 9. *Let $z_1, z_2, \cdots$ and z_n be LNNs that are sorted into two groups: F_1 and F_2. In F_1, the elements are divided into t clusters $P_1, P_2, \cdots, P_t$, which satisfy $P_x \cap P_y = \varnothing$, $x \neq y$ and $\overset{t}{\underset{r=1}{\cup}} P_r = F_1$; in F_2, the elements are irrelevant to any element. The weighted form of the LNGPBM operator of the LNNs $z_1, z_2, \cdots$ and z_n is defined as follows:*

$$LNGWPBM^{p,q}(z_1, z_2, \cdots, z_n) = \left(\frac{n-|F_2|}{n}\left(\frac{1}{t}\overset{t}{\underset{r=1}{\oplus}}\left(\frac{1}{\sum_{i=1}^{h_r}\omega_i}\overset{h_r}{\underset{i=1}{\oplus}} \omega_i z_i^p \otimes\left(\frac{1}{\sum_{\substack{j=1\\ j\neq i}}^{h_r}\omega_j}\overset{h_r}{\underset{\substack{j=1\\ j\neq i}}{\oplus}} \omega_j z_j^q\right)\right)\right)^{\frac{p}{p+q}} \oplus \frac{|F_2|}{n}\left(\frac{1}{\sum_{i=1}^{|F_2|}\omega_i}\overset{|F_2|}{\underset{i=1}{\oplus}} \omega_i z_i^p\right)\right)^{\frac{1}{p}} \tag{18}$$

where $z_i = (l_{\alpha_i}, l_{\beta_i}, l_{\gamma_i})$ and $\alpha_i, \beta_i, \gamma_i \in [0, s]$ $(i = 1, 2, \cdots, n)$; ω_i is the weight of input argument z_i meeting $\omega_i \in [0, 1]$ and $\sum_{i=1}^{n} \omega_i = 1$; $p, q \geq 0$ and $p + q > 0$; $|F_2|$ denotes the number of elements in F_2; h_r indicates the number of elements in partition P_r; and $\sum_{r=1}^{t} h_r = n - |F_2|$. Then, we call it a linguistic neutrosophic generalized weighted PBM (LNGWPBM) operator.

Theorem 2. *Let $z_1, z_2, \cdots$ and z_n be LNNs, where $z_i = \left(l_{\alpha_i}, l_{\beta_i}, l_{\gamma_i}\right)$ and $\alpha_i, \beta_i, \gamma_i \in [0, s]$ $(i = 1, 2, \cdots, n)$, and let the weight of input argument z_i be ω_i, where $\omega_i \in [0, 1]$ and $\sum\limits_{i=1}^{n} \omega_i = 1$. Then, the synthesized result of the LNGWPBM operator of the LNNs $z_1, z_2, \cdots$ and z_n is still a LNN, which is shown as follows:*

$$LNGWPBM^{pq}(z_1, z_2, \cdots, z_n) = \left(l_{s\left(1-\left(\left(\left(\prod\limits_{r=1}^{t}\left(1-(1-K_\alpha)^{\frac{p}{p+q}}\right)\right)^{\frac{1}{t}}\right)^{\frac{n-|F_2|}{n}}\right)\left(\left(\left(\prod\limits_{i=1}^{|F_2|}(1-(\alpha_i/s)^p)^{\omega_i}\right)^{\frac{1}{\sum_{i=1}^{|F_2|}\omega_i}}\right)^{\frac{|F_2|}{n}}\right)\right)^{\frac{1}{p}}}, \right. \tag{19}$$

$$\left. l_{s-s\left(1-\left(\left(\prod\limits_{r=1}^{t}\left(1-(1-K_\beta)^{\frac{p}{p+q}}\right)\right)^{\frac{1}{t}}\right)^{\frac{n-|F_2|}{n}}\left(\left(\prod\limits_{i=1}^{|F_2|}(1-(1-\beta_i/s)^p)^{\omega_i}\right)^{\frac{1}{\sum_{i=1}^{|F_2|}\omega_i}}\right)^{\frac{|F_2|}{n}}\right)^{\frac{1}{p}}}, l_{s-s\left(1-\left(\left(\prod\limits_{r=1}^{t}\left(1-(1-K_\gamma)^{\frac{p}{p+q}}\right)\right)^{\frac{1}{t}}\right)^{\frac{n-|F_2|}{n}}\left(\left(\prod\limits_{i=1}^{|F_2|}(1-(1-\gamma_i/s)^p)^{\omega_i}\right)^{\frac{1}{\sum_{i=1}^{|F_2|}\omega_i}}\right)^{\frac{|F_2|}{n}}\right)^{\frac{1}{p}}} \right)$$

where $K_\alpha = \left(\prod\limits_{i=1}^{h_r}\left(1-\left(1-(1-(\alpha_i/s)^p)^{\omega_i}\right)\left(1-\left(\prod\limits_{\substack{j=1\\j\neq i}}^{h_r}\left(1-(\alpha_j/s)^q\right)^{\omega_j}\right)^{\frac{1}{\sum_{j=1, j\neq i}^{h_r}\omega_j}}\right)\right)\right)^{\frac{1}{\sum_{i=1}^{h_r}\omega_i}}$,

$$K_\beta = \left(\prod\limits_{i=1}^{h_r}\left(1-\left(1-(1-(1-\beta_i/s)^p)^{\omega_i}\right)\left(1-\left(\prod\limits_{\substack{j=1\\j\neq i}}^{h_r}\left(1-(1-\beta_j/s)^q\right)^{\omega_j}\right)^{\frac{1}{\sum_{j=1, j\neq i}^{h_r}\omega_j}}\right)\right)\right)^{\frac{1}{\sum_{i=1}^{h_r}\omega_i}}, \text{ and}$$

$$K_\gamma = \left(\prod\limits_{i=1}^{h_r}\left(1-\left(1-(1-(1-\gamma_i/s)^p)^{\omega_i}\right)\left(1-\left(\prod\limits_{\substack{j=1\\j\neq i}}^{h_r}\left(1-(1-\gamma_j/s)^q\right)^{\omega_j}\right)^{\frac{1}{\sum_{j=1, j\neq i}^{h_r}\omega_j}}\right)\right)\right)^{\frac{1}{\sum_{i=1}^{h_r}\omega_i}}.$$

Proof. Along the lines of Theorem 1, we also process the groups F_1 and F_2 separately, and then combine them to prove.

(i) The processing of F_1:

Firstly, we successively use Formulas (7), (6), and (4) to get the following formula:

$$\mathop{\oplus}_{\substack{j=1\\ j\neq i}}^{h_r}\omega_j z_j^q=\left(l_{s-s\prod\limits_{\substack{j=1\\ j\neq i}}^{h_r}\left(1-(\alpha_j/s)^q\right)^{\omega_j}},l_{s\prod\limits_{\substack{j=1\\ j\neq i}}^{h_r}\left(1-(1-\beta_j/s)^q\right)^{\omega_j}},l_{s\prod\limits_{\substack{j=1\\ j\neq i}}^{h_r}\left(1-(1-\gamma_j/s)^q\right)^{\omega_j}}\right),$$

Then, we have:

$$\frac{1}{\sum\limits_{\substack{j=1\\ j\neq i}}^{h_r}\omega_j}\mathop{\oplus}_{\substack{j=1\\ j\neq i}}^{h_r}\omega_j z_j^q=\left(l_{s-s\left(\prod\limits_{\substack{j=1\\ j\neq i}}^{h_r}\left(1-(\alpha_j/s)^q\right)^{\omega_j}\right)^{\frac{1}{\sum\limits_{\substack{j=1\\ j\neq i}}^{h_r}\omega_j}}},l_{s\left(\prod\limits_{\substack{j=1\\ j\neq i}}^{h_r}\left(1-(1-\beta_j/s)^q\right)^{\omega_j}\right)^{\frac{1}{\sum\limits_{\substack{j=1\\ j\neq i}}^{h_r}\omega_j}}},l_{s\left(\prod\limits_{\substack{j=1\\ j\neq i}}^{h_r}\left(1-(1-\gamma_j/s)^q\right)^{\omega_j}\right)^{\frac{1}{\sum\limits_{\substack{j=1\\ j\neq i}}^{h_r}\omega_j}}}\right).$$

Since $\omega_i z_i^p=\left(l_{s-s\left(1-(\alpha_i/s)^p\right)^{\omega_i}},l_{s\left(1-(1-\beta_i/s)^p\right)^{\omega_i}},l_{s\left(1-(1-\gamma_i/s)^p\right)^{\omega_i}}\right)$, we can get:

$$\omega_i z_i^p\otimes\left(\frac{1}{\sum\limits_{\substack{j=1\\ j\neq i}}^{h_r}\omega_j}\mathop{\oplus}_{\substack{j=1\\ j\neq i}}^{h_r}\omega_j z_j^q\right)=$$

$$\left(l_{s\left(1-\left(1-(\alpha_i/s)^p\right)^{\omega_i}\right)\left(1-\left(\prod\limits_{\substack{j=1\\ j\neq i}}^{h_r}\left(1-(\alpha_j/s)^q\right)^{\omega_j}\right)^{\frac{1}{\sum\limits_{\substack{j=1\\ j\neq i}}^{h_r}\omega_j}}\right)},l_{s-s\left(1-\left(1-(1-\beta_i/s)^p\right)^{\omega_i}\right)\left(1-\left(\prod\limits_{\substack{j=1\\ j\neq i}}^{h_r}\left(1-(1-\beta_j/s)^q\right)^{\omega_j}\right)^{\frac{1}{\sum\limits_{\substack{j=1\\ j\neq i}}^{h_r}\omega_j}}\right)},l_{s-s\left(1-\left(1-(1-\gamma_i/s)^p\right)^{\omega_i}\right)\left(1-\left(\prod\limits_{\substack{j=1\\ j\neq i}}^{h_r}\left(1-(1-\gamma_j/s)^q\right)^{\omega_j}\right)^{\frac{1}{\sum\limits_{\substack{j=1\\ j\neq i}}^{h_r}\omega_j}}\right)}\right)$$

$$\Rightarrow\mathop{\oplus}_{i=1}^{h_r}\omega_i z_i^p\otimes\left(\frac{1}{\sum\limits_{\substack{j=1\\ j\neq i}}^{h_r}\omega_j}\mathop{\oplus}_{\substack{j=1\\ j\neq i}}^{h_r}\omega_j z_j^q\right)=$$

$$\left(l_{s\left(1-\prod\limits_{i=1}^{h_r}\left(1-\left(1-\left(1-(\alpha_i/s)^p\right)^{\omega_i}\right)\left(1-\left(\prod\limits_{\substack{j=1\\ j\neq i}}^{h_r}\left(1-(\alpha_j/s)^q\right)^{\omega_j}\right)^{\frac{1}{\sum\limits_{\substack{j=1\\ j\neq i}}^{h_r}\omega_j}}\right)\right)\right)},l_{s\prod\limits_{i=1}^{h_r}\left(1-\left(1-\left(1-(1-\beta_i/s)^p\right)^{\omega_i}\right)\left(1-\left(\prod\limits_{\substack{j=1\\ j\neq i}}^{h_r}\left(1-(1-\beta_j/s)^q\right)^{\omega_j}\right)^{\frac{1}{\sum\limits_{\substack{j=1\\ j\neq i}}^{h_r}\omega_j}}\right)\right)},l_{s\prod\limits_{i=1}^{h_r}\left(1-\left(1-\left(1-(1-\gamma_i/s)^p\right)^{\omega_i}\right)\left(1-\left(\prod\limits_{\substack{j=1\\ j\neq i}}^{h_r}\left(1-(1-\gamma_j/s)^q\right)^{\omega_j}\right)^{\frac{1}{\sum\limits_{\substack{j=1\\ j\neq i}}^{h_r}\omega_j}}\right)\right)}\right)$$

Hence, the following equation is established in the light of the upper:

$$\overline{H}=\frac{1}{\sum_{i=1}^{h_r}\omega_i}\mathop{\oplus}_{i=1}^{h_r}\omega_i z_i^p\otimes\left(\frac{1}{\sum\limits_{\substack{j=1\\ j\neq i}}^{h_r}\omega_j}\mathop{\oplus}_{\substack{j=1\\ j\neq i}}^{h_r}\omega_j z_j^q\right)=\left(l_{s-s\left(\prod\limits_{i=1}^{h_r}\left(1-\left(1-\left(1-(\alpha_i/s)^p\right)^{\omega_i}\right)\left(1-\left(\prod\limits_{\substack{j=1\\ j\neq i}}^{h_r}\left(1-(\alpha_j/s)^q\right)^{\omega_j}\right)^{\frac{1}{\sum\limits_{\substack{j=1\\ j\neq i}}^{h_r}\omega_j}}\right)\right)\right)^{\frac{1}{\sum_{i=1}^{h_r}\omega_i}}},\right.$$

$$l_{s\left(\prod_{i=1}^{h_r}\left(1-\left(1-\left(1-\left(1-\beta_i/s\right)^p\right)^{\omega_i}\right)\left(1-\left(\prod_{\substack{j=1\\j\neq i}}^{h_r}\left(1-\left(1-\beta_j/s\right)^q\right)^{\omega_j}\right)^{\frac{1}{\sum_{\substack{j=1\\j\neq i}}^{h_r}\omega_j}}\right)\right)\right)^{\frac{1}{\sum_{i=1}^{h_r}\omega_i}}}, l_{s\left(\prod_{i=1}^{h_r}\left(1-\left(1-\left(1-\left(1-\gamma_i/s\right)^p\right)^{\omega_i}\right)\left(1-\left(\prod_{\substack{j=1\\j\neq i}}^{h_r}\left(1-\left(1-\gamma_j/s\right)^q\right)^{\omega_j}\right)^{\frac{1}{\sum_{\substack{j=1\\j\neq i}}^{h_r}\omega_j}}\right)\right)\right)^{\frac{1}{\sum_{i=1}^{h_r}\omega_i}}}\Bigg)$$

Since the expression $\overline{H}$ is too long, we suppose:

$$K_\alpha = \left(\prod_{i=1}^{h_r}\left(1-\left(1-\left(1-(\alpha_i/s)^p\right)^{\omega_i}\right)\left(1-\left(\prod_{\substack{j=1\\j\neq i}}^{h_r}\left(1-(\alpha_j/s)^q\right)^{\omega_j}\right)^{\frac{1}{\sum_{\substack{j=1\\j\neq i}}^{h_r}\omega_j}}\right)\right)\right)^{\frac{1}{\sum_{i=1}^{h_r}\omega_i}},$$

$$K_\beta = \left(\prod_{i=1}^{h_r}\left(1-\left(1-\left(1-\left(1-\beta_i/s\right)^p\right)^{\omega_i}\right)\left(1-\left(\prod_{\substack{j=1\\j\neq i}}^{h_r}\left(1-\left(1-\beta_j/s\right)^q\right)^{\omega_j}\right)^{\frac{1}{\sum_{\substack{j=1\\j\neq i}}^{h_r}\omega_j}}\right)\right)\right)^{\frac{1}{\sum_{i=1}^{h_r}\omega_i}},$$

$$K_\gamma = \left(\prod_{i=1}^{h_r}\left(1-\left(1-\left(1-\left(1-\gamma_i/s\right)^p\right)^{\omega_i}\right)\left(1-\left(\prod_{\substack{j=1\\j\neq i}}^{h_r}\left(1-\left(1-\gamma_j/s\right)^q\right)^{\omega_j}\right)^{\frac{1}{\sum_{\substack{j=1\\j\neq i}}^{h_r}\omega_j}}\right)\right)\right)^{\frac{1}{\sum_{i=1}^{h_r}\omega_i}},$$

Then, the expression $\overline{H}$ can be written as $\overline{H} = \left(l_{s\times(1-K_\alpha)}, l_{s\times K_\beta}, l_{s\times K_\gamma}\right)$. Next, we can get the below expression:

$$A' = \frac{1}{t}\bigoplus_{r=1}^{t}\left(\frac{1}{\sum_{i=1}^{h_r}\omega_i}\bigoplus_{i=1}^{h_r}\omega_i z_i^p \otimes\left(\frac{1}{\sum_{\substack{j=1\\j\neq i}}^{h_r}\omega_j}\bigoplus_{\substack{j=1\\j\neq i}}^{h_r}\omega_j z_j^q\right)\right)^{\frac{p}{p+q}} = \frac{1}{t}\bigoplus_{r=1}^{t}\left(\overline{H}\right)^{\frac{p}{p+q}} = \left(l_{s-s\left(\prod_{r=1}^{t}\left(1-(1-K_\alpha)^{\frac{p}{p+q}}\right)\right)^{\frac{1}{t}}}, l_{s\left(\prod_{r=1}^{t}\left(1-(1-K_\beta)^{\frac{p}{p+q}}\right)\right)^{\frac{1}{t}}}, l_{s\left(\prod_{r=1}^{t}\left(1-(1-K_\gamma)^{\frac{p}{p+q}}\right)\right)^{\frac{1}{t}}}\right)$$

(ii) The processing of F_2:

Based on the operational laws of LNNs, it is easy to obtain:

$$B' = \frac{1}{\sum_{i=1}^{|F_2|}\omega_i}\bigoplus_{i=1}^{|F_2|}\omega_i z_i^p = \left(l_{s-s\left(\prod_{i=1}^{|F_2|}\left(1-(\alpha_i/s)^p\right)^{\omega_i}\right)^{\frac{1}{\sum_{i=1}^{|F_2|}\omega_i}}}, l_{s\left(\prod_{i=1}^{|F_2|}\left(1-\left(1-\beta_i/s\right)^p\right)^{\omega_i}\right)^{\frac{1}{\sum_{i=1}^{|F_2|}\omega_i}}}, l_{s\left(\prod_{i=1}^{|F_2|}\left(1-\left(1-\gamma_i/s\right)^p\right)^{\omega_i}\right)^{\frac{1}{\sum_{i=1}^{|F_2|}\omega_i}}}\right)$$

Finally, we compute the synthesized result of the *LNGWPBM* operator:

$$\left(\frac{n-|F_2|}{n}A' \oplus \frac{|F_2|}{n}B'\right)^{\frac{1}{p}} = \left(l_{s\left(1-\left(\left(\prod_{r=1}^{t}\left(1-(1-K_\alpha)^{\frac{p}{p+q}}\right)\right)^{\frac{1}{t}}\right)^{\frac{n-|F_2|}{n}}\left(\left(\prod_{i=1}^{|F_2|}(1-(\alpha_i/s)^p)^{\omega_i}\right)^{\frac{1}{\sum_{i=1}^{|F_2|}\omega_i}}\right)^{\frac{|F_2|}{n}}\right)^{\frac{1}{p}}}, \right.$$

$$\left. l_{s-s\left(1-\left(\left(\prod_{r=1}^{t}\left(1-(1-K_\beta)^{\frac{p}{p+q}}\right)\right)^{\frac{1}{t}}\right)^{\frac{n-|F_2|}{n}}\left(\left(\prod_{i=1}^{|F_2|}(1-(1-\beta_i/s)^p)^{\omega_i}\right)^{\frac{1}{\sum_{i=1}^{|F_2|}\omega_i}}\right)^{\frac{|F_2|}{n}}\right)^{\frac{1}{p}}}, l_{s-s\left(1-\left(\left(\prod_{r=1}^{t}\left(1-(1-K_\gamma)^{\frac{p}{p+q}}\right)\right)^{\frac{1}{t}}\right)^{\frac{n-|F_2|}{n}}\left(\left(\prod_{i=1}^{|F_2|}(1-(1-\gamma_i/s)^p)^{\omega_i}\right)^{\frac{1}{\sum_{i=1}^{|F_2|}\omega_i}}\right)^{\frac{|F_2|}{n}}\right)^{\frac{1}{p}}} \right)$$

That proves that Formula (19) is kept. Then, we prove that the aggregated result of Formula (19) is an LNN. It is easy to prove the following inequalities:

$$0 \le s\left(1-\left(\left(\left(\prod_{r=1}^{t}\left(1-(1-K_\alpha)^{\frac{p}{p+q}}\right)\right)^{\frac{1}{t}}\right)^{\frac{n-|F_2|}{n}}\right)\left(\left(\left(\prod_{i=1}^{|F_2|}(1-(\alpha_i/s)^p)^{\omega_i}\right)^{\frac{1}{\sum_{i=1}^{|F_2|}\omega_i}}\right)^{\frac{|F_2|}{n}}\right)\right)^{\frac{1}{p}} \le s,$$

$$0 \le s-s\left(1-\left(\left(\prod_{r=1}^{t}\left(1-(1-K_\beta)^{\frac{p}{p+q}}\right)\right)^{\frac{1}{t}}\right)^{\frac{n-|F_2|}{n}}\left(\left(\prod_{i=1}^{|F_2|}(1-(1-\beta_i/s)^p)^{\omega_i}\right)^{\frac{1}{\sum_{i=1}^{|F_2|}\omega_i}}\right)^{\frac{|F_2|}{n}}\right)^{\frac{1}{p}} \le s,$$

and $0 \le s-s\left(1-\left(\left(\prod_{r=1}^{t}\left(1-(1-K_\gamma)^{\frac{p}{p+q}}\right)\right)^{\frac{1}{t}}\right)^{\frac{n-|F_2|}{n}}\left(\left(\prod_{i=1}^{|F_2|}(1-(1-\gamma_i/s)^p)^{\omega_i}\right)^{\frac{1}{\sum_{i=1}^{|F_2|}\omega_i}}\right)^{\frac{|F_2|}{n}}\right)^{\frac{1}{p}} \le s.$

Firstly, we prove $0 \le K_\alpha \le 1$, $0 \le K_\beta \le 1$, and $0 \le H_\gamma \le 1$.

Based on the previous conditions such as $\alpha_j \in [0,s]$, $p \ge 0$, $q \ge 0$, $\omega_i \in [0,1]$, and so on, we can get $0 \le (1-(\alpha_i/s)^p)^{\omega_i} \le 1$ and $0 \le \left(\prod_{\substack{j=1\\ j\ne i}}^{h_r}\left(1-(\alpha_j/s)^q\right)^{\omega_j}\right)^{\frac{1}{\sum_{\substack{j=1\\ j\ne i}}^{h_r}\omega_j}} \le 1$, which can deduce the following inequality:

$$0 \le \left(1-\left(1-(\alpha_i/s)^p\right)^{\omega_i}\right)\left(1-\left(\prod_{\substack{j=1\\ j\neq i}}^{h_r}\left(1-(\alpha_j/s)^q\right)^{\omega_j}\right)^{\frac{1}{\sum_{\substack{j=1\\ j\neq i}}^{h_r}\omega_j}}\right) \le 1.$$

So, we can easily obtain $0 \le \left(\prod_{i=1}^{h_r}\left(1-\left(1-\left(1-(\alpha_i/s)^p\right)^{\omega_i}\right)\left(1-\left(\prod_{\substack{j=1\\ j\neq i}}^{h_r}\left(1-(\alpha_j/s)^q\right)^{\omega_j}\right)^{\frac{1}{\sum_{\substack{j=1\\ j\neq i}}^{h_r}\omega_j}}\right)\right)\right)^{\frac{1}{\sum_{i=1}^{h_r}\omega_i}} \le 1,$

i.e., $0 \le K_\alpha \le 1$.

Similarly, we also have $0 \le K_\beta \le 1$ and $0 \le H_\gamma \le 1$.

Next, we put the first to prove $0 \le s\left(1-\left(\left(\left(\prod_{r=1}^{t}\left(1-(1-K_\alpha)^{\frac{p}{p+q}}\right)\right)^{\frac{1}{t}}\right)^{\frac{n-|F_2|}{n}}\right)\left(\left(\left(\prod_{i=1}^{|F_2|}\left(1-(\alpha_i/s)^p\right)^{\omega_i}\right)^{\frac{1}{\sum_{i=1}^{|F_2|}\omega_i}}\right)^{\frac{|F_2|}{n}}\right)\right)^{\frac{1}{p}} \le s.$

According to $0 \le K_\alpha \le 1$, $p+q>0$ and $t>0$, we can illustrate $0 \le \left(\prod_{r=1}^{t}\left(1-(1-K_\alpha)^{\frac{p}{p+q}}\right)\right)^{\frac{1}{t}} \le 1$

and $0 \le \left(\prod_{i=1}^{|F_2|}\left(1-(\alpha_i/s)^p\right)^{\omega_i}\right)^{\frac{1}{\sum_{i=1}^{|F_2|}\omega_i}} \le 1$, which can deduce the following inequality:

$$0 \le \left(\left(\left(\prod_{r=1}^{t}\left(1-(1-K_\alpha)^{\frac{p}{p+q}}\right)\right)^{\frac{1}{t}}\right)^{\frac{n-|F_2|}{n}}\right)\left(\left(\left(\prod_{i=1}^{|F_2|}\left(1-(\alpha_i/s)^p\right)^{\omega_i}\right)^{\frac{1}{\sum_{i=1}^{|F_2|}\omega_i}}\right)^{\frac{|F_2|}{n}}\right) \le 1$$

Then, we find that $0 \le s\left(1-\left(\left(\left(\prod_{r=1}^{t}\left(1-(1-K_\alpha)^{\frac{p}{p+q}}\right)\right)^{\frac{1}{t}}\right)^{\frac{n-|F_2|}{n}}\right)\left(\left(\left(\prod_{i=1}^{|F_2|}\left(1-(\alpha_i/s)^p\right)^{\omega_i}\right)^{\frac{1}{\sum_{i=1}^{|F_2|}\omega_i}}\right)^{\frac{|F_2|}{n}}\right)\right)^{\frac{1}{p}} \le s.$

Likewise, we can illustrate $0 \le s - s\left(1-\left(\left(\prod_{r=1}^{t}\left(1-(1-K_\beta)^{\frac{p}{p+q}}\right)\right)^{\frac{1}{t}}\right)^{\frac{n-|F_2|}{n}}\left(\left(\prod_{i=1}^{|F_2|}\left(1-(1-\beta_i/s)^p\right)^{\omega_i}\right)^{\frac{1}{\sum_{i=1}^{|F_2|}\omega_i}}\right)^{\frac{|F_2|}{n}}\right)^{\frac{1}{p}} \le s$

and $0 \le s - s\left(1-\left(\left(\prod_{r=1}^{t}\left(1-(1-K_\gamma)^{\frac{p}{p+q}}\right)\right)^{\frac{1}{t}}\right)^{\frac{n-|F_2|}{n}}\left(\left(\prod_{i=1}^{|F_2|}\left(1-(1-\gamma_i/s)^p\right)^{\omega_i}\right)^{\frac{1}{\sum_{i=1}^{|F_2|}\omega_i}}\right)^{\frac{|F_2|}{n}}\right)^{\frac{1}{p}} \le s.$

Therefore, Theorem 2 is kept. □

In the following, we demonstrate the desired properties of the proposed *LNGWPBM* operator:

(1) **Monotonicity**: If $z_i = (l_{\alpha_i}, l_{\beta_i}, l_{\gamma_i})$ $(i = 1, 2, \cdots, n)$ and $y_i = (l_{\delta_i}, l_{\eta_i}, l_{\sigma_i})$ $(i = 1, 2, \cdots, n)$ are any two sets of LNNs, they satisfy the conditions $\alpha_i \ge \delta_i$, $\beta_i \le \eta_i$ and $\gamma_i \le \sigma_i$, then:

$$LNGWPBM^{p,q}(z_1, z_2, \cdots, z_n) \ge LNGWPBM^{p,q}(y_1, y_2, \cdots, y_n).$$

Proof. Similar to the monotonicity property of the *LNGPBM* operator, we also suppose that $LNGWPBM^{p,q}(z_1, z_2, \cdots, z_n) = z = (l_\alpha, l_\beta, l_\gamma)$ and $LNGWPBM^{p,q}(y_1, y_2, \cdots, y_n) = y = (l_\delta, l_\eta, l_\sigma)$. Then:

$$\alpha = s\left(1 - \left(\left(\left(\prod_{r=1}^{t}\left(1-(1-K_\alpha)^{\frac{p}{p+q}}\right)\right)^{\frac{1}{t}}\right)^{\frac{n-|F_2|}{n}}\right)\left(\left(\left(\prod_{i=1}^{|F_2|}(1-(\alpha_i/s)^p)^{\omega_i}\right)^{\frac{1}{\sum_{i=1}^{|F_2|}\omega_i}}\right)^{\frac{|F_2|}{n}}\right)\right)^{\frac{1}{p}}$$

$$\delta = s\left(1 - \left(\left(\left(\prod_{r=1}^{t}\left(1-(1-K_\delta)^{\frac{p}{p+q}}\right)\right)^{\frac{1}{t}}\right)^{\frac{n-|F_2|}{n}}\right)\left(\left(\left(\prod_{i=1}^{|F_2|}(1-(\delta_i/s)^p)^{\omega_i}\right)^{\frac{1}{\sum_{i=1}^{|F_2|}\omega_i}}\right)^{\frac{|F_2|}{n}}\right)\right)^{\frac{1}{p}},$$

$$\beta = s - s\left(1 - \left(\left(\prod_{r=1}^{t}\left(1-(1-K_\beta)^{\frac{p}{p+q}}\right)\right)^{\frac{1}{t}}\right)^{\frac{n-|F_2|}{n}}\left(\left(\prod_{i=1}^{|F_2|}(1-(1-\beta_i/s)^p)^{\omega_i}\right)^{\frac{1}{\sum_{i=1}^{|F_2|}\omega_i}}\right)^{\frac{|F_2|}{n}}\right)^{\frac{1}{p}},$$

$$\eta = s - s\left(1 - \left(\left(\prod_{r=1}^{t}\left(1-(1-K_\eta)^{\frac{p}{p+q}}\right)\right)^{\frac{1}{t}}\right)^{\frac{n-|F_2|}{n}}\left(\left(\prod_{i=1}^{|F_2|}(1-(1-\eta_i/s)^p)^{\omega_i}\right)^{\frac{1}{\sum_{i=1}^{|F_2|}\omega_i}}\right)^{\frac{|F_2|}{n}}\right)^{\frac{1}{p}},$$

$$\gamma = s - s\left(1 - \left(\left(\prod_{r=1}^{t}\left(1-(1-K_\gamma)^{\frac{p}{p+q}}\right)\right)^{\frac{1}{t}}\right)^{\frac{n-|F_2|}{n}}\left(\left(\prod_{i=1}^{|F_2|}(1-(1-\gamma_i/s)^p)^{\omega_i}\right)^{\frac{1}{\sum_{i=1}^{|F_2|}\omega_i}}\right)^{\frac{|F_2|}{n}}\right)^{\frac{1}{p}}, \text{ and}$$

$$\sigma = s - s\left(1 - \left(\left(\prod_{r=1}^{t}\left(1-(1-K_\sigma)^{\frac{p}{p+q}}\right)\right)^{\frac{1}{t}}\right)^{\frac{n-|F_2|}{n}}\left(\left(\prod_{i=1}^{|F_2|}(1-(1-\sigma_i/s)^p)^{\omega_i}\right)^{\frac{1}{\sum_{i=1}^{|F_2|}\omega_i}}\right)^{\frac{|F_2|}{n}}\right)^{\frac{1}{p}}.$$

In order to prove this property, we need to compute their score function values $C(z)$ and $C(y)$, and their accuracy values $A(z)$ and $A(y)$ to compare their synthesized result, i.e., $z \geq y$. Firstly, on the basis of the condition $\alpha_i \geq \delta_i$, $\beta_i \leq \eta_i$ and $\gamma_i \leq \sigma_i$, we can get the compared result of their truth-membership degrees, indeterminacy-membership degrees, and falsity-membership degrees, respectively.

(i) The comparison of the truth-membership degrees:

Based on $\alpha_i \geq \delta_i$, we can get:

$$K_\alpha = \left(\prod_{i=1}^{h_r} \left(1 - \left(1 - \left(1 - (\alpha_i/s)^p\right)^{\omega_i}\right) \left(1 - \left(\prod_{\substack{j=1 \\ j \neq i}}^{h_r} \left(1 - (\alpha_j/s)^q\right)^{\omega_j} \right)^{\frac{1}{\sum_{\substack{j=1 \\ j \neq i}}^{h_r} \omega_j}} \right) \right) \right)^{\frac{1}{\sum_{i=1}^{h_r} \omega_i}}$$

$$\leq \left(\prod_{i=1}^{h_r} \left(1 - \left(1 - \left(1 - (\delta_i/s)^p\right)^{\omega_i}\right) \left(1 - \left(\prod_{\substack{j=1 \\ j \neq i}}^{h_r} \left(1 - (\delta_j/s)^q\right)^{\omega_j} \right)^{\frac{1}{\sum_{\substack{j=1 \\ j \neq i}}^{h_r} \omega_j}} \right) \right) \right)^{\frac{1}{\sum_{i=1}^{h_r} \omega_i}} = K_\delta$$

$$\Rightarrow \left(\left(\prod_{r=1}^{t} \left(1 - (1 - K_\alpha)^{\frac{p}{p+q}}\right) \right)^{\frac{1}{t}} \right)^{\frac{n-|F_2|}{n}} \leq \left(\left(\prod_{r=1}^{t} \left(1 - (1 - K_\delta)^{\frac{p}{p+q}}\right) \right)^{\frac{1}{t}} \right)^{\frac{n-|F_2|}{n}}$$

$$\text{and} \left(\left(\prod_{i=1}^{|F_2|} \left(1 - (\alpha_i/s)^p\right)^{\omega_i} \right)^{\frac{1}{\sum_{i=1}^{|F_2|} \omega_i}} \right)^{\frac{|F_2|}{n}} \leq \left(\left(\prod_{i=1}^{|F_2|} \left(1 - (\delta_i/s)^p\right)^{\omega_i} \right)^{\frac{1}{\sum_{i=1}^{|F_2|} \omega_i}} \right)^{\frac{|F_2|}{n}}.$$

In accordance with the upper two inequalities, we have:

$$s\left(1 - \left(\left(\left(\prod_{r=1}^{t}\left(1-(1-K_\alpha)^{\frac{p}{p+q}}\right)\right)^{\frac{1}{t}}\right)^{\frac{n-|F_2|}{n}}\right)\left(\left(\left(\prod_{i=1}^{|F_2|}\left(1-(\alpha_i/s)^p\right)^{\omega_i}\right)^{\frac{1}{\sum_{i=1}^{|F_2|}\omega_i}}\right)^{\frac{|F_2|}{n}}\right)\right)^{\frac{1}{p}} \geq s\left(1 - \left(\left(\left(\prod_{r=1}^{t}\left(1-(1-K_\delta)^{\frac{p}{p+q}}\right)\right)^{\frac{1}{t}}\right)^{\frac{n-|F_2|}{n}}\right)\left(\left(\left(\prod_{i=1}^{|F_2|}\left(1-(\delta_i/s)^p\right)^{\omega_i}\right)^{\frac{1}{\sum_{i=1}^{|F_2|}\omega_i}}\right)^{\frac{|F_2|}{n}}\right)\right)^{\frac{1}{p}}$$

That is, $\alpha \geq \delta$.

(ii) The comparison of indeterminacy-membership degrees and falsity-membership degrees, respectively:

Based on $\beta_i \leq \eta_i$ and $\gamma_i \leq \sigma_i$, we can also obtain $\beta \leq \eta$ and $\gamma \leq \sigma$, which is similar to the process of the truth-membership degrees.

Thus, it can be obtained that $C(z) = \frac{2s+\alpha-\beta-\gamma}{3s} \geq \frac{2s+\delta-\eta-\sigma}{3s} = C(y)$. In the following, we discuss two cases.

(i) If $C(z) > C(y)$, then $z > y$ according to Definition 2.
(ii) If $C(z) = C(y)$, then $(\alpha - \gamma) - \beta = (\delta - \sigma) - \eta$. Since $\alpha - \gamma \geq \delta - \sigma$ in the light of $\alpha \geq \delta$ and $\gamma \leq \sigma$, now we assume $\alpha - \gamma > \delta - \sigma$, then $\beta > \eta$, which is in contradiction with the previous proof $\beta \leq \eta$. So, we can conclude that $\alpha - \gamma = \delta - \sigma$. That is $A(z) = \frac{\alpha-\gamma}{s} = \frac{\delta-\sigma}{s} = A(y)$, which testifies $z = y$.

In conclusion, the synthesized result is $z \geq y$, which explains $LNGWPBM^{p,q}(z_1, z_2, \cdots, z_n) \geq LNGWPBM^{p,q}(y_1, y_2, \cdots, y_n)$. □

(2) **Boundedness**: Let $z_i = (l_{\alpha_i}, l_{\beta_i}, l_{\gamma_i})$ $(i = 1, 2, \cdots, n)$ be any set of LNNs, then:

$$LNGWPBM^{p,q}\left(\min_i z_i, \min_i z_i, \cdots, \min_i z_i\right) \leq LNGWPBM^{p,q}(z_1, z_2, \cdots, z_n) \leq LNGWPBM^{p,q}\left(\max_i z_i, \max_i z_i, \cdots, \max_i z_i\right).$$

Based on the monotonicity property of the *LNGWPBM* operator, it is easy to prove, and the detailed process is omitted here.

Based on the character of F_2, some special cases are discussed about the *LNGPBM* operator, as shown in the following.

(1) When $|F_2| = 0$, all of the arguments belong to the group F_1, and are divided into t partitions; then, the proposed *LNGWPBM* operator is simplified as the linguistic neutrosophic weighted *PBM* (*LNWPBM*) operator:

$$LNGWPBM^{p,q}(z_1,z_2,\cdots,z_n) = \frac{1}{t}\mathop{\oplus}_{r=1}^{t}\left(\frac{1}{\sum_{i=1}^{h_r}\omega_i}\mathop{\oplus}_{i=1}^{h_r}\omega_i z_i^p \otimes\left(\frac{1}{\sum_{j=1, j\neq i}^{h_r}\omega_j}\mathop{\oplus}_{j=1, j\neq i}^{h_r}\omega_j z_j^q\right)\right)^{\frac{1}{p+q}} = LNWPBM^{p,q}(z_1,z_2,\cdots,z_n)$$

(2) When $|F_2| = 0$ and $t = 1$, all of the arguments belong to the same partition, i.e., $h_r = n$; then, the proposed *LNGWPBM* operator is translated into the linguistic neutrosophic normalized weighted *BM* (*LNNWBM*) operator:

$$LNGWPBM^{p,q}(z_1,z_2,\cdots,z_n) = \left(\frac{1}{\sum_{i=1}^{h_r}\omega_i}\mathop{\oplus}_{i=1}^{h_r}\omega_i z_i^p \otimes\left(\frac{1}{\sum_{j=1, j\neq i}^{h_r}\omega_j}\mathop{\oplus}_{j=1, j\neq i}^{h_r}\omega_j z_j^q\right)\right)^{\frac{1}{p+q}} = \left(\frac{1}{\sum_{i=1}^{n}\omega_i}\mathop{\oplus}_{i=1}^{n}\omega_i z_i^p \otimes\left(\frac{1}{\sum_{j=1, j\neq i}^{n}\omega_j}\mathop{\oplus}_{j=1, j\neq i}^{n}\omega_j z_j^q\right)\right)^{\frac{1}{p+q}}$$

$$= \left(\mathop{\oplus}_{i=1}^{n}\omega_i z_i^p \otimes\left(\frac{1}{1-\omega_i}\mathop{\oplus}_{j=1, j\neq i}^{n}\omega_j z_j^q\right)\right)^{\frac{1}{p+q}} = LNNWBM^{p,q}(z_1,z_2,\cdots,z_n).$$

(3) When $|F_2| = n$, there is no element in group F_1 and all of the elements are independent; then, the proposed *LNGWPBM* operator reduces to the linguistic neutrosophic power root weighted mean (*LNPRWM*) operator:

$$LNGWPBM^{p,q}(z_1,z_2,\cdots,z_n) = \left(\frac{1}{\sum_{i=1}^{|F_2|}\omega_i}\mathop{\oplus}_{i=1}^{|F_2|}\omega_i z_i^p\right)^{\frac{1}{p}} = \left(\frac{1}{\sum_{i=1}^{n}\omega_i}\mathop{\oplus}_{i=1}^{n}\omega_i z_i^p\right)^{\frac{1}{p}} = \left(\mathop{\oplus}_{i=1}^{n}\omega_i z_i^p\right)^{\frac{1}{p}} = LNPRWM^{p}(z_1,z_2,\cdots,z_n)$$

Moreover, we can also get some special cases by distributing different values to the parameters p and q.

(1) When $q \to 0$, the *LNGWPBM* operator is translated into the *LNPRWM* operator, as described in the previous discussion. Since there are no inner connections in group F_1, all of the elements are unrelated.

(2) When $p = 1$ and $q \to 0$, the *LNGWPBM* operator becomes the *LNNWAA* operator, as defined by Fang and Ye [29]:

$$LNGWPBM^{p=1,q\to 0}(z_1,z_2,\cdots,z_n) = \left(\mathop{\oplus}_{i=1}^{n}\omega_i z_i^p\right)^{\frac{1}{p}} = \mathop{\oplus}_{i=1}^{n}\omega_i z_i = LNNWAA(z_1,z_2,\cdots,z_n).$$

(3) When $p = 2$ and $q \to 0$, the *LNGWPBM* operator is transformed into the linguistic neutrosophic square root weighted mean (*LNSRWM*) operator, which is shown as follows:

$$LNGWPBM^{p=2,q\to 0}(z_1,z_2,\cdots,z_n) = \left(\mathop{\oplus}_{i=1}^{n}\omega_i z_i^p\right)^{\frac{1}{p}} = \left(\mathop{\oplus}_{i=1}^{n}\omega_i z_i^2\right)^{\frac{1}{2}} = LNSRWA(z_1,z_2,\cdots,z_n).$$

(4) When $p = q = 1$, the *LNGWPBM* operator is simplified as the simplest form of the *LNGWPBM* operator, which is shown as follows:

$$LNGWPBM^{p=1,q=1}(z_1, z_2, \cdots, z_n) = \left(\frac{n-|F_2|}{n} \left(\frac{1}{t} \underset{r=1}{\overset{t}{\oplus}} \left(\frac{1}{\sum_{i=1}^{h_r} \omega_i} \underset{i=1}{\overset{h_r}{\oplus}} \omega_i z_i \otimes \left(\frac{1}{\sum_{j=1, j\neq i}^{h_r} \omega_j} \underset{j=1, j\neq i}{\overset{h_r}{\oplus}} \omega_j z_j \right) \right) \right)^{\frac{1}{2}} \oplus \frac{|F_2|}{n} \left(\frac{1}{\sum_{i=1}^{|F_2|} \omega_i} \underset{i=1}{\overset{|F_2|}{\oplus}} \omega_i z_i \right) \right)$$

It is often used to simplify the calculation in a problem with different weights.

4. A Novel MAGDM Method by the Introduced *LNGWPBM* Operator

In this section, we develop a novel MAGDM method based on the proposed *LNGWPBM* operator to address the kind of problems where the attributes are sorted into two groups: one group contains several clusters where the attributes are relevant in same cluster, but independent in different clusters, and another contains the attributes that are irrelevant to any other attribute. Firstly, we put this kind of problem in a nutshell. Then, we detail the procedures of the proposed method to solve the above problems.

Suppose $X = \{X_1, X_2, \cdots, X_m\}$ is a set of alternatives, and $G = \{G_1, G_2, \cdots, G_n\}$ is a set of attributes, ω_j is the weight of the attribute $G_j (j = 1, 2, \cdots, n)$, where $0 \le \omega_j \le 1$ $(j = 1, 2, \cdots, n)$, $\sum_{j=1}^{n} \omega_j = 1$. Experts D_k $(k = 1, 2, \cdots, d)$ can use the LNNs to judge the alternative X_i for attribute G_j and denote it as $z_{ij}^k = \left(l_{\alpha_{ij}^k}, l_{\beta_{ij}^k}, l_{\gamma_{ij}^k} \right)$ in a linguistic term set $L = (l_0, l_1, \cdots, l_s)$, which meets $\alpha_{ij}^k, \beta_{ij}^k, \gamma_{ij}^k \in [0, s]$, and s is an even number. The experts' weight vector is $\pi = (\pi_1, \pi_2, \cdots, \pi_d)^T$ satisfying with $0 \le \pi_k \le 1$ $(k = 1, 2, \cdots, d)$, $\sum_{k=1}^{d} \pi_k = 1$. Thus, we form the evaluation values given by expert D_k into a decision matrix $Z^k = \left[z_{ij}^k \right]_{m \times n}$ $(k = 1, 2, \cdots, d)$.

We further hypothesize that the set of attributes $G = \{G_1, G_2, \cdots, G_n\}$ is sorted into two groups: F_1 and F_2. In F_1, the attributes are divided into t clusters $P_1, P_2, \cdots, P_t$, which satisfies $P_x \cap P_y = \varnothing, x \neq y$ and $\bigcup_{r=1}^{t} P_r = F_1$. It means that the group F_1 contains several clusters, where the attributes are relevant in same cluster, but independent in different clusters; in F_2, the attributes are irrelevant to any attribute. Afterwards, we decide the priority of alternatives according to the information provided above.

The procedures of the proposed method are designed as follows.

Step 1. Normalize the LNNs.

Since the attributes generally fall into two types, the corresponding attribute values have the two types. In order to achieve normalization, we generally transform the cost attribute values into benefit attribute values. First of all, we assume that $Y^k = \left[y_{ij}^k \right]_{m \times n}$ is the normalized matrix of $Z^k = \left[z_{ij}^k \right]_{m \times n}$, where $y_{ij}^k = \left(l_{\delta_{ij}^k}, l_{\eta_{ij}^k}, l_{\sigma_{ij}^k} \right)$, $1 \le i \le m$, $1 \le j \le n$, and $1 \le k \le d$. Then, the standardizing method is described in the following [7]:

(1) For benefit attribute values:

$$y_{ij}^k = z_{ij}^k = \left(l_{\alpha_{ij}^k}, l_{\beta_{ij}^k}, l_{\gamma_{ij}^k} \right) \tag{20}$$

(2) For cost attribute values:

$$y_{ij}^k = z_{ij}^k = \left(l_{s-\alpha_{ij}^k}, l_{s-\beta_{ij}^k}, l_{s-\gamma_{ij}^k} \right) \tag{21}$$

Step 2. Calculate the collective decision information by the *LNGWPBM* operator fixed with $|F_2| = 0$ and $t = 1$ (i.e., the *LNNWBM* operator discussed in Section 3.2), because there is no need to divide the experts into different clusters. Then, we can get the unfolding form:

$$y_{ij} = \left(l_{\delta_{ij}}, l_{\eta_{ij}}, l_{\sigma_{ij}} \right) = LNNWBM^{p,q}\left(y_{ij}^1, y_{ij}^2, \cdots, y_{ij}^k \right) = \left(\overset{d}{\underset{k=1}{\oplus}} \pi_k \left(y_{ij}^k \right)^p \otimes \left(\frac{1}{1-\pi_k} \underset{\substack{h=1 \\ h \neq k}}{\overset{d}{\oplus}} \pi_h \left(y_{ij}^h \right)^q \right) \right)^{\frac{1}{p+q}}$$

$$= \left(l_{s\left(1-\prod_{k=1}^{d}\left(1-\left(1-\left(1-\left(\delta_{ij}^k/s\right)^p\right)^{\pi_k}\right)\left(1-\left(\prod_{\substack{h=1 \\ h\neq k}}^{d}\left(1-\left(\delta_{ij}^h/s\right)^q\right)^{\pi_h}\right)^{\frac{1}{1-\pi_k}}\right)\right)\right)^{\frac{1}{p+q}}}, l_{s-s\left(1-\prod_{k=1}^{d}\left(1-\left(1-\left(1-\left(1-\eta_{ij}^k/s\right)^p\right)^{\pi_k}\right)\left(1-\left(\prod_{\substack{h=1 \\ h\neq k}}^{d}\left(1-\left(1-\eta_{ij}^h/s\right)^q\right)^{\pi_h}\right)^{\frac{1}{1-\pi_k}}\right)\right)\right)^{\frac{1}{p+q}}}, l_{s-s\left(1-\prod_{k=1}^{d}\left(1-\left(1-\left(1-\left(1-\sigma_{ij}^k/s\right)^p\right)^{\pi_k}\right)\left(1-\left(\prod_{\substack{h=1 \\ h\neq k}}^{d}\left(1-\left(1-\sigma_{ij}^h/s\right)^q\right)^{\pi_h}\right)^{\frac{1}{1-\pi_k}}\right)\right)\right)^{\frac{1}{p+q}}} \right) \tag{22}$$

where $1 \leq i \leq m, 1 \leq j \leq n, p, q \geq 0$, and $p + q > 0$.

Step 3. Compute the comprehensive value of each alternative based on the *LNGWPBM* operator; the unfolding form is detailed in the following:

$$y_i = \left(l_{\delta_i}, l_{\eta_i}, l_{\sigma_i} \right) = LNGWPBM^{p,q}(y_{i1}, y_{i2}, \cdots, y_{in}) = \left(\frac{n-|F_2|}{n} \left(\frac{1}{t} \overset{t}{\underset{r=1}{\oplus}} \left(\frac{1}{\sum_{j=1}^{h_r} \omega_j} \overset{h_r}{\underset{j=1}{\oplus}} \omega_j y_{ij}^p \otimes \left(\frac{1}{\sum_{\substack{g=1 \\ g\neq j}}^{h_r} \omega_g} \underset{\substack{g=1 \\ g\neq j}}{\overset{h_r}{\oplus}} \omega_g y_{ig}^q \right) \right) \right)^{\frac{p}{p+q}} \oplus \frac{|F_2|}{n} \left(\frac{1}{\sum_{j=1}^{|F_2|} \omega_j} \overset{|F_2|}{\underset{j=1}{\oplus}} \omega_j y_{ij}^p \right) \right)^{\frac{1}{p}}$$

$$= \left(l_{s\left(1-\left(\left(\prod_{r=1}^{t}\left(1-\left(1-K_\alpha\right)^{\frac{p}{p+q}}\right)\right)^{\frac{1}{t}}\right)^{\frac{n-|F_2|}{n}}\left(\left(\prod_{j=1}^{|F_2|}\left(1-\left(\alpha_{ij}/s\right)^p\right)^{\omega_j}\right)^{\frac{1}{\sum_{j=1}^{|F_2|}\omega_j}}\right)^{\frac{|F_2|}{n}}\right)^{\frac{1}{p}}}, l_{s-s\left(1-\left(\left(\prod_{r=1}^{t}\left(1-\left(1-K_\beta\right)^{\frac{p}{p+q}}\right)\right)^{\frac{1}{t}}\right)^{\frac{n-|F_2|}{n}}\left(\left(\prod_{j=1}^{|F_2|}\left(1-\left(1-\beta_{ij}/s\right)^p\right)^{\omega_j}\right)^{\frac{1}{\sum_{j=1}^{|F_2|}\omega_j}}\right)^{\frac{|F_2|}{n}}\right)^{\frac{1}{p}}}, l_{s-s\left(1-\left(\left(\prod_{r=1}^{t}\left(1-\left(1-K_\gamma\right)^{\frac{p}{p+q}}\right)\right)^{\frac{1}{t}}\right)^{\frac{n-|F_2|}{n}}\left(\left(\prod_{j=1}^{|F_2|}\left(1-\left(1-\gamma_{ij}/s\right)^p\right)^{\omega_j}\right)^{\frac{1}{\sum_{j=1}^{|F_2|}\omega_j}}\right)^{\frac{|F_2|}{n}}\right)^{\frac{1}{p}}} \right)$$

$$K_\alpha = \left(\prod_{j=1}^{h_r} \left(1 - \left(1 - \left(1 - (\alpha_{ij}/s)^p \right)^{\omega_j} \right) \left(1 - \left(\prod_{\substack{g=1 \\ g\neq j}}^{h_r} \left(1 - (\alpha_{ig}/s)^q \right)^{\omega_g} \right)^{\frac{1}{\sum_{\substack{g=1 \\ g\neq j}}^{h_r} \omega_g}} \right) \right) \right)^{\frac{1}{\sum_{j=1}^{h_r} \omega_j}},$$

$$K_\beta = \left(\prod_{j=1}^{h_r} \left(1 - \left(1 - \left(1 - (1-\beta_{ij}/s)^p \right)^{\omega_j} \right) \left(1 - \left(\prod_{\substack{g=1 \\ g\neq j}}^{h_r} \left(1 - (1-\beta_{ig}/s)^q \right)^{\omega_g} \right)^{\frac{1}{\sum_{\substack{g=1 \\ g\neq j}}^{h_r} \omega_g}} \right) \right) \right)^{\frac{1}{\sum_{j=1}^{h_r} \omega_j}},$$

$$K_\gamma = \left(\prod_{j=1}^{h_r} \left(1 - \left(1 - \left(1 - (1-\gamma_{ij}/s)^p \right)^{\omega_j} \right) \left(1 - \left(\prod_{\substack{g=1 \\ g\neq j}}^{h_r} \left(1 - (1-\gamma_{ig}/s)^q \right)^{\omega_g} \right)^{\frac{1}{\sum_{\substack{g=1 \\ g\neq j}}^{h_r} \omega_g}} \right) \right) \right)^{\frac{1}{\sum_{j=1}^{h_r} \omega_j}}. \tag{23}$$

where $1 \leq i \leq m$, $p, q \geq 0$, and $p + q > 0$; $|F_2|$ denotes the number of attributes in F_2, h_r indicates the number of attributes in cluster P_r, and $\sum_{r=1}^{t} h_r = n - |F_2|$.

Step 4. Calculate the score value $C(y_i)$ and the accuracy value $A(y_i)$ of the synthesized evaluation value y_i in the light of Definitions 2 and 3, where $1 \leq i \leq m$.

Step 5. Compare the obtained score values $C(y_1)$, $C(y_2)$, …, and $C(y_m)$ based on Definition 4. The larger the value of $C(y_i)$, the more front the order of alternative X_i, where $1 \leq i \leq m$. If the value of $C(y_i)$ is the same, then compare the obtained accuracy values $A(y_1)$, $A(y_2)$, …, and $A(y_m)$ to determine the ranking orders of alternatives.

Step 6. Ends.

5. A Practical Application on Selecting Green Suppliers

In this section, we use a realistic example to illustrate the effectiveness and advantage of the proposed MAGDM method by the proposed *LNGWPBM* operator.

Example 2. *The example is about the selection of green suppliers. A car manufacturer wants to choose parts, and there are four alternative green suppliers expressed by* $\{X_1, X_2, X_3, X_4\}$, *which can be seen as evaluation objects. The car manufacturer establishes seven criteria to assess the four green suppliers and the measured evaluation criteria* $G = \{G_1, G_2, G_3, G_4, G_5, G_6, G_7\}$ *are shown as follows: price (* G_1*), green degree (* G_2 *), quality (* G_3*), service level (* G_4*), environment for development (* G_5*), response time (* G_6*), and innovation ability (* G_7*). Their weight vector is* $\omega = (0.1, 0.2, 0.2, 0.1, 0.1, 0.2, 0.1)^T$. *Since the green degree shows the influence degree of the green suppliers on the environment and resources, the criterion* G_2 *has nothing to do with the other criteria. Besides, according to the interrelationship patterns, we are able to divide the other evaluation criteria into three partition structures:* $P_1 = \{G_1, G_3\}$, $P_2 = \{G_4, G_6\}$, *and* $P_3 = \{G_5, G_7\}$. *The car manufacturer assembled a panel of three related principals to conduct field explorations and surveys in depth, so that the optimal green supplier can be selected. We use* D_k $(k = 1, 2, 3)$ *to denote each related principal and their weight vector* $\pi = (0.4, 0.3, 0.3)^T$. *On the basis of their investigation, professional knowledge and experience, every related principal* D_k $(k = 1, 2, 3)$ *needs to assess each green supplier* X_i $(i = 1, 2, 3, 4)$ *under each evaluation criterion* G_j $(j = 1, 2, \cdots, 7)$ *by using scores or linguistic information directly. Suppose the linguistic term set* $L = \{l_0, l_1, l_2, l_3, l_4, l_5, l_6, l_7, l_8\}$, *which expresses, from left to right: extremely low, very low, low, slightly low, medium, slightly high, high, very high, and extremely high, respectively. The corresponding relationships between score and LV are detailed in Table 1 [7]. Therefore, we can unify evaluation information with LNNs to depict the fuzziness and uncertainty of evaluation criteria. Finally, these related principals' evaluation information constructs the three following decision matrices* $Z^k = [z_{ij}^k]_{m \times n}$ $(k = 1, 2, 3)$ *described in Tables 2–4, where* z_{ij}^k *can be depicted as* $\left(l_{\alpha_{ij}^k}, l_{\beta_{ij}^k}, l_{\gamma_{ij}^k}\right)$.

Table 1. The corresponding relationships between score and linguistic values (LV).

Score	0~19	20~29	30~39	40~49	50~59	60~69	70~79	80~89	90~100
Evaluation	extremely low	very low	low	slightly low	medium	slightly high	high	very high	very high
Linguistic value	l_0	l_1	l_2	l_3	l_4	l_5	l_6	l_7	l_8

Table 2. Evaluation matrix Z^1 given by the related principal D_1.

	G_1	G_2	G_3	G_4	G_5	G_6	G_7
X_1	(l_4, l_4, l_3)	(l_3, l_5, l_1)	(l_6, l_3, l_4)	(l_7, l_1, l_2)	(l_4, l_1, l_3)	(l_2, l_1, l_3)	(l_4, l_4, l_3)
X_2	(l_4, l_3, l_2)	(l_5, l_4, l_2)	(l_5, l_3, l_2)	(l_6, l_3, l_1)	(l_5, l_4, l_3)	(l_2, l_7, l_2)	(l_2, l_4, l_1)
X_3	(l_5, l_1, l_2)	(l_3, l_1, l_1)	(l_7, l_1, l_2)	(l_7, l_1, l_2)	(l_4, l_6, l_1)	(l_4, l_3, l_3)	(l_1, l_4, l_1)
X_4	(l_3, l_4, l_3)	(l_6, l_3, l_3)	(l_6, l_4, l_2)	(l_5, l_1, l_1)	(l_5, l_2, l_2)	(l_3, l_1, l_2)	(l_4, l_6, l_3)

Table 3. Evaluation matrix Z^2 given by the related principal D_2.

	G_1	G_2	G_3	G_4	G_5	G_6	G_7
X_1	(l_3, l_5, l_2)	(l_3, l_1, l_4)	(l_5, l_2, l_3)	(l_6, l_2, l_1)	(l_5, l_1, l_3)	(l_3, l_1, l_2)	(l_3, l_2, l_3)
X_2	(l_5, l_1, l_2)	(l_4, l_4, l_3)	(l_6, l_1, l_1)	(l_7, l_2, l_2)	(l_7, l_4, l_4)	(l_3, l_3, l_4)	(l_3, l_1, l_1)
X_3	(l_4, l_3, l_1)	(l_3, l_5, l_1)	(l_7, l_4, l_3)	(l_5, l_3, l_1)	(l_6, l_1, l_2)	(l_4, l_1, l_2)	(l_2, l_3, l_3)
X_4	(l_3, l_4, l_3)	(l_5, l_2, l_2)	(l_7, l_2, l_4)	(l_7, l_3, l_4)	(l_4, l_2, l_1)	(l_2, l_3, l_4)	(l_4, l_4, l_3)

Table 4. Evaluation matrix Z^3 given by the related principal D_3.

	G_1	G_2	G_3	G_4	G_5	G_6	G_7
X_1	(l_4, l_1, l_2)	(l_4, l_2, l_3)	(l_6, l_3, l_2)	(l_6, l_1, l_4)	(l_6, l_3, l_1)	(l_3, l_4, l_5)	(l_4, l_1, l_2)
X_2	(l_5, l_3, l_4)	(l_5, l_4, l_3)	(l_5, l_1, l_2)	(l_5, l_3, l_5)	(l_5, l_3, l_3)	(l_2, l_1, l_2)	(l_3, l_2, l_1)
X_3	(l_3, l_1, l_2)	(l_3, l_1, l_1)	(l_7, l_1, l_3)	(l_5, l_2, l_2)	(l_4, l_1, l_1)	(l_3, l_2, l_3)	(l_2, l_2, l_1)
X_4	(l_4, l_1, l_4)	(l_4, l_2, l_3)	(l_5, l_3, l_5)	(l_6, l_1, l_5)	(l_7, l_2, l_4)	(l_3, l_1, l_2)	(l_3, l_2, l_1)

5.1. The Evaluation Procedures

[Step 1] Normalize the LNNs in the evaluation matrix. Since the price (G_1) and the response time (G_6) belong to the cost attributes, we need to transform the corresponding LNNs of the attributes G_1 and G_6 into the benefit attributes values according to Formula (21) in the evaluation matrices Z^k $(k = 1, 2, 3)$. The normalized matrices are $Y^k = \left[y_{ij}^k\right]_{4\times 7}$ $(k = 1, 2, 3)$, which are displayed in Tables 5–7.

Table 5. The normalized matrix Y^1.

	G_1	G_2	G_3	G_4	G_5	G_6	G_7
X_1	(l_4, l_4, l_5)	(l_3, l_5, l_1)	(l_6, l_3, l_4)	(l_7, l_1, l_2)	(l_4, l_1, l_3)	(l_6, l_7, l_5)	(l_4, l_4, l_3)
X_2	(l_4, l_5, l_6)	(l_5, l_4, l_2)	(l_5, l_3, l_2)	(l_6, l_3, l_1)	(l_5, l_4, l_3)	(l_6, l_1, l_6)	(l_2, l_4, l_1)
X_3	(l_3, l_7, l_6)	(l_3, l_1, l_1)	(l_7, l_1, l_2)	(l_7, l_1, l_2)	(l_4, l_6, l_1)	(l_4, l_5, l_5)	(l_1, l_4, l_1)
X_4	(l_5, l_4, l_5)	(l_6, l_3, l_3)	(l_6, l_4, l_2)	(l_5, l_1, l_1)	(l_5, l_2, l_2)	(l_5, l_7, l_6)	(l_4, l_6, l_3)

Table 6. The normalized matrix Y^2.

	G_1	G_2	G_3	G_4	G_5	G_6	G_7
X_1	(l_5, l_3, l_6)	(l_3, l_1, l_4)	(l_5, l_2, l_3)	(l_6, l_2, l_1)	(l_5, l_1, l_3)	(l_5, l_7, l_6)	(l_3, l_2, l_3)
X_2	(l_3, l_7, l_6)	(l_4, l_4, l_3)	(l_6, l_1, l_1)	(l_7, l_2, l_2)	(l_7, l_4, l_4)	(l_5, l_5, l_4)	(l_3, l_1, l_1)
X_3	(l_4, l_5, l_7)	(l_3, l_5, l_1)	(l_7, l_4, l_3)	(l_5, l_3, l_1)	(l_6, l_1, l_2)	(l_4, l_7, l_6)	(l_2, l_3, l_3)
X_4	(l_5, l_4, l_5)	(l_5, l_2, l_2)	(l_7, l_2, l_4)	(l_7, l_3, l_4)	(l_4, l_2, l_1)	(l_6, l_5, l_4)	(l_4, l_4, l_3)

Table 7. The normalized matrix Y^3.

	G_1	G_2	G_3	G_4	G_5	G_6	G_7
X_1	(l_4, l_7, l_6)	(l_4, l_2, l_3)	(l_6, l_3, l_2)	(l_6, l_1, l_4)	(l_6, l_3, l_1)	(l_5, l_4, l_3)	(l_4, l_1, l_2)
X_2	(l_3, l_5, l_4)	(l_5, l_4, l_3)	(l_5, l_1, l_2)	(l_5, l_3, l_5)	(l_5, l_3, l_3)	(l_6, l_7, l_6)	(l_3, l_2, l_1)
X_3	(l_5, l_7, l_6)	(l_3, l_1, l_1)	(l_7, l_1, l_3)	(l_5, l_2, l_2)	(l_4, l_1, l_1)	(l_5, l_6, l_5)	(l_2, l_2, l_1)
X_4	(l_4, l_7, l_4)	(l_4, l_2, l_3)	(l_5, l_3, l_5)	(l_6, l_1, l_5)	(l_7, l_2, l_4)	(l_5, l_7, l_6)	(l_3, l_2, l_1)

[Step 2] Calculate the collective decision information by the *LNNWBM* operator in Formula (22). In order to reduce the complexity of computing, we fix $p = q = 1$.

As an example, we can calculate the collective decision value y_{11}, and the below is its calculative process:

$$
\begin{aligned}
y_{11} &= LNNWBM^{p,q}(y_{11}^1, y_{11}^2, y_{11}^3) = \\
&= \left(l_{s\left(1-\prod_{k=1}^{3}\left(1-\left(1-\left(1-(\delta_{11}^k/s)\right)^{\pi_k}\right)\left(1-\left(\prod_{h=1,h\neq k}^{3}\left(1-(\delta_{11}^h/s)\right)^{\pi_h}\right)^{\frac{1}{1-\pi_k}}\right)\right)\right)^{\frac{1}{2}}},\ l_{s-s\left(1-\prod_{k=1}^{3}\left(1-\left(1-(\eta_{11}^k/s)^{\pi_k}\right)\left(1-\left(\prod_{h=1,h\neq k}^{3}(\eta_{11}^h/s)^{\pi_h}\right)^{\frac{1}{1-\pi_k}}\right)\right)\right)^{\frac{1}{2}}},\ l_{s-s\left(1-\prod_{k=1}^{3}\left(1-\left(1-(\sigma_{11}^k/s)^{\pi_k}\right)\left(1-\left(\prod_{h=1,h\neq k}^{3}(\sigma_{11}^h/s)^{\pi_h}\right)^{\frac{1}{1-\pi_k}}\right)\right)\right)^{\frac{1}{2}}} \right) \\
&= (l_{4.5685}, l_{4.3818}, l_{5.5541})
\end{aligned}
$$

The other collective decision values y_{ij} are shown in the following:

$$y_{21} = (l_{3.5344}, l_{5.5586}, l_{5.2871});\ y_{31} = (l_{4.1699}, l_{6.4201}, l_{6.2805});\ y_{41} = (l_{4.9669}, l_{4.7841}, l_{4.5176});$$

$$y_{12} = (l_{3.4824}, l_{2.2406}, l_{2.1420});\ y_{22} = (l_{4.9669}, l_{3.7793}, l_{2.3128});$$
$$y_{32} = (l_{3.1424}, l_{1.5104}, l_{0.7488});\ y_{42} = (l_{5.3832}, l_{2.0407}, l_{2.3634});$$

$$y_{13} = (l_{6.0202}, l_{2.3634}, l_{2.7300});\ y_{23} = (l_{5.6366}, l_{1.2948}, l_{1.3362});$$
$$y_{33} = (l_{7.2512}, l_{1.3489}, l_{2.3128});\ y_{43} = (l_{6.4202}, l_{2.7300}, l_{3.2926});$$

$$y_{14} = (l_{6.7107}, l_{0.9833}, l_{1.7986});\ y_{24} = (l_{6.4202}, l_{2.3634}, l_{1.9954});$$
$$y_{34} = (l_{6.1444}, l_{1.5443}, l_{1.3362});\ y_{44} = (l_{6.3515}, l_{1.1760}, l_{2.8122});$$

$$y_{15} = (l_{5.2700}, l_{1.1760}, l_{1.9394});\ y_{25} = (l_{6.0606}, l_{3.4315}, l_{3.0331});$$
$$y_{35} = (l_{4.9485}, l_{2.0139}, l_{0.9833});\ y_{45} = (l_{5.7982}, l_{1.6889}, l_{1.7986});$$

$$y_{16} = (l_{5.6872}, l_{6.1687}, l_{4.5018});\ y_{26} = (l_{6.0202}, l_{4.0018}, l_{5.2871});$$
$$y_{36} = (l_{4.5685}, l_{5.9320}, l_{5.2005});\ y_{46} = (l_{5.6366}, l_{6.4201}, l_{5.2871});$$

$$y_{17} = (l_{3.8863}, l_{1.9465}, l_{2.3634});\ y_{27} = (l_{2.7306}, l_{1.9465}, l_{0.7488});$$
$$y_{37} = (l_{1.6456}, l_{2.7300}, l_{1.1760});\ y_{47} = (l_{3.8863}, l_{3.8560}, l_{1.9394}).$$

[Step 3] According to Formula (23), we can get the comprehensive value y_i of each alternative X_i $(i = 1, 2, 3, 4)$ (suppose $p = q = 1$); the results are shown below:

$$y_1 = (l_{5.5760}, l_{0.0007}, l_{0.0016});\ y_2 = (l_{5.4447}, l_{0.0019}, l_{0.0009});$$
$$y_3 = (l_{5.0616}, l_{0.0036}, l_{0.0004});\ y_4 = (l_{5.8684}, l_{0.0046}, l_{0.0017}).$$

[Step 4] According to Formula (2), we can obtain the score values $C(y_1)$, $C(y_2)$, $C(y_3)$, and $C(y_4)$ of the comprehensive values y_1, y_2, y_3, and y_4, respectively, which are displayed as follows:

$$C(y_1) = 0.8989; C(y_2) = 0.8934; C(y_3) = 0.8774; C(y_4) = 0.9109.$$

[Step 5] Since $C(y_4) > C(y_1) > C(y_2) > C(y_3)$, which is based on Definition 4, we can see that the ranking order of the alternatives X_1, X_2, X_3 and X_4 is: $X_4 \succ X_1 \succ X_2 \succ X_3$, where the most suitable alternative is X_4.

According to the upper computation of the proposed method, we can find that the most suitable green supplier is X_4, the second is X_1, and the worst is X_2 or X_3. So, we recommend that the car manufacturer choose green supplier X_4.

5.2. Exploration of the Parameters' Influence

In the above steps, we fix parameters p and q with 1, but we can easily find that the parameters p and q play an important role in the procedures of the proposed method, based on the *LNGWPBM* operator. When we change the values of parameters p and q, the integration results are usually different, so that the ranking order may be changed accordingly. Table 8 shows the ranking orders of the green suppliers when we assign the parameters p and q to different values. Then, we further explore the influence of parameters p and q on the ranking order.

Table 8. Ranking orders of the green suppliers under different values of the parameters p and q.

Parameters p and q	Score Value y_i $(i = 1, 2, 3, 4)$	Ranking Orders
$p = 1, q = 1$	$C(y_1) = 0.8989; C(y_2) = 0.8934;$ $C(y_3) = 0.8774; C(y_4) = 0.9109.$	$X_4 \succ X_1 \succ X_2 \succ X_3$
$p = 1, q = 0.01$	$C(y_1) = 0.8775; C(y_2) = 0.8796;$ $C(y_3) = 0.8742; C(y_4) = 0.8921.$	$X_4 \succ X_2 \succ X_1 \succ X_3$
$p = 0.01, q = 1$	$C(y_1) = 0.9977; C(y_2) = 0.9969;$ $C(y_3) = 0.9951; C(y_4) = 0.9984.$	$X_4 \succ X_1 \succ X_2 \succ X_3$
$p = 1, q = 2$	$C(y_1) = 0.9019; C(y_2) = 0.8954;$ $C(y_3) = 0.8765; C(y_4) = 0.9136.$	$X_4 \succ X_1 \succ X_2 \succ X_3$
$p = 1, q = 5$	$C(y_1) = 0.9009; C(y_2) = 0.8993;$ $C(y_3) = 0.8840; C(y_4) = 0.9140.$	$X_4 \succ X_1 \succ X_2 \succ X_3$
$p = 1, q = 10$	$C(y_1) = 0.9028; C(y_2) = 0.9073;$ $C(y_3) = 0.9007; C(y_4) = 0.9200.$	$X_4 \succ X_2 \succ X_1 \succ X_3$
$p = 2, q = 1$	$C(y_1) = 0.8835; C(y_2) = 0.8800;$ $C(y_3) = 0.8694; C(y_4) = 0.8938.$	$X_4 \succ X_1 \succ X_2 \succ X_3$
$p = 5, q = 1$	$C(y_1) = 0.8792; C(y_2) = 0.8790;$ $C(y_3) = 0.8828; C(y_4) = 0.8872.$	$X_4 \succ X_3 \succ X_1 \succ X_2$
$p = 10, q = 1$	$C(y_1) = 0.8914; C(y_2) = 0.8923;$ $C(y_3) = 0.9061; C(y_4) = 0.8999.$	$X_3 \succ X_4 \succ X_2 \succ X_1$
$p = 2, q = 2$	$C(y_1) = 0.8846; C(y_2) = 0.8788;$ $C(y_3) = 0.8634; C(y_4) = 0.8938.$	$X_4 \succ X_1 \succ X_2 \succ X_3$
$p = 5, q = 5$	$C(y_1) = 0.8734; C(y_2) = 0.8683;$ $C(y_3) = 0.8583; C(y_4) = 0.8780.$	$X_4 \succ X_1 \succ X_2 \succ X_3$
$p = 9, q = 9$	$C(y_1) = 0.8788; C(y_2) = 0.8781;$ $C(y_3) = 0.8659; C(y_4) = 0.8820.$	$X_4 \succ X_1 \succ X_2 \succ X_3$
$p = 10, q = 10$	$C(y_1) = 0.8806; C(y_2) = 0.8809;$ $C(y_3) = 0.8677; C(y_4) = 0.8835.$	$X_4 \succ X_2 \succ X_1 \succ X_3$

From Table 8, it is easy to find that the bigger the value of parameter p or q is, the more chaotic the ranking order. Let's explain with an example. When $p = 1$ $q = 10$, the ranking order is $X_4 \succ X_2 \succ X_1 \succ X_3$; when $p = 5$ $q = 1$, the ranking order is $X_4 \succ X_3 \succ X_1 \succ X_2$; however, when $p = 10$ $q = 1$, the ranking order is $X_3 \succ X_4 \succ X_2 \succ X_1$. So, it's hard to get the regularity of arrangements under this situation. However, when the parameters p and q are equal and less than 10, the ranking orders are relatively stable, and the best green supplier is X_4, and the worst is X_3.

Generally, the bigger the values of parameters p and q, the more complex the calculation becomes, and the more the interrelations between the attributes are emphasized. DMs usually choose the right parameters p and q according to their preferences. However, there is a special case, i.e., $q = 0$, and the proposed method cannot reflect inner connections between attributes, which is similar to another case such as $|F_2| = n$. Hence, this is not in conformity with this example, and we only allow q to be close to 0 infinitely when discussing. When $p = 1$ $q = 0.01$, the ranking order is $X_4 \succ X_2 \succ X_1 \succ X_3$, and the best green supplier is still X_4. Therefore, in real decision making, we generally recommend that the parameter values be 1 from a practical point of view, which is not only intuitionistic and simple, but is also able to consider the inner connections between attributes.

5.3. Comparison with Other Existing Methods

In this subsection, in order to illustrate the validity and advantage of the proposed MAGDM method related to the *LNGWPBM* operator, we plan to compare it with Fang and Ye's MAGDM method [29], which is related to the *LNNWAA* operator, and Liang et al.'s MAGDM method [7], which is about improving classical TOPSIS with LNNs based on Example 2; their ranking results are displayed in Table 9.

Table 9. A comparison of the ranking results of the alternatives for different multiple attribute group decision-making (MAGDM) methods for Example 2. *LNGWPBM*: linguistic neutrosophic generalized weighted partitioned Bonferroni mean operator.

Methods	Score Values	Ranking Orders
Fang and Ye's MAGDM method [29] by the *LNNWAA* operator	$C_1 = 0.6403, C_2 = 0.6508,$ $C_3 = 0.6748, C_4 = 0.6300.$	$X_3 \succ X_2 \succ X_1 \succ X_4$
Liang et al.'s MAGDM method [7] by improving classical TOPSIS	No	$X_3 \succ X_2 \succ X_1 \succ X_4$
Our proposed MAGDM method by the *LNGWPBM* operator (when $p = q = 1$)	$C_1 = 0.8989, C_2 = 0.8934,$ $C_3 = 0.8774, C_4 = 0.9109.$	$X_4 \succ X_1 \succ X_2 \succ X_3$
Our proposed MAGDM method by the *LNGWPBM* operator (when $p = 1, q = 0$ and $\mid F_2 \mid = 7$)	$C_1 = 0.6403, C_2 = 0.6508,$ $C_3 = 0.6748, C_4 = 0.6300.$	$X_3 \succ X_2 \succ X_1 \succ X_4$

Note: C_i is the abbreviation of the score value $C(y_i)$ of the collective decision information y_i $(i = 1, 2, 3, 4)$, respectively.

(1) Since Fang and Ye's method [29] by the *LNNWAA* operator can only address the MAGDM problems where the attributes are not associated with each other, in order to complete the comparison between it and our proposed MAGDM method by the *LNGWPBM* operator, we suppose that the seven attributes are independent of each other in Example 2, i.e., $\mid F_2 \mid =7$; then, we compare their ranking results. In terms of Table 9, we can find that the ranking order of Fang and Ye's method [29] by the *LNNWAA* operator is consistent with the one of our proposed MAGDM method by the *LNGWPBM* operator (when $p = 1$, $q = 0$ and $\mid F_2 \mid =7$), which is $X_3 \succ X_2 \succ X_1 \succ X_4$. However, the ranking order of Fang and Ye's method [13] by the *LNNWAA* operator has a great difference from the one of our proposed MAGDM method by the *LNGWPBM* operator (when $p = q = 1$); even the best alternatives are not the same. In the following, we explain the reason for the ranking results.

Fang and Ye's method [29] by the *LNNWAA* operator cannot capture inner connections between attributes. In this practical application about the selection of green suppliers, if our assumption is that the attributes have nothing to do with any other attribute, then, i.e., $\mid F_2 \mid = 7$. Besides, we take $p = 1$, $q = 0$ to make the *LNNWBM* operator become the *LNNWAA* operator in integrating the evaluation information given by DMs, which is consistent with step 1 in Fang and Ye's method [29]. Then, the ranking result of Fang and Ye's method [29] by the *LNNWAA* operator should be consistent with the one of our proposed MAGDM method by the *LNGWPBM* operator (when $p = 1, q = 0$ and $\mid F_2 \mid = 7$). By using the two methods to deal with Example 2, respectively, we find that the ranking result of Fang and Ye's method [29] by the *LNNWAA* operator is equal to the one of our proposed MAGDM method by the *LNGWPBM* operator (when $p = 1, q = 0$ and $\mid F_2 \mid = 7$), which is $X_3 \succ X_2 \succ X_1 \succ X_4$. Therefore, this can explain that our proposed MAGDM method is tried and true. However, the ranking result of Fang and Ye's method [29] by the *LNNWAA* operator has a great difference from the one of our proposed MAGDM method by the *LNGWPBM* operator (when $p = q = 1$); even the best alternatives are not the same. In Example 2, we can find that inner connections exist between the attributes and a special condition where the criterion G_2 has nothing to do with other criteria, i.e., $\mid F_2 \mid = 1$; this can be solved by the *LNGWPBM* operator well, but the *LNNWAA* operator does not have the same ability. It is easy to compute in our proposed MAGDM method; we assume $p = q = 1$, and then the ranking result by the *LNGWPBM* operator is $X_4 \succ X_1 \succ X_2 \succ X_3$, which is very different from Fang and Ye's method [29] by the *LNNWAA* operator. In addition, DMs can choose the right value of the parameters p and q according to the actual decision-making situation and their personal preferences, so our proposed MAGDM method is universal and elastic. Meanwhile, Fang and Ye's method [29] can only solve the MAGDM problems with independent attributes, and is not suitable for this kind of question, such as in Example 2. Therefore, our proposed MAGDM method by the *LNGWPBM* operator is more workable and elastic than Fang and Ye's method [29] by the *LNNWAA* operator.

(2) From Table 9, we find that the ranking result of Liang et al.'s MAGDM method [7] by improving classical TOPSIS is the same as the one of our proposed MAGDM method by the *LNGWPBM* operator (when $p = 1$, $q = 0$ and $|F_2|$=7), which is $X_3 \succ X_2 \succ X_1 \succ X_4$; however, it is inconsistent with the one of our proposed MAGDM method by the *LNGWPBM* operator (when $p = q = 1$). Then, we elaborate what leads to the ranking results.

Liang et al.'s MAGDM method [7] uses the *LNNWAA* operator to integrate the evaluation information given by DMs, and then adopts the extended TOPSIS model to rank the alternatives. To compare our proposed MAGDM method with Liang et al.'s MAGDM method [7], we also take $p = 1$, $q = 0$, and $|F_2| = 7$ similar to in the previous analysis; so, the ranking result of Liang et al.'s MAGDM method [7] should be consistent with the one of our proposed MAGDM method. It is important to note that when using Liang et al.'s MAGDM method [7] to solve Example 2, we use the weights of the attributes given in Example 2. By calculating separately, the ranking result of Liang et al.'s MAGDM method [7] by improving classical TOPSIS is the same as the one of our proposed MAGDM method by the *LNGWPBM* operator (when $p = 1$, $q = 0$ and $|F_2| = 7$), which is $X_3 \succ X_2 \succ X_1 \succ X_4$. This proves the validity of our proposed MAGDM method again. However, Liang et al.'s MAGDM method [7] cannot integrate evaluation information, and does not reflect inner connections between attributes, while our proposed MAGDM method by the *LNGWPBM* operator can easily achieve these two points. Furthermore, in the extended TOPSIS model used by Liang et al.'s MAGDM method [7], the correlation coefficient cannot guarantee that the best solution should have the closest distance from the positive ideal solution and the farthest distance from the negative ideal solution, simultaneously [49]. At the same time, Liang et al.'s MAGDM method [7] by improving the classical TOPSIS model neglects DMs' utilities or preferences, whereas our proposed MAGDM method can draw attention to the influence of DMs' utilities or preferences on the final results, and select the appropriate parameters p and q. When $p = q = 1$, the ranking result of our proposed MAGDM method is $X_4 \succ X_1 \succ X_2 \succ X_3$, which is even the opposite result of Liang et al.'s MAGDM method [7]. Therefore, our proposed MAGDM method is more appropriate and effective than Liang et al.'s MAGDM method [7] in solving the problem, such as in Example 2.

(3) To further interpret the effectiveness of our proposed MAGDM method by the *LNGWPBM* operator, we use our proposed MAGDM method to solve the illustrative examples in [29] and [7], and compare our proposed MAGDM method by the *LNGWPBM* operator with Fang and Ye's MAGDM method [29] by the *LNNWAA* operator and Liang et al.'s MAGDM method [7] by improving the classical TOPSIS model. Of course, because the attributes are independent of each other in these two illustrative examples, we still fix with $p = 1$, $q = 0$ and $|F_2| = n$, where n denotes the numbers of the attributes. By applying our proposed MAGDM method to these two illustrative examples, we can find that the ranking result of our proposed MAGDM method is consistent with that of Fang and Ye's MAGDM method [29] and Liang et al.'s MAGDM method [7], respectively, which are detailed in Tables 10 and 11. This further illustrate the effectiveness of our proposed MAGDM method by the *LNGWPBM* operator.

Table 10. A ranking comparison of the alternatives for different MAGDM methods for example described by Fang and Ye in [29].

Methods	Score Values	Ranking Order
Fang and Ye's MAGDM method [29] by the *LNNWAA* operator	$C_1 = 0.7528$, $C_2 = 0.7777$, $C_3 = 0.7613$, $C_4 = 0.8060$.	$X_4 \succ X_2 \succ X_3 \succ X_1$
Our proposed MAGDM method by the *LNGWPBM* operator (when $p = 1, q = 0$ and $\mid F_2 \mid = 3$)	$C_1 = 0.7528$, $C_2 = 0.7777$, $C_3 = 0.7613$, $C_4 = 0.8060$.	$X_4 \succ X_2 \succ X_3 \succ X_1$

Note: C_i is abbreviation of score value $C(y_i)$ of the collective decision information y_i $(i = 1,2,3,4)$, respectively.

Table 11. A ranking comparison of the alternatives for different MAGDM methods for example described by Liang et al. in [7].

Methods	Score Values	Ranking Order
Liang et al.'s MAGDM method [7] by improving classical TOPSIS	No	$X_4 \succ X_2 \succ X_3 \succ X_1$
Our proposed MAGDM method by the *LNGWPBM* operator (when $p = 1, q = 0$ and $\mid F_2 \mid = 5$)	$C_1 = 0.4941$, $C_2 = 0.7901$, $C_3 = 0.6495$, $C_4 = 0.7925$.	$X_4 \succ X_2 \succ X_3 \succ X_1$

Note: C_i is abbreviation of score value $C(y_i)$ of the collective decision information y_i $(i = 1,2,3,4)$, respectively.

In the following, we compare the desirable properties of our proposed MAGDM method with the ones of Fang and Ye's MAGDM method [29] and Liang et al.'s MAGDM method [7] to go even further in the advantages of our proposed MAGDM method. Table 12 describes the final comparison results.

Table 12. A comparison of the properties for different MAGDM methods. DM: decision makers.

Methods / Properties	Fang and Ye's MAGDM Method [29] by the *LNNWAA* Operator	Liang et al.'s MAGDM Method [7] by Improving Classical TOPSIS	Our proposed MAGDM Method by the *LNGWPBM* Operator
Integrate evaluation information	Yes	No	Yes
Reflect DMs' preferences	No	No	Yes
Consider inner relations between attributes in the same cluster	No	No	Yes
Consider the clusters of the input arguments	No	No	Yes

From Table 12, the following conclusions are drawn:

(1) Our proposed MAGDM method and Fang and Ye's MAGDM method [29] can integrate evaluation information, while Liang et al.'s MAGDM method [7] cannot do this and only rank the alternatives by comparing the relative closeness of the positive ideal alternative and the negative ideal alternative.

(2) Although Fang and Ye's MAGDM method [29] can integrate evaluation information, it ignores DMs' preferences, and does not capture the inherent relation pattern between attributes. Besides, Liang et al.'s MAGDM method [7] also cannot reflect DMs' preferences and the inherent relation patterns between attributes.

(3) Our proposed MAGDM method contains regulatory factors that are determined by DMs' preferences, and considers the clusters of the input arguments and the inner relations between the attributes in the same cluster. So, our proposed MAGDM method can effectively address the problems with the heterogeneous relationship among attributes. However, the other two methods do not have these advantages, which show that the application scopes of the two methods are relatively narrow.

In summary, the contrastive analysis further illustrates the validity and merit of our proposed MAGDM method, compared with Fang and Ye's MAGDM method [29] and Liang et al.'s MAGDM method [7].

6. Conclusions

The GPBM operator can model the average of the respective satisfaction of the independent and dependent inputs, and is an extended form of the *PBM* operator, the arithmetic mean operator, and the BM operator. Its merit is to capture the heterogeneous relationship among attributes where all of the attributes are sorted into two groups: F_1 and F_2. In F_1, the elements are divided into several clusters, and the members have inherent connections in the same cluster, but independence in different clusters; in F_2, the elements do not belong to any cluster of the correlated input arguments in F_1. Besides, LNNs can depict the qualitative information more appropriately than the SNNs, and are also an extension of the LIFNs. However, now, based on LNNs, we yet have not seen any studies addressing the MAGDM problems with the heterogeneous relationships among attributes. Therefore, in order to fill this gap, we have expanded the GPBM operator to adapt the linguistic neutrosophic environment, and have proposed the *LNGPBM* operator in this paper. At the same time, its desired properties and special cases have been discussed. Moreover, aiming at the condition where different attributes have different weights in practical applications, we also have introduced its weighted version, namely the *LNGWPBM* operator, including discussing its desired properties and special cases. Then, based on the developed *LNGWPBM* operator, we have developed a novel MAGDM method with LNNs to solve the MAGDM problems with the heterogeneous relationship among attributes. By comparing with Fang and Ye's MAGDM method [29] and Liang et al.'s MAGDM method [7], we find that the developed MAGDM method is more valid and general for solving the MAGDM problems with co-dependent attributes. This is because the developed MAGDM method can intuitively and realistically depict qualitative information and reflect the heterogeneous relationship among attributes. In further research, our developed operators can be improved by considering the unknown weights, objective data, or other forms of information, such as unbalanced linguistic information [50]. Besides, we can apply our developed operators to the other practices such as medical diagnosis, clustering analysis, pattern recognition, discordance analysis, and so on.

Author Contributions: P.L. proposed the *LNGPBM* and *LNGWPBM* operators and investigated their properties, and Y.W. provided the calculation and comparative analysis of examples. We wrote the paper together.

Acknowledgments: This work is supported by the National Natural Science Foundation of China (Nos. 71471172 and 71271124), the Special Funds of Taishan Scholars Project of Shandong Province (No. ts201511045) and the Humanities and Social Sciences Research Project of Ministry of Education of China (17YJA630065 and 17YJC630077).

Conflicts of Interest: The authors declare no conflict of interest.

References

1. Boran, F.E.; Genç, S.; Kurt, M.; Akay, D. A multi-criteria intuitionistic fuzzy group decision making for supplier selection with TOPSIS method. *Expert Syst. Appl.* **2009**, *36*, 11363–11368. [CrossRef]
2. Liu, P.; Wang, P. Some interval-valued intuitionistic fuzzy Schweizer–Sklar power aggregation operators and their application to supplier selection. *Int. J. Syst. Sci.* **2018**. [CrossRef]
3. Sanayei, A.; Mousavi, S.F.; Yazdankhah, A. Group decision making process for supplier selection with VIKOR under fuzzy environment. *Expert Syst. Appl.* **2010**, *37*, 24–30. [CrossRef]
4. Banerjee, D.; Dutta, B.; Guha, D.; Goh, M. Generalized partition Bonferroni mean for multi-attribute group decision making problems based on interval data. *IEEE Trans. Syst. Man Cybern. Syst.* **2017**, 1–18, in press.
5. Herrera, F.; Herrera-Viedma, E. A model of consensus in group decision making under linguistic assessments. *Fuzzy Sets Syst.* **1996**, *78*, 73–87. [CrossRef]
6. Herrera, F.; Herrera-Viedma, E. Linguistic decision analysis: Steps for solving decision problems under linguistic information. *Fuzzy Sets Syst.* **2000**, *115*, 67–82. [CrossRef]

7. Liang, W.; Zhao, G.; Wu, H. Evaluating investment risks of metallic mines using an extended TOPSIS method with linguistic neutrosophic numbers. *Symmetry* **2017**, *9*, 149. [CrossRef]
8. Liao, X.; Li, Y.; Lu, B. A model for selecting an ERP system based on linguistic information processing. *Inf. Syst.* **2007**, *32*, 1005–1017. [CrossRef]
9. Wang, T.; Liu, J.; Li, J.; Niu, C. An integrating OWA–TOPSIS framework in intuitionistic fuzzy settings for multiple attribute decision making. *Comput. Ind. Eng.* **2016**, *98*, 185–194. [CrossRef]
10. Yager, R.R. Multicriteria decision making with ordinal/linguistic intuitionistic fuzzy sets for mobile apps. *IEEE Trans. Fuzzy Syst.* **2016**, *24*, 590–599. [CrossRef]
11. Zhang, Z.; Chu, X. Fuzzy group decision making for multi-format and multi-granularity linguistic judgments in quality function deployment. *Expert Syst. Appl.* **2009**, *36*, 9150–9158. [CrossRef]
12. Zadeh, L.A. The concept of a linguistic variable and its application to approximate reasoning Part I. *Inf. Sci.* **1975**, *8*, 199–249. [CrossRef]
13. Bordogna, G.; Fedrizzi, M.; Pasi, G. A linguistic modelling of consensus in group decision making based on OWA operator. *IEEE Trans. Syst. Man Cybern. Syst.* **1997**, *27*, 126–133. [CrossRef]
14. Pei, Z.; Shi, P. Fuzzy risk analysis based on linguistic aggregation operators. *Int. J. Innov. Comput. Inf. Control* **2011**, *7*, 7105–7118.
15. Xu, Z. A method based on linguistic aggregation operators for group decision making with linguistic preference relations. *Inf. Sci.* **2004**, *166*, 19–30. [CrossRef]
16. Atanassov, K.T. Intuitionistic fuzzy sets. *Fuzzy Sets Syst.* **1986**, *20*, 87–96. [CrossRef]
17. Zadeh, L.A. Fuzzy sets. *Inf. Control* **1965**, *8*, 338–356. [CrossRef]
18. Montajabiha, M. An extended PROMETHE II multi-criteria group decision making technique based on intuitionistic fuzzy logic for sustainable energy planning. *Group Decis. Negot.* **2016**, *25*, 221–244. [CrossRef]
19. Peng, J.; Yeh, W.; Lai, T.; Hsu, C. The incorporation of the Taguchi and the VIKOR methods to optimize multi-response problems in intuitionistic fuzzy environments. *J. Chin. Inst. Eng.* **2015**, *38*, 897–907. [CrossRef]
20. Qin, Q.; Liang, F.; Li, L. A TODIM-based multi-criteria group decision making with triangular intuitionistic fuzzy numbers. *Appl. Soft Comput.* **2017**, *55*, 93–107. [CrossRef]
21. Xu, Z. Intuitionistic fuzzy aggregation operators. *IEEE Trans. Fuzzy Syst.* **2007**, *15*, 1179–1187.
22. Xu, Z.; Yager, R.R. Some geometric aggregation operators based on intuitionistic fuzzy sets. *Int. J. Gen. Syst.* **2006**, *35*, 417–433. [CrossRef]
23. Xu, Z.; Yager, R.R. Intuitionistic fuzzy Bonferroni means. *IEEE Trans. Syst. Man Cybern. Part B Cybern.* **2011**, *41*, 568–578.
24. Chen, Z.C.; Liu, P.H.; Pei, Z. An approach to multiple attribute group decision making based on linguistic intuitionistic fuzzy numbers. *Int. J. Comput. Intell. Syst.* **2015**, *8*, 747–760. [CrossRef]
25. Liu, P.; Liu, J.; Merigó, J.M. Partitioned Heronian means based on linguistic intuitionistic fuzzy numbers for dealing with multi-attribute group decision making. *Appl. Soft Comput.* **2018**, *62*, 395–422. [CrossRef]
26. Liu, P.; Wang, P. Some improved linguistic intuitionistic fuzzy aggregation operators and their applications to multiple-attribute decision making. *Int. J. Inf. Technol. Decis. Mak.* **2017**, *16*, 817–850. [CrossRef]
27. Peng, H.; Wang, J.; Cheng, P. A linguistic intuitionistic multi-criteria decision-making method based on the Frank Heronian mean operator and its application in evaluating coal mine safety. *Int. J. Mach. Learn. Cybern.* **2017**. [CrossRef]
28. Zhang, H.; Peng, H.; Wang, J.; Wang, J.Q. An extended outranking approach for multi-criteria decision-making problems with linguistic intuitionistic fuzzy numbers. *Appl. Soft Comput.* **2017**, *59*, 462–474. [CrossRef]
29. Fang, Z.B.; Ye, J. Multiple attribute group decision-making method based on linguistic neutrosophic numbers. *Symmetry* **2017**, *9*, 111. [CrossRef]
30. Smarandache, F. *A Unifying Field in Logics. Neutrosophy: Neutrosophic Probability, Set and Logic*; American Research Press: Rehoboth, NM, USA, 1999.
31. Wang, H.; Smarandache, F.; Zhang, Y.; Sunderraman, R. Single valued neutrosophic sets. In Proceedings of the 10th International Conference on Fuzzy Theory and Technology, Salt Lake City, UT, USA, 21–26 July 2005.
32. Wang, H.; Smarandache, F.; Zhang, Y.Q.; Sunderraman, R. *Interval Neutrosophic Sets and Logic: Theory and Applications in Computing*; Hexis: Phoenix, AZ, USA, 2005.
33. Ye, J. A multicriteria decision-making method using aggregation operators for simplified neutrosophic sets. *J. Intell. Fuzzy Syst.* **2014**, *26*, 2459–2466.

34. Baušys, R.; Juodagalvienė, B. Garage location selection for residential house by WASPAS-SVNS method. *J. Civ. Eng. Manag.* **2017**, *23*, 421–429. [CrossRef]
35. Baušys, R.; Zavadskas, E.K. Multicriteria decision making approach by vikor under interval neutrosophic set environment. *Econ. Comput. Econ. Cybern. Stud. Res.* **2015**, *49*, 33–48.
36. Baušys, R.; Zavadskas, E.K.; Kaklauskas, A. Application of neutrosophic set to multi-criteria decision making by copras. *Econ. Comput. Econ. Cybern. Stud. Res.* **2015**, *49*, 91–105.
37. Kazimieras Zavadskas, E.; Baušys, R.; Lazauskas, M. Sustainable assessment of alternative sites for the construction of a waste incineration plant by applying WASPAS method with single-valued neutrosophic set. *Sustainability* **2015**, *7*, 15923–15936. [CrossRef]
38. Ma, H.; Zhu, H.; Hu, Z.; Li, K.; Tang, W. Time-aware trustworthiness ranking prediction for cloud services using interval neutrosophic set and ELECTRE. *Knowl.-Based Syst.* **2017**, *138*, 27–45. [CrossRef]
39. Peng, X.; Liu, C. Algorithms for neutrosophic soft decision making based on EDAS, new similarity measure and level soft set. *J. Intell. Fuzzy Syst.* **2017**, *32*, 955–968. [CrossRef]
40. Zavadskas, E.K.; Bausys, R.; Juodagalviene, B.; Garnyte-Sapranavicienec, I. Model for residential house element and material selection by neutrosophic MULTIMOORA method. *Eng. Appl. Artif. Intell.* **2017**, *64*, 315–324. [CrossRef]
41. Zavadskas, E.K.; Bausys, R.; Kaklauskas, A.; Ubarte, I.; Kuzminske, A.; Gudiene, N. Sustainable market valuation of buildings by the single-valued neutrosophic MAMVA method. *Appl. Soft Comput.* **2017**, *57*, 74–87. [CrossRef]
42. Liu, P.; Zhang, X. Research on the supplier selection of a supply chain based on entropy weight and improved ELECTRE-III method. *Int. J. Prod. Res.* **2011**, *49*, 637–646. [CrossRef]
43. Jiang, Y.; Liang, X.; Liang, H. An I-TODIM method for multi-attribute decision making with interval numbers. *Soft Comput.* **2017**, *21*, 5489–5506. [CrossRef]
44. Bonferroni, C. Sulle medie multiple di potenze. *Bollettino dell' Unione Matematica Italiana* **1950**, *5*, 267–270.
45. Maclaurin, C. A second letter to Martin Folkes, Esq.; concerning the roots of equations, with the demonstartion of other rules in algebra. *Philos. Trans. R. Soc. Lond. Ser. A* **1729**, *36*, 59–96.
46. Hara, T.; Uchiyama, M.; Takahasi, S.-E. A refinement of various mean inequalities. *J. Inequal. Appl.* **1998**, *2*, 387–395. [CrossRef]
47. Wang, J.Q.; Yang, Y.; Li, L. Multi-criteria decision-making method based on single-valued neutrosophic linguistic Maclaurin symmetric mean operators. *Neural Comput. Appl.* **2016**. [CrossRef]
48. Dutta, B.; Guha, D. Partitioned Bonferroni mean based on linguistic 2-tuple for dealing with multi-attribute group decision making. *Appl. Soft Comput.* **2015**, *37*, 166–179. [CrossRef]
49. Hadi-Vencheh, A.; Mirjaberi, M. Fuzzy inferior ratio method for multiple attribute decision making problems. *Inf. Sci.* **2014**, *277*, 263–272. [CrossRef]
50. Cabrerizo, F.J.; Al-Hmouz, R.; Morfeq, A.; Balamash, A.S.; Martinez, M.A.; Herrera-Viedma, E. Soft consensus measures in group decision making using unbalanced fuzzy linguistic information. *Soft Comput.* **2017**, *21*, 3037–3050. [CrossRef]

symmetry

MDPI

Article

Neutrosophic Hesitant Fuzzy Subalgebras and Filters in Pseudo-BCI Algebras

Songtao Shao [1], Xiaohong Zhang [2,3,*], Chunxin Bo [1] and Florentin Smarandache [4]

1 College of Information Engineering, Shanghai Maritime University, Shanghai 201306, China; 201740310005@stu.shmtu.edu.cn (S.S); 201640311001@stu.shmtu.edu.cn (C.B)
2 Department of Mathematics, School of Arts and Sciences, Shaanxi University of Science & Technology, Xi'an 710021, China
3 Department of Mathematics, College of Arts and Sciences, Shanghai Maritime University, Shanghai 201306, China
4 Department of Mathematics, University of New Mexico, Gallup, NM 87301, USA; smarand@unm.edu
* Correspondence: zhangxiaohong@sust.edu.cn or zhangxh@shmtu.edu.cn

Received: 29 April 2018; Accepted: 14 May 2018; Published: 18 May 2018

Abstract: The notions of the neutrosophic hesitant fuzzy subalgebra and neutrosophic hesitant fuzzy filter in pseudo-BCI algebras are introduced, and some properties and equivalent conditions are investigated. The relationships between neutrosophic hesitant fuzzy subalgebras (filters) and hesitant fuzzy subalgebras (filters) is discussed. Five kinds of special sets are constructed by a neutrosophic hesitant fuzzy set, and the conditions for the two kinds of sets to be filters are given. Moreover, the conditions for two kinds of special neutrosophic hesitant fuzzy sets to be neutrosophic hesitant fuzzy filters are proved.

Keywords: pseudo-BCI algebra; hesitant fuzzy set; neutrosophic set; filter

1. Introduction

G. Georgescu and A. Iogulescu presented pseudo-BCKalgebras, which was an extension of the famous BCK algebra theory. In [1], the notion of the pseudo-BCI algebra was introduced by W.A. Dudek and Y.B. Jun. They investigated some properties of pseudo-BCI algebras. In [2], Y.B. Jun et al. presented the concept of the pseudo-BCI ideal in pseudo-BCI algebras and researched its characterizations. Then, some classes of pseudo-BCI algebras and pseudo-ideals (filters) were studied; see [3–14].

In 1965, Zadeh introduced fuzzy set theory [15]. In the study of modern fuzzy logic theory, algebraic systems played an important role, such as [16–22]. In 2010, Torra introduced hesitant fuzzy set theory [23]. The hesitant fuzzy set was a useful tool to express peoples' hesitancy in real life, and uncertainty problems were resolved. Furthermore, hesitant fuzzy sets have been applied to decision making and algebraic systems [24–31]. As a generalization of fuzzy set theory, Smarandache introduced neutrosophic set theory [32]; the neutrosophic set theory is a useful tool to deal with indeterminate and inconsistent decision information [33,34]. The neutrosophic set includes the truth membership, indeterminacy membership and falsity membership. Then, Wang et al. [35,36] introduced the interval neutrosophic set and single-valued neutrosophic set. Ye [37] introduced the single-valued neutrosophic hesitant fuzzy set as an extension of the single-valued neutrosophic set and hesitant fuzzy set. Recently, the neutrosophic triplet structures were introduced and researched [38–40].

In this paper, some preliminary concepts in pseudo-BCI algebras, hesitant fuzzy set theory and neutrosophic set theory are briefly reviewed in Section 2. In Section 3, the notion of neutrosophic hesitant fuzzy subalgebras in pseudo-BCI algebras is introduced. The relationships between neutrosophic hesitant fuzzy subalgebras and hesitant fuzzy subalgebras are investigated. Five kinds of special sets are constructed. Some properties are studied. Third, the two kinds of sets to be filters are given. In Section 4, the concept of neutrosophic hesitant fuzzy filters in pseudo-BCI algebras is proposed. The equivalent conditions of the neutrosophic hesitant fuzzy filters in the construction of hesitant fuzzy filters are given. The conditions for two kinds of special neutrosophic hesitant fuzzy sets to be neutrosophic hesitant fuzzy filters are given.

2. Preliminaries

Let us review some fundamental notions of pseudo-BCI algebra and interval-valued hesitant fuzzy filter in this section.

Definition 1. *([13]) A pseudo-BCI algebra is a structure $(X; \rightarrow, \hookrightarrow, 1)$, where "$\rightarrow$" and "$\hookrightarrow$" are binary operations on X and "1" is an element of X, verifying the axioms: $\forall x, y, z \in X$,*

(1) $(y \rightarrow z) \rightarrow ((z \rightarrow x) \hookrightarrow (y \rightarrow x)) = 1, (y \hookrightarrow z) \hookrightarrow ((z \hookrightarrow x) \rightarrow (y \hookrightarrow x)) = 1;$
(2) $x \rightarrow ((x \rightarrow y) \hookrightarrow y) = 1, x \hookrightarrow ((x \hookrightarrow y) \rightarrow y) = 1;$
(3) $x \rightarrow x = 1;$
(4) $x \rightarrow y = y \rightarrow x = 1 \Longrightarrow x = y;$
(5) $x \rightarrow y = 1 \Longleftrightarrow x \hookrightarrow y = 1.$

If $(X; \rightarrow, \hookrightarrow, 1)$ is a pseudo-BCI algebra satisfying $\forall x, y \in X, x \rightarrow y = x \hookrightarrow y$, then $(X; \rightarrow, 1)$ is a BCI algebra. If $(X; \rightarrow, \hookrightarrow, 1)$ is a pseudo-BCI algebra satisfying $\forall x \in X, x \rightarrow 1 = 1$, then $(X; \rightarrow, \hookrightarrow, 1)$ is a pseudo-BCK algebra.

Remark 1. *([1]) In any pseudo-BCI algebra $(X; \rightarrow, \hookrightarrow)$, we can define a binary relation '$\leq$' by putting:*

$$x \leq y \text{ if and only if } x \rightarrow y \text{ (or } x \hookrightarrow y\text{)}.$$

Proposition 1. *([13]) Let $(X; \rightarrow, \hookrightarrow)$ be a pseudo-BCI algebra, then X satisfies the following properties, $\forall x, y, z \in X$,*

(1) $1 \leq x \Rightarrow x = 1;$
(2) $x \leq y \Rightarrow y \rightarrow z \leq x \rightarrow z, y \hookrightarrow z \leq x \hookrightarrow z;$
(3) $x \leq y, y \leq z \Rightarrow x \leq z;$
(4) $x \hookrightarrow (y \rightarrow z) = y \rightarrow (x \hookrightarrow z);$
(5) $x \leq y \rightarrow z \Rightarrow y \leq x \hookrightarrow z;$
(6) $x \rightarrow y \leq (z \rightarrow x) \rightarrow (z \rightarrow y), x \hookrightarrow y \leq (z \hookrightarrow x) \hookrightarrow (z \hookrightarrow y);$
(7) $x \leq y \Rightarrow z \rightarrow x \leq z \rightarrow y, z \hookrightarrow x \leq z \hookrightarrow y;$
(8) $1 \rightarrow x = x, 1 \hookrightarrow x = x;$
(9) $((y \rightarrow x) \hookrightarrow x) \rightarrow x = y \rightarrow x, ((y \hookrightarrow x) \rightarrow x) \hookrightarrow x = y \hookrightarrow x;$
(10) $x \rightarrow y \leq (y \rightarrow x) \hookrightarrow 1, x \hookrightarrow y \leq (y \hookrightarrow x) \rightarrow 1;$
(11) $(x \rightarrow y) \rightarrow 1 = (x \rightarrow 1) \hookrightarrow (y \hookrightarrow 1), (x \hookrightarrow y) \hookrightarrow 1 = (x \hookrightarrow 1) \rightarrow (y \rightarrow 1);$
(12) $x \rightarrow 1 = x \hookrightarrow 1.$

Definition 2. *([13]) A subset F of a pseudo-BCI algebra X is called a filter of X if it satisfies:*
(F1) $1 \in F;$
(F2) $x \in F, x \rightarrow y \in F \Rightarrow y \in F;$
(F3) $x \in F, x \hookrightarrow y \in F \Rightarrow y \in F.$

Definition 3. *([1]) By a pseudo-BCI subalgebra of a pseudo-BCI algebra X, we mean a subset S of X that satisfies $\forall x, y \in S, x \rightarrow y \in S, x \hookrightarrow y \in S$.*

Definition 4. *([12]) A pseudo-BCK algebra is called a type-2 positive implicative if it satisfies:*

$$x \rightarrow (y \hookrightarrow z) = (x \rightarrow y) \hookrightarrow (x \rightarrow z),$$
$$x \hookrightarrow (y \rightarrow z) = (x \hookrightarrow y) \rightarrow (x \hookrightarrow z).$$

If X is a type-2 positive implicative pseudo-BCK algebra, then $x \rightarrow y = x \hookrightarrow y$ *for all* $x \in X$.

Definition 5. *([23]) Let X be a reference set. A hesitant fuzzy set A on X is defined in terms of a function* $h_A(x)$ *that returns a subset of* $[0,1]$ *when it is applied to X, i.e.,*

$$A = \{(x, h_A(x)) | x \in X\}.$$

where $h_A(x)$ *is a set of some different values in* $[0,1]$*, representing the possible membership degrees of the element* $x \in X$. $h_A(x)$ *is called a hesitant fuzzy element, a basis unit of the hesitant fuzzy set.*

Example 1. *Let* $X = \{a, b, c\}$ *be a reference set,* $h_A(a) = [0.1, 0.2]$, $h_A(b) = [0.3, 0.6]$, $h_A(c) = [0.7, 0.8]$. *Then, A is considered as a hesitant fuzzy set,*

$$A = \{(a, [0.1, 0.2]), (b, [0.3, 0.6]), (c, [0.7, 0.8])\}.$$

Definition 6. *([13]) A fuzzy set* $\mu : X \rightarrow [0,1]$ *is called a fuzzy pseudo-filter (fuzzy filter) of a pseudo-BCI algebra X if it satisfies:*

(FF1) $\mu(1) \geq \mu(x), \forall x \in X$;

(FF2) $\mu(y) \geq \mu(x \rightarrow y) \wedge \mu(x), \forall x, y \in X$;

(FF3) $\mu(y) \geq \mu(x \hookrightarrow y) \wedge \mu(x), \forall x, y \in X$.

Definition 7. *([32]) Let X be a non-empty fixed set, a neutrosophic set A on X is defined as:*

$$A = \{(x, T_A(x), I_A(x), F_A(x)) | x \in X\},$$

where $T_A(x), I_A(x), F_A(x) \in [0,1]$*, denoting the truth, indeterminacy and falsity membership degree of the element* $x \in X$*, respecting, and satisfying the limit:* $0 \leq T_A(x) + I_A(x) + F_A(x) \leq 3$.

Definition 8. *([34]) Let X be a fixed set; a neutrosophic hesitant fuzzy set N on X is defined as*

$$N = \{(x, \tilde{t}_N(x), \tilde{i}_N(x), \tilde{f}_N(x)) | x \in X\},$$

in which $\tilde{t}_N(x), \tilde{i}_N(x), \tilde{f}_N(x) \in P([0,1])$*, denoting the possible truth membership hesitant degrees, indeterminacy membership hesitant degrees and falsity membership hesitant degrees of* $x \in X$ *to the set N, respectively, with the conditions* $0 \leq \delta, \gamma, \eta \leq 1$ *and* $0 \leq \delta^+ + \gamma^+ + \eta^+ \leq 3$*, where* $\gamma \in \tilde{t}_N(x)$, $\delta \in \tilde{i}_N(x)$, $\eta \in \tilde{f}_N(x)$, $\gamma^+ \in \bigcup_{\gamma \in \tilde{t}_N(x)} max\{\gamma\}$, $\delta^+ \in \bigcup_{\delta \in \tilde{i}_N(x)} max\{\delta\}$, $\eta^+ \in \bigcup_{\eta \in \tilde{f}_N(x)} max\{\eta\}$ *for* $x \in X$.

Example 2. *Let* $X = \{a, b, c\}$ *be a reference set,* $h_A(a) = ([0.4, 0.5], [0.1, 0.2], [0.2, 0.4])$, $h_A(b) = ([0.5, 0.6], \{0.2, 0.3\}, [0.3, 0.4])$, $h_A(c) = ([0.5, 0.8], [0.2, 0.4], \{0.3, 0.5\})$. *Then, A is considered as a neutrosophic hesitant fuzzy set,*

$$A = \{(a, [0.4, 0.5], [0.1, 0.2], [0.2, 0.4]), (b, [0.5, 0.6], \{0.2, 0.3\}, [0.3, 0.4]), (c, [0.5, 0.8], [0.2, 0.4], \{0.3, 0.5\})\}.$$

Conveniently, $N(x) = \{\tilde{t}_N(x), \tilde{i}_N(x), \tilde{f}_N(x)\}$ is called a neutrosophic hesitant fuzzy element, which is denoted by the simplified symbol $N(x) = \{\tilde{t}_N, \tilde{i}_N, \tilde{f}_N\}$.

Definition 9. *([34]) Let* $N_1 = \{\tilde{t}_{N_1}, \tilde{i}_{N_1}, \tilde{f}_{N_1}\}$ *and* $N_2 = \{\tilde{t}_{N_2}, \tilde{i}_{N_2}, \tilde{f}_{N_2}\}$ *be two neutrosophic hesitant fuzzy sets, then:*

$$N_1 \cup N_2 = \{\tilde{t}_{N_1} \cup \tilde{t}_{N_2}, \tilde{i}_{N_1} \cap \tilde{i}_{N_2}, \tilde{f}_{N_1} \cap f_{N_2}\};$$
$$N_1 \cap N_2 = \{\tilde{t}_{N_1} \cap \tilde{t}_{N_2}, \tilde{i}_{N_1} \cup \tilde{i}_{N_2}, \tilde{f}_{N_1} \cup f_{N_2}\}.$$

3. Neutrosophic Hesitant Fuzzy Subalgebras of Pseudo-BCI Algebras

In the following, let X be a pseudo-BCI algebra, unless otherwise specified.

Definition 10. *A hesitant fuzzy set $A = \{(x, h_A(x)) | x \in X\}$ is called a hesitant fuzzy pseudo-subalgebra (hesitant fuzzy subalgebra) of X if it satisfies:*

(HFS2) $h_A(x) \cap h_A(y) \subseteq h_A(x \to y)$, $\forall x, y \in X$;

(HFS3) $h_A(x) \cap h_A(y) \subseteq h_A(x \hookrightarrow y)$, $\forall x, y \in X$.

Definition 11. *A neutrosophic hesitant fuzzy set $N = \{(x, \tilde{t}_N(x), \tilde{i}_N(x), \tilde{f}_N(x)) | x \in X\}$ is called a neutrosophic hesitant fuzzy pseudo-subalgebra (neutrosophic hesitant fuzzy subalgebra) of X if it satisfies:*

(1) $\tilde{t}_N(x) \cap \tilde{t}_N(y) \subseteq \tilde{t}_N(x \to y)$, $\tilde{t}_N(x) \cap \tilde{t}_N(y) \subseteq \tilde{t}_N(x \hookrightarrow y)$, $\forall x, y \in X$;

(2) $\tilde{i}_N(x) \cup \tilde{i}_N(y) \supseteq \tilde{i}_N(x \to y)$, $\tilde{i}_N(x) \cup \tilde{i}_N(y) \supseteq \tilde{i}_N(x \hookrightarrow y)$, $\forall x, y \in X$;

(3) $\tilde{f}_N(x) \cup \tilde{f}_N(y) \supseteq \tilde{f}_N(x \to y)$, $\tilde{f}_N(x) \cup \tilde{f}_N(y) \supseteq \tilde{f}_N(x \hookrightarrow y)$, $\forall x, y \in X$.

Example 3. *Let $X = \{a, b, c, d, 1\}$ with two binary operations in Tables 1 and 2.*

Table 1. $\to$.

$\to$	a	b	c	d	1
a	1	c	1	1	1
b	d	1	1	1	1
c	d	c	1	1	1
d	c	c	c	1	1
1	a	b	c	d	1

Table 2. $\hookrightarrow$.

$\hookrightarrow$	a	b	c	d	1
a	1	d	1	1	1
b	d	1	1	1	1
c	d	d	1	1	1
d	c	b	c	1	1
1	a	b	c	d	1

Then, $(X; \to, \hookrightarrow, 1)$ is a pseudo-BCI algebra. Let:

$$N = \{(1, [0,1], \{0, \tfrac{1}{16}\}, [0, \tfrac{1}{6}]), (a, [\tfrac{1}{3}, \tfrac{1}{4}], [0, \tfrac{1}{2}], [0, \tfrac{5}{6}]), (b, [0, \tfrac{1}{2}], [0, \tfrac{2}{3}], [0, \tfrac{2}{3}]), \\ (c, [\tfrac{1}{3}, \tfrac{2}{3}], [0, \tfrac{1}{6}], [0, \tfrac{1}{5}]), (d, [\tfrac{1}{3}, 1], [0, \tfrac{1}{3}], [0, \tfrac{1}{5}])\}.$$

then, N is a neutrosophic hesitant fuzzy subalgebra of X.

Considering three hesitant fuzzy sets $H_{\tilde{t}_N}, H_{\tilde{i}_N}, H_{\tilde{f}_N}$ by:

$$H_{\tilde{t}_N} = \{(x, \tilde{t}_N(x)) | x \in X\}, H_{\tilde{i}_N} = \{(x, 1 - \tilde{i}_N(x)) | x \in X\}, H_{\tilde{f}_N} = \{(x, 1 - \tilde{f}_N(x)) | x \in X\}.$$

Therefore, $H_{\tilde{t}_N}$ is called a generated hesitant fuzzy set by function $\tilde{t}_N(x)$; $H_{\tilde{i}_N}$ is called a generated hesitant fuzzy set by function $\tilde{i}_N(x)$; $H_{\tilde{f}_N}$ is called a generated hesitant fuzzy set by function $\tilde{f}_N(x)$.

Theorem 1. *Let $N = \{(x, \tilde{t}_N(x), \tilde{i}_N(y), \tilde{f}_N(x)) | x \in X\}$ be a neutrosophic hesitant fuzzy set on X. Then, N is a neutrosophic hesitant fuzzy subalgebra of X if and only if it satisfies the conditions: $\forall x \in X$, $H_{\tilde{t}_N}$ and $H_{\tilde{i}_N}$, $H_{\tilde{f}_N}$ are hesitant fuzzy subalgebras of X.*

Proof. Necessity: (i) By Definition 10 and Definition 11, we can obtain that $H_{\tilde{t}_N}$ is a hesitant fuzzy subalgebra of X.

(ii) $\forall x, y \in X$, $(1-\tilde{i}_N(x)) \cap (1-\tilde{i}_N(y)) = 1-(\tilde{i}_N(x) \cup \tilde{i}_N(y)) \subseteq 1-\tilde{i}_N(x \to y), (1-\tilde{i}_N(x)) \cap (1-\tilde{i}_N(y)) = 1-(\tilde{i}_N(x) \cup \tilde{i}_N(y)) \subseteq 1-\tilde{i}_N(x \hookrightarrow y)$.

Similarly, $(1-\tilde{f}_N(x)) \cap (1-\tilde{f}_N(y)) \subseteq 1-\tilde{f}_N(x \to y), (1-\tilde{f}_N(x)) \cap (1-\tilde{f}_N(y)) \subseteq 1-\tilde{f}_N(x \to y)$. Therefore, $\forall x \in X$, $H_{\tilde{i}_N} = \{(x, 1-\tilde{i}(x)) | x \in X\}$ and $H_{\tilde{f}_N} = \{(x, 1-\tilde{f}_N(x)) | x \in X\}$ are hesitant fuzzy subalgebras of X.

Sufficiency: (i) Let $x, y \in H_{\tilde{t}_N}$. Obviously, $\tilde{t}_N(x) \cap \tilde{t}_N(y) \subseteq \tilde{t}_N(x \to y), \tilde{t}_N(x) \cap \tilde{t}_N(y) \subseteq \tilde{t}_N(x \hookrightarrow y)$.

(ii) Let $x, y \in H_{\tilde{i}_N}$. By Definition 10, we have $(1-\tilde{i}_N(x)) \cap (1-\tilde{i}_N(y)) \subseteq 1-\tilde{i}_N(x \to y), (1-\tilde{i}_N(x)) \cap (1-\tilde{i}_N(y)) \subseteq 1-\tilde{i}_N(x \to y)$, thus $\tilde{i}_N(x) \cup \tilde{i}_N(y) \supseteq \tilde{i}_N(x \to y), \tilde{i}_N(x) \cup \tilde{i}_N(y) \supseteq \tilde{i}_N(x \hookrightarrow y)$.

Similarly, Let $x, y \in H_{\tilde{f}_N}$; we have $\tilde{f}_N(x) \cup \tilde{f}_N(y) \supseteq \tilde{f}_N(x \to y), \tilde{f}_N(x) \cup \tilde{f}_N(y) \supseteq \tilde{f}(x \hookrightarrow y)$.

That is, N is a neutrosophic hesitant fuzzy subalgebra of X. □

Theorem 2. *Let $N = \{(x, \tilde{t}_N(x), \tilde{i}_N(x), \tilde{f}_N(x)) | x \in X\}$ be a neutrosophic hesitant fuzzy set on X. Then, the following conditions are equivalent:*

(1) $N = \{(x, \tilde{t}_N(x), \tilde{i}_N(x), \tilde{f}_N(x)) | x \in X\}$ is a neutrosophic hesitant fuzzy subalgebra of X;

(2) $\forall \lambda_1, \lambda_2, \lambda_3 \in P([0,1])$, the nonempty hesitant fuzzy level sets $H_{\tilde{t}_N}(\lambda_1), H_{\tilde{i}_N}(\lambda_2), H_{\tilde{f}_N}(\lambda_3)$ are subalgebras of X, where $P([0,1])$ is the power set of $[0,1]$,

$$H_{\tilde{t}_N}(\lambda_1) = \{x \in X | \lambda_1 \subseteq \tilde{t}_N(x)\},$$
$$H_{\tilde{i}_N}(\lambda_2) = \{x \in X | \lambda_2 \subseteq 1-\tilde{i}_N(x)\},$$
$$H_{\tilde{f}_N}(\lambda_3) = \{x \in X | \lambda_3 \subseteq 1-\tilde{f}_N(x)\}.$$

Proof. (1)⇒(2) Suppose $H_{\tilde{t}_N}(\lambda_1), H_{\tilde{i}_N}(\lambda_2), H_{\tilde{f}_N}(\lambda_3)$ are nonempty sets. If $x, y \in H_{\tilde{t}_N}(\lambda_1)$, then $\lambda_1 \subseteq \tilde{t}_N(x), \lambda_1 \subseteq \tilde{t}_N(y)$. Since N is a neutrosophic hesitant fuzzy subalgebra of X, by Definition 11, we can obtain:

$$\lambda_1 \subseteq \tilde{t}_N(x) \cap \tilde{t}_N(y) \subseteq \tilde{t}_N(x \to y), \lambda_1 \subseteq \tilde{t}_N(x) \cap \tilde{t}_N(y) \subseteq \tilde{t}_N(x \hookrightarrow y);$$

then $x \to y, x \hookrightarrow y \in H_{\tilde{t}_N}(\lambda_1)$, $H_{\tilde{t}_N}(\lambda_1)$ is a subalgebra of X.

If $x, y \in H_{\tilde{i}_N}(\lambda_2)$, then $\lambda_2 \subseteq 1-\tilde{i}_N(x), \lambda_2 \subseteq 1-\tilde{i}_N(y)$. Since N is a neutrosophic hesitant fuzzy subalgebra of X, by Definition 11, we can obtain:

$$\lambda_2 \subseteq (1-\tilde{i}_N(x)) \cap (1-\tilde{i}_N(y)) = 1-(\tilde{i}_N(x) \cup \tilde{i}_N(y)) \subseteq 1-\tilde{i}_N(x \to y),$$
$$\lambda_2 \subseteq (1-\tilde{i}_N(x)) \cap (1-\tilde{i}_N(y)) = 1-(\tilde{i}_N(x) \cup \tilde{i}_N(y)) \subseteq 1-\tilde{i}_N(x \hookrightarrow y);$$

Thus, $x \to y, x \hookrightarrow y \in H_{\tilde{i}_N}(\lambda_2)$, $H_{\tilde{i}_N}(\lambda_2)$ is a subalgebra of X.

Similarly, we can obtain then that $H_{\tilde{f}_N}(\lambda_3)$ is a subalgebra of X.

(2)⇒(1) Suppose that $H_{\tilde{t}_N}(\lambda_1), H_{\tilde{i}_N}(\lambda_2), H_{\tilde{f}_N}(\lambda_3)$ are nonempty subalgebras of X, $\forall \lambda_1, \lambda_2, \lambda_3 \in P([0,1])$. Let $x, y \in X$ with $\tilde{t}_N(x) = \mu_1, \tilde{t}_N(y) = \mu_2$. Let $\mu_1 \cap \mu_2 = \lambda_1$. Therefore, we have $x, y \in H_X^{(1)}(\lambda_1)$. Since $H_X^{(1)}(\lambda_1)$ is a subalgebra, we can obtain $x \to y, x \hookrightarrow y \in H_{\tilde{t}_N}(\lambda_1)$. Hence, we can obtain:

$$\tilde{t}_N(x) \cap \tilde{t}_N(y) \subseteq \tilde{t}_N(x \to y), \tilde{t}_N(x) \cap \tilde{t}_N(y) \subseteq \tilde{t}_N(x \hookrightarrow y);$$

Let $x, y \in X$ with $\tilde{i}(x) = \mu_3, \tilde{i}(y) = \mu_4$. Let $(1-\mu_3) \cap (1-\mu_4) = \lambda_2$. Then, we have $x, y \in H_{\tilde{i}_N}(\lambda_2)$. Since $H_{\tilde{i}_N}(\lambda_2)$ is a subalgebra, we can obtain $x \to y, x \hookrightarrow y \in H_{\tilde{f}_N}(\lambda_2)$. Hence, we can obtain $(1-\tilde{i}_N(x)) \cap (1-\tilde{i}_N(y)) = 1-(\tilde{i}_N(x) \cup \tilde{i}_N(y)) = \lambda_2 \subseteq 1-\tilde{i}_N(x \to y)$, $(1-\tilde{i}_N(x)) \cap (1-\tilde{i}_N(y)) = 1-(\tilde{i}_N(x) \cup \tilde{i}_N(y)) = \lambda_2 \subseteq 1-\tilde{i}_N(x \hookrightarrow y)$. Then, we have $\tilde{i}_N(x) \cup \tilde{i}_N(y) \supseteq \tilde{i}_N(x \to y)$, $\tilde{i}_N(x) \cup \tilde{i}_N(y) \supseteq \tilde{i}_N(x \hookrightarrow y)$.

Similarly, let $x, y \in X$ with $\tilde{f}_N(x) = \mu_5, \tilde{f}_N(y) = \mu_6$; we can obtain $\tilde{f}_N(x) \cup \tilde{f}_N(y) \supseteq \tilde{f}_N(x \to y)$, $\tilde{f}_N(x) \cup \tilde{f}_N(y) \supseteq \tilde{f}_N(x \hookrightarrow y)$.

Thus, N is a neutrosophic hesitant fuzzy subalgebra of X. □

Definition 12. *Let* $N = \{(x, \tilde{t}_N(x), \tilde{i}_N(x), \tilde{f}_N(x)) | x \in X\}$ *be a neutrosophic hesitant fuzzy set on* X. $X_N^{(1)}(a^k, b)$, $X_N^{(2)}(a^k, b)$, $X_N^{(3)}(a^k, b)$, $X_N^{(4)}(a^k, b)$, $X_N^{(5)}(a)$ *are called generated subsets by* N: $\forall a, b \in X, k \in \mathbb{N}$,

$$\begin{aligned}
X_N^{(1)}(a^k, b) = \{x \in X | &\tilde{t}_N(a^k * (b * x)) = \tilde{t}_N(1),\\
&\tilde{i}_N(a^k * (b * x)) = \tilde{i}_N(1), \tilde{f}_N(a^k * (b * x)) = \tilde{f}_N(1)\};\\
X_N^{(2)}(a^k, b) = \{x \in X | &\tilde{t}_N(a^k \to (b \hookrightarrow x)) = \tilde{t}_N(1),\\
&\tilde{i}_N(a^k \to (b \hookrightarrow x)) = \tilde{t}_N(1), \tilde{f}_N(a^k \to (b \hookrightarrow x)) = \tilde{f}_N(1)\};\\
X_N^{(3)}(a^k, b) = \{x \in X | &\tilde{t}_N(a^k \hookrightarrow (b \to x)) = \tilde{t}_N(1),\\
&\tilde{i}_N(a^k \hookrightarrow (b \to x)) = \tilde{t}_N(1), \tilde{f}_N(a^k \hookrightarrow (b \to x)) = \tilde{f}_N(1)\};\\
X_N^{(4)}(a^k, b) = \{x \in X | &\tilde{t}_N(a^k \to (b \to x)) = \tilde{t}_N(1),\\
&\tilde{i}_N(a^k \to (b \to x)) = \tilde{i}_N(1), \tilde{f}_N(a^k \to (b \to x)) = \tilde{f}_N(1),\\
&\tilde{t}_N(a^k \hookrightarrow (b \hookrightarrow x)) = \tilde{t}_N(1), \tilde{i}_N(a^k \hookrightarrow (b \hookrightarrow x)) = \tilde{i}_N(1), \tilde{f}_N(a^k \hookrightarrow (b \hookrightarrow x)) = \tilde{f}_N(1)\};\\
X_N^{(5)}(a) = \{x \in X | &\tilde{t}_N(a) \subseteq \tilde{t}_N(x),\\
&\tilde{i}_N(a) \supseteq \tilde{i}_N(x), \tilde{f}_N(a) \supseteq \tilde{f}_N(x)\}.
\end{aligned}$$

where "a" appears "k" times, "$*$" represents any binary operation "$\to$" or "$\hookrightarrow$" on X,

$$\begin{aligned}
a^k * (b * x) &= a * (a * (\cdots (a * (b * x)) \cdots));\\
a^k \to (b \hookrightarrow x)) &= a \to (a \to (\cdots (a \to (b \hookrightarrow x)) \cdots));\\
a^k \hookrightarrow (b \to x)) &= a \hookrightarrow (a \hookrightarrow (\cdots (a \hookrightarrow (b \to x)) \cdots));\\
a^k \to (b \to x)) &= a \to (a \to (\cdots (a \to (b \to x)) \cdots));\\
a^k \hookrightarrow (b \hookrightarrow x) &= a \hookrightarrow (a \hookrightarrow (\cdots (a \hookrightarrow (b \hookrightarrow x)) \cdots)).
\end{aligned}$$

Theorem 3. *Let* $N = \{(x, \tilde{t}_N(x), \tilde{i}_N(x), \tilde{f}_N(x)) | x \in X\}$ *be a neutrosophic hesitant fuzzy set on* X. *If* N *satisfies the following conditions:*

(1) $\tilde{t}_N(x) \subseteq \tilde{t}_N(1), \tilde{t}_N(x \hookrightarrow y) = \tilde{t}_N(x) \cup \tilde{t}_N(y), \forall x, y \in X$;

(2) $\tilde{i}_N(x) \supseteq \tilde{i}_N(1), \tilde{i}_N(x \hookrightarrow y) = \tilde{i}_N(x) \cap \tilde{i}_N(y), \forall x, y \in X$;

(3) $\tilde{f}_N(x) \supseteq \tilde{f}_N(1), \tilde{f}_N(x \hookrightarrow y) = \tilde{f}_N(x) \cap \tilde{f}_N(y), \forall x, y \in X$;

then $X_N^{(1)}(a^k, b) = X, k \in \mathbb{N}$.

Proof. By Proposition 1, we can obtain $\forall x \in X$,

$$\begin{aligned}
&\tilde{t}_N(a^k * (b * x) = \tilde{t}_N(1 \hookrightarrow (a^k * (b * x)))\\
=&\tilde{t}_N(1) \cup \tilde{t}_N(a^k * (b * x))) = \tilde{t}_N(1).\\
&\tilde{i}_N(a^k * (b * x)) = \tilde{i}_N(1 \hookrightarrow (a^k * (b * x)))\\
=&\tilde{i}_N(1) \cap \tilde{t}_N(a^k * (b * x))) = \tilde{i}_N(1).\\
&\tilde{f}_N(a^k * (b * x)) = \tilde{f}_N(1 \hookrightarrow (a^k * (b * x)))\\
=&\tilde{f}_N(1) \cap \tilde{t}_N(a^k * (b * x))) = \tilde{f}_N(1).
\end{aligned}$$

Thus, $x \in X_N^{(1)}(a^k, b)$, $X \subseteq X_N^{(1)}(a^k, b)$.

Conversely, it is easy to check that $X_N^{(1)}(a^k, b) \subseteq X$.

Finally, we can obtain $X = X_N^{(1)}(a^k, b)$. □

Corollary 1. *Let* $N = \{(x, \tilde{t}_N(x), \tilde{i}_N(x), \tilde{f}_N(x)) | x \in X\}$ *be a neutrosophic hesitant fuzzy set on* X. *If* N *satisfies the following conditions:*

(1) $\tilde{t}_N(x) \subseteq \tilde{t}_N(1), \tilde{t}_N(x \to y) = \tilde{t}_N(x) \cup \tilde{t}_N(y), \forall x, y \in X$;

(2) $\tilde{i}_N(x) \supseteq \tilde{i}_N(1), \tilde{i}_N(x \rightarrow y) = \tilde{i}_N(x) \cap \tilde{i}_N(y), \forall x, y \in X$;
(3) $\tilde{f}_N(x) \supseteq \tilde{f}_N(1), \tilde{f}_N(x \rightarrow y) = \tilde{f}_N(x) \cap \tilde{f}_N(y), \forall x, y \in X$;
then $X_N^{(1)}(a^k, b) = X, k \in \mathbb{N}$.

Theorem 4. *Let* $N = \{(x, \tilde{t}_N(x), \tilde{i}_N(x), \tilde{f}_N(x)) | x \in X\}$ *be a neutrosophic hesitant fuzzy set on X. N satisfies the following conditions:*
(1) $\tilde{t}_N(1) \supseteq \tilde{t}_N(x), \tilde{i}_N(1) \subseteq \tilde{i}_N(x), \tilde{f}_N(1) \subseteq \tilde{f}_N(x), \forall x \in X$;
(2) $x \hookrightarrow y = 1 \Rightarrow \tilde{t}_N(x) \subseteq \tilde{t}_N(y), \tilde{i}_N(x) \supseteq \tilde{i}_N(y), \tilde{f}_N(x) \supseteq \tilde{f}_N(y), \forall x, y \in X$.
If $\forall a, b, c \in X, k \in \mathbb{N}, b \leq c$, *then* $X_N^{(2)}(a^k, c) \subseteq X_N^{(2)}(a^k, b)$.

Proof: Let $x \in X_N^{(2)}(a^k, c)$. If $b \leq c$, by Proposition 1, we can obtain:

$$\begin{aligned} \tilde{t}_N(1) &= \tilde{t}_N(a^k \rightarrow (c \hookrightarrow x)) \\ &= \tilde{t}_N(c \hookrightarrow (a^k \rightarrow x)) \\ &\subseteq \tilde{t}_N(b \hookrightarrow (a^k \rightarrow x)) \\ &= \tilde{t}_N(a^k \rightarrow (b \hookrightarrow x)). \end{aligned}$$

Similarly, we can obtain:

$$\tilde{i}_N(a^k \rightarrow (b \hookrightarrow x)) \subseteq \tilde{i}_N(a^k \rightarrow (c \hookrightarrow x)) \subseteq \tilde{i}_N(1);$$
$$\tilde{f}_N(a^k \rightarrow (b \hookrightarrow x)) \subseteq \tilde{f}_N(a^k \rightarrow (c \hookrightarrow x)) \subseteq \tilde{f}_N(1).$$

That is, $x \in X_N^{(2)}(a^k, b), X_N^{(2)}(a^k, c) \subseteq X_N^{(2)}(a^k, b)$.

Corollary 2. *Let* $N = \{(x, \tilde{t}_N(x), \tilde{i}_N(x), \tilde{f}_N(x)) | x \in X\}$ *be a neutrosophic hesitant fuzzy set on X. N satisfies the following conditions:*
(1) $\tilde{t}_N(1) \supseteq \tilde{t}_N(x), \tilde{i}_N(1) \subseteq \tilde{i}_N(x), \tilde{f}_N(1) \subseteq \tilde{f}_N(x), \forall x \in X$;
(2) $x \rightarrow y = 1 \Rightarrow \tilde{t}_N(x) \subseteq \tilde{t}_N(y), \tilde{i}_N(x) \supseteq \tilde{i}_N(y), \tilde{f}_N(x) \supseteq \tilde{f}_N(y), \forall x, y \in X$.
If $\forall a, b, c \in X, k \in \mathbb{N}, b \leq c$, *then* $X_N^{(3)}(a^k, c) \subseteq X_N^{(3)}(a^k, b)$.

The following example shows that $X_N^{(4)}(a^k, b)$ may not be a filter of X.

Example 4. *Let* $X = \{a, b, c, d, 1\}$ *with two binary operations in Tables 3 and 4.*

Table 3. $\rightarrow$.

$\rightarrow$	a	b	c	d	1
a	1	1	1	1	1
b	d	1	1	1	1
c	d	c	1	1	1
d	c	c	c	1	1
1	a	b	c	d	1

Table 4. $\hookrightarrow$.

$\hookrightarrow$	a	b	c	d	1
a	1	d	1	1	1
b	d	1	1	1	1
c	d	d	1	1	1
d	c	b	c	1	1
1	a	b	c	d	1

Then, $(X; \rightarrow, \hookrightarrow, 1)$ *is a pseudo-BCI algebra. Let:*

$$N = \{(1, [0,1], [\tfrac{1}{6}, \tfrac{1}{5}], [0, \tfrac{1}{5}]), (a, [\tfrac{1}{3}, \tfrac{1}{4}], [0, \tfrac{5}{6}], [0, \tfrac{3}{4}]), (b, [0, \tfrac{1}{2}], [\tfrac{1}{6}, \tfrac{3}{4}], [0, \tfrac{1}{3}]), (c, [\tfrac{1}{3}, \tfrac{2}{3}], [0, \tfrac{3}{5}], [0, \tfrac{1}{4}]), (d, [\tfrac{1}{3}, 1], [\tfrac{1}{6}, \tfrac{1}{3}], [0, \tfrac{5}{6}])\}.$$

then $X_N^{(4)}(c,d) = \{a,c,d,1\}$ *is not a filter of X. Since* $c \to b = c \in X_N^{(4)}(c,d)$, *but* $b \notin X_N^{(4)}(c,d)$.

Theorem 5. *Let* $N = \{(x, \tilde{t}_N(x), \tilde{i}_N(x), \tilde{f}_N(x)) | x \in X\}$ *be a neutrosophic hesitant fuzzy set on X. Let X be a type-2 positive implicative pseudo-BCK algebra. If functions* $\tilde{t}_N(x), \tilde{i}_N(x)$ *and* $\tilde{f}_N(x)$ *are injective, then* $X_N^{(4)}(a^k, b)$ *is a filter of X for all* $a, b \in X, k \in \mathbb{N}$.

Proof. (1) If X is a pseudo-BCK algebra, then by Definition 1 and Proposition 1, we can obtain $1 \in X_N^{(4)}(a^k, b)$.

(2) Let $x, y \in X$ with $x, x \to y \in X_N^{(4)}(a^k, b)$. Thus, $a^k \hookrightarrow (b \hookrightarrow x) = 1, a^k \hookrightarrow (b \hookrightarrow (x \to y)) = 1$. Since functions $\tilde{t}_N, \tilde{i}_N$ and $\tilde{f}_N$ are injective, by Definition 5, we have:

$$\begin{aligned}
\tilde{t}_N(1) &= \tilde{t}_N(a^k \hookrightarrow (b \hookrightarrow (x \to y))) \\
&= \tilde{t}_N(a^k \hookrightarrow ((b \hookrightarrow x) \to (b \hookrightarrow y))) \\
&= \tilde{t}_N((a^k \hookrightarrow (b \hookrightarrow x)) \to (a^k \hookrightarrow (b \hookrightarrow y))) \\
&= \tilde{t}_N(1 \to (a^k \hookrightarrow (b \hookrightarrow y))) \\
&= \tilde{t}_N(a^k \hookrightarrow ((b \hookrightarrow y)).
\end{aligned}$$

Similarly, we can obtain $\tilde{i}_N(a^k \hookrightarrow ((b \hookrightarrow y)) = \tilde{i}_N(1), \tilde{f}_N(a^k \hookrightarrow ((b \hookrightarrow y)) = \tilde{f}_N(1)$. Thus, we have $y \in X_N^{(4)}(a^k, b)$.

(3) Similarly, let $x, y \in X$ with $x, x \hookrightarrow y \in X_N^{(4)}(a^k, b)$; we have $y \in X_N^{(4)}(a^k, b)$. □

This means that $X_N^{(4)}(a^k, b)$ is a filter of X for all $a, b \in X, k \in \mathbb{N}$.

Theorem 6. *Let* $N = \{(x, \tilde{t}_N(x), \tilde{i}_N(x), \tilde{f}_N(x)) | x \in X)\}$ *be a neutrosophic hesitant fuzzy set on X. Let X be a type-2 positive implicative pseudo-BCK algebra. If functions* $\tilde{t}_N(x), \tilde{i}_N(x)$ *and* $\tilde{f}_N(x)$ *satisfy the following identifies:* $\forall x, y \in X$,

(1) $\tilde{t}_N(x) \subseteq \tilde{t}_N(1), \tilde{i}_N(x) \supseteq i_N(1), \tilde{f}_N(x) \supseteq f_N(1)$;

(2) $\tilde{t}_N(x \to y) = \tilde{t}_N(x) \cap \tilde{t}_N(y), \tilde{i}_N(x \to y) = \tilde{i}_N(x) \cup \tilde{i}_N(y), \tilde{f}_N(x \to y) = \tilde{f}_N(x) \cup \tilde{f}_N(y)$;

(3) $\tilde{t}_N(x \hookrightarrow y) = \tilde{t}_N(x) \cap \tilde{t}_N(y), \tilde{i}_N(x \hookrightarrow y) = \tilde{i}_N(x) \cup \tilde{i}_N(y), \tilde{f}_N(x \hookrightarrow y) = \tilde{f}_N(x) \cup \tilde{f}_N(y)$;

then $X_N^{(4)}(a^k, b)$ *is a filter of X for all* $a, b \in X, k \in \mathbb{N}$.

Proof. (1) If X is a pseudo-BCK algebra, by Definition 1 and Proposition 1, $1 \in X_N^{(4)}(a^k, b)$.

(2) Let $x, y \in X$ with $x, x \to y \in X_N^{(4)}(a^k, b)$. We have $\tilde{t}_N(a^k \hookrightarrow (b \hookrightarrow x)) = \tilde{t}_N(1), \tilde{t}_N(a^k \hookrightarrow (b \hookrightarrow (x \to y))) = \tilde{t}_N(1)$. By Definition 5, we have:

$$\begin{aligned}
\tilde{t}_N(1) &= \tilde{t}_N(a^k \hookrightarrow (b \hookrightarrow (x \to y))) \\
&= \tilde{t}_N(a^k \hookrightarrow ((b \hookrightarrow x) \to (b \hookrightarrow y))) \\
&= \tilde{t}_N((a^k \hookrightarrow (b \hookrightarrow x)) \to (a^k \hookrightarrow (b \hookrightarrow y))) \\
&= \tilde{t}_N(a^k \hookrightarrow (b \hookrightarrow x)) \cap \tilde{t}(a^k \hookrightarrow (b \hookrightarrow y)) \\
&= \tilde{t}_N(1) \cap \tilde{t}(a^k \hookrightarrow (b \hookrightarrow y)) \\
&= \tilde{t}_N(a^k \hookrightarrow (b \hookrightarrow y)).
\end{aligned}$$

Similarly, we can obtain $\tilde{i}_N(a^k \hookrightarrow (b \hookrightarrow y)) = \tilde{i}_N(1), \tilde{f}_N(a^k \hookrightarrow (b \hookrightarrow y)) = \tilde{f}_N(1)$. Thus, we have $y \in X_N^{(4)}(a^k, b)$.

(3) Similarly, let $x, y \in X$ with $x, x \hookrightarrow y \in X_N^{(4)}(a^k, b)$; we have $y \in X_N^{(4)}(a^k, b)$.

This means that $X_N^{(4)}(a^k, b)$ is a filter of X for all $a, b \in X, k \in \mathbb{N}$. □

Theorem 7. *Let $N = \{(x, \tilde{t}_N(x), \tilde{i}_N(x), \tilde{f}_N(x)) | x \in X)\}$ be a neutrosophic hesitant fuzzy set on X and F be a filter of X. If functions $\tilde{t}_N(x), \tilde{i}_N(x)$ and $\tilde{f}_N(x)$ are injective, then $\bigcup X_N^{(4)}(a^k, b) = F$ for all $a, b \in F, k \in \mathbb{N}$.*

Proof. (1) Let $x \in \bigcup X_N^{(4)}(a^k, b)$. By Definition 12, we have $\tilde{t}_N(a \to (a^{k-1} \to (b \to x))) = \tilde{t}_N(1), \tilde{i}_N(a \to (a^{k-1} \to (b \to x))) = \tilde{i}_N(1), \tilde{f}_N(a \to (a^{k-1} \to (b \to x))) = \tilde{f}_N(1)$. Since F is a filter of X and $\tilde{t}_N, \tilde{i}_N, \tilde{f}_N$ are injective, thus we can obtain $a \to (a^{k-1} \to (b \to x)) = 1$ and $a^{k-1} \to (b \to x) \in F$. Continuing, we can obtain $b \to x \in F$. Since $b \in F$, thus $x \in F, \bigcup X_N^{(4)}(a^k, b) \subseteq F$.

(2) Let $x \in F$. When $a = 1, b = x$, we can obtain $\tilde{t}_N(1^k \to (x \to x)) = \tilde{t}_N(1^k \hookrightarrow (x \hookrightarrow x)) = \tilde{t}_N(1)$. Similarly, we have $\tilde{i}_N(1^k \to (x \to x)) = \tilde{i}_N(1^k \hookrightarrow (x \hookrightarrow x)) = \tilde{i}_N(1), \tilde{f}_N(1^k \to (x \to x)) = \tilde{f}_N(1^k \hookrightarrow (x \hookrightarrow x)) = \tilde{f}_N(1)$. Thus, we have $F \subseteq \bigcup X_N^{(4)}(a^k, b)$.

This means that $\bigcup X_N^{(4)}(a^k, b) = F$ for all $a, b \in F, k \in \mathbb{N}$. □

Theorem 8. *Let $N = \{(x, \tilde{t}_N(x), \tilde{i}_N(x), \tilde{f}_N(x)) | x \in X)\}$ be a neutrosophic hesitant fuzzy set on X.*

(1) If $X_N^{(5)}(a)$ is a filter of X, then N satisfies: $\forall x, y \in X$,

(i) $\tilde{t}_N(a) \subseteq \tilde{t}_N(x \to y) \cap \tilde{t}_N(x), \tilde{i}_N(a) \supseteq \tilde{i}_N(x \to y) \cup \tilde{i}_N(x), \tilde{f}_N(a) \supseteq \tilde{f}_N(x \to y) \cup \tilde{f}_N(x) \Rightarrow \tilde{t}_N(a) \subseteq \tilde{t}_N(y), \tilde{i}_N(a) \supseteq \tilde{i}_N(y), \tilde{f}_N(a) \supseteq \tilde{f}_N(y)$;

(ii) $\tilde{t}_N(a) \subseteq \tilde{t}_N(x \hookrightarrow y) \cap \tilde{t}_N(x), \tilde{i}_N(a) \supseteq \tilde{i}_N(x \hookrightarrow y) \cup \tilde{i}_N(x), \tilde{f}_N(a) \supseteq \tilde{f}_N(x \hookrightarrow y) \cup \tilde{f}_N(x) \Rightarrow \tilde{t}_N(a) \subseteq \tilde{t}_N(y), \tilde{i}_N(a) \supseteq \tilde{i}_N(y), \tilde{f}_N(a) \supseteq \tilde{f}_N(y)$.

(2) If N satisfies Conditions (i), (ii) and $\tilde{t}_N(x) \subseteq \tilde{t}_N(1), \tilde{i}_N(x) \supseteq \tilde{i}_N(1), \tilde{f}_N(x) \supseteq \tilde{f}_N(1)$ for all $x, y \in X$, then $X_N^{(5)}(a)$ is a filter of X.

Proof. (1) (i) Let $x, y \in X$ with $\tilde{t}_N(a) \subseteq \tilde{t}_N(x \to y) \cap \tilde{t}_N(x), \tilde{i}_N(a) \supseteq \tilde{i}_N(x \to y) \cup \tilde{i}_N(x), \tilde{f}_N(a) \supseteq \tilde{f}_N(x \to y) \cup \tilde{f}_N(x)$; we have $x \in X_N^{(5)}(a), x \to y \in X_N^{(5)}(a)$. Since $X_N^{(5)}(a)$ is a filter, thus we can have $y \in X_N^{(5)}(a), \tilde{t}_N(a) \subseteq \tilde{t}_N(y), \tilde{i}_N(a) \supseteq \tilde{i}_N(y), \tilde{f}_N(a) \supseteq \tilde{f}_N(y)$.

(ii) Similarly, we know that (ii) is correct.

(2) Since $\tilde{t}_N(x) \subseteq \tilde{t}_N(1), \tilde{i}_N(x) \supseteq \tilde{i}_N(1), \tilde{f}_N(x) \supseteq \tilde{f}_N(1)$ for all $x \in X$, thus $1 \in X_N^{(5)}(a)$. Let $x, y \in X$ with $x, x \to y \in X_N^{(5)}(a)$; we can obtain $\tilde{t}_N(a) \subseteq \tilde{t}_N(x), \tilde{t}_N(a) \subseteq \tilde{t}_N(x \to y), \tilde{i}_N(a) \supseteq \tilde{i}_N(x), \tilde{i}_N(a) \supseteq \tilde{i}_N(x \to y), \tilde{f}_N(a) \supseteq \tilde{f}_N(x), \tilde{f}_N(a) \supseteq \tilde{f}_N(x \to y)$. By Condition (i), we have $\tilde{t}_N(a) \subseteq \tilde{t}_N(y), \tilde{i}_N(a) \supseteq \tilde{i}_N(y), \tilde{f}_N(a) \supseteq \tilde{f}_N(y)$. Thus, we can obtain $y \in X_N^{(5)}(a)$. Similarly, let $x, y \in X$ with $x, x \hookrightarrow y \in X_N^{(5)}(a)$, by Condition (1)(ii); we can obtain $y \in X_N^{(5)}(a)$.

This means that $X_N^{(5)}(a)$ is a filter of X. □

4. Neutrosophic Hesitant Fuzzy Filters of Pseudo-BCI Algebras

In the following, let X be a pseudo-BCI algebra, unless otherwise specified.

Definition 13. *([22]) A hesitant fuzzy set $A = \{(x, h_A(x)) | x \in X\}$ is called a hesitant fuzzy pseudo-filter (briefly, hesitant fuzzy filter) of X if it satisfies:*

(HFF1) $h_A(x) \subseteq h_A(1), \forall x \in X$;

(HFF2) $h_A(x) \cap h_A(x \to y) \subseteq h_A(y), \forall x, y \in X$;

(HFF3) $h_A(x) \cap h_A(x \hookrightarrow y) \subseteq h_A(y), \forall x, y \in X$.

Definition 14. *A neutrosophic hesitant fuzzy set $N = \{(x, \tilde{t}_N(x), \tilde{i}_N(x), \tilde{f}_N(x)) | x \in X\}$ is called a neutrosophic hesitant fuzzy pseudo-filter (neutrosophic hesitant fuzzy filter) of X if it satisfies:*

(NHFF1) $\tilde{t}_N(x) \subseteq \tilde{t}_N(1), \tilde{i}_N(x) \supseteq \tilde{i}_N(1), \tilde{f}_N(x) \supseteq \tilde{f}_N(1), \forall x \in X$;

(NHFF2) $\tilde{t}_N(x \to y) \cap \tilde{t}_N(x) \subseteq \tilde{t}_N(y), \tilde{i}_N(x \to y) \cup \tilde{i}_N(x) \supseteq \tilde{i}_N(y), \tilde{f}_N(x \to y) \cup \tilde{f}_N(x) \supseteq \tilde{f}_N(y), \forall x, y \in X$;

(NHFF3) $\tilde{t}_N(x \hookrightarrow y) \cap \tilde{t}_N(x) \subseteq \tilde{t}_N(y), \tilde{i}_N(x \hookrightarrow y) \cup \tilde{i}_N(x) \supseteq \tilde{i}_N(y), \tilde{f}_N(x \hookrightarrow y) \cup \tilde{f}_N(x) \supseteq \tilde{f}_N(y)$, $\forall x, y \in X$.

A neutrosophic hesitant fuzzy set $N = \{(x, \tilde{t}_N(x), \tilde{i}_N(x), \tilde{f}_N(x)) | x \in X)\}$ is called a neutrosophic hesitant fuzzy closed filter of X if it is a neutrosophic hesitant fuzzy filter such that:

$$\tilde{t}_N(x \to 1) \supseteq \tilde{t}_N(x), \tilde{i}_N(x \to 1) \subseteq \tilde{i}_N(x), \tilde{f}_N(x \to 1) \subseteq \tilde{f}_N(x).$$

Example 5. *Let* $X = \{a, b, c, d, 1\}$ *with two binary operations in Tables 5 and 6. Then,* $(X; \to, \hookrightarrow, 1)$ *is a pseudo-BCI algebra. Let:*

$$N = \{(1, [0,1], [0, \tfrac{3}{7}], [0, \tfrac{1}{10}]), (a, [0, \tfrac{1}{4}], [0, \tfrac{3}{4}], [0, \tfrac{1}{2}]), (b, [0, \tfrac{1}{4}], [0, \tfrac{3}{4}], [0, \tfrac{1}{2}]), (c, [0, \tfrac{1}{3}], [0, \tfrac{3}{5}], [0, \tfrac{1}{4}]), (d, [0, \tfrac{3}{4}]), [0, \tfrac{3}{6}], [0, \tfrac{1}{5}])\}.$$

Then, N is a neutrosophic hesitant fuzzy filter of X.

Table 5. $\to$.

$\to$	a	b	c	d	1
a	1	1	1	1	1
b	c	1	1	1	1
c	a	b	1	d	1
d	b	b	c	1	1
1	a	b	c	d	1

Table 6. $\hookrightarrow$.

$\hookrightarrow$	a	b	c	d	1
a	1	1	1	1	1
b	d	1	1	1	1
c	b	b	1	d	1
d	a	b	c	1	1
1	a	b	c	d	1

Theorem 9. *Let* $N = \{(x, \tilde{t}_N(x), \tilde{i}_N(y), \tilde{f}_N(x)) | x \in X\}$ *be a neutrosophic hesitant fuzzy set on X. Then, N is a neutrosophic hesitant fuzzy filter of X if and only if it satisfies the following conditions:* $\forall x \in X$, $H_{\tilde{t}_N}$, $H_{\tilde{i}_N}$, $H_{\tilde{f}_N}$ *are hesitant fuzzy filters of X.*

Proof. Necessity: If N is a neutrosophic hesitant fuzzy filter:

(1) Obviously, $H_{\tilde{t}_N}$ is a hesitant fuzzy filter of X.

(2) By Definition 14, we have $(1 - \tilde{i}_N(x)) \subseteq (1 - \tilde{i}_N(1)), 1 - (\tilde{i}_N(x) \cup \tilde{i}_N(x \to y)) = (1 - \tilde{i}_N(x)) \cap (1 - \tilde{i}_N(x \to y)) \subseteq (1 - \tilde{i}_N(y))$; similarly, by Definition 14, we have $(1 - \tilde{i}_N(x)) \cap (1 - \tilde{i}_N(x \hookrightarrow y)) \subseteq (1 - \tilde{i}_N(y))$. Thus, $H_{\tilde{i}_N}$ is hesitant fuzzy filter of X.

(3) Similarly, we have that $H_{\tilde{f}_N}$ is a hesitant fuzzy filter of X.

Sufficiency: If $H_{\tilde{t}_N}$, $H_{\tilde{i}_N}$, $H_{\tilde{f}_N}$ are hesitant fuzzy filters of X. It is easy to prove that $\tilde{t}_N(x)$, $\tilde{i}_N(x)$, $\tilde{f}_N(x)$ satisfies Definition 14. Therefore, $N = \{(x, \tilde{t}_N(x), \tilde{i}_N(x), \tilde{f}_N(x)) | x \in X\}$ is a neutrosophic hesitant fuzzy filter of X. □

Theorem 10. *Let* $N = \{(x, \tilde{t}_N(x), \tilde{i}_N(x), \tilde{f}_N(x)) | x \in X\}$ *be a neutrosophic hesitant fuzzy set on X. Then, the following are equivalent:*

(1) $N = \{(x, \tilde{t}_N(x), \tilde{i}_N(x), \tilde{f}_N(x)) | x \in X\}$ *is a neutrosophic hesitant fuzzy filter of X;*

(2) $\forall \lambda_1, \lambda_2, \lambda_3 \in P([0,1])$, the nonempty hesitant fuzzy level sets $H_{\tilde{t}_N}(\lambda_1), H_{\tilde{i}_N}(\lambda_2), H_{\tilde{f}_N}(\lambda_3)$ are filters of X, where $P([0,1])$ is the power set of $[0,1]$,

$$\begin{aligned} H_{\tilde{t}_N}(\lambda_1) &= \{x \in X \mid \lambda_1 \subseteq \tilde{t}_N(x)\}; \\ H_{\tilde{i}_N}(\lambda_2) &= \{x \in X \mid \lambda_2 \subseteq 1 - \tilde{i}_N(x)\}; \\ H_{\tilde{f}_N}(\lambda_3) &= \{x \in X \mid \lambda_3 \subseteq 1 - \tilde{f}_N(x)\}. \end{aligned}$$

Proof. (1)$\Rightarrow$(2) (i) Suppose $H_{\tilde{t}_N}(\lambda_1) \neq \emptyset$. Let $x \in H_{\tilde{t}_N}(\lambda_1)$, then $\lambda_1 \subseteq \tilde{t}_N(x)$. Since N is a neutrosophic hesitant fuzzy filter of X, by Definition 14, we have $\lambda_1 \subseteq \tilde{t}_N(x) \subseteq \tilde{t}_N(1)$. Thus, $1 \in H_{\tilde{t}_N}(\lambda_1)$.

Let $x, y \in X$ with $x, x \to y \in H_{\tilde{t}_N}(\lambda_1)$, then $\lambda_1 \subseteq \tilde{t}_N(x), \lambda_1 \subseteq \tilde{t}_N(x \to y)$. Since N is a neutrosophic hesitant fuzzy filter of X, by Definition 14, we have $\lambda_1 \subseteq \tilde{t}_N(x \to y) \cap \tilde{t}_N(x) \subseteq \tilde{t}_N(y)$. Thus $y \in H_{\tilde{t}_N}(\lambda_1)$. Similarly, let $x, y \in X$ with $x, x \hookrightarrow y \in H_{\tilde{t}_N}(\lambda_1)$. We have $y \in H_{\tilde{t}_N}(\lambda_1)$.

Thus, we can obtain that $H_{\tilde{t}_N}(\lambda_1)$ is a filter of X.

(ii) Suppose $H_{\tilde{i}_N}(\lambda_2) \neq \emptyset$. Let $x \in H_{\tilde{i}_N}(\lambda_2)$, then $\lambda_2 \subseteq 1 - \tilde{i}_N(x)$. Since N is a neutrosophic hesitant fuzzy filter of X, we have $\tilde{i}_N(1) \subseteq \tilde{i}_N(x)$. Thus, $\lambda_2 \subseteq 1 - \tilde{i}_N(x) \subseteq 1 - \tilde{i}_N(1), 1 \in H_{\tilde{i}_N}(\lambda_2)$.

Let $x, y \in X$ with $x, x \to y \in H_{\tilde{i}_N}(\lambda_2)$, then $\lambda_2 \subseteq 1 - \tilde{i}_N(x), \lambda_2 \subseteq 1 - \tilde{i}_N(x \to y)$. Since N is a neutrosophic hesitant fuzzy filter of X, we have $\tilde{i}_N(x \to y) \cup \tilde{i}_N(x) \supseteq \tilde{i}_N(y)$. Thus, $1 - (\tilde{i}_N(x \to y) \cup \tilde{i}_N(x)) = (1 - \tilde{i}_N(x \to y)) \cap (1 - \tilde{i}_N(x)) \subseteq (1 - \tilde{i}_N(y)), \lambda_2 \subseteq (1 - \tilde{i}_N(y)), y \in H_{\tilde{i}_N}(\lambda_2)$. Similarly, let $x, y \in X$ with $x, x \hookrightarrow y \in H_{\tilde{i}_N}(\lambda_2)$. We have $y \in H_{\tilde{i}_N}(\lambda_2)$.

Thus, we can obtain that $H_{\tilde{i}_N}(\lambda_2)$ is a filter of X.

(iii) We have that $H_{\tilde{f}_N}(\lambda_3)$ is a filter of X. The progress of proof is similar to (ii).

(2)$\Rightarrow$(1) Suppose $H_{\tilde{t}_N}(\lambda_1) \neq \emptyset, H_{\tilde{i}_N}(\lambda_2) \neq \emptyset, H_{\tilde{f}_N}(\lambda_3) \neq \emptyset$ for all $\lambda_1, \lambda_2, \lambda_3 \in P([0,1])$.

(i') Let $x \in X$ with $\tilde{t}_N(x) = \mu_1$. Let $\lambda_1 = \mu_1$. Since $H_{\tilde{t}_N}(\lambda_1)$ is a filter of X, we have $1 \in H_{\tilde{t}_N}(\lambda_1)$. Thus, $\lambda_1 = \mu_1 = \tilde{t}_N(x) \subseteq \tilde{t}_N(1)$.

Let $x, y \in X$ with $\tilde{t}_N(x) = \mu_1, \tilde{t}_N(x \to y) = \mu_4$. Let $\mu_1 \cap \mu_4 = \lambda_1$. Since $H_{\tilde{t}_N}(\lambda_1)$ is a filter of X for all $\lambda_1 \in P([0,1])$, we have $y \in H_{\tilde{t}_N}(\lambda_1)$. Thus, $\lambda_1 = \tilde{t}_N(x) \cap \tilde{t}_N(x \to y) \subseteq \tilde{t}_N(y)$.

Similarly, let $x, y \in X$ with $\tilde{t}_N(x) = \mu_1, \tilde{t}_N(x \hookrightarrow y) = \mu_4'$. We can obtain $\tilde{t}_N(x \hookrightarrow y) \cap \tilde{t}_N(x) \subseteq \tilde{t}_N(y)$.

(ii') Let $x \in X$ with $\tilde{i}_N(x) = \mu_2$. Let $\lambda_2 = 1 - \mu_2$. Since $H_{\tilde{i}_N}(\lambda_2)$ is a filter of X for all $\lambda_2 \in P([0,1])$, we have $1 \in H_{\tilde{i}_N}(\lambda_2), \lambda_2 \subseteq 1 - \tilde{i}_N(1)$. Thus, $1 - \lambda_2 = \mu_2 = \tilde{i}_N(x) \supseteq \tilde{i}_N(1)$.

Let $x, y \in X$ with $\tilde{i}_N(x) = \mu_2, \tilde{i}_N(x \to y) = \mu_5$. Let $(1 - \mu_2) \cap (1 - \mu_5) = \lambda_2$. Since $H_{\tilde{i}_N}(\lambda_2)$ is a filter of X for all $\lambda_2 \in P([0,1])$, we have $y \in H_{\tilde{i}_N}(\lambda_2), \lambda_2 \subseteq 1 - \tilde{i}_N(y)$. Thus, $\lambda_2 = (1 - \mu_2) \cap (1 - \mu_5) = (1 - \tilde{i}_N(x)) \cap (1 - \tilde{i}_N(x \to y)) = 1 - (\tilde{i}_N(x) \cup \tilde{i}_N(x \to y)) \subseteq (1 - \tilde{i}_N(y)), \tilde{i}_N(x) \cup \tilde{i}_N(x \to y) \supseteq \tilde{i}_N(y)$.

Similarly, let $x, y \in X$ with $\tilde{i}_N(x) = \mu_2, \tilde{i}_N(x \hookrightarrow y) = \mu_5'$; we have $\tilde{i}_N(x) \cup \tilde{i}_N(x \hookrightarrow y) \supseteq \tilde{i}_N(y)$.

(iii') Similarly, we can obtain $\tilde{f}_N(x) \supseteq \tilde{f}_N(1), \tilde{f}_N(x) \cup \tilde{f}_N(x \to y) \supseteq \tilde{f}_N(y), \tilde{f}_N(x) \cup \tilde{f}_N(x \hookrightarrow y) \supseteq \tilde{f}_N(y)$.

Therefore, $N = \{(x, \tilde{t}_N(x), \tilde{i}_N(x), \tilde{f}_N(x)) \mid x \in X\}$ is a neutrosophic hesitant fuzzy filter of X. □

Definition 15. *$N = \{(x, \tilde{t}_N(x), \tilde{i}_N(x), \tilde{f}_N(x)) \mid x \in X\}$ is a neutrosophic hesitant fuzzy set on X. Define a neutrosophic hesitant fuzzy set $N^* = \{(x, \tilde{t}_N^*(x), \tilde{i}_N^*(x), \tilde{f}_N^*(x)) \mid x \in X\}$ by:*

$$\tilde{t}_N^* : X \Longrightarrow P([0,1]), x \mapsto \begin{cases} \tilde{t}_N(x), & x \in H_{\tilde{t}_N}(\lambda_1) \\ \varphi_1, & x \notin H_{\tilde{t}_N}(\lambda_1) \end{cases}$$

$$\tilde{i}_N^* : X \Longrightarrow P([0,1]), x \mapsto \begin{cases} \tilde{i}_N(x), & x \in H_{\tilde{i}_N}(\lambda_2) \\ 1 - \varphi_2, & x \notin H_{\tilde{i}_N}(\lambda_2) \end{cases}$$

$$\tilde{f}_N^* : X \Longrightarrow P([0,1]), x \mapsto \begin{cases} \tilde{f}_N(x), & x \in H_{\tilde{f}_N}(\lambda_3) \\ 1 - \varphi_3, & x \notin H_{\tilde{f}_N}(\lambda_3) \end{cases}$$

where $\lambda_1, \lambda_2, \lambda_3, \varphi_1, \varphi_2, \varphi_3 \in P([0,1])$, $\varphi_1 \subseteq \lambda_1, \varphi_2 \subseteq \lambda_2, \varphi_3 \subseteq \lambda_3$. *Then,* N^* *is called a generated neutrosophic hesitant fuzzy set by hesitant fuzzy level sets* $H_{\tilde{t}_N}(\lambda_1), H_{\tilde{i}_N}(\lambda_2)$ *and* $H_{\tilde{f}_N}(\lambda_3)$.

Theorem 11. *Let* $N = \{(x, \tilde{t}_N(x), \tilde{i}_N(x), \tilde{f}_N(x)) | x \in X\}$ *be a neutrosophic hesitant fuzzy filter of X. Then,* N^* *is a neutrosophic hesitant fuzzy filter of X.*

Proof. (1) If N is a neutrosophic hesitant fuzzy filter of X, by Theorem 10, we know that $H_{\tilde{t}_N}(\lambda_1), H_{\tilde{i}_N}(\lambda_2), H_{\tilde{f}_N}(\lambda_3)$ are filters of X. Thus, $1 \in H_{\tilde{t}_N}(\lambda_1), 1 \in H_{\tilde{i}_N}(\lambda_2), 1 \in H_{\tilde{f}_N}(\lambda_3)$, $\tilde{t}_N^*(1) = \tilde{t}_N(1) \supseteq \tilde{t}_N^*(x), \tilde{i}_N^*(1) = \tilde{i}_N(1) \subseteq \tilde{i}_N^*(x), \tilde{f}_N^*(1) = \tilde{f}_N(1) \subseteq \tilde{f}_N^*(x), \forall x \in X$

(2) (i) Let $x, y \in X$ with $x, x \to y \in H_{\tilde{t}_N}(\lambda_1)$. By Theorem 9, Theorem 10 and Definition 15, we know $\lambda_1 \subseteq \tilde{t}_N^*(x \to y) \cap \tilde{t}_N^*(x) = \tilde{t}_N(x \to y) \cap \tilde{t}_N(x) \subseteq \tilde{t}_N(y) = \tilde{t}_N^*(y)$.

Let $x, y \in X$ with $x, x \to y \in H_{\tilde{i}_N}(\lambda_2)$. By Theorem 9, Theorem 10 and Definition 15, we know $\lambda_2 \subseteq (1 - \tilde{i}_N^*(x \to y)) \cap (1 - \tilde{t}_N^*(x)) = (1 - \tilde{i}_N(x \to y)) \cap (1 - \tilde{t}_N(x)) = 1 - (\tilde{i}_N(x \to y) \cup \tilde{i}_N(x)) \subseteq 1 - \tilde{i}_N(y) = 1 - \tilde{t}_N^*(y)$. Thus, we have $1 - \lambda_2 \supseteq \tilde{i}_N^*(x \to y) \cup \tilde{i}_N^*(x) = \tilde{i}_N(x \to y) \cup \tilde{i}_N(x) \supseteq i_N(y) = \tilde{i}_N^*(y)$.

Similarly, let $x, y \in X$ with $x, x \to y \in H_{\tilde{f}_N}(\lambda_3)$; we have $1 - \lambda_3 \supseteq \tilde{f}_N^*(x \to y) \cup \tilde{f}_N^*(x) = \tilde{f}_N(x \to y) \cup \tilde{f}_N(x) \supseteq f_N(y) = \tilde{f}_N^*(y)$.

(ii) Let $x, y \in X$ with $x \notin H_{\tilde{t}_N}(\lambda_1)$ or $x \to y \notin H_{\tilde{t}_N}(\lambda_1)$. By Definition 15, we have $\tilde{t}_N^*(x) = \varphi_1$ or $\tilde{t}_N^*(x \to y) = \varphi_1$. Thus, we can obtain $\tilde{t}_N^*(x) \cap \tilde{t}_N^*(x \to y) = \varphi_1 \subseteq \tilde{t}_N^*(y)$.

Let $x, y \in X$ with $x \notin H_{\tilde{i}_N}(\lambda_2)$ or $x \to y \notin H_{\tilde{i}_N}(\lambda_2)$. By Definition 15, we have $\tilde{i}_N^*(x) = 1 - \varphi_2$ or $\tilde{i}_N^*(x \to y) = 1 - \varphi_2$. Since $1 - \lambda_2 \subseteq 1 - \varphi_2$; thus, we can obtain $\tilde{i}_N^*(x) \cup \tilde{i}_N^*(x \to y) = 1 - \varphi_2 \supseteq \tilde{t}_N^*(y)$.

Similarly, let $x, y \in X$ with $x \notin H_{\tilde{f}_N}(\lambda_3)$ or $x \to y \notin H_{\tilde{f}_N}(\lambda_3)$; we have $\tilde{f}^*(x) \cup \tilde{f}^*(x \to y) = 1 - \varphi_3 \supseteq \tilde{f}^*(y)$.

(3) We can obtain $\tilde{t}^*(x) \cap \tilde{t}^*(x \hookrightarrow y) \subseteq \tilde{t}^*(y), \tilde{i}^*(x) \cup \tilde{i}^*(x \hookrightarrow y) \supseteq \tilde{i}^*(y), \tilde{f}^*(x) \cup \tilde{f}^*(x \hookrightarrow y) \supseteq \tilde{f}^*(y)$. The process of proof is similar to (2).

Thus N^* is a neutrosophic hesitant fuzzy filter of X. □

Theorem 12. *Let* $N = \{(x, \tilde{t}_N(x), \tilde{i}_N(x), \tilde{f}_N(x)) | x \in X\}$ *be a neutrosophic hesitant fuzzy filter of X. Then, N satisfies the following properties,* $\forall x, y, z \in X$,

(1) $x \leq y \Rightarrow \tilde{t}_N(x) \subseteq \tilde{t}_N(y), \tilde{i}_N(x) \supseteq \tilde{i}_N(y), \tilde{f}_N(x) \supseteq \tilde{f}_N(y)$;

(2) $\tilde{t}_N(x \to z) \supseteq \tilde{t}_N(x \to (y \hookrightarrow z)) \cap \tilde{t}_N(y), \tilde{t}_N(x \hookrightarrow z) \supseteq \tilde{t}_N(x \hookrightarrow (y \to z)) \cap \tilde{t}_N(y)$;
$\tilde{i}_N(x \to z) \subseteq \tilde{i}_N(x \to (y \hookrightarrow z)) \cup \tilde{i}_N(y), \tilde{i}_N(x \hookrightarrow z) \subseteq \tilde{i}_N(x \hookrightarrow (y \to z)) \cup \tilde{i}_N(y)$;
$\tilde{f}_N(x \to z) \subseteq \tilde{f}_N(x \to (y \hookrightarrow z)) \cup \tilde{f}_N(y), \tilde{f}_N(x \hookrightarrow z) \subseteq \tilde{f}_N(x \hookrightarrow (y \to z)) \cup \tilde{f}_N(y)$;

(3) $\tilde{t}_N((x \to y) \hookrightarrow y) \supseteq \tilde{t}_N(x), \tilde{t}_N((x \hookrightarrow y) \to y) \supseteq \tilde{t}_N(x)$;
$\tilde{i}_N((x \to y) \hookrightarrow y) \subseteq \tilde{i}_N(x), \tilde{i}_N((x \hookrightarrow y) \to y) \subseteq \tilde{i}_N(x)$;
$\tilde{f}_N((x \to y) \hookrightarrow y) \subseteq \tilde{f}_N(x), \tilde{f}_N((x \hookrightarrow y) \to y) \subseteq \tilde{f}_N(x)$;

(4) $z \leq x \to y \Rightarrow \tilde{t}_N(x) \cap \tilde{t}_N(z) \subseteq \tilde{t}_N(y), \tilde{i}_N(x) \cup \tilde{i}_N(z) \supseteq \tilde{i}_N(y), \tilde{f}_N(x) \cup \tilde{f}_N(z) \supseteq \tilde{f}_N(y)$;
$z \leq x \hookrightarrow y \Rightarrow \tilde{t}_N(x) \cap \tilde{t}_N(z) \subseteq \tilde{t}_N(y), \tilde{i}_N(x) \cup \tilde{i}_N(z) \supseteq \tilde{i}_N(y), \tilde{f}_N(x) \cup \tilde{f}_N(z) \supseteq \tilde{f}_N(y)$.

Proof. (1) Let $x, y \in X$ with $x \leq y$. By Proposition 1, we know $x \to y = 1$ (or $x \hookrightarrow y = 1$). If N is a neutrosophic hesitant fuzzy filter of X, by Definition 14, we have $\tilde{t}_N(x) = \tilde{t}_N(1) \cap \tilde{t}_N(x) = \tilde{t}_N(x \to y) \cap \tilde{t}_N(x) \subseteq \tilde{t}_N(y)$ ($\tilde{t}_N(x) = \tilde{t}_N(1) \cap \tilde{t}_N(x) = \tilde{t}_N(x \hookrightarrow y) \cap \tilde{t}_N(x) \subseteq \tilde{t}_N(y)$). Thus, $\tilde{t}_N(x) \subseteq \tilde{t}_N(y)$.

Similarly, we have $\tilde{i}_N(x) \supseteq \tilde{i}_N(y), \tilde{f}_N(x) \supseteq \tilde{f}_N(y)$.

(2) By Proposition 1, Definition 14, we know, $\forall x, y, z \in X$,

$$\tilde{t}_N(x \to z) \supseteq \tilde{t}_N(y \hookrightarrow (x \to z)) \cap \tilde{t}_N(y) = \tilde{t}_N(x \to (y \hookrightarrow z)) \cap \tilde{t}_N(y),$$
$$\tilde{t}_N(x \hookrightarrow z) \supseteq \tilde{t}_N(y \to (x \hookrightarrow z)) \cap \tilde{t}_N(y) = \tilde{t}_N(x \hookrightarrow (y \to z)) \cap \tilde{t}_N(y).$$

Similarly, we have, $\forall x, y, z \in X$:

$$\tilde{i}_N(x \to z) \subseteq \tilde{i}_N(x \to (y \hookrightarrow z)) \cup \tilde{i}_N(y), \tilde{i}_N(x \hookrightarrow z) \subseteq \tilde{i}_N(x \hookrightarrow (y \to z)) \cup \tilde{i}_N(y);$$
$$\tilde{f}_N(x \to z) \subseteq \tilde{f}_N(x \to (y \hookrightarrow z)) \cup \tilde{f}_N(y), \tilde{f}_N(x \hookrightarrow y) \subseteq \tilde{f}_N(x \hookrightarrow (y \to z)) \cup \tilde{f}_N(y).$$

(3) By Definition 1 and Definition 14, with regard to the function $\tilde{t}_N(x)$, we can obtain, $\forall x, y \in X$,

$$\begin{aligned}\tilde{t}_N((x \to y) \hookrightarrow y) &\supseteq \tilde{t}_N(x \to ((x \to y) \hookrightarrow y)) \cap \tilde{t}_N(x) \\ &= \tilde{t}_N((x \to y) \hookrightarrow (x \to y)) \cap \tilde{t}_N(x) \\ &= \tilde{t}_N(1) \cap \tilde{t}_N(x) \\ &= \tilde{t}_N(x).\end{aligned}$$

Similarly, we have $\tilde{t}_N((x \hookrightarrow y) \to y) \supseteq \tilde{t}_N(x)$.
With regard to the function $\tilde{i}_N(x)$, we can obtain, $\forall x, y \in X$,

$$\begin{aligned}\tilde{i}_N((x \to y) \hookrightarrow y) &\subseteq \tilde{i}_N(x \to ((x \to y) \hookrightarrow y)) \cup \tilde{i}_N(x) \\ &= \tilde{i}_N((x \to y) \hookrightarrow (x \to y)) \cup \tilde{i}_N(x) \\ &= \tilde{i}_N(1) \cup \tilde{i}_N(x) \\ &= \tilde{i}_N(x).\end{aligned}$$

Similarly, we have $\tilde{i}_N((x \hookrightarrow y) \to y) \subseteq \tilde{i}_N(x)$.
Similarly, with regard to the function $\tilde{f}_N(x)$, we can obtain $\tilde{f}_N((x \to y) \hookrightarrow y) \subseteq \tilde{f}_N(x)$, $\tilde{f}_N((x \hookrightarrow y) \to y) \subseteq \tilde{f}_N(x)$.

(4) Let $x, y, z \in X$ with $z \leq x \to y$. By Remark 1 and Definition 14, we can obtain:

$$\begin{aligned}\tilde{t}_N(x) \cap \tilde{t}_N(z) &= \tilde{t}_N(x) \cap (\tilde{t}_N(1) \cap \tilde{t}_N(z)) \\ &= \tilde{t}_N(x) \cap (\tilde{t}_N(z \hookrightarrow (x \to y)) \cap \tilde{t}_N(z)) \\ &\subseteq \tilde{t}_N(x) \cap \tilde{t}_N(x \to y), \\ &\subseteq \tilde{t}_N(y). \\ \tilde{i}_N(x) \cup \tilde{i}_N(z) &= \tilde{i}_N(x) \cup (\tilde{i}_N(1) \cup \tilde{i}_N(z)) \\ &= \tilde{i}_N(x) \cup (\tilde{i}_N(z \to (x \to y)) \cup \tilde{i}_N(z)) \\ &\supseteq \tilde{i}_N(x) \cup \tilde{i}_N(x \to y), \\ &\supseteq \tilde{i}_N(y).\end{aligned}$$

Similarly, we can obtain $\tilde{f}_N(x) \cup \tilde{f}_N(z) \supseteq \tilde{f}_N(y)$.

Let $x, y, z \in X$ with $z \leq x \hookrightarrow y$. We can obtain $\tilde{t}_N(x) \cap \tilde{t}_N(z) \subseteq \tilde{t}_N(y)$, $\tilde{i}_N(x) \cup \tilde{i}_N(z) \supseteq \tilde{i}_N(y)$, $\tilde{f}_N(x) \cup \tilde{f}_N(z) \supseteq \tilde{f}_N(y)$. The process of the proof is similar to the above. □

Theorem 13. *A neutrosophic hesitant fuzzy set $N = \{(x, \tilde{t}_N(x), \tilde{i}_N(x), \tilde{f}_N(x)) | x \in X)\}$ is a neutrosophic hesitant fuzzy filter of X if and only if hesitant fuzzy sets $H_{\tilde{t}_N}, H_{\tilde{i}_N}, H_{\tilde{f}_N}$ satisfy the following conditions, respectively.*

(1) $\tilde{t}_N(x) \subseteq \tilde{t}_N(1)$, $\tilde{t}_N(x \to (y \hookrightarrow z)) \cap \tilde{t}_N(y) \subseteq \tilde{t}_N(x \to z)$, $\tilde{t}_N(x \hookrightarrow (y \to z)) \cap \tilde{t}_N(y) \subseteq \tilde{t}_N(x \hookrightarrow z)$, $\forall x, y, z \in X$;

(2) $\tilde{i}_N(x) \supseteq \tilde{i}_N(1)$, $\tilde{i}_N(x \to (y \hookrightarrow z)) \cup \tilde{i}_N(y) \supseteq \tilde{i}_N(x \to z)$, $\tilde{i}_N(x \hookrightarrow (y \to z)) \cup \tilde{i}_N(y) \supseteq \tilde{i}_N(x \hookrightarrow z)$, $\forall x, y, z \in X$;

(3) $\tilde{f}_N(x) \supseteq \tilde{f}_N(1)$, $\tilde{f}_N(x \to (y \hookrightarrow z)) \cup \tilde{f}_N(y) \supseteq \tilde{f}_N(x \to z)$, $\tilde{f}_N(x \hookrightarrow (y \to z)) \cup \tilde{f}_N(y) \supseteq \tilde{f}_N(x \hookrightarrow z)$, $\forall x, y, z \in X$.

Proof. Necessity: By Theorem 9, Theorem 12 and Definition 14, (1)$\sim$(3) holds.

Sufficiency: (1) $\forall x, y, z \in X$, by Proposition 1, we can obtain $\tilde{t}_N(y) = \tilde{t}_N(1 \to y) \supseteq \tilde{t}_N(1 \to (x \hookrightarrow y)) \cap \tilde{t}_N(x) = \tilde{t}_N(x \hookrightarrow y) \cap \tilde{t}_N(x)$ and $\tilde{t}_N(y) = \tilde{t}_N(1 \hookrightarrow y) \supseteq \tilde{t}_N(1 \hookrightarrow (x \to y)) \cap \tilde{t}_N(x) = \tilde{t}_N(x \to y) \cap \tilde{t}_N(x)$. We have $\tilde{i}_N(x) \supseteq \tilde{i}_N(1)$ for all $x \in X$. Thus, $H_{\tilde{t}_N}$ is a hesitant fuzzy filter of X.

(2) $\forall x, y, z \in X$, by Proposition 1, we can obtain $\tilde{i}_N(y) = \tilde{i}_N(1 \to y) \subseteq \tilde{i}_N(1 \to (x \hookrightarrow y)) \cup \tilde{i}_N(x) = \tilde{i}_N(x \hookrightarrow y) \cup \tilde{i}_N(x)$; thus, we have $(1 - \tilde{i}_N(x \hookrightarrow y)) \cap (1 - \tilde{i}_N(x)) \subseteq (1 - \tilde{i}_N(y))$.

Similarly, we can have $(1-\tilde{i}_N(x \to y)) \cap (1-\tilde{i}_N(x)) \subseteq (1-\tilde{i}_N(y))$.

It is easy to obtain $(1-\tilde{i}_N(x)) \subseteq (1-\tilde{t}_N(1))$ for all $x \in X$. Thus, $H_{\tilde{i}_N}$ is a hesitant fuzzy filter of X.

(3) We have that $H_{\tilde{f}_N}$ is a hesitant fuzzy filter of X. The process of the proof is similar (2).

Therefore, $H_{\tilde{t}_N}, H_{\tilde{i}_N}, H_{\tilde{f}_N}$ are hesitant fuzzy filters of X. By Theorem 9, we know that N is a neutrosophic hesitant fuzzy filter of X. □

Theorem 14. *Let $N = \{(x, \tilde{t}_N(x), \tilde{i}_N(x), \tilde{f}_N(x)) | x \in X)\}$ be a neutrosophic hesitant fuzzy filter of X. Then:*

$$\prod_{k=1}^{n} x_k \to y = 1 \Rightarrow \tilde{t}_N(y) \supseteq \bigcap_{k=1}^{n} \tilde{t}_N(x_k), \tilde{i}_N(y) \subseteq \bigcup_{i=k}^{n} \tilde{i}_N(x_k), \tilde{f}_N(y) \subseteq \bigcup_{k=1}^{n} \tilde{f}_N(x_k).$$

where $n \in \mathbb{N}$,

$$\prod_{k=1}^{n} x_k \to y = x_n \to (x_{n-1} \to (\cdots (x_1 \to y) \cdots)).$$

Proof. If N is a neutrosophic hesitant fuzzy filter of X:

(i) By Theorem 12, we know that $\tilde{t}_N(x_1) \subseteq \tilde{t}_N(y), \tilde{i}_N(x_1) \supseteq \tilde{i}_N(y), \tilde{f}_N(x_1) \supseteq \tilde{f}_N(y)$ for $n = 1$.

(ii) By Theorem 12, we know that $\tilde{t}_N(x_2) \subseteq \tilde{t}_N(x_1 \to y), \tilde{i}_N(x_2) \supseteq \tilde{i}_N(x_1 \to y), \tilde{f}_N(x_2) \supseteq \tilde{f}_N(x_1 \to y)$ for $n = 2$. By Definition 14, we have $\tilde{t}_N(x_1) \cap \tilde{t}_N(x_1 \to y) \subseteq \tilde{t}_N(y), \tilde{i}_N(x_1) \cup \tilde{i}_N(x_1 \to y) \supseteq \tilde{i}_N(y), \tilde{f}_N(x_1) \cup \tilde{f}_N(x_1 \to y) \supseteq \tilde{f}_N(y)$. Thus, $\tilde{t}_N(x_1) \cap \tilde{t}_N(x_2) \subseteq \tilde{t}_N(y), \tilde{i}_N(x_1) \cup \tilde{i}_N(x_2) \supseteq \tilde{i}_N(y), \tilde{f}_N(x_1) \cup \tilde{f}_N(x_2) \supseteq \tilde{f}_N(y)$.

(iii) Suppose that the above formula is true for $n = j$; thus, $\prod_{k=1}^{j} x_k \to y = 1, \forall x_j, \cdots, x_1, y \in X$, and we can obtain $\bigcap_{k=1}^{j} \tilde{t}_N(x_k) \subseteq \tilde{t}_N(y), \bigcup_{k=1}^{j} \tilde{i}_N(x_k) \supseteq \tilde{i}_N(y), \bigcup_{k=1}^{j} \tilde{f}_N(x_k) \supseteq \tilde{f}_N(y)$. Therefore, suppose that $\prod_{k=1}^{j+1} x_k \to y = 1, \forall x_{j+1}, \cdots, x_1, y \in X$, then we have $\bigcap_{k=2}^{j+1} \tilde{t}_N(x_k) \subseteq \tilde{t}_N(x_1 \to y), \bigcup_{k=2}^{j+1} \tilde{i}_N(x_k) \supseteq \tilde{i}_N(x_1 \to y), \bigcup_{k=2}^{j+1} \tilde{f}_N(x_k) \supseteq \tilde{f}_N(x_1 \to y)$. By Definition 14, we can obtain:

$$\tilde{t}_N(y) \supseteq \tilde{t}_N(x_1) \cap \tilde{t}_N(x_1 \to y) \supseteq \tilde{t}_N(x_1) \cap (\bigcap_{k=2}^{j+1} \tilde{t}_N(x_k)) = \bigcap_{k=1}^{j+1} \tilde{t}_N(x_k),$$
$$\tilde{i}_N(y) \subseteq \tilde{i}_N(x_1) \cup \tilde{i}_N(x_1 \to y) \subseteq \tilde{i}_N(x_1) \cup (\bigcup_{k=2}^{j+1} \tilde{i}_N(x_k)) = \bigcup_{k=1}^{j+1} \tilde{i}_N(x_k),$$
$$\tilde{f}_N(y) \subseteq \tilde{f}_N(x_1) \cup \tilde{f}_N(x_1 \to y) \subseteq \tilde{f}_N(x_1) \cup (\bigcup_{k=2}^{j+1} \tilde{f}_N(x_k)) = \bigcup_{k=1}^{j+1} \tilde{f}_N(x_k),$$

which complete the proof. □

Corollary 3. *Let $N = \{(x, \tilde{t}_N(x), \tilde{i}_N(x), \tilde{f}_N(x))) | x \in X)\}$ be a neutrosophic hesitant fuzzy filter of X. Then:*

$$\prod_{k=1}^{n} x_k * y = 1 \Rightarrow \tilde{t}_N(y) \supseteq \bigcap_{k=1}^{n} \tilde{t}_N(x_k), \tilde{i}_N(y) \subseteq \bigcup_{k=1}^{n} \tilde{i}_N(x_k), \tilde{f}_N(y) \subseteq \bigcup_{k=1}^{n} \tilde{f}_N(x_k).$$

where "$*$" represents any binary operation "$\to$" or "$\hookrightarrow$" on X, $n \in \mathbb{N}$,

$$\prod_{k=1}^{n} x_k * y = x_n * (x_{n-1} * (\cdots (x_1 * y) \cdots)).$$

Theorem 15. *Let $N = \{(x, \tilde{t}_N(x), \tilde{i}_N(x), \tilde{f}_N(x)) | x \in X)\}$ be a neutrosophic hesitant fuzzy filter of X and X be a pseudo-BCK algebra, then N is a neutrosophic hesitant fuzzy subalgebra of X.*

Proof. If $N = \{(x, \tilde{t}_N(x), \tilde{i}_N(x), \tilde{f}_N(x)) | x \in X)\}$ is a neutrosophic hesitant fuzzy filter of X, then we can obtain $\forall x, y \in X$,

$$\begin{aligned}
\tilde{t}_N(x \to y) &\supseteq \tilde{t}_N(y \hookrightarrow (x \to y)) \cap \tilde{t}_N(y) \\
&= \tilde{t}_N(x \to (y \hookrightarrow y)) \cap \tilde{t}_N(y) \\
&= \tilde{t}_N(x \to 1) \cap \tilde{t}_N(y) \\
&\supseteq \tilde{t}_N(x) \cap \tilde{t}_N(y). \\
\tilde{i}_N(x \to y) &\subseteq \tilde{i}_N(y \hookrightarrow (x \to y)) \cup \tilde{i}_N(y) \\
&= \tilde{i}_N(x \to (y \hookrightarrow y)) \cup \tilde{i}_N(y) \\
&= \tilde{i}_N(x \to 1) \cup \tilde{i}_N(y) \\
&\subseteq \tilde{i}_N(x) \cup \tilde{i}_N(y).
\end{aligned}$$

$$\begin{aligned}
\tilde{f}_N(x \to y) &\subseteq \tilde{f}_N(y \hookrightarrow (x \to y)) \cup \tilde{f}(y) \\
&= \tilde{f}_N(x \to (y \hookrightarrow y)) \cup \tilde{f}_N(y) \\
&= \tilde{f}_N(x \to 1) \cup \tilde{f}_N(y) \\
&\subseteq \tilde{f}_N(x) \cup \tilde{f}_N(y).
\end{aligned}$$

Similarly, we can obtain $\tilde{t}_N(x \hookrightarrow y) \supseteq \tilde{t}_N(x) \cap \tilde{t}_N(y), \tilde{i}_N(x \hookrightarrow y) \subseteq \tilde{i}_N(x) \cup \tilde{i}_N(y), \tilde{f}_N(x \hookrightarrow y) \subseteq \tilde{f}_N(x) \cup \tilde{f}_N(y)$. Thus, N is a neutrosophic hesitant fuzzy subalgebra of X. □

Theorem 16. *Let $N = \{(x, \tilde{t}_N(x), \tilde{i}_N(x), \tilde{f}_N(x)) | x \in X)\}$ be a neutrosophic hesitant fuzzy closed filter of X. Then, N is a neutrosophic hesitant fuzzy subalgebra of X.*

Proof. The process of proof is similar to Theorem 15. □

If $N = \{(x, \tilde{t}_N(x), \tilde{i}_N(x), \tilde{f}_N(x)) | x \in X)\}$ is a neutrosophic hesitant fuzzy subalgebra of X, then N may not be a neutrosophic hesitant fuzzy filter of X.

Example 6. *Let $X = \{a, b, c, d, 1\}$ with two binary operations in Tables 1 and 2. Then, $(X; \to, \hookrightarrow, 1)$ is a pseudo-BCI algebra. N is a neutrosophic hesitant fuzzy subalgebra of X. However, N is not a neutrosophic hesitant fuzzy filter of X. Since $\tilde{t}(b \to a) \cap \tilde{t}(b) = [\frac{1}{3}, \frac{1}{2}]$, $\tilde{t}(a) = [\frac{1}{3}, \frac{1}{4}]$, we cannot obtain $\tilde{t}(b \to a) \cap \tilde{t}(b) \subseteq \tilde{t}(a)$.*

Definition 16. *$N = \{(x, \tilde{t}_N(x), \tilde{i}_N(x), \tilde{f}_N(x)) | x \in X)\}$ is a neutrosophic hesitant fuzzy set on X. Define a neutrosophic hesitant fuzzy set $N^{(a,b)} = \{(x, \tilde{t}_N^{(a,b)}(x), \tilde{i}_N^{(a,b)}(x), \tilde{f}_N^{(a,b)}(x)) | x \in X\}$ by $\forall a, b \in X$,*

$$\tilde{t}_N^{(a,b)} : X \Longrightarrow P([0,1]), x \mapsto \begin{cases} \psi_1, & a \to (b \to x) = 1, a \hookrightarrow (b \hookrightarrow x) = 1; \\ \psi_2, & otherwise: \end{cases}$$

$$\tilde{i}_N^{(a,b)} : X \Longrightarrow P([0,1]), x \mapsto \begin{cases} \psi_3, & a \to (b \to x) = 1, a \hookrightarrow (b \hookrightarrow x) = 1; \\ \psi_4, & otherwise: \end{cases}$$

$$\tilde{f}_N^{(a,b)} : X \Longrightarrow P([0,1]), x \mapsto \begin{cases} \psi_5, & a \to (b \to x) = 1, a \hookrightarrow (b \hookrightarrow x) = 1; \\ \psi_6, & otherwise: \end{cases}$$

where $\psi_1, \psi_2, \psi_3, \psi_4, \psi_5, \psi_6 \in P([0,1])$, $\psi_1 \supseteq \psi_2, \psi_3 \subseteq \psi_4, \psi_5 \subseteq \psi_6$. Then, $N^{(a,b)}$ is called a generated neutrosophic hesitant fuzzy set.

A generated neutrosophic hesitant fuzzy set $N^{(a,b)}$ may not be a neutrosophic hesitant fuzzy filter of X.

Example 7. *Let* $X = \{a, b, c, d, 1\}$ *with two binary operations in Tables 1 and 2. Then,* $(X; \rightarrow, \hookrightarrow, 1)$ *is a pseudo-BCI algebra. N is a neutrosophic hesitant fuzzy set of X. However,* $N^{(a,b)}$ *is not a neutrosophic hesitant fuzzy filter of X. Since* $\tilde{t}^{(1,a)}(a \rightarrow b) \cap \tilde{t}^{(1,a)}(a) = [0,1]$, $\tilde{t}^{(1,a)}(b) = [\frac{1}{3}, \frac{2}{3}]$, *we cannot obtain* $\tilde{t}^{(1,a)}(a \rightarrow b) \cap \tilde{t}^{(1,a)}(a) \subseteq \tilde{t}^{(1,a)}(b)$.

Theorem 17. *Let X be a pseudo-BCK algebra. If X is a type-2 positive implicative pseudo-BCK algebra, then* $N^{(a,b)}$ *is a neutrosophic hesitant fuzzy filter of X for all* $a, b \in X$.

Proof. If X is a pseudo-BCK algebra, (1) by Definition 1 and Proposition 1, we can obtain $a \rightarrow (b \rightarrow 1) = 1$ $(a \hookrightarrow (b \hookrightarrow 1) = 1)$. $\tilde{t}_N^{(a,b)}(1) = \psi_1 \supseteq \tilde{t}_N^{(a,b)}(x), \tilde{i}_N^{(a,b)}(1) = \psi_3 \subseteq \tilde{i}_N^{(a,b)}(x), \tilde{f}_N^{(a,b)}(1) = \psi_5 \subseteq \tilde{f}_N^{(a,b)}(x)$ for all $x \in X$.

(2) (i) Let $x, y \in X$ with $a \rightarrow (b \rightarrow x) \neq 1$ or $a \hookrightarrow (b \hookrightarrow x) \neq 1$ or $a \rightarrow (b \rightarrow (x \rightarrow y)) \neq 1$ or $a \hookrightarrow (b \hookrightarrow (x \rightarrow y)) \neq 1$. Thus, we can obtain:

$$\begin{gathered}
\tilde{t}_N^{(a,b)}(x) \cap \tilde{t}_N^{(a,b)}(x \rightarrow y) = \psi_2 \subseteq \tilde{t}_N^{(a,b)}(y), \tilde{t}_N^{(a,b)}(x) \cap \tilde{t}_N^{(a,b)}(x \hookrightarrow y) = \psi_2 \subseteq \tilde{t}_N^{(a,b)}(y); \\
\tilde{i}_N^{(a,b)}(x) \cup \tilde{i}_N^{(a,b)}(x \rightarrow y) = \psi_4 \supseteq \tilde{i}_N^{(a,b)}(y), \tilde{i}_N^{(a,b)}(x) \cup \tilde{i}_N^{(a,b)}(x \hookrightarrow y) = \psi_4 \supseteq \tilde{i}_N^{(a,b)}(y); \\
\tilde{f}_N^{(a,b)}(x) \cup \tilde{f}_N^{(a,b)}(x \rightarrow y) = \psi_6 \supseteq \tilde{f}_N^{(a,b)}(y), \tilde{f}_N^{(a,b)}(x) \cup \tilde{f}_N^{(a,b)}(x \hookrightarrow y) = \psi_6 \supseteq \tilde{f}_N^{(a,b)}(y).
\end{gathered}$$

(ii) Let $x, y \in X$ with $a \rightarrow (b \rightarrow x) = 1$, $a \hookrightarrow (b \hookrightarrow x) = 1$ and $a \rightarrow (b \rightarrow (x \rightarrow y)) = 1$, $a \hookrightarrow (b \hookrightarrow (x \hookrightarrow y)) = 1$. Then, by Proposition 1 and Definition 4, we can obtain:

$$\begin{aligned}
&\tilde{t}_N^{(a,b)}(a \hookrightarrow (b \hookrightarrow y)) \\
=&\tilde{t}_N^{(a,b)}(1 \rightarrow (a \hookrightarrow (b \hookrightarrow y))) \\
=&\tilde{t}_N^{(a,b)}((a \hookrightarrow (b \hookrightarrow x)) \rightarrow (a \hookrightarrow (b \hookrightarrow y))) \\
=&\tilde{t}_N^{(a,b)}(a \hookrightarrow ((b \hookrightarrow x) \rightarrow (b \hookrightarrow y))) \\
=&\tilde{t}_N^{(a,b)}(a \hookrightarrow (b \hookrightarrow (x \rightarrow y))) \\
=&\tilde{t}_N^{(a,b)}(1).
\end{aligned}$$

$$\begin{aligned}
&\tilde{t}_N^{(a,b)}(a \rightarrow (b \rightarrow y)) \\
=&\tilde{t}_N^{(a,b)}(1 \hookrightarrow (a \rightarrow (b \rightarrow y))) \\
=&\tilde{t}_N^{(a,b)}(((a \rightarrow (b \rightarrow x)) \hookrightarrow (a \rightarrow (b \rightarrow y))) \\
=&\tilde{t}_N^{(a,b)}(a \rightarrow ((b \rightarrow x) \hookrightarrow (b \rightarrow y))) \\
=&\tilde{t}_N^{(a,b)}(a \rightarrow (b \rightarrow (x \hookrightarrow y))) \\
=&\tilde{t}_N^{(a,b)}(1).
\end{aligned}$$

Therefore, we can obtain,

$$\tilde{t}_N^{(a,b)}(y) = \psi_1 = \tilde{t}_N^{(a,b)}(x) \cap \tilde{t}_N^{(a,b)}(x \rightarrow y), \tilde{t}_N^{(a,b)}(y) = \psi_1 = \tilde{t}_N^{(a,b)}(x) \cap \tilde{t}_N^{(a,b)}(x \hookrightarrow y).$$

Similarly, we can obtain,

$$\begin{gathered}
\tilde{i}_N^{(a,b)}(y) = \psi_3 = \tilde{i}_N^{(a,b)}(x) \cup \tilde{i}_N^{(a,b)}(x \rightarrow y), \tilde{i}_N^{(a,b)}(y) = \psi_3 = \tilde{i}_N^{(a,b)}(x) \cup \tilde{i}_N^{(a,b)}(x \hookrightarrow y); \\
\tilde{f}_N^{(a,b)}(y) = \psi_5 = \tilde{f}_N^{(a,b)}(x) \cup \tilde{f}_N^{(a,b)}(x \rightarrow y), \tilde{f}_N^{(a,b)}(y) = \psi_5 = \tilde{f}_N^{(a,b)}(x) \cup \tilde{f}_N^{(a,b)}(x \hookrightarrow y).
\end{gathered}$$

This means that $N^{(a,b)}$ is a neutrosophic hesitant fuzzy filter of X. □

Example 8. *Let* $X = \{a, b, c, d, 1\}$ *with two binary operations in Tables 7 and 8. Then,* $(X; \rightarrow, \hookrightarrow, 1)$ *is a type-2 positive implicative pseudo-BCI algebra. Let N be a neutrosophic hesitant fuzzy set. We take b, c as* *an example; thus, we have* $\{b,c,d,1\}$ *satisfy* $d \to (c \to x) = 1, d \hookrightarrow (c \hookrightarrow x) = 1$. *Let* $\psi_1 = [0.1,0.4]$, $\psi_2 = [0.2,0.3]$, $\psi_3 = [0.4,0.5]$, $\psi_4 = [0.3,0.6]$, $\psi_5 = [0.2,0.8]$, $\psi_6 = [0.1,0.9]$,

$$N^{(d,c)} = \{(1,\psi_1,\psi_3,\psi_5),(a,\psi_2,\psi_4,\psi_6),(b,\psi_1,\psi_3,\psi_5),(c,\psi_1,\psi_3,\psi_5),(e,\psi_1,\psi_3,\psi_5)\} = \{(1,[0.1,0.4],[0.4,0.5],[0.2,0.8]),(a,[0.2,0.3],[0.3,0.6],[0.1,0.9]),(b,[0.1,0.4],[0.4,0.5],[0.2,0.8]),(c,[0.1,0.4],[0.4,0.5],[0.2,0.8]),(d,[0.1,0.4],[0.4,0.5],[0.2,0.8])\}.$$

Then, we can obtain that $N^{(d,c)}$ *is a neutrosophic hesitant fuzzy filter of X.*

Table 7. $\to$.

$\to$	a	b	c	d	1
a	1	b	c	d	1
b	a	1	1	1	1
c	a	d	1	d	1
d	a	b	c	1	1
1	a	b	c	d	1

Table 8. $\hookrightarrow$.

$\hookrightarrow$	a	b	c	d	1
a	1	b	c	d	1
b	a	1	1	1	1
c	a	d	1	d	1
d	a	b	c	1	1
1	a	b	c	d	1

Theorem 18. *Let* $N = \{(x,\tilde{t}_N(x),\tilde{i}(x),\tilde{f}(x))|x \in X\}$ *be a neutrosophic hesitant fuzzy filter of X. Then,* $X_N^{(5)}(a) = \{x|\tilde{t}_N(a) \subseteq \tilde{t}_N(x), \tilde{i}_N(a) \supseteq \tilde{i}_N(x), \tilde{f}_N(a) \supseteq \tilde{f}_N(x)\}$ *is a filter of X for all* $a \in X$.

Proof. (1) Let $x,y \in X$ with $x, x \to y \in X_N^5(a)$. Then, we have $\tilde{t}_N(a) \subseteq \tilde{t}_N(x), \tilde{t}_N(a) \subseteq \tilde{t}_N(x \to y)$. Since $N = \{(x,\tilde{t}_N(x),\tilde{i}_N(x),\tilde{f}_N(x))|x \in X\}$ is a neutrosophic hesitant fuzzy filter, thus we have $\tilde{t}_N(a) \subseteq \tilde{t}_N(x) \cap \tilde{t}_N(x \to y) \subseteq \tilde{t}_N(y) \subseteq \tilde{t}_N(1)$. Similarly, we can get $\tilde{i}_N(a) \supseteq \tilde{i}_N(x) \cup \tilde{i}(x \to y) \supseteq \tilde{i}_N(y) \supseteq \tilde{i}_N(1), \tilde{f}_N(a) \supseteq \tilde{f}_N(x) \cup \tilde{f}_N(x \to y) \supseteq \tilde{f}_N(y) \supseteq \tilde{f}_N(1)$.

(2) Similarly, let $x,y \in X$ with $x, x \hookrightarrow y \in X_N^{(5)}(a)$; we have $\tilde{t}_N(a) \subseteq \tilde{t}_N(x) \cap \tilde{t}_N(x \hookrightarrow y) \subseteq \tilde{t}_N(y) \subseteq \tilde{t}_N(1), \tilde{i}_N(a) \supseteq \tilde{i}_N(x) \cup \tilde{i}_N(x \hookrightarrow y) \supseteq \tilde{i}_N(y) \supseteq \tilde{i}_N(1), \tilde{f}_N(a) \supseteq \tilde{f}_N(x) \cup \tilde{f}_N(x \hookrightarrow y) \supseteq \tilde{f}_N(y) \supseteq \tilde{f}_N(1)$.

This means that $X_N^{(5)}(a)$ satisfies the conditions of Definition 2 (F1), (F2) and (F3); $X_N^{(5)}(a)$ is a filter of X. □

Example 9. *Let* $X = \{a,b,c,d,1\}$ *with two binary operations in Tables 5 and 6. Then,* $(X;\to,\hookrightarrow,1)$ *is a pseudo-BCI algebra. Let:*

$$N = \{(1,[0,1],[0,\tfrac{3}{7}],[0,\tfrac{1}{10}]),(a,[0,\tfrac{1}{4}],[0,\tfrac{3}{4}],[0,\tfrac{1}{2}]),(b,[0,\tfrac{1}{4}],[0,\tfrac{3}{4}],[0,\tfrac{1}{2}]),(c,[0,\tfrac{1}{3}],[0,\tfrac{3}{5}],[0,\tfrac{1}{4}]),(d,[0,\tfrac{3}{4}]),[0,\tfrac{3}{6}],[0,\tfrac{1}{5}])\}.$$

Then, N is a neutrosophic hesitant fuzzy filter of X. Let $X_N^{(5)}(c) = \{c,d,1\}$. *It is easy to get that* $X_N^{(5)}(a)$ *is a filter.*

5. Conclusions

In this paper, the neutrosophic hesitant fuzzy set theory was applied to pseudo-BCI algebra, and the neutrosophic hesitant fuzzy subalgebras (filters) in pseudo-BCI algebras were developed. The relationships between neutrosophic hesitant fuzzy subalgebras (filters) and hesitant fuzzy subalgebras (filters) was discussed, and some properties were demonstrated. In future work, different types of neutrosophic hesitant fuzzy filters will be defined and discussed.

Author Contributions: All authors have contributed equally to this paper.

Acknowledgments: This work was supported by the National Natural Science Foundation of China (Grant Nos. 61573240, 61473239).

Conflicts of Interest: The authors declare no conflicts of interest.

References

1. Dudek, W.A.; Jun, Y.B. Pseudo-BCI algebras. *East Asian Math. J.* **2008**, *24*, 187–190.
2. Jun, Y.B.; Kim, H.S.; Neggers, J. On pseudo-BCI ideals of pseudo-*BCI* algebras. *Mat. Vesn.* **2006**, *58*, 39–46.
3. Ahn, S.S.; Ko, J.M. Rough fuzzy ideals in BCK/BCI-algebras. *J. Comput. Anal. Appl.* **2018**, *25*, 75–84.
4. Huang, Y. BCI-algebra. In *Science Press*; Publishing House: Beijing, China, 2006.
5. Jun, Y.B.; Sun, S.A. Hesitant fuzzy set theory applied to BCK/BCI-algebras. *J. Comput. Anal. Appl.* **2016**, *20*, 635–646.
6. Lim, C.R.; Kim, H.S. Rough ideals in BCK/BCI-algebras. *Bull. Pol. Acad. Math.* **2003**, *51*, 59–67.
7. Meng, J.; Jun, Y.B. *BCK-Algebras*; Kyungmoon Sa Co.: Seoul, Korea, 1994.
8. Zhang, X.H. Fuzzy commutative filters and fuzzy closed filters in pseudo-BCI algebras. *J. Comput. Inf. Syst.* **2014**, *10*, 3577–3584.
9. Zhang, X.H. On some fuzzy filters in pseudo-BCI algebras. *Sci. World J.* **2014**, *2014*. [CrossRef]
10. Zhang, X.H.; Jun, Y.B. Anti-grouped pseudo-BCI algebras and anti-grouped filters. *Fuzzy Syst. Math.* **2014**, *28*, 21–33.
11. Zhang, X.H.; Park, Choonkil; Wu, S.P. Soft set theoretical approach to pseudo-BCI algebras. *J. Intell. Fuzzy Syst.* **2018**, *34*, 559–568. [CrossRef]
12. Jun, Y.B. Characterizations of pseudo-BCK algebras. *Sci. Math. Jpn.* **2002**, *57*, 265–270.
13. Zhang, X.H. Fuzzy Anti-grouped Filters and Fuzzy normal Filters in Pseudo-*BCI* Algebras. *J. Intell. Fuzzy Syst.* **2017**, *33*, 1767–1774. [CrossRef]
14. Zhang, X.H. Fuzzy 1-type and 2-type positive implicative filters of pseudo-BCK algebras. *J. Intell. Fuzzy Syst.* **2015**, *28*, 2309–2317. [CrossRef]
15. Zadeh, L.A. Fuzzy sets. *Inf. Control.* **1965**, *8*, 338–353. [CrossRef]
16. Hajek, P. Observations on non-commutative fuzzy logic. *Soft Comput.* **2003**, *8*, 38–43. [CrossRef]
17. Pei, D. Fuzzy logic and algebras on residuated latties. *South. Asian Bull. Math.* **2004**, *28*, 519–531.
18. Wu, W.Z.; Mi, J.S.; Zhang, W.X. Generalized fuzzy rough sets. *Inf. Sci.* **2003**, *152*, 263–282. [CrossRef]
19. Zadeh, L.A. Toward a theory of fuzzy information granulation and its centrality in human reasoning and fuzzy logic. *Fuzzy Sets Syst.* **1997**, *90*, 111–127. [CrossRef]
20. Zhang, X.H. Fuzzy logic and algebraic analysis. In *Science Press*; Publishing House: Beijing, China, 2008.
21. Zhang, X.H.; Dudek, W.A. Fuzzy BIK+- logic and non-commutative fuzzy logics. *Fuzzy Syst. Math.* **2009**, *23*, 8–20.
22. Bo, C.X; Zhang, X.H.; Shao, S.T.; Park, Choonkil. The lattice generated by hesitant fuzzy filters in pseudo-BCI algebras. *J. Intell. Fuzzy Syst.* **2018**, *In press*
23. Torra, V. Hesitant fuzzy sets. *Int. J. Intell. Syst.* **2010**, *25*, 529–539. [CrossRef]
24. Faizi, S.; Rashid, T.; Salabun, W.; Zafar, S. Decision Making with Uncertainty Using Hesitant Fuzzy Sets. *Int. J. Fuzzy Syst.* **2017**, *20*, 1–11. [CrossRef]
25. Torra, V.; Narukawa, Y. On hesitant fuzzy sets and decision. In *Proceedings of the 18th IEEE International Conference on Fuzzy Systems, Jeju Island, Korea, 20–24 August 2009*; Publishing House: Jeju Island, Korea, 2009; pp. 1378–1382.
26. Wang, F.Q.; Li, X.; Chen, X.H. Hesitant fuzzy soft set and its applications in multicriteria decision making. *J. Appl. Math.* **2014**, *2014*. [CrossRef]
27. Wei, G. Hesitant fuzzy prioritized operators and their application to multiple attribute decision making. *Knowl. Based Syst.* **2012**, *31*, 176–182. [CrossRef]
28. Xia, M.; Xu, Z.S. Hesitant fuzzy information aggregation in decision making. *Int. J. Approx. Reason.* **2011**, *52*, 395–407. [CrossRef]
29. Xu, Z.S.; xia, M. Distance and similarity measures for hesitant fuzzy sets. *Inf. Sci.* **2011**, *181*, 2128–2138. [CrossRef]

30. Alcantud J C R, Torra V. Decomposition theorems and extension principles for hesitant fuzzysets. *Inf. Fusion* **2018**, *41*, 48–56. [CrossRef]
31. Wang Z.X, Li J. Correlation coefficients of probabilistic hesitant fuzzy elements and their applications to evaluation of the alternatives. *Symmetry* **2017**, *9*, 259. [CrossRef]
32. Smarandache, F. A unifying field in logics neutrosophy: neutrosophic probability, set and logic. *Mult. Valued Log.* **1999**, *8*, 489–503.
33. Peng, J.; Wang, J.; Wu, X. Multi-Valued neutrosophic sets and power aggregation operators with their applications in multi-criteria group decision-making problems. *Int. J. Comput. Int. Syst.* **2015**, *8*, 345–363. [CrossRef]
34. Ye, J. A multicriteria decision-making method using aggregation operators for simplified neutrosophic sets. *J. Intell. Fuzzy Syst.* **2014**, *26*, 2459–2466.
35. Wang, H.; Smarandache, F.; Zhang, Y.Q.; Sunderraman, R. Interval Neutrosophic Sets and Logic: Theory and Applications in Computing. *arXiv* **2005**, arXiv:cs/0505014. [CrossRef]
36. Wang, H.; Smarandache, F.; Sunderraman, R. Single-valued neutrosophic sets. *Rev. Air Force Acad.* **2013**, *17*, 10–13.
37. Ye, J. Multiple-attribute decision-making method under a single-valued neutrosophic hesitant fuzzy environment. *J. Intell. Syst.* **2014**, *24*, 23–36. [CrossRef]
38. Smarandache, F.; Ali, M. Neutrosophic triplet group. *Neur. Comput. Appl.* **2018**, *29*, 595–601. [CrossRef]
39. Zhang, X.H.; Smarandache, F.; Liang X.L. Neutrosophic duplet semi-group and cancellable neutrosophic triplet groups. *Symmetry* **2017**, *9*, 275. 9110275. [CrossRef]
40. Zhang, X.H.; Bo, C.X.; Smarandache, F.; Dai, J.H. New inclusion relation of neutrosophic sets with applications and related lattice structure. *Int. J. Mach. Learn. Cyben.* **2018**. [CrossRef]

symmetry

MDPI

Article

Neutrosophic Weighted Support Vector Machines for the Determination of School Administrators Who Attended an Action Learning Course Based on Their Conflict-Handling Styles

Muhammed Turhan [1], Dönüş Şengür [1,*], Songül Karabatak [1], Yanhui Guo [2] and Florentin Smarandache [3]

[1] Department of Education, University of Firat at Elazig, 23119 Elazig, Turkey; muhammedturhan66@gmail.com (M.T.); s_halici@hotmail.com (S.K.)

[2] Department of Computer Science, University of Illinois at Springfield, Springfield, IL 62703, USA; guoyanhui@gmail.com

[3] Department of Mathematics, University of New Mexico, 705 Gurley Ave., Gallup, NM 87301, USA; fsmarandache@gmail.com

* Correspondence: kdksengur@gmail.com

Received: 30 March 2018; Accepted: 18 May 2018; Published: 20 May 2018

Abstract: In the recent years, school administrators often come across various problems while teaching, counseling, and promoting and providing other services which engender disagreements and interpersonal conflicts between students, the administrative staff, and others. Action learning is an effective way to train school administrators in order to improve their conflict-handling styles. In this paper, a novel approach is used to determine the effectiveness of training in school administrators who attended an action learning course based on their conflict-handling styles. To this end, a Rahim Organization Conflict Inventory II (ROCI-II) instrument is used that consists of both the demographic information and the conflict-handling styles of the school administrators. The proposed method uses the Neutrosophic Set (NS) and Support Vector Machines (SVMs) to construct an efficient classification scheme neutrosophic support vector machine (NS-SVM). The neutrosophic c-means (NCM) clustering algorithm is used to determine the neutrosophic memberships and then a weighting parameter is calculated from the neutrosophic memberships. The calculated weight value is then used in SVM as handled in the Fuzzy SVM (FSVM) approach. Various experimental works are carried in a computer environment out to validate the proposed idea. All experimental works are simulated in a MATLAB environment with a five-fold cross-validation technique. The classification performance is measured by accuracy criteria. The prediction experiments are conducted based on two scenarios. In the first one, all statements are used to predict if a school administrator is trained or not after attending an action learning program. In the second scenario, five independent dimensions are used individually to predict if a school administrator is trained or not after attending an action learning program. According to the obtained results, the proposed NS-SVM outperforms for all experimental works.

Keywords: action learning; school administrator; SVM; neutrosophic classification

---

1. Introduction

Support Vector Machine (SVM) is a widely used supervised classifier, which has provided better achievements than traditional classifiers in many pattern recognition applications in the last two decades [1]. SVM is also known as a kernel-based learning algorithm where the input features are transformed into a high-dimensional feature space to increment the class separability of the input features. Then SVM seeks a separating optimal hyperplane that maximizes the margin between two classes in high-dimensional feature space [2]. Maximizing the margin is an optimization problem which can be solved using the Lagrangian multiplier [2]. In addition, some of the input features, which are called support vectors, can also be used to determine the optimal hyperplane [2].

Although SVM outperforms many classification applications, in some applications, some of the input data points may not be truly classified [3]. This misclassification may arise due to noises or other conditions. To handle such a problem, Lin et al. proposed Fuzzy SVMs (FSVMs), in which a fuzzy membership is assigned to each input data point [3]. Thus, a robust SVM architecture is constructed by combining the fuzzy memberships into the learning of the decision surface. Another fuzzy-based improved SVMs approach was proposed by Wang et al. The authors applied it to a credit risk analysis of consumer lending [4]. Ilhan et al. proposed a hybrid method where a genetic algorithm (GA) and SVM were used to predict Single Nucleotide Polymorphisms (SNP) [5]. In other words, GA was used to select the optimum C and γ parameters in order to predict the SNP. The authors also used a particle swarm optimization (PSO) algorithm to optimize C and γ parameters of SVMs. Peng et al. proposed an improved SVM for heterogeneous datasets [6]. To do so, the authors used a mapping procedure to map nominal features to another space via the minimization of the predicted generalization errors. Ju et al. proposed neutrosophic logic to improve the efficiency of the SVMs classifier (N-SVM) [7]. More specifically, the proposed N-SVM approach was applied to image segmentation. The authors used the diverse density support vector machine (DD-SVM) to improve its efficiency with neutrosophic set theory [8]. Almasi et al. proposed a new fuzzy SVM method, which was based on an optimization method [9]. The proposed method simultaneously generated appropriate fuzzy memberships and solved the model selection problem for the SVM family in linear/nonlinear and separable/non-separable classification problems. In Reference [10], Tang et al. proposed a novel fuzzy membership function for linear and nonlinear FSVMs. The structural information of two classes in the input space and in the feature space was used for the calculation of the fuzzy memberships. Wu et al. used an artificial immune system (AIS) in the optimization of SVMs [11]. The authors used the AIS algorithm to optimize the C and γ parameters of SVMs and developed an efficient scheme called AISSVM. Chen et al. optimized the parameters of the SVM by using the artificial bee colony (ABC) approach [12]. Specifically, the authors used an enhanced ABC algorithm where cat chaotic mapping initialization and current optimum were used to improve the ABC approach. Zhao et al. used an ant colony algorithm (ACA) to improve the efficiency of SVMs [13]. The ACA optimization method was used to select the kernel function parameter and soft margin constant C penalty parameter. Guraksin et al. used particle swarm optimization (PSO) to tune SVM parameters to improve its efficiency [14]. The improved SVM approach was applied to a bone age determination system.

In this paper, a new approach is proposed: Neutrosophic SVM (NS-SVM). The neutrosophic set (NS) is defined as the generalization of the fuzzy set [15]. NS is quite effective in dealing with outliers and noises. The noises and outlier samples in a dataset can be treated as a kind of indeterminacy. NS has been successfully applied for indeterminate information processing, and demonstrates advantages to deal with the indeterminacy information of data [16–18]. NS employs three memberships to measure the degree of truth (T), indeterminacy (I), and falsity (F) of each dataset. The neutrosophic c-means (NCM) algorithm is used to produce T, I, and F memberships [16,17]. In recent years, school administrators often come across various problems while teaching, counseling, and promoting and providing other services which engender disagreements and interpersonal conflicts between students, the administrative staff, and others. Action learning is an effective way to train school administrators in order to improve their conflict-handling styles. To this end, the developed NS-SVM approach is applied to determine the effectiveness of training in school administrators who attended an action learning course based on their conflict-handling styles. A Rahim Organization Conflict Inventory II (ROCI-II) instrument is used that consists of both the demographic information and the conflict-handling styles of the school administrators. A five-fold cross-validation test is applied to evaluate the proposed method. The classification accuracy is calculated for performance measure. The proposed method is also compared with SVM and FSVM.

The paper is organized as follows. In the next section, a summarization of the present works on this topic is given. The proposed NS-SVM is introduced in Section 3. Section 4 gives the experimental work and results. We conclude the paper in Section 5.

2. Related Works

As mentioned earlier, there have been a number of presented works about the feature weighting for improving the efficiency of classifiers. To this end, Akbulut et al. proposed an NS-based Extreme Learning Machine (ELM) approach for imbalanced data classification [18]. They initially employed an NS-based clustering algorithm to assign a weight for each input data point and then the obtained weights were linked to the ELM formulation to improve its efficiency. In the experiments, the proposed scheme highly improved the classification accuracy. Ju et al. proposed a similar work and applied it to improve image segmentation performance [7]. The authors opted to construct the NS weights based on the formulations given in Reference [7]. The obtained weights were then used in SVM equations. In other words, the authors used the DD-SVM to improve its efficiency with neutrosophic logic. Guo et al. proposed an unsupervised approach for data clustering [16]. The authors combined NS theory in an unsupervised data clustering which can be seen as a weighting procedure. Thus, the indeterminate data points were also considered in the classification process more efficiently. An NS-based k-NN approach was proposed by Akbulut et al. [19]. The authors used the NS memberships to improve the classification performance of the k-NN classifier. The proposed scheme calculated the NS memberships based on a supervised neutrosophic c-means (NCM) algorithm. A final belonging membership U was calculated from the NS triples. A final voting scheme as given in fuzzy k-NN was considered for class label determination. Budak et al. proposed an NS-based efficient Hough transform [20]. The authors initially transferred the Hough space into the NS space by calculating the NS membership triples. An indeterminacy filtering was constructed where the neighborhood information was used to remove the indeterminacy in the spatial neighborhood of the neutrosophic Hough space. The potential peaks were detected based on thresholding on the neutrosophic Hough space, and these peak locations were then used to detect the lines in the image domain.

3. Proposed Neutrosophic Set Support Vector Machines (NS-SVM)

In this section, we briefly introduce the theories of SVM and NS. The readers may refer to related references for detailed information [1,3]. Then, the proposed neutrosophic set support vector machine is presented in detail below.

3.1. Support Vector Machine (SVM)

SVM is an important and efficient supervised classification algorithm [1,2]. Given a set of N training data points $\{(x_i, y_i)_{n=1}^{N}\}$ where x_i is a multidimensional feature vector and $y_i \in \{-1, 1\}$ is the corresponding label, an SVM models a decision boundary between classes of training data as a separating hyperplane. SVM aims to find an optimal solution by maximizing the margin around the separating hyperplane, which is equivalent to minimizing $||w||$ with the constraint:

$$y_i(w.x_i + b) \geq 1 \tag{1}$$

SVM employs non-linear mapping to transform the input data into a higher dimensional space. Thus, the hyperplane can be found in the higher dimensional space with a maximum margin as:

$$w.\varphi(x) + b = 0 \tag{2}$$

such that for each data sample $(\varphi(x_i), y_i)$:

$$y_i(w.\varphi(x_i) + b) \geq 1, \qquad i = 1, \ldots, N. \tag{3}$$

when the input dataset is not linearly separable, then the soft margin is allowed by defining N non-negative variables, denoted by $\xi = (\xi_1, \xi_2, \ldots, \xi_N)$, such that the constraint for each sample in Equation (3) is rewritten as:

$$y_i(w.\varphi(x_i) + b) \geq 1 - \xi_i, \qquad i = 1, \ldots, N \tag{4}$$

where the optimal hyperplane is determined as;

$$\text{minimum}\left(\frac{1}{2}w^2 + C\sum_{i=1}^{N}\xi_i\right) \tag{5}$$

$$\text{subjected to } y_i(w.\varphi(x_i) + b) \geq 1 - \xi_i, \qquad i = 1, \ldots, N \tag{6}$$

where C is a constant parameter that tunes the balance between the maximum margin and the minimum classification error.

3.2. Neutrosophic c-Means Clustering

In this section, a weighting function is defined by samples using the neutrosophic c-means (NCM) clustering. Let A $= \{A_1, A_2, , \ldots., A_m\}$ be a set of alternatives in the neutrosophic set. A sample A_i is represented as $\{\mathrm{T}(A_i), \mathrm{I}(A_i), \mathrm{F}(A_i)\}/A_i$, where $\mathrm{T}(A_i)$, $\mathrm{I}(A_i)$ and $\mathrm{F}(A_i)$ are the membership values to the true, indeterminate, and false sets. $\mathrm{T}(A_i)$ is used to measure the belonging degree of the sample to the center of the labeled class, $\mathrm{I}(A_i)$ for indiscrimination degree between two classes, and $\mathrm{F}(A_i)$ for the belonging degree to the outliers.

The NCM clustering overcomes the disadvantages of handling indeterminate points in other algorithms [16]. Here we improve the NCM by only computing neutrosophic memberships to the true and indeterminate sets based on the samples' distribution.

Using NCM, the truth and indeterminacy memberships are defined as:

$$K = \left[\frac{1}{\varpi_1}\sum_{j=1}^{C}(x_i - c_j)^{-\frac{2}{m-1}} + \frac{1}{\varpi_2}(x_i - \bar{c}_{imax})^{-\left(\frac{2}{m-1}\right)} + \frac{1}{\varpi_3}\delta^{-\left(\frac{2}{m-1}\right)}\right] \tag{7}$$

$$T_{ij} = \frac{K}{\varpi_1}(x_i - c_j)^{-\left(\frac{2}{m-1}\right)} \tag{8}$$

$$I_i = \frac{K}{\varpi_2}(x_i - \bar{c}_{imax})^{-\left(\frac{2}{m-1}\right)} \tag{9}$$

where T_{ij} and I_i are the true and indeterminacy membership values of point i, and the cluster center is denoted as c_j. $\bar{c}_{\mathrm{imax}}$ is obtained from indexes of the largest and second largest value of T_{ij}. ϖ_1, ϖ_2, and ϖ_3 are constant weights. T_{ij} and I_i are updated at each iteration until $\left|T_{ij}^{(k+1)} - T_{ij}^{(k)}\right| < \varepsilon$, where ε is a termination criterion.

3.3. Proposed Neutrosophic Set Support Vector Machine (NS-SVM)

In the fuzzy support vector machine (FSVM), a membership g_i is assigned for each input data point $\{(x_i, y_i)_{n=1}^{N}\}$, where $0 < g_i < 1$ [3]. As g_i and ξ_i shows the membership and the error of SVM for input data point x_i , respectively, the term $g_i\xi_i$ shows the measure of error with different weighting. Thus, the optimal hyperplane problem can be re-solved as;

$$\text{minimum}\left(\frac{1}{2}w^2 + C\sum_{i=1}^{N}g_i\xi_i\right) \tag{10}$$

$$\text{subjected to } y_i(w.\varphi(x_i) + b) \geq 1 - \xi_i, \qquad i = 1, \ldots, N \tag{11}$$

In the proposed method, a weighting function is defined in the NS based on the memberships to truth and indeterminacy and then used to remove the effect of indeterminacy information for classification.

$$g_{Ni} = \sum_{j=1}^{C} T_{ij} \cdot I_i \tag{12}$$

Then we use the newly defined weight function g_{Ni} to replace the weight function in Equation (4), and an optimization procedure is employed to minimize the cost function as:

$$\text{minimum}\left(\frac{1}{2}w^2 + C\sum_{i=1}^{N} g_{Ni} \cdot \xi_i\right) \tag{13}$$

$$\text{subjected to } y_i(w.\varphi(x_i) + b) \geq 1 - \xi_i, \qquad i = 1, \ldots, N \tag{14}$$

Finally, the support vectors are identified and their weights are obtained for classification. The semantic algorithm of the proposed method is given as:

Input: Labeled training dataset.

Output: Predicted class labels.

Step 1: Calculate the cluster centers according to the labeled dataset and employ NCM algorithm to determine NS memberships T and I for each data point.

Step 2: Calculate g_{Ni} by using T and I components according to Equation (8).

Step 3: Optimize NS-SVM by minimizing the cost function according to Equation (9).

Step 4: Calculate the labels of test data.

4. Experimental Work and Results

In this study, a new approach NS-SVM is proposed and applied to determine if an action learning experience resulted in school administrators being more productive in their conflict-management skills [21]. To this end, an experimental organization was constructed where 38 administrators from various schools in Elazig/Turkey were administered a pre-test and a post-test of the Rahim Organization Conflict Inventory II (ROCI-II) [22]. The pre-test was applied to the administrators before the action learning experience and the post-test was applied after the action learning experience. The ROCI-II contains 28 scale items. These scale items are grouped into five dimensions: integrating, obliging, dominating, avoiding, and compromising. The dataset, which was used in this work, is given in Appendix A. The MATLAB software is used in construction of the NS-SVM approach. In the evaluation of the proposed method, a five-fold cross-validation test is used and the mean accuracy value is recorded. During the experimental work, two different scenarios are considered. In the first one, all 28 scale items are used to determine the trained and non-trained school administrators. In the second scenario, each dimension of ROCI-II is used to determine trained and non-trained administrators in order to determine the relationship between the dimensions and the trained and non-trained school administrators. The NS-SVM parameter C is searched in the range of $[10^{-3}, 10^{2}]$ at a step size of 10^{-1}. In addition, for NCM the following parameters are chosen: $\varepsilon = 10^{-3}$, $\varpi_1 = 0.75$, $\varpi_2 = 0.125$, $\varpi_3 = 0.125$, which were obtained from trial and error. The δ parameter of NCM method is also searched in the range of $\{2^{-10}, 2^{-8}, \ldots, 2^{8}, 2^{10}\}$. The dataset is normalized with zero mean and unit variance. Table 1 shows the obtained accuracy scores for the first scenario. The obtained results are further compared with FSVM and other SVM types such as Linear, Quadratic, Cubic, Fine Gaussian, Medium Gaussian, and Coarse Gaussian SVMs.

As seen in Table 1, 81.2% accuracy is obtained with the proposed NS-SVM method, which is the highest among all compared classifier types. The second highest accuracy, 76.9%, is obtained by the FSVM method. An accuracy score of 73.7% is produced by both linear and medium Gaussian SVM methods. In addition, quadratic and cubic SVM techniques produce 68.4% accuracy scores. An accuracy score of 63.2% is obtained by the coarse Gaussian SVM method and finally, the worst accuracy score, 48.7%, is obtained by the fine Gaussian SVM method. Generally speaking, contributing memberships as weighting to SVM highly increases the efficiency. Both FSVM and NS-SVM produce better results than traditional SVM methods. The experimental results that cover the second scenario are given in Tables 2–6. Table 2 shows the obtained accuracy scores when the integrating dimension is used as input. The integrating dimension has six scale items.

Table 1. Prediction accuracies for the first scenario. The bold case shows the highest accuracy. SVM: Support Vector Machines; FSVM: Fuzzy Support Vector Machines; NS-SVM: Neutrosophic Support Vector Machines.

Classifier Type	Accuracy (%)
Linear SVM	73.7
Quadratic SVM	68.4
Cubic SVM	68.4
Fine Gaussian SVM	48.7
Medium Gaussian SVM	73.7
Coarse Gaussian SVM	63.2
FSVM	76.9
NS-SVM	**81.2**

As seen in Table 2, the highest accuracy score, 80.3%, is obtained by the proposed method. This score is 4% better than that achieved by FSVM. The FSVM method produces a 76.3% accuracy score, which is the second highest. Linear and medium Gaussian SVM methods produce 73.7% accuracy scores, which are the third highest. In addition, linear and medium Gaussian SVM methods achieve the best accuracy among the ordinary SVM techniques. It is worth mentioning that cubic SVM has the lowest accuracy score, with an achievement of 53.9%.

Table 2. Prediction accuracies for the second scenario. The integrating dimension is used as input. The bold case shows the highest accuracy.

Classifier Type	Accuracy (%)
Linear SVM	73.7
Quadratic SVM	57.9
Cubic SVM	53.9
Fine Gaussian SVM	60.5
Medium Gaussian SVM	73.7
Coarse Gaussian SVM	67.1
FSVM	76.3
NS-SVM	**80.3**

Table 3 shows the achievements obtained when the obliging dimension is used as input to the classifiers. The obliging dimension covers five scale items and 73.8% accuracy score, which is the highest, obtained by the NS-SVM method. FSVM also produces a 71.3% accuracy score, which is the second-best achievement. The worst accuracy score is obtained by quadratic SVM, for which the accuracy score is 50.0%. One important inference from Table 3 is that ordinary SVM techniques produce almost similar achievements, while weighting with memberships highly improves the accuracy.

The dominating dimension also covers five scale items and the produced results are shown in Table 4. As seen in Table 4, the highest accuracy, 70.0%, is produced by the proposed NS-SVM method. In addition, the second-best accuracy score, 65.0%, is obtained by the FSVM method. The linear SVM obtains 59.2% accuracy, which is the third highest accuracy score. When one considers the ordinary SVM's achievements, an obvious improvement can be seen easily that is achieved by the NS-SVM method.

Table 3. Prediction accuracies for the second scenario. The obliging dimension is used as input. The bold case shows the highest accuracy.

Classifier Type	Accuracy (%)
Linear SVM	61.8
Quadratic SVM	50.0
Cubic SVM	51.3
Fine Gaussian SVM	52.6
Medium Gaussian SVM	61.8
Coarse Gaussian SVM	55.3
FSVM	71.3
NS-SVM	**73.8**

Table 4. Prediction accuracies for the second scenario. The dominating dimension is used as input. The bold case shows the highest accuracy.

Classifier Type	Accuracy (%)
Linear SVM	59.2
Quadratic SVM	57.9
Cubic SVM	52.6
Fine Gaussian SVM	55.3
Medium Gaussian SVM	52.6
Coarse Gaussian SVM	55.3
FSVM	65.0
NS-SVM	**70.0**

The avoiding dimension covers six scale items and the produced results are given in Table 5. As one evaluates the obtained results given in Table 5, it can be observed that the avoiding dimension is not efficient enough in discriminating trained and non-trained participants. In other words, the ordinary SVM techniques do not achieve better accuracy scores. Among them, the highest accuracy, 53.9%, is produced by the cubic SVM method. On the other hand, both FSVM and the proposed NS-SVM methods produce better accuracy scores, with achievements of 63.8% and 66.3%, respectively. Once more, the best accuracy is obtained by the proposed NS-SVM method.

Table 5. Prediction accuracies for the second scenario. The avoiding dimension is used as input. The bold case shows the highest accuracy.

Classifier Type	Accuracy (%)
Linear SVM	50.0
Quadratic SVM	43.4
Cubic SVM	53.9
Fine Gaussian SVM	48.7
Medium Gaussian SVM	44.7
Coarse Gaussian SVM	42.1
FSVM	63.8
NS-SVM	**66.3**

Finally, the compromising dimension covers six scale items and the produced results are given in Table 6. As seen in Table 6, the compromising dimension is quite efficient in the determination of trained and non-trained participants, where better accuracy scores are visible when compared with the avoiding dimension's accuracy scores. A 75.0% accuracy score, the highest among all methods, is obtained by NS-SVM. A 73.8% accuracy score is obtained by the FSVM method. The highest third accuracy score is produced by medium Gaussian SVM.

Table 6. Prediction accuracies for the second scenario. The compromising dimension is used as input. The bold case shows the highest accuracy.

Classifier Type	Accuracy (%)
Linear SVM	67.1
Quadratic SVM	67.1
Cubic SVM	57.9
Fine Gaussian SVM	65.8
Medium Gaussian SVM	71.1
Coarse Gaussian SVM	68.4
FSVM	73.8
NS-SVM	**75.0**

We further analyze the results obtained from the first scenario by considering a statistical measure and the running time. To this end, the f-measure metric was considered. The f-measure calculates the weighted harmonic mean of recall and precision [23]. The results are tabulated in Table 7.

Table 7. Calculated f-measure and running times for the first scenario. The bold cases show the better achievements.

Classifier Type	f-Measure (%)	Time (s)
Linear SVM	73.50	0.314
Quadratic SVM	68.50	0.129
Cubic SVM	68.50	0.122
Fine Gaussian SVM	48.50	0.119
Medium Gaussian SVM	71.00	0.130
Coarse Gaussian SVM	61.00	0.129
FSVM	76.50	0.089
NS-SVM	**80.00**	**0.065**

In Table 7, the best f-measure achievement score, 80.00%, was achieved by the proposed NS-SVM method. The second-best f-measure score, 76.50%, was produced by FSVM. The other SVM techniques also produced reasonable f-measure scores when their accuracy achievements were considered (Table 1). In addition, the running time of the proposed method was less than those of the other SVM methods. The proposed method achieved its process at 0.065 s. In other words, this running time is almost half the running times of the non-weighted SVM methods. Thus, it is evident that the proposed NS-SVM performed more accurate results in a very short time, demonstrating its efficiency.

5. Conclusions

In this paper, neutrosophic set theory and SVM is used to construct an efficient classification approach called NS-SVM. It is then applied to an educational problem. More specifically, the determination of the effectiveness of training in school administrators who attended an action learning course based on their conflict-handling styles is achieved. To this end, a ROCI-II instrument is used that consists of both the demographic information and the conflict-handling styles of the school administrators. Six various SVM approaches and FSVM are used in performance comparison. The experimental works are carried out with a five-fold cross-validation technique and the classification accuracy is measured to evaluate the performance of the proposed NS-SVM approach. The experiments are conducted based on two scenarios. In the first one, all statements are used to predict if a school administrator is trained or not after attending an action learning program. In the second scenario, five independent dimensions are used individually to predict if a school administrator is educated or not after attending an action learning program. According to the obtained results, the first scenario achieves the best performance with the NS-SVM method, resulting in an accuracy score of 81.2%. In addition, for all experiments in the second scenario, the proposed NS-SVM achieves the highest accuracy scores as given in Tables 2–6. Furthermore,

FSVM achieved the second highest accuracy scores for all experiments that are handled in scenarios 1 and 2. This situation shows that embedding the membership degrees into the SVM method highly improves its discriminatory ability. To further analyze the efficiency of the proposed method, we used the f-measure test and the running times of the methods. The proposed NS-SVM yielded the highest f-measure score. In addition, the running time of the proposed method was much less than those of the traditional SVM techniques.

This study revealed important results for both educational research and determining the effectiveness of educational practices. First, this research showed that the NS-SVM technique can be used in pre-test and post-test comparisons in experimental educational research. In addition, this study demonstrated that the effectiveness levels of training courses can be determined by examining the NS-SVM discrimination accuracy of individuals who attended training courses compared to those who did not.

Author Contributions: M.T., D.Ş., S.K., Y.G. and F.S. conceived and worked together to achieve this work.

Conflicts of Interest: The authors declare no conflict of interest.

Appendix A

The dataset was used in the experimental works is given in Figure A1. The features are in the columns and the last column shows the class labels. Moreover, the rows show the number of samples.

This dataset was originally constructed based on the questionnaire that was based on the ROCI-II instrument [24]. As mentioned earlier, the ROCI-II instrument contains 28 scale items which are grouped into five dimensions; integrating (six scale items, Features 1–6), obliging (five scale items, Features 7–11), dominating (five scale items, Features 12–16), avoiding (six scale items, Features 17–22), and compromising (six scale items, Features 23–28). The school administrators were asked to fill out this questionnaire by assigning a five-point Likert scale (1–5) for each feature before and after a action learning course. Thus, 76 questionnaires were obtained. In scenario 1, the 28 scale items were used in the prediction of trained and non-trained school administrators and in scenario 2, each dimension of the ROCI-II instrument was used to predict trained and non-trained school administrators.

Samples	Feature1	Feature2	Feature3	Feature4	Feature5	Feature6	Feature7	Feature8	Feature9	Feature10	Feature11	Feature12	Feature13	Feature14	Feature15	Feature16	Feature17	Feature18	Feature19	Feature20	Feature21	Feature22	Feature23	Feature24	Feature25	Feature26	Feature27	Feature28	Labels
1	5	5	5	5	5	5	4	1	3	3	4	2	1	4	4	2	1	1	2	2	4	5	4	4	4	4	4	1	1
2	5	4	5	5	5	5	5	1	3	4	5	3	1	5	5	4	4	2	2	1	1	5	3	3	4	3	5	3	1
3	3	4	4	4	4	4	4	3	4	4	4	1	1	4	4	4	2	2	4	2	4	2	4	4	4	4	4	4	1
4	5	4	4	4	5	3	4	3	3	3	3	4	4	4	4	4	3	2	4	4	1	3	4	4	5	4	3	4	1
5	5	5	4	4	5	5	4	2	2	5	4	2	2	2	4	2	4	2	2	1	4	5	4	5	5	5	5	3	1
6	4	4	4	4	4	4	4	1	2	1	4	2	1	3	3	3	2	1	1	1	3	5	3	4	4	4	4	1	1
7	4	5	5	4	3	4	3	2	3	3	3	1	1	3	4	3	2	1	2	1	5	5	4	4	4	4	5	3	1
8	4	5	4	4	4	4	3	2	1	3	3	1	1	1	4	2	3	3	3	1	1	4	4	4	4	3	4	1	1
9	4	4	4	4	4	4	4	1	1	4	4	1	1	5	4	2	2	2	1	1	3	5	3	4	5	4	4	2	1
10	4	4	5	5	4	5	3	3	3	4	5	1	1	5	4	3	3	4	1	1	3	5	4	4	4	4	4	2	1
11	5	5	4	4	4	4	4	2	3	3	3	2	1	2	4	1	5	1	1	1	1	4	5	4	4	4	4	1	1
12	3	4	3	4	3	3	4	1	3	3	3	2	2	4	4	4	2	2	2	1	2	5	4	5	4	3	3	3	1
13	4	4	4	2	4	4	4	1	1	3	3	1	2	3	4	4	3	2	2	1	2	4	4	5	3	3	4	3	1
14	5	4	4	3	4	4	5	1	4	3	4	2	1	4	4	4	5	3	1	4	4	5	5	4	5	4	4	2	1
15	4	4	4	4	5	4	4	1	3	1	4	2	1	3	4	1	3	3	1	1	1	5	5	4	4	4	4	3	1
16	5	5	5	2	3	4	4	1	2	3	3	1	1	5	3	3	1	1	1	1	3	4	4	4	4	4	4	2	1
17	4	4	4	3	4	4	3	1	3	2	3	1	1	1	4	2	2	3	3	2	3	4	4	4	5	3	4	1	1
18	4	5	4	4	5	5	4	1	3	3	4	2	1	4	4	5	5	3	4	2	5	5	5	4	5	3	4	2	1
19	4	4	4	4	5	3	4	1	3	2	3	1	1	3	4	3	5	4	5	2	4	4	4	5	4	4	4	2	1
20	4	5	4	4	5	5	3	2	3	3	5	1	1	3	4	2	2	1	4	1	3	5	3	4	5	4	4	3	1
21	4	4	5	5	5	5	4	4	4	5	5	3	1	4	5	4	2	2	4	1	4	4	4	4	5	5	5	4	1
22	4	4	4	4	3	4	4	4	4	3	3	1	3	4	4	3	4	5	5	1	4	5	5	5	4	4	5	1	1
23	4	5	4	3	3	3	3	1	2	2	4	1	1	2	5	3	1	1	1	1	2	1	4	4	4	3	4	4	1
24	4	4	4	4	4	4	4	1	3	2	4	1	1	2	5	4	2	2	4	4	1	1	4	5	4	4	2	3	1
25	5	3	5	3	3	4	4	3	3	3	4	3	4	4	4	4	4	4	2	1	4	5	4	4	4	4	4	4	1
26	5	5	5	5	5	5	5	1	3	1	5	1	1	1	4	2	4	2	1	1	1	5	4	5	3	5	5	1	1
27	4	4	4	4	4	5	4	2	3	3	4	2	1	1	4	3	4	3	2	1	3	5	4	5	5	4	4	1	1
28	5	5	4	4	4	4	4	3	3	5	4	3	3	3	4	4	4	5	5	1	4	4	5	5	3	5	5	4	1
29	4	4	4	4	4	4	3	2	3	3	4	2	3	4	5	3	3	2	2	2	3	5	4	5	4	4	4	3	1
30	5	5	5	3	4	4	4	1	3	3	4	2	3	4	4	4	5	3	2	1	4	2	4	3	5	4	5	2	1
31	3	2	3	3	3	3	2	2	2	3	2	2	2	3	4	3	3	2	3	2	5	3	3	3	2	2	2	3	1
32	5	3	3	3	3	3	4	2	1	4	3	2	2	4	4	4	3	1	4	4	5	3	3	3	4	3	3	4	1
33	3	3	1	3	4	3	2	2	1	3	3	2	2	3	3	2	3	2	4	1	4	3	3	3	1	1	2	2	1
34	3	2	4	3	3	4	1	1	1	3	1	2	2	3	5	1	1	1	4	1	1	3	2	3	3	3	3	1	1
35	4	3	4	3	5	3	4	1	1	3	3	2	2	4	5	5	3	1	3	1	1	1	3	4	4	4	4	5	1
36	3	4	4	4	3	3	3	2	3	4	3	2	2	1	5	1	3	3	4	2	4	3	3	4	4	3	3	4	1
37	4	3	3	4	3	3	3	1	3	3	3	3	2	4	5	4	3	3	4	3	5	3	3	3	3	3	3	2	1
38	4	4	3	4	3	3	4	2	2	2	3	3	4	4	4	5	3	5	3	1	3	4	4	3	4	3	4	3	1
39	5	4	4	5	4	4	4	2	1	4	4	1	1	4	5	1	4	1	1	1	4	5	4	5	4	5	5	2	2
40	5	5	5	5	5	5	4	1	1	3	1	1	1	1	5	1	1	1	1	1	1	5	2	5	5	5	5	1	2
41	4	4	4	2	4	4	3	2	3	4	4	1	2	2	4	4	3	2	4	2	4	4	4	4	4	4	4	4	2
42	4	4	4	2	3	4	3	2	3	4	4	1	2	2	5	4	3	2	4	2	4	4	4	4	4	4	4	4	2
43	5	4	5	4	5	4	4	2	2	5	4	1	1	2	4	2	4	1	2	1	4	4	5	5	5	5	5	4	2
44	5	4	4	4	4	5	4	1	3	3	4	2	2	4	5	3	2	2	3	1	3	1	3	4	4	3	3	2	2
45	4	4	4	5	4	5	4	5	3	3	4	1	1	2	5	1	3	1	1	1	5	5	4	4	4	4	4	3	2
46	4	4	5	4	4	4	4	1	3	3	4	1	1	4	4	4	4	4	2	2	4	4	4	5	4	5	5	3	2
47	5	5	5	5	5	5	5	1	1	4	5	1	1	4	5	3	3	2	2	2	2	5	3	5	5	5	5	3	2
48	3	4	5	5	5	5	4	1	2	5	5	1	1	1	2	1	3	1	1	1	5	5	4	4	5	5	5	1	2
49	4	4	4	4	4	5	3	1	2	3	4	1	1	2	4	2	2	1	1	1	1	5	4	4	4	4	4	1	2
50	4	4	4	4	5	5	4	1	3	3	3	1	1	4	4	3	2	1	1	1	4	5	5	4	5	4	4	1	2
51	4	4	4	4	4	4	4	2	3	3	4	2	2	4	4	4	4	4	2	2	4	4	4	5	4	4	4	3	2
52	5	5	4	5	5	5	5	1	3	3	5	2	1	4	5	3	5	5	4	4	5	5	5	5	3	4	4	3	2
53	5	4	5	5	5	5	4	1	1	3	4	1	3	2	5	1	1	2	5	1	1	1	5	5	5	5	5	3	2
54	4	3	4	2	2	2	3	1	3	2	3	1	1	4	3	1	2	2	3	2	1	2	3	3	5	4	4	2	2
55	5	5	5	5	5	5	4	1	2	2	2	1	1	1	4	1	1	1	3	1	2	3	4	5	5	5	5	1	2
56	5	5	5	5	5	5	4	1	3	4	5	1	1	3	3	5	5	3	3	2	5	5	5	5	5	4	5	3	2
57	4	5	5	5	5	4	4	1	3	4	4	1	1	3	3	3	5	3	4	2	5	5	4	5	5	5	5	3	2
58	5	5	5	5	5	5	3	2	3	4	5	1	1	3	3	2	3	3	4	1	4	5	4	5	5	5	5	3	2
59	4	5	5	5	5	5	5	3	3	5	5	2	1	3	4	4	3	3	4	1	4	5	4	5	5	5	5	4	2
60	5	5	5	5	4	5	4	3	3	4	4	1	2	3	4	3	4	5	4	1	4	5	5	5	5	5	5	2	2
61	4	5	5	4	4	4	4	1	2	3	5	1	1	2	4	3	2	1	1	1	3	3	4	5	5	4	5	4	2
62	4	4	5	5	4	5	4	1	3	3	5	1	1	2	4	4	2	2	3	3	2	3	4	5	5	5	3	3	2
63	5	4	5	4	4	5	4	2	3	3	4	2	3	3	4	4	4	4	2	1	5	5	4	5	4	5	5	4	2
64	5	5	5	5	5	5	5	1	3	3	5	1	1	1	4	2	4	2	1	1	2	5	5	5	4	5	5	3	2
65	5	5	4	5	5	5	4	1	3	4	5	2	1	1	4	3	4	3	2	1	4	5	5	5	5	5	5	3	2
66	5	5	5	5	5	5	4	2	3	5	5	2	2	2	4	4	4	5	3	1	5	5	5	5	4	5	5	4	2
67	5	5	5	5	5	5	3	1	3	4	4	2	2	2	4	3	4	2	3	2	4	5	4	5	5	5	5	3	2
68	5	5	5	4	5	5	4	1	3	4	5	2	2	2	4	4	5	3	2	1	5	3	4	4	5	5	5	2	2
69	3	3	3	4	4	4	3	2	2	3	3	1	2	3	4	3	4	2	3	2	4	4	3	3	2	3	2	3	2
70	5	4	4	4	4	4	5	1	1	4	4	1	1	4	4	4	4	1	4	4	4	4	4	4	5	4	4	4	2
71	4	4	1	4	4	4	3	2	1	3	4	2	1	3	3	2	3	2	4	1	4	4	5	4	1	1	2	2	2
72	4	3	5	5	5	5	1	1	1	3	1	1	1	3	5	1	1	1	4	1	1	5	3	4	4	4	4	1	2
73	5	4	5	5	5	4	5	1	1	3	4	1	2	4	5	5	1	1	3	1	1	1	4	5	5	5	5	5	2
74	4	5	5	5	5	4	4	2	3	4	3	1	1	1	5	1	4	3	4	2	4	5	4	5	5	4	4	4	2
75	5	4	4	4	4	4	4	1	3	3	4	3	2	4	5	4	4	3	4	3	4	2	4	4	4	3	3	2	2
76	5	5	4	4	3	4	5	2	2	2	4	3	4	4	4	5	5	5	3	1	2	5	5	5	5	4	5	3	2

Figure A1. The dataset that was used in the experimental works.

References

1. Vapnik, V.N. *The Nature of Statistical Learning Theory*; Springer: New York, NY, USA, 1995.
2. Burges, C. A tutorial on support vector machines for pattern recognition. *Data Min. Knowl. Discov.* **1998**, *2*, 121–167. [CrossRef]
3. Lin, C.F.; Wang, S.D. Fuzzy support vector machine. *IEEE Trans. Neural Netw.* **2002**, *13*, 464–471. [PubMed]
4. Wang, Y.; Wang, S.; Lai, K.K. A new fuzzy support vector machine to evaluate credit risk. *IEEE Trans. Fuzzy Syst.* **2005**, *13*, 820–831. [CrossRef]
5. Ilhan, İ.; Tezel, G. A genetic algorithm–support vector machine method with parameter optimization for selecting the tag SNPs. *J. Biomed. Inform.* **2013**, *46*, 328–340. [CrossRef] [PubMed]
6. Peng, S.; Hu, Q.; Chen, Y.; Dang, J. Improved support vector machine algorithm for heterogeneous data. *Pattern Recognit.* **2015**, *48*, 2072–2083. [CrossRef]
7. Ju, W.; Cheng, H.D. A Novel Neutrosophic Logic SVM (N-SVM) and Its Application to Image Categorization. *New Math. Natural Comput.* **2013**, *9*, 27–42. [CrossRef]
8. Smarandache, F.A. *A Unifying Field in Logics: Neutrosophic Logic*; Neutrosophy, Neutrosophic Set, Neutrosophic Probability; American Research Press: Santa Fe, NM, USA, 2003.
9. Almasi, O.N.; Rouhani, M. A new fuzzy membership assignment and model selection approach based on dynamic class centers for fuzzy SVM family using the firefly algorithm. *Turk. J. Electr. Eng. Comput. Sci.* **2016**, *24*, 1797–1814. [CrossRef]
10. Tang, W.M. Fuzzy SVM with a new fuzzy membership function to solve the two-class problems. *Neural Process. Lett.* **2011**, *34*, 209. [CrossRef]
11. Wu, W.-J.; Lin, S.-W.; Moon, W.K. An artificial immune system-based support vector machine approach for classifying ultrasound breast tumor images. *J. Digit. Imaging* **2015**, *28*, 576–585. [CrossRef] [PubMed]
12. Chen, G.; Zhang, X.; Wang, Z.J.; Li, F. An enhanced artificial bee colony-based support vector machine for image-based fault detection. *Math. Probl. Eng.* **2015**, *2015*, 638926. [CrossRef]
13. Zhao, B.; Qi, Y. Image classification with ant colony based support vector machine. In Proceedings of the IEEE 2011 30th Chinese Control Conference (CCC), Yantai, China, 22–24 July 2011.
14. Güraksın, G.E.; Haklı, H.; Uğuz, H. Support vector machines classification based on particle swarm optimization for bone age determination. *Appl. Soft Comput.* **2014**, *24*, 597–602. [CrossRef]
15. Smarandache, F. A Unifying Field in Logics: Neutrosophic Logic. Neutrosophic Probability, Neutrosophic Set. In Proceedings of the 2000 Western Section Meeting (Meeting #951), Santa Barbara, CA, USA, 11–12 March 2000; Volume 951, pp. 11–12.
16. Guo, Y.; Şengür, A. NCM: Neutrosophic c-means clustering algorithm. *Pattern Recognit.* **2015**, *48*, 2710–2724. [CrossRef]
17. Guo, Y.; Şengür, A. NECM: Neutrosophic evidential c-means clustering algorithm. *Neural Comput. Appl.* **2015**, *26*, 561–571. [CrossRef]
18. Akbulut, Y.; Şengür, A.; Guo, Y.; Smarandache, F. A Novel Neutrosophic Weighted Extreme Learning Machine for Imbalanced Data Set. *Symmetry* **2017**, *9*, 142. [CrossRef]
19. Akbulut, Y.; Sengur, A.; Guo, Y.; Smarandache, F. NS-k-NN: Neutrosophic Set-Based k-Nearest Neighbors classifier. *Symmetry* **2017**, *9*, 179. [CrossRef]
20. Budak, Ü.; Guo, Y.; Şengür, A.; Smarandache, F. Neutrosophic Hough Transform. *Axioms* **2017**, *6*, 35. [CrossRef]
21. Marquardt, M.J. *Optimizing the Power of Action Learning*; Davies-Black Publishing: Palo Alto, CA, USA, 2004.
22. Rahim, M.A. A measure of styles of handling interpersonal conflict. *Acad. Manag. J.* **1983**, *26*, 368–376.
23. Sengur, A.; Guo, Y. Color texture image segmentation based on neutrosophic set and wavelet transformation. *Comput. Vis. Image Underst.* **2011**, *115*, 1134–1144. [CrossRef]
24. Gümüşeli, A.İ. *İzmir Ortaöğretim Okulları Yöneticilerinin Öğretmenler İle Aralarındaki Çatışmaları Yönetme Biçimleri*; A.Ü. Sosyal Bilimler Enstitüsü, Yayımlanmamış Doktora Tezi: Ankara, Turkey, 1994.

symmetry

MDPI

Article

New Operations of Totally Dependent-Neutrosophic Sets and Totally Dependent-Neutrosophic Soft Sets

Xiaohong Zhang [1,2,*], Chunxin Bo [3], Florentin Smarandache [4] and Choonkil Park [5]

1 School of Arts and Sciences, Shaanxi University of Science and Technology, Xi'an 710021, China
2 College of Arts and Sciences, Shanghai Maritime University, Shanghai 201306, China
3 College of Information Engineering, Shanghai Maritime University, Shanghai 201306, China; 201640311001@stu.shmtu.edu.cn
4 Department of Mathematics, University of New Mexico, Gallup, NM 87301, USA; smarand@unm.edu
5 Department of Mathematics, Hanyang University, Seoul 04763, Korea; baak@hanyang.ac.kr
* Correspondence: zhangxiaohong@sust.edu.cn or zhangxh@shmtu.edu.cn

Received: 29 April 2018; Accepted: 28 May 2018; Published: 30 May 2018

Abstract: The purpose of the paper is to study new algebraic operations and fundamental properties of totally dependent-neutrosophic sets and totally dependent-neutrosophic soft sets. First, the in-coordination relationships among the original inclusion relations of totally dependent-neutrosophic sets (called type-1 and typ-2 inclusion relations in this paper) and union (intersection) operations are analyzed, and then type-3 inclusion relation of totally dependent-neutrosophic sets and corresponding type-3 union, type-3 intersection, and complement operations are introduced. Second, the following theorem is proved: all totally dependent-neutrosophic sets (based on a certain universe) determined a generalized De Morgan algebra with respect to type-3 union, type-3 intersection, and complement operations. Third, the relationships among the type-3 order relation, score function, and accuracy function of totally dependent-neutrosophic sets are discussed. Finally, some new operations and properties of totally dependent-neutrosophic soft sets are investigated, and another generalized De Morgan algebra induced by totally dependent-neutrosophic soft sets is obtained.

Keywords: neutrosophic set; soft set; totally dependent-neutrosophic set; totally dependent-neutrosophic soft set; generalized De Morgan algebra

---

1. Introduction

In the real world, uncertainty exists universally, so uncertainty becomes the research object of many branches of science. In order to express and deal with uncertainty, many mathematical tools and methods have been put forward, for example, probability theory, fuzzy set theory [1], intuitionistic fuzzy set [2], and soft set theory [3], and these theories have been widely used in many fields [4–17].

As a general framework, F. Smarandache proposed the concept of a neutrosophic set to deal with incomplete, indeterminate, and inconsistent decision information [18]. A neutrosophic set includes truth membership, falsity membership, and indeterminacy membership. In this paper, we only discuss single-valued neutrosophic sets [19]. Recently, the neutrosophic set theory has been applied to many scientific fields (see [20–25]).

In 2006, F. Smarandache introduced, for the first time, the degree of dependence (and consequently the degree of independence) between the components of the fuzzy set, and also between the components of the neutrosophic set [26]. In 2016, the refined neutrosophic set was generalized to the degree of dependence or independence of subcomponets [26]. In this paper, we will discuss a special kind of neutrosophic set, that is, a totally dependent-neutrosophic set. A neutrosophic set A on the universe X is called totally dependent if T_A, I_A, F_A are 100% dependent, that is $T_A(x) + I_A(x) + F_A(x) \leq 1$ for any x in X.

It should be noted that a "totally dependent-neutrosophic set" is also known as a picture fuzzy set (see [27–29]) or standard neutrosophic set (see [30]). But, F. Smarandache, for the first time, used the name "totally dependent", so this name will be used from the beginning of this article.

This paper tried to prove the new ordering relation on D^* that is given in paper [29] (it is named as type-3 ordering relation in this paper) as a partial ordering relation and consider some new operations on totally dependent-neutrosophic sets and totally dependent-neutrosophic soft sets. In Section 2, we first review some basic notions of intuitionistic fuzzy sets, fuzzy soft sets, totally dependent-neutrosophic sets, and so on. Moreover, we analyze the in-coordination relationships among the original inclusion relations of totally dependent-neutrosophic sets (picture fuzzy sets), called type-1 inclusion relation, and type-2 inclusion relation in this paper; and union (intersection) operations. In Section 3, we prove that the type-3 ordering relation is a partial ordering relation and D^* makes up a lattice about type-3 intersection and type-3 union relations. In Section 4, new algebraic operations (called type-3 union and type-3 intersection) of totally dependent-neutrosophic sets are given with their operations rules. Additionally, we point out that all totally dependent-neutrosophic sets on a certain universe make up generalized De Morgan algebra about the type-3 intersection operation, type-3 union operation, and complement operation. In Section 5, we study some new operations and properties of totally dependent-neutrosophic soft sets (that is, picture fuzzy soft sets) and show that, for the appointed parameter set, totally dependent-neutrosophic soft sets over a certain universe make up generalized De Morgan algebra about type-3 intersection, type-3 union, and complement operations.

2. Preliminaries and Motivation

2.1. Some Basic Concepts

We will now review several basic concepts of intuitionistic fuzzy sets, fuzzy soft sets, standard neutrosophic sets (picture fuzzy sets), and so on.

Definition 1 [2]. *Let X be a nonempty set (universe). An intuitionistic fuzzy set A on X is an object of the form:*

$$A = \{(x, \mu_A(x), \nu_A(x)) \mid x \in X\},$$

where $\mu_A(x), \nu_A(x) \in [0,1]$, $\mu_A(x) + \nu_A(x) \leq 1$ for all x in X. $\mu_A(x) \in [0,1]$ is named the "degree of membership of x in A", and $\nu_A(x)$ is named the "degree of non-membership of x in A".

Definition 2 [6]. *Assume that F(U) is the set of all fuzzy sets on U, and E is a set of parameters, $A \subseteq E$. If F is a mapping given by $F{:}A \to F(U)$, then the pair $\langle F, A\rangle$ is known as a fuzzy soft set over U.*

Definition 3 [26,27]. *Let X be a nonempty set (universe). A totally dependent-neutrosophic set (or picture fuzzy set) A on X is an object of the form:*

$$A = \{(x, \mu_A(x), \eta_A(x), \nu_A(X)) \mid x \in X\},$$

where $\mu_A(x), \eta_A(x), \nu_A(x) \in [0,1]$, $\mu_A(x) + \eta_A(x) + \nu_A(x) \leq 1$, for all x in X. $\mu_A(x)$ is named as the "degree of positive membership of x in A", $\eta_A(x)$ is named as the "degree of neutral membership of x in A", and $\nu_A(x)$ is named the "degree of negative membership of x in A".

Let $TDNS(X)$ denote the set of all totally dependent-neutrosophic sets (or picture fuzzy sets) on X.

Definition 4 [26,28]. *Assume that U is an initial universe set and E is a set of parameters, $A \subseteq E$. If F is a mapping given by $F{:}A \to TDNS(U)$, then the pair (F, A) is called a totally dependent-neutrosophic soft set (or picture fuzzy soft set) over U.*

Obviously, a totally dependent-neutrosophic soft set (*TDNSSs*) is a mapping from parameters to *TDNS*(*U*). It is a parameterized family of totally dependent-neutrosophic sets of *U*. Clearly, $\forall e \in A$, $F(e)$ can be written as a totally dependent-neutrosophic set such that:

$$F(e) = \left\{ \left(x, \mu_{F(e)}(x), \eta_{F(e)}(x), \nu_{F(e)}(x) \right) | x \in U \right\},$$

where $\mu_{F(e)}(x)$, $\eta_{F(e)}(x)$, and $\nu_{F(e)}(x)$ are the positive membership, neutral membership, and negative membership functions, respectively.

Remark 1. *For inclusion relation and basic algebraic operations of totally dependent-neutrosophic sets (or picture fuzzy sets), we can define them as special simple-valued neutrosophic sets, that is (see [18,19], we will call them type-1 operations): for every two totally dependent-neutrosophic sets (TDNSs) A and B,*

(1) $A \subseteq_1 B$ *if* $\forall x \in X$, $\mu_A(x) \leq \mu_B(x)$, $\eta_A(x) \geq \eta_B(x)$, $\nu_A(x) \geq \nu_B(x)$;
(2) $A = B$ *if* $A \subseteq_1 B$ *and* $B \subseteq_1 A$;
(3) $A \cup_1 B = \{(x, max(\mu_A(x), \mu_B(x)), min(\eta_A(x), \eta_B(x)), min(\nu_A(x), \nu_B(x))) \mid x \in X\}$;
(4) $A \cap_1 B = \{(x, min(\mu_A(x), \mu_B(x)), max(\eta_A(x), \eta_B(x)), max(\nu_A(x), \nu_B(x))) \mid x \in X\}$;
(5) $co(A) = A^c = \{(x, \nu_A(x), \eta_A(x), \mu_A(x)) \mid x \in X\}$.

In [27], inclusion relation and basic algebraic operations of totally dependent-neutrosophic sets (or picture fuzzy sets) are defined using another approach, and we will call them type-2 operations.

Definition 5 [27]. For every two totally dependent-neutrosophic sets (TDNSs) A and B, type-2 inclusion relation, union, intersection operations, and the complement operation are defined as follows:

(1) $A \subseteq_2 B$ *if* $\forall x \in X$, $\mu_A(x) \leq \mu_B(x)$, $\eta_A(x) \leq \eta_B(x)$, $\nu_A(x) \geq \nu_B(x)$;
(2) $A = B$ *if* $A \subseteq_2 B$ *and* $B \subseteq_2 A$;
(3) $A \cup_2 B = \{(x, max(\mu_A(x), \mu_B(x)), min(\eta_A(x), \eta_B(x)), min(\nu_A(x), \nu_B(x))) \mid x \in X\}$;
(4) $A \cap_2 B = \{(x, min(\mu_A(x), \mu_B(x)), min(\eta_A(x), \eta_B(x)), max(\nu_A(x), \nu_B(x))) \mid x \in X\}$;
(5) $co(A) = A^c = \{(x, \nu_A(x), \eta_A(x), \mu_A(x)) \mid x \in X\}$.

Remark 2. *It should be noted that the type-2 operations here (for totally dependent-neutrosophic sets) are not the same as in the literature [25] (for neutrosophic sets).*

Proposition 1 [27]. *For every TDNS's A, B, and C, the following assertions are true:*

(1) *If* $A \subseteq_2 B$ *and* $B \subseteq_2 C$, *then* $A \subseteq_2 C$;
(2) $(A^c)^c = A$;
(3) $A \cap_2 B = B \cap_2 A$, $A \cup_2 B = B \cup_2 A$;
(4) $(A \cap_2 B) \cap_2 C = A \cap_2 (B \cap_2 C)$, $(A \cup_2 B) \cup_2 C = A \cup_2 (B \cup_2 C)$;
(5) $(A \cap_2 B) \cup_2 C = (A \cup_2 C) \cap_2 (B \cup_2 C)$, $(A \cup_2 B) \cap_2 C = (A \cap_2 C) \cup_2 (B \cap_2 C)$;
(6) $(A \cap_2 B)^c = A^c \cup_2 B^c$, $(A \cup_2 B)^c = A^c \cap_2 B^c$.

Definition 6 [31]. *Assume that* $\alpha = (\mu_\alpha, \eta_\alpha, \nu_\alpha, \rho_\alpha)$ *is a totally dependent-neutrosophic number (picture fuzzy number), where* $\mu_\alpha + \eta_\alpha + \nu_\alpha \leq 1$ *and* $\rho_\alpha = 1 - \mu_\alpha - \eta_\alpha - \nu_\alpha$. *The mapping* $S(\alpha) = \mu_\alpha - \nu_\alpha$ *is called the score function, and the mapping* $H(\alpha) = \mu_\alpha + \eta_\alpha + \nu_\alpha$ *is called the accuracy function, where* $S(\alpha) \in [-1, 1]$, $H(\alpha) \in [0, 1]$. *Moreover, for any two totally dependent-neutrosophic numbers (picture fuzzy number)* α *and* β,

(1) *when* $S(\alpha) > S(\beta)$, *we say that* α *is superior to* β, *and it is expressed by* $\alpha \succ \beta$;
(2) *when* $S(\alpha) = S(\beta)$, *then*

(i) *when* $H(\alpha) = H(\beta)$, *we say that* α *is equivalent to* β, *and it is expressed by* $\alpha \sim \beta$;

(ii) *when* $H(\alpha) > H(\beta)$, *we say that* α *is superior to* β, *and it is expressed by* $\alpha \succ \beta$.

Definition 7 [32]. *Let* $(M, \vee, \wedge, ^-, 0, 1)$ *be a universal algebra. Then* $(M, \vee, \wedge, ^-, 0, 1)$ *is called a generalized De Morgan algebra (or GM-algebra), if* $(M, \vee, \wedge, 0, 1)$ *is a bounded lattice and the unary operation satisfies the identities:*

(1) $(x^-)^- = x$;

(2) $(x \wedge y)^- = x^- \vee y^-$;

(3) $1^- = 0$.

2.2. On Inclusion Relations of Totally Dependent-Neutrosophic Sets (Picture Fuzzy Sets)

In Ref. [29], the set D^* is defined as follows:

$$D^* = \left\{x = (x_1, x_2, x_3)\middle| x \in [0,1]^3, x_1 + x_2 + x_3 \leq 1\right\}.$$

When $x \in D^*$, it is denoted by $x = (x_1, x_2, x_3)$, that is, the first component of x is expressed by x_1, the second component of x is expressed by x_2, and the third component of x is expressed by x_3. Moreover, the units of D^* are expressed by $1_{D^*} = (1, 0, 0)$ and $0_{D^*} = (0, 0, 1)$, respectively.

It can easily be seen that a totally dependent-neutrosophic set

$$A = \{(x, \mu_A(x), \eta_A(x), \nu_A(x)) | x \in X\},$$

can be regarded as a D^*-fuzzy set, that is, a mapping of:

$$A : X \to D^* : x \to (\mu_A(x), \eta_A(x), \nu_A(x)).$$

By Definition 5(1), the original inclusion relation of totally dependent-neutrosophic sets is built on the following order relation on D^* (it is named a type-2 inclusion relation in this paper):

$$\forall x, y \in D^*, x \leq_2 y \Leftrightarrow (x_1 \leq y_1) \wedge (x_2 \leq y_2) \wedge (x_3 \geq y_3).$$

The above "$\wedge$" denotes "and". Then,

$A \subseteq_2 B$ if and only if $(\forall x \in X)$ $(\mu_A(x), \eta_A(x), \nu_A(x)) \leq_2 (\mu_B(x), \eta_B(x), \nu_B(x))$.

Accordingly, type-2 union, intersection, and complement operations in Definition 5 are denoted as the following:

$$\begin{aligned} A \cup_2 B &= \{(\max(\mu_A(x), \mu_B(x)), \min(\eta_A(x), \eta_B(x)), \min(\nu_A(x), \nu_B(x))) \mid x \in X\} \\ &= \{(\mu_A(x), \eta_A(x), \nu_A(x)) \vee_2 (\mu_B(x), \eta_B(x), \nu_B(x)) \mid x \in X\}; \end{aligned}$$

$$\begin{aligned} A \cap_2 B &= \{(\min(\mu_A(x), \mu_B(x)), \min(\eta_A(x), \eta_B(x)), \max(\nu_A(x), \nu_B(x))) \mid x \in X\} \\ &= \{(\mu_A(x), \eta_A(x), \nu_A(x)) \wedge_2 (\mu_B(x), \eta_B(x), \nu_B(x)) \mid x \in X\}; \end{aligned}$$

$$A^{c_2} = \{ (\nu_A(x), \eta_A(x), \mu_A(x)) | x \in X\} = \left\{ (\mu_A(x), \eta_A(x), \nu_A(x))^{c_2} \middle| x \in X\right\}.$$

Now, we discuss the in-coordination relationships among type-2 inclusion relations of totally dependent-neutrosophic sets and type-2 union (intersection) operations. Consider the following examples.

Example 1. *Let* $x = (0.3, 0.4, 0.1)$, $y = (0.4, 0.3, 0.2) \in D^*$. *Then,*

$$x \vee_2 y = (0.4, 0.3, 0.1),\ x \wedge_2 y = (0.3, 0.3, 0.2).$$

Therefore, $x \nleq_2 x \vee_2 y$. *This means that* $x \vee_1 y$ *is not an upper bound of* x *and* y. *Moreover,*

$$x \vee_2 (x \wedge_2 y) = (0.3, 0.3, 0.1) \neq x.$$

It follows that the absorption law is not true for $\vee_2$ *and* $\wedge_2$.

Example 2. *Let* $x = (0.3, 0.4, 0.2)$, $y = (0.4, 0.35, 0.1) \in D^*$. *Then,*

$$x \leq_2 y,\ x \vee_2 y = (0.4, 0.35, 0.1) \neq y.$$

This means that $x \leq_2 y \nRightarrow x \vee_2 y = y$.

The above examples show that the type-2 inclusion relation of totally dependent-neutrosophic sets is inconsistent with the union and intersection operations. Now, we introduce a new inclusion of totally dependent-neutrosophic sets.

Definition 8. *Assume that A and B are two totally dependent-neutrosophic sets on X. Then, the relation* $\subseteq_3$ *defined as the following is called a type-3 inclusion relation:* $A \subseteq_3 B$ *when and only when*

$$\text{for all } x \text{ in } X,\ (\mu_A(x) < \mu_B(x),\ \nu_A(x) \geq \nu_B(x)), \text{ or } (\mu_A(x) = \mu_B(x),\ \nu_A(x) > \nu_B(x)),$$
$$\text{or } (\mu_A(x) = \mu_B(x),\ \nu_A(x) = \nu_B(x) \text{ and } \eta_A(x) \leq \eta_B(x)).$$

It should be noted that the relation $\subseteq_3$ is built on the following order relation on D^* (see [29], it is named a type-3 order relation):

$$x \leq_3 y \Leftrightarrow ((x_1 < y_1) \wedge (x_3 \geq y_3)) \vee ((x_1 = y_1) \wedge (x_3 > y_3)) \vee ((x_1 = y_1) \wedge (x_3 = y_3) \wedge (x_2 \leq y_2)).$$

Here, "$\wedge$" represents the logic and operation, and "$\vee$" represents the logic or operation.

Remark 3. *To avoid confusion, type-3 order relation on* D^* *is represented by the symbol "*$\leq_3$*". The strict proof process of the basic properties of the order relation "*$\leq_3$*" is not given in the literature [29], and these proofs are presented in this article (see next section). In addition, if* $x \leq_3 y$ *and* $y \leq_3 x$ *are not true for* $x, y \in D^*$, *then* x *is not comparable to* y, *which is expressed by* $x||_{\leq_3} y$.

3. On Type-3 Ordering Relation

In this section, we first prove that $(D^*, \leq_3)$ is a partial ordered set. Then, we prove that D^* makes up a lattice through type-3 intersection and type-3 union operations.

Proposition 2. *Let* $D^* = \left\{ x = (x_1, x_2, x_3) \middle| x \in [0,1]^3, x_1 + x_2 + x_3 \leq 1 \right\}$. *Then* $(D^*, \leq_3)$ *is a partial ordered set.*

Proof. Suppose $x, y, z \in D^*$.

(1) By the definition of $\leq_3$, we have $x \leq_3 x$.
(2) Assume that $x \leq_3 y$ and $y \leq_3 x$, then

Case 1: $x_1 < y_1$ and $x_3 \geq y_3$. According to the definition of $y \leq_3 x$, we can get $x_1 \geq y_1$, which is contradictory.

Case 2: $x_1 = y_1$ and $x_3 > y_3$. According to the definition of $y \leq_3 x$, we can get $x_3 \leq y_3$, which is contradictory.

Case 3: $x_1 = y_1$ and $x_3 = y_3$. From $x \leq_3 y$, we have $x_2 \leq y_2$; also from $y \leq_3 x$, we have $x_2 \geq y_2$. Thus, $x_2 = y_2$.

It follows that, $(x \leq_3 y$ and $y \leq_3 x) \Rightarrow (x_1 = y_1, x_2 = y_2$ and $x_3 = y_3) \Rightarrow x = y$.

(3) Assume that $x \leq_3 y$, $y \leq_3 z$, then

Case 1: $(x_1 < y_1, x_3 \geq y_3)$ and $(y_1 < z_1, y_3 \geq z_3)$. It follows that $x_1 < z_1$ and $x_3 \geq z_3$, thus $x \leq_3 z$.

Case 2: $(x_1 = y_1, x_3 > y_3)$ and $(y_1 = z_1, y_3 > z_3)$. It follows that $x_1 = z_1$ and $x_3 > z_3$, thus $x \leq_3 z$.

Case 3: $(x_1 = y_1, x_3 = y_3, x_2 \leq y_2)$ and $(y_1 = z_1, y_3 = z_3, y_2 \leq z_2)$. It follows that $x_1 = z_1$, $x_3 = z_3$ and $x_2 \leq z_2$, thus $x \leq_3 z$.

Case 4: $(x_1 < y_1, x_3 \geq y_3)$ and $(y_1 = z_1, y_3 \geq z_3)$. It follows that $x_1 < z_1$ and $x_3 \geq z_3$, thus $x \leq_3 z$.

Case 5: $(x_1 < y_1, x_3 \geq y_3)$ and $(y_1 = z_1, y_3 = z_3, y_2 \leq z_2)$. It follows that $x_1 < z_1$, $x_3 \geq z_3$ and $y_2 \leq z_2$, thus $x \leq_3 z$.

Case 6: $(x_1 = y_1, x_3 \geq y_3)$ and $(y_1 < z_1, y_3 \geq z_3)$. It follows that $x_1 < z_1$ and $x_3 \geq z_3$, thus $x \leq_3 z$.

Case 7: $(x_1 = y_1, x_3 \geq y_3)$ and $(y_1 = z_1, y_3 = z_3, y_2 \leq z_2)$. It follows that $x_1 = z_1$, $x_3 \geq z_3$ and $y_2 \leq z_2$, thus $x \leq_3 z$.

Case 8: $(x_1 = y_1, x_3 = y_3, x_2 \leq y_2)$ and $(y_1 < z_1, y_3 \geq z_3)$. It follows that $x_1 < z_1$, $x_3 \geq z_3$ and $x_2 \leq y_2$, thus $x \leq_3 z$.

Case 9: $(x_1 = y_1, x_3 = y_3, x_2 \leq y_2)$ and $(y_1 = z_1, y_3 > z_3)$, then $x_1 = z_1$, $x_3 > z_3$ and $x_2 \leq y_2$, thus $x \leq_3 z$.

It follows that, $(x \leq_3 y$ and $y \leq_3 z) \Rightarrow x \leq_3 z$.

Therefore, $(D^*, \leq_3)$ is a partial ordered set. □

Remark 4. *It is important to note that the set D^* in this paper is different from the set D^* of the literature [25]. It follows that the corresponding type-3 intersection, type-3 union, and type-3 complement operations in this paper and related operations in [25] are not the same, respectively. So, the relevant results of this paper are not the direct inference of the results in [25] (although the research ideas are similar), and the readers must pay attention to it.*

Proposition 3. *Two operations are defined on D^* as follows: $\forall x, y \in D^*$,*

$$x \wedge_3 y = \begin{cases} x, & \text{when} x \leq_3 y \\ y, & \text{when} y \leq_3 x \\ (\min(x_1, y_1), 1 - \min(x_1, y_1) - \max(x_3, y_3), \max(x_3, y_3)), & \text{otherwise} \end{cases}$$

$$x \vee_3 y = \begin{cases} y, & \text{when} x \leq_3 y \\ x, & \text{when} y \leq_3 x \\ (\max(x_1, y_1), 0, \min(x_3, y_3)), & \text{otherwise} \end{cases}$$

Then, $x \wedge_3 y = \inf(x, y)$, $x \vee_3 y = \sup(x, y)$, and $(D^*, \leq_3)$ is a lattice.

Proof. Suppose that $x \leq_3 y$ or $y \leq_3 x$, then, by the definition of "$\wedge_3$", $x \wedge_3 y$ is the largest lower bound of x, y, i.e., $x \wedge_3 y = \inf(x, y)$. Moreover, suppose that $x \leq_3 y$ or $y \leq_3 x$, then $x \vee_3 y$ is the smallest upper bound of x, y, i.e., $x \vee_3 y = \sup(x, y)$.

Next, assume that $x||_{\leq_3}y$. Then, from the definitions of "$\wedge_3$" and "$\vee_3$", we have:

$$x \wedge_3 y = (\min(x_1, y_1), 1 - \min(x_1, y_1) - \max(x_3, y_3), \max(x_3, y_3)),$$

$$x \vee_3 y = (\max(x_1, y_1), 0, \min(x_3, y_3)).$$

(1) To prove $x \wedge_3 y = \inf(x, y)$: Let

$$z = (z_1, z_2, z_3) = (\min(x_1, y_1), 1 - \min(x_1, y_1) - \max(x_3, y_3), \max(x_3, y_3)).$$

Then, $x_1 \geq \min(x_1, y_1) = z_1, x_3 \leq \max(x_3, y_3) = z_3$.
If $x_1 > z_1$ and $x_3 \leq z_3$, then $z \leq_3 x$.
If $x_1 = z_1$ and $x_3 < z_3$, then $z \leq_3 x$.
If $x_1 = z_1$ and $x_3 = z_3$, then $y_1 \geq x_1, y_3 \leq x_3$, and $x \leq_3 y$ or $y \leq_3 x$. This is contradictory to the assumed condition $x||_{\leq_3}y$.

Therefore, $z \leq_3 x$. In the same way, we can obtain $z \leq_3 y$. That is, z is a lower bound of x and y.
The next goal is to prove that z is the largest lower bound of x and y.
Suppose $a = (a_1, a_2, a_3) \in D^*$ such that $a \leq_3 x$ and $a \leq_3 y$.

Case 1: $(a_1 < x_1, a_3 \geq x_3)$ and $(a_1 < y_1, a_3 \geq y_3)$. It follows that $a_1 < \min(x_1, y_1) = z_1$ and $a_3 \geq \max(x_3, y_3) = z_3$, thus $a \leq_3 z$.
Case 2: $(a_1 = x_1, a_3 > x_3)$ and $(a_1 = y_1, a_3 > y_3)$. It follows that $a_1 = \min(x_1, y_1) = z_1$ and $a_3 > \max(x_3, y_3) = z_3$, thus $a \leq_3 z$.
Case 3: $(a_1 = x_1, a_3 = x_3, a_2 \leq x_2)$ and $(a_1 = y_1, a_3 = y_3, a_2 \leq y_2)$. It follows that $a_1 = \min(x_1, y_1) = z_1, a_3 = \max(x_3, y_3) = z_3$ and $a_2 \leq \min(x_2, y_2)$. Since $a_1 + a_2 + a_3 \leq 1$, so $a_2 \leq 1 - \min(x_1, y_1) - \max(x_3, y_3) = z_2$, thus $a \leq_3 z$.
Case 4: $(a_1 = x_1, a_3 > x_3)$ and $(a_1 < y_1, a_3 \geq y_3)$. It follows that $a_1 \leq \min(x_1, y_1)$ and $a_3 \geq \max(x_3, y_3)$. If $(a_1 < \min(x_1, y_1), a_3 \geq \max(x_3, y_3))$ or $(a_1 = \min(x_1, y_1), a_3 > \max(x_3, y_3))$, then $a \leq_3 z$; If $a_1 = \min(x_1, y_1)$ and $a_3 = \max(x_3, y_3)$, from this and the hidden condition $a_1 + a_2 + a_3 \leq 1$, we get $a_2 \leq 1 - \min(x_1, y_1) - \max(x_3, y_3) = z_2$, hence $a \leq_3 z$.
Case 5: $(a_1 < x_1, a_3 \geq x_3)$ and $(a_1 = y_1, a_3 > y_3)$. It follows that $a_1 \leq \min(x_1, y_1) = z_1$ and $a_3 \geq \max(x_3, y_3) = z_3$. Similar to Case 4, we can get $a \leq_3 z$.
Case 6: $(a_1 < x_1, a_3 \geq x_3)$ and $(a_1 = y_1, a_3 = y_3, a_2 \leq y_2)$. It follows that $y_1 < x_1$ and $y_3 \geq x_3$, so $y \leq_3 x$, it is a contradiction with hypothesis $x||_{\leq_3}y$.
Case 7: $(a_1 = x_1, a_3 > x_3)$ and $(a_1 = y_1, a_3 = y_3, a_2 \leq y_2)$. It follows that $y_1 = x_1$ and $y_3 > x_3$, so $y \leq_3 x$, which is a contradiction with hypothesis $x||_{\leq_3}y$.
Case 8: $(a_1 = x_1, a_3 = x_3, a_2 \leq x_2)$ and $(a_1 < y_1, a_3 \geq y_3)$. It follows that $x_1 < y_1$ and $x_3 \geq y_3$, so $x \leq_3 y$, which is a contradiction with hypothesis $x||_{\leq_3}y$.
Case 9: $(a_1 = x_1, a_3 = x_3, a_2 \leq x_2)$ and $(a_1 = y_1, a_3 > y_3)$. It follows that $x_1 = y_1$ and $x_3 > y_3$, so $x \leq_3 y$, which is a contradiction with hypothesis $x||_{\leq_3}y$.

Hence, $a \leq_3 z$. That is, $z = (\min(x_1, y_1), 1 - \min(x_1, y_1) - \max(x_3, y_3), \max(x_3, y_3))$ is the largest lower bound of x, y.

(2) To prove $x \vee_3 y = \sup(x, y)$: Let

$$w = (w_1, w_2, w_3) = (\max(x_1, y_1), 0, \min(x_3, y_3)).$$

Then $x_1 \leq \max(x_1, y_1) = w_1, x_3 \geq \min(x_3, y_3) = w_3$.
If $x_1 < w_1$ and $x_3 \geq w_3$, then $x \leq_3 w$.
If $x_1 = w_1$ and $x_3 > w_3$, then $x \leq_3 w$.
If $x_1 = w_1$ and $x_3 = w_3$, then $y_1 \leq x_1, y_3 \geq x_3$, so $y \leq_3 x$ or $x \leq_3 y$, which is contradictory to the assumed condition $x||_{\leq_3}y$. Thus, $x \leq_3 w$.

In the same way, we can obtain $y \leq_3 w$. Hence, w is an upper bound of x and y.
The next goal is to prove that w is the smallest upper bound of x and y.
Assume $a = (a_1, a_2, a_3) \in D^*$ such that $x \leq_3 a, y \leq_3 a$.

Case 1: $(x_1 < a_1, x_3 \geq a_3)$ and $(y_1 < a_1, y_3 \geq a_3)$. It follows that $a_1 > \max(x_1, y_1) = w_1$ and $a_3 \leq \min(x_3, y_3) = w_3$. Thus, $w \leq_3 a$.

Case 2: $(a_1 = x_1, x_3 > a_3)$ and $(a_1 = y_1, y_3 > a_3)$. It follows that $a_1 = \max(x_1, y_1) = w_1$, $a_3 < \min(x_3, y_3) = w_3$, thus $w \leq_3 a$.

Case 3: $(a_1 = x_1, a_3 = x_3, x_2 \leq a_2)$ and $(a_1 = y_1, a_3 = y_3, y_2 \leq a_2)$. It follows that $a_1 = \max(x_1, y_1) = w_1$, $a_3 = \min(x_3, y_3) = w_3$ and $a_2 \geq \max(x_2, y_2) \geq 0$, thus $w \leq_3 a$.

Case 4: $(a_1 = x_1, x_3 > a_3)$ and $(y_1 < a_1, y_3 \geq a_3)$. It follows that $a_1 \geq \max(x_1, y_1) = w_1$ and $a_3 \leq \min(x_3, y_3) = w_3$. If $(a_1 > \max(x_1, y_1) = w_1, a_3 \leq \min(x_3, y_3) = w_3)$ or $((a_1 = \max(x_1, y_1) = w_1, a_3 < \min(x_3, y_3) = w_3)$, then $w \leq {}_3a$; If $a_1 = \max(x_1, y_1) = w_1$ and $a_3 = \min(x_3, y_3) = w_3$, according the hidden condition $a_2 \geq 0$, we can get $w \leq_3 a$.

Case 5: $(x_1 < a_1, x_3 \geq a_3)$ and $(a_1 = y_1, y_3 > a_3)$. It follows that $a_1 \geq \max(x_1, y_1) = w_1$ and $a_3 \leq \min(x_3, y_3) = w_3$, similar to Case 4, so we can get $w \leq_3 a$.

Case 6: $(x_1 < a_1, x_3 \geq a_3)$ and $(a_1 = y_1, a_3 = y_3, y_2 \leq a_2)$. It follows that $x_1 < y_1$ and $x_3 \geq y_3$, so $x \leq_3 y$, which is a contradiction with hypothesis $x||_{\leq_3}y$.

Case 7: $(a_1 = x_1, x_3 > a_3)$ and $(a_1 = y_1, a_3 = y_3, y_2 \leq a_2)$. It follows that $y_1 = x_1$ and $y_3 < x_3$, so $x \leq_3 y$, which is a contradiction with hypothesis $x||_{\leq_3}y$.

Case 8: $(a_1 = x_1, a_3 = x_3, x_2 \leq a_2)$ and $(y_1 < a_1, y_3 \geq a_3)$. It follows that $y_1 < x_1$ and $y_3 \geq x_3$, so $y \leq_3 x$, which is a contradiction with hypothesis $x||_{\leq_3}y$.

Case 9: $(a_1 = x_1, a_3 = x_3, x_2 \leq a_2)$ and $(a_1 = y_1, y_3 > a_3)$. It follows that $y_1 = x_1$ and $y_3 > x_3$, so $y \leq_3 x$, which is a contradiction with hypothesis $x||_{\leq_3}y$.

Hence, $w \leq_3 a$. That is, $w = (\max(x_1, y_1), 0, \min(x_3, y_3))$ is the smallest upper bound of x, y. Integrating (1) and (2), $x \wedge_3 y = \inf(x, y)$, $x \vee_3 y = \sup(x, y)$, and $(D^*, \leq_3)$ is a lattice. □

4. New Operations and Properties of Totally Dependent-Neutrosophic Sets (Picture Fuzzy Sets)

In this section, we investigate the properties of the type-3 inclusion relation of totally dependent- neutrosophic sets, and give some new operations named type-3 union, type-3 intersection, and type-3 complement of totally dependent-neutrosophic sets and study their basic properties. Moreover, we discuss the relationship between type-3 ordering relation $\leq_3$ and the rank of totally dependent-neutrosophic sets determined by score function and accuracy function (see Definition 6).

For any totally dependent-neutrosophic sets A and B on X, applying Definition 8, we see that:

$$A \subseteq_3 B \text{ if and only if } (\mu_A(x), \eta_A(x), \nu_A(x)) \leq_3 (\mu_B(x), \eta_B(x), \nu_B(x)), \forall x \in X.$$

From this, using Proposition 2, we can get the following proposition.

Proposition 4. *If A, B, and C are totally dependent-neutrosophic sets on X, then*

(1) $A \subseteq_3 A$;

(2) $(A \subseteq_3 B, B \subseteq_3 A) \Rightarrow A = B$;

(3) $(A \subseteq_3 B, B \subseteq_3 C) \Rightarrow A \subseteq_3 C$.

Definition 9. *Assume that A and B are totally dependent-neutrosophic sets on X. The operations defined as follows are called a type-3 union, type-3 intersection, and type-3 complement, respectively:*

(1) $$(A \cup_3 B)(x) = \begin{cases} (\mu_A(x), \eta_A(x), \nu_A(x)), & \text{if} (\mu_B(x), \eta_B(x), \nu_B(x)) \leq_3 (\mu_A(x), \eta_A(x), \nu_A(x)) \\ (\mu_B(x), \eta_B(x), \nu_B(x)), & \text{if } (\mu_A(x), \eta_A(x), \nu_A(x)) \leq_3 (\mu_B(x), \eta_B(x), \nu_B(x)) \\ (\max(\mu_A(x), \mu_B(x)), 0, \min(\nu_A(x), \nu_B(x))), & \text{otherwise} \end{cases}$$

(2) $$(A \cap_3 B)(x) = \begin{cases} (\mu_A(x), \eta_A(x), \nu_A(x)), & \text{if } (\mu_A(x), \eta_A(x), \nu_A(x)) \leq_3 (\mu_B(x), \eta_B(x), \nu_B(x)) \\ (\mu_B(x), \eta_B(x), \nu_B(x)), & \text{if} (\mu_B(x), \eta_B(x), \nu_B(x)) \leq_3 (\mu_A(x), \eta_A(x), \nu_A(x)) \\ (\min(\mu_A(x), \mu_B(x)), 1 - \min(\mu_A(x), \mu_B(x)) - \max(\nu_A(x), \nu_B(x)), \max(\nu_A(x), \nu_B(x))), & \text{otherwise} \end{cases}$$

(3) $A^{c_3} = \{(x, \nu_A(x), 1 - \mu_A(x) - \eta_A(x) - \nu_A(x), \mu_A(x)) \mid x \in X\}$.

By Definition 9 and Proposition 3, we have:

Proposition 5. *If A and B are totally dependent-neutrosophic sets on X, then*

(1) $A \cup_3 B = \{(\mu_A(x), \eta_A(x), \nu_A(x)) \vee_3 (\mu_B(x), \eta_B(x), \nu_B(x)) \mid x \in X\}$;
(2) $A \cap_3 B = \{(\mu_A(x), \eta_A(x), \nu_A(x)) \wedge_3 (\mu_B(x), \eta_B(x), \nu_B(x)) \mid x \in X\}$.

Proposition 6. *If A, B, and C are totally dependent-neutrosophic sets on X, then*

(1) $A \cap_3 A = A, A \cup_3 A = A$;
(2) $A \cap_3 B = B \cap_3 A, A \cup_3 B = B \cup_3 A$;
(3) $(A \cap_3 B) \cap_3 C = A \cap_3 (B \cap_3 C), (A \cup_3 B) \cup_3 C = A \cup_3 (B \cup_3 C)$;
(4) $A \cap_3 (B \cup_3 A) = A, A \cup_3 (B \cap_3 A) = A$;
(5) $A \subseteq_3 B \Leftrightarrow A \cup_3 B = B; A \subseteq_3 B \Leftrightarrow A \cap_3 B = A$.

By Definition 9(3), we have:

Proposition 7. *For any totally dependent-neutrosophic sets on X,* $(A^{c_3})^{c_3} = A$.

Proposition 8. *If A and B are totally dependent-neutrosophic sets on X, then:*

(1) $(A \cap_3 B)^{c_3} = A^{c_3} \cup_3 B^{c_3}$;
(2) $(A \cup_3 B)^{c_3} = A^{c_3} \cap_3 B^{c_3}$.

Proof. By Definition 9(3), we have:

$$A^{c_3} = \{(x, \nu_A(x), 1 - \mu_A(x) - \eta_A(x) - \nu_A(x), \mu_A(x)) | x \in X\}$$

$$B^{c_3} = \{(x, \nu_B(x), 1 - \mu_B(x) - \eta_B(x) - \nu_B(x), \mu_B(x)) | x \in X\}$$

(1) If $B \subseteq_3 A$, then:

Case 1: $\mu_B(x) < \mu_A(x)$ and $\nu_B(x) \geq \nu_A(x)$. It follows that $A^{c_3} \subseteq_3 B^{c_3}$. Thus $(A \cap_3 B)^{c_3} = A^{c_3} \cup_3 B^{c_3}$.
Case 2: $\mu_B(x) = \mu_A(x)$ and $\nu_B(x) > \nu_A(x)$. It follows that $A^{c_3} \subseteq_3 B^{c_3}$. Thus $(A \cap_3 B)^{c_3} = A^{c_3} \cup_3 B^{c_3}$.
Case 3: $\mu_B(x) < \mu_A(x)$, $\nu_B(x) = \nu_A(x)$ and $\eta_B(x) \leq \eta_A(x)$. Then $1 - \mu_A(x) - \eta_A(x) - \nu_A(x) \leq 1 - \mu_B(x) - \eta_B(x) - \nu_B(x)$. Thus $A^{c_3} \subseteq_3 B^{c_3}$, and $(A \cap_3 B)^{c_3} = B^{c_3} = A^{c_3} \cup_3 B^{c_3}$.

Similarly, if $A \subseteq_3 B$, then $(A \cap_3 B)^{c_3} = A^{c_3} \cup_3 B^{c_3}$.
If neither $B \subseteq_3 A$ nor $A \subseteq_3 B$, then:

$$A \cap_3 B = \{(x, \min(\mu_A(x), \mu_B(x), 1 - \min(\mu_A(x), \mu_B(x)) - \max(\nu_A(x), \nu_B(x)), \max(\nu_A(x), \nu_B(x))) | x \in X\},$$

$$A \cup_3 B = \{(x, \max(\mu_A(x), \mu_B(x)), 0, \min(\nu_A(x), \nu_B(x))) | x \in X\}.$$

Thus

$$(A \cap_3 B)^{c_3} = \{(x, \max(\nu_A(x), \nu_B(x)), 0, \min(\mu_A(x), \mu_B(x))) | x \in X\}.$$

$$A^{c_3} \cup_3 B^{c_3} = \{(x, \max(\nu_A(x), \nu_B(x)), 0, \min(\mu_A(x), \mu_B(x))) | x \in X\}.$$

Hence $(A \cap_3 B)^{c_3} = A^{c_3} \cup_3 B^{c_3}$.

(2) By (1) and Proposition 7, we can get that $(A \cup_3 B)^{c_3} = A^{c_3} \cap_3 B^{c_3}$. □

Theorem 1. *Let TDNS(X) be the set of all totally dependent-neutrosophic sets on X, and*

$$0_{TDNS} = \{(x, 0, 0, 1) \mid x \in X\},\ 1_{TDNS} = \{(x, 1, 0, 0) \mid x \in X\}.$$

Then, $(TDNS(X), \cup_3, \cap_3, {}^{c}{}_3, 0_{TDNS}, 1_{TDNS})$ is a GM-algebra (i.e., generalized De Morgan algebra).

Proof. Applying Proposition 6–8 and Definition 7, we can see that $(TDNS(X), \cup_3, \cap_3, {}^{c}{}_3, 0_{TDNS}, 1_{TDNS})$ is a GM-algebra.

We can verify that the distributive law is not true for $(TDNS(X), \cup_3, \cap_3, {}^{c}{}_3, 0_{TDNS}, 1_{TDNS})$, that is, it is not a De Morgan algebra. □

Example 3. *Let X = {a, b} and*

$$A = \{(a, 0.1, 0.4, 0.3), (b, 0, 0, 1)\},\ B = \{(a, 0.3, 0.1, 0.5), (b, 0, 0, 1)\},\ C = \{(a, 0.2, 0.2, 0.4), (b, 1, 0, 0)\}.$$

Then:

$$(A \cap_3 B) \cup_3 C = \{(a, 0.2, 0.2, 0.4), (b, 1, 0, 0)\};\ (A \cup_3 C) \cap_3 (B \cup_3 C) = \{(a, 0.2, 0.4, 0.4), (b, 1, 0, 0)\};$$

$$(A \cup_3 B) \cap_3 C = \{(a, 0.2, 0.2, 0.4), (b, 0, 0, 1)\};\ (A \cap_3 C) \cup_3 (B \cap_3 C) = \{(a, 0.2, 0, 0.4), (b, 0, 0, 1)\}.$$

Therefore,

$$(A \cap_3 B) \cup_3 C \neq (A \cup_3 C) \cap_3 (B \cup_3 C),\ (A \cup_3 B) \cap_3 C \neq (A \cap_3 C) \cup_3 (B \cap_3 C).$$

Proposition 9. *If $\alpha = (\mu_\alpha, \eta_\alpha, \nu_\alpha, \rho_\alpha)$, $\beta = (\mu_\beta, \eta_\beta, \nu_\beta, \rho_\beta)$ are two totally dependent-neutrosophic numbers, and $(\mu_\alpha, \eta_\alpha, \nu_\alpha) \leq_3 (\mu_\beta, \eta_\beta, \nu_\beta)$, then $\alpha \prec \beta$ or $\alpha \sim \beta$.*

Proof. Assume $(\mu_\alpha, \eta_\alpha, \nu_\alpha) \leq_3 (\mu_\beta, \eta_\beta, \nu_\beta)$. By the definition of type-3 order relation $\leq_3$, we have

Case 1: $(\mu_\alpha < \mu_\beta, \nu_\alpha \geq \nu_\beta)$ or $(\mu_\alpha = \mu_\beta, \nu_\alpha > \nu_\beta)$. It follows that $S(\alpha) = \mu_\alpha - \nu_\alpha < \mu_\beta - \nu_\beta = S(\beta)$. Thus, $\alpha \prec \beta$.

Case 2: $(\mu_\alpha = \mu_\beta, \nu_\alpha = \nu_\beta$ and $\eta_\alpha < \eta_\beta$. It follows that $S(\alpha) = S(\beta)$, $H(\alpha) < H(\beta)$. Thus, $\alpha \prec \beta$.

Case 3: $\mu_\alpha = \mu_\beta, \nu_\alpha = \nu_\beta$ and $\eta_\alpha = \eta_\beta$. It follows that $S(\alpha) = S(\beta)$, $H(\alpha) = H(\beta)$. Thus, $\alpha \sim \beta$.

Therefore, the proof is completed. □

Example 4. *Let $\alpha = (0.3, 0.4, 0.1, 0.2)$ and $\beta = (0.5, 0.2, 0.2, 0.1)$ be two totally dependent-neutrosophic numbers, then $S(\alpha) < S(\beta)$, $\alpha \prec \beta$, but $(0.3, 0.4, 0.1) ||_{\leq_3} (0.5, 0.2, 0.2)$. That is, the inverse of Proposition 9 is not true.*

5. New Operations and Properties of Totally Dependent-Neutrosophic Soft Sets

In this section, we investigate some new operations on totally dependent-neutrosophic soft sets (picture fuzzy soft sets), including type-3 intersection (type-3 union, type-3 complement).

The notions of intersection, union, and complement of totally dependent-neutrosophic (picture fuzzy) soft sets are introduced in [28]. To avoid confusion, these operations are called a type-2 intersection, type-2 union, and type-2 complement in this paper.

Remark 5. *Note that, for type-1 operations of totally dependent-neutrosophic soft sets (picture fuzzy soft sets), we denote the operations in neutrosophic soft sets (see [33,34]), that is, the corresponding operations of totally dependent-neutrosophic soft sets, as a kind of special neutrosophic soft sets.*

Definition 10 [28]. *The type-2 complement of a totally dependent-neutrosophic soft set (F, A) over U is denoted as* $(F, A)^{c_2}$ *and is defined by* $(F, A)^{c_2} = (F^{c_2}, A)$*, where* $F^{c_2}: A \to SNS(U)$ *is a mapping given by:*

$$F^{c_2}(e) = (F(e))^{c_2} = \left\{ \left(x, \nu_{F(e)}(x), \eta_{F(e)}(x), \mu_{F(e)}(x)\right) \middle| x \in U \right\}, \forall e \in A.$$

Definition 11 [28]. *The type-2 intersection of two totally dependent-neutrosophic soft sets (F, A) and (G, B) over a common universe U is a totally dependent-neutrosophic soft set (H, C), where* $C = A \cup B$ *and for all* $e \in C$,

$$H(e) = \begin{cases} F(e), & \text{if } e \in A - B \\ G(e), & \text{if } e \in B - A \\ F(e) \cap_2 G(e), & \text{if } e \in A \cap B \end{cases}$$

That is, $\forall e \in A \cap B$, $H(e) = \{(x, \min(\mu_{F(e)}(x), \mu_{G(e)}(x)), \min(\eta_{F(e)}(x), \eta_{G(e)}(x)), \max(\nu_{F(e)}(x), \nu_{G(e)}(x) \mid x \in U\}$*. This relation is denoted by* $(F, A) \cap_2 (G, B) = (H, C)$*.*

Definition 12 [28]. *The type-2 union of two totally dependent-neutrosophic soft sets (F, A) and (G, B) over a common universe U is a totally dependent-neutrosophic soft set (H, C), where* $C = A \cup B$ *and for all* $e \in C$,

$$H(e) = \begin{cases} F(e), & \text{if } e \in A - B \\ G(e), & \text{if } e \in B - A \\ F(e) \cup_2 G(e), & \text{if } e \in A \cap B \end{cases}$$

That is, $\forall e \in A \cap B$, $H(e) = \{(x, \max(\mu_{F(e)}(x), \mu_{G(e)}(x)), \min(\eta_{F(e)}(x), \eta_{G(e)}(x)), \min(\nu_{F(e)}(x), \nu_{G(e)}(x) \mid x \in U\}$*. This relation is denoted by* $(F, A) \cup_2 (G, B) = (H, C)$*.*

Now, we discuss the type-3 complement, type-3 intersection, and type-3 union of totally dependent-neutrosophic soft sets. First, we introduce type-3 inclusion relation on totally dependent neutrosophic soft sets.

Definition 13. *Let (F, A) and (G, B) be two totally dependent-neutrosophic soft sets over U. Then, (F, A) is called a totally dependent-neutrosophic soft subset of (G, B), denoted by* $(F, A) \subseteq_3 (G, B)$*, if:*

(1) $A \subseteq B$;

(2) $\forall e \in A$, $F(e) \subseteq_3 G(e)$*, that is,* $\forall x \in U$, $(\mu_{F(e)}(x) < \mu_{G(e)}(x), \nu_{F(e)}(x) \geq \nu_{G(e)}(x))$*, or* $(\mu_{F(e)}(x) = \mu_{G(e)}(x), \nu_{F(e)}(x) > \nu_{G(e)}(x))$*, or* $(\mu_{F(e)}(x) = \mu_{G(e)}(x), \nu_{F(e)}(x) = \nu_{G(e)}(x)$ *and* $\eta_{F(e)}(x) \leq \eta_{G(e)}(x))$*.*

Example 5. *Let* $U = \{x_1, x_2, x_3, x_4\}$ *and* $E = \{e_1, e_2, e_3, e_4, e_5\}$*. Suppose that (F, A) and (G, B) are two SNSSs over U,* $A = \{e_1, e_2\}$, $B = \{e_1, e_2, e_5\}$ *and*

$$(F, A) = \begin{pmatrix} & e_1 & e_2 \\ x_1 & (0.1, 0.2, 0.5) & (0.2, 0.1, 0.6) \\ x_2 & (0.4, 0.2, 0.3) & (0.2, 0.2, 0.5) \\ x_3 & (0.2, 0.3, 0.4) & (0.1, 0.4, 0.2) \\ x_4 & (0.3, 0.3, 0.2) & (0.4, 0.0, 0.5) \end{pmatrix},$$

$$(G, B) = \begin{pmatrix} & e_1 & e_2 & e_5 \\ x_1 & (0.3, 0.2, 0.4) & (0.2, 0.2, 0.4) & (0.7, 0.1, 0.2) \\ x_2 & (0.6, 0.0, 0.3) & (0.2, 0.1, 0.3) & (0.5, 0.3, 0.1) \\ x_3 & (0.3, 0.4, 0.2) & (0.1, 0.4, 0.2) & (0.8, 0.0, 0.1) \\ x_4 & (0.5, 0.2, 0.2) & (0.4, 0.1, 0.5) & (0.2, 0.5, 0.1) \end{pmatrix}$$

Then, (F, A) is a totally dependent-neutrosophic soft subset of (G, B).

Definition 14. *Let (F, A) and (G, B) be two totally dependent-neutrosophic soft sets over U. (F, A) and (G, B) are said totally dependent-neutrosophic soft equals, denoted (F, A) = (G, B), if (F, A)* $\subseteq_3$ *(G, B) and (G, B)* $\subseteq_3$ *(F, A).*

By Proposition 4, we know that $(F, A) = (G, B) \Leftrightarrow A = B$ and $F(e) = G(e)$, $\forall e \in A$.

Definition 15. *Let (F, A) be a totally dependent-neutrosophic soft set over U. Type-3 complement of (F, A) is denoted as* $(F, A)^{c_3}$ *and is defined by* $(F, A)^{c_3} = (F^{c_3}, A)$*, where* $F^{c_3}\colon A \to TDNS(U)$ *is a mapping given by:*

$$F^{c_3}(e) = (F(e))^{c_3} = \left\{ \left(x, \nu_{F(e)}(x), 1 - \mu_{F(e)}(x) - \eta_{F(e)}(x) - \nu_{F(e)}(x), \mu_{F(e)}(x)\right) \middle| x \in U \right\}, \forall e \in A.$$

Definition 16. *Type-3 union of two totally dependent-neutrosophic soft sets (F, A) and (G, B) over U can be defined as* $(F, A) \cup_3 (G, B) = (H, C)$*, where* $C = A \cup B$*, and* $\forall e \in C$*,*

$$H(e) = \begin{cases} F(e), & \text{if } e \in A - B \\ G(e), & \text{if } e \in B - A \\ F(e) \cup_3 G(e), & \text{if } e \in A \cap B \end{cases}$$

Example 6. *Let* $U = \{x_1, x_2, x_3, x_4\}$*,* $E = \{e_1, e_2, e_3, e_4, e_5\}$*,* $A = \{e_1, e_2\}$*,* $B = \{e_1, e_3, e_5\}$*, and*

$$(F, A) = \begin{pmatrix} & e_1 & e_2 \\ x_1 & (0.1, 0.2, 0.6) & (0.4, 0.2, 0.3) \\ x_2 & (0.2, 0.1, 0.1) & (0.3, 0.1, 0.6) \\ x_3 & (0.7, 0.3, 0.0) & (0.5, 0.2, 0.3) \\ x_4 & (0.4, 0.0, 0.3) & (0.8, 0.0, 0.1) \end{pmatrix},$$

$$(G, B) = \begin{pmatrix} & e_1 & e_3 & e_5 \\ x_1 & (0.4, 0.3, 0.2) & (0.2, 0.7, 0.1) & (0.6, 0.1, 0.2) \\ x_2 & (0.7, 0.1, 0.1) & (0.8, 0.0, 0.2) & (0.3, 0.2, 0.4) \\ x_3 & (0.3, 0.5, 0.1) & (0.2, 0.4, 0.2) & (0.5, 0.3, 0.0) \\ x_4 & (0.6, 0.1, 0.2) & (0.4, 0.3, 0.1) & (0.1, 0.1, 0.6) \end{pmatrix}.$$

Then, $(F, A) \cup_3 (G, B) = (H, C)$*, where* $C = A \cup B = \{e_1, e_2, e_3, e_5\}$ *and*

$$(F, A) \cup_3 (G, B) = (H, C) =$$

$$\begin{pmatrix} & e_1 & e_2 & e_3 & e_5 \\ x_1 & (0.4, 0.3, 0.2) & (0.4, 0.2, 0.3) & (0.2, 0.7, 0.1) & (0.6, 0.1, 0.2) \\ x_2 & (0.7, 0.1, 0.1) & (0.7, 0.1, 0.1) & (0.3, 0.1, 0.6) & (0.3, 0.2, 0.4) \\ x_3 & (0.7, 0.3, 0.0) & (0.5, 0.2, 0.3) & (0.2, 0.4, 0.2) & (0.5, 0.3, 0.0) \\ x_4 & (0.6, 0.1, 0.2) & (0.8, 0.0, 0.1) & (0.4, 0.3, 0.1) & (0.1, 0.1, 0.6) \end{pmatrix}.$$

Definition 17. *Assume that* $A, B \subseteq E$ *and* $A \cap B \neq \emptyset$*. Type-3 intersection of two totally dependent-neutrosophic soft sets (F, A) and (G, B) over U can be defined as* $(F, A) \cap_3 (G, B) = (H, C)$*, where* $C = A \cap B$*, and* $\forall e \in C$*,* $H(e) = F(e) \cap_3 F(e)$*.*

Example 7. *Consider the totally dependent-neutrosophic soft sets (F, A), (G, B) in Example 6. We have (F, A) $\cap_3$ (G, B) = (H, C), where C = A $\cap$ B = $\{e_1\}$ and*

$$(F,A) \cap_3 (G,B) = (H,C) = \begin{pmatrix} & e_1 \\ x_1 & (0.1, 0.2, 0.6) \\ x_2 & (0.2, 0.1, 0.1) \\ x_3 & (0.3, 0.5, 0.1) \\ x_4 & (0.4, 0.0, 0.3) \end{pmatrix}.$$

Proposition 10. *If (F, A), (G, B), and (H, C) are totally dependent-neutrosophic soft sets over U, then:*

(1) $\left((F,A)^{c_3}\right)^{c_3} = (F,A)$;
(2) $(F, A) \cup_3 (F, A) = (F, A)$, $(F, A) \cap_3 (F, A) = (F, A)$;
(3) $(F, A) \cup_3 (G, B) = (G, B) \cup_3 (F, A)$, $(F, A) \cap_3 (G, B) = (G, B) \cap_3 (F, A)$;
(4) $((F, A) \cup_3 (G, B)) \cup_3 (H, C) = (F, A) \cup_3 ((G, B) \cup_3 (H, C))$;
(5) $((F, A) \cap_3 (G, B)) \cap_3 (H, C) = (F, A) \cap_3 ((G, B) \cap_3 (H, C))$, *when* $A \cap B \cap C \neq \emptyset$.

Proof. (1) It is easy to verify from Proposition 7 and Definition 15.
(2) and (3) It is obvious from Definitions 16 and 17.
(4) The proof is similar to Proposition 3.9 in [15].
(5) The proof is similar to Proposition 3.10 in [15].
□

Proposition 11. *If (F, A) and (G, A) are two totally dependent-neutrosophic soft sets over U, then:*

(1) $((F, A) \cup_3 (G, A))^{c3} = (F, A)^{c3} \cap_3 (G, A)^{c3}$;
(2) $((F, A) \cap_3 (G, A))^{c3} = (F, A)^{c3} \cup_3 (G, A)^{c3}$.

Proof.

(1) Assume that $(F, A) \cup_3 (G, A) = (H, A)$ and $(F, A)^{c3} \cap_3 (G, A)^{c3} = (I, A)$. Then:

$\forall e \in A$, $H(e) = F(e) \cup_3 G(e)$ (by Definition 16);

$\forall e \in A$, $I(e) = F^{c3}(e) \cap_3 G^{c3}(e) = (F(e))^{c3} \cap_3 (G(e))^{c3} = (F(e) \cup_3 G(e))^{c3}$
(by Definitions 15, 17 and Proposition 8).

Thus $\forall e \in A$, $H^{c3}(e) = (H(e))^{c3} = (F(e) \cup_3 G(e))^{c3} = I(e)$. Since $((F, A) \cup_3 (G, A))^{c3} = (H^{c3}, A)$, it follows that $((F, A) \cup_3 (G, A))^{c3} = (H^{c3}, A) = (I, A) = (F, A)^{c3} \cap_3 (G, A)^{c3}$.

(2) By (1), and using Proposition 10(1) we can get $((F, A) \cap_3 (G, A))^{c3} = (F, A)^{c3} \cup_3 (G, A)^{c3}$. □

Proposition 12. *If (F, A) and (G, A) are two totally dependent-neutrosophic soft sets over U, then:*

(1) $(F, A) \cap_3 ((F, A) \cup_3 (G, A)) = (F, A)$;
(2) $(F, A) \cup_3 ((F, A) \cap_3 (G, A)) = (F, A)$;
(3) $(F, A) \subseteq_3 (G, A) \Leftrightarrow (F, A) \cup_3 (G, A) = (G, A)$;
(4) $(F, A) \subseteq_3 (G, A) \Leftrightarrow (F, A) \cap_3 (G, A) = (F, A)$.

Proof.

(1) Assume that $(F, A) \cup_3 (G, A) = (H, A)$ and $(F, A) \cap_3 ((F, A) \cup_3 (G, A)) = (I, A)$. Then:

$$\forall e \in A, H(e) = F(e) \cup_3 G(e) \text{ (by Definition 16);}$$

$$\forall e \in A, I(e) = F(e) \cap_3 H(e) = F(e) \cap_3 (F(e) \cup_3 G(e)) = F(e)$$
$$\text{(by Definition 17 and Proposition 6(3)).}$$

Thus, $\forall e \in A, I(e) = F(e)$, and it follows that $(F, A) \cap_3 ((F, A) \cup_3 (G, A)) = (F, A)$.

(2) The proof is similar to (1).

(3) From Proposition 6(4) and Definition 16, we get that $(F, A) \subseteq_3 (G, A) \Leftrightarrow (F, A) \cup_3 (G, A) = (G, A)$.

(4) From Proposition 6(4) and Definition 17, we get that $(F, A) \subseteq_3 (G, A) \Leftrightarrow (F, A) \cap_3 (G, A) = (F, A)$. □

Theorem 2. *Let TDNSS(U, A) be the set of all totally dependent-neutrosophic soft sets over universe U with a fixed parameter set A. Denote:*

$$(0_{TDNSS}, A) \in TDNSS(U, A);\ \forall e \in A,\ 0_{TDNSS}(e) = \{(x, 0, 0, 1) | x \in U\};$$

$$(1_{TDNSS}, A) \in TDNSS(U, A);\ \forall e \in A,\ 1_{TDNSS}(e) = \{(x, 1, 0, 0) | x \in U\}.$$

Then $(TDNSS(U, A), \cup_3, \cap_3, {}^{c3}, (0_{TDNSS}, A), (1_{TDNSS}, A))$ *is a GM-algebra.*

Proof. By Definition 13, we have $(F, A) \subseteq_3 (1_{TDNSS}, A)$ and $(0_{TDNSS}, A) \subseteq_3 (F, A), \forall (F, A) \in TDNSS(U, A)$.

By Propositions 10 and 12, we know that $(TDNSS(U, A), \cup_3, \cap_3, (0_{TDNSS}, A), (1_{TDNSS}, A))$ is a bounded lattice. Therefore, by Propositions 10(1), 12 and Definition 7, we get that $(TDNSS(U, A), \cup_3, \cap_3, {}^{c3}, (0_{TDNSS}, A), (1_{TDNSS}, A))$ is a GM-algebra. □

We can verify that the distributive law (with respect to $\cup_3$ and $\cap_3$) in $TDNSS(U, A)$ is not satified.

Example 8. *Let* $U = \{x_1, x_2, x_3, x_4\}$, $E = \{e_1, e_2, e_3, e_4\}$, *and* $A = \{e_1, e_2\}$. *Suppose that (F, A), (G, A), and (H, A) are totally dependent-neutrosophic soft sets over U, and*

$$F(e_1) = \{(x_1, 0.1, 0.4, 0.3), (x_2, 0.4, 0.2, 0.3), (x_3, 0.2, 0.3, 0.4), (x_4, 0.3, 0.3, 0.2)\},$$
$$F(e_2) = \{(x_1, 0.1, 0.1, 0.1), (x_2, 0.2, 0.2, 0.2), (x_3, 0.3, 0.3, 0.3), (x_4, 0.4, 0.4, 0.2)\};$$

$$G(e_1) = \{(x_1, 0.3, 0.1, 0.5), (x_2, 0.6, 0.0, 0.3), (x_3, 0.3, 0.4, 0.2), (x_4, 0.5, 0.2, 0.2)\},$$
$$G(e_2) = \{(x_1, 0.2, 0.2, 0.4), (x_2, 0.2, 0.1, 0.3), (x_3, 0.1, 0.4, 0.2), (x_4, 0.4, 0.1, 0.5)\};$$

$$H(e_1) = \{(x_1, 0.2, 0.2, 0.4), (x_2, 0.2, 0.1, 0.1), (x_3, 0.7, 0.3, 0.0), (x_4, 0.4, 0.0, 0.3)\},$$
$$H(e_2) = \{(x_1, 0.4, 0.2, 0.3), (x_2, 0.3, 0.1, 0.6), (x_3, 0.5, 0.2, 0.3), (x_4, 0.8, 0.0, 0.1)\}.$$

Then:

$$((F, A) \cup_3 (G, A)) \cap_3 (H, A) = \begin{pmatrix} & e_1 & e_2 \\ x_1 & (0.2, 0.2, 0.4) & (0.2, 0.5, 0.3) \\ x_2 & (0.2, 0.5, 0.3) & (0.2, 0.2, 0.6) \\ x_3 & (0.3, 0.4, 0.2) & (0.3, 0.4, 0.3) \\ x_4 & (0.4, 0.0, 0.3) & (0.4, 0.4, 0.2) \end{pmatrix},$$

$$((F, A) \cap_3 (H, A)) \cup_3 ((G, A) \cap_3 (H, A) = \begin{pmatrix} & e_1 & e_2 \\ x_1 & (0.2, 0.0, 0.4) & (0.2, 0.0, 0.3) \\ x_2 & (0.2, 0.5, 0.3) & (0.2, 0.2, 0.6) \\ x_3 & (0.3, 0.4, 0.2) & (0.3, 0.3, 0.3) \\ x_4 & (0.4, 0.0, 0.3) & (0.4, 0.4, 0.2) \end{pmatrix};$$

$$((F,A)\cap_3(G,A))\cup_3(H,A)=\begin{pmatrix} & e_1 & e_2 \\ x_1 & (0.2,0.2,0.4) & (0.4,0.2,0.3) \\ x_2 & (0.4,0.0,0.1) & (0.3,0.0,0.3) \\ x_3 & (0.7,0.3,0.0) & (0.5,0.2,0.3) \\ x_4 & (0.4,0.0,0.2) & (0.8,0.0,0.1) \end{pmatrix},$$

$$((F,A)\cup_3(H,A))\cap_3((G,A)\cup_3(H,A)=\begin{pmatrix} & e_1 & e_2 \\ x_1 & (0.2,0.4,0.4) & (0.4,0.2,0.3) \\ x_2 & (0.4,0.0,0.1) & (0.3,0.0,0.3) \\ x_3 & (0.7,0.3,0.0) & (0.5,0.2,0.3) \\ x_4 & (0.4,0.0,0.2) & (0.8,0.0,0.1) \end{pmatrix}.$$

Hence

$$((F,A)\cup_3(G,A))\cap_3(H,A)\neq((F,A)\cap_3(H,A))\cup_3((G,A))\cap_3(H,A));$$

$$((F,A)\cap_3(G,A))\cup_3(H,A)\neq((F,A)\cup_3(H,A))\cap_3((G,A))\cup_3(H,A)).$$

6. Conclusions

In this paper, we prove that the type-3 ordering relation $\leq_3$ is a partial ordering relation and D^* makes up a lattice with respect to type-3 intersection and type-3 union operations. Then, we give some new operations of totally dependent-neutrosophic (picture fuzzy) sets and totally dependent-neutrosophic (picture fuzzy) soft sets, and their properties are presented. At the same time, we point out all of the totally dependent-neutrosophic (picture fuzzy) sets on X make up generalized De Morgan algebra with respect to type-3 intersection, type-3 union, and type-3 complement operations. Moreover, we prove that for appointed parameter sets, all of the totally dependent-neutrosophic (picture fuzzy) soft sets over U can also generate a generalized De Morgan algebra based on type-3 algebraic operations.

It can be seen from the results of this paper that the type-3 algebraic operations of the totally dependent-neutrosophic (picture fuzzy) sets have good properties and a new property which is different from the fuzzy sets and the intuitionistic fuzzy sets (because the distribution law is not established). This theoretically shows that although the totally dependent-neutrosophic (picture fuzzy) set is a generalization of the fuzzy set and intuitionistic fuzzy set, it has different characteristics. In fact, the type-1 and type-2 algebraic operations of totally dependent-neutrosophic (picture fuzzy) sets (including the order relations, see Remark 1 and Definition 5) simply imitate the corresponding operations of the intuitionistic fuzzy sets, which cannot truly reflect the original idea of totally dependent-neutrosophic (picture fuzzy) sets. For example, for type-2 inclusion relation Definition 5, $A\subseteq_2 B$ if ($\forall x\in X$, $\mu_A(x)\leq\mu_B(x)$, $\eta_A(x)\leq\eta_B(x)$, $\nu_A(x)\geq\nu_B(x)$), which means that the first two membership functions (μ, η) have the same effect, but the three membership functions in the original definition of neutrosophic sets are completely independent, which is incongruous. For the type-1 inclusion relation, there is a similar problem. From Definition 8, we know that the type-3 inclusion relation has overcome this defect.

As further research topics, we will discuss the applications of the new algebraic operations in multiple attribute decision making and uncertainty reasoning. At the same time, it is a meaningful topic for reviewers to suggest developing new directions, such as drawing on new ideas in [35].

Author Contributions: X.Z. and C.B. initiated the research and wrote the paper; F.S. and C.P. supervised the research work and provided helpful suggestions.

Acknowledgments: This research was funded by the National Natural Science Foundation of China (Grant Numbers. 61573240, 61473239).

Conflicts of Interest: The authors declare no conflicts of interest.

References

1. Zadeh, L.A. Fuzzy sets. *Inf. Control* **1965**, *8*, 338–353. [CrossRef]
2. Atanassov, K.T. Intuitionistic fuzzy sets. *Fuzzy Sets Syst.* **1986**, *20*, 87–96. [CrossRef]
3. Molodtsov, D.A. Soft set theory-first results. *Comput. Math. Appl.* **1999**, *37*, 19–31. [CrossRef]
4. Zhang, X.H.; Pei, D.W.; Dai, J.H. *Fuzzy Mathematics and Rough Set Theory*; Tsinghua University Press: Beijing, China, 2013.
5. Zhang, X.H. Fuzzy anti-grouped filters and fuzzy normal filters in pseudo-BCI algebras. *J. Intell. Fuzzy Syst.* **2017**, *33*, 1767–1774. [CrossRef]
6. Roy, A.R.; Maji, P.K. A fuzzy soft set theoretic approach to decision making problems. *J. Comput. Math. Appl.* **2007**, *203*, 412–418. [CrossRef]
7. Jun, Y.B. Soft BCK/BCI-algebras. *Comput. Math. Appl.* **2008**, *56*, 1408–1413. [CrossRef]
8. Feng, F.; Jun, Y.B.; Liu, X.; Li, L. An adjustable approach to fuzzy soft set based decision making. *J. Comput. Appl. Math.* **2010**, *234*, 10–20. [CrossRef]
9. Zhan, J.M.; Ali, M.I.; Mehmood, N. On a novel uncertain soft set model: Z-soft fuzzy rough set model and corresponding decision making methods. *Appl. Soft Comput.* **2017**, *56*, 446–457. [CrossRef]
10. Zhan, J.M.; Zhu, K.Y. A novel soft rough fuzzy set: Z-soft rough fuzzy ideals of hemirings and corresponding decision making. *Soft Comput.* **2017**, *21*, 1923–1936. [CrossRef]
11. Ma, X.L.; Zhan, J.M.; Ali, M.I.; Mehmood, N. A survey of decision making methods based on two classes of hybrid soft set models. *Artif. Intell. Rev.* **2018**, *49*, 511–529. [CrossRef]
12. Zhan, J.M.; Alcantud, J.C.R. A novel type of soft rough covering and its application to multicriteria group decision making. *Artif. Intell. Rev.* **2018**. [CrossRef]
13. Qin, K.Y.; Zhao, H. Lattice structures of fuzzy soft sets. In *Advanced Intelligent Computing Theories and Applications*; Springer: Berlin, Germany, 2010; pp. 126–133.
14. Yang, H.L.; Guo, Z.L. Kernels and closures of soft set relations and soft relation mapping. *Comput. Math. Appl.* **2011**, *61*, 651–662. [CrossRef]
15. Li, Z.W.; Zheng, D.W.; Hao, J. L-fuzzy soft sets based on complete Boolean lattices. *Comput. Math. Appl.* **2012**, *64*, 2558–2574. [CrossRef]
16. Zhang, X.H.; Wang, W.S. Lattice-valued interval soft sets-A general frame of many soft models. *J. Intell. Fuzzy Syst.* **2014**, *26*, 1311–1321.
17. Zhang, X.H.; Park, C.; Wu, S.P. Soft set theoretical approach to pseudo-BCI algebras. *J. Intell. Fuzzy Syst.* **2018**, *34*, 559–568. [CrossRef]
18. Smarandache, F. *A Unifying Field in Logics: Neutrosophic Logic, Neutrosophy, Neutrosophic Set, Neutrosophic Probability*; American Research Press: Rehoboth, DE, USA, 1999.
19. Wang, H.; Smarandache, F.; Sunderraman, R. Single-valued neutrosophic sets. *Rev. Air Force Acad.* **2013**, *17*, 10–13.
20. Ye, J. Multicriteria decision-making method using the correlation coefficient under single-valued neutrosophic environment. *Int. J. Gen. Syst.* **2013**, *42*, 386–394. [CrossRef]
21. Ye, J. Similarity measures between interval neutrosophic sets and their multicriteria decision-making method. *J. Intell. Fuzzy Syst.* **2014**, *26*, 165–172.
22. Liu, P.D.; Tang, G.L. Some power generalized aggregation operators based on the interval neutrosophic numbers and their application to decision making. *J. Intell. Fuzzy. Syst.* **2015**, *30*, 2517–2528. [CrossRef]
23. Liu, P.D.; Teng, F. Multiple attribute decision making method based on normal neutrosophic generalized weighted power averaging operator. *Int. J. Mach. Learn. Cybern.* **2015**, 1–13. [CrossRef]
24. Zhang, X.H.; Smarandache, F.; Liang, X.L. Neutrosophic duplet semi-group and cancellable neutrosophic triplet groups. *Symmetry* **2017**, *9*, 275. [CrossRef]
25. Zhang, X.H.; Bo, C.X.; Smarandache, F.; Dai, J.H. New inclusion relation of neutrosophic sets with applications and related lattice structrue. *Int. J. Mach. Learn. Cybern.* **2018**. [CrossRef]
26. Smarandache, F. Degree of dependence and independence of the (sub)components of fuzzy set and neutrosophic set. *Neutrosophic Sets Syst.* **2016**, *11*, 95–97.
27. Cuong, B.C. *Picture Fuzzy Sets-First Results, Part 1 and Part 2, Seminar "Neuro-Fuzzy Systems with Applications"*; Preprint 03/2013 and Preprint 04/2013; Institute of Mathematics: Hanoi, Vietnam, 2013.
28. Cuong, B.C. Picture fuzzy sets. *J. Comput. Sci. Cybern.* **2014**, *30*, 409–420.

29. Cuong, B.C.; Hai, P.V. Some fuzzy logic operators for picture fuzzy sets. In Proceedings of the IEEE Seventh International Conference on Knowledge and Systems Engineering, Ho Chi Minh City, Vietnam, 8–10 October 2015; pp. 132–137.
30. Cuong, B.C.; Phong, P.H.; Smarandache, F. Standard neutrosophic soft theory: Some first results. *Neutrosophic Sets Syst.* **2016**, *12*, 80–91.
31. Wang, C.; Zhou, X.; Tu, H.; Tao, S. Some geometric aggregation operators based on picture fuzzy sets and their application in multiple attribute decision making. *Ital. J. Pure Appl. Math.* **2017**, *37*, 477–492.
32. Daniel, S. Free non-distributive morgan-stone algebars. *N. Z. J. Math.* **1996**, *25*, 85–94.
33. Maji, P.K. Neutrosophic soft set. *Ann. Fuzzy Math. Inform.* **2013**, *5*, 157–168.
34. Şahin, R.; Küçük, A. On similarity and entropy of neutrosophic soft sets. *J. Intell. Fuzzy Syst.* **2014**, *27*, 2417–2430.
35. Bucolo, M.; Fortuna, L.; La Rosa, M. Network self-organization through "small-worlds" topologies. *Chaos Solitons Fractals* **2001**, *14*, 1059–1064. [CrossRef]

symmetry

MDPI

Article

Some Results on the Graph Theory for Complex Neutrosophic Sets

Shio Gai Quek [1,*], Said Broumi [2], Ganeshsree Selvachandran [3], Assia Bakali [4], Mohamed Talea [2] and Florentin Smarandache [5]

1 A-Level Academy, UCSI College KL Campus, Lot 12734, Jalan Choo Lip Kung, Taman Taynton View, Cheras 56000, Kuala Lumpur, Malaysia
2 Laboratory of Information Processing, Faculty of Science Ben M'Sik, University Hassan II, B.P 7955 Sidi Othman, Casablanca, Morocco; broumisaid78@gmail.com (S.B.); taleamohamed@yahoo.fr (M.T.)
3 Department of Actuarial Science and Applied Statistics, Faculty of Business & Information Science, UCSI University, Jalan Menara Gading, Cheras 56000, Kuala Lumpur, Malaysia; ganeshsree86@yahoo.com
4 Ecole Royale Navale, Boulevard Sour Jdid, B.P 16303 Casablanca, Morocco; assiabakali@yahoo.fr
5 Department of Mathematics, University of New Mexico, 705 Gurley Avenue, Gallup, NM 87301, USA; fsmarandache@gmail.com or smarand@unm.edu
* Correspondence: quekshg@yahoo.com or quekshg@ucsicollege.edu.my; Tel.: +60-17-5008359

Received: 25 February 2018; Accepted: 21 May 2018; Published: 31 May 2018

Abstract: Fuzzy graph theory plays an important role in the study of the symmetry and asymmetry properties of fuzzy graphs. With this in mind, in this paper, we introduce new neutrosophic graphs called complex neutrosophic graphs of type 1 (abbr. CNG1). We then present a matrix representation for it and study some properties of this new concept. The concept of CNG1 is an extension of the generalized fuzzy graphs of type 1 (GFG1) and generalized single-valued neutrosophic graphs of type 1 (GSVNG1). The utility of the CNG1 introduced here are applied to a multi-attribute decision making problem related to Internet server selection.

Keywords: complex neutrosophic set; complex neutrosophic graph; fuzzy graph; matrix representation

1. Introduction

Smarandache [1] introduced a new theory called neutrosophic theory, which is basically a branch of philosophy that focuses on the origin, nature, and scope of neutralities and their interactions with different ideational spectra. On the basis of neutrosophy, Smarandache defined the concept of a neutrosophic set (NS) which is characterized by a degree of truth membership T, a degree of indeterminacy membership I, and a degree of falsity membership F. The concept of neutrosophic set theory generalizes the concept of classical sets, fuzzy sets by Zadeh [2], intuitionistic fuzzy sets by Atanassov [3], and interval-valued fuzzy sets by Turksen [4]. In fact, this mathematical tool is apt for handling problems related to imprecision, indeterminacy, and inconsistency of data. The indeterminacy component present in NSs is independent of the truth and falsity membership values. To make it more convenient to apply NSs to real-life scientific and engineering problems, Smarandache [1] proposed the single-valued neutrosophic set (SVNS) as a subclass of neutrosophic sets. Later on, Wang et al. [5] presented the set-theoretic operators and studied some of the properties of SVNSs. The NS model and its generalizations have been successfully applied in many diverse areas, and these can be found in [6].

Graphs are among the most powerful and convenient tools to represent information involving the relationship between objects and concepts. In crisp graphs, two vertices are either related or not related to one another so, mathematically, the degree of relationship is either 0 or 1. In fuzzy graphs on the other hand, the degree of relationship takes on values from the interval [0, 1]. Subsequently, Shannon and Atanassov [7] defined the concept of intuitionistic fuzzy graphs (IFGs) using five types of Cartesian products. The concept fuzzy graphs and their extensions have a common property that each edge must have a membership value of less than, or equal to, the minimum membership of the nodes it connects.

In the event that the description of the object or their relations or both is indeterminate and inconsistent, it cannot be handled by fuzzy, intuitionistic fuzzy, bipolar fuzzy, vague, or interval-valued fuzzy graphs. For this reason, Smarandache [8] proposed the concept of neutrosophic graphs based on the indeterminacy (*I*) membership values to deal with such situations. Smarandache [9,10] then gave another definition for neutrosophic graph theory using the neutrosophic truth-values (*T*, *I*, *F*) and constructed three structures of neutrosophic graphs: neutrosophic edge graphs, neutrosophic vertex graphs and neutrosophic vertex-edge graphs. Subsequently, Smarandache [11] proposed new versions of these neutrosophic graphs, such as the neutrosophic off graph, neutrosophic bipolar graph, neutrosophic tripolar graph, and neutrosophic multipolar graph. Presently, works on neutrosophic vertex-edge graphs and neutrosophic edge graphs are progressing rapidly. Broumi et al. [12] combined the SVNS model and graph theory to introduce certain types of SVNS graphs (SVNG), such as strong SVNG, constant SVNG, and complete SVNG, and proceeded to investigate some of the properties of these graphs with proofs and examples. Broumi et al. [13] then introduced the concept of neighborhood degree of a vertex and closed neighborhood degree of a vertex in SVNG as a generalization of the neighborhood degree of a vertex and closed neighborhood degree of a vertex found in fuzzy graphs and intuitionistic fuzzy graphs. In addition, Broumi et al. [14] proved a necessary and sufficient condition for a SVNG to be an isolated SVNG.

Recently, Smarandache [15] initiated the idea of the removal of the edge degree restriction for fuzzy graphs, intuitionistic fuzzy graphs and SVNGs. Samanta et al. [16] proposed a new concept called generalized fuzzy graphs (GFG) and defined two types of GFG. Here the authors also studied some of the major properties of GFGs, such as the completeness and regularity of GFGs, and verified the results. In [16], the authors claim that fuzzy graphs and their extensions are limited to the representations of only certain systems, such as social networks. Broumi et al. [17] then discussed the removal of the edge degree restriction of SVNGs and presented a new class of SVNG, called generalized SVNG of type 1, which is an extension of generalized fuzzy graphs of type 1 proposed in [16]. Since the introduction of complex fuzzy sets (CFSs) by Ramot et al. in [18], several new extensions of CFSs have been proposed in literature [19–25]. The latest model related to CFS is the complex neutrosophic set (CNS) model which is a combination of CFSs [18] and complex intuitionistic fuzzy sets [21] proposed by Ali and Smarandache [26]. The CNS model is defined by three complex-valued membership functions which represent the truth, indeterminate, and falsity components. Therefore, a complex-valued truth membership function is a combination of the traditional truth membership function with the addition of the phase term. Similar to fuzzy graphs, complex fuzzy graphs (CFG) introduced by Thirunavukarasu et al. [27] have a common property that each edge must have a membership value of less than or equal to the minimum membership of the nodes it connects.

In this paper, we extend the research works mentioned above, and introduce the novel concept of type 1 complex neutrosophic graphs (CNG1) and a matrix representation of CNG1. To the best of our knowledge, there is no research on CNGs in the literature at present. We also present an investigation pertaining to the symmetric properties of CNG1 in this paper. In the study of fuzzy graphs, a symmetric fuzzy graph refers to a graph structure with one edge (i.e., two arrows on opposite directions) or no edges, whereas an asymmetric fuzzy graph refers to a graph structure with no arcs or only one arc present between any two vertices. Motivated by this, we have dedicated an entire section in this paper (Section 7) to study the symmetric properties of our proposed CNG1.

The remainder of this paper is organized as follows: in Section 2, we review some basic concepts about NSs, SVNSs, CNSs, and generalized SVNGs of type 1; in Section 3, the formal definition of CNG1 is introduced and supported with illustrative examples; in Section 4 a representation matrix of CNG1 is introduced; some advanced theoretical results pertaining to our CNG1 is presented in Section 5, followed by an investigation on the shortest CNG1 in Section 6; the symmetric properties of ordinary simple CNG1 is presented in Section 7; and Section 8 outlines the conclusion of this paper and suggests directions for future research. This is followed by the acknowledgments and the list of references.

2. Preliminaries

In this section, we present brief overviews of NSs, SVNSs, SVNGs, and generalize fuzzy graphs that are relevant to the present work. We refer the readers to [1,5,17,18,27] for further information related to these concepts.

The key feature that distinguishes the NS from the fuzzy and intuitionistic fuzzy set (IFS) models is the presence of the indeterminacy membership function. In the NS model the indeterminacy membership function is independent from the truth and falsity membership functions, and we are able to tell the exact value of the indeterminacy function. In the fuzzy set model this indeterminacy function is non-existent, whereas in the IFS model, the value of the indeterminacy membership function is dependent on the values of the truth and falsity membership functions. This is evident from the structure of the NS model in which $T + I + F \leq 3$, whereas it is $T + F \leq 1$ and $I = 1 - T - F$ in the IFS model. This structure is reflective of many real-life situations, such as in sports (wining, losing, draw), voting (for, against, abstain from voting), and decision-making (affirmative, negative, undecided), in which the proportions of one outcome is independent of the others. The NS model is able to model these situations more accurately compared to fuzzy sets and IFSs as it is able to determine the degree of indeterminacy from the truth and falsity membership function more accurately, whereas this distinction cannot be done when modelling information using the fuzzy sets and IFSs. Moreover, the NS model has special structures called neutrosophic oversets and neutrosophic undersets that were introduced by Smarandache in [11], in which the values of the membership functions can exceed 1 or be below 0, in order to cater to special situations. This makes the NS more flexible compared to fuzzy sets and IFSs, and gives it the ability to cater to a wider range of applications. The flexibility of this model and its ability to clearly distinguish between the truth, falsity, and indeterminacy membership functions served as the main motivation to study a branch of graph theory of NSs in this paper. We refer the readers to [28,29] for more information on the degree of dependence and independence of neutrosophic sets, and [11] for further information on the concepts of neutrosophic oversets and undersets.

Definition 1 [1]. *Let X be a space of points and let $x \in X$. A neutrosophic set $A \in X$ is characterized by a truth membership function T, an indeterminacy membership function I, and a falsity membership function F. The values of T, I, F are real standard or nonstandard subsets of $]^-0,\ 1^+[$, and T, I, $F : X \to]^-0,\ 1^+[$. A neutrosophic set can therefore be represented as:*

$$A = \{(x, T_A(x), I_A(x), F_A(x)) : x \in X\} \tag{1}$$

Since T, I, $F \in [0,\ 1]$, the only restriction on the sum of T, I, F is as given below:

$$^-0 \ \leq T_A(x) + I_A(x) + F_A(x) \leq 3^+ \tag{2}$$

Although theoretically the NS model is able to handle values from the real standard or non-standard subsets of $]^-0,\ 1^+[$, it is often unnecessary or computationally impractical to use values from this non-standard range when dealing with real-life applications. Most problems in engineering, and computer science deal with values from the interval [0, 1] instead of the interval $]^-0,\ 1^+[$, and this led to the introduction of the single-valued neutrosophic set (SVNS) model in [5]. The SVNS model is a special case of the general NS model in which the range of admissible values are from the standard interval of [0, 1], thereby making it more practical to be used to deal with most real-life problems. The formal definition of the SVNS model is given in Definition 2.

Definition 2 [5]. *Let X be a space of points (objects) with generic elements in X denoted by x. A single-valued neutrosophic set A (SVNS A) is characterized by a truth-membership function $T_A(x)$,* *an indeterminacy-membership function* $I_A(x)$*, and a falsity-membership function* $F_A(x)$*. For each point* $x \in X$*,* $T_A(x),\ I_A(x),\ F_A(x) \in [0,\ 1]$*. The SVNS A can therefore be written as:*

$$A = \{(x, T_A(x), I_A(x), F_A(x)) : x \in X\} \tag{3}$$

Definition 3 [26]. *Denote* $i = \sqrt{-1}$*. A complex neutrosophic set A defined on a universe of discourse X, which is characterized by a truth membership function* $T_A(x)$*, an indeterminacy-membership function* $I_A(x)$*, and a falsity-membership function* $F_A(x)$ *that assigns a complex-valued membership grade to* $T_A(x),\ I_A(x),\ F_A(x)$ *for any* $x \in X$*. The values of* $T_A(x),\ I_A(x),\ F_A(x)$ *and their sum may be any values within a unit circle in the complex plane and is therefore of the form* $T_A(x) = p_A(x)e^{i\mu_A(x)}$*,* $I_A(x) = q_A(x)e^{i\nu_A(x)}$*, and* $F_A(x) = r_A(x)e^{i\omega_A(x)}$*. All the amplitude and phase terms are real-valued and* $p_A(x),\ q_A(x),\ r_A(x) \in [0,\ 1]$*, whereas* $\mu_A(x),\ \nu_A(x),\ \omega_A(x) \in (0,\ 2\pi]$*, such that the condition:*

$$0 \le p_A(x) + q_A(x) + r_A(x) \le 3 \tag{4}$$

is satisfied. A complex neutrosophic set A can thus be represented in set form as:

$$A = \{\langle x,\ T_A(x) = a_T,\ I_A(x) = a_I,\ F_A(x) = a_F\rangle : x \in X\}, \tag{5}$$

where $T_A : X \to \{a_T : a_T \in C,\ |a_T| \le 1\}$*,* $I_A : X \to \{a_I : a_I \in C,\ |a_I| \le 1\}$*,* $F_A : X \to \{a_F : a_F \in C,\ |a_F| \le 1\}$*, and also:*

$$|T_A(x) + I_A(x) + F_A(x)| \le 3. \tag{6}$$

Definition 4 [26]. *Let* $A = \{(x, T_A(x), I_A(x), F_A(x)) : x \in X\}$ *and* $B = \{(x, T_B(x), I_B(x), F_B(x)) : x \in X\}$ *be two CNSs in X. The union and intersection of A and B are as defined below.*

(i) *The union of A and B, denoted as* $A\cup_N B$*, is defined as:*

$$A\cup_N B = \{(x, T_{A\cup B}(x), I_{A\cup B}(x), F_{A\cup B}(x)) : x \in X\}, \tag{7}$$

where $T_{A\cup B}(x), I_{A\cup B}(x), F_{A\cup B}(x)$ *are given by:*

$$\begin{aligned} T_{A\cup B}(x) &= \max(p_A(x),\ p_B(x)).e^{i\mu_{A\cup B}(x)}, \\ I_{A\cup B}(x) &= \min(q_A(x),\ q_B(x)).e^{i\nu_{A\cup B}(x)}, \\ F_{A\cup B}(x) &= \min(r_A(x),\ r_B(x)).e^{i\omega_{A\cup B}(x)}. \end{aligned}$$

(ii) *The intersection of A and B, denoted as* $A\cap_N B$*, is defined as:*

$$A\cap_N B = \{(x, T_{A\cap B}(x), I_{A\cap B}(x), F_{A\cap B}(x)) : x \in X\}, \tag{8}$$

where $T_{A\cap B}(x), I_{A\cap B}(x), F_{A\cap B}(x)$ *are given by:*

$$\begin{aligned} T_{A\cap B}(x) &= \min(p_A(x),\ p_B(x)).e^{i\mu_{A\cap B}(x)}, \\ I_{A\cap B}(x) &= \max(q_A(x),\ q_B(x)).e^{i\nu_{A\cap B}(x)}, \\ F_{A\cap B}(x) &= \max(r_A(x),\ r_B(x)).e^{i\omega_{A\cap B}(x)}. \end{aligned}$$

The union and the intersection of the phase terms of the complex truth, falsity and indeterminacy membership functions can be calculated from, but not limited to, any one of the following operations:

(a) Sum:

$$\begin{aligned}\mu_{A\cup B}(x) &= \mu_A(x) + \mu_B(x),\\ \nu_{A\cup B}(x) &= \nu_A(x) + \nu_B(x),\\ \omega_{A\cup B}(x) &= \omega_A(x) + \omega_B(x).\end{aligned}$$

(b) Max:

$$\begin{aligned}\mu_{A\cup B}(x) &= \max(\mu_A(x), \mu_B(x)),\\ \nu_{A\cup B}(x) &= \max(\nu_A(x), \nu_B(x)),\\ \omega_{A\cup B}(x) &= \max(\omega_A(x), \omega_B(x)).\end{aligned}$$

(c) Min:

$$\begin{aligned}\mu_{A\cup B}(x) &= \min(\mu_A(x), \mu_B(x)),\\ \nu_{A\cup B}(x) &= \min(\nu_A(x), \nu_B(x)),\\ \omega_{A\cup B}(x) &= \min(\omega_A(x), \omega_B(x)).\end{aligned}$$

(d) "The game of winner, neutral, and loser":

$$\begin{aligned}\mu_{A\cup B}(x) &= \begin{cases}\mu_A(x) & if \quad p_A > p_B\\ \mu_B(x) & if \quad p_B > p_A\end{cases},\\ \nu_{A\cup B}(x) &= \begin{cases}\nu_A(x) & if \quad q_A < q_B\\ \nu_B(x) & if \quad q_B < q_A\end{cases},\\ \omega_{A\cup B}(x) &= \begin{cases}\omega_A(x) & if \quad r_A < r_B\\ \omega_B(x) & if \quad r_B < r_A\end{cases}.\end{aligned}$$

Definition 5 [17]. *Let the following statements hold:*

(a) *V is a non-void set.*
(b) *$\breve{\rho}_T$, $\breve{\rho}_I$, $\breve{\rho}_F$ are three functions, each from V to $[0, 1]$.*
(c) *$\breve{\omega}_T$, $\breve{\omega}_I$, $\breve{\omega}_F$ are three functions, each from $V \times V$ to $[0, 1]$.*
(d) *$\breve{\rho} = (\breve{\rho}_T, \breve{\rho}_I, \breve{\rho}_F)$ and $\breve{\omega} = (\breve{\omega}_T, \breve{\omega}_I, \breve{\omega}_F)$.*

Then the structure $\breve{\xi} = \langle V, \breve{\rho}, \breve{\omega}\rangle$ is said to be a generalized single valued neutrosophic graph of type 1 (GSVNG1).

Remark 1.

(i) *$\breve{\rho}$ depends on $\breve{\rho}_T$, $\breve{\rho}_I$, $\breve{\rho}_F$ and $\breve{\omega}$ depends on $\breve{\omega}_T$, $\breve{\omega}_I$, $\breve{\omega}_F$. Hence there are seven mutually independent parameters in total that make up a CNG1: V, $\breve{\rho}_T, \breve{\rho}_I, \breve{\rho}_F, \breve{\omega}_T, \breve{\omega}_I, \breve{\omega}_F$.*
(ii) *For each $x \in V$, x is said to be a vertex of $\breve{\xi}$. The entire set V is thus called the vertex set of $\breve{\xi}$.*
(iii) *For each $u, v \in V$, (u, v) is said to be a directed edge of $\breve{\xi}$. In particular, (v, v) is said to be a loop of $\breve{\xi}$*
(iv) *For each vertex: $\breve{\rho}_T(v), \breve{\rho}_I(v), \breve{\rho}_F(v)$ are called the truth-membership value, indeterminate membership value, and false-membership value, respectively, of that vertex v. Moreover, if $\breve{\rho}_T(v) = \breve{\rho}_I(v) = \breve{\rho}_F(v) = 0$, then v is said to be a void vertex.*
(v) *Likewise, for each edge (u, v) : $\breve{\omega}_T(u, v)$, $\breve{\omega}_I(u, v)$, $\breve{\omega}_F(u, v)$ are called the truth-membership value, indeterminate-membership value, and false-membership value, respectively of that directed edge (u, v). Moreover, if $\breve{\omega}_T(u, v) = \breve{\omega}_I(u, v) = \breve{\omega}_F(u, v) = 0$, then (u, v) is said to be a void directed edge.*

Here we shall restate the concept of complex fuzzy graph of type 1. Moreover, for all the remaining parts of this paper, we shall denote the set $\{z \in \mathbb{C}\colon |z| \leq 1\}$ as O_1.

Definition 6 [27]. *Let the following statements hold:*

(a) *V is a non-void set.*
(b) $\dot{\rho}$ *is a function from V to* O_1.
(c) $\dot{\omega}$ *is a function from* $V \times V$ *to* O_1.

Then:

(i) *the structure* $\dot{\xi} = \langle V, \dot{\rho}, \dot{\omega} \rangle$ *is said to be a complex fuzzy graph of type 1 (abbr. CFG1).*
(ii) *For each* $x \in V$, *x is said to be a vertex of* $\dot{\xi}$. *The entire set V is thus called the vertex set of* $\dot{\xi}$.
(iii) *For each* $u, v \in V$, (u, v) *is said to be a directed edge of* $\dot{\xi}$. *In particular,* (v, v) *is said to be a loop of* $\dot{\xi}$.

3. Complex Neutrosophic Graphs of Type 1

By using the concept of complex neutrosophic sets [26], the concept of complex fuzzy graph of type 1 [27], and the concept of generalized single valued neutrosophic graph of type 1 [17], we define the concept of *complex neutrosophic graph of type* 1 as follows:

Definition 7. *Let the following statements hold:*

(a) *V is a non-void set.*
(b) ρ_T, ρ_I, ρ_F *are three functions, each from V to* O_1.
(c) $\omega_T, \omega_I, \omega_F$ *are three functions, each from* $V \times V$ *to* O_1.
(d) $\rho = (\rho_T, \rho_I, \rho_F)$ *and* $\omega = (\omega_T, \omega_I, \omega_F)$.

Then the structure $\xi = \langle V, \rho, \omega \rangle$ *is said to be a complex neutrosophic graph of type 1 (abbr. CNG1).*

Remark 2. ρ *depends on* ρ_T, ρ_I, ρ_F, *and* ω *depends on* $\omega_T, \omega_I, \omega_F$. *Hence there are seven mutually independent parameters in total that make up a CNG1: V,* $\rho_T, \rho_I, \rho_F, \omega_T, \omega_I, \omega_F$. *Furthermore, in analogy to a GSVNG1:*

(i) *For each* $x \in V$, *x is said to be a vertex of* ξ. *The entire set V is thus called the vertex set of* ξ.
(ii) *For each* $u, v \in V$, (u, v) *is said to be a directed edge of* ξ. *In particular,* (v, v) *is said to be a loop of* ξ.
(iii) *For each vertex:* $\rho_T(v), \rho_I(v), \rho_F(v)$ *are called the complex truth, indeterminate, and falsity membership values, respectively, of the vertex v. Moreover, if* $\rho_T(v) = \rho_I(v) = \rho_F(v) = 0$, *then v is said to be a void vertex.*
(iv) *Likewise, for each directed edge* (u, v) : $\omega_T(u, v)$, $\omega_I(u, v)$, $\omega_F(u, v)$ *are called the complex truth, indeterminate and falsity membership value, of the directed edge* (u, v). *Moreover, if* $\omega_T(u, v) = \omega_I(u, v) = \omega_F(u, v) = 0$, *then* (u, v) *is said to be a void directed edge.*

For the sake of brevity, we shall denote $\omega(u, v) = (\omega_T(u, v), \omega_I(u, v), \omega_F(u, v))$ and $\rho(v) = (\rho_T(v), \rho_I(v), \rho_F(v))$ for all the remaining parts of this paper.

As mentioned, CNG1 is generalized from both GSVNG1 and CFG1. As a result, we have ω_T, ω_I and ω_T being functions themselves. This further implies that $\omega_T(u, v)$, $\omega_I(u, v)$ and $\omega_T(u, v)$ can only be single values from O_1. In particular, $\omega_T(v, v)$, $\omega_I(v, v)$, and $\omega_T(v, v)$ can only be single values.

As a result, each vertex v in a CNG1 possess a single, undirected loop, whether void or not. And each of the two distinct vertices u, v in a CNG1 possess *two* directed edges, resulting from (u, v) and (v, u), whether void or not.

Recall that in classical graph theory, we often deal with ordinary (or undirected) graphs, and also simple graphs. To further relate our CNG1 with it, we now proceed with the following definition.

Definition 8. *Let* $\xi = \langle V, \rho, \omega \rangle$ *be a CNG1.*

(a) *If* $\omega(a, b) = \omega(b, a)$, *then* $\{a, b\} = \{(a, b), (b, a)\}$ *is said to be an (ordinary) edge of* ξ. *Moreover,* $\{a, b\}$ *is said to be a void (ordinary) edge if both* (a, b) *and* (b, a) *are void.*

(b) *If* $\omega(u,v) = \omega(v,u)$ *holds for all* $u, v \in V$, *then* ξ *is said to be ordinary (or undirected), otherwise it is said to be directed.*

(c) *If all the loops of* ξ *are void, then* ξ *is said to be simple.*

Definition 9. *Let* $\xi = \langle V, \rho, \omega \rangle$ *be an ordinary CNG1. If for all* $u, v \in V$ *with* $u \neq v$, *there exist non-void edges* $\{u = w_1, w_2\}, \{w_2, w_3\}, \ldots, \{w_{n-1}, w_n = v\}$ *for some* $n \geq 2$, *then* ξ *is said to be connected.*

Definition 10. *Let* $\xi = \langle V, \rho, \omega \rangle$ *be an ordinary CNG1. Let* $u, v \in V$. *Then:*

(a) $\{u, v\}$ *is said to be adjacent to* u *(and to* v*).*

(b) u *(and* v *as well) is said to be an end-point of* $\{u, v\}$.

We now discuss a real life scenario that can only be represented by a CNG1.

3.1. The Scenario

Note: All the locations mentioned are fictional

Suppose there is a residential area in Malaysia with four families: a, b, c, d. All of them have Internet access. In other words, they are Internet clients, which will access the Internet servers from around the world (including those servers located within Malaysia) depending on which website they are visiting.

If they access the internet on their own, the outcomes can be summarized as given in the Table 1 and Figure 1.

Table 1. The outcomes of individuals, for Scenario 3.1.

Activities	a	b	c	d
Some members will seek excitement (e.g., playing online games)	Happens on 80% of the day, and those will be connecting towards 0° (because that server is located in China ⊛)	Happens on 70% of the day, and those will be connecting towards 30°	Happens on 90% of the day, and those will be connecting towards 120°	Happens on 80% of the day, and those will be connecting towards 250°
Some members will want to surf around (e.g., online shopping)	Happens on 50% of the day, and those will be connecting towards 130° (because that server is located in Australia ⊛)	Happens on 60% of the day, and those will be connecting towards 180°	Happens on 20% of the day, and those will be connecting towards 340°	Happens on 40% of the day, and those will be connecting towards 200°
Some members will need to relax (e.g., listening to music)	Happens on 20% of the day, and those will be connecting towards 220° (because that server is located in Sumatra ⊛, Indonesia)	Happens on 30% of the day, and those will be connecting towards 200°	Happens on 50% of the day, and those will be connecting towards 40°	Happens on 10% of the day, and those will be connecting towards 110°

(⊛) as illustrated in Figure 1.

Moreover, the following (unordered) pairs of the four families are close friends or relatives:

$$\{a,b\}, \{a,c\}, \{a,d\}, \{b,d\}.$$

Thus, each pair of family mentioned (e.g., $\{a,b\}$) may invite one another for a visit, accessing the Internet as one team. In particular:

(i) When $\{a,b\}$ or $\{a,d\}$ access the internet together, they will simply search for "a place of common interest". This is regardless of who initiates the invitation.

(ii) a and c rarely meet. Thus, each time they do, everyone (especially the children) will be so excited that they would like to try something fresh, so all will seek excitement and connect towards to a local broadcasting server at 240° to watch soccer matches (that server will take care of which country to connect to) for the entire day. This is also regardless of who initiates the visitation.

(iii) The size and the wealth of d far surpasses b. Thus, it would always be d who invites b to their house, never the other way, and during the entire visit, members of b will completely behave like members of d and, therefore, will visit the same websites as d.

Denote the first term of the ordered pair (u, v) as the family who initiates the invitation, and the second term as family who receives the invitation and visit the other family. The outcomes of the seven possible teams (a, b), (a, c), (a, d), (b, a), (c, a), (d, a), (d, b) are, thus, summarized by Table 2.

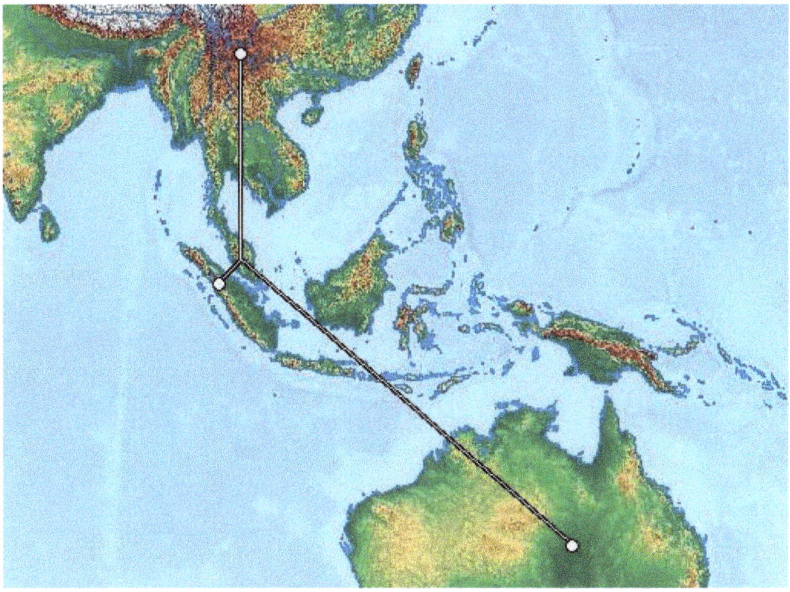

Figure 1. The illustration of the servers' relative positions using a public domain map, for Scenario 3.1.

Table 2. The outcomes of teams in pairs, for the scenario.

Activities	(a,b),(b,a)	(a,c),(c,a)	(a,d),(d,a)	(d,b)
Some members will seek excitement	Happens on 80% of the day, and those will be connecting towards 15°	Happens on the entire day, all will be connecting towards 240°	Happens on 80% of the day, and those will be connecting towards 305°	Happens on 80% of the day, and those will be connecting towards 250°
Some members will want to surf around	Happens on 60% of the day, and those will be connecting towards 155°	Does not happen	Happens on 50% of the day, and those will be connecting towards 165°	Happens on 40% of the day, and those will be connecting towards 200°
Some members will need to relax	Happens on 30% of the day, and those will be connecting towards 210°	Does not happen	Happens on 50% of the day, and those will be connecting towards 40°	Happens on 10% of the day, and those will be connecting towards 110°

On the other hand, $\{c, b\}$ and $\{d, c\}$ are mutual strangers. So c and b will visit each other. The same goes to d and c.

3.2. *Representation of the Scenario with CNG1*

We now follow all the steps from (a) to (e) in Definition 7, to represent the scenario with a particular CNG1.

(a) Take $V_0 = \{a, b, c, d\}$.

(b) In accordance with the scenario, define the three functions on V_0: ρ_T, ρ_I, ρ_F, as illustrated in Table 3.

Table 3. $k(v)$, where k represents any of the 3 functions on V_0 ρ_T, ρ_I, ρ_F, for the scenario. Also mentioned in Section 4.2.

k \ v	a	b	c	d
ρ_T	$0.8e^{i2\pi}$	$0.7e^{i\frac{\pi}{6}}$	$0.9e^{i\frac{2\pi}{3}}$	$0.8e^{i\frac{25\pi}{18}}$
ρ_I	$0.5e^{i\frac{13\pi}{18}}$	$0.6e^{i\pi}$	$0.2e^{i\frac{17\pi}{9}}$	$0.4e^{i\frac{10\pi}{9}}$
ρ_F	$0.2e^{i\frac{11\pi}{9}}$	$0.3e^{i\frac{10\pi}{9}}$	$0.5e^{i\frac{2\pi}{9}}$	$0.1e^{i\frac{11\pi}{18}}$

(c) In accordance with the scenario, define the three functions ω_T, ω_I, ω_F, as illustrated in Tables 4–6.

Table 4. The outcomes of $\omega_T(u, v)$, for the scenario. Also mentioned in Section 4.2.

u \ v	a	b	c	d
a	0	$0.8e^{i\frac{\pi}{12}}$	$1e^{i\frac{4\pi}{3}}$	$0.8e^{i\frac{61\pi}{36}}$
b	$0.8e^{i\frac{\pi}{12}}$	0	0	0
c	$1e^{i\frac{4\pi}{3}}$	0	0	0
d	$0.8e^{i\frac{61\pi}{36}}$	$0.8e^{i\frac{25\pi}{18}}$	0	0

Table 5. The outcomes of $\omega_I(u, v)$, for the scenario. Also mentioned in Section 4.2.

u \ v	a	b	c	d
a	0	$0.6e^{i\frac{31\pi}{36}}$	0	$0.5e^{i\frac{33\pi}{36}}$
b	$0.6e^{i\frac{31\pi}{36}}$	0	0	0
c	0	0	0	0
d	$0.5e^{i\frac{33\pi}{36}}$	$0.4e^{i\frac{10\pi}{9}}$	0	0

Table 6. The outcomes of $\omega_F(u, v)$, for the scenario. Also mentioned in Section 4.2.

u \ v	a	b	c	d
a	0	$0.3e^{i\frac{7\pi}{6}}$	0	$0.5e^{i\frac{2\pi}{9}}$
b	$0.3e^{i\frac{7\pi}{6}}$	0	0	0
c	0	0	0	0
d	$0.5e^{i\frac{2\pi}{9}}$	$0.1e^{i\frac{11\pi}{18}}$	0	0

(d) By statement (d) from Definition 7, let $\rho_0 = (\rho_T, \rho_I, \rho_F)$, and $\omega_0 = (\omega_T, \omega_I, \omega_F)$. We have now formed a CNG1 $\langle V_0, \rho_0, \omega_0 \rangle$.

One of the way of representing the entire $\langle V_0, \rho_0, \omega_0 \rangle$ is by using a diagram that is analogous with graphs as in classical graph theory, as shown in the Figure 2.

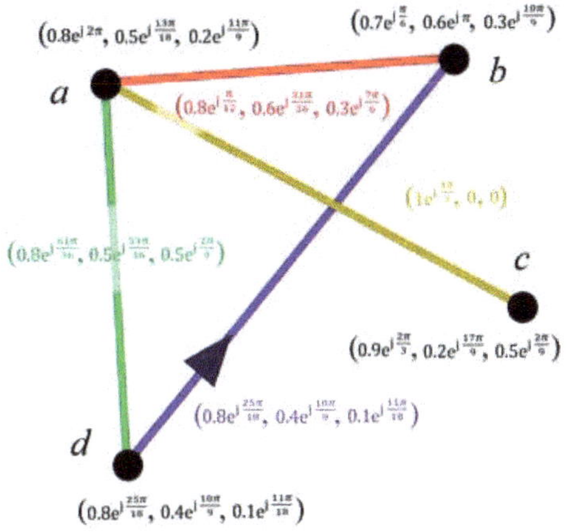

Figure 2. A diagram representing $\langle V_0, \rho_0, \omega_0 \rangle$, for the scenario.

In other words, only the non-void edges (whether directed or ordinary) and vertices are to be drawn in such a diagram.

Hence, we have shown how a CGN1 can be constructed from a data set with four homes. The same concept mentioned can certainly be used on a larger dataset, such as one with thousands of locations and thousands of homes, which will result in a more complicated diagram being generated. However, one will definitely require computer algebraic systems, such as SAGE, to process the data and to display the data in diagram form.

Additionally, recall that, in classical graph theory, a graph can be represented by an adjacency matrix, for which the entries are either a positive integer (connected) or 0 (not connected).

This motivates us to represent CNG1 using a matrix as well, in a similar manner. Nonetheless, instead of a single value that is either 0 or 1, we have *three* values to deal with: $\omega_T, \omega_I, \omega_F$, with each of them capable of being anywhere in O_1. Moreover, each of the vertices themselves also contain ρ_T, ρ_I, ρ_F, which must be taken into account as well.

4. Representation of a CNG1 by an Adjacency Matrix

4.1. Two Methods of Representation

In this section, we discuss the representation of CNG1 in two ways, which are both analogous to the one encountered in classical literature.

Let $\xi = \langle V, \rho, \omega \rangle$ be a CNG1 where vertex set $V = \{v_1, v_2, \ldots, v_n\}$ (i.e., CNG1 has finite vertices). We first form an $n \times n$ matrix as shown:

$$\mathbf{M} = \left[\mathbf{a}_{i,j}\right]_n = \begin{pmatrix} \mathbf{a}_{1,1} & \mathbf{a}_{1,2} & \cdots & \mathbf{a}_{1,n} \\ \mathbf{a}_{2,1} & \mathbf{a}_{2,2} & & \mathbf{a}_{2,n} \\ \vdots & & \ddots & \vdots \\ \mathbf{a}_{n,1} & \mathbf{a}_{n,2} & \cdots & \mathbf{a}_{n,n} \end{pmatrix},$$

where $\mathbf{a}_{i,j} = \omega(v_i, v_j)$ for all i, j.

In other words, each element of the matrix $\mathbf{M}$ is itself an ordered set of three elements, instead of just a number of either 0 or 1 in the classical literature.

Remark 3. *Since ξ can only possess undirected loops, we decided not to multiply the main diagonal elements of* $\mathbf{M}$ *by 2, as seen in adjacency matrices for graphs classical literature (2 for undirected, 1 for directed, 0 for void).*

Meanwhile, also recall that each of the vertices in ξ contains ρ_T, ρ_I, ρ_F, which must be taken into account as well.

Thus, we form another matrix **K** as shown:

$$\mathbf{K} = [\mathbf{k}_i]_{n,1} = \begin{pmatrix} \mathbf{k}_1 \\ \mathbf{k}_2 \\ \vdots \\ \mathbf{k}_n \end{pmatrix}, \text{where } \mathbf{k}_i = \rho(v_i) \text{for all} i.$$

To accomplish one of our methods of representing the entire ξ, we, therefore, augment the matrix **K** with **M**, forming the *adjacency matrix of* CNG1, **[K | M]**, as shown:

$$[\mathbf{K}|\mathbf{M}] = \begin{pmatrix} \mathbf{k}_1 & \mathbf{a}_{1,1} & \mathbf{a}_{1,2} & \cdots & \mathbf{a}_{1,n} \\ \mathbf{k}_2 & \mathbf{a}_{2,1} & \mathbf{a}_{2,2} & & \mathbf{a}_{2,n} \\ & \vdots & & \ddots & \vdots \\ \mathbf{k}_n & \mathbf{a}_{n,1} & \mathbf{a}_{n,2} & \cdots & \mathbf{a}_{n,n} \end{pmatrix},$$

where $\mathbf{a}_{i,j} = \omega(v_i, v_j)$, and $\mathbf{k}_i = \rho(v_i)$, for all i, j.

Although **[K | M]** is an $n \times (n + 1)$ matrix and therefore not a square, this representation will save us another *separate* ordered set to represent the ρ_T, ρ_I, ρ_F values of the vertices themselves.

Sometimes it is more convenient to separately deal with each of the three kinds of membership values for both edges and vertices. As a result, here we provide another method of representing the entire ξ: using three $n \times (n + 1)$ matrices, $[\mathbf{K}|\mathbf{M}]_T$, $[\mathbf{K}|\mathbf{M}]_I$, and $[\mathbf{K}|\mathbf{M}]_F$, each derived from $[\mathbf{K}|\mathbf{M}]$ by taking only one kind of the membership values from its elements:

$$[\mathbf{K}|\mathbf{M}]_T = [\mathbf{K}_T|\mathbf{M}_T] \begin{pmatrix} \rho_T(v_1) & \omega_T(v_1, v_1) & \omega_T(v_1, v_2) & \cdots & \omega_T(v_1, v_n) \\ \rho_T(v_2) & \omega_T(v_2, v_1) & \omega_T(v_2, v_2) & & \omega_T(v_2, v_n) \\ & \vdots & & \ddots & \vdots \\ \rho_T(v_n) & \omega_T(v_n, v_1) & \omega_T(v_n, v_2) & \cdots & \omega_T(v_n, v_n) \end{pmatrix},$$

$$[\mathbf{K}|\mathbf{M}]_I = [\mathbf{K}_I|\mathbf{M}_I] \begin{pmatrix} \rho_I(v_1) & \omega_I(v_1, v_1) & \omega_I(v_1, v_2) & \cdots & \omega_I(v_1, v_n) \\ \rho_I(v_2) & \omega_I(v_2, v_1) & \omega_I(v_2, v_2) & & \omega_I(v_2, v_n) \\ & \vdots & & \ddots & \vdots \\ \rho_I(v_n) & \omega_I(v_n, v_1) & \omega_I(v_n, v_2) & \cdots & \omega_I(v_n, v_n) \end{pmatrix},$$

$$[\mathbf{K}|\mathbf{M}]_F = [\mathbf{K}_F|\mathbf{M}_F] \begin{pmatrix} \rho_F(v_1) & \omega_F(v_1, v_1) & \omega_F(v_1, v_2) & \cdots & \omega_F(v_1, v_n) \\ \rho_F(v_2) & \omega_F(v_2, v_1) & \omega_F(v_2, v_2) & & \omega_F(v_2, v_n) \\ & \vdots & & \ddots & \vdots \\ \rho_F(v_n) & \omega_F(v_n, v_1) & \omega_F(v_n, v_2) & \cdots & \omega_F(v_n, v_n) \end{pmatrix}.$$

$[\mathbf{K}|\mathbf{M}]_T$, $[\mathbf{K}|\mathbf{M}]_I$, and $[\mathbf{K}|\mathbf{M}]_F$ shall, thus, be called, respectively, the *truth-adjacency matrix*, the *indeterminate-adjacency matrix*, and the *false-adjacency matrix* of ξ.

Remark 4. *If* $[\mathbf{K}|\mathbf{M}]_I = [\mathbf{K}|\mathbf{M}]_F = [0]_{n,n+1}$, $\mathbf{K}_T = [1]_{n,1}$, *all the entries of* $\mathbf{M}_T$ *are either 1 or 0, then* ξ *is reduced to a graph in classical literature. Furthermore, if that* $\mathbf{M}_T$ *is symmetrical and with main diagonal elements being zero, then* ξ *is further reduced to a simple ordinary graph in the classical literature.*

Remark 5. *If* $[\mathbf{K}|\mathbf{M}]_I = [\mathbf{K}|\mathbf{M}]_F = [0]_{n,n+1}$, *and all the entries of* $[\mathbf{K}|\mathbf{M}]_T$ *are real values from the interval [0, 1], then* ξ *is reduced to a generalized fuzzy graph type 1 (GFG1).*

Remark 6. *If all the entries of* $[\mathbf{K}|\mathbf{M}]_T$, $[\mathbf{K}|\mathbf{M}]_I$, *and* $[\mathbf{K}|\mathbf{M}]_F$ *are real values from the interval* [0, 1], *then* ξ *is reduced to a generalized single valued neutrosophic graphs of type 1 (GSVNG1).*

Remark 7. *If* $\mathbf{M}_T$, $\mathbf{M}_I$, *and* $\mathbf{M}_F$ *are symmetric matrices, then* ξ *is ordinary.*

4.2. Illustrative Example

For the sake of brevity, we now give representation for our example for the scenario in 3.1 by the latter method using three matrices: $[\mathbf{K}|\mathbf{M}]_T$, $[\mathbf{K}|\mathbf{M}]_I$, and $[\mathbf{K}|\mathbf{M}]_F$:

$$[\mathbf{K}|\mathbf{M}]_T = \begin{pmatrix} 0.8e^{i2\pi} & 0 & 0.8e^{i\frac{\pi}{12}} & 1e^{i\frac{4\pi}{3}} & 0.8e^{i\frac{61\pi}{36}} \\ 0.7e^{i\frac{\pi}{6}} & 0.8e^{i\frac{\pi}{12}} & 0 & 0 & 0 \\ 0.9e^{i\frac{2\pi}{3}} & 1e^{i\frac{4\pi}{3}} & 0 & 0 & 0 \\ 0.8e^{i\frac{25\pi}{18}} & 0.8e^{i\frac{61\pi}{36}} & 0.8e^{i\frac{25\pi}{18}} & 0 & 0 \end{pmatrix}$$

$$[\mathbf{K}|\mathbf{M}]_I = \begin{pmatrix} 0.5e^{i\frac{13\pi}{18}} & 0 & 0.6e^{i\frac{31\pi}{36}} & 0 & 0.5e^{i\frac{33\pi}{36}} \\ 0.6e^{i\pi} & 0.6e^{i\frac{31\pi}{36}} & 0 & 0 & 0 \\ 0.2e^{i\frac{17\pi}{9}} & 0 & 0 & 0 & 0 \\ 0.4e^{i\frac{10\pi}{9}} & 0.5e^{i\frac{33\pi}{36}} & 0.4e^{i\frac{10\pi}{9}} & 0 & 0 \end{pmatrix}$$

$$[\mathbf{K}|\mathbf{M}]_F = \begin{pmatrix} 0.2e^{i\frac{11\pi}{9}} & 0 & 0.3e^{i\frac{7\pi}{6}} & 0 & 0.5e^{i\frac{2\pi}{9}} \\ 0.3e^{i\frac{10\pi}{9}} & 0.3e^{i\frac{7\pi}{6}} & 0 & 0 & 0 \\ 0.5e^{i\frac{2\pi}{9}} & 0 & 0 & 0 & 0 \\ 0.1e^{i\frac{11\pi}{18}} & 0.5e^{i\frac{2\pi}{9}} & 0.1e^{i\frac{11\pi}{18}} & 0 & 0 \end{pmatrix}$$

As in Section 3, we have shown how a matrix representation of a CNG1 with $|V| = 4$ can be constructed. Likewise, the same concept mentioned can certainly be used on a larger CNG1 but, again, one will definitely require computer algebraic systems, such as SAGE to process the data and to display such a matrix representation.

5. Some Theoretical Results on Ordinary CNG1

We now discuss some theoretical results that follows from the definition of ordinary CNG1, as well as its representation with adjacency matrix. Since we are concerned about ordinary CNG1, all the edges that we will be referring to are ordinary edges.

Definition 11. *Let* $\xi = \langle V, \rho, \omega\rangle$ *be an ordinary CNG1. Let* $V = \{v_1, v_2, \ldots, v_n\}$ *be the vertex set of* ξ*. Then, for each i, the resultant degree of* v_i*, denoted as* $D(v_i)$*, is defined to be the ordered set* $(D_T(v_i), D_I(v_i), D_F(v_i))$*, for which:*

(a) $D_T(v_i) = \sum_{r=1}^{n} \omega_T(v_i, v_r) + \omega_T(v_i, v_i)$,

(b) $D_I(v_i) = \sum_{r=1}^{n} \omega_I(v_i, v_r) + \omega_I(v_i, v_i)$,

(c) $D_F(v_i) = \sum_{r=1}^{n} \omega_F(v_i, v_r) + \omega_F(v_i, v_i)$.

Remark 8. *In analogy to classical graph theory, each undirected loop has both its ends connected to the same vertex, so is counted twice.*

Remark 9. *Each of the values of* $D_T(v_i)$, $D_I(v_i)$, *and* $D_F(v_i)$ *need not be an integer as in a classical graph.*

Definition 12. *Let* $\xi = \langle V, \rho, \omega\rangle$ *be an ordinary CNG1. Let* $V = \{v_1, v_2, \ldots, v_n\}$ *be the vertex set of* ξ*. Then, the resultant amount of edges of* ξ*, denoted as* E_ξ*, is defined to be the ordered set* (E_T, E_I, E_F) *for which:*

(a) $E_T = \sum_{\{r,s\}\subseteq\{1,2,\ldots,n\}} \omega_T(v_r, v_s)$,

(b) $E_I = \sum_{\{r,s\}\subseteq\{1,2,\ldots,n\}} \omega_I(v_r, v_s)$,

(c) $E_F = \sum_{\{r,s\}\subseteq\{1,2,\ldots,n\}} \omega_F(v_r, v_s)$.

Remark 10. *As in classical graph theory, each edge is counted only once, as shown by $\{r, s\} \subseteq \{1, 2, \ldots, n\}$ in the expression. For example, if $\omega_T(v_a, v_b)$ is added, we will not add $\omega_T(v_b, v_a)$ again since $\{a, b\} = \{b, a\}$.*

Remark 11. *Each of the values of E_T, E_I and E_F need not be an integer as in a classical graph. As a result, we call it the "amount" of edges, instead of the "number" of edges as in the classical literature.*

For each vertex v_i, just because $D(v_i)$ equals 0, that does not mean that all the edges connect to v_i are void. It could be two distinct edges $\{v_i, v_1\}$ and $\{v_i, v_2\}$ with $\omega_T(v_i, v_1) = -\omega_T(v_i, v_2)$, $\omega_I(v_i, v_1) = -\omega_I(v_i, v_2)$ and $\omega_F(v_i, v_1) = -\omega_F(v_i, v_2)$ (i.e., equal in magnitude, but opposite in phase). The same goes to the value of E_ξ. This differs from the classical theory of graphs and, therefore, it motivates us to look at a CNG1 in yet another approach. We, thus, further define the following:

Definition 13. *Let $\xi = \langle V, \rho, \omega\rangle$ be an ordinary CNG1. Let $V = \{v_1, v_2, \ldots, v_n\}$ be the vertex set of ξ. Then, for each i, the absolute degree of v_i, denoted as $|D|(v_i)$, is defined to be the ordered set $(|D|_T(v_i), |D|_I(v_i), |D|_F(v_i))$, for which:*

(a) $|D|_T(v_i) = \sum_{r=1}^{n} |\omega_T(v_i, v_r)| + |\omega_T(v_i, v_i)|$,

(b) $|D|_I(v_i) = \sum_{r=1}^{n} |\omega_I(v_i, v_r)| + |\omega_I(v_i, v_i)|$,

(c) $|D|_F(v_i) = \sum_{r=1}^{n} |\omega_F(v_i, v_r)| + |\omega_F(v_i, v_i)|$.

Definition 14. *Let $\xi = \langle V, \rho, \omega\rangle$ be an ordinary CNG1. Let $V = \{v_1, v_2, \ldots, v_n\}$ be the vertex set of ξ. Then, the absolute amount of edges of ξ, denoted as $|E|_\xi$, is defined to be the ordered set $(|E|_T, |E|_I, |E|_F)$ for which:*

(a) $|E|_T = \sum_{\{r,s\}\subseteq\{1,2,\ldots,n\}} |\omega_T(v_r, v_s)|$,

(b) $|E|_I = \sum_{\{r,s\}\subseteq\{1,2,\ldots,n\}} |\omega_I(v_r, v_s)|$,

(c) $|E|_F = \sum_{\{r,s\}\subseteq\{1,2,\ldots,n\}} |\omega_F(v_r, v_s)|$.

On the other hand, sometimes we are particularly concerned about the number of non-void edges in an ordinary CNG1. In other words, we just want to know how many edges $\{v_i, v_j\}$ with:

$$\omega(v_i, v_j) \neq (0, 0, 0).$$

Instead of a mere visual interpretation, we must however form a precise definition as follows:

Definition 15. *Let $\xi = \langle V, \rho, \omega\rangle$ be an ordinary CNG1. Let $V = \{v_1, v_2, \ldots, v_n\}$ to be the vertex set of ξ. Then, the number of non-void edges of ξ, denoted as M_ξ, is defined to be the cardinality of the set:*

$$\{\ \{v_i, v_j\} \subseteq V \mid \omega(v_i, v_j) \neq (0, 0, 0)\ \}.$$

Definition 16. *Let $\xi = \langle V, \rho, \omega\rangle$ be an ordinary CNG1. Let $V = \{v_1, v_2, \ldots, v_n\}$ to be the vertex set of ξ. Then, the number of vertices of ξ, denoted as N_ξ, is defined to be the cardinality of the set V itself.*

Remark 12. *In this paper, we often deal with both* M_ξ *and* N_ξ *at the same time. Thus, we will not denote* N_ξ *as* $|V|$.

Remark 13. *By Definition 7, V is non-void, so* $N_\xi \geq 1$ *follows.*

Lemma 1. *Let* $\xi = \langle V,\ \rho,\ \omega \rangle$ *be an ordinary CNG1. Let* $V = \{v_1,\ v_2, \ldots,\ v_n\}$ *be the vertex set of* ξ*. Then, for each i:*

(a) $D_T(v_i) = \sum\limits_{r=1}^{n} \omega_T(v_r, v_i) + \omega_T(v_i, v_i)$,

(b) $D_I(v_i) = \sum\limits_{r=1}^{n} \omega_I(v_r, v_i) + \omega_I(v_i, v_i)$,

(c) $D_F(v_i) = \sum\limits_{r=1}^{n} \omega_F(v_r, v_i) + \omega_F(v_i, v_i)$.

Proof. Since ξ is ordinary, $\omega(v_r, v_i) = \omega(v_i, v_r)$ for all i and r. The lemma thus follows. □

Lemma 2. *Let* $\xi = \langle V,\ \rho,\ \omega \rangle$ *to be an ordinary CNG1. If* ξ *is simple. then, for each i:*

(a) $D_T(v_i) = \sum\limits_{r\in\{1,2,\ldots,n\}-\{i\}} \omega_T(v_i, v_r)$,

(b) $D_I(v_i) = \sum\limits_{r\in\{1,2,\ldots,n\}-\{i\}} \omega_I(v_i, v_r)$,

(c) $D_F(v_i) = \sum\limits_{r\in\{1,2,\ldots,n\}-\{i\}} \omega_F(v_i, v_r)$.

Proof. Since ξ is simple, $\omega(v_i, v_i) = (0, 0, 0)$ for all i. The lemma thus follows. □

Lemma 3. *Let* $\xi = \langle V,\ \rho,\ \omega \rangle$ *be an ordinary CNG1. Let* $V = \{v_1,\ v_2, \ldots,\ v_n\}$ *be the vertex set of* ξ*. Then, for each i:*

(a) $|D|_T(v_i) = \sum\limits_{r=1}^{n} |\omega_T(v_r, v_i)| + |\omega_T(v_i, v_i)|$,

(b) $|D|_I(v_i) = \sum\limits_{r=1}^{n} |\omega_I(v_r, v_i)| + |\omega_I(v_i, v_i)|$,

(c) $|D|_F(v_i) = \sum\limits_{r=1}^{n} |\omega_F(v_r, v_i)| + |\omega_F(v_i, v_i)|$.

Proof. The arguments are similar to Lemma 1. □

Lemma 4. *Let* $\xi = \langle V,\ \rho,\ \omega \rangle$ *be an ordinary CNG1. If* ξ *is simple. then, for each i:*

(a) $|D|_T(v_i) = \sum\limits_{r\in\{1,2,\ldots,n\}-\{i\}} |\omega_T(v_i, v_r)|$,

(b) $|D|_I(v_i) = \sum\limits_{r\in\{1,2,\ldots,n\}-\{i\}} |\omega_I(v_i, v_r)|$,

(c) $|D|_F(v_i) = \sum\limits_{r\in\{1,2,\ldots,n\}-\{i\}} |\omega_F(v_i, v_r)|$.

Proof. The arguments are similar to Lemma 2. □

Lemma 5. *Let* $\xi = \langle V,\ \rho,\ \omega \rangle$ *be an ordinary CNG1. Then* $\sum\limits_{r=1}^{n} |D|(v_r) = (0, 0, 0)$ *if and only if* $|D|(v_i) = (0, 0, 0)$ *for all i.*

Proof . Without loss of generality, since $|D|_T(v_i) = \sum_{r=1}^{n} |\omega_T(v_i, v_r)| + |\omega_T(v_i, v_i)|$ by Definition 13, it is always a non-negative real number. Thus, in order that $\sum_{r=1}^{n} |D|_T(v_r) = 0$, there can be only one possibility: all $|D|_T(v_i)$ must be zero. □

Remark 14. *A similar statement does not hold for the resultant degree.*

We now proceed with two of our theorems which both serve as generalizations of the well-known theorem in classical literature:

"For an ordinary graph, the sum of the degree of all its vertices is always twice the number of its edges".

Theorem 1. *Let $\xi = \langle V, \rho, \omega \rangle$ be an ordinary CNG1. Then $\sum_{r=1}^{n} D(v_r) = 2E_\xi$.*

Proof . As $D(v_i) = (D_T(v_i), D_I(v_i), D_F(v_i))$ for all i, and $E_\xi = (E_T, E_I, E_F)$. Without loss of generality, it suffices to prove that $2E_T = \sum_{r=1}^{n} D_T(v_r)$:

$$E_T = \sum_{\{r,s\} \subseteq \{1,2,\ldots,n\}} \omega_T(v_r, v_s) = \sum_{\substack{\{r,s\} \subseteq \{1,2,\ldots,n\} \\ r \neq s}} \omega_T(v_r, v_s) + \sum_{r=1}^{n} \omega_T(v_r, v_r).$$

Since $\{r, s\} = \{s, r\}$ for all s and r, it follows that:

$$\begin{aligned} 2E_T &= 2\sum_{\substack{\{r,s\} \subseteq \{1,2,\ldots,n\} \\ r \neq s}} \omega_T(v_r, v_s) + 2\sum_{r=1}^{n} \omega_T(v_r, v_r) \\ &= \sum_{\substack{r \in \{1,2,\ldots,n\} \\ s \in \{1,2,\ldots,n\} \\ r \neq s}} \omega_T(v_r, v_s) + 2\sum_{r=1}^{n} \omega_T(v_r, v_r) \\ &= \sum_{\substack{r \in \{1,2,\ldots,n\} \\ s \in \{1,2,\ldots,n\}}} \omega_T(v_r, v_s) + \sum_{r=1}^{n} \omega_T(v_r, v_r) \\ &= \sum_{r=1}^{n} \sum_{s=1}^{n} \omega_T(v_r, v_s) + \sum_{r=1}^{n} \omega_T(v_r, v_r) \\ &= \sum_{r=1}^{n} \left(\sum_{s=1}^{n} \omega_T(v_r, v_s) + \omega_T(v_r, v_r) \right) \\ &= \sum_{r=1}^{n} D_T(v_r). \end{aligned}$$

This completes the proof. □

Theorem 2. *Let $\xi = \langle V, \rho, \omega \rangle$ be an ordinary CNG1. Then $\sum_{r=1}^{n} |D|(v_r) = 2|E|_\xi$.*

Proof . The arguments are similar to Theorem 1 and can be easily proven by replacing all the terms $\omega_T(v_i, v_j)$ with $|\omega_T(v_i, v_j)|$. □

Lemma 6. *Let $\xi = \langle V, \rho, \omega \rangle$ be an ordinary CNG1, with $\sum_{r=1}^{n} D(v_r) = (0, 0, 0)$. If $M_\xi > 0$, then $M_\xi \geq 2$.*

Proof . By Theorem 1, $\sum_{r=1}^{n} D(v_r) = 2E_\xi$, so $E_\xi = (0, 0, 0)$ as well.

If only one edge is non-void, then $\omega(v_{r_0}, v_{s_0}) \neq (0,0,0)$ only for one particular set $\{r_0, s_0\}$. This implies that:

$$E_T = \sum_{\{r,s\}\subseteq\{1,2,\ldots,n\}} \omega_T(v_r, v_s) = \omega_T(v_{r_0}, v_{s_0}),$$
$$E_I = \sum_{\{r,s\}\subseteq\{1,2,\ldots,n\}} \omega_I(v_r, v_s) = \omega_I(v_{r_0}, v_{s_0}),$$
$$E_F = \sum_{\{r,s\}\subseteq\{1,2,\ldots,n\}} \omega_F(v_r, v_s) = \omega_F(v_{r_0}, v_{s_0}),$$

which contradicts the statement that $E_\xi = (0,0,0)$. □

Since $M_\xi \geq 2$, one may have thought either M_ξ or N_ξ must be even. However, this is proven to be false, even by letting ξ to be simple and by letting $D(v) = (0,0,0)$ for all i, as shown by the following counter-example (Figure 3):

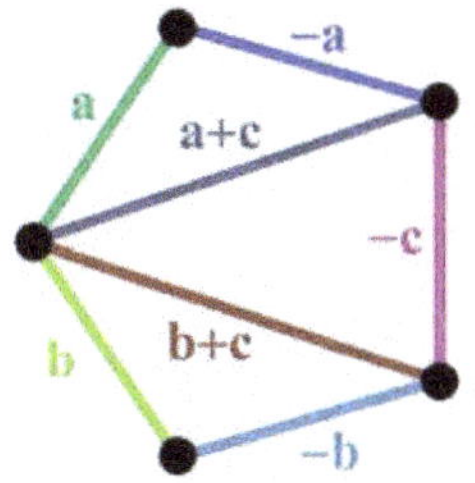

Figure 3. A counterexample, showing that M_ξ or N_ξ need not be even. $\mathbf{a} = \left(\frac{1}{5}e^{\ell 2\pi}, \frac{1}{5}e^{\ell\frac{4}{3}\pi}, \frac{1}{5}e^{\ell\frac{2}{3}\pi}\right)$, $\mathbf{b} = \left(\frac{1}{5}e^{\ell\frac{4}{3}\pi}, \frac{1}{5}e^{\ell\frac{2}{3}\pi}, \frac{1}{5}e^{\ell 2\pi}\right)$, $\mathbf{c} = \left(\frac{1}{5}e^{\ell\frac{2}{3}\pi}, \frac{1}{5}e^{\ell 2\pi}, \frac{1}{5}e^{\ell\frac{4}{3}\pi}\right)$.

for which $M_\xi = 7$, $N_\xi = 5$, and with all vertices being end-points of some edges. Moreover, such a result is not related to the value of $\rho(v)$ for any of the vertex v.

This motivates to consider what is the least possible values of M_ξ and N_ξ, for the special case of an ordinary ξ being simple, with $D(v) = (0,0,0)$ and $\rho(v) = (1,0,0)$ for all of its vertices v.

6. The Shortest CNG1 of Certain Conditions

We now proceed with the following definitions.

Definition 17. *Let $\xi = \langle V, \rho, \omega\rangle$ be an ordinary CNG1. ξ is said to be net if all of the following are satisfied:*

(a) *ξ is simple.*
(b) *ξ is connected.*
(c) *for all $v \in V$, $D(v) = (0,0,0)$ and $\rho(v) = (1,0,0)$.*

Furthermore, ξ is said to be *trivial* if the entire ξ consist of one single vertex v with $\rho(v) = (1,0,0)$. On the other hand, ξ is said to be *gross* if it is not net.

Lemma 7. *Let $\xi = \langle V, \rho, \omega\rangle$ be a non-trivial net CNG1. Then each vertex must have least two non-void edges adjacent to it.*

Proof. Let $v \in V$. Since $N_\xi \geq 2$ and ξ is connected, there must exist a non-void edge $\{v, u\}$ for some $u \in V - \{v\}$.

If $\{v, u\}$ is the only non-void edge adjacent to v, then $D(v) = \omega(v, u) \neq (0,0,0)$. This a contradiction. □

Theorem 3. *Let $\xi = \langle V, \rho, \omega \rangle$ be a non-trivial net CNG1. Then $M_\xi \geq 4$. Moreover, two of those non-void edges must be $\{a, b\}$ and $\{a, c\}$, for some mutually distinct vertices a, b, c.*

Proof . Since $N_\xi \geq 2$ and ξ is connected, non-void edge(s) must exist, so $M_\xi > 0$. Furthermore, $D(v) = (0,0,0)$ for all $v \in V$ would imply $\sum_{v \in V} D(v) = (0,0,0)$. $M_\xi \geq 2$ now follows by Lemma 6.

Let a be an end-point of some of those non-void edges. From Lemma 7, we conclude that at least two non-void edges must be adjacent to a.

Since ξ is simple, it now follows that those 2 non-void edges must be $\{a, b\}$ and $\{a, c\}$, with a, b, c being 3 mutually distinct vertices of ξ.

If $M_\xi = 2$:

$\{a, b\}$ and $\{a, c\}$ are therefore the only two non-void edges. By Lemma 7, both $\{a, b\}$ and $\{a, c\}$ must be adjacent to b. This is a contradiction.

If $M_\xi = 3$:

There can only be one more non-void edges besides $\{a, b\}$ and $\{a, c\}$.

By Lemma 7: b must be an end-point of another non-void edge besides $\{a, b\}$; and c must also be an end-point of another non-void edge besides $\{a, c\}$.

We now deduce that the third non-void edge must therefore be adjacent to both b and c. This yields Figure 4:

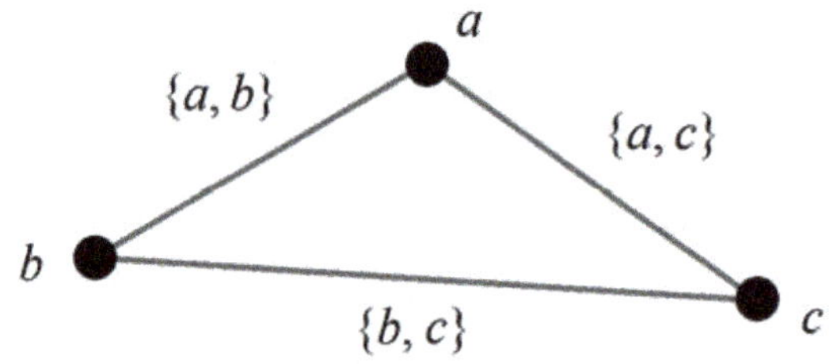

Figure 4. The triangle formed when $\{a, b\}$, $\{a, c\}$ and $\{b, c\}$ are all non-void. Mentioned in Theorem 3, 6.

Since $\{a, b\}$ and $\{c, a\}$ are non void, $\omega(a, b) = \mathbf{k} = -\omega(c, a)$ for some $\mathbf{k} \neq (0,0,0)$.

Since $\{b, c\}$ is adjacent to both b and c, $\omega(b, c) = \mathbf{k} = -\mathbf{k}$. This is again a contradiction.

$M_\xi \geq 4$ now follows. □

Theorem 4. *Let $\xi = \langle V, \rho, \omega \rangle$ be a non-trivial net CNG1. Then $M_\xi \geq 4$ and $N_\xi \geq 4$.*

Proof . By Theorem 3, $M_\xi \geq 4$, and two of those non-void edges must be $\{a, b\}$ and $\{a, c\}$, for some mutually distinct vertices a, b, c.

Suppose $N_\xi < 4$. Since ξ is simple, the maximum possible number of edges (whether it is void or not) is $3 + \frac{3}{2}(3 - 3) = 3 < 4$, which is a contradiction. $N_\xi \geq 4$ now follows. □

Theorem 5. *The smallest non-trivial net CNG1 must be of the structure in Figure 5:*

Proof . Let $\xi = \langle V, \rho, \omega \rangle$ be a non-trivial net CNG1. By Theorem 4, $M_\xi \geq 4$ and $N_\xi \geq 4$. By Theorem 3, two of those non-void edges must be $\{a, b\}$ and $\{a, c\}$, with a, b, c being three mutually distinct vertices of ξ.

Consider the scenario where $M_\xi = 4$ and $N_\xi = 4$ (i.e., the least possible number).

If the edge $\{b, c\}$ is non-void, then we would have formed Figure 4, as mentioned in the proof of Theorem 3

That leaves us with only one vertex d and only one extra non-void edge being adjacent to d. This is a contradiction.

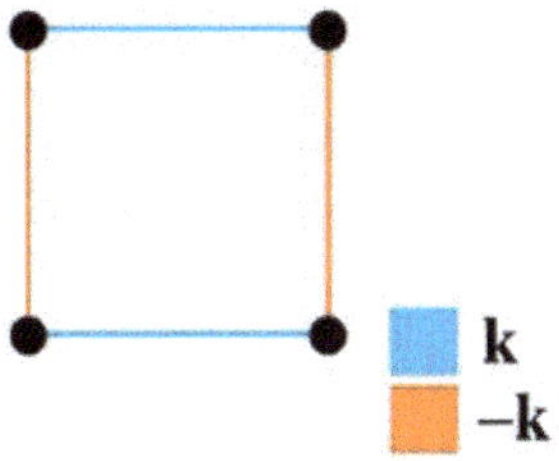

Figure 5. The smallest non-trivial net CNG1. Mentioned in Theorem 5 and Example 4. $\mathbf{k} \neq (0,0,0)$.

There is now only one choice left: both the edges $\{d, b\}$ and $\{d, c\}$ must be non-void. This gives rise to the following structure in Figure 6:

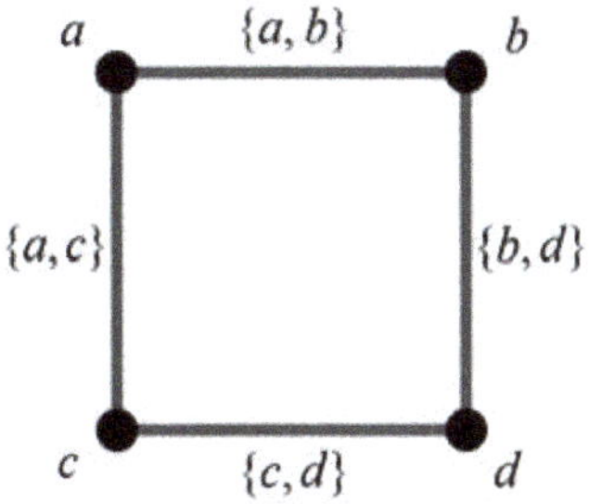

Figure 6. The only choices left because eace vertex must have at least 2 adjacent non void edges. Mentioned in Theorem 5, 6.

Without loss of generality, let $\omega(a, b) = \mathbf{k}$. Then both $\omega(b, d) = -\mathbf{k}$ and $\omega(a, c) = -\mathbf{k}$ must follow, leaving us with $\omega(c, d) = \mathbf{k}$ as the only valid option.

We are therefore left with the only way of assigning ω as shown by the theorem. □

Lemma 8. *Let $\xi = \langle V, \rho, \omega \rangle$ be a non-trivial net CNG1. Then $M_\xi \geq N_\xi$.*

Proof . Every single non-void edge is connected to two vertices. Thus, if we count the total number of adjacent non-void edges for each vertex, and then summing the results for all the vertices together, the result will be $2M_\xi$ (note: this paragraph is analogous to classical graph theory).

By Lemma 7, each vertex must have at least two non-void edges connect to it. We now have $2M_\xi \geq 2N_\xi$, so $M_\xi \geq N_\xi$ follows. □

Theorem 6. *Let $\xi = \langle V, \rho, \omega \rangle$ be a non-trivial net CNG1 with both M_ξ and N_ξ being odd numbers. Then $M_\xi \geq 7$ and $N_\xi \geq 5$.*

Proof. Let $\xi = V, \rho, \omega$ be a non-trivial net CNG1. By Theorem 4, $M_\xi \geq 4$ and $N_\xi \geq 4$. By Theorem 3, two of those non-void edges must be $\{a, b\}$ and $\{a, c\}$, for some a, b and c being three mutually distinct vertices of ξ.

Since both M_ξ and N_ξ are odd, it follows that $M_\xi \geq 5$ and $N_\xi \geq 5$. So in addition to a, b, c, there exist another 2 vertices d, e.

Consider the scenario where $M_\xi = 5$ and $N_\xi = 5$ (i.e., the least possible number).

Case 1. Suppose the edge $\{b,c\}$ is non-void. Then we would have formed Figure 4, as mentioned in the proof of Theorem 3.

That leaves us with two vertices d and e, and two extra non-void edge, which both of them must be adjacent to d. Even if $\{d,e\}$ is non-void, the other non-void edge adjacent to d cannot possibly be $\{d,e\}$ itself. Therefore, we have, at most, one non-void edge being adjacent to e. This is a contradiction.

Case 2. Without loss of generality, suppose the edges $\{b,d\}$ and $\{c,d\}$ are non-void. Then we would have formed Figure 6, as mentioned in the proof of Theorem 5.

That leaves us with only one vertex e and only one extra edge being adjacent to e, which is, again, a contradiction.

Case 3. Without loss of generality, suppose the edges $\{b,d\}$ and $\{c,e\}$ are non-void. Then, besides $\{b,d\}$, another edge must be adjacent to d. Likewise, besides $\{c,e\}$, another edge must be adjacent to e. Since we are left with one edge, it must, therefore, be $\{d,e\}$. This gives rise to the following structure in Figure 7:

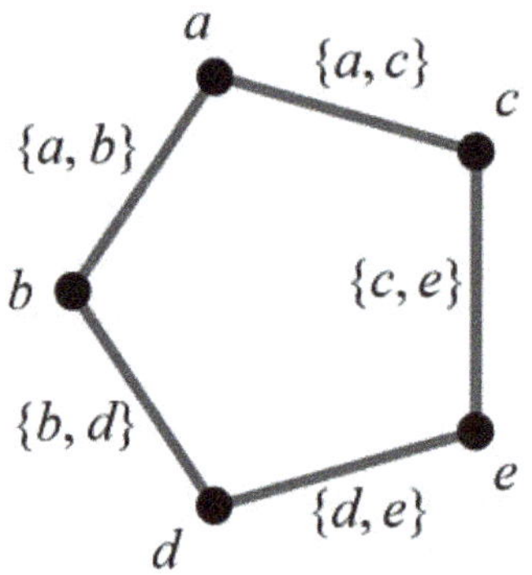

Figure 7. The only choice left for the case of 5 non-void edges connecting to 5 vertices. Mentioned in Theorem 6.

Without loss of generality, let $\omega(a,b) = \mathbf{k}$. Then both $\omega(b,d) = -\mathbf{k}$ and $\omega(a,c) = -\mathbf{k}$ must follow, leaving us with both $\omega(c,e) = \mathbf{k}$ and $\omega(d,e) = \mathbf{k}$.

We have, thus, arrived at $D(e) = 2\mathbf{k} \neq (0,0,0)$, again a contradiction.

Hence, it is either $M_\xi > 5$ or $N_\xi > 5$.

Since both M_ξ and N_ξ are odd, either one of the following must hold:

(a) $M_\xi \geq 7$ and $N_\xi \geq 7$.
(b) $M_\xi = 7$ and $N_\xi = 5$.
(c) $M_\xi = 5$ and $N_\xi = 7$.

Furthermore, by Lemma 8, $M_\xi \geq N_\xi$. Hence (c) will not occur, which implies that $M_\xi \geq 7$ and $N_\xi \geq 5$. This completes the proof. □

Theorem 7. *The smallest non-trivial net CNG1 ξ, with both M_ξ and N_ξ being odd numbers, must be of the structure as shown in Figure 8:*

Proof . Let $\xi = \langle V,\ \rho,\ \omega\rangle$ be a non-trivial net CNG1 with both M_ξ and N_ξ being odd numbers. Then $M_\xi \geq 7$ and $N_\xi \geq 5$.

Consider the scenario where $M_\xi = 7$ and $N_\xi = 5$ (i.e., the least possible number).

Since ξ is an ordinary CNG1, each vertex must have 5 edges adjacent to it (whether void or not). Since ξ is simple, one of the five edges for each vertex, which is a loop, must be void. As a result, we now conclude that each vertex must have at most 4 non-void edges adjacent to it.

On the other hand, by Lemma 7, each vertex must have at least two non-void edges adjacent to it.

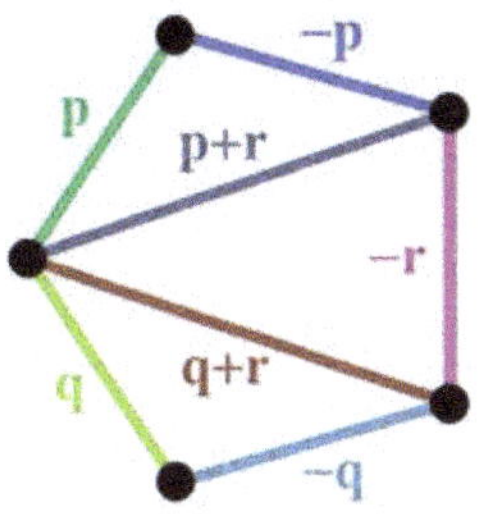

Figure 8. The smallest non-trivial net CNG1, with both M_ξ and N_ξ being odd numbers. . Mentioned in Theorem 7 and Example 5. $\mathbf{p}+\mathbf{q}+\mathbf{r}=(0,0,0)$; $|\mathbf{p}+\mathbf{r}|, |\mathbf{q}+\mathbf{r}| \leq 1$.

Since every single non-void edge is adjacent to two vertices. Thus, if we count the total number of adjacent non-void edges for each vertex, and then summing the results for all the vertices together, the result will be $7 \times 2 = 14$ (note: this paragraph is analogous to classical graph theory).

Hence, the set representing the number of non-void edges adjacent to each of the five vertices, must be one of the following:

(a) $\{2,3,3,3,3\}$ (most "widely spread" possibility)
(b) $\{2,2,3,3,4\}$
(c) $\{2,2,2,4,4\}$ (most "concentrated" possibility)

We now consider each the three cases:

Case 1. $\{2,3,3,3,3\}$

Without loss of generality:

$$\begin{array}{c}\text{Let } a \text{ be that one vertex which is an end-point to only 2} \\ \text{non-void edges } \{a,b\} \text{ and } \{a,c\}. \text{ (i.e., } \{a,d\}, \{a,e\} \text{ are void) (Figure 9).}\end{array} \tag{9}$$

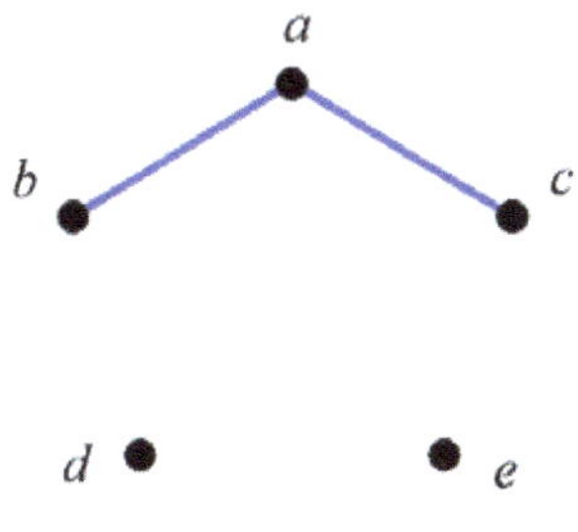

Figure 9. The 2 non-void edges $\{a,b\}$ and $\{a,c\}$, for all the 3 cases of Theorem 7.

Then, each one among b, c, d, e must be an end-point of three non-void edges.

Besides $\{d,a\}$ and $\{d,d\}$, which are both void, there are three more edges adjacent to d: $\{d,b\}$, $\{d,c\}$, $\{d,e\}$.

Since d is an end-point of exactly three non-void edges, we conclude that:

$$\{d,b\},\{d,c\},\{d,e\} \text{ areallnon} - \text{void} \tag{10}$$

Similarly, besides $\{e,a\}$ and $\{e,e\}$, which are both void, there are three more edges adjacent to e: $\{e,b\}$, $\{e,c\}$, $\{e,d\}$.

Since e is also an end-point of exactly three non-void edges, we conclude that:

$$\{e,b\},\{e,c\},\{e,d\} \text{ areallnon} - \text{void} \tag{11}$$

From (10) and (11), we conclude that:

$$\{d,b\},\{d,c\},\{e,b\},\{e,c\},\{d,e\} = \{e,d\} \text{ areallnon} - \text{void} \tag{12}$$

From (9) and (12), we have obtained all the seven non-void edges:

$$\{d,b\},\{d,c\},\{e,b\},\{e,c\},\{d,e\},\{a,b\},\{a,c\}.$$

Hence, $\{b,c\}$, $\{a,d\}$, $\{a,e\}$ must be all void. We, thus, obtain the following structure (Figure 10):

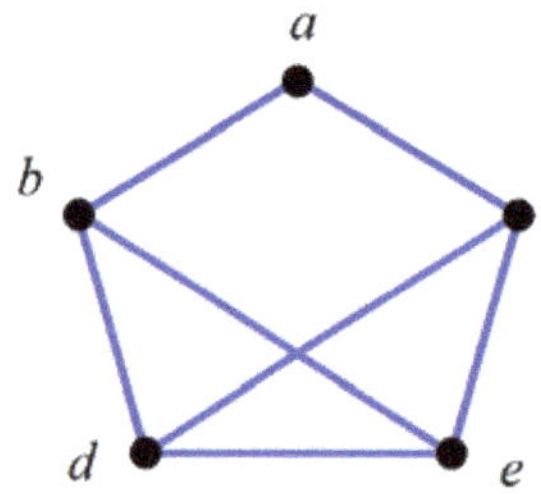

Figure 10. The only possible way of connection for $\{2,3,3,3,3\}$.

Let $\omega(a,b)$ = $\mathbf{p}$, $\omega(b,d)$ = $\mathbf{q}$, $\omega(b,e)$ = $\mathbf{r}$, $\omega(c,d)$ = $\mathbf{s}$, $\omega(c,e)$ = $\mathbf{t}$, $\omega(d,e)$ = $\mathbf{u}$.

Since $\{a,b\}$ and $\{a,c\}$ are the only two non-void edges adjacent to a, we now have $\omega(a,c) = -\mathbf{p}$ (Figure 11).

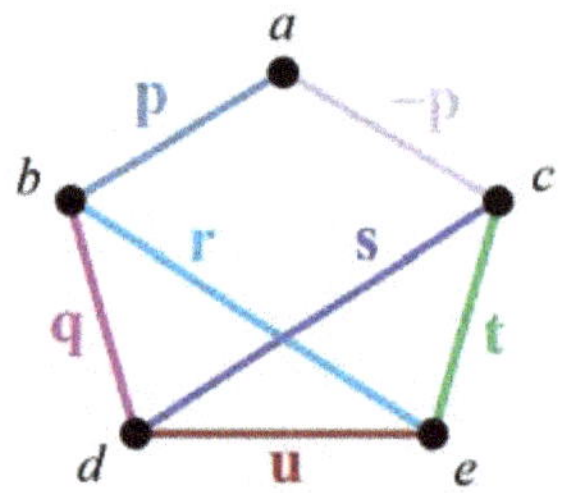

Figure 11. The labeling of the non-void edges for $\{2,3,3,3,3\}$.

We how have: $\mathbf{r}+\mathbf{q}+\mathbf{p} = \mathbf{s}+\mathbf{t}-\mathbf{p} = \mathbf{s}+\mathbf{q}+\mathbf{u} = \mathbf{r}+\mathbf{t}+\mathbf{u} = (0,0,0)$.

This further implies that: $\mathbf{r}+\mathbf{q}+\mathbf{p}+\mathbf{s}+\mathbf{t}-\mathbf{p} = \mathbf{s}+\mathbf{q}+\mathbf{u}+\mathbf{r}+\mathbf{t}+\mathbf{u} = (0,0,0)$.

Therefore, $\mathbf{q}+\mathbf{r}+\mathbf{s}+\mathbf{t} = \mathbf{q}+\mathbf{r}+\mathbf{s}+\mathbf{t}+2\mathbf{u}$, which implies that $\mathbf{u} = (0,0,0)$. This is a contradiction.

Case 2. $\{2,2,3,3,4\}$

Without loss of generality:

$$\text{Let } a \text{ be that one vertex which is an end-point to only two non-void edges } \{a,b\} \text{ and } \{a,c\}. \text{ (i. e., } \{a,d\}, \{a,e\} \text{ are void)} \tag{13}$$

as shown in Figure 9.

Since $\{a,d\}$, $\{a,e\}$ are void, both d and e cannot be that vertex which is an end-point to four non-void edges.

By symmetry, fix b to be that vertex which is an end-point to four non-void edges. Then:

$$\{b,a\},\{b,c\},\{b,d\},\{b,e\} \text{ areallnon} - \text{void.} \tag{14}$$

From (13) and (14), we have now arrived at the following structure (Figure 12):

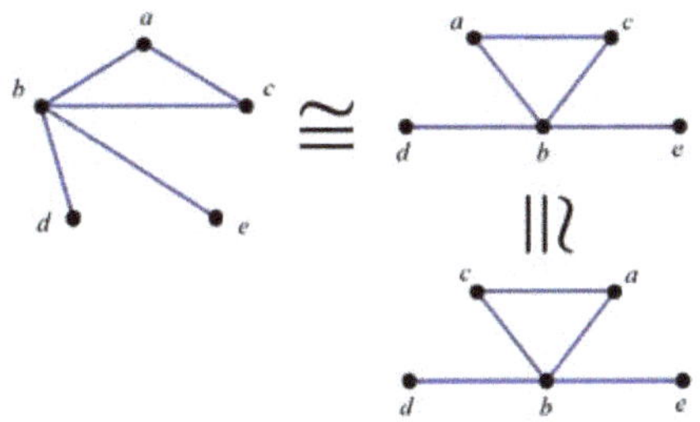

Figure 12. The first 5 non-void edges for $\{2,2,3,3,4\}$.

Suppose $\{d,e\}$ is void. Then exactly one out of $\{d,c\}$ and $\{d,a\}$ must be non-void. Similarly, exactly one out of $\{e,c\}$ and $\{e,a\}$ must be non-void. By symmetry and the rules of graph isomorphism, fix $\{d,a\}$ to be non-void, then a would have been an end-point of three non-void edges: $\{d,a\}$, $\{b,a\}$, $\{c,a\}$. So $\{e,a\}$ must be void and, therefore, $\{e,c\}$ is non-void. We, thus, obtain the following structure (Figure 13):

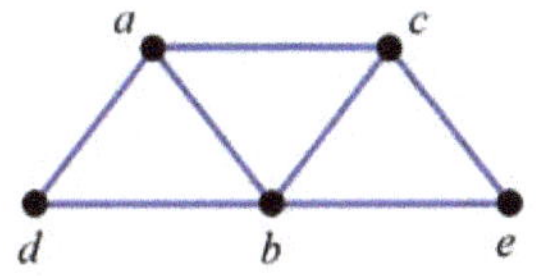

Figure 13. The only possible way of connection for $\{2,2,3,3,4\}$, if $\{d,e\}$ is void.

Suppose $\{d,e\}$ is non-void. Then we now arrived at the following structure (Figure 14):

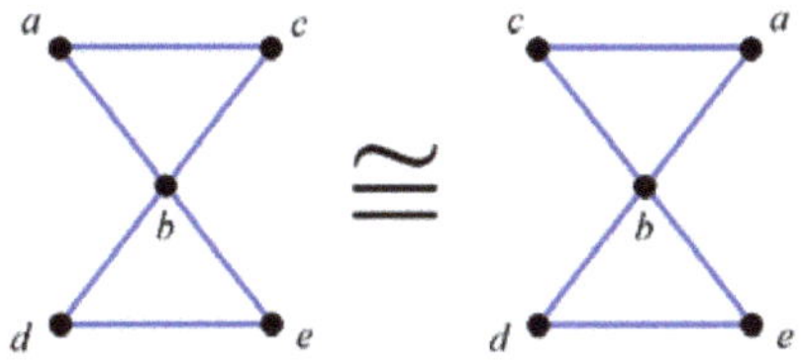

Figure 14. The first 6 non-void edges for $\{2,2,3,3,4\}$, for the case of non-void $\{d,e\}$.

By symmetry, fix a to be a vertex which is an end-point of three non-void edges. Then exactly one edge out of $\{a,d\}$ and $\{a,e\}$ must be non-void. By the rules of graph isomorphism, we can fix $\{a,d\}$ to be non-void. Again we obtain the following structure (Figure 15):

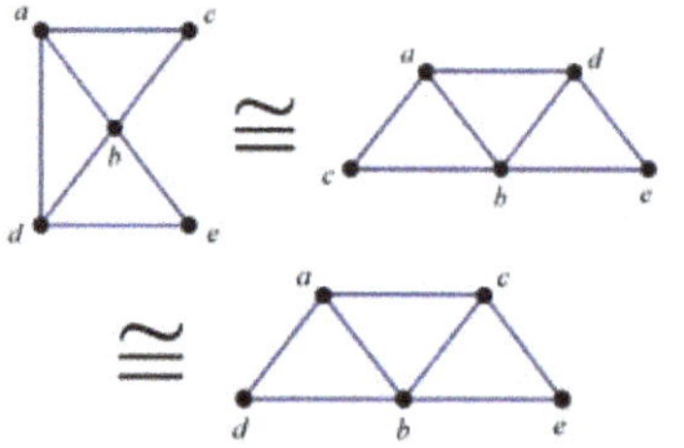

Figure 15. The only possible way of connection for $\{2,2,3,3,4\}$, if $\{d,e\}$ is non-void.

Let $\omega(a,b) = \mathbf{g}$, $\omega(c,b) = \mathbf{h}$, $\omega(a,c) = \mathbf{k}$, $\omega(b,d) = \mathbf{p}$, $\omega(b,e) = \mathbf{q}$.

Since $\{a,d\}$ and $\{b,d\}$ are the only two non-void edges adjacent to d, we now have $\omega(a,d) = -\mathbf{p}$.

Likewise, since $\{c,e\}$ and $\{b,e\}$ are the only two non-void edges adjacent to e, we now have $\omega(c,e) = -\mathbf{q}$ (Figure 16).

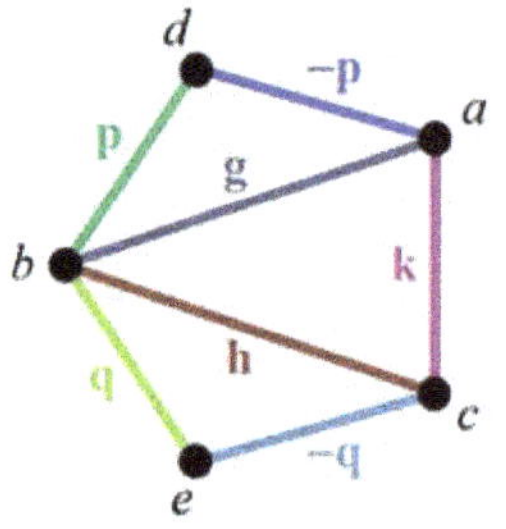

Figure 16. The labeling of the non-void edges for $\{2,2,3,3,4\}$.

We how have: $\mathbf{p}+\mathbf{q}+\mathbf{g}+\mathbf{h} = \mathbf{g}+\mathbf{k}-\mathbf{p} = \mathbf{h}+\mathbf{k}-\mathbf{q} = (0,0,0)$.

Therefore, $\mathbf{g} = \mathbf{p}-\mathbf{k}$, $\mathbf{h} = \mathbf{q}-\mathbf{k}$. As a result: $\mathbf{p}+\mathbf{q}+\mathbf{p}-\mathbf{k}+\mathbf{q}-\mathbf{k} = 2\mathbf{p}+2\mathbf{q}-2\mathbf{k} = (0,0,0)$, which implies $\mathbf{p}+\mathbf{q}-\mathbf{k} = (0,0,0)$.

Denote $-\mathbf{k} = \mathbf{r}$. Then $\mathbf{g} = \mathbf{p}+\mathbf{r}$, $\mathbf{h} = \mathbf{q}+\mathbf{r}$, and $\mathbf{p}+\mathbf{q}+\mathbf{r} = (0,0,0)$ follows. We have, thus, formed the structure as mentioned in this theorem.

Case 3. $\{2,2,2,4,4\}$

Without loss of generality:

Let a be one of that two vertices which is an end-point to four non-void edges. Then:

$$\{a,b\},\{a,c\},\{a,d\},\{a,e\} \text{ arenon} - \text{void}. \tag{15}$$

Let b be the other one vertices which is also an end-point to four non-void edges. Then:

$$\{b,a\},\{b,c\},\{b,d\},\{b,e\} \text{ arenon} - \text{void}. \tag{16}$$

From (15) and (16), we have obtained the seven non-void edges:

$$\{a,c\},\{a,d\},\{a,e\},\{b,c\},\{b,d\},\{b,e\},\{a,b\}.$$

Hence, $\{c,d\}$, $\{c,e\}$, $\{d,e\}$ are all void. We, thus, obtain the following structure (Figure 17):

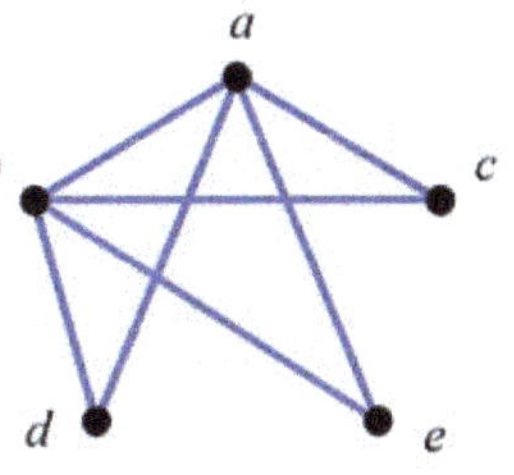

Figure 17. The only possible way of connection for $\{2,2,2,4,4\}$.

Let $\omega(b,d) = \mathbf{p}$, $\omega(b,e) = \mathbf{q}$, $\omega(b,c) = \mathbf{r}$, $\omega(b,a) = \mathbf{s}$,

Since $\{a,d\}$ and $\{b,d\}$ are the only two non-void edges adjacent to d, we now have = $-\mathbf{p}$.

Since $\{a,e\}$ and $\{b,e\}$ are the only two non-void edges adjacent to e, we now have $\omega(a,e) = -\mathbf{q}$.

Since $\{a,c\}$ and $\{b,c\}$ are the only two non-void edges adjacent to c, we now have $\omega(a,c) = -\mathbf{r}$ (Figure 18).

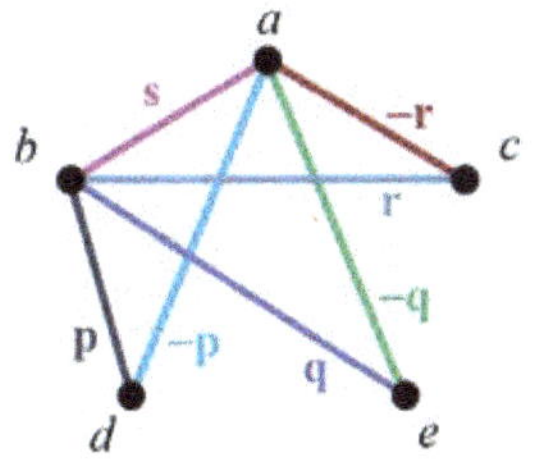

Figure 18. The labeling of the non-void edges for $\{2,2,2,4,4\}$.

We how have: $\mathbf{s}+\mathbf{p}+\mathbf{q}+\mathbf{r} = \mathbf{s}-\mathbf{p}-\mathbf{q}-\mathbf{r} = (0,0,0)$.

This further implies that: $\mathbf{s}+\mathbf{p}+\mathbf{q}+\mathbf{r}+\mathbf{s}-\mathbf{p}-\mathbf{q}-\mathbf{r} = (0,0,0)$.

We now have $2\mathbf{s} = (0,0,0)$, which implies that $\mathbf{s} = (0,0,0)$. This is a contradiction.

Our proof is now complete. □

Note that both 5 and 7 are not divisible even by 3, the next prime number after 2. This yields the following corollary:

Corollary 1. *The smallest non-trivial net CNG1 ξ, with both M_ξ and N_ξ not divisible by 2 or 3, must also be of the structure as shown in Figure:*

7. Symmetric Properties of Ordinary Simple CNG1

Definition 18. *Let V and W be two non-void sets. Let $\xi = \langle V, \rho, \omega\rangle$ and $\zeta = \langle W, \varsigma, \psi\rangle$ be two ordinary CNG1s. If $V = W$, $\rho = \varsigma$ and $\omega = \psi$, then ξ and ζ are said to be equal, and shall be denoted by $\xi \equiv \zeta$.*

Definition 19. *Let V and W be a non-void set. Let $\xi = \langle V, \rho, \omega\rangle$ and $\zeta = \langle W, \varsigma, \psi\rangle$ be two ordinary CNG1s. If there exist a bijection $f : V \to W$ such that:*

(a) $\rho(u) = \varsigma(f(u))$ *for all* $u \in V$.
(b) $\omega(u,v) = \psi(f(u), f(v))$ *for all* $u, v \in V$.

Then:

(i) *Such f is said to be an isomorphism from ξ to ζ, and we shall denote such case by $f[\xi] \equiv \zeta$.*
(ii) *ξ and ζ are said to be isomorphic, and shall be denoted by $\xi \cong \zeta$.*

Remark 15. *As both ξ and ζ are ordinary, $\omega(u,v) = \omega(v,u)$ and $\psi(f(u), f(v)) = \psi(f(v), f(u))$ follow for all $u, v \in V$.*

Example 1. *Consider $\xi_0 = V_0, \rho_0, \omega_0$ and $\zeta_0 = W_0, \varsigma_0, \psi_0$ as follows:*

$V_0 = \{v_1, v_2, v_3, v_4, v_5, v_6\}$. $W_0 = \{w_1, w_2, w_3, w_4, w_5, w_6\}$.
$\rho_0(v_1) = \mathbf{p}, \rho_0(v_2) = \mathbf{q}, \rho_0(v_3) = \mathbf{t}, \varsigma_0(w_1) = \mathbf{t}, \varsigma_0(w_3) = \mathbf{p}, \varsigma_0(w_4) = \mathbf{s},$
$\rho_0(v_4) = \rho_0(v_6) = \mathbf{r}, \rho_0(v_5) = \mathbf{s}$. $\varsigma_0(w_2) = \varsigma_0(w_4) = \mathbf{r}, \varsigma_0(w_6) = \mathbf{q}$.
$\omega_0(v_1, v_2) = \omega_0(v_2, v_1) = \omega_0(v_2, v_2) = \mathbf{a}, \psi_0(w_1, w_2) = \psi_0(w_2, w_1) = \mathbf{d},$
$\omega_0(v_1, v_3) = \omega_0(v_3, v_1) = \mathbf{c}, \psi_0(w_1, w_3) = \psi_0(w_3, w_1) = \mathbf{c},$
$\omega_0(v_2, v_3) = \omega_0(v_3, v_2) = \mathbf{b}, \psi_0(w_1, w_4) = \psi_0(w_4, w_1) = \mathbf{e},$
$\omega_0(v_3, v_4) = \omega_0(v_4, v_3) = \mathbf{d}, \psi_0(w_4, w_5) = \psi_0(w_5, w_4) = \mathbf{f},$
$\omega_0(v_3, v_5) = \omega_0(v_5, v_3) = \mathbf{e}, \psi_0(w_1, w_6) = \psi_0(w_6, w_1) = \mathbf{b},$
$\omega_0(v_5, v_6) = \omega_0(v_6, v_5) = \mathbf{f}, \psi_0(w_3, w_6) = \psi_0(w_6, w_3) = \psi_0(w_6, w_6) = \mathbf{a},$
otherwise, $\omega_0(u, v) = (0,0,0)$. *otherwise*, $\psi_0(w, v) = (0,0,0)$.

Moreover, $|\{\mathbf{p}, \mathbf{q}, \mathbf{r}, \mathbf{s}, \mathbf{t}\}| = 5$ *and* $|\{\mathbf{a}, \mathbf{b}, \mathbf{c}, \mathbf{d}, \mathbf{e}, \mathbf{f}\}| = 6$ (Figure 19).

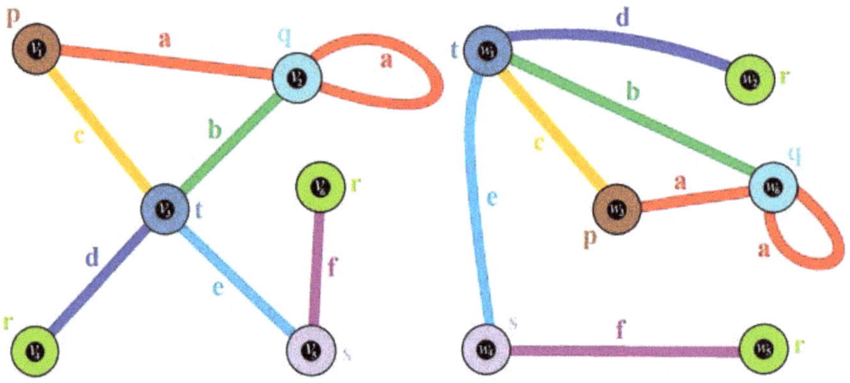

Figure 19. Two isomorphic CNG1's, as mentioned in Example 1.

Thus, we define the bijection $f_0 : V \to W$ as:

$$f_0(v_1) = v_3, f_0(v_2) = v_6, f_0(v_3) = v_1, f_0(v_4) = v_2, f_0(v_5) = v_4, f_0(v_6) = v_5.$$

It now follows that f_0 is an isomorphism from ξ_0 to ζ_0, so $\xi_0 \cong \zeta_0$. Still, $\xi_0 \zeta_0$ in accordance with Definition 18.

In all the following passages of this paper, let $\mathscr{I} : V \to V$ be the identity mapping from V to itself.

Like classical graph theory, whenever $\zeta \equiv \xi$, $\mathscr{I}$ is an isomorphism from ξ to ξ itself in accordance with Definition 19. It is, therefore, motivational to investigate if there are other non-identity bijections from V to itself, which is also an isomorphism from ξ to ξ itself. Additionally, recall that, in classical graph theory, an isomorphism from a graph to itself will be called an automorphism on that graph. Thus, we proceed with the following definition:

Definition 20. *Let V be a non-void set. Let $\xi = \langle V,\ \rho,\ \omega \rangle$ be an ordinary CNG1's. Let $\mathscr{f} : V \rightarrow V$ be a bijection such that:*

(a) $\rho(u) = \rho(\mathscr{f}(u))$ *for all* $u \in V$.
(b) $\omega(u,v) = \omega(\mathscr{f}(u), \mathscr{f}(v))$ *for all* $u, v \in V$.

Then $\mathscr{f}$ is said to be an automorphism of ξ.

Remark 16. *As ξ is ordinary, $\omega(u,v) = \omega(v,u)$ follows for all $u, v \in V$.*

Remark 17. *Just because $\rho(u) = \rho(\mathscr{f}(u))$ and $\omega(u,v) = \omega(\mathscr{f}(u), \mathscr{f}(v))$, does not mean that $u = \mathscr{f}(u)$ or $v = \mathscr{f}(v)$.*

Remark 18. *$\mathscr{I}$ is thus called the trivial automorphism of ξ.*

Example 2. *Consider* $\xi_1 = \langle V_1,\ \rho_1,\ \omega_1 \rangle$ as shown in Figure 20:
$V_1 = \{a, b, c, d\}$. $\rho_1(a) = \rho_1(b) = \rho_1(d) = \mathbf{p}$, $\rho_1(c) = \mathbf{q}$.
$\omega_1(a,c) = \omega_1(c,a) = \mathbf{h}$, $\omega_1(b,c) = \omega_1(c,b) = \omega_1(d,c) = \omega_1(c,d) = \mathbf{g}$, *otherwise,* $\omega_1(u,v) = (0,0,0)$.
$|\{\mathbf{p}, \mathbf{q}\}| = |\{\mathbf{g}, \mathbf{h}\}| = 2$.

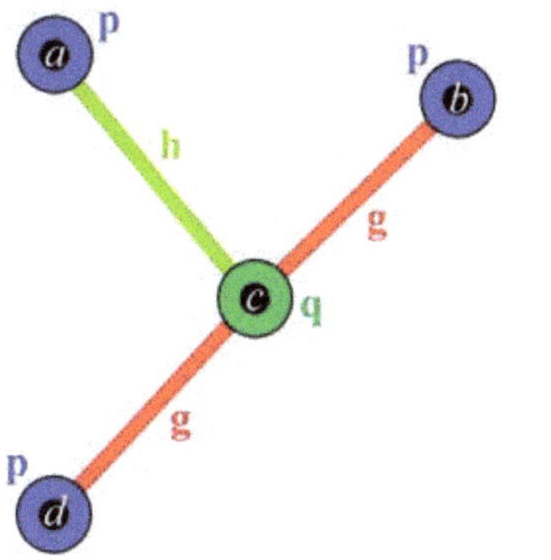

Figure 20. ξ_1 as mentioned in Example 2.

Let $\mathscr{f}_1, \mathscr{b}_1, \mathscr{h}_1 : V_1 \rightarrow V_1$ be three bijections defined as follows:

(a) $\mathscr{f}_1(c) = a, \mathscr{f}_1(a) = d, \mathscr{f}_1(d) = c, \mathscr{f}_1(b) = b$.
(b) $\mathscr{b}_1(b) = a, \mathscr{b}_1(d) = d, \mathscr{b}_1(c) = c, \mathscr{b}_1(a) = b$.
(c) $\mathscr{h}_1(b) = d, \mathscr{h}_1(a) = a, \mathscr{h}_1(c) = c, \mathscr{h}_1(d) = b$.

Then:

(i) *$\mathscr{f}_1$ is an isomorphism from V_1 to the following ordinary CNG1 (Figure 21).*

which is not equal to ξ_1 in accordance with Definition 18. *$\mathscr{f}_1$ is therefore not an automorphism of ξ_1.*

(ii) *$\mathscr{b}_1$ is an isomorphism from V_1 to the following ordinary CNG1 (Figure 22).*

which is also not equal to ξ_1 in accordance with Definition 18. *Likewise $\mathscr{b}_1$ is, therefore, not an automorphism of ξ_1.*

(iii) *$\mathscr{h}_1$ is an isomorphism from V_1 to itself and, therefore, it is an automorphism of ξ_1.* Note that, *even if $\mathscr{h}_1(b) = d$ and $\mathscr{h}_1(d) = b$, as $\rho_1(b) = \rho_1(d) = \mathbf{p}$ and $\omega_1(b,c) = \omega_1(d,c) = \mathbf{g}$, so $\mathscr{h}_1[\xi_1] \equiv \xi_1$ still holds.*

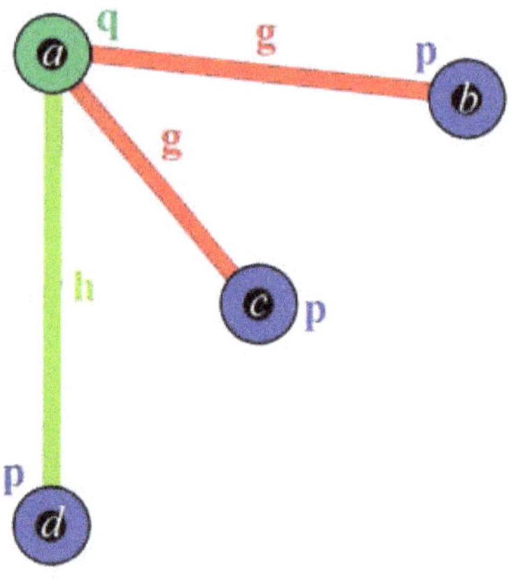

Figure 21. This is not an automorphism of ξ_1 as mentioned in Example 2.

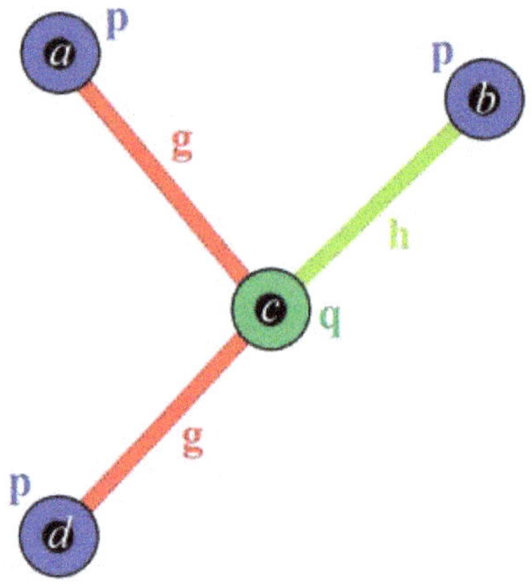

Figure 22. This is not an automorphism of ξ_1 as mentioned in Example 2.

Definition 21. *Let $\xi = \langle V,\ \rho,\ \omega \rangle$ be an ordinary simple CNG1. ξ is said to be total symmetric if, for all $\{u_1, v_1\}, \{u_2, v_2\} \subseteq V$, with $|\{u_1, v_1\}| = |\{u_2, v_2\}|$, there exist an automorphism of ξ, ℓ, such that $u_2 = \ell(u_1)$, $v_2 = \ell(v_1)$.*

Remark 19. *In other words, $\{u_1, v_1\}, \{u_2, v_2\}$ can either be two edges, or two vertices as when $u_1 = v_1$ and $u_2 = v_2$.*

Example 3. *With this definition, the following CNG1 (Figure 23) is, thus, totally-symmetric.*

However, unlike symmetry of classical graphs, the concept of total symmetry takes all the edges into account, whether void or not. As a result, the following graph (Figure 24), though looks familiar to the classical literature, is not totally-symmetric.

As a result, the concept of total-symmetry in ordinary simple CNG1 proves even more stringent than the concept of symmetry in classical ordinary simple graphs. Additionally, recall that edges and vertices in CNG1 have three membership values instead of only 0 (disconnected, void) and 1 (connected). To give more characterization of symmetry among ordinary simple CNG1, we now proceed with the following definitions.

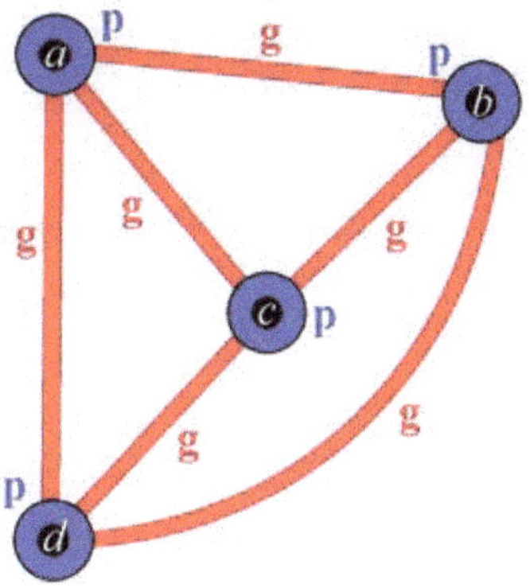

Figure 23. A totally-symmetric CNG1, as mentioned in Example 3.

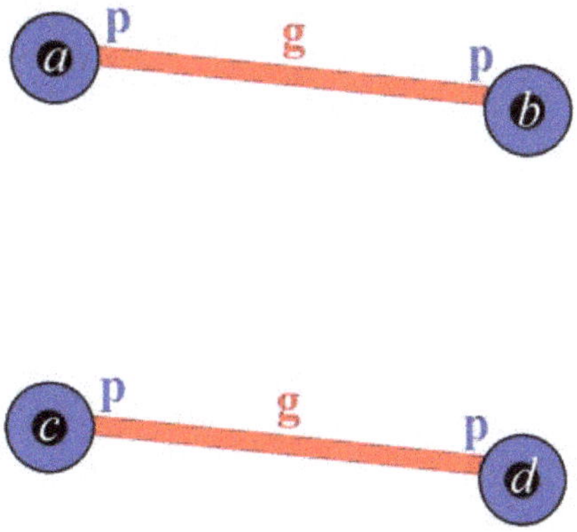

Figure 24. This graph is not totally-symmetric. Mentioned in Example 3.

Definition 22. *Let $\xi = \langle V, \rho, \omega \rangle$ be an ordinary simple CNG1. ξ is said to be strong edge-wise symmetric (abbr. SES) if: For all $\{u_1, v_1\}, \{u_2, v_2\} \subseteq V$ with both $\omega(u_1, v_1)$ and $\omega(u_2, v_2)$ non-void, there exist an automorphism f of ξ, such that $u_2 = f(u_1)$, $v_2 = f(v_1)$.*

Remark 20. *As ξ is simple, it follows that $|\{u_1, v_1\}| = |\{u_2, v_2\}| = 2$.*

Remark 21. *An ordinary simple CNG1 with all edges being void is classified as strong edge-wise symmetric as well.*

Definition 23. *Let $\xi = \langle V, \rho, \omega \rangle$ be an ordinary simple CNG1. ξ is said to be strong point-wise(or vertex-wise) symmetric (abbr. SPS) if: For all $u_1, u_2 \in V$ with both $\rho(u_1)$ and $\rho(u_2)$ non-void, there exists an automorphism f of ξ, such that $u_2 = f(u_1)$.*

Remark 22. *An ordinary simple CNG1 with all vertices being void is classified as strong point-wise symmetric as well.*

Definition 24. *Let $\xi = \langle V, \rho, \omega\rangle$ be an ordinary simple CNG1. ξ is said to be strong symmetric (abbr. SS) if it is both strong edge-wise symmetric and strong point-wise symmetric.*

Definition 25. *Let $\xi = \langle V, \rho, \omega\rangle$ be an ordinary simple CNG1. ξ is said to be weak edge-wise symmetric (abbr. wES) if: For all $\{u_1, v_1\}, \{u_2, v_2\} \subseteq V$ with $\omega(u_1, v_1) = \omega(u_2, v_2) \neq (0, 0, 0)$, there exists an automorphism $\mathscr{f}$ of ξ, such that $u_2 = \mathscr{f}(u_1)$, $v_2 = \mathscr{f}(v_1)$. Otherwise, ξ is said to be edge-wise asymmetric (abbr. EA).*

Remark 23. *Again, as ξ is simple, it follows that $|\{u_1, v_1\}| = |\{u_2, v_2\}| = 2$.*

Remark 24. *An ordinary simple CNG1 with all non-void edges having different membership value is classified as weak edge-wise symmetric as well.*

Definition 26. *Let $\xi = V, \rho, \omega$ be an ordinary simple CNG1. ξ is said to be weak point-wise (or vertex-wise) symmetric (abbr. wPS) if: For all $u_1, u_2 \in V$ with $\rho(u_1) = \rho(u_2) \neq (0, 0, 0)$, there exists an automorphism $\mathscr{f}$ of ξ, such that $u_2 = \mathscr{f}(u_1)$. Otherwise, ξ is said to be point-wise asymmetric (abbr. PA).*

Remark 25. *An ordinary simple CNG1 with all non-void vertices having different membership value is classified as weak point-wise symmetric as well.*

Definition 27. *Let $\xi = \langle V, \rho, \omega\rangle$ be an ordinary simple CNG1. ξ is said to be asymmetric if it is both edge-wise asymmetric and point-wise asymmetric.*

Based on the definition, we now state such symmetric properties on the smallest non-trivial net CNG1, as mentioned in Theorem 5, as well as the smallest non-trivial net CNG1 with both M_ξ and N_ξ being odd numbers, as mentioned in Theorem 7.

Example 4. *With regards to the structure of* Figure 5, *as mentioned in Theorem 5, with $\rho(a) = \rho(b) = \rho(c) = \rho(d) = (1, 0, 0)$.*

Consider the following three automorphisms $\mathscr{f}, \mathscr{g}, \mathscr{h}$ of $\xi_{4,4}$:

(a) $\mathscr{f}(a) = b, \mathscr{f}(b) = a, \mathscr{f}(c) = d, \mathscr{f}(d) = c,$
(b) $\mathscr{g}(a) = c, \mathscr{g}(b) = d, \mathscr{g}(c) = a, \mathscr{g}(d) = b,$
(c) $\mathscr{h}(a) = d, \mathscr{h}(b) = c, \mathscr{h}(c) = b, \mathscr{h}(d) = a,$

together with $\mathscr{I}$, the trivial automorphism of ξ.

As a result, $\xi_{4,4}$ is thus strong point-wise symmetric (SPS) and weak edge-wise symmetric (wES).

Example 5. *With regards to the structure of Figure 8, as mentioned in Theorem in Theorem 7, with $\mathbf{p} + \mathbf{q} + \mathbf{r} = (0, 0, 0)$; $|\mathbf{p} + \mathbf{r}|$, $|\mathbf{q} + \mathbf{r}| \leq 1$; and $\rho(a) = \rho(b) = \rho(c) = \rho(d) = \rho(e) = (1, 0, 0)$.*

In this case, as non-void vertices having different membership values, only one automorphism of $\xi_{5,7}$, which is the identity mapping $\mathscr{I} : V \to V$ where $\mathscr{I}(v) = v$ for all $v \in V$. As $\mathscr{I}(a) \neq b$, $\xi_{5,7}$ is, thus, point-wise asymmetric (PA). It is, nonetheless, weak edge-wise symmetric (wES).

We now give an example of CNG1 which is asymmetric (i.e., both edge-wise and point-wise).

Example 6. *$\tilde{\xi} = \langle V, \rho, \omega\rangle$ has the structure as shown in Figure 25:*

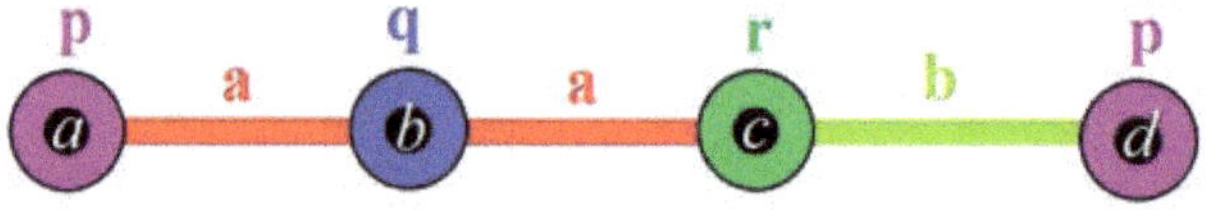

Figure 25. A $\tilde{\xi}$ which is both point-wise asymmetric and edge asymmetric, as mentioned in Example 6.

with $|\{\mathbf{p},\mathbf{q},\mathbf{r}\}| = 3, |\{\mathbf{a},\mathbf{b}\}| = 2$.

Only the trivial automorphism $\mathcal{J}$ *can be formed. As* $\mathcal{J}(a) \neq d$, $\tilde{\xi}$ *is point-wise asymmetric. Moreover, as* $\mathcal{J}(a) \neq b$, $\tilde{\xi}$ *is edge asymmetric.*

We end this section by giving a conjecture, which shall be dealt with in our future work:

Conjecture 1. *The smallest non-trivial net CNG1* ξ*, with both* M_ξ *and* N_ξ *being odd numbers, and is both SPS and wES, must be of the structure as shown in Figure 26:*

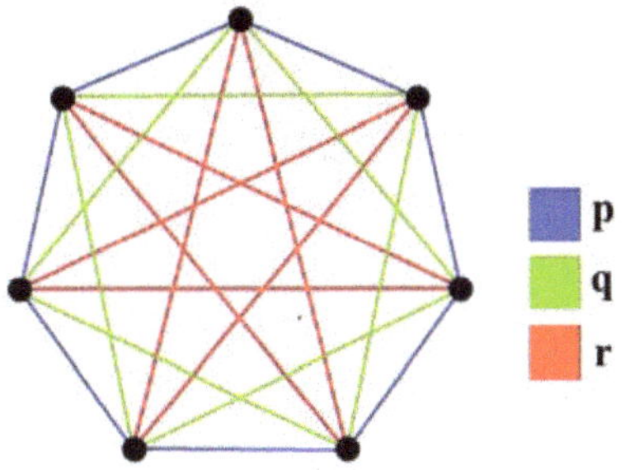

Figure 26. ξ for conjecture 1. $\mathbf{p}+\mathbf{q}+\mathbf{r} = (0,0,0)$.

8. Conclusions

In this article, we presented a new concept of the neutrosophic graph called complex neutrosophic graphs of type 1 (CNG1), and also proceeded to present a matrix representation of it.

The strength of CNG1 lies in the presence of both magnitude and direction for the parameters involved, as has been illustrated in Section 3. As the parameters have directions, even when the resultant degree of a vertex is zero, the edges to that vertex need not necessarily be void. Thus the concept of CNG1 may also be used in engineering, such as in metal frameworks, for example in the construction of power lines, so that even when the beams are under tension, the resultant force at a point (possibly being a cornerstone) joining all those beams are zero.

The concept of CNG1 can also be applied to the case of bipolar complex neutrosophic graphs (BCNG1). We have plans to expand on this interesting concept in the near future, and plan to study the concept of completeness, regularity, and CNGs of type 2.

As we can see in Section 6, when the choices of M_ξ and N_ξ becomes more restrictive, the smallest non-trivial net CNG1 ξ increases in complexity. This makes us wonder what will be the smallest non-trivial net CNG1 ξ in the case when both M_ξ and N_ξ are not divisible by all primes up to 5 (7, 11, etc.), as well as whether their symmetric properties, as outlined in Section 7. However, the proof of such cases will become much more tedious and, therefore, we would have to utilize computer programs, such as MATLAB and SAGE, in order to find those non-trivial net CNG1 ξ. Therefore our future research in this area involves plans to deal with those non-trivial net CNG1 ξ. We are motivated by the interest to know if there exist some general patterns or general theorems governing such smallest non-trivial net CNG1 as M_ξ and N_ξ become more restrictive.

We are currently working on developing a more in-depth theoretical framework concerning the symmetric properties of CNG1, and have plans to extend this to other types of fuzzy graphs in the future. We are also motivated by the works presented in [30–32], and look forward to extending our work to other generalizations of neutrosophic sets, such as interval complex neutrosophic sets, and apply the work in medical imaging problems and recommender systems.

Author Contributions: Conceptualization, S.G.Q., S.B., A.B., and M.T.; Methodology, S.B.; Software, S.G.Q.; Validation, G.S., S.B. and S.G.Q.; Formal Analysis, G.S. and F.S.; Writing-Original Draft Preparation, S.B.; Writing-Review & Editing, G.S. and F.S.; Visualization, S.G.Q.; Supervision, G.S. and F.S.

Funding: This research was funded by the Ministry of Education, Malaysia under grant no. FRGS/1/2017/STG06/UCSI/03/1 and UCSI University, Malaysia under grant no. Proj-In-FOBIS-014.

Acknowledgments: The authors would like to thank the Editor-in-Chief and the anonymous reviewers for their valuable comments and suggestions.

Conflicts of Interest: The authors declare no conflict of interest.

References

1. Smarandache, F. *Neutrosophic Probability, Set, and Logic*; ProQuest Information & Learning: Ann Arbor, MI, USA, 1998; 105p.
2. Zadeh, L.A. Fuzzy sets. *Inform. Control* **1965**, *8*, 338–353. [CrossRef]
3. Atanassov, K.T. Intuitionistic fuzzy sets. *Fuzzy Sets Syst.* **1986**, *20*, 87–96. [CrossRef]
4. Turksen, I. Interval valued fuzzy sets based on normal forms. *Fuzzy Sets Syst.* **1986**, *20*, 191–210. [CrossRef]
5. Wang, H.; Smarandache, F.; Zhang, Y.; Sunderraman, R. Single valued neutrosophic sets. *Multispace Multistruct.* **2010**, *4*, 410–413.
6. Neutrosophic Set Theory. Available online: http://fs.gallup.unm.edu/NSS/ (accessed on 1 December 2017).
7. Shannon, A.; Atanassov, K. A first step to a theory of the intuitionistic fuzzy graphs. In Proceedings of the First Workshop on Fuzzy Based Expert Systems, Sofia, Bulgaria, 28–30 September 1994; Akov, D., Ed.; 1994; pp. 59–61.
8. Smarandache, F. *Symbolic Neutrosophic Theory*; Europanova Asbl: Brussels, Belgium, 2015.
9. Smarandache, F. Refined literal indeterminacy and the multiplication law of sub-indeterminacies. *Neutrosophic Sets Syst.* **2015**, *9*, 58–63.
10. Smarandache, F. Types of Neutrosophic Graphs and neutrosophic Algebraic Structures together with their Applications in Technology. In Proceedings of the Seminar UniversitateaTransilvania din Brasov, Facultatea de Design de ProdussiMediu, Brasov, Romania, 6 June 2015.
11. Smarandache, F. (Ed.) Neutrosophic overset, neutrosophic underset, neutrosophic offset. In *Similarly for Neutrosophic Over-/Under-/Off-Logic, Probability, and Statistics*; Pons Editions: Brussels, Belgium, 2016; pp. 158–168.
12. Broumi, S.; Talea, M.; Bakali, A.; Smarandache, F. Single valued neutrosophic graphs. *J. New Theory* **2016**, *10*, 86–101.
13. Broumi, S.; Talea, M.; Smarandache, F.; Bakali, A. Single valued neutrosophic graphs: Degree, order and size. In Proceedings of the IEEE International Conference on Fuzzy Systems, Vancouver, BC, Canada, 24–29 July 2016; IEEE: Piscataway, NJ, USA, 2016; pp. 2444–2451.
14. Broumi, S.; Bakali, A.; Talea, M.; Hassan, A.; Smarandache, F. Isolated single valued neutrosophic graphs. *Neutrosophic Sets Syst.* **2016**, *11*, 74–78.
15. Smarandache, F. (Ed.) Nidus idearum. In *Scilogs, III: Viva la Neutrosophia*; Pons asbl: Brussels, Belgium, 2017; ISBN 978-1-59973-508-5.
16. Samanta, S.; Sarkar, B.; Shin, D.; Pal, M. Completeness and regularity of generalized fuzzy graphs. *SpringerPlus* **2016**, *5*, 1–14. [CrossRef] [PubMed]
17. Broumi, S.; Bakali, A.; Talea, M.; Hassan, A.; Smarandache, F. Generalized single valued neutrosophic graphs of first type. In Proceedings of the 2017 IEEE International Conference on Innovations in Intelligent Systems and Applications (INISTA), Gdynia Maritime University, Gdynia, Poland, 3–5 July 2017; pp. 413–420.
18. Ramot, D.; Menahem, F.; Gideon, L.; Abraham, K. Complex Fuzzy Set. *IEEE Trans. Fuzzy Syst.* **2002**, *10*, 171–186. [CrossRef]
19. Selvachandran, G.; Maji, P.K.; Abed, I.E.; Salleh, A.R. Complex vague soft sets and its distance measures. *J. Intell. Fuzzy Syst.* **2016**, *31*, 55–68. [CrossRef]
20. Selvachandran, G.; Maji, P.K.; Abed, I.E.; Salleh, A.R. Relations between complex vague soft sets. *Appl. Soft Comp.* **2016**, *47*, 438–448. [CrossRef]
21. Kumar, T.; Bajaj, R.K. On complex intuitionistic fuzzy soft sets with distance measures and entropies. *J. Math.* **2014**, *2014*, 1–12. [CrossRef]
22. Alkouri, A.; Salleh, A. Complex intuitionistic fuzzy sets. *AIP Conf. Proc.* **2012**, *1482*, 464–470.
23. Selvachandran, G.; Hafeed, N.A.; Salleh, A.R. Complex fuzzy soft expert sets. *AIP Conf. Proc.* **2017**, *1830*, 1–8. [CrossRef]

24. Selvachandran, G.; Salleh, A.R. Interval-valued complex fuzzy soft sets. *AIP Conf. Proc.* **2017**, *1830*, 1–8. [CrossRef]
25. Greenfield, S.; Chiclana, F.; Dick, S. Interval-valued Complex Fuzzy Logic. In Proceedings of the IEEE International Conference on Fuzzy Systems, Vancouver, BC, Canada, 24–29 July 2016; IEEE: Piscataway, NJ, USA, 2016; pp. 1–6.
26. Ali, M.; Smarandache, F. Complex neutrosophic set. *Neural Comp. Appl.* **2017**, *28*, 1817–1834. [CrossRef]
27. Thirunavukarasu, P.; Suresh, R.; Viswanathan, K.K. Energy of a complex fuzzy graph. *Int. J. Math. Sci. Eng. Appl.* **2016**, *10*, 243–248.
28. Smarandache, F. A geometric interpretation of the neutrosophic set—A generalization of the intuitionistic fuzzy set. In Proceedings of the IEEE International Conference on Granular Computing, National University of Kaohsiung, Kaohsiung, Taiwan, 8–10 November 2011; Hong, T.P., Kudo, Y., Kudo, M., Lin, T.Y., Chien, B.C., Wang, S.L., Inuiguchi, M., Liu, G.L., Eds.; IEEE Computer Society; IEEE: Kaohsiung, Taiwan, 2011; pp. 602–606.
29. Smarandache, F. Degree of dependence and independence of the (sub) components of fuzzy set and neutrosophic set. *Neutrosophic Sets Syst.* **2016**, *11*, 95–97.
30. Ali, M.; Son, L.H.; Khan, M.; Tung, N.T. Segmentation of dental X-ray images in medical imaging using neutrosophic orthogonal matrices. *Expert Syst. Appl.* **2018**, *91*, 434–441. [CrossRef]
31. Ali, M.; Dat, L.Q.; Son, L.H.; Smarandache, F. Interval complex neutrosophic set: Formulation and applications in decision-making. *Int. J. Fuzzy Syst.* **2018**, *20*, 986–999. [CrossRef]
32. Ali, M.; Son, L.H.; Thanh, N.D.; Minh, N.V. A neutrosophic recommender system for medical diagnosis based on algebraic neutrosophic measures. *Appl. Soft Comp.* **2017**. [CrossRef]

symmetry

MDPI

Article

A Classical Group of Neutrosophic Triplet Groups Using $\{Z_{2p}, \times\}$

Vasantha Kandasamy W.B. [1], **Ilanthenral Kandasamy** [1,*] and **Florentin Smarandache** [2]

1 School of Computer Science and Engineering, VIT, Vellore 632014, India; vasantha.wb@vit.ac.in
2 Department of Mathematics, University of New Mexico, 705 Gurley Avenue, Gallup, NM 87301, USA; smarand@unm.edu
* Correspondence: ilanthenral.k@vit.ac.in

Received: 15 May 2018; Accepted: 25 May 2018; Published: 1 June 2018

Abstract: In this paper we study the neutrosophic triplet groups for $a \in Z_{2p}$ and prove this collection of triplets $(a, neut(a), anti(a))$ if trivial forms a semigroup under product, and semi-neutrosophic triplets are included in that collection. Otherwise, they form a group under product, and it is of order $(p-1)$, with $(p+1, p+1, p+1)$ as the multiplicative identity. The new notion of pseudo primitive element is introduced in Z_{2p} analogous to primitive elements in Z_p, where p is a prime. Open problems based on the pseudo primitive elements are proposed. Here, we restrict our study to Z_{2p} and take only the usual product modulo $2p$.

Keywords: neutrosophic triplet groups; semigroup; semi-neutrosophic triplets; classical group of neutrosophic triplets; S-semigroup of neutrosophic triplets; pseudo primitive elements

1. Introduction

Fuzzy set theory was introduced by Zadeh in [1] and was generalized to the Intuitionistic Fuzzy Set (IFS) by Atanassov [2]. Real-world, uncertain, incomplete, indeterminate, and inconsistent data were presented philosophically as a neutrosophic set by Smarandache [3], who also studied the notion of neutralities that exist in all problems. Many [4–7] have studied neutralities in neutrosophic algebraic structures. For more about this literature and its development, refer to [3–10].

It has not been feasible to relate this neutrosophic set to real-world problems and the engineering discipline. To implement such a set, Wang et al. [11] introduced a Single-Valued Neutrosophic Set (SVNS), which was further developed into a Double Valued Neutrosophic Set (DVNS) [12] and a Triple Refined Indeterminate Neutrosophic Set (TRINS) [13]. These sets are capable of dealing with the real world's indeterminate data, and fuzzy sets and IFSs are not.

Smarandache [14] presents recent developments in neutrosophic theories, including the neutrosophic triplet, the related triplet group, the neutrosophic duplet, and the duplet set. The new, innovative, and interesting notion of the neutrosophic triplet group, which is a group of three elements, was introduced by Florentin Smarandache and Ali [10]. Since then, neutrosophic triplets have been a field of interest that many researchers have worked on [15–22]. In [21], cancellable neutrosophic triplet groups were introduced, and it was proved that it coincides with the group. The paper also discusses weak neutrosophic duplets in BCI algebras. Notions such as the neutrosophic triplet coset and its connection with the classical coset, neutrosophic triplet quotient groups, and neutrosophic triplet normal subgroups were defined and studied by [20].

Using the notion of neutrosophic triplet groups introduced in [10], which is different from classical groups, several interesting structural properties are developed and defined in this paper. Here, we study the neutrosophic triplet groups using only $\{Z_{2p}, \times\}$, p is a prime and the operation $\times$ is product modulo $2p$. The properties as a neutrosophic triplet group under the inherited operation $\times$

is studied. This leads to the definition of a semi-neutrosophic triplet. However, it has been proved that semi-neutrosophic triplets form a semigroup under $\times$, but the neutrosophic triplet groups, which are nontrivial and are not semi-neutrosophic triplets, form a classical group of neutrosophic triplets under $\times$.

This paper is organized into five sections. Section 2 provides basic concepts. In Section 3, we study neutrosophic triplets in the case of Z_{2p}, where p is an odd prime. Section 4 defines the semi-neutrosophic triplet and shows several interesting properties associated with the classical group of neutrosophic triplets. The final section provides the conclusions and probable applications.

2. Basic Concepts

We recall here basic definitions from [10].

Definition 1. *Consider $(S, \times)$ to be a nonempty set with a closed binary operation. S is called a neutrosophic triplet set if for any $x \in S$ there will exist a neutral of x called $neut\,(x)$, which is different from the algebraic unitary element (classical), and an opposite of x called $anti\,(x)$, with both $neut\,(x)$ and $anti\,(x)$ belonging to S such that*

$$x * neut\,(x) = neut\,(x) * x = x$$

and

$$x * anti\,(x) = anti\,(x) * x = neut\,(x)\,.$$

The elements x, $neut\,(x)$, and $anti\,(x)$ are together called a neutrosophic triplet group, denoted by $(x, neut\,(x)\,, anti\,(x))$.

$neut\,(x)$ denotes the neutral of x. x is the first coordinate of a neutrosophic triplet group and not a neutrosophic triplet. y is the second component, denoted by $neut\,(x)$, of a neutrosophic triplet if there are elements x and $z \in S$ such that $x * y = y * x = x$ and $x * z = z * x = y$. Thus, (x, y, z) is the neutrosophic triplet.

We know that $(neut\,(x)\,, neut\,(x)\,, neut\,(x))$ is a neutrosophic triplet group. Let $\{S, *\}$ be the neutrosophic triplet set. If $(S, *)$ is well defined and for all $x, y \in S$, $x * y \in S$, and $(x * y) * z = x * (y * z)$ for all $x, y, z \in S$, then $\{S, *\}$ is defined as the neutrosophic triplet group. Clearly, $\{S, *\}$ is not a group in the classical sense.

In the following section, we define the notion of a semi-neutrosophic triplet, which is different from neutrosophic duplets and the classical group of neutrosophic triplets of $\{Z_{2p}, \times\}$, and derive some of its interesting properties.

3. The Classical Group of Neutrosophic Triplet Groups of $\{Z_{2p}, \times\}$ and Its Properties

Here we define the classical group of neutrosophic triplets using $\{Z_{2p}, \times\}$, where p is an odd prime. The collection of all nontrivial neutrosophic triplet groups forms a classical group under the usual product modulo $2p$, and the order of that group is $p - 1$. We also derive interesting properties of such groups.

We will first illustrate this situation with some examples.

Example 1. *Let $S = \{Z_{22}, \times\}$ be the semigroup under $\times$ modulo 22. Clearly, 11 and 12 are the only idempotents or neutral elements of Z_{22}. The idempotent $11 \in Z_{22}$ yields only a trivial neutrosophic triplet $(11, 11, 11)$ for $11 \times 21 = 11$, where 21 is a unit in Z_{22}. The other nontrivial neutrosophic triplets associated with the neutral element 12 are $H = \{(2, 12, 6)\,, (6, 12, 2)\,, (4, 12, 14)\,, (14, 12, 4)\,, (16, 12, 20)\,, (20, 12, 16)\,, (12, 12, 12)\,, (10, 12, 10)\,, (8, 12, 18)\,, (18, 12, 8)\}$. It is easily verified that $\{H, \times\}$ is a classical group of order 10 under component-wise multiplication modulo 22, with $(12, 12, 12)$ as the identity element. $(12, 12, 12) \times (12, 12, 12) = (12, 12, 12)$ product modulo 22. Likewise,*

$$(2, 12, 6) \times (2, 12, 6) = (4, 12, 14)\,,$$

$$
\begin{gathered}
\textit{and}\,(2,12,6)\times(4,12,14)=(8,12,18)\,;\\
(2,12,6)\times(8,12,18)=(16,12,20)\,,\\
\textit{and}\,(2,12,6)\times(16,12,20)=(10,12,10)\,;\\
(10,12,10)\times(2,12,6)=(20,12,16)\,,\\
\textit{and}\,(2,12,6)\times(20,12,16)=(18,12,8)\,;\\
(2,12,6)\times(18,12,8)=(14,12,4)\,,\\
\textit{and}\,(2,12,6)\times(14,12,4)=(6,12,2)\,;\\
(6,12,2)\times(2,12,6)=(12,12,12)\,,\\
\textit{and}\,(2,12,6)^{10}=(12,12,12)\,.
\end{gathered}
$$

Thus, H is a cyclic group of order 10.

Example 2. *Let* $S=\{Z_{14},\times\}$ *be the semigroup under product modulo* 14. *The neutral elements or idempotents of* Z_{14} *are* 7 *and* 8. *The neutrosophic triplets are*

$$H=\{(2,8,4)\,,(4,8,2)\,,(6,8,6)\,,(10,8,12)\,,(12,8,10)\,,(8,8,8)\},$$

associated with the neutral element 8. *H is a classical group of order* 6. *Clearly,*

$$
\begin{gathered}
(10,8,12)\times(10,8,12)=(2,8,4),\\
(10,8,12)\times(2,8,4)=(6,8,6),\\
(10,8,12)\times(6,8,6)=(4,8,2),\\
(10,8,12)\times(4,8,2)=(12,8,10),\,\textit{and}\\
(10,8,12)\times(12,8,10)=(8,8,8).
\end{gathered}
$$

Thus, H is generated by $(10,8,12)$ *as* $(10,8,12)^6=(8,8,8)$, *and* $(8,8,8)$ *is the multiplicative identity of the classical group of neutrosophic triplets.*

Example 3. *Let* $S=\{Z_{38},\times\}$ *be the semigroup under product modulo* 38. $19,20\in Z_{38}$ *are the idempotents of* Z_{38}.

$$
\begin{gathered}
H=\{(2,20,10)\,,(10,20,2)\,,(4,20,24)\,,(24,20,4)\,,(20,20,20)\,,(8,20,12)\,,\\
(12,20,8)\,,(16,20,6)\,,(6,20,16)\,,(32,20,22)\,,(22,20,32)\,,(18,20,18)\,,\\
(34,20,14)\,,(14,20,34)\,,(26,20,28)\,,(28,20,26)\,,(30,2036)\,,(36,20,30)\}
\end{gathered}
$$

is the classical group of neutrosophic triplets with $(20,20,20)$ *as the identity element of H.*

In view of all these example, we have the following results.

Theorem 1. *Every semigroup* $\{Z_{2p},\times\}$, *where p is an odd prime, has only two idempotents:* p *and* $p+1$.

Proof. Clearly, p is a prime of the form $2n+1$ in Z_{2p}.

$$
\begin{aligned}
p^2=(2n+1)^2&=4n^2+4n+1\\
&=4n^2+2n+2n+1\\
&=4n^2+2n+p\\
&=2n\,(2n+1)+p\\
&=2np+p\\
&=p.
\end{aligned}
$$

Thus, p is an idempotent in Z_{2p}. Consider $p+1 \in Z_{2p}$:

$$\begin{aligned}(p+1)^2 &= p^2+2p+1\\ &= p^2+1\\ &= p+1 \quad \text{as} \quad p^2 = p.\end{aligned}$$

Thus, p and $p+1$ are the only idempotents of Z_{2p}. In fact, Z_{2p} has no other nontrivial idempotent.

Let $x \in Z_{2p}$ be an idempotent. This implies that x must be even as all odd elements other than p are units.

Let $x = 2n$ (where n is an integer), and $2 < n < p-1$ such that $x^2 = 4n^2 = x = 2n$, which implies that $2n\,(2n-1) = 0$.

This is zero only if $2n-1 = p$ as $2n-1$ is odd. Otherwise, $2n = 0$, which is not possible, as n is even and n is not equal to 0, $x \neq 0$, so $2n-1 = p$. That is, $x = 2n = p+1$ is the only possibility. Otherwise, $x = 0$, which is a contradiction.

Thus, Z_{2p} has only two idempotents, p and $p+1$. □

Theorem 2. *Let G = $\{Z_{2p}, \times\}$, where p is an odd prime, be the semigroup under $\times$, product modulo 2p.*

1. *If $a \in Z_{2p}$ has neut (a) and anti (a), then a is even.*
2. *The only nontrivial neutral element is $p+1$ for all a, which contributes to neutrosophic triplet groups in G.*

Proof. Let a in G be such that $a \times neut\,(a) = a$ if a is odd and $a \neq p$. Then a^{-1} exists in Z_{2p} and we have $neut\,(a) = 1$, but $neut\,(a) \neq 1$ by definition. Hence the result is true.

Further, we know $neut\,(a) \times neut\,(a) = neut\,(a)$, that is $neut\,(a)$ is an idempotent. This is possible if and only if $a = p+1$ or p.

Clearly, $a = p$ is ruled out because $ap = 0$ for all even a in Z_{2p}, hence the claim.

Thus, $neut\,(a) = p+1$ is the only neutral element for all relevant a in Z_{2p}. □

Definition 2. *Let $\{Z_{2p}, \times\}$ be the semigroup under multiplication modulo 2p, where p is an odd prime. $H = \{(a, neut\,(a), anti\,(a))\,|a \in 2Z_{2p} \setminus \{0\}\}$. $\{H, \times\}$ is the collection of all neutrosophic triplet groups. H has the multiplicative identity $(p+1, p+1, p+1)$ under the component-wise product modulo 2p. H is defined as the classical group of neutrosophic triplets.*

We have already given examples of them. It is important to mention this definition is valid only for Z_{2p} under the product modulo $2p$ where p is an odd prime.

Example 4. *Let $S = \{Z_{46}, \times\}$ be the semigroup under product modulo 46. Let*

$$\begin{gathered}H = \{(24,24,24), (2,24,12), (12,24,2), (4,24,6), (6,24,4), (8,24,26),\\ (26,24,8), (16,24,36), (36,24,16), (32,24,18), (18,24,32), (22,24,22),\\ (10,24,30), (14,24,28), (28,24,14), (30,24,10), (20,24,38), (38,24,20),\\ (34,24,44), (44,24,34), (40,24,42), (42,24,40)\}\end{gathered}$$

be the classical group of neutrosophic triplets, with $(24,24,24)$ as the identity under $\times$. $o\,(H) = 22$.

In view of all of this, we have to define the following for Z_{2p}.

Definition 3. *Let $\{Z_{2p}, \times\}$ be the semigroup under product modulo 2p, where p is an odd prime. Let $K = \{2, 4, \ldots, 2p-2\}$ be the set of all even elements of Z_{2p}. For $p+1 \in K$, $x \times p+1 = x, \forall\, x \in K$. There also exists a $y \in K$ such that $y^{p-1} = p+1$. We define this y as the pseudo primitive element of $K \subseteq Z_{2p}$.*

Note: We can define pseudo primitive elements only for Z_{2p} where p is an odd prime and not for any Z_n, where n is an even integer that is analogous to primitive elements in Z_p, where p is a prime.

We will illustrate this situation with some examples.

Example 5. *Let $\{Z_6, \times\}$ be the modulo semigroup. For $K = \{2,4\}$, 2 is the pseudo primitive element of $K \subseteq Z_6$.*

Example 6. *Let $\{Z_{14}, \times\}$ be the modulo semigroup under product $\times$, modulo 14. Consider $K = \{2,4,6,8,10,12\} \subseteq Z_{14}$. Then 10 is the pseudo primitive element of $K \subseteq Z_{14}$.*

Example 7. *Let $\{Z_{34}, \times\}$ be the semigroup under product modulo integer 34. 10 is the pseudo primitive element of $K = \{2,4,6,8,10,12,14,16,18,20,22,24,26,28,30,32\} \subseteq Z_{34}$.*

Similarly, for $\{Z_{38}, \times\}$, 10 is the pseudo primitive element of $K = 2Z_{38} \setminus \{0\} \subseteq Z_{38}$.

However, in the case of Z_{22}, Z_{58}, and Z_{26}, 2 is the pseudo primitive element for these semigroups.

We leave it as an open problem to find the number of such pseudo primitive elements of $K = \{2,4,6,\ldots,2(p-1)\}$ of Z_{2p}.

We have the following theorem.

Theorem 3. *Let $S = \{Z_{2p}, \times\}$ be the semigroup under product modulo $2p$, where p is an odd prime.*

1. *$K = \{2,4,\ldots,2p-2\} \subseteq Z_{2p}$ has a pseudo primitive element $x \in K$ with $x^{p-1} = p+1$, where $p+1$ is the multiplicative identity of K.*
2. *K is a cyclic group under $\times$ of order $p-1$ generated by that x, and $p+1$ is the identity element of K.*
3. *S is a Smarandache semigroup.*

Proof. Consider Z_{2p}, where p is an odd prime. Let $K = \{2,4,6,\ldots,2p-2\} \subseteq Z_{2p}$. For any $x \in K$, $(p+1)x = px + x = x$ is$px = 0(mod\ 2p)$, where x is even. Thus, $p+1$ is the identity element of Z_{2p}. There is a $x \in K$ such that $x^{p-1} = p+1$ using the principle of $2p \equiv 0$, where x is even. This x is the pseudo primitive element of K.

This $x \in K$ proves part (2) of the claim.

Since K is a group under $\times$ and $K \subseteq \{Z_{2p}, \times\}$, by the definition of Smarandache semigroup [4], S is an S-semigroup, so (3) is true. □

Next, we prove that the following theorem for our research pertains to the classical group of neutrosophic triplets and their structure.

Theorem 4. *Let $S = \{Z_{2p}, \times\}$ be the semigroup. Then*

$$H = \{(a, neut(a), anti(a))\ |a \in 2Z_{2p} \setminus \{0\}\},$$

is the classical group of neutrosophic triplets, which is cyclic and of the order $p-1$.

Proof. Clearly, from the earlier theorem, $K = 2Z_{2p} \setminus \{0\}$ is a cyclic group of the order $p-1$, and $p+1$ acts as the identity element of K.

$H = \{(a, neut(a), anti(a))\ |a \in K\}$ is a neutrosophic triplet groups collection and $neut(a) = p+1$ acts as the identity and is the unique element (neutral element) for all $a \in K$.

$(neut(a), neut(a), neut(a)) = (p+1, p+1, p+1)$ acts as the unique identity element of every neutrosophic triplet group h in H.

Since $K \subseteq Z_{2p} \setminus \{0\}$ is a cyclic group of order $p-1$ with $p+1$ as the identity element of K, we have $H = \{(a, neut\,(a)\,, anti\,(a))\ |a \in K\}$, to be cyclic. If $x \in K$ is such that $x^{p-1} = p+1$, then that neutrosophic triplet group element $(x, p+1, anti(x))$ in H will generate H as a cyclic group of order $p-1$ as $a \times anti(a) = neut(a)$.

Hence, H is a cyclic group of order $p-1$. □

Next, we proceed to describe the semi-neutrosophic triplets in the following section.

4. Semi-Neutrosophic Triplets and Their Properties

In this section, we define the notion of semi-neutrosophic triplet groups and trivial neutrosophic triplet groups and show some interesting results.

Example 8. *Let* $\{Z_{26}, \times\} = S$ *be the semigroup under product modulo 26.*

We see that $13 \in Z_{26}$ *is an idempotent, but* $13 \times 25 = 13$*, where 25 is a unit of* Z_{26}*. Therefore, for this 25, we cannot find* $anti(13)$*, but* $13 \times 13 = 13$ *is an idempotent, and* $(13, 13, 13)$ *is a neutrosophic triplet group. We do not accept it as a neutrosophic triplet, as it cannot yield any other nontrivial triplet other than* $(13, 13, 13)$.

Further, the authors of [10] defined $(0, 0, 0)$ as a trivial neutrosophic triplet group.

Definition 4. *Let* $S = \{Z_{2p}, \times\}$ *be the semigroup under product modulo* $2p$. $p \in Z_{2p}$ *is an idempotent of* Z_{2p}*. However,* p *is not a neutrosophic triplet group as* $p \times (2p-1) = 2p - p = p$*. Hence,* $(p, neut(p), anti(p)) = (p, p, p)$ *is defined as a semi-neutrosophic triplet group.*

Proposition 1. *Let* $S = \{Z_{2p}, \times\}$ *be the semigroup under product modulo* $2p$. (p, p, p) *is the semi-neutrosophic triplet group of* Z_{2p}.

Proof. This is obvious from the definition and the fact $p^2 = p$ in Z_{2p} under product modulo $2p$. □

Example 9. *Let* $S = \{Z_{46}, \times\}$ *be the semigroup under product modulo 46.* $T = \{(23, 23, 23), (0, 0, 0)\}$ *is the semi-neutrosophic triplet group and the zero neutrosophic triplet group. Clearly,* T *is a semigroup under* $\times$*, and* T *is defined as the semigroup of semi-neutrosophic triplet groups of order two as* $(23, 23, 23) \times (23, 23, 23) = (23, 23, 23)$. $K = \{(a, neut\,(a), anti\,(a))\, | a \in 2Z_{46} \setminus \{0\} = \{2, 4, 6, 8, 10, 12, 14, 16, \ldots, 42, 44\}\}$ *is a classical group of neutrosophic triplets.*

Let $P = \langle K \cup T \rangle = K \cup T$. For every $x \in K$ and for every $y \in T$, $x \times y = y \times x = (0, 0, 0)$.

Thus, P is a semigroup under product, and P is defined as the semigroup of neutrosophic triplets.

Further, we define T as the annihilating neutrosophic triplet semigroup of the classical group of neutrosophic triplets.

Definition 5. *Let* $S = \{Z_{2p}, \times\}$*, where* p *is an odd prime, be the semigroup under product modulo* $2p$*. Let* $K = \{(a, neut\,(a), anti\,(a))\, | a \in 2Z_{2p} \setminus \{0\}, \times\}$ *be the classical group of neutrosophic triplets. Let* $T = \{(p, p, p), (0, 0, 0)\}$ *be the semigroup of semi-neutrosophic triplets (as a minomer, we call the trivial neutrosophic triplet* $(0, 0, 0)$ *as a semi-neutrosophic triplet). Clearly,* $\langle T \cup K \rangle = T \cup K = P$ *is defined as the semigroup of neutrosophic triplets with* $o\,(P) = o\,(T) + o\,(K) = p - 1 + 2 = p + 1$.

Further, T *is defined as the annihilating semigroup of the classical group of neutrosophic triplets* K.

We have seen examples of classical group of neutrosophic triplets, and we have defined and studied this only for Z_{2p} under the product modulo $2p$ for every odd prime p.

In the following section, we identify open problems and probable applications of these concepts.

5. Discussions and Conclusions

This paper studies the neutrosophic triplet groups introduced by [10] only in the case of $\{Z_{2p}, \times\}$, where p is an odd prime, under product modulo $2p$. We have proved the triplets of Z_{2p} are contributed only by elements in $2Z_{2p} \setminus \{0\} = \{2, 4, \ldots, 2p-2\}$, and these triplets under product form a group of order $p-1$, defined as the classical group of neutrosophic triplets.

Further, the notion of pseudo primitive element is defined for elements $K_1 = 2Z_{2p} \setminus \{0\} = \{2, 4, 6, \ldots, 2p-2\} \subseteq Z_{2p}$. This K_1 is a cyclic group of order $p-1$ with $p+1$ as its multiplicative identity. Based on this,

$$K = \{(a, neut(a), anti(a)) \,|a \in K_1, \times\}$$

is proved to be a cyclic group of order $p-1$.

We suggest the following problems:

1. How many pseudo primitive elements are there in $\{Z_{2p}, \times\}$, where p is an odd prime?
2. Can $\{Z_n, \times\}$, where n is any composite number different from $2p$, have pseudo primitive elements? If so, which idempotent serves as the identity?

For future research, one can apply the proposed neutrosophic triplet group to SVNS and develop it for the case of DVNS or TRINS. These neutrosophic triplet groups can be applied to problems where $neut(a)$ and $anti(a)$ are fixed once a is chosen, and vice versa. It can be realized as a special case of Single Valued Neutrosophic Sets (SVNSs) where neutral is always fixed. For every a in K_1, the other factor $anti(a)$ is automatically fixed, thereby eliminating the arbitrariness in determining $anti(a)$; however, there is only one case in which $a = anti(a)$. The set $2Z_{2p} \setminus \{0\}$ can be used to model this sort of problem and thereby reduce the arbitrariness in determining $anti(a)$, which is an object of future study.

Author Contributions: The contributions of the authors are roughly equal.

Acknowledgments: The authors would like to thank the reviewers for their reading of the manuscript and their many insightful comments and suggestions.

Conflicts of Interest: The authors declare no conflict of interest.

Abbreviations

The following abbreviations are used in this manuscript:

SVNS	Single Valued Neutrosophic Set
DVNS	Double Valued Neutrosophic Set
TRINS	Triple Refined Indeterminate Neutrosophic Set
IFS	Intuitionistic Fuzzy Set

References

1. Zadeh, L.A. Fuzzy sets. *Inf. Control* **1965**, *8*, 338–353. [CrossRef]
2. Atanassov, K.T. Intuitionistic fuzzy sets. *Fuzzy Sets Syst.* **1986**, *20*, 87–96. [CrossRef]
3. Smarandache, F. *A Unifying Field in Logics: Neutrosophic Logic. Neutrosophy, Neutrosophic Set, Neutrosophic Probability and Statistics*; American Research Press: Rehoboth, DE, USA, 2005; ISBN 978-1-59973-080-6.
4. Vasantha, W.B. *Smarandache Semigroups*; American Research Press: Rehoboth, MA, USA, 2002; ISBN 978-1-931233-59-4.
5. Vasantha, W.B.; Smarandache, F. *Basic Neutrosophic Algebraic Structures and Their Application to Fuzzy and Neutrosophic Models*; Hexis: Phoenix, AZ, USA, 2004; ISBN 978-1-931233-87-X.
6. Vasantha, W.B.; Smarandache, F. *N-Algebraic Structures and SN-Algebraic Structures*; Hexis: Phoenix, AZ, USA, 2005; ISBN 978-1-931233-05-5.
7. Vasantha, W.B.; Smarandache, F. *Some Neutrosophic Algebraic Structures and Neutrosophic N-Algebraic Structures*; Hexis: Phoenix, AZ, USA, 2006; ISBN 978-1-931233-15-2.
8. Smarandache, F. Neutrosophic set-a generalization of the intuitionistic fuzzy set. In Proceedings of the 2006 IEEE International Conference on Granular Computing, Atlanta, GA, USA, 10–12 May 2006; pp. 38–42.
9. Smarandache, F. Operators on Single-Valued Neutrosophic Oversets, Neutrosophic Undersets, and Neutrosophic Offsets. *J. Math. Inf.* **2016**, *5*, 63–67. [CrossRef]

10. Smarandache, F.; Ali, M. Neutrosophic triplet group. *Neural Comput. Appl.* **2018**, *29*, 595–601. [CrossRef]
11. Wang, H.; Smarandache, F.; Zhang, Y.; Sunderraman, R. Single valued neutrosophic sets. *Review* **2010**, *1*, 10–15.
12. Kandasamy, I. Double-Valued Neutrosophic Sets, their Minimum Spanning Trees, and Clustering Algorithm. *J. Intell. Syst.* **2018**, *27*, 163–182. [CrossRef]
13. Kandasamy, I.; Smarandache, F. Triple Refined Indeterminate Neutrosophic Sets for personality classification. In Proceedings of the 2016 IEEE Symposium Series on Computational Intelligence (SSCI), Athens, Greece, 6–9 December 2016; pp. 1–8.
14. Smarandache, F. *Neutrosophic Perspectives: Triplets, Duplets, Multisets, Hybrid Operators, Modal Logic, Hedge Algebras and Applications*, 2nd ed.; Pons Publishing House: Brussels, Belgium, 2017; ISBN 978-1-59973-531-3.
15. Sahin, M.; Abdullah, K. Neutrosophic triplet normed space. *Open Phys.* **2017**, *15*, 697–704. [CrossRef]
16. Smarandache, F. Hybrid Neutrosophic Triplet Ring in Physical Structures. *Bull. Am. Phys. Soc.* **2017**, *62*, 17.
17. Smarandache, F.; Ali, M. Neutrosophic Triplet Field used in Physical Applications. In Proceedings of the 18th Annual Meeting of the APS Northwest Section, Pacific University, Forest Grove, OR, USA, 1–3 June 2017.
18. Smarandache, F.; Ali, M. Neutrosophic Triplet Ring and its Applications. In Proceedings of the 18th Annual Meeting of the APS Northwest Section, Pacific University, Forest Grove, OR, USA, 1–3 June 2017.
19. Zhang, X.H.; Smarandache, F.; Liang, X.L. Neutrosophic Duplet Semi-Group and Cancellable Neutrosophic Triplet Groups. *Symmetry* **2017**, *9*, 275–291. [CrossRef]
20. Bal, M.; Shalla, M.M.; Olgun, N. Neutrosophic Triplet Cosets and Quotient Groups. *Symmetry* **2017**, *10*, 126–139. [CrossRef]
21. Zhang, X.H.; Smarandache, F.; Ali, M.; Liang, X.L. Commutative neutrosophic triplet group and neutro-homomorphism basic theorem. *Ital. J. Pure Appl. Math.* **2017**, in press.
22. Vasantha, W.B.; Kandasamy, I.; Smarandache, F. *Neutrosophic Triplet Groups and Their Applications to Mathematical Modelling*; EuropaNova: Brussels, Belgium, 2017; ISBN 978-1-59973-533-7.

symmetry

MDPI

Article

Exponential Aggregation Operator of Interval Neutrosophic Numbers and Its Application in Typhoon Disaster Evaluation

Ruipu Tan [1,2], Wende Zhang [3,4,*] and Shengqun Chen [2]

1 School of Economics and Management, Fuzhou University, Fuzhou 350116, China; tanruipu123@163.com
2 College of Electronics and Information Science, Fujian Jiangxia University, Fuzhou 350108, China; csq@fjjxu.edu.cn
3 Institute of Information Management, Fuzhou University, Fuzhou 350116, China
4 Information Construction Office, Fuzhou University, Fuzhou 350116, China
* Correspondence: zhangwd@fzu.edu.cn; Tel.: +86-0591-2286-6169

Received: 18 April 2018; Accepted: 24 May 2018; Published: 1 June 2018

Abstract: In recent years, typhoon disasters have occurred frequently and the economic losses caused by them have received increasing attention. This study focuses on the evaluation of typhoon disasters based on the interval neutrosophic set theory. An interval neutrosophic set (INS) is a subclass of a neutrosophic set (NS). However, the existing exponential operations and their aggregation methods are primarily for the intuitionistic fuzzy set. So, this paper mainly focus on the research of the exponential operational laws of interval neutrosophic numbers (INNs) in which the bases are positive real numbers and the exponents are interval neutrosophic numbers. Several properties based on the exponential operational law are discussed in detail. Then, the interval neutrosophic weighted exponential aggregation (INWEA) operator is used to aggregate assessment information to obtain the comprehensive risk assessment. Finally, a multiple attribute decision making (MADM) approach based on the INWEA operator is introduced and applied to the evaluation of typhoon disasters in Fujian Province, China. Results show that the proposed new approach is feasible and effective in practical applications.

Keywords: neutrosophic sets (NSs); interval neutrosophic numbers (INNs); exponential operational laws of interval neutrosophic numbers; interval neutrosophic weighted exponential aggregation (INWEA) operator; multiple attribute decision making (MADM); typhoon disaster evaluation

1. Introduction

Natural hazards attract worldwide attention. Typhoons are one of the main natural hazards in the world. When a typhoon makes landfall, the impacted coastal areas experience torrential rain, strong winds, storm surges, and other weather-related disasters [1]. Typhoons can cause extremely serious harm, frequently generating heavy economic losses and personnel casualty [2]. In the last 50 years, economic damage from typhoon disasters around the coastal regions of China has increased dramatically. The Yearbook of Tropical Cyclones in China shows that from 2000 to 2014, on average, typhoon disasters caused economic losses of 45.784 billion yuan (RMB), 244 deaths, and affected 37.77 million people per year [3]. Effective evaluation of typhoon disasters can improve the typhoon disaster management efficacy, preventing or reducing disaster loss. Furthermore, precise evaluation of typhoon disasters is critical to the timely allocation and delivery of aid and materials to the disaster area. Therefore, in-depth studies of typhoon disaster evaluation are of great value.

The evaluation of typhoon disasters is a popular research topic in disaster management. Researchers have made contributions to this topic from several different perspectives [1]. Wang et al. [4]

proposed a typhoon disaster evaluation model based on an econometric and input-output joint model to evaluate the direct and indirect economic loss caused by typhoon disasters for related industrial departments. Zhang et al. [5] proposed a typhoon disaster evaluation model for the rubber plantations of Hainan Island which is based on extension theory. Lou et al. [6] adopted a back-propagation neural network method to evaluate typhoon disasters, and a real case in Zhejiang Province of China was studied in detail. Lu et al. [7] used the multi-dimensional linear dependence model to evaluate typhoon disaster losses in China. Yu et al. [1] and Lin [8] asserted that establishing a decision support system is crucial to improving data analysis capabilities for decision makers.

Since the influencing factors of the typhoon disasters are completely hard to describe accurately, the typhoon disasters may include economic loss and environmental damage. Taking economic loss for example, it includes many aspects such as the building's collapse, the number and extent of damage to housing, and the affected local economic conditions [1]. Therefore, it is impossible to describe the economic loss precisely because the estimation is based on incomplete and indeterminate data. Therefore, fuzzy set (FS) and intuitionistic fuzzy set (IFS) have been used for typhoon disaster assessment in recent years. Li et al. [9] proposed evaluating typhoon disasters with a method that applied an extension of the Technique for Order of Preference by Similarity to Ideal Solution (TOPSIS) method with intuitionistic fuzzy theory. Ma [10] proposed a fuzzy synthetic evaluation model for typhoon disasters. Chen et al. [11] provided an evaluation model based on a discrete Hopfield neural network. Yu et al. [1] studied typhoon disaster evaluation in Zhejiang Province, China, using new generalized intuitionistic fuzzy aggregation operators. He [12] proposed a typhoon disaster assessment method based on Dombi hesitant fuzzy information aggregation operators. However, this review reveals that the application of the neutrosophic sets theory in typhoon disaster assessment has yet to be examined. We believe that neutrosophic sets (NSs) offer a powerful technique to enhance typhoon disaster assessment.

Neutrosophic sets can express and handle incomplete, indeterminant, and inconsistent information. NSs were originally defined by Smarandache [13,14], who added an independent indeterminacy-membership on the basis of IFS. Neutrosophic sets are a generalization of set theories including the classic set, the fuzzy set [15] and the intuitionistic fuzzy set [16]. Neutrosophic sets are characterized by a truth-membership function (T), an indeterminacy-function (I), and a falsity-membership function (F). This theory is very important in many application areas because indeterminacy is quantified explicitly and the three primary functions are all independent. Since Smarandache's initial proposal of NSs in 1998, the concept has attracted broad attention and achieved several successful implementations. For example, Wang et al. [17] proposed single-valued neutrosophic sets (SVNSs), a type of NS. Ye [18] introduced simplified neutrosophic sets (SNSs) and defined the operational laws of SNSs, as well as some aggregation operators. Wang et al. [19] and Peng et al. [20] defined multi-valued neutrosophic sets and the multi-valued neutrosophic number, as well as proposing the application of the TODIM (a Portuguese acronym of interactive and multi-criteria decision making) method in a multi-valued neutrosophic number environment. Wang et al. [21] proposed interval neutrosophic sets (INSs) along with their set-theoretic operators and Zhang et al. [22] proposed an improved weighted correlation coefficient measure for INSs for use in multi-criteria decision making. Ye [23] offered neutrosophic hesitant fuzzy sets with single-valued neutrosophic sets. Tian et al. [24] defined simplified neutrosophic linguistic sets, which combine the concepts of simplified neutrosophic sets and linguistic term sets, and have enabled great progress in describing linguistic information. Biswas [25] and Ye [26] defined the trapezoidal fuzzy neutrosophic number, and applied it to multi-criteria decision making. Deli [27] defined the interval valued neutrosophic soft set (ivn-soft set), which is a combination of an interval valued neutrosophic set and a soft set, and then applied the concept as a decision making method. Broumi et al. [28–30] combined the neutrosophic sets and graph theory to introduce various types of neutrosophic graphs.

When Smarandache proposed the concept of NSs [13], he also introduced some basic NS operations rules. Ye [16] defined some basic operations of simplified neutrosophic sets. Wang et al. [21]

defined some basic operations of interval neutrosophic sets, including "containment", "complement", "intersection", "union", "difference", "addition", "Scalar multiplication" and "Scalar division". Based on these operations, Liu et al. [31] proposed a simplified neutrosophic correlated averaging (SNCA) operator and a simplified neutrosophic correlated geometric (SNCG) operator for multiple attribute group decision making. Ye [32] and Zhang et al. [33] introduced interval neutrosophic number ordered weighted aggregation operators, the interval neutrosophic number weighted averaging (INNWA) operator, and the interval neutrosophic number weighted geometric (INNWG) operator for multi-criteria decision making. Liu et al. [34] proposed a single-valued neutrosophic normalized weighted Bonferroni mean (SVNNWBM) operator and analyzed its properties. Ye [35] proposed interval neutrosophic uncertain linguistic variables, and further proposed the interval neutrosophic uncertain linguistic weighted arithmetic averaging (INULWAA) and the interval neutrosophic uncertain linguistic weighted arithmetic averaging (INULWGA) operator. Peng et al. [36] introduced multi-valued neutrosophic sets (MVNSs) and proposed the multi-valued neutrosophic power weighted average (MVNPWA) operator and the multi-valued neutrosophic power weighted geometric (MVNPWG) operator. A trapezoidal neutrosophic number weighted arithmetic averaging (TNNWAA) operator and a trapezoidal neutrosophic number weighted geometric averaging (TNNWGA) operator have also been proposed and applied to multiple attribute decision making (MADM) with trapezoidal neutrosophic numbers [26]. Tan et al. [37] proposed the trapezoidal fuzzy neutrosophic number ordered weighted arithmetic averaging (TFNNOWAA) operator and the trapezoidal fuzzy neutrosophic number hybrid weighted arithmetic averaging (TFNNHWAA) operator for multiple attribute group decision making. Sahin [38] proposed generalized prioritized weighted aggregation operators, including the normal neutrosophic generalized prioritized weighted averaging (NNGPWA) operator and the normal neutrosophic generalized prioritized weighted geometric (NNGPWG) operator for normal neutrosophic multiple attribute decision making.

As the study of the NS theory has expanded in both depth and scope, effective aggregation and handling of neutrosophic number information have become increasingly imperative. In response, many techniques for aggregating neutrosophic number information have been developed [18,26,31–38]. However, an important operational law is lacking, we are unable to handle information aggregation in which the bases are positive real numbers and the exponents are neutrosophic numbers. For example, when decision makers determine the attribute importance under a complex decision environment, the attribute weights are characterized by incompleteness, uncertainty, and inconsistency, while the attribute values are real numbers. In the existing literature about exponential operational laws and exponential aggregation operator, Gou et al. [39] introduced a new exponential operational law about intuitionistic fuzzy numbers (IFNs), in which the bases are positive real numbers and the exponents are IFNs. Gou et al. [40] defined exponential operational laws of interval intuitionistic fuzzy numbers (IIFNs), in which the bases are positive real numbers and the exponents are IFNs. Lu et al. [41] defined new exponential operations of single-valued neutrosophic numbers (NNs), in which the bases are positive real numbers, and the exponents are single-valued NNs. In addition, they also proposed the single-valued neutrosophic weighted exponential aggregation (SVNWEA) operator and the SVNWEA operator-based decision making method. Sahin [42] proposed two new operational laws in which the bases are positive real numbers and interval numbers, respectively; the exponents in both operational laws are simplified neutrosophic numbers (SNNs), and they introduce the simplified neutrosophic weighted exponential aggregation (SNWEA) operator and the dual simplified neutrosophic weighted exponential aggregation (DSNWEA) operator for multi-criteria decision making. Unfortunately, to date, there are not enough theoretical and applied researches on the exponential operational laws and exponential aggregation operators of interval neutrosophic numbers [43]. This is what we need to do. In order to perfect the existing neutrosophic aggregation methods, we further enriched the theoretical research of the exponential operational laws of interval neutrosophic numbers (INNs) and the applied research of the corresponding interval neutrosophic aggregation method based on [43]. In this paper, we discussed in detail several properties of the exponential operation laws of interval neutrosophic numbers, in which the bases are positive real numbers and the exponents are interval neutrosophic numbers. Then, we investigated in detail several properties of the interval neutrosophic weighted exponential aggregation (INWEA) operator, and applied the operator to aggregate assessment information to obtain comprehensive evaluation value. Additionally, a MADM method based on the INWEA operator is proposed. In the MADM problem, the attribute values in the decision matrix are expressed as positive real numbers and the attribute weights are expressed as INNs. Although traditional aggregation operators of INNs cannot address the above decision problem, the exponential aggregation operators of INNs can effectively resolve this issue.

The remainder of this paper is organized as follows: Section 2 briefly introduces some basic definitions dealing with NSs, INSs and so on. Section 3 discusses the exponential operational properties of INSs and INNs in detail. Moreover, this paper investigates in detail the properties of the interval neutrosophic exponential aggregation (INWEA) operator in Section 4. After that, a MADM method based on the INWEA operator is given in Section 5. Section 6 uses a typhoon disaster evaluation example to illustrate the applicability of the exponential operational laws and the information aggregation method proposed in Sections 3 and 4. Finally, in Section 7, the conclusions are drawn.

2. Preliminaries

In this section, we review some basic concepts related to neutrosophic sets, single-valued neutrosophic sets, and interval neutrosophic sets. We will also introduce the operational rules.

Definition 1 [13]. *Let X be a space of points (objects), with a generic element in X denoted by x. A neutrosophic set (NS) A in X is characterized by a truth-membership function $T_A(x)$, an indeterminacy-membership function $I_A(x)$, and a falsity-membership function $F_A(x)$. The function $T_A(x)$, $I_A(x)$ and $F_A(x)$ are real standard or nonstandard subsets of $]0^-,1^+[$, i.e., $T_A(x): X \to]0^-,1^+[$, $I_A(x): X \to]0^-,1^+[$, and $F_A(x): X \to]0^-,1^+[$. Therefore, the sum of $T_A(x)$, $I_A(x)$ and $F_A(x)$ satisfies the condition $0^- \leq \sup T_A(x) + \sup I_A(x) + \sup F_A(x) \leq 3^+$.*

Definition 2 [21]. *Let X be a space of point (objects) with generic elements in X denoted by x. An interval neutrosophic set (INS) $\widetilde{A}$ in X is characterized by a truth-membership function$\widetilde{T}_{\widetilde{A}}(x)$, an indeterminacy-membership function $\widetilde{I}_{\widetilde{A}}(x)$, and a falsity-membership function $\widetilde{F}_{\widetilde{A}}(x)$. There are $\widetilde{T}_{\widetilde{A}}(x)$, $\widetilde{I}_{\widetilde{A}}(x)$, $\widetilde{F}_{\widetilde{A}}(x) \subseteq [0,1]$ for each point x in X. Thus, an INS $\widetilde{A}$ can be denoted by*

$$\begin{aligned}\widetilde{A} &= \{< x, \widetilde{T}_{\widetilde{A}}(x), \widetilde{I}_{\widetilde{A}}(x), \widetilde{F}_{\widetilde{A}}(x) > | x \in X\} \\ &= \{< x, [\inf T_{\widetilde{A}}(x), \sup T_{\widetilde{A}}(x)], [\inf I_{\widetilde{A}}(x), \sup I_{\widetilde{A}}(x)], [\inf F_{\widetilde{A}}(x), \sup F_{\widetilde{A}}(x)] > | x \in X\}.\end{aligned} \tag{1}$$

Then, the sum of $\widetilde{T}_{\widetilde{A}}(x)$, $\widetilde{I}_{\widetilde{A}}(x)$, and $\widetilde{F}_{\widetilde{A}}(x)$ satisfies the condition of $0 \leq \sup T_{\widetilde{A}}(x) + \sup I_{\widetilde{A}}(x) + \sup F_{\widetilde{A}}(x) \leq 3$.

For convenience, we can use $a = < [T^L, T^U], [I^L, I^U], [F^L, F^U] >$ to represent an interval neutrosophic number (INN) in an INS.

Definition 3 [33]. *Let $a_1 = < [T_1^L, T_1^U], [I_1^L, I_1^U], [F_1^L, F_1^U] >$ and $a_2 = < [T_2^L, T_2^U], [I_2^L, I_2^U], [F_2^L, F_2^U] >$ be two INNs and $\lambda > 0$. Then, the operational rules are defined as follows:*

1. $a_1 \oplus a_2 = \langle [T_1^L + T_2^L - T_1^L \cdot T_2^L, T_1^U + T_2^U - T_1^U \cdot T_2^U], [I_1^L \cdot I_2^L, I_1^U \cdot I_2^U], [F_1^L \cdot F_2^L, F_1^U \cdot F_2^U] \rangle$;
2. $a_1 \otimes a_2 = \langle [T_1^L \cdot T_2^L, T_1^U \cdot T_2^U], [I_1^L + I_2^L - I_1^L \cdot I_2^L, I_1^U + I_2^U - I_1^U \cdot I_2^U], [F_1^L + F_2^L - F_1^L \cdot F_2^L, F_1^U + F_2^U - F_1^U \cdot F_2^U] \rangle$;
3. $\lambda a_1 = \left\langle \left[1 - (1 - T_1^L)^\lambda, 1 - (1 - T_1^U)^\lambda\right], \left[(I_1^L)^\lambda, (I_1^U)^\lambda\right], \left[(F_1^L)^\lambda, (F_1^U)^\lambda\right] \right\rangle$;
4. $a_1^\lambda = \left\langle \left[(T_1^L)^\lambda, (T_1^U)^\lambda\right], \left[1 - (1 - I_1^L)^\lambda, 1 - (1 - I_1^U)^\lambda\right], \left[1 - (1 - F_1^L)^\lambda, 1 - (1 - F_1^U)^\lambda\right] \right\rangle$.

Furthermore, for any three INNs $a_1 =< [T_1^L, T_1^U], [I_1^L, I_1^U], [F_1^L, F_1^U] >, a_2 =< [T_2^L, T_2^U], [I_2^L, I_2^U], [F_2^L, F_2^U] >$, $a_3 =< [T_3^L, T_3^U], [I_3^L, I_3^U], [F_3^L, F_3^U] >$ and any real numbers $\lambda, \lambda_1 > 0, \lambda_2 > 0$, then, there are the following properties:

1. $a_1 \oplus a_2 = a_2 \oplus a_1$;
2. $a_1 \otimes a_2 = a_2 \otimes a_1$;
3. $\lambda(a_1 \oplus a_2) = \lambda a_2 \oplus \lambda a_1$;
4. $(a_1 \otimes a_2)^\lambda = a_1^\lambda \oplus a_2^\lambda$;
5. $\lambda_1 a_1 + \lambda_2 a_1 = (\lambda_1 + \lambda_2) a_1$;
6. $a^{\lambda_1} \otimes a^{\lambda_2} = a^{(\lambda_1 + \lambda_2)}$;
7. $(a_1 \oplus a_2) \oplus a_3 = a_1 \oplus (a_2 \oplus a_3)$;
8. $(a_1 \otimes a_2) \otimes a_3 = a_1 \otimes (a_2 \otimes a_3)$.

Definition 4 [44]. *Let $a =< [T^L, T^U], [I^L, I^U], [F^L, F^U] >$ be an INN, a score function S of an interval neutrosophic value, based on the truth-membership degree, indeterminacy-membership degree, and falsity-membership degree is defined by*

$$S(a) = \frac{2 + T^L + T^U - 2I^L - 2I^U - F^L - F^U}{4} \tag{2}$$

where $S(a) \in [-1, 1]$.

Definition 5. *Let $a =< [T^L, T^U], [I^L, I^U], [F^L, F^U] >$ be an INN. Then an accuracy function A of an interval neutrosophic value, based on the truth-membership degree, indeterminacy-membership degree, and falsity-membership degree is defined by*

$$A(a) = \frac{1}{2}\left(T^L + T^U - I^U\left(1 - T^U\right) - I^L\left(1 - T^L\right) - F^U\left(1 - I^L\right) - F^L\left(1 - I^U\right)\right), \tag{3}$$

where $A(a) \in [-1, 1]$.

Definition 6. *Let $a_1 =< [T_1^L, T_1^U], [I_1^L, I_1^U], [F_1^L, F_1^U] >$, and $a_2 =< [T_2^L, T_2^U], [I_2^L, I_2^U], [F_2^L, F_2^U] >$ be two INNs, then the ranking method is defined by*

1. *If $S(a_1) > S(a_2)$, then $a_1 > a_2$;*
2. *If $S(a_1) = S(a_2)$, and then $A(a_1) = A(a_2)$, then $a_1 > a_2$.*

Definition 7 [33]. *Let $a_j (j = 1, 2, \cdots, n)$ be a collection of INNs, and $\omega = (\omega_1, \omega_2, \cdots, \omega_n)^T$ be the weight vector of $a_j (j = 1, 2, \cdots, n)$, with $\omega_j \in [0, 1]$, and $\sum_{j=1}^{n} \omega_j = 1$. Then the interval neutrosophic number weighted averaging (INNWA) operator of dimension n is defined by*

$$\begin{aligned} &INNWA(a_1, a_2, \cdots, a_n) = \omega_1 a_1 + \omega_2 a_2 + \cdots \omega_n a_n = \sum_{j=1}^{n} \omega_j a_j \\ &=< [1 - \prod_{j=1}^{n} (1 - T_j^L)^{\omega_j}, 1 - \prod_{j=1}^{n} (1 - T_j^U)^{\omega_j}], [\prod_{j=1}^{n} (I_j^L)^{\omega_j}, \prod_{j=1}^{n} (I_j^U)^{\omega_j}], [\prod_{j=1}^{n} (F_j^L)^{\omega_j}, \prod_{j=1}^{n} (F_j^U)^{\omega_j}] >. \end{aligned} \tag{4}$$

Definition 8 [33]. *Let $a_j (j = 1, 2, \cdots, n)$ be a collection of INNs, and $\omega = (\omega_1, \omega_2, \cdots, \omega_n)^T$ be the weight vector of $a_j (j = 1, 2, \cdots, n)$, with $\omega_j \in [0, 1]$, and $\sum_{j=1}^{n} \omega_j = 1$. Then the interval neutrosophic number weighted geometric (INNWG) operator of dimension n is defined by*

$$
\begin{aligned}
&INNWG(a_1, a_2, \cdots, a_n) = a_1^{\omega_1} \otimes a_2^{\omega_2} \otimes \cdots a_n^{\omega_n} = \prod_{j=1}^{n} a_j^{\omega_j} \\
&=< [\prod_{j=1}^{n} (T_j^L)^{\omega_j}, \prod_{j=1}^{n} (T_j^U)^{\omega_j}], [1 - \prod_{j=1}^{n} (1 - I_j^L)^{\omega_j}, 1 - \prod_{j=1}^{n} (1 - I_j^U)^{\omega_j}], [1 - \prod_{j=1}^{n} (1 - F_j^L)^{\omega_j}, 1 - \prod_{j=1}^{n} (1 - F_j^U)^{\omega_j}] > .
\end{aligned}
\tag{5}
$$

3. The Exponential Operational Laws of INSs and INNs

As a supplement, we discussed in detail several properties of the exponential operational laws about INSs and INNs, respectively, in which the bases are positive real numbers and the exponents are INSs or INNs.

Lu and Ye [41] and Ye [43] introduced the exponential operations of SVNSs as follows:

Definition 9 [41]. *Let $A = \{\langle x, T_A(x), I_A(x), F_A(x)\rangle | x \in U\}$ be a SVNS in a universe of discourse X. Then an exponential operational law of the SVNS A is defined as*

$$
\lambda^A = \begin{cases} \left\{\left\langle x, \lambda^{1-T_A(x)}, 1 - \lambda^{I_A(x)}, 1 - \lambda^{F_A(x)} \right\rangle | x \in X\right\}, & \lambda \in (0,1), \\ \left\{\left\langle x, \left(\frac{1}{\lambda}\right)^{1-T_A(x)}, 1 - \left(\frac{1}{\lambda}\right)^{I_A(x)}, 1 - \left(\frac{1}{\lambda}\right)^{F_A(x)} \right\rangle | x \in X\right\}, & \lambda \geq 1. \end{cases}
\tag{6}
$$

Definition 10 [43]. *Let X be a fixed set, $\widetilde{A} = \{< x, \widetilde{T}_{\widetilde{A}}(x), \widetilde{I}_{\widetilde{A}}(x), \widetilde{F}_{\widetilde{A}}(x) > | x \in X\}$ be an INS, then we can define the exponential operational law of INSs as:*

$$
\lambda^{\widetilde{A}} = \begin{cases} \left\{\left\langle x, \left[\lambda^{1-\inf T_{\widetilde{A}}(x)}, \lambda^{1-\sup T_{\widetilde{A}}(x)}\right], \left[1 - \lambda^{\inf I_{\widetilde{A}}(x)}, 1 - \lambda^{\sup I_{\widetilde{A}}(x)}\right], \left[1 - \lambda^{\inf F_{\widetilde{A}}(x)}, 1 - \lambda^{\sup F_{\widetilde{A}}(x)}\right] \right\rangle | x \in X\right\}, & \lambda \in (0,1), \\ \left\{\left\langle x, \left[\left(\frac{1}{\lambda}\right)^{1-\inf T_{\widetilde{A}}(x)}, \left(\frac{1}{\lambda}\right)^{1-\sup T_{\widetilde{A}}(x)}\right], \left[1 - \left(\frac{1}{\lambda}\right)^{\inf I_{\widetilde{A}}(x)}, 1 - \left(\frac{1}{\lambda}\right)^{\sup I_{\widetilde{A}}(x)}\right], \left[1 - \left(\frac{1}{\lambda}\right)^{\inf F_{\widetilde{A}}(x)}, 1 - \left(\frac{1}{\lambda}\right)^{\sup F_{\widetilde{A}}(x)}\right] \right\rangle | x \in X\right\}, & \lambda \geq 1. \end{cases}
\tag{7}
$$

Theorem 1. *The value of $\lambda^{\widetilde{A}}$ is an INS.*

Proof.

(1) Let $\lambda \in (0, 1)$, and $\widetilde{A} = \{< x, \widetilde{T}_{\widetilde{A}}(x), \widetilde{I}_{\widetilde{A}}(x), \widetilde{F}_{\widetilde{A}}(x) > | x \in X\}$ be an INS, where $\widetilde{T}_{\widetilde{A}}(x) \subseteq [0, 1]$, $\widetilde{I}_{\widetilde{A}}(x) \subseteq [0, 1]$ and $\widetilde{F}_{\widetilde{A}}(x) \subseteq [0, 1]$ with the condition: $0 \leq \sup T_{\widetilde{A}}(x) + \sup I_{\widetilde{A}}(x) + \sup F_{\widetilde{A}}(x) \leq 3$. So we can get $\left[\lambda^{1-\inf T_{\widetilde{A}}(x)}, \lambda^{1-\sup T_{\widetilde{A}}(x)}\right] \subseteq [0, 1]$, $\left[1 - \lambda^{\inf I_{\widetilde{A}}(x)}, 1 - \lambda^{\sup I_{\widetilde{A}}(x)}\right] \subseteq [0, 1]$ and $\left[1 - \lambda^{\inf F_{\widetilde{A}}(x)}, 1 - \lambda^{\sup F_{\widetilde{A}}(x)}\right] \subseteq [0, 1]$. Then, we get $0 \leq \lambda^{1-\sup T_{\widetilde{A}}(x)} + 1 - \lambda^{\sup I_{\widetilde{A}}(x)} + 1 - \lambda^{\sup F_{\widetilde{A}}(x)} \leq 3$. So $\lambda^{\widetilde{A}}$ is an INS.

(2) Let $\lambda \in (0, 1)$, and $0 \leq \frac{1}{\lambda} \leq 1$, it is easy to proof that $\lambda^{\widetilde{A}}$ is an INS.

Combining (1) and (2), it follows that the value of $\lambda^{\widetilde{A}}$ is an INS. Similarly, we propose an operational law for an INN. □

Definition 11 [43]. *Let $a =< [T^L, T^U], [I^L, I^U], [F^L, F^U] >$ be an INN, then the exponential operational law of the INN a is defined as follows:*

$$
\lambda^a = \begin{cases} \left\langle \left[\lambda^{1-T^L}, \lambda^{1-T^U}\right], \left[1 - \lambda^{I^L}, 1 - \lambda^{I^U}\right], \left[1 - \lambda^{F^L}, 1 - \lambda^{F^U}\right] \right\rangle, & \lambda \in (0,1), \\ \left\langle \left[\left(\frac{1}{\lambda}\right)^{1-T^L}, \left(\frac{1}{\lambda}\right)^{1-T^U}\right], \left[1 - \left(\frac{1}{\lambda}\right)^{I^L}, 1 - \left(\frac{1}{\lambda}\right)^{I^U}\right], \left[1 - \left(\frac{1}{\lambda}\right)^{F^L}, 1 - \left(\frac{1}{\lambda}\right)^{F^U}\right] \right\rangle, & \lambda \geq 1. \end{cases}
\tag{8}
$$

It is obvious that λ^a is also an INN. Let us consider the following example.

Example 1. *Let* $a =< [0.4, 0.6], [0.1, 0.3], [0.2, 0.4] >$ *be an INN, and* $\lambda_1 = 0.3$ *and* $\lambda_2 = 2$ *are two real numbers. Then, according to Definition 11, we obtain*

$$\begin{aligned}\lambda_1^a &= 0.3^{<[0.4,0.6],[0.1,0.3],[0.2,0.4]>} = \left\langle \left[0.3^{1-0.4}, 0.3^{1-0.6}\right], \left[1-0.3^{0.1}, 1-0.3^{0.3}\right], \left[1-0.3^{0.2}, 1-0.3^{0.4}\right] \right\rangle \\ &= \left\langle \left[0.3^{0.6}, 0.3^{0.4}\right], \left[1-0.3^{0.1}, 1-0.3^{0.3}\right], \left[1-0.3^{0.2}, 1-0.3^{0.4}\right] \right\rangle \\ &= \langle [0.4856, 0.6178], [0.1134, 0.3032], [0.2140, 0.3822] \rangle.\end{aligned}$$

$$\begin{aligned}\lambda_2^a &= 2^{<[0.4,0.6],[0.1,0.3],[0.2,0.4]>} = \left\langle \left[\left(\tfrac{1}{2}\right)^{1-0.4}, \left(\tfrac{1}{2}\right)^{1-0.6}\right], \left[1-\left(\tfrac{1}{2}\right)^{0.1}, 1-\left(\tfrac{1}{2}\right)^{0.3}\right], \left[1-\left(\tfrac{1}{2}\right)^{0.2}, 1-\left(\tfrac{1}{2}\right)^{0.4}\right] \right\rangle \\ &= \left\langle \left[0.5^{0.6}, 0.5^{0.4}\right], \left[1-0.5^{0.1}, 1-0.5^{0.3}\right], \left[1-0.5^{0.2}, 1-0.5^{0.4}\right] \right\rangle \\ &= \langle [0.6598, 0.7579], [0.0670, 0.1877], [0.1294, 0.2421] \rangle.\end{aligned}$$

Here, when $T^L = T^U$, $I^L = I^U$ and $F^L = F^U$, the exponential operational law for INNs is equal to the exponential operational law of SVNNs [41]. When $0^- \leq T^U + I^U + F^U \leq 1$, the exponential operational law for INNs is equivalent to the exponential operational law of IIFNs [40]. When $T^L = T^U$, $I^L = I^U$, $F^L = F^U$ and $0^- \leq T^U + I^U + F^U \leq 1$, the exponential operational law for INSs is equivalent to the exponential operational law of IFNs [39]. So the exponential operational laws of INNs is a more generalized representation, and the exponential operational laws of SVNNs, IIFNs and IFNs are special cases.

Next, we investigate in detail some basic properties of the exponential operational laws of INNs. We notice that when $\lambda \in (0, 1)$, the operational process and the form of λ^a are similar to the case when $\lambda \geq 1$. So, below we only discuss the case when $\lambda \in (0, 1)$.

Theorem 2. *Let* $a_i =< [T_i^L, T_i^U], [I_i^L, I_i^U], [F_i^L, F_i^U] > (i = 1, 2)$ *be two INNs,* $\lambda \in (0, 1)$ *, then*

(1) $\lambda^{a_1} \oplus \lambda^{a_2} = \lambda^{a_2} \oplus \lambda^{a_1}$;

(2) $\lambda^{a_1} \otimes \lambda^{a_2} = \lambda^{a_2} \otimes \lambda^{a_1}$.

Proof. By Definition 3 and Definition 11, we have

(1)
$$\begin{aligned}&\lambda^{a_1} \oplus \lambda^{a_2} \\ &= \left\langle \left[\lambda^{1-T_1^L}, \lambda^{1-T_1^U}\right], \left[1-\lambda^{I_1^L}, 1-\lambda^{I_1^U}\right], \left[1-\lambda^{F_1^L}, 1-\lambda^{F_1^U}\right] \right\rangle \oplus \left\langle \left[\lambda^{1-T_2^L}, \lambda^{1-T_2^U}\right], \left[1-\lambda^{I_2^L}, 1-\lambda^{I_2^U}\right], \left[1-\lambda^{F_2^L}, 1-\lambda^{F_2^U}\right] \right\rangle \\ &= \left\langle \begin{array}{l} \left[\lambda^{1-T_1^L} + \lambda^{1-T_2^L} - \lambda^{1-T_1^L} \cdot \lambda^{1-T_2^L}, \lambda^{1-T_1^U} + \lambda^{1-T_2^U} - \lambda^{1-T_1^U} \cdot \lambda^{1-T_2^U}\right], \\ \left[\left(1-\lambda^{I_1^L}\right) \cdot \left(1-\lambda^{I_2^L}\right), \left(1-\lambda^{I_1^U}\right) \cdot \left(1-\lambda^{I_2^U}\right)\right], \left[\left(1-\lambda^{F_1^L}\right) \cdot \left(1-\lambda^{F_2^L}\right), \left(1-\lambda^{F_1^U}\right) \cdot \left(1-\lambda^{F_2^U}\right)\right] \end{array} \right\rangle \\ &= \lambda^{a_2} \oplus \lambda^{a_1}.\end{aligned}$$

(2)
$$\begin{aligned}&\lambda^{a_1} \otimes \lambda^{a_2} \\ &= \left\langle \left[\lambda^{1-T_1^L}, \lambda^{1-T_1^U}\right], \left[1-\lambda^{I_1^L}, 1-\lambda^{I_1^U}\right], \left[1-\lambda^{F_1^L}, 1-\lambda^{F_1^U}\right] \right\rangle \oplus \left\langle \left[\lambda^{1-T_2^L}, \lambda^{1-T_2^U}\right], \left[1-\lambda^{I_2^L}, 1-\lambda^{I_2^U}\right], \left[1-\lambda^{F_2^L}, 1-\lambda^{F_2^U}\right] \right\rangle \\ &= \left\langle \begin{array}{l} \left[\lambda^{1-T_1^L} \cdot \lambda^{1-T_2^L}, \lambda^{1-T_1^U} \cdot \lambda^{1-T_2^U}\right], \\ \left[\left(1-\lambda^{I_1^L}\right) + \left(1-\lambda^{I_2^L}\right) - \left(1-\lambda^{I_1^L}\right) \cdot \left(1-\lambda^{I_2^L}\right), \left(1-\lambda^{I_1^U}\right) + \left(1-\lambda^{I_2^U}\right) - \left(1-\lambda^{I_1^U}\right) \cdot \left(1-\lambda^{I_2^U}\right)\right], \\ \left[\left(1-\lambda^{F_1^L}\right) + \left(1-\lambda^{F_2^L}\right) - \left(1-\lambda^{F_1^L}\right) \cdot \left(1-\lambda^{F_2^L}\right), \left(1-\lambda^{F_1^U}\right) + \left(1-\lambda^{F_2^U}\right) - \left(1-\lambda^{F_1^U}\right) \cdot \left(1-\lambda^{F_2^U}\right)\right] \end{array} \right\rangle \\ &= \lambda^{a_2} \otimes \lambda^{a_1}. \square\end{aligned}$$

Theorem 3. *Let* $a_i =< [T_i^L, T_i^U], [I_i^L, I_i^U], [F_i^L, F_i^U] > (i = 1, 2, 3)$ *be three INNs,* $\lambda \in (0, 1)$*, then*

(1) $(\lambda^{a_1} \oplus \lambda^{a_2}) \oplus \lambda^{a_3} = \lambda^{a_1} \oplus (\lambda^{a_2} \oplus \lambda^{a_3})$;

(2) $(\lambda^{a_1} \otimes \lambda^{a_2}) \otimes \lambda^{a_3} = \lambda^{a_1} \otimes (\lambda^{a_2} \otimes \lambda^{a_3})$.

Proof. By Definition 3 and Definition 11, we have

$$
\begin{aligned}
&(\lambda^{a_1} \oplus \lambda^{a_2}) \oplus \lambda^{a_3} \\
&= \left\langle \begin{array}{l} \left[\lambda^{1-T_1^L} + \lambda^{1-T_2^L} - \lambda^{1-T_1^L} \cdot \lambda^{1-T_2^L}, \lambda^{1-T_1^U} + \lambda^{1-T_2^U} - \lambda^{1-T_1^U} \cdot \lambda^{1-T_2^U}\right], \\ \left[\left(1-\lambda^{I_1^L}\right) \cdot \left(1-\lambda^{I_2^L}\right), \left(1-\lambda^{I_1^U}\right) \cdot \left(1-\lambda^{I_2^U}\right)\right], \\ \left[\left(1-\lambda^{F_1^L}\right) \cdot \left(1-\lambda^{F_2^L}\right), \left(1-\lambda^{F_1^U}\right) \cdot \left(1-\lambda^{F_2^U}\right)\right] \end{array} \right\rangle \\
&\oplus \left\langle \left[\lambda^{1-T_3^L}, \lambda^{1-T_3^U}\right], \left[1-\lambda^{I_3^L}, 1-\lambda^{I_3^U}\right], \left[1-\lambda^{F_3^L}, 1-\lambda^{F_3^U}\right] \right\rangle \\
&= \left\langle \begin{array}{l} \left[\begin{array}{l} \left(\lambda^{1-T_1^L} + \lambda^{1-T_2^L} - \lambda^{1-T_1^L} \cdot \lambda^{1-T_2^L}\right) + \lambda^{1-T_3^L} - \left(\lambda^{1-T_1^L} + \lambda^{1-T_2^L} - \lambda^{1-T_1^L} \cdot \lambda^{1-T_2^L}\right) \cdot \lambda^{1-T_3^L}, \\ \left(\lambda^{1-T_1^U} + \lambda^{1-T_2^U} - \lambda^{1-T_1^U} \cdot \lambda^{1-T_2^U}\right) + \lambda^{1-T_3^U} - \left(\lambda^{1-T_1^U} + \lambda^{1-T_2^U} - \lambda^{1-T_1^U} \cdot \lambda^{1-T_2^U}\right) \cdot \lambda^{1-T_3^U} \end{array}\right], \\ \left[\left(1-\lambda^{I_1^L}\right) \cdot \left(1-\lambda^{I_2^L}\right) \cdot \left(1-\lambda^{I_3^L}\right), \left(1-\lambda^{I_1^U}\right) \cdot \left(1-\lambda^{I_2^U}\right) \cdot \left(1-\lambda^{I_3^U}\right)\right], \\ \left[\left(1-\lambda^{F_1^L}\right) \cdot \left(1-\lambda^{F_2^L}\right) \cdot \left(1-\lambda^{F_3^L}\right), \left(1-\lambda^{F_1^U}\right) \cdot \left(1-\lambda^{F_2^U}\right) \cdot \left(1-\lambda^{F_3^U}\right)\right] \end{array} \right\rangle \\
(1) \quad &= \left\langle \begin{array}{l} \left[\begin{array}{l} \lambda^{1-T_1^L} + \lambda^{1-T_2^L} + \lambda^{1-T_3^L} - \lambda^{1-T_1^L} \cdot \lambda^{1-T_2^L} - \lambda^{1-T_1^L} \cdot \lambda^{1-T_3^L} - \lambda^{1-T_2^L} \cdot \lambda^{1-T_3^L} + \lambda^{1-T_1^L} \cdot \lambda^{1-T_2^L} \cdot \lambda^{1-T_3^L}, \\ \lambda^{1-T_1^U} + \lambda^{1-T_2^U} + \lambda^{1-T_3^U} - \lambda^{1-T_1^U} \cdot \lambda^{1-T_2^U} - \lambda^{1-T_1^U} \cdot \lambda^{1-T_3^U} - \lambda^{1-T_2^U} \cdot \lambda^{1-T_3^U} + \lambda^{1-T_1^U} \cdot \lambda^{1-T_2^U} \cdot \lambda^{1-T_3^U} \end{array}\right], \\ \left[\left(1-\lambda^{I_1^L}\right) \cdot \left(1-\lambda^{I_2^L}\right) \cdot \left(1-\lambda^{I_3^L}\right), \left(1-\lambda^{I_1^U}\right) \cdot \left(1-\lambda^{I_2^U}\right) \cdot \left(1-\lambda^{I_3^U}\right)\right], \\ \left[\left(1-\lambda^{F_1^L}\right) \cdot \left(1-\lambda^{F_2^L}\right) \cdot \left(1-\lambda^{F_3^L}\right), \left(1-\lambda^{F_1^U}\right) \cdot \left(1-\lambda^{F_2^U}\right) \cdot \left(1-\lambda^{F_3^U}\right)\right] \end{array} \right\rangle \\
&= \left\langle \begin{array}{l} \left[\begin{array}{l} \lambda^{1-T_1^L} + \left(\lambda^{1-T_2^L} + \lambda^{1-T_3^L} - \lambda^{1-T_2^L} \cdot \lambda^{1-T_3^L}\right) - \lambda^{1-T_1^L} \cdot \left(\lambda^{1-T_2^L} + \lambda^{1-T_3^L} - \lambda^{1-T_2^L} \cdot \lambda^{1-T_3^L}\right), \\ \lambda^{1-T_1^U} + \left(\lambda^{1-T_2^U} + \lambda^{1-T_3^U} - \lambda^{1-T_2^U} \cdot \lambda^{1-T_3^U}\right) - \lambda^{1-T_1^U} \cdot \left(\lambda^{1-T_2^U} + \lambda^{1-T_3^U} - \lambda^{1-T_2^U} \cdot \lambda^{1-T_3^U}\right) \end{array}\right], \\ \left[\left(1-\lambda^{I_1^L}\right) \cdot \left(1-\lambda^{I_2^L}\right) \cdot \left(1-\lambda^{I_3^L}\right), \left(1-\lambda^{I_1^U}\right) \cdot \left(1-\lambda^{I_2^U}\right) \cdot \left(1-\lambda^{I_3^U}\right)\right], \\ \left[\left(1-\lambda^{F_1^L}\right) \cdot \left(1-\lambda^{F_2^L}\right) \cdot \left(1-\lambda^{F_3^L}\right), \left(1-\lambda^{F_1^U}\right) \cdot \left(1-\lambda^{F_2^U}\right) \cdot \left(1-\lambda^{F_3^U}\right)\right] \end{array} \right\rangle \\
&= \left\langle \left[\lambda^{1-T_1^L}, \lambda^{1-T_1^U}\right], \left[1-\lambda^{I_1^L}, 1-\lambda^{I_1^U}\right], \left[1-\lambda^{F_1^L}, 1-\lambda^{F_1^U}\right] \right\rangle \oplus \\
&\quad \left\langle \begin{array}{l} \left[\lambda^{1-T_2^L} + \lambda^{1-T_3^L} - \lambda^{1-T_2^L} \cdot \lambda^{1-T_3^L}, \lambda^{1-T_2^U} + \lambda^{1-T_3^U} - \lambda^{1-T_2^U} \cdot \lambda^{1-T_3^U}\right], \\ \left[\left(1-\lambda^{I_2^L}\right) \cdot \left(1-\lambda^{I_3^L}\right), \left(1-\lambda^{I_2^U}\right) \cdot \left(1-\lambda^{I_3^U}\right)\right], \left[\left(1-\lambda^{F_2^L}\right) \cdot \left(1-\lambda^{F_3^L}\right), \left(1-\lambda^{F_2^U}\right) \cdot \left(1-\lambda^{F_3^U}\right)\right] \end{array} \right\rangle \\
&= \lambda^{a_1} \oplus (\lambda^{a_2} \oplus \lambda^{a_3}).
\end{aligned}
$$

$$
\begin{aligned}
&(\lambda^{a_1} \otimes \lambda^{a_2}) \otimes \lambda^{a_3} \\
&= \left\langle \begin{array}{l} \left[\lambda^{1-T_1^L} \cdot \lambda^{1-T_2^L}, \lambda^{1-T_1^U} \cdot \lambda^{1-T_2^U}\right], \\ \left[\left(1-\lambda^{I_1^L}\right) + \left(1-\lambda^{I_2^L}\right) - \left(1-\lambda^{I_1^L}\right) \cdot \left(1-\lambda^{I_2^L}\right), \left(1-\lambda^{I_1^U}\right) + \left(1-\lambda^{I_2^U}\right) - \left(1-\lambda^{I_1^U}\right) \cdot \left(1-\lambda^{I_2^U}\right)\right], \\ \left[\left(1-\lambda^{F_1^L}\right) + \left(1-\lambda^{F_2^L}\right) - \left(1-\lambda^{F_1^L}\right) \cdot \left(1-\lambda^{F_2^L}\right), \left(1-\lambda^{F_1^U}\right) + \left(1-\lambda^{F_2^U}\right) - \left(1-\lambda^{F_1^U}\right) \cdot \left(1-\lambda^{F_2^U}\right)\right] \end{array} \right\rangle \\
&\otimes \left\langle \left[\lambda^{1-T_3^L}, \lambda^{1-T_3^U}\right], \left[1-\lambda^{I_3^L}, 1-\lambda^{I_3^U}\right], \left[1-\lambda^{F_3^L}, 1-\lambda^{F_3^U}\right] \right\rangle \\
&= \left\langle \begin{array}{l} \left[\lambda^{1-T_1^L} \cdot \lambda^{1-T_2^L} \cdot \lambda^{1-T_3^L}, \lambda^{1-T_1^U} \cdot \lambda^{1-T_2^U} \cdot \lambda^{1-T_3^U}\right], \\ \left[\begin{array}{l} \left(1-\lambda^{I_1^L}\right) + \left(1-\lambda^{I_2^L}\right) - \left(1-\lambda^{I_1^L}\right) \cdot \left(1-\lambda^{I_2^L}\right) + \left(1-\lambda^{I_3^L}\right) - \left(\left(1-\lambda^{I_1^L}\right) + \left(1-\lambda^{I_2^L}\right) - \left(1-\lambda^{I_1^L}\right) \cdot \left(1-\lambda^{I_2^L}\right)\right) \cdot \left(1-\lambda^{I_3^L}\right), \\ \left(1-\lambda^{I_1^U}\right) + \left(1-\lambda^{I_2^U}\right) - \left(1-\lambda^{I_1^U}\right) \cdot \left(1-\lambda^{I_2^U}\right) + \left(1-\lambda^{I_3^U}\right) - \left(\left(1-\lambda^{I_1^U}\right) + \left(1-\lambda^{I_2^U}\right) - \left(1-\lambda^{I_1^U}\right) \cdot \left(1-\lambda^{I_2^U}\right)\right) \cdot \left(1-\lambda^{I_3^U}\right) \end{array}\right], \\ \left[\begin{array}{l} \left(1-\lambda^{F_1^L}\right) + \left(1-\lambda^{F_2^L}\right) - \left(1-\lambda^{F_1^L}\right) \cdot \left(1-\lambda^{F_2^L}\right) + \left(1-\lambda^{F_3^L}\right) - \left(\left(1-\lambda^{F_1^L}\right) + \left(1-\lambda^{F_2^L}\right) - \left(1-\lambda^{F_1^L}\right) \cdot \left(1-\lambda^{F_2^L}\right)\right) \cdot \left(1-\lambda^{F_3^L}\right), \\ \left(1-\lambda^{F_1^U}\right) + \left(1-\lambda^{F_2^U}\right) - \left(1-\lambda^{F_1^U}\right) \cdot \left(1-\lambda^{F_2^U}\right) + \left(1-\lambda^{F_3^U}\right) - \left(\left(1-\lambda^{F_1^U}\right) + \left(1-\lambda^{F_2^U}\right) - \left(1-\lambda^{F_1^U}\right) \cdot \left(1-\lambda^{F_2^U}\right)\right) \cdot \left(1-\lambda^{F_3^U}\right) \end{array}\right] \end{array} \right\rangle \\
(2) \quad &= \left\langle \begin{array}{l} \left[\lambda^{1-T_1^L} \cdot \lambda^{1-T_2^L} \cdot \lambda^{1-T_3^L}, \lambda^{1-T_1^U} \cdot \lambda^{1-T_2^U} \cdot \lambda^{1-T_3^U}\right], \\ \left[\begin{array}{l} \left(1-\lambda^{I_1^L}\right) + \left(1-\lambda^{I_2^L}\right) + \left(1-\lambda^{I_3^L}\right) - \left(1-\lambda^{I_1^L}\right) \cdot \left(1-\lambda^{I_2^L}\right) - \left(1-\lambda^{I_1^L}\right) \cdot \left(1-\lambda^{I_3^L}\right) - \left(1-\lambda^{I_2^L}\right) \cdot \left(1-\lambda^{I_3^L}\right) + \left(1-\lambda^{I_1^L}\right) \cdot \left(1-\lambda^{I_2^L}\right) \cdot \left(1-\lambda^{I_3^L}\right), \\ \left(1-\lambda^{I_1^U}\right) + \left(1-\lambda^{I_2^U}\right) + \left(1-\lambda^{I_3^U}\right) - \left(1-\lambda^{I_1^U}\right) \cdot \left(1-\lambda^{I_2^U}\right) - \left(1-\lambda^{I_1^U}\right) \cdot \left(1-\lambda^{I_3^U}\right) - \left(1-\lambda^{I_2^U}\right) \cdot \left(1-\lambda^{I_3^U}\right) + \left(1-\lambda^{I_1^U}\right) \cdot \left(1-\lambda^{I_2^U}\right) \cdot \left(1-\lambda^{I_3^U}\right) \end{array}\right], \\ \left[\begin{array}{l} \left(1-\lambda^{F_1^L}\right) + \left(1-\lambda^{F_2^L}\right) + \left(1-\lambda^{F_3^L}\right) - \left(1-\lambda^{F_1^L}\right) \cdot \left(1-\lambda^{F_2^L}\right) - \left(1-\lambda^{F_1^L}\right) \cdot \left(1-\lambda^{F_3^L}\right) + \left(1-\lambda^{F_2^L}\right) \cdot \left(1-\lambda^{F_3^L}\right) - \left(1-\lambda^{F_1^L}\right) \cdot \left(1-\lambda^{F_2^L}\right) \cdot \left(1-\lambda^{F_3^L}\right), \\ \left(1-\lambda^{F_1^U}\right) + \left(1-\lambda^{F_2^U}\right) + \left(1-\lambda^{F_3^U}\right) - \left(1-\lambda^{F_1^U}\right) \cdot \left(1-\lambda^{F_2^U}\right) - \left(1-\lambda^{F_1^U}\right) \cdot \left(1-\lambda^{F_3^U}\right) + \left(1-\lambda^{F_2^U}\right) \cdot \left(1-\lambda^{F_3^U}\right) - \left(1-\lambda^{F_1^U}\right) \cdot \left(1-\lambda^{F_2^U}\right) \cdot \left(1-\lambda^{F_3^U}\right) \end{array}\right] \end{array} \right\rangle \\
&= \left\langle \begin{array}{l} \left[\lambda^{1-T_1^L} \cdot \lambda^{1-T_2^L} \cdot \lambda^{1-T_3^L}, \lambda^{1-T_1^U} \cdot \lambda^{1-T_2^U} \cdot \lambda^{1-T_3^U}\right], \\ \left[\begin{array}{l} \left(1-\lambda^{I_1^L}\right) + \left(1-\lambda^{I_2^L}\right) + \left(1-\lambda^{I_3^L}\right) - \left(1-\lambda^{I_2^L}\right) \cdot \left(1-\lambda^{I_3^L}\right) - \left(1-\lambda^{I_1^L}\right) \cdot \left(\left(1-\lambda^{I_2^L}\right) + \left(1-\lambda^{I_3^L}\right) - \left(1-\lambda^{I_2^L}\right) \cdot \left(1-\lambda^{I_3^L}\right)\right), \\ \left(1-\lambda^{I_1^U}\right) + \left(1-\lambda^{I_2^U}\right) + \left(1-\lambda^{I_3^U}\right) - \left(1-\lambda^{I_2^U}\right) \cdot \left(1-\lambda^{I_3^U}\right) - \left(1-\lambda^{I_1^U}\right) \cdot \left(\left(1-\lambda^{I_2^U}\right) + \left(1-\lambda^{I_3^U}\right) - \left(1-\lambda^{I_2^U}\right) \cdot \left(1-\lambda^{I_3^U}\right)\right) \end{array}\right], \\ \left[\begin{array}{l} \left(1-\lambda^{F_1^L}\right) + \left(1-\lambda^{F_2^L}\right) + \left(1-\lambda^{F_3^L}\right) - \left(1-\lambda^{F_2^L}\right) \cdot \left(1-\lambda^{F_3^L}\right) - \left(1-\lambda^{F_1^L}\right) \cdot \left(\left(1-\lambda^{F_2^L}\right) + \left(1-\lambda^{F_3^L}\right) - \left(1-\lambda^{F_2^L}\right) \cdot \left(1-\lambda^{F_3^L}\right)\right), \\ \left(1-\lambda^{F_1^U}\right) + \left(1-\lambda^{F_2^U}\right) + \left(1-\lambda^{F_3^U}\right) - \left(1-\lambda^{F_2^U}\right) \cdot \left(1-\lambda^{F_3^U}\right) - \left(1-\lambda^{F_1^U}\right) \cdot \left(\left(1-\lambda^{F_2^U}\right) + \left(1-\lambda^{F_3^U}\right) - \left(1-\lambda^{F_2^U}\right) \cdot \left(1-\lambda^{F_3^U}\right)\right) \end{array}\right] \end{array} \right\rangle \\
&= \lambda^{a_1} \otimes (\lambda^{a_2} \otimes \lambda^{a_3}). \square
\end{aligned}
$$

Theorem 4. *Let* $a =< [T^L, T^U], [I^L, I^U], [F^L, F^U] >$ *and* $a_i =< [T_i^L, T_i^U], [I_i^L, I_i^U], [F_i^L, F_i^U] >$ $(i = 1, 2)$ *be three INNs,* $\lambda \in (0, 1)$*,* $k, k_1, k_2 > 0$*, then*

(1) $k(\lambda^{a_1} \oplus \lambda^{a_2}) = k\lambda^{a_1} \oplus k\lambda^{a_2}$;
(2) $(\lambda^{a_1} \otimes \lambda^{a_2})^k = (\lambda^{a_2})^k \otimes (\lambda^{a_1})^k$;
(3) $k_1\lambda^a \oplus k_2\lambda^a = (k_1 + k_2)\lambda^a$;
(4) $(\lambda^a)^{k_1} \otimes (\lambda^a)^{k_2} = (\lambda^a)^{k_1+k_2}$;
(5) $(\lambda_1)^a \otimes (\lambda_2)^a = (\lambda_1\lambda_2)^a$.

Proof. By Definition 3 and Definition 11, we have

(1)

$$
\begin{aligned}
&k(\lambda^{a_1} \oplus \lambda^{a_2}) \\
&= k\left\langle \begin{array}{l} \left[\lambda^{1-T_1^L} + \lambda^{1-T_2^L} - \lambda^{1-T_1^L} \cdot \lambda^{1-T_2^L}, \lambda^{1-T_1^U} + \lambda^{1-T_2^U} - \lambda^{1-T_1^U} \cdot \lambda^{1-T_2^U}\right], \\ \left[\left(1-\lambda^{I_1^L}\right) \cdot \left(1-\lambda^{I_2^L}\right), \left(1-\lambda^{I_1^U}\right) \cdot \left(1-\lambda^{I_2^U}\right)\right], \left[\left(1-\lambda^{F_1^L}\right) \cdot \left(1-\lambda^{F_2^L}\right), \left(1-\lambda^{F_1^U}\right) \cdot \left(1-\lambda^{F_2^U}\right)\right] \end{array} \right\rangle \\
&= \left\langle \begin{array}{l} \left[1-\left(1-\left(\lambda^{1-T_1^L} + \lambda^{1-T_2^L} - \lambda^{1-T_1^L} \cdot \lambda^{1-T_2^L}\right)\right)^k, 1-\left(1-\left(\lambda^{1-T_1^U} + \lambda^{1-T_2^U} - \lambda^{1-T_1^U} \cdot \lambda^{1-T_2^U}\right)\right)^k\right], \\ \left[\left(\left(1-\lambda^{I_1^L}\right) \cdot \left(1-\lambda^{I_2^L}\right)\right)^k, \left(\left(1-\lambda^{I_1^U}\right) \cdot \left(1-\lambda^{I_2^U}\right)\right)^k\right], \left[\left(\left(1-\lambda^{F_1^L}\right) \cdot \left(1-\lambda^{F_2^L}\right)\right)^k, \left(\left(1-\lambda^{F_1^U}\right) \cdot \left(1-\lambda^{F_2^U}\right)\right)^k\right] \end{array} \right\rangle \\
&= \left\langle \left[1-(1-\lambda^{1-T_1^L})^k, 1-(1-\lambda^{1-T_1^U})^k\right], \left[(1-\lambda^{I_1^L})^k, (1-\lambda^{I_1^U})^k\right], \left[(1-\lambda^{F_1^L})^k, (1-\lambda^{F_1^U})^k\right] \right\rangle \\
&\oplus \left\langle \left[1-(1-\lambda^{1-T_2^L})^k, 1-(1-\lambda^{1-T_2^U})^k\right], \left[(1-\lambda^{I_2^L})^k, (1-\lambda^{I_2^U})^k\right], \left[(1-\lambda^{F_2^L})^k, (1-\lambda^{F_2^U})^k\right] \right\rangle \\
&= k\lambda^{a_1} \oplus k\lambda^{a_2}.
\end{aligned}
$$

(2)

$$
\begin{aligned}
&(\lambda^{a_1} \otimes \lambda^{a_2})^k \\
&= \left\langle \begin{array}{l} \left[\lambda^{1-T_1^L} \cdot \lambda^{1-T_2^L}, \lambda^{1-T_1^U} \cdot \lambda^{1-T_2^U}\right], \\ \left[\left(1-\lambda^{I_1^L}\right) + \left(1-\lambda^{I_2^L}\right) - \left(1-\lambda^{I_1^L}\right) \cdot \left(1-\lambda^{I_2^L}\right), \left(1-\lambda^{I_1^U}\right) + \left(1-\lambda^{I_2^U}\right) - \left(1-\lambda^{I_1^U}\right) \cdot \left(1-\lambda^{I_2^U}\right)\right], \\ \left[\left(1-\lambda^{F_1^L}\right) + \left(1-\lambda^{F_2^L}\right) - \left(1-\lambda^{F_1^L}\right) \cdot \left(1-\lambda^{F_2^L}\right), \left(1-\lambda^{F_1^U}\right) + \left(1-\lambda^{F_2^U}\right) - \left(1-\lambda^{F_1^U}\right) \cdot \left(1-\lambda^{F_2^U}\right)\right] \end{array} \right\rangle^k \\
&= \left\langle \begin{array}{l} \left[(\lambda^{1-T_1^L} \cdot \lambda^{1-T_2^L})^k, (\lambda^{1-T_1^U} \cdot \lambda^{1-T_2^U})^k\right], \\ \left[1-(1-\left(1-\lambda^{I_1^L}\right) - \left(1-\lambda^{I_2^L}\right) + \left(1-\lambda^{I_1^L}\right) \cdot \left(1-\lambda^{I_2^L}\right))^k, 1-(1-\left(1-\lambda^{I_1^U}\right) - \left(1-\lambda^{I_2^U}\right) + \left(1-\lambda^{I_1^U}\right) \cdot \left(1-\lambda^{I_2^U}\right))^k\right], \\ \left[1-(1-\left(1-\lambda^{F_1^L}\right) - \left(1-\lambda^{F_2^L}\right) + \left(1-\lambda^{F_1^L}\right) \cdot \left(1-\lambda^{F_2^L}\right))^k, 1-(1-\left(1-\lambda^{F_1^U}\right) - \left(1-\lambda^{F_2^U}\right) + \left(1-\lambda^{F_1^U}\right) \cdot \left(1-\lambda^{F_2^U}\right))^k\right] \end{array} \right\rangle \\
&= \left\langle \begin{array}{l} \left[(\lambda^{1-T_2^L})^k \cdot (\lambda^{1-T_1^L})^k, (\lambda^{1-T_2^U})^k \cdot (\lambda^{1-T_1^U})^k\right], \\ \left[\begin{array}{l} \left(1-(1-\left(1-\lambda^{I_2^L}\right))^k\right) + \left(1-(1-\left(1-\lambda^{I_1^L}\right))^k\right) - \left(1-(1-\left(1-\lambda^{I_2^L}\right))^k\right) \cdot \left(1-(1-\left(1-\lambda^{I_1^L}\right))^k\right), \\ \left(1-(1-\left(1-\lambda^{I_2^U}\right))^k\right) + \left(1-(1-\left(1-\lambda^{I_1^U}\right))^k\right) - \left(1-(1-\left(1-\lambda^{I_2^U}\right))^k\right) \cdot \left(1-(1-\left(1-\lambda^{I_1^U}\right))^k\right) \end{array}\right], \\ \left[\begin{array}{l} \left(1-(1-\left(1-\lambda^{F_2^L}\right))^k\right) + \left(1-(1-\left(1-\lambda^{F_1^L}\right))^k\right) - \left(1-(1-\left(1-\lambda^{F_2^L}\right))^k\right) \cdot \left(1-(1-\left(1-\lambda^{F_1^L}\right))^k\right), \\ \left(1-(1-\left(1-\lambda^{F_2^U}\right))^k\right) + \left(1-(1-\left(1-\lambda^{F_1^U}\right))^k\right) - \left(1-(1-\left(1-\lambda^{F_2^U}\right))^k\right) \cdot \left(1-(1-\left(1-\lambda^{F_1^U}\right))^k\right) \end{array}\right] \end{array} \right\rangle \\
&= (\lambda^{a_2})^k \otimes (\lambda^{a_1})^k.
\end{aligned}
$$

(3)

$$
\begin{aligned}
&k_1\lambda^a \oplus k_2\lambda^a \\
&= k_1\left\langle \left[\lambda^{1-T^L}, \lambda^{1-T^U}\right], \left[1-\lambda^{I^L}, 1-\lambda^{I^U}\right], \left[1-\lambda^{F^L}, 1-\lambda^{F^U}\right] \right\rangle \\
&\quad \oplus k_2\left\langle \left[\lambda^{1-T^L}, \lambda^{1-T^U}\right], \left[1-\lambda^{I^L}, 1-\lambda^{I^U}\right], \left[1-\lambda^{F^L}, 1-\lambda^{F^U}\right] \right\rangle \\
&= \left\langle \left[1-(1-\lambda^{1-T^L})^{k_1}, 1-(1-\lambda^{1-T^U})^{k_1}\right], \left[(1-\lambda^{I^L})^{k_1}, (1-\lambda^{I^U})^{k_1}\right], \left[(1-\lambda^{F^L})^{k_1}, (1-\lambda^{F^U})^{k_1}\right] \right\rangle \\
&\quad \oplus \left\langle \left[1-(1-\lambda^{1-T^L})^{k_2}, 1-(1-\lambda^{1-T^U})^{k_2}\right], \left[(1-\lambda^{I^L})^{k_2}, (1-\lambda^{I^U})^{k_2}\right], \left[(1-\lambda^{F^L})^{k_2}, (1-\lambda^{F^U})^{k_2}\right] \right\rangle \\
&= \left\langle \begin{array}{l} \left[\begin{array}{l} 1-(1-\lambda^{1-T^L})^{k_1} + 1-(1-\lambda^{1-T^L})^{k_2} - \left(1-(1-\lambda^{1-T^L})^{k_1}\right) \cdot \left(1-(1-\lambda^{1-T^L})^{k_2}\right), \\ 1-(1-\lambda^{1-T^U})^{k_1} + 1-(1-\lambda^{1-T^U})^{k_2} - \left(1-(1-\lambda^{1-T^U})^{k_1}\right) \cdot \left(1-(1-\lambda^{1-T^U})^{k_2}\right) \end{array}\right], \\ \left[(1-\lambda^{I^L})^{k_1} \cdot (1-\lambda^{I^L})^{k_2}, (1-\lambda^{I^U})^{k_1} \cdot (1-\lambda^{I^U})^{k_2}\right], \left[(1-\lambda^{F^L})^{k_1} \cdot (1-\lambda^{F^L})^{k_2}, (1-\lambda^{F^U})^{k_1} \cdot (1-\lambda^{F^U})^{k_2}\right] \end{array} \right\rangle \\
&= \left\langle \begin{array}{l} \left[\begin{array}{l} 1-(1-\lambda^{1-T^L})^{k_2}(1-\lambda^{1-T^L})^{k_1}, \\ 1-(1-\lambda^{1-T^U})^{k_2}(1-\lambda^{1-T^U})^{k_1} \end{array}\right], \\ \left[(1-\lambda^{I^L})^{k_1} \cdot (1-\lambda^{I^L})^{k_2}, (1-\lambda^{I^U})^{k_1} \cdot (1-\lambda^{I^U})^{k_2}\right], \left[(1-\lambda^{F^L})^{k_1} \cdot (1-\lambda^{F^L})^{k_2}, (1-\lambda^{F^U})^{k_1} \cdot (1-\lambda^{F^U})^{k_2}\right] \end{array} \right\rangle \\
&= (k_1+k_2)\lambda^a.
\end{aligned}
$$

$$
\begin{aligned}
&(\lambda^a)^{k_1} \otimes (\lambda^a)^{k_2} \\
&= \left\langle \left[\lambda^{1-T^L}, \lambda^{1-T^U}\right], \left[1-\lambda^{I^L}, 1-\lambda^{I^U}\right], \left[1-\lambda^{F^L}, 1-\lambda^{F^U}\right] \right\rangle^{k_1} \\
&\otimes \left\langle \left[\lambda^{1-T^L}, \lambda^{1-T^U}\right], \left[1-\lambda^{I^L}, 1-\lambda^{I^U}\right], \left[1-\lambda^{F^L}, 1-\lambda^{F^U}\right] \right\rangle^{k_2} \\
&= \left\langle \left[(\lambda^{1-T^L})^{k_1}, (\lambda^{1-T^U})^{k_1}\right], \left[1-(1-(1-\lambda^{I^L}))^{k_1}, 1-(1-(1-\lambda^{I^U}))^{k_1}\right], \left[1-(1-(1-\lambda^{F^L}))^{k_1}, 1-(1-(1-\lambda^{F^U}))^{k_1}\right] \right\rangle \\
&\otimes \left\langle \left[(\lambda^{1-T^L})^{k_2}, (\lambda^{1-T^U})^{k_2}\right], \left[1-(1-(1-\lambda^{I^L}))^{k_2}, 1-(1-(1-\lambda^{I^U}))^{k_2}\right], \left[1-(1-(1-\lambda^{F^L}))^{k_2}, 1-(1-(1-\lambda^{F^U}))^{k_2}\right] \right\rangle \\
&= \left\langle \begin{array}{l} \left[(\lambda^{1-T^L})^{k_1} \cdot (\lambda^{1-T^L})^{k_2}, (\lambda^{1-T^U})^{k_1} \cdot (\lambda^{1-T^U})^{k_2}\right], \\
\left[\begin{array}{l} 1-(1-(1-\lambda^{I^L}))^{k_1} + 1-(1-(1-\lambda^{I^L}))^{k_2} - \left(1-(1-(1-\lambda^{I^L}))^{k_1}\right) \cdot \left(1-(1-(1-\lambda^{I^L}))^{k_2}\right), \\ 1-(1-(1-\lambda^{I^U}))^{k_1} + 1-(1-(1-\lambda^{I^U}))^{k_2} - \left(1-(1-(1-\lambda^{I^U}))^{k_1}\right) \cdot \left(1-(1-(1-\lambda^{I^U}))^{k_2}\right) \end{array}\right], \\
\left[\begin{array}{l} 1-(1-(1-\lambda^{F^L}))^{k_1} + 1-(1-(1-\lambda^{F^L}))^{k_2} - \left(1-(1-(1-\lambda^{F^L}))^{k_1}\right) \cdot \left(1-(1-(1-\lambda^{F^L}))^{k_2}\right), \\ 1-(1-(1-\lambda^{F^U}))^{k_1} + 1-(1-(1-\lambda^{F^U}))^{k_2} - \left(1-(1-(1-\lambda^{F^U}))^{k_1}\right) \cdot \left(1-(1-(1-\lambda^{F^U}))^{k_2}\right) \end{array}\right] \end{array} \right\rangle \\
&= \left\langle \begin{array}{l} \left[(\lambda^{1-T^L})^{k_1} \cdot (\lambda^{1-T^L})^{k_2}, (\lambda^{1-T^U})^{k_1} \cdot (\lambda^{1-T^U})^{k_2}\right], \\ \left[1-(\lambda^{I^L})^{k_2}(\lambda^{I^L})^{k_1}, 1-(\lambda^{I^U})^{k_2}(\lambda^{I^U})^{k_1}\right], \\ \left[1-(\lambda^{F^L})^{k_2}(\lambda^{F^L})^{k_1}, 1-(\lambda^{F^U})^{k_2}(\lambda^{F^U})^{k_1}\right] \end{array} \right\rangle \\
&= (\lambda^a)^{k_1+k_2}.
\end{aligned} \tag{4}
$$

$$
\begin{aligned}
&(\lambda_1)^a \otimes (\lambda_2)^a \\
&= \left\langle \left[\lambda_1^{1-T^L}, \lambda_1^{1-T^U}\right], \left[1-\lambda_1^{I^L}, 1-\lambda_1^{I^U}\right], \left[1-\lambda_1^{F^L}, 1-\lambda_1^{F^U}\right] \right\rangle \\
&\otimes \left\langle \left[\lambda_2^{1-T^L}, \lambda_2^{1-T^U}\right], \left[1-\lambda_2^{I^L}, 1-\lambda_2^{I^U}\right], \left[1-\lambda_2^{F^L}, 1-\lambda_2^{F^U}\right] \right\rangle \\
&= \left\langle \begin{array}{l} \left[\lambda_1^{1-T^L} \cdot \lambda_2^{1-T^L}, \lambda_1^{1-T^U} \cdot \lambda_2^{1-T^U}\right], \\ \left[1-\lambda_1^{I^L} + 1-\lambda_2^{I^L} - \left(1-\lambda_1^{I^L}\right) \cdot \left(1-\lambda_2^{I^L}\right), 1-\lambda_1^{I^U} + 1-\lambda_2^{I^U} - \left(1-\lambda_1^{I^U}\right) \cdot \left(1-\lambda_2^{I^U}\right)\right], \\ \left[1-\lambda_1^{F^L} + 1-\lambda_2^{F^L} - \left(1-\lambda_1^{F^L}\right) \cdot \left(1-\lambda_2^{F^L}\right), 1-\lambda_1^{F^U} + 1-\lambda_2^{F^U} - \left(1-\lambda_1^{F^U}\right) \cdot \left(1-\lambda_2^{F^U}\right)\right] \end{array} \right\rangle \\
&= \left\langle \begin{array}{l} \left[\lambda_1^{1-T^L} \cdot \lambda_2^{1-T^L}, \lambda_1^{1-T^U} \cdot \lambda_2^{1-T^U}\right], \\ \left[1-\lambda_1^{I^L}\lambda_2^{I^L}, 1-\lambda_1^{I^U}\lambda_2^{I^U}\right], \\ \left[1-\lambda_1^{F^L}\lambda_2^{F^L}, 1-\lambda_1^{F^U}\lambda_2^{F^U}\right] \end{array} \right\rangle \\
&= \left\langle \left[(\lambda_1\lambda_2)^{1-T^L}, (\lambda_1\lambda_2)^{1-T^U}\right], \left[1-(\lambda_1\lambda_2)^{I^L}, 1-(\lambda_1\lambda_2)^{I^U}\right], \left[1-(\lambda_1\lambda_2)^{F^L}, 1-(\lambda_1\lambda_2)^{F^U}\right] \right\rangle \\
&= (\lambda_1\lambda_2)^a. \square
\end{aligned} \tag{5}
$$

Theorem 5. *Let* $a =< [T^L, T^U], [I^L, I^U], [F^L, F^U] >$ *be an INN. If* $\lambda_1 \geq \lambda_2$*, then one can obtain* $(\lambda_1)^a \geq (\lambda_2)^a$ *for* $\lambda_1, \lambda_2 \in (0,1)$*, and* $(\lambda_1)^a \leq (\lambda_2)^a$ *for* $\lambda_1, \lambda_2 \geq 1$.

Proof. When $\lambda_1 \geq \lambda_2$ and $\lambda_1, \lambda_2 \in (0,1)$, based on Definition 11, we can obtain

$$
\begin{aligned}
(\lambda_1)^a &= \left\langle \left[\lambda_1^{1-T^L}, \lambda_1^{1-T^U}\right], \left[1-\lambda_1^{I^L}, 1-\lambda_1^{I^U}\right], \left[1-\lambda_1^{F^L}, 1-\lambda_1^{F^U}\right] \right\rangle, \\
(\lambda_2)^a &= \left\langle \left[\lambda_2^{1-T^L}, \lambda_2^{1-T^U}\right], \left[1-\lambda_2^{I^L}, 1-\lambda_2^{I^U}\right], \left[1-\lambda_2^{F^L}, 1-\lambda_2^{F^U}\right] \right\rangle,
\end{aligned}
$$

Since $\lambda_1 \geq \lambda_2$, then $\lambda_1^{1-T^L} \geq \lambda_2^{1-T^L}, \lambda_1^{1-T^U} \geq \lambda_2^{1-T^U}$, and $1-\lambda_1^{I^L} \leq 1-\lambda_2^{I^L}, 1-\lambda_1^{I^U} \leq 1-\lambda_2^{I^U}$, and $1-\lambda_1^{F^L} \geq 1-\lambda_2^{F^L}, 1-\lambda_1^{F^U} \geq 1-\lambda_2^{F^U}$.

$$
\begin{aligned}
S((\lambda_1)^a) &= \frac{2+\lambda_1^{1-T^L}+\lambda_1^{1-T^U}-2\left(1-\lambda_1^{I^L}\right)-2\left(1-\lambda_1^{I^U}\right)-\left(1-\lambda_1^{F^L}\right)-\left(1-\lambda_1^{F^U}\right)}{4} \\
&= \frac{\lambda_1^{1-T^L}+\lambda_1^{1-T^U}+2\lambda_1^{I^L}+2\lambda_1^{I^U}+\lambda_1^{F^L}+\lambda_1^{F^U}-4}{4},
\end{aligned}
$$

$$
\begin{aligned}
S((\lambda_2)^a) &= \frac{2+\lambda_2^{1-T^L}+\lambda_2^{1-T^U}-2\left(1-\lambda_2^{I^L}\right)-2\left(1-\lambda_2^{I^U}\right)-\left(1-\lambda_2^{F^L}\right)-\left(1-\lambda_2^{F^U}\right)}{4} \\
&= \frac{\lambda_2^{1-T^L}+\lambda_2^{1-T^U}+2\lambda_2^{I^L}+2\lambda_2^{I^U}+\lambda_2^{F^L}+\lambda_2^{F^U}-4}{4},
\end{aligned}
$$

$$
\begin{aligned}
& S((\lambda_1)^a) - S((\lambda_2)^a) \\
&= \frac{\lambda_1^{1-T^L}+\lambda_1^{1-T^U}+2\lambda_1^{I^L}+2\lambda_1^{I^U}+\lambda_1^{F^L}+\lambda_1^{F^U}-4}{4} - \frac{\lambda_2^{1-T^L}+\lambda_2^{1-T^U}+2\lambda_2^{I^L}+2\lambda_2^{I^U}+\lambda_2^{F^L}+\lambda_2^{F^U}-4}{4} \\
&= \frac{\left(\lambda_1^{1-T^L}-\lambda_2^{1-T^L}\right)+\left(\lambda_1^{1-T^U}-\lambda_2^{1-T^U}\right)+\left(2\lambda_1^{I^L}-2\lambda_2^{I^L}\right)+\left(2\lambda_1^{I^U}-2\lambda_2^{I^U}\right)+\left(\lambda_1^{F^L}-\lambda_2^{F^L}\right)+\left(\lambda_1^{F^U}-\lambda_2^{F^U}\right)}{4} \\
&\geq 0.
\end{aligned}
$$

Then $S((\lambda_1)^a) \geq S((\lambda_2)^a)$, $(\lambda_1)^a \geq (\lambda_2)^a$.

Then, when $\lambda_1, \lambda_2 \geq 1$ and $\lambda_1 \geq \lambda_2$, we can know $0 \leq \frac{1}{\lambda_1} \leq \frac{1}{\lambda_2} \leq 1$. As discussed above, we can obtain $(\lambda_1)^a \leq (\lambda_2)^a$. This completes the proof. □

In what follows, let us take a look at some special values about λ^a:

(1) If $\lambda = 1$, then $\lambda^a = \langle [1,1], [0,0], [0,0] \rangle = \langle 1,0,0 \rangle$;
(2) If $a = \langle [1,1], [0,0], [0,0] \rangle = \langle 1,0,0 \rangle$, then $\lambda^a = \langle [1,1], [0,0], [0,0] \rangle = \langle 1,0,0 \rangle$;
(3) If $a = \langle [0,0][1,1], [1,1], \rangle = \langle 0,1,1 \rangle$, then

$$\lambda^a = \langle [\lambda, \lambda], [1-\lambda, 1-\lambda], [1-\lambda, 1-\lambda] \rangle$$

4. Interval Neutrosophic Weighted Exponential Aggregation (INWEA) Operator

Aggregation operators have been commonly used to aggregate the evaluation information in decision making. Here, we utilize the INNs rather than real numbers as weight of criterion, which is more comprehensive and reasonable. In this section, we introduced the interval neutrosophic weighted exponential aggregation (INWEA) operator. Furthermore, some characteristics of the proposed aggregation operator, such as boundedness and monotonicity are discussed in detail.

Definition 12. *Let $a_i =< [T_i^L, T_i^U], [I_i^L, I_i^U], [F_i^L, F_i^U] > (i = 1, 2, \cdots, n)$ be a collection of INNs, and $\lambda_i \in (0,\ 1)$ $(i = 1, 2, \cdots, n)$ be the collection of real numbers, and let INWEA: $\Theta^n \to \Theta$. If*

$$INWEA(a_1, a_2, \cdots, a_n) = \lambda_1{}^{a_1} \otimes \lambda_2{}^{a_2} \otimes \cdots \otimes \lambda_n{}^{a_n}. \tag{9}$$

Then the function INWEA is called an interval neutrosophic weighted exponential aggregation (INWEA) operator, where a_i $(i = 1, 2, \cdots, n)$ are the exponential weighting vectors of attribute values $\lambda_i (i = 1, 2, \cdots, n)$.

Theorem 6 [43]. *Let $a_i =< [T_i^L, T_i^U], [I_i^L, I_i^U], [F_i^L, F_i^U] > (i = 1, 2, \cdots, n)$ be a collection of INNs, the aggregated value by using the INWEA operator is also an INN, where*

$$
\begin{aligned}
& INWEA(a_1, a_2, \cdots, a_n) \\
&= \begin{cases} \left\langle \left[\prod_{i=1}^n \lambda_i^{1-T_i^L}, \prod_{i=1}^n \lambda_i^{1-T_i^U}\right], \left[1-\prod_{i=1}^n \lambda_i^{I_i^L}, 1-\prod_{i=1}^n \lambda_i^{I_i^U}\right], \left[1-\prod_{i=1}^n \lambda_i^{F_i^L}, 1-\prod_{i=1}^n \lambda_i^{F_i^U}\right] \right\rangle, & \lambda_i \in (0,1) \\ \left\langle \left[\prod_{i=1}^n \left(\frac{1}{\lambda_i}\right)^{1-T_i^L}, \prod_{i=1}^n \left(\frac{1}{\lambda_i}\right)^{1-T_i^U}\right], \left[1-\prod_{i=1}^n \left(\frac{1}{\lambda_i}\right)^{I_i^L}, 1-\prod_{i=1}^n \left(\frac{1}{\lambda_i}\right)^{I_i^U}\right], \left[1-\prod_{i=1}^n \left(\frac{1}{\lambda_i}\right)^{F_i^L}, 1-\prod_{i=1}^n \left(\frac{1}{\lambda_i}\right)^{F_i^U}\right] \right\rangle, & \lambda_i \geq 1 \end{cases}
\end{aligned} \tag{10}
$$

and a_i $(i = 1, 2, \cdots, n)$ are the exponential weights of λ_i $(i = 1, 2, \cdots, n)$.

Proof. By using mathematical induction, we can prove the Equation (10).

(1) When $n = 2$, we have

$$
\begin{aligned}
&INWEA(a_1,a_2) = \lambda_1{}^{a_1} \otimes \lambda_2{}^{a_2} \\
&= \left\langle \left[\lambda_1^{1-T_1^L}, \lambda_1^{1-T_1^U}\right], \left[1-\lambda_1^{I_1^L}, 1-\lambda_1^{I_1^U}\right], \left[1-\lambda_1^{F_1^L}, 1-\lambda_1^{F_1^U}\right] \right\rangle \\
&\otimes \left\langle \left[\lambda_2^{1-T_2^L}, \lambda_2^{1-T_2^U}\right], \left[1-\lambda_2^{I_2^L}, 1-\lambda_2^{I_2^U}\right], \left[1-\lambda_2^{F_2^L}, 1-\lambda_2^{F_2^U}\right] \right\rangle \\
&= \left\langle \begin{array}{l} \left[\lambda_1^{1-T_1^L} \cdot \lambda_2^{1-T_2^L}, \lambda_1^{1-T_1^U} \cdot \lambda_2^{1-T_2^U}\right], \\ \left[1-\lambda_1^{I_1^L} + 1-\lambda_2^{I_2^L} - \left(1-\lambda_1^{I_1^L}\right) \cdot \left(1-\lambda_2^{I_2^L}\right), 1-\lambda_1^{I_1^U} + 1-\lambda_2^{I_2^U} - \left(1-\lambda_1^{I_1^U}\right) \cdot \left(1-\lambda_2^{I_2^U}\right)\right], \\ \left[1-\lambda_1^{F_1^L} + 1-\lambda_2^{F_2^L} - \left(1-\lambda_1^{F_1^L}\right) \cdot \left(1-\lambda_2^{F_2^L}\right), 1-\lambda_1^{F_1^U} + 1-\lambda_2^{F_2^U} - \left(1-\lambda_1^{F_1^U}\right) \cdot \left(1-\lambda_2^{F_2^U}\right)\right] \end{array} \right\rangle \\
&= \left\langle \left[\prod_{i=1}^{2} \lambda_i^{1-T_i^L}, \prod_{i=1}^{2} \lambda_i^{1-T_i^U}\right], \left[1-\prod_{i=1}^{2} \lambda_i^{I_i^L}, 1-\prod_{i=1}^{2} \lambda_i^{I_i^U}\right], \left[1-\prod_{i=1}^{2} \lambda_i^{F_i^L}, 1-\prod_{i=1}^{2} \lambda_i^{F_i^U}\right] \right\rangle.
\end{aligned}
\tag{11}
$$

(2) When $n = k$, according to Equation (10) there is the following formula:

$$
\begin{aligned}
&INWEA(a_1,a_2,\cdots,a_k) \\
&= \left\langle \left[\prod_{i=1}^{k} \lambda_i^{1-T_i^L}, \prod_{i=1}^{k} \lambda_i^{1-T_i^U}\right], \left[1-\prod_{i=1}^{k} \lambda_i^{I_i^L}, 1-\prod_{i=1}^{k} \lambda_i^{I_i^U}\right], \left[1-\prod_{i=1}^{k} \lambda_i^{F_i^L}, 1-\prod_{i=1}^{k} \lambda_i^{F_i^U}\right] \right\rangle.
\end{aligned}
\tag{12}
$$

When $n = k + 1$, we have the following results based on the operational rules of Definition 3 and combining (2) and (3).

$$
\begin{aligned}
&INWEA(a_1,a_2,\cdots,a_k,a_{k+1}) \\
&= \left\langle \left[\prod_{i=1}^{k} \lambda_i^{1-T_i^L}, \prod_{i=1}^{k} \lambda_i^{1-T_i^U}\right], \left[1-\prod_{i=1}^{k} \lambda_i^{I_i^L}, 1-\prod_{i=1}^{k} \lambda_i^{I_i^U}\right], \left[1-\prod_{i=1}^{k} \lambda_i^{F_i^L}, 1-\prod_{i=1}^{k} \lambda_i^{F_i^U}\right] \right\rangle \otimes a_{k+1} \\
&= \left\langle \left[\prod_{i=1}^{k} \lambda_i^{1-T_i^L}, \prod_{i=1}^{k} \lambda_i^{1-T_i^U}\right], \left[1-\prod_{i=1}^{k} \lambda_i^{I_i^L}, 1-\prod_{i=1}^{k} \lambda_i^{I_i^U}\right], \left[1-\prod_{i=1}^{k} \lambda_i^{F_i^L}, 1-\prod_{i=1}^{k} \lambda_i^{F_i^U}\right] \right\rangle \\
&\otimes \left\langle \left[\lambda_{k+1}^{1-T_{k+1}^L}, \lambda_{k+1}^{1-T_{k+1}^U}\right], \left[1-\lambda_{k+1}^{I_{k+1}^L}, 1-\lambda_{k+1}^{I_{k+1}^U}\right], \left[1-\lambda_{k+1}^{F_{k+1}^L}, 1-\lambda_{k+1}^{F_{k+1}^U}\right] \right\rangle \\
&= \left\langle \left[\prod_{i=1}^{n} \lambda_i^{1-T_i^L}, \prod_{i=1}^{n} \lambda_i^{1-T_i^U}\right], \left[1-\prod_{i=1}^{n} \lambda_i^{I_i^L}, 1-\prod_{i=1}^{n} \lambda_i^{I_i^U}\right], \left[1-\prod_{i=1}^{n} \lambda_i^{F_i^L}, 1-\prod_{i=1}^{n} \lambda_i^{F_i^U}\right] \right\rangle. \square
\end{aligned}
$$

Therefore, for the above results we determine that Equation (10) holds for any n. Thus, the proof is completed. When $\lambda_i \geq 1$, and $0 < \frac{1}{\lambda_i} \leq 1$, we can also obtain

$$
\begin{aligned}
&INWEA(\alpha_1,\alpha_2,\cdots,\alpha_n) \\
&= \left\langle \left[\prod_{i=1}^{n} \left(\frac{1}{\lambda_i}\right)^{1-T_i^L}, \prod_{i=1}^{n} \left(\frac{1}{\lambda_i}\right)^{1-T_i^U}\right], \left[1-\prod_{i=1}^{n} \left(\frac{1}{\lambda_i}\right)^{I_i^L}, 1-\prod_{i=1}^{n} \left(\frac{1}{\lambda_i}\right)^{I_i^U}\right], \left[1-\prod_{i=1}^{n} \left(\frac{1}{\lambda_i}\right)^{F_i^L}, 1-\prod_{i=1}^{n} \left(\frac{1}{\lambda_i}\right)^{F_i^U}\right] \right\rangle.
\end{aligned}
$$

and the aggregated value is an INN.

Here, we discuss the relationship between the $INWEA$ operator and other exponential aggregation operators. When $T^L = T^U$, $I^L = I^U$ and $F^L = F^U$, the $INWEA$ operator of INNs is equivalent to the $SVNWEA$ operator of SVNNs [41].

$$
INWEA(a_1,a_2,\cdots,a_n) = \left\langle \left[\prod_{i=1}^{n} \lambda_i^{1-T_i}\right], \left[1-\prod_{i=1}^{n} \lambda_i^{I_i}\right], \left[1-\prod_{i=1}^{n} \lambda_i^{F_i}\right] \right\rangle = SVNWEA(a_1,a_2,\cdots,a_n)
$$

When $0^- \leq T^U + I^U + F^U \leq 1$, the $INWEA$ operator of INNs is equivalent to the $IIFWEA$ operator of IIFNs [40]. When $T^L = T^U$, $I^L = I^U$, $F^L = F^U$ and $0^- \leq T^U + I^U + F^U \leq 1$, the $INWEA$ operator of INNs is equivalent to the $IFWEA$ operator of IFNs [39]. So the $INWEA$ operator of INNs is a more generalized representation, and the other exponential aggregation operators of SVNNs, IIFNs and IFNs are special cases.

Theorem 7. *The INWEA operator has the following properties:*

(1) *Boundedness: Let* $a_i = < [T_i^L, T_i^U], [I_i^L, I_i^U], [F_i^L, F_i^U] >$ $(i = 1, 2, \cdots, n)$ *be a collection of INNs, and let* $a_{\min} = < [\min_i T_i^L, \min_i T_i^U], [\max_i I_i^L, \max_i I_i^U], [\max_i F_i^L, \max_i F_i^U] >$, $a_{\max} = < [\max_i T_i^L, \max_i T_i^U], [\min_i I_i^L, \min_i I_i^U], [\min_i F_i^L, \min_i F_i^U] >$ *for* $i = 1, 2, \cdots, n$,

$$a^- = \mathrm{INWEA}(a_{\min}, a_{\min}, \cdots, a_{\min})$$
$$= \left\langle \left[\prod_{i=1}^n \lambda_i^{1-\min_i T_i^L}, \prod_{i=1}^n \lambda_i^{1-\min_i T_i^U}\right], \left[1 - \prod_{i=1}^n \lambda_i^{\max_i I_i^L}, 1 - \prod_{i=1}^n \lambda_i^{\max_i I_i^U}\right], \left[1 - \prod_{i=1}^n \lambda_i^{\max_i F_i^L}, 1 - \prod_{i=1}^n \lambda_i^{\max_i F_i^U}\right] \right\rangle,$$

$$a^+ = \mathrm{INWEA}(a_{max}, a_{max}, \cdots, a_{max})$$
$$= \left\langle \left[\prod_{i=1}^n \lambda_i^{1-\max_i T_i^L}, \prod_{i=1}^n \lambda_i^{1-\max_i T_i^U}\right], \left[1 - \prod_{i=1}^n \lambda_i^{\min_i I_i^L}, 1 - \prod_{i=1}^n \lambda_i^{\min_i I_i^U}\right], \left[1 - \prod_{i=1}^n \lambda_i^{\min_i F_i^L}, 1 - \prod_{i=1}^n \lambda_i^{\min_i F_i^U}\right] \right\rangle,$$

Then $a^- \leq \mathrm{INWEA}(a_1, a_2, \cdots, a_n) \leq a^+$.

Proof. For any i, we have $\min_i T_i^L \leq T_i^L \leq \max_i T_i^L$, $\min_i T_i^U \leq T_i^U \leq \max_i T_i^U$, $\min_i I_i^L \leq I_i^L \leq \max_i I_i^L$, $\min_i I_i^U \leq I_i^U \leq \max_i I_i^U$, $\min_i F_i^L \leq F_i^L \leq \max_i F_i^L$, $\min_i F_i^U \leq F_i^U \leq \max_i F_i^U$.

$$\prod_{i=1}^n \lambda_i^{1-T_i^L} \geq \prod_{i=1}^n \lambda_i^{1-\min_i T_i^L}, \prod_{i=1}^n \lambda_i^{1-T_i^U} \geq \prod_{i=1}^n \lambda_i^{1-\min_i T_i^U},$$
$$1 - \prod_{i=1}^n \lambda_i^{I_i^L} \leq 1 - \prod_{i=1}^n \lambda_i^{\max_i I_i^L}, 1 - \prod_{i=1}^n \lambda_i^{I_i^U} \leq 1 - \prod_{i=1}^n \lambda_i^{\max_i I_i^U},$$
$$1 - \prod_{i=1}^n \lambda_i^{F_i^L} \leq 1 - \prod_{i=1}^n \lambda_i^{\max_i F_i^L}, 1 - \prod_{i=1}^n \lambda_i^{F_i^U} \leq 1 - \prod_{i=1}^n \lambda_i^{\max_i F_i^U},$$
$$\prod_{i=1}^n \lambda_i^{1-T_i^L} \leq \prod_{i=1}^n \lambda_i^{1-\mathit{mxa}_i T_i^L}, \prod_{i=1}^n \lambda_i^{1-T_i^U} \leq \prod_{i=1}^n \lambda_i^{1-\mathit{mxa}_i T_i^U},$$
$$1 - \prod_{i=1}^n \lambda_i^{I_i^L} \geq 1 - \prod_{i=1}^n \lambda_i^{\min_i I_i^L}, 1 - \prod_{i=1}^n \lambda_i^{I_i^U} \geq 1 - \prod_{i=1}^n \lambda_i^{\min_i I_i^U},$$
$$1 - \prod_{i=1}^n \lambda_i^{F_i^L} \geq 1 - \prod_{i=1}^n \lambda_i^{\min_i F_i^L}, 1 - \prod_{i=1}^n \lambda_i^{F_i^U} \geq 1 - \prod_{i=1}^n \lambda_i^{\min_i F_i^U},$$

Let $\mathrm{INWEA}(a_1, a_2, \cdots, a_n) = a$, $a^- = < [T^{L-}, T^{U-}], [I^{L-}, I^{U-}], [F^{L-}, F^{U-}] >$, and $a^+ = < [T^{L+}, T^{U+}], [I^{L+}, I^{U+}], [F^{L+}, F^{U+}] >$, then based on the score function, where

$$\begin{aligned} S(a) &= \frac{2+\prod_{i=1}^n \lambda_i^{1-T_i^L} + \prod_{i=1}^n \lambda_i^{1-T_i^U} - 2\left(1-\prod_{i=1}^n \lambda_i^{I_i^L}\right) - 2\left(1-\prod_{i=1}^n \lambda_i^{I_i^U}\right) - \left(1-\prod_{i=1}^n \lambda_i^{F_i^L}\right) - \left(1-\prod_{i=1}^n \lambda_i^{F_i^U}\right)}{4} \\ &\geq \frac{2+\prod_{i=1}^n \lambda_i^{1-\min_i T_i^L} + \prod_{i=1}^n \lambda_i^{1-\min_i T_i^U} - 2\left(1-\prod_{i=1}^n \lambda_i^{\max_i I_i^L}\right) - 2\left(1-\prod_{i=1}^n \lambda_i^{\max_i I_i^U}\right) - \left(1-\prod_{i=1}^n \lambda_i^{\max_i F_i^L}\right) - \left(1-\prod_{i=1}^n \lambda_i^{\max_i F_i^U}\right)}{4} \\ &= S(\alpha^-), \end{aligned}$$

$$\begin{aligned} S(a) &= \frac{2+\prod_{i=1}^n \lambda_i^{1-T_i^L} + \prod_{i=1}^n \lambda_i^{1-T_i^U} - 2\left(1-\prod_{i=1}^n \lambda_i^{I_i^L}\right) - 2\left(1-\prod_{i=1}^n \lambda_i^{I_i^U}\right) - \left(1-\prod_{i=1}^n \lambda_i^{F_i^L}\right) - \left(1-\prod_{i=1}^n \lambda_i^{F_i^U}\right)}{4} \\ &\geq \frac{2+\prod_{i=1}^n \lambda_i^{1-\max_i T_i^L} + \prod_{i=1}^n \lambda_i^{1-\max_i T_i^U} - 2\left(1-\prod_{i=1}^n \lambda_i^{\min_i I_i^L}\right) - 2\left(1-\prod_{i=1}^n \lambda_i^{\min_i I_i^U}\right) - \left(1-\prod_{i=1}^n \lambda_i^{\min_i F_i^L}\right) - \left(1-\prod_{i=1}^n \lambda_i^{\min_i F_i^U}\right)}{4} \\ &= S(\alpha^+). \square \end{aligned}$$

In what follows, we discuss three cases:

(I) If $S(a^-) < S(a) < S(a^+)$, then $a^- < \text{INWEA}(a_1, a_2, \cdots, a_n) < a^+$ holds obviously.

(II) If $S(a) = S(a^-)$, then there is

$T^L + T^U - 2I^L - 2I^U - F^L - F^U = T^{L-} + T^{U-} - 2I^{L-} - 2I^{U-} - F^{L-} - F^{U-}$. Thus, we can obtain $T^L = T^{L-}$, $T^U = T^{U-}$, $I^L = I^{L-}$, $I^U = I^{U-}$, $F^L = F^{L-}$, $F^U = F^{U-}$. Hence, there is

$$\begin{aligned} A(a) &= \tfrac{1}{2}(T^L + T^U - I^U(1 - T^U) - I^L(1 - T^L) - F^U(1 - I^L) - F^L(1 - I^U)) \\ &= \tfrac{1}{2}(T^{L-} + T^{U-} - I^{U-}(1 - T^{U-}) - I^{L-}(1 - T^{L-}) - F^{U-}(1 - I^{L-}) - F^{L-}(1 - I^{U-})) \\ &= A(a^-). \end{aligned}$$

So we have $\text{INWEA}(a_1, a_2, \cdots, a_n) = a^-$.

(III) If $S(a) = S(a^+)$, then there is

$T^L + T^U - 2I^L - 2I^U - F^L - F^U = T^{L+} + T^{U+} - 2I^{L+} - 2I^{U+} - F^{L+} - F^{U+}$. Thus, we can obtain $T^L = T^{L+}$, $T^U = T^{U+}$, $I^L = I^{L+}$, $I^U = I^{U+}$, $F^L = F^{L+}$, $F^U = F^{U+}$. Hence, there is

$$\begin{aligned} A(a) &= \tfrac{1}{2}(T^L + T^U - I^U(1 - T^U) - I^L(1 - T^L) - F^U(1 - I^L) - F^L(1 - I^U)) \\ &= \tfrac{1}{2}(T^{L+} + T^{U+} - I^{U+}(1 - T^{U+}) - I^{L+}(1 - T^{L+}) - F^{U+}(1 - I^{L+}) - F^{L+}(1 - I^{U+})) \\ &= A(a^+). \end{aligned}$$

Hence, we have $\text{INWEA}(a_1, a_2, \cdots, a_n) = a^+$.
Based on the above three cases, there is $a^- \le \text{INWEA}(a_1, a_2, \cdots, a_n) \le a^+$.

(2) *Monotonity: Let* $a_i = < [T_i^L, T_i^U], [I_i^L, I_i^U], [F_i^L, F_i^U] >$ $(i = 1, 2, \cdots, n)$ *and* $a_i^* = < [T_i^{L*}, T_i^{U*}], [I_i^{L*}, I_i^{U*}], [F_i^{L*}, F_i^{U*}] >$ *be two collections of INNs. If* $a_i \le a_i^*$, *then* $\text{INWEA}(a_1, a_2, \cdots, a_n) \le \text{INWEA}(a_1^*, a_2^*, \cdots, a_n^*)$.

Proof. Let $a = \text{INWEA}(a_1, a_2, \cdots, a_n) = \left\langle \left[\prod_{i=1}^n \lambda_i^{1-T_i^L}, \prod_{i=1}^n \lambda_i^{1-T_i^U}\right], \left[1 - \prod_{i=1}^n \lambda_i^{I_i^L}, 1 - \prod_{i=1}^n \lambda_i^{I_i^U}\right], \left[1 - \prod_{i=1}^n \lambda_i^{F_i^L}, 1 - \prod_{i=1}^n \lambda_i^{F_i^U}\right] \right\rangle$, $a^* = \text{INWEA}(a_1^*, a_2^*, \cdots, a_n^*) = \left\langle \left[\prod_{i=1}^n \lambda_i^{1-T_i^{L*}}, \prod_{i=1}^n \lambda_i^{1-T_i^{U*}}\right], \left[1 - \prod_{i=1}^n \lambda_i^{I_i^{L*}}, 1 - \prod_{i=1}^n \lambda_i^{I_i^{U*}}\right], \left[1 - \prod_{i=1}^n \lambda_i^{F_i^{L*}}, 1 - \prod_{i=1}^n \lambda_i^{F_i^{U*}}\right] \right\rangle$,

If $a_i \le a_i^*$, then $T_i^L \le T_i^{L*}$, $T_i^U \le T_i^{U*}$, $I_i^L \ge I_i^{L*}$, $I_i^U \ge I_i^{U*}$, $F_i^L \ge F_i^{L*}$, $F_i^U \ge F_i^{U*}$ for any i. So we have $\prod_{i=1}^n \lambda_i^{1-T_i^L} \le \prod_{i=1}^n \lambda_i^{1-T_i^{L*}}$, $\prod_{i=1}^n \lambda_i^{1-T_i^U} \le \prod_{i=1}^n \lambda_i^{1-T_i^{U*}}$, $1 - \prod_{i=1}^n \lambda_i^{I_i^L} \ge 1 - \prod_{i=1}^n \lambda_i^{I_i^{L*}}$, $1 - \prod_{i=1}^n \lambda_i^{I_i^U} \ge 1 - \prod_{i=1}^n \lambda_i^{I_i^{U*}}$, $1 - \prod_{i=1}^n \lambda_i^{F_i^L} \ge 1 - \prod_{i=1}^n \lambda_i^{F_i^{L*}}$, $1 - \prod_{i=1}^n \lambda_i^{I_i^U} \ge 1 - \prod_{i=1}^n \lambda_i^{I_i^{U*}}$.

Thus,

$$\begin{aligned} & S(a) \\ &= \frac{2 + \prod_{i=1}^n \lambda_i^{1-T_i^L} + \prod_{i=1}^n \lambda_i^{1-T_i^U} - 2\left(1 - \prod_{i=1}^n \lambda_i^{I_i^L}\right) - 2\left(1 - \prod_{i=1}^n \lambda_i^{I_i^U}\right) - \left(1 - \prod_{i=1}^n \lambda_i^{F_i^L}\right) - \left(1 - \prod_{i=1}^n \lambda_i^{F_i^U}\right)}{4} \\ &\le \frac{2 + \prod_{i=1}^n \lambda_i^{1-T_i^{L*}} + \prod_{i=1}^n \lambda_i^{1-T_i^{U*}} - 2\left(1 - \prod_{i=1}^n \lambda_i^{I_i^{L*}}\right) - 2\left(1 - \prod_{i=1}^n \lambda_i^{I_i^{U*}}\right) - \left(1 - \prod_{i=1}^n \lambda_i^{F_i^{L*}}\right) - \left(1 - \prod_{i=1}^n \lambda_i^{F_i^{U*}}\right)}{4} \\ &= S(\alpha^*). \end{aligned}$$

Hence, there are the following two cases:

(1) If $S(a) < S(a^*)$, then we can get $\text{INWEA}(a_1, a_2, \cdots, a_n) < \text{INWEA}(a_1^*, a_2^*, \cdots, a_n^*)$;

(2) If $S(a) = S(a^*)$, then

$$\begin{aligned} & \prod_{i=1}^n \lambda_i^{1-T_i^L} + \prod_{i=1}^n \lambda_i^{1-T_i^U} - 2\left(1 - \prod_{i=1}^n \lambda_i^{I_i^L}\right) - 2\left(1 - \prod_{i=1}^n \lambda_i^{I_i^U}\right) - \left(1 - \prod_{i=1}^n \lambda_i^{F_i^L}\right) - \left(1 - \prod_{i=1}^n \lambda_i^{F_i^U}\right) \\ &= \prod_{i=1}^n \lambda_i^{1-T_i^{L*}} + \prod_{i=1}^n \lambda_i^{1-T_i^{U*}} - 2\left(1 - \prod_{i=1}^n \lambda_i^{I_i^{L*}}\right) - 2\left(1 - \prod_{i=1}^n \lambda_i^{I_i^{U*}}\right) - \left(1 - \prod_{i=1}^n \lambda_i^{F_i^{L*}}\right) - \left(1 - \prod_{i=1}^n \lambda_i^{F_i^{U*}}\right). \end{aligned}$$

Therefore, by the condition $T_i^L \leq T_i^{L*}$, $T_i^U \leq T_i^{U*}$, $I_i^L \geq I_i^{L*}$, $I_i^U \geq I_i^{U*}$, $F_i^L \geq F_i^{L*}$, $F_i^U \geq F_i^{U*}$ for any i, we can get

$\prod_{i=1}^{n}\lambda_i^{1-T_i^L} = \prod_{i=1}^{n}\lambda_i^{1-T_i^{L*}}$, $\prod_{i=1}^{n}\lambda_i^{1-T_i^U} = \prod_{i=1}^{n}\lambda_i^{1-T_i^{U*}}$, $1-\prod_{i=1}^{n}\lambda_i^{I_i^L} = 1-\prod_{i=1}^{n}\lambda_i^{I_i^{L*}}$, $1-\prod_{i=1}^{n}\lambda_i^{I_i^U} = 1-\prod_{i=1}^{n}\lambda_i^{I_i^{U*}}$, $1-\prod_{i=1}^{n}\lambda_i^{F_i^L} = 1-\prod_{i=1}^{n}\lambda_i^{F_i^{L*}}$, $1-\prod_{i=1}^{n}\lambda_i^{I_i^U} = 1-\prod_{i=1}^{n}\lambda_i^{I_i^{U*}}$.

Thus,

$$\begin{aligned}
&A(a)\\
&= \tfrac{1}{2}\Big(\prod_{i=1}^{n}\lambda_i^{1-T_i^L} + \prod_{i=1}^{n}\lambda_i^{1-T_i^U} - \Big(1-\prod_{i=1}^{n}\lambda_i^{I_i^U}\Big)\Big(1-\prod_{i=1}^{n}\lambda_i^{1-T_i^U}\Big) - \Big(1-\prod_{i=1}^{n}\lambda_i^{I_i^L}\Big)\Big(1-\prod_{i=1}^{n}\lambda_i^{1-T_i^L}\Big) - \Big(1-\prod_{i=1}^{n}\lambda_i^{F_i^U}\Big)\Big(1-\Big(1-\prod_{i=1}^{n}\lambda_i^{I_i^L}\Big)\Big) - \Big(1-\prod_{i=1}^{n}\lambda_i^{F_i^L}\Big)\Big(1-\Big(1-\prod_{i=1}^{n}\lambda_i^{I_i^U}\Big)\Big)\Big)\\
&= \tfrac{1}{2}\Big(\prod_{i=1}^{n}\lambda_i^{1-T_i^{L*}} + \prod_{i=1}^{n}\lambda_i^{1-T_i^{U*}} - \Big(1-\prod_{i=1}^{n}\lambda_i^{I_i^{U*}}\Big)\Big(1-\prod_{i=1}^{n}\lambda_i^{1-T_i^{U*}}\Big) - \Big(1-\prod_{i=1}^{n}\lambda_i^{I_i^{L*}}\Big)\Big(1-\prod_{i=1}^{n}\lambda_i^{1-T_i^{L*}}\Big) - \Big(1-\prod_{i=1}^{n}\lambda_i^{F_i^{U*}}\Big)\Big(1-\Big(1-\prod_{i=1}^{n}\lambda_i^{I_i^{L*}}\Big)\Big) - \Big(1-\prod_{i=1}^{n}\lambda_i^{F_i^{L*}}\Big)\Big(1-\Big(1-\prod_{i=1}^{n}\lambda_i^{I_i^{U*}}\Big)\Big)\Big)\\
&= A(a^*).
\end{aligned}$$

Therefore, $\text{INWEA}(a_1, a_2, \cdots, a_n) = \text{INWEA}(a_1^*, a_2^*, \cdots, a_n^*)$.

Based on (1) and (2), there is $\text{INWEA}(a_1, a_2, \cdots, a_n) \leq \text{INWEA}(a_1^*, a_2^*, \cdots, a_n^*)$. □

5. Multiple Attribute Decision Making Method Based on the INWEA Operator

To better understand the new operational law and the new operational aggregation operator, we will address some MADM problems, where the attribute weights will be expressed as INNs, and the attribute values for alternatives are represented as positive real numbers. So, we establish a MADM method.

In MADM problems, let $X = \{x_1, x_2, \cdots x_m\}$ be a discrete set of m alternatives, and $C = \{c_1, c_2, \cdots c_n\}$ be the set of n attributes. The evaluation values of attribute $c_j (j = 1, 2, \cdots, n)$ for alternative $x_i (i = 1, 2, \cdots, m)$ is expressed by a positive real number $\lambda_{ij} \in (0, 1), (i = 1, 2, \cdots, m, j = 1, 2, \cdots, n)$. So, the decision matrix $R = (\lambda_{ij})_{m\times n}$ can be given. The INN $a_j =< [T_j^L, T_j^U], [I_j^L, I_j^U], [F_j^L, F_j^U] >$ is represented as the attribute weight of the $c_j (j = 1, 2, \cdots, n)$, here $[T_j^L, T_j^U] \subseteq [0, 1]$ indicates the degree of certainty of the attribute c_j supported by the experts, $[I_j^L, I_j^U] \subseteq [0, 1]$ indicates the degree of uncertainty of the attribute c_j supported by the experts, and $[F_j^L, F_j^U] \subseteq [0, 1]$ indicates the negative degree of the attribute c_j supported by the experts. Then, we can rank the alternatives and obtain the best alternatives based on the given information; the specific steps are as follows:

Step 1 Utilize the INWEA operator $d_i = INWEA(a_1, a_2, \cdots, a_m)$ $(i = 1, 2, \cdots, m; j = 1, 2, \cdots n)$ to aggregate the characteristic λ_{ij} of the alternative x_i.

Step 2 Utilize the score function to calculate the scores $S(d_i)$ $(i = 1, 2, \cdots, m)$ of the alternatives x_i $(i = 1, 2, \cdots, m)$.

Step 3 Utilize the scores $S(d_i)$ $(i = 1, 2, \cdots, m)$ to rank and select the alternatives x_i $(i = 1, 2, \cdots, n)$, if the two scores $S(d_i)$ and $S(d_j)$ are equal, then we need to calculate the accuracy degrees $A(d_i)$ and $A(d_j)$ of the overall criteria values d_i and d_j, then we rank the alternatives x_i and x_j by using $A(d_i)$ and $A(d_j)$.

Step 4 End.

6. Typhoon Disaster Evaluation Based on Neutrosophic Information

6.1. Illustrative Example

In China, typhoons are among the most serious types of natural disasters. They primarily impact the eastern coastal regions of China, where the population is extremely dense, the economy is highly developed, and social wealth is notably concentrated. Fujian Province is one of the most severely impacted typhoon disaster areas in both local and global contexts, routinely enduring substantial economic losses caused by typhoon disasters. For example, in 2017, a total of 208,900 people in 59 counties of Fujian Province were affected by the successive landings of twin typhoons No. 9 "Nassa" and No. 10 "Haicang." There were 434 collapsed houses and 273,300 people were urgently displaced; 26.73 thousand hectares of crops were affected, 101.9 thousand hectares affected, and 2.19 thousand hectares were lost. The typhoon also led to the cancellation of 507 Fujian flights and 139 trains. According to incomplete statistics, the total direct economic loss was 966 million yuan (RNB). We examine the problem of typhoon disaster evaluation in Fujian Province.

We will use several indices to evaluate the typhoon disaster effectively. The assessment indicators $C = \{c_1, c_2, c_3, c_4\}$ include economic loss c_1, social impact c_2, environmental damage c_3, and other impact c_4 proposed by Yu [1]. Several experts are responsible for this assessment, and the evaluation information is expressed by positive real numbers and INNs. The assessment decision matrix based on this is constructed $R = (\lambda_{ij})_{m\times n}$(see Table 1), and the λ_{ij} is positive real numbers. The λ_{ij} in the matrix indicates the degree of damage to the city in the typhoon. The data between 0 and 1 is used to indicate the degree of disaster received. 0 means that the city is basically unaffected by disasters, 0.2 means that the extent of the disaster is relatively small, 0.4 means that the extent of the disaster is middle, 0.6 means that the degree of disaster is slightly larger, 0.8 means the extent of the disaster is relatively large. 1 means that the extent of the disaster is extremely large. The rest of the data located in the middle of the two data indicates that the extent of the disaster is between the two. The interval neutrosophic weights $\omega_1, \omega_2, \omega_3, \omega_4$ for the four attributes voted by experts. Take ω_1 as an example to explain its meaning, $[0.6, 0.8]$ indicates the degree of certainty of the attribute c_1 supported by the experts is between 0.6 and 0.8, $[0.2, 0.4]$ indicates the uncertainty of the expert's support for attribute c_1 is between 0.2 and 0.4, and $[0.1, 0.2]$ indicates the negative degree of expert's support for attribute c_1 is between 0.1 and 0.2.

$$\omega_1 =< [0.6,0.8],[0.2,0.4],[0.1,0.2] >, \quad \omega_2 =< [0.5,0.9],[0.2,0.5],[0.1,0.3] >$$
$$\omega_3 =< [0.4,0.7],[0.3,0.6],[0.3,0.5] >, \quad \omega_4 =< [0.2,0.4],[0.4,0.8],[0.6,0.7] > .$$

Table 1. Decision matrix.

Attributes / Cities	c_1	c_2	c_3	c_4
Nanping (NP)	0.2	0.2	0.2	0.2
Ningde (ND)	0.9	0.8	0.7	0.4
Sanming (SM)	0.2	0.2	0.2	0.2
Fuzhou (FZ)	0.8	0.5	0.5	0.3
Putian (PT)	0.7	0.7	0.6	0.3
Longyan (LY)	0.4	0.3	0.3	0.2
Quanzhou (QZ)	0.3	0.4	0.2	0.3
Xiamen (XM)	0.3	0.3	0.3	0.2
Zhangzhou (ZZ)	0.6	0.5	0.8	0.3

According to Section 5, Typhoon disaster evaluation using the MADM model contains the following steps:

Step 1 Using the $INWEA$ operator defined by equation (10) to aggregate all evaluation information to obtain a comprehensive assessment value d_i for each city as follows:

When $i = 2$, we can get

$$\begin{aligned}
d_2^{ND} &= INWEA(\alpha_1, \alpha_2, \cdots, \alpha_n) \\
&= \left\langle \left[\prod_{i=1}^{4} \lambda_i^{1-T_i^L}, \prod_{i=1}^{4} \lambda_i^{1-T_i^U}\right], \left[1 - \prod_{i=1}^{4} \lambda_i^{I_i^L}, 1 - \prod_{i=1}^{4} \lambda_i^{I_i^U}\right], \left[1 - \prod_{i=1}^{4} \lambda_i^{F_i^L}, 1 - \prod_{i=1}^{4} \lambda_i^{F_i^U}\right] \right\rangle \\
&=< \left[0.9^{(1-0.6)} \times 0.8^{(1-0.5)} \times 0.7^{(1-0.4)} \times 0.4^{(1-0.2)}, 0.9^{(1-0.8)} \times 0.8^{(1-0.9)} \times 0.7^{(1-0.7)} \times 0.4^{(1-0.4)}\right], \\
&\quad \left[1 - 0.9^{0.2} \times 0.8^{0.2} \times 0.7^{0.3} \times 0.4^{0.4}, 1 - 0.9^{0.4} \times 0.8^{0.5} \times 0.7^{0.6} \times 0.4^{0.8}\right], \\
&\quad \left[1 - 0.9^{0.1} \times 0.8^{0.1} \times 0.7^{0.3} \times 0.4^{0.6}, 1 - 0.9^{0.2} \times 0.8^{0.3} \times 0.7^{0.5} \times 0.4^{0.7}\right], \\
&=< [0.333, 0.497], [0.417, 0.667], [0.498, 0.597] >
\end{aligned}$$

In a similar way, we can get

$d_1^{NP} = < [0.025, 0.145], [0.830, 0.975], [0.830, 0.935] >$,

$d_2^{ND} = < [0.333, 0.497], [0.417, 0.667], [0.498, 0.597] >$,

$d_3^{SM} = < [0.025, 0.145], [0.830, 0.975], [0.830, 0.935] >$,

$d_4^{FZ} = < [0.163, 0.352], [0.582, 0.837], [0.640, 0.764] >$,

$d_5^{PT} = < [0.204, 0.374], [0.540, 0.796], [0.612, 0.721] >$,

$d_6^{LY} = < [0.051, 0.196], [0.760, 0.949], [0.785, 0.897] >$,

$d_7^{QZ} = < [0.057, 0.215], [0.751, 0.943], [0.758, 0.885] >$,

$d_8^{XM} = < [0.045, 0.185], [0.774, 0.955], [0.791, 0.903] >$,

$d_9^{ZZ} = < [0.192, 0.383], [0.546, 0.808], [0.597, 0.718] >$.

Step 2 Using Definition 4 to calculate the score function value of the comprehensive assessment value d_i for each city as follows:

$S(d_1^{NP}) = -0.801$, $S(d_2^{ND}) = -0.109$, $S(d_3^{SM}) = -0.801$, $S(d_4^{FZ}) = -0.432$,

$S(d_5^{PT}) = -0.357$, $S(d_6^{LY}) = -0.714$, $S\left(d_7^{QZ}\right) = -0.690$, $S(d_8^{XM}) = -0.730$,

$S(d_9^{ZZ}) = -0.360$.

Step 3 According to Definition 6, the ranking order of the nine cities is $d_2^{ND} \succ d_5^{PT} \succ d_9^{ZZ} \succ d_4^{FZ} \succ d_7^{QZ} \succ d_6^{LY} \succ d_8^{XM} \succ d_3^{SM} \sim d_1^{NP}$. The ranking results of the cities are shown in Figure 1.

Step 4 End.

6.2. Comparative Analysis Based on Different Sorting Methods

To illustrate the stability of the ranking results, the degree of possibility-based ranking method proposed in [33,45] is used in this paper. We obtain the matrix of degrees of possibility of the comprehensive assessment values of nine cities as follows:

$$P = \begin{array}{c} \\ NP \\ ND \\ SM \\ FZ \\ PT \\ LY \\ QZ \\ XM \\ ZZ \end{array} \begin{array}{c} \begin{array}{ccccccccc} NP & ND & SM & FZ & PT & LY & QZ & XM & ZZ \end{array} \\ \left[\begin{array}{ccccccccc} 0.500 & 0.000 & 0.500 & 0.000 & 0.000 & 0.344 & 0.302 & 0.371 & 0.000 \\ 1.000 & 0.500 & 1.000 & 0.944 & 0.854 & 1.000 & 1.000 & 1.000 & 0.843 \\ 0.500 & 0.000 & 0.500 & 0.000 & 0.000 & 0.344 & 0.302 & 0.371 & 0.000 \\ 1.000 & 0.056 & 1.000 & 0.500 & 0.402 & 0.913 & 0.862 & 0.943 & 0.406 \\ 1.000 & 0.146 & 1.000 & 0.598 & 0.500 & 1.000 & 0.980 & 1.000 & 0.501 \\ 0.656 & 0.000 & 0.656 & 0.087 & 0.000 & 0.500 & 0.456 & 0.527 & 0.000 \\ 0.698 & 0.000 & 0.698 & 0.138 & 0.020 & 0.544 & 0.500 & 0.570 & 0.038 \\ 0.629 & 0.000 & 0.629 & 0.057 & 0.000 & 0.473 & 0.430 & 0.500 & 0.000 \\ 1.000 & 0.157 & 1.000 & 0.594 & 0.499 & 1.000 & 0.962 & 0.100 & 0.500 \end{array}\right] \end{array}$$

Here, $ND \succ PT \succ ZZ \succ FZ \succ QZ \succ LY \succ XM \succ SM \sim NP$. The ranking order of the nine cities is also $d_2^{ND} \succ d_5^{PT} \succ d_9^{ZZ} \succ d_4^{FZ} \succ d_7^{QZ} \succ d_6^{LY} \succ d_8^{XM} \succ d_3^{SM} \sim d_1^{NP}$. The ranking results of the cities are shown in Figure 2. As can be seen from the above results, the two sorting results are the same.

6.3. Comparative Analysis of Different Aggregation Operators

In order to illustrate the rationality and predominance of the proposed method, we compare this method with other methods [33]. The comparative analysis is shown in Table 2 and Figure 3.

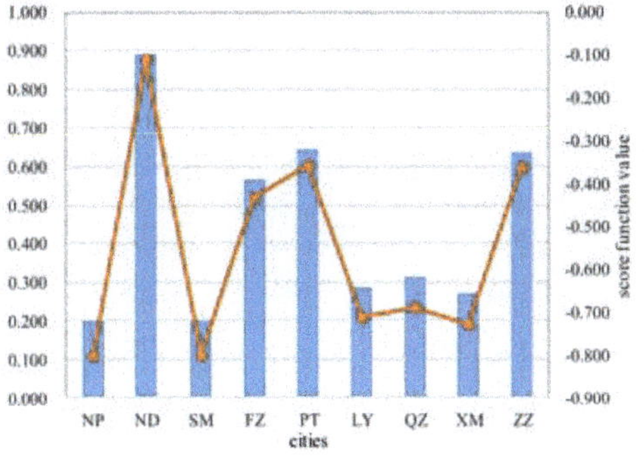

Figure 1. Ranking results based on the score function.

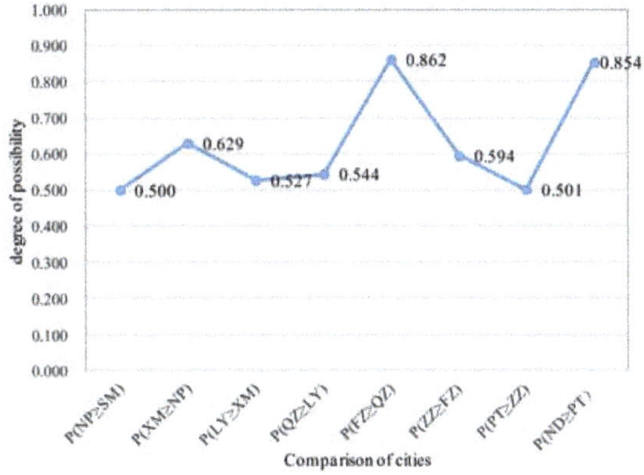

Figure 2. Ranking results based on the degree of possibility.

Table 2. Comparative analysis of different aggregation operators.

Different Aggregation Method	Ranking Result
INWEA operator of our method	$d_2^{ND} \succ d_5^{PT} \succ d_9^{ZZ} \succ d_4^{FZ} \succ d_7^{QZ} \succ d_6^{LY} \succ d_8^{XM} \succ d_3^{SM} \sim d_1^{NP}$
INNWA operator of [33]	$d_2^{ND} \succ d_5^{PT} \succ d_4^{FZ} \succ d_9^{ZZ} \succ d_6^{LY} \succ d_7^{QZ} \succ d_8^{XM} \succ d_3^{SM} \sim d_1^{NP}$

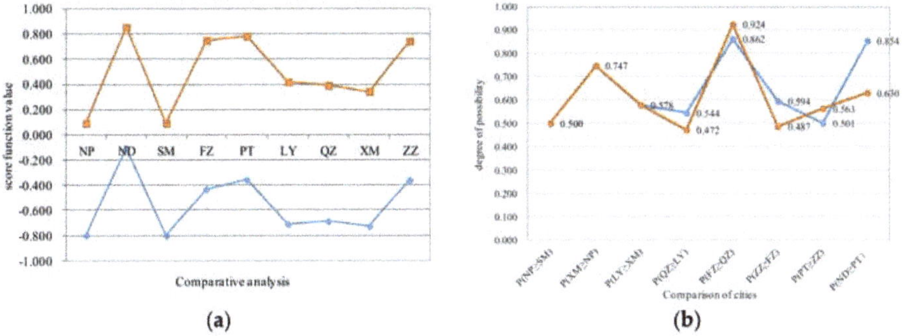

Figure 3. Comparative analysis of different aggregation operators. (**a**) Ranking results of two operators based on the score function; (**b**) Ranking results of two operators based on the possibility degree.

First in Step 1, using the $INNWA$ operator proposed by [33] instead of the $INWEA$ operator to aggregate all evaluation information to obtain a comprehensive assessment value d_i for each city, then using Definition 4 to calculate the score function value of d_i as follows:

$$S(d_1^{NP}) = 0.092,\ S(d_2^{ND}) = 0.853,\ S(d_3^{SM}) = 0.092,\ S(d_4^{FZ}) = 0.750,\ S(d_5^{PT}) = 0.782,\ S(d_6^{LY}) = 0.418,$$
$$S\left(d_7^{QZ}\right) = 0.389,\ S(d_8^{XM}) = 0.340,\ S(d_9^{ZZ}) = 0.742.$$

Here, we compare and analyze several aggregation methods to illustrate the advantages of the proposed method.

(1) Can be seen from Table 2 and Figure 3, the two ranking results based on the $INWEA$ operator and the $INNWA$ operator are different. The main reason is that the positions and meanings of the attribute values and the attribute weights are different. For the $INWEA$ operator, its bases are positive real numbers and the exponents are interval neutrosophic numbers. It can deal with the decision making problem, in which attribute values are positive real numbers, and the attribute weights are interval neutrosophic numbers. However, the $INNWA$ operator is just the opposite. It needs to exchange the roles of the attribute values and the attribute weights because its bases are interval neutrosophic numbers and its exponents are positive real numbers. Therefore, it cannot be used to solve the typhoon disaster assessment problem in this paper, and the second ranking results in Table 2 and Figure 3 are unreasonable.

(2) Compared with the existing $SVNWAA$ operator introduced in an SVNN environment [41], our method is a more generalized representation, and the $SVNWAA$ operator is a special case. When the upper limit and lower limit of the INNs are the same, the $INWEA$ operator is equivalent to the $SVNWEA$ operator.

(3) Compared with the existing $IIFWEA$ operator of IIFNs [40] and the $IFWEA$ operator of IFNs [39], our method uses interval neutrosophic weights, which include truth degree, falsity degree, and indeterminacy degree, and can deal with the indeterminate, incomplete, and inconsistent problems. However, the $IIFWEA$ operator and $IFWEA$ operator use intuitionistic fuzzy weights, which only contain truth degree and falsity degree, and cannot handle the assessment problem in this paper. Since IFN and IIFN are only special cases of interval NN, our exponential aggregation operator is the extension of the existing exponential operators [39–41].

7. Conclusions

In this paper, a typhoon disaster evaluation approach based on exponential aggregation operators of interval neutrosophic numbers under the neutrosophic fuzzy environment, is proposed. First, this paper introduced the exponential operational laws of INSs and INNs, which are a useful supplement to the existing neutrosophic fuzzy aggregation techniques. Then, we investigated a series of properties of these operational laws. Next, we discussed in detail some favorable properties of the interval neutrosophic weighted exponential aggregation (INWEA) operator. Finally, we applied the proposed decision making method successfully to the evaluation of typhoon disaster assessment. The research in this paper will be helpful to deepen the study of typhoon disaster evaluation and improve decision making for disaster reduction and disaster prevention. In addition, it provides methodological guidance for the handling of typhoon disasters and can improve the government's ability to effectively improve disaster reduction. In future research, we will expand the proposed method and apply it to other natural disaster assessment problems. We will continue to study related theories of exponential aggregation operators in a neutrosophic fuzzy environment and their application in typhoon disaster assessment. The authors will also study the related theory of single-valued neutrosophic sets, interval-valued neutrosophic sets, bipolar neutrosophic sets, neutrosophic hesitant fuzzy sets, multi-valued neutrosophic sets, simplified neutrosophic linguistic sets, and their applications in typhoon disaster evaluation problems.

Author Contributions: All three authors contribute to this article, and the specific contribution is as follows: the idea and mathematical model were put forward by R.T., she also wrote the paper. W.Z. analyzed the existing work of the research problem and collected relevant data. Submission and review of the paper are the responsibility of S.C.

Funding: This research was funded by [Fujian Federation Of Social Science Circles] grant number [FJ2016C028], [The Education Department Of Fujian Province] grant numbers [JAT160556,JAT170113], [National Planning Office of Philosophy and Social Science] grant number [17CGL058], [Ministry of Education of the People's Republic of China] grant number [16YJC630008], [Fuzhou University] grant number [BPZD1601].

Acknowledgments: This research is supported by the Fujian Province Social Science Planning Project of China (No. FJ2016C028), Education and Scientific Research Projects of Young and Middle-aged Teachers of Fujian Province (No. JAT160556, JAT170113), Projects of the National Social Science Foundation of China (No. 17CGL058), Social Sciences Foundation of the Ministry of Education of China (No. 16YJC630008), and Fuzhou University Funding Project (No. BPZD1601).

Conflicts of Interest: The authors declare no conflicts of interest.

References

1. Yu, D.J. Intuitionistic fuzzy theory based typhoon disaster evaluation in Zhejiang Province, China: A comparative Perspective. *Nat. Hazards* **2015**, *75*, 2559–2576. [CrossRef]
2. Knutson, T.R.; Tuleya, R.E.; Kurihara, Y. Simulated increase of hurricane intensities in a CO_2-warmed climate. *Science* **1998**, *279*, 1018–1020. [CrossRef] [PubMed]
3. Zhang, Y.; Fan, G.F.; He, Y.; Cao, L.J. Risk assessment of typhoon disaster for the Yangtze River Delta of China. *Geomat. Nat. Hazards Risk* **2017**, *8*, 1580–1591. [CrossRef]
4. Wang, G.; Chen, R.; Chen, J. Direct and indirect economic loss assessment of typhoon disasters based on EC and IO joint model. *Nat. Hazards* **2017**, *87*, 1751–1764. [CrossRef]
5. Zhang, J.; Huang, H.; Che, X.; Zhang, M. Evaluation of Typhoon Disaster Losses of Hainan Island Rubber Plantation. In Proceedings of the 7th Annual Meeting of Risk Analysis Council of China Association for Disaster Prevention, Changsha, China, 4–6 November 2016; pp. 251–256.
6. Lou, W.P.; Chen, H.Y.; Qiu, X.F.; Tang, Q.Y.; Zheng, F. Assessment of economic losses from tropical cyclone disasters based on PCA-BP. *Nat. Hazards* **2012**, *60*, 819–829. [CrossRef]
7. Lu, C.L.; Chen, S.H. Multiple linear interdependent models (Mlim) applied to typhoon data from China. *Theor. Appl. Climatol.* **1998**, *61*, 143–149.
8. Yang, A.; Sui, G.; Tang, D.; Lin, J.; Chen, H. An intelligent decision support system for typhoon disaster management. *J. Comput. Inf. Syst.* **2012**, *8*, 6705–6712.
9. Li, C.; Zhang, L.; Zeng, S. Typhoon disaster evaluation based on extension of intuitionistic fuzzy TOPSIS method in Zhejiang Province of China. *Int. J. Earth Sci. Eng.* **2015**, *8*, 1031–1035.
10. Ma, Q. A Fuzzy Synthetic Evaluation Model for Typhoon Disaster. *Meteorol. Mon.* **2008**, *34*, 20–25. (In Chinese)
11. Chen, S.H.; Liu, X.Q. Typhoon disaster evaluation model based on discrete Hopfield neural network. *J. Nat. Disasters* **2011**, *20*, 47–52.
12. He, X.R. Typhoon disaster assessment based on Dombi hesitant fuzzy information aggregation operators. *Nat. Hazards* **2018**, *90*, 1153–1175. [CrossRef]
13. Smarandache, F. *Neutrosophy: Neutrosophic Probability, Set, and Logic*; American Research Press: Rehoboth, DE, USA, 1998.
14. Smarandache, F. Neutrosophic set—A generalization of intuitionistic fuzzy set. *IEEE Int. Conf. Granul. Comput.* **2006**, *1*, 38–42.
15. Zadeh, L.A. Fuzzy sets. *Inf. Control* **1965**, *8*, 338–353. [CrossRef]
16. Atanassov, K.T. Intuitionistic fuzzy sets. *Fuzzy Sets Syst.* **1986**, *20*, 87–96. [CrossRef]
17. Wang, H.B.; Smarandache, F.; Zhang, Y.Q. Single valued neutrosophic sets. *Multisp. Multistruct.* **2010**, *4*, 410–413.
18. Ye, J. A multicriteria decision-making method using aggregation operators for simplified neutrosophic sets. *Int. J. Fuzzy Syst.* **2014**, *26*, 2459–2466.

19. Wang, J.J.; Li, X.E. TODIM method with multi-valued neutrosophic sets. *Control Decis.* **2015**, *30*, 1139–1142. (In Chinese)
20. Peng, J.J.; Wang, J.Q. Multi-valued Neutrosophic Sets and its Application in Multi-criteria Decision-making Problems. *Neutrosophic Sets Syst.* **2015**, *10*, 3–17. [CrossRef]
21. Wang, H.; Smarandache, F.; Zhang, Y.Q.; Sunderraman, R. *Interval Neutrosophic Sets and Logic: Theory and Applications in Computing*; Hexis: Phoenix, AZ, USA, 2005.
22. Zhang, H.Y.; Ji, P.J.; Wang, J.Q.; Chen, X.H. Improved Weighted Correlation Coefficient based on Integrated Weight for Interval Neutrosophic Sets and its Application in Multi-criteria Decision Making Problems. *Int. J. Comput. Intell. Syst.* **2015**, *8*, 1027–1043. [CrossRef]
23. Ye, J. Multiple-attribute decision-making method under a single-valued neutrosophic hesitant fuzzy environment. *J. Intell. Syst.* **2014**, *24*, 23–36. [CrossRef]
24. Tian, Z.P.; Wang, J.; Zhang, H.Y.; Chen, X.H.; Wang, J.Q. Simplified neutrosophic linguistic normalized weighted Bonferroni mean operator and its application to multi-criteria decision-making problems. *Filomat* **2015**, *30*, 3339–3360. [CrossRef]
25. Biswas, P.; Pramanik, S.; Giri, B.C. Cosine Similarity Measure Based Multi-attribute Decision-making with Trapezoidal Fuzzy Neutrosophic Numbers. *Neutrosophic Sets Syst.* **2014**, *8*, 46–56.
26. Ye, J. Trapezoidal neutrosophic set and its applicaion to multiple attribute decision-making. *Neural Comput. Appl.* **2015**, *26*, 1157–1166. [CrossRef]
27. Deli, I. Interval-valued neutrosophic soft sets and its decision making. *Int. J. Mach. Learn. Cybern.* **2017**, *8*, 665–676. [CrossRef]
28. Broumi, S.; Talea, M.; Bakali, A.; Smarandache, F. Single Valued Neutrosophic Graphs. *J. New Theory* **2016**, *10*, 86–101.
29. Broumi, S.; Smarandache, F.; Talea, M.; Bakali, A. An Introduction to Bipolar Single Valued Neutrosophic Graph Theory. *Appl. Mech. Mater.* **2016**, *841*, 184–191. [CrossRef]
30. Broumi, S.; Talea, M.; Bakali, A.; Smarandache, F. ON Strong Interval Valued Neutrosophic graphs. *Crit. Rev.* **2016**, *XII*, 49–71.
31. Liu, C.F.; Luo, Y.S. Correlated aggregation operators for simplified neutrosophic set and their application in multi-attribute group decision making. *J. Intell. Fuzzy Syst.* **2016**, *30*, 1755–1761. [CrossRef]
32. Ye, J. Multiple attribute decision-making method based on the possibility degree ranking method and ordered weighted aggregation operators of interval neutrosophic numbers. *J. Intell. Fuzzy Syst.* **2015**, *28*, 1307–1317.
33. Zhang, H.Y.; Wang, J.Q.; Chen, X.H. Interval neutrosophic sets and their application in multicriteria decision making problems. *Sci. World J.* **2014**, *2014*, 645653. [CrossRef] [PubMed]
34. Liu, P.D.; Wang, Y.M. Multiple attribute decision-making method based on single-valued neutrosophic normalized weighted Bonferroni mean. *Neural Comput. Appl.* **2014**, *25*, 2001–2010. [CrossRef]
35. Ye, J. Multiple attribute group decision making based on interval neutrosophic uncertain linguistic variables. *Int. J. Mach. Learn. Cybern.* **2017**, *8*, 837–848. [CrossRef]
36. Peng, J.J.; Wang, J.Q.; Wu, X.H.; Wang, J.; Chen, X.H. Multi-valued Neutrosophic Sets and Power Aggregation Operators with Their Applications in Multi-criteria Group Decision-making Problems. *Int. J. Comput. Intell. Syst.* **2015**, *8*, 345–363. [CrossRef]
37. Tan, R.; Zhang, W. Multiple attribute group decision making methods based on trapezoidal fuzzy neutrosophic numbers. *J. Intell. Fuzzy Syst.* **2017**, *33*, 2547–2564. [CrossRef]
38. Şahin, R. Normal neutrosophic multiple attribute decision making based on generalized prioritized aggregation operators. *Neural Comput. Appl.* **2017**, 1–21. [CrossRef]
39. Gou, X.J.; Xu, Z.S.; Lei, Q. New operational laws and aggregation method of intuitionistic fuzzy information. *J. Intell. Fuzzy Syst.* **2016**, *30*, 129–141. [CrossRef]
40. Gou, X.J.; Xu, Z.S.; Lei, Q. Exponential operations of interval-valued intuitionistic fuzzy numbers. *Int. J. Mach. Learn. Cybern.* **2016**, *7*, 501–518. [CrossRef]
41. Lu, Z.; Ye, J. Exponential Operations and an Aggregation Method for Single-Valued Neutrosophic Numbers in Decision Making. *Information* **2017**, *8*, 62. [CrossRef]

42. Şahin, R.; Liu, P. Some approaches to multi criteria decision making based on exponential operations of simplified neutrosophic numbers. *J. Intell. Fuzzy Syst.* **2017**, *32*, 2083–2099. [CrossRef]
43. Ye, J. Exponential operations and aggregation operators of interval neutrosophic sets and their decision making methods. *Springerplus* **2016**, *5*, 1488.
44. Şahin, R. Multi-criteria neutrosophic decision making method based on score and accuracy functions under neutrosophic environment. *Comput. Sci.* **2014**, 1–9. [CrossRef]
45. Xu, Z. On method for uncertain multiple attribute decision making problems with uncertain multiplicative preference information on alternatives. *Fuzzy Optim. Decis. Mak.* **2005**, *4*, 131–139. [CrossRef]

symmetry

MDPI

Article

Multi-Criteria Decision-Making Method Based on Simplified Neutrosophic Linguistic Information with Cloud Model

Jian-Qiang Wang [1], Chu-Quan Tian [1], Xu Zhang [1], Hong-Yu Zhang [1,*] and Tie-Li Wang [2,*]

1 School of Business, Central South University, Changsha 410083, China; jqwang@csu.edu.cn (J.-Q.W.); chuquantian@163.com (C.-Q.T.); qing6707@126.com (X.Z.)

2 Management School, University of South China, Hengyang 421001, China

* Correspondence: Hyzhang@csu.edu.cn (H.-Y.Z.); wangtieli@usc.edu.cn (T.-L.W.)

Received: 25 April 2018; Accepted: 29 May 2018; Published: 1 June 2018

Abstract: This study introduces simplified neutrosophic linguistic numbers (SNLNs) to describe online consumer reviews in an appropriate manner. Considering the defects of studies on SNLNs in handling linguistic information, the cloud model is used to convert linguistic terms in SNLNs to three numerical characteristics. Then, a novel simplified neutrosophic cloud (SNC) concept is presented, and its operations and distance are defined. Next, a series of simplified neutrosophic cloud aggregation operators are investigated, including the simplified neutrosophic clouds Maclaurin symmetric mean (SNCMSM) operator, weighted SNCMSM operator, and generalized weighted SNCMSM operator. Subsequently, a multi-criteria decision-making (MCDM) model is constructed based on the proposed aggregation operators. Finally, a hotel selection problem is presented to verify the effectiveness and validity of our developed approach.

Keywords: simplified neutrosophic linguistic numbers; cloud model; Maclaurin symmetric mean; multi-criteria decision-making

1. Introduction

Nowadays, multi-criteria decision-making (MCDM) problems are attracting more and more attention. Lots of studies suggest that it is difficult to describe decision information completely because the information is usually inconsistent and indeterminate in real-life problems. To address this issue, Smarandache [1] put forward neutrosophic sets (NSs). Now, NSs have been applied to many fields and extended to various forms. Wang et al. [2] presented the concept of single-valued neutrosophic sets (SVNSs) and demonstrated its application, Ye [3] proposed several kinds of projection measures of SVNSs, and Ji et al. [4] proposed Bonferroni mean aggregation operators of SVNSs. Wang et al. [5] used interval numbers to extend SVNSs, and proposed the interval-valued neutrosophic set (IVNS). Ye [6] introduced trapezoidal neutrosophic sets (TrNSs), and proposed a series of trapezoidal neutrosophic aggregation operators. Liang et al. [7] introduced the preference relations into TrNSs. Peng et al. [8] combined the probability distribution with NSs to propose the probability multi-valued neutrosophic sets. Wu et al. [9] further extended this set to probability hesitant interval neutrosophic sets. All of the aforementioned sets are the descriptive tools of quantitative information.

Zhang et al. [10] proposed a method of using NSs to describe online reviews posted by consumers. For example, a consumer evaluates a hotel with the expressions: 'the location is good', 'the service is neither good nor bad', and 'the room is in a mess'. Obviously, there is active, neutral, and passive information in this review. According to the NS theory, such review information can be characterized by employing truth, neutrality, and falsity degrees. This information presentation method has been proved to be feasible [11]. However, in practical online reviews, the consumer usually gives a comprehensive evaluation before posting the text reviews. NSs can describe the text reviews, but they cannot represent the comprehensive evaluation. To deal with this issue, many scholars have studied the combination of NSs and linguistic term sets [12,13]. The semantic of linguistic term set provides precedence on a qualitative level, and such precedence is more sensitive for decision-makers than a common ranking due to the expression of absolute benchmarks [14–16]. Based on the concepts of NSs and linguistic term sets, Ye [17] proposed interval neutrosophic linguistic sets (INLSs) and interval neutrosophic linguistic numbers (INLNs). Then, many interval neutrosophic linguistic MCDM approaches were developed [18,19]. Subsequently, Tian et al. [20] introduced the concepts of simplified neutrosophic linguistic sets (SNLSs) and simplified neutrosophic linguistic numbers (SNLNs). Wang et al. [21] proposed a series of simplified neutrosophic linguistic Maclaurin symmetric mean aggregation operators and developed a MCDM method. The existed studies on SNLNs simply used the linguistic functions to deal with linguistic variables in SNLNs. This strategy is simple, but it cannot effectively deal with qualitative information because it ignores the randomness of linguistic variables.

The cloud model is originally proposed by Li [22] in the light of probability theory and fuzzy set theory. It characterizes the randomness and fuzziness of a qualitative concept rely on three numerical characters and makes the conversion between qualitative concepts and quantitative values becomes effective. Since the introduction of the cloud model, many scholars have conducted lots of studies and applied it to various fields [23–25], such as hotel selection [26], data detection [27], and online recommendation algorithms [28]. Currently, the cloud model is considered as the best way to handle linguistic information and it is used to handle multiple qualitative decision-making problems [29–31], such as linguistic intuitionistic problems [32] and Z-numbers problems [33]. Considering the effectiveness of the cloud model in handling qualitative information, we utilize the cloud model to deal with linguistic terms in SNLNs. In this way, we propose a new concept by combining SNLNs and cloud model to solve real-life problems.

The aggregation operator is one of the most important tool of MCDM method [34–37]. Maclaurin symmetric mean (MSM) operator, defined by Maclaurin [38], possess the prominent advantage of summarizing the interrelations among input variables lying between the maximum value and minimum value. The MSM operator can not only take relationships among criteria into account, but it can also improve the flexibility of aggregation operators in application by adding parameters. Since the MSM operator was proposed, it has been expanded to various fuzzy sets [39–43]. For example, Liu and Zhang [44] proposed many MSM operators to deal with single-valued trapezoidal neutrosophic information, Ju et al. [45] proposed a series of intuitionistic linguistic MSM aggregation operators, and Yu et al. [46] proposed the hesitant fuzzy linguistic weighted MSM operator.

From the above analysis, the motivation of this paper is presented as follows:

1. The cloud model is a reliable tool for dealing with linguistic information, and it has been successfully applied to handle multifarious linguistic problems, such as probabilistic linguistic decision-making problems. The existing studies have already proved the effectiveness and feasibility of using the cloud model to process linguistic information. In view of this, this paper introduces the cloud model to process linguistic evaluation information involved in SNLNs.
2. As an efficient and applicable aggregation operator, MSM not only takes into account the correlation among criteria, but also adjusts the scope of the operator through the transformation of parameters. Therefore, this paper aims to accommodate the MSM operator to simplified neutrosophic linguistic information environments.

The remainder of this paper is organized as follows. Some basic definitions are introduced in Section 2. In Section 3, we propose a new concept of SNCs and the corresponding operations and distance. In Section 4, we propose some simplified neutrosophic cloud aggregation operators. In Section 5, we put forward a MCDM approach in line with the proposed operators. Then, in Section 6, we provide a practical example concerning hotel selection to verify the validity of the developed method. In Section 7, a conclusion is presented.

2. Preliminaries

This section briefly reviews some basic concepts, including linguistic term sets, linguistic scale function, NSs, SNSs, and cloud model, which will be employed in the subsequent analyses.

2.1. Linguistic Term Sets and Linguistic Scale Function

Definition 1 ([47]). *Let $H = \{h_\tau | \tau = 1, 2, \cdots, 2t+1, t \in N^*\}$ be a finite and totally ordered discrete term set, where N^* is a set of positive integers, and h_τ is interpreted as the representation of a linguistic variable. Then, the following properties should be satisfied:*

(1) *The linguistic term set is ordered: $h_\tau < h_\upsilon$ if and only if $\tau < \upsilon$, where $(h_\tau, h_\upsilon \in H)$;*
(2) *If a negation operator exists, then $neg(h_\tau) = h_{(2t+1-\tau)}$ $(\tau, \upsilon = 1, 2, \cdots, 2t+1)$.*

Definition 2 ([48]). *Let $h_\tau \in H$ be a linguistic term. If $\theta_\tau \in [0,1]$ is a numerical value, then the linguistic scale function f that conducts the mapping from h_τ to θ_τ $(\tau = 1, 2, \cdots, 2t+1)$ can be defined as*

$$f : s_\tau \to \theta_\tau \ (\tau = 1, 2, \cdots, 2t+1), \tag{1}$$

where $0 \le \theta_1 < \theta_2 < \cdots < \theta_{2t+1} \le 1$.

Based on the existed studies, three types of linguistic scale functions are described as

$$f_1(h_x) = \theta_x = \frac{x}{2t}, \quad (x = 1, 2, \cdots, 2t+1), \quad \theta_x \in [0,1]; \tag{2}$$

$$f_2(h_y) = \theta_y = \begin{cases} \frac{\alpha^t - \alpha^{t-y}}{2\alpha^t - 2}, & (y = 1, 2, \cdots, t+1), \\ \frac{\alpha^t + \alpha^{y-t} - 2}{2\alpha^t - 2}, & (y = t+2, t+3, \cdots, 2t+1); \end{cases} \tag{3}$$

$$f_3(h_z) = \theta_z = \begin{cases} \frac{t^\beta - (t-z)^\beta}{2t^\beta}, & (z = 1, \cdots, t+1), \\ \frac{t^\gamma + (z-t)^\gamma}{2t^\gamma}, & (z = t+2, \cdots, 2t+1). \end{cases} \tag{4}$$

2.2. SNSs and SNLSs

Definition 3 ([1]). *Let X be a space of points (objects), and x be a generic element in X. A NS A in X is characterized by a truth-membership function $T_A(x)$, a indeterminacy-membership function $I_A(x)$, and a falsity-membership function $F_A(x)$. $T_A(x)$, $I_A(x)$, and $F_A(x)$ are real standard or nonstandard subsets $]0^-, 1^+[$. That is, $T_A(x) : x \to]0^-, 1^+[$, $I_A(x) : x \to]0^-, 1^+[$, and $F_A(x) : x \to]0^-, 1^+[$. There is no restriction on the sum of $T_A(x)$, $I_A(x)$, and $F_A(x)$, so $0^- \le \sup T_A(x) + \sup I_A(x) + \sup F_A(x) \le 3^+$.*

In fact, NSs are very difficult for application without specification. Given this, Ye [34] introduced SNSs by reducing the non-standard intervals of NSs into a kind of standard intervals.

Definition 4 ([17]). *Let X be a space of points with a generic element x. Then, an SNS B in X can be defined as $B = \{(x, T_B(x), I_B(x), F_B(x)) | x \in X\}$, where $T_B(x) : X \to [0,1]$, $I_B(x) : X \to [0,1]$, and $F_B(x) : X \to [0,1]$. In addition, the sum of $T_B(x)$, $I_B(x)$, and $F_B(x)$ satisfies $0 \le T_B(x) + I_B(x) + F_B(x) \le 3$. For simplicity, B can be denoted as $B = \langle T_B(x), I_B(x), F_B(x) \rangle$, which is a subclass of NSs.*

Definition 5 ([20]). *Let X be a space of points with a generic element x, and $H = \{h_\tau | \tau = 1, 2, \cdots, 2t+1, t \in N^*\}$ be a linguistic term set. Then an SNLS C in X is defined as $C = \{\langle x, h_C(x), (T_C(x), I_C(x), F_C(x)) \rangle | x \in X\}$, where $h_C(x) \in H$, $T_C(x) \in [0,1]$, $I_C(x) \in [0,1]$,*

$F_C(x) \in [0,1]$ and $0 \leq T_C(x) + I_C(x) + F_C(x) \leq 3$ for any $x \in X$. In addition, $T_C(x)$, $I_C(x)$, and $F_C(x)$ represent the degree of truth-membership, indeterminacy-membership, and falsity-membership of the element x in X to the linguistic term $h_C(x)$, respectively. For simplicity, a SNLN is expressed as $\langle h_C(x), (T_C(x), I_C(x), F_C(x))\rangle$.

2.3. The Cloud Model

Definition 6 ([22]). *Let U be a universe of discourse and T be a qualitative concept in U. $x \in U$ is a random instantiation of the concept T, and x satisfies $x \sim N\left(Ex, (En^*)^2\right)$, where $En^* \sim N(En, He^2)$, and the degree of certainty that x belongs to the concept T is defined as*

$$\mu = e^{-\frac{(x-Ex)^2}{2(En^*)^2}},$$

then the distribution of x in the universe U is called a normal cloud, and the cloud C is presented as $C = (Ex, En, He)$.

Definition 7 ([33]). *Let $M(Ex_1, En_1, He_1)$ and $N(Ex_2, En_2, He_2)$ be two clouds, then the operations between them are defined as*

(1) $M + N = \left(Ex_1 + Ex_2, \sqrt{En_1{}^2 + En_2{}^2}, \sqrt{He_1{}^2 + He_2{}^2}\right)$;

(2) $M - N = \left(Ex_1 - Ex_2, \sqrt{En_1{}^2 + En_2{}^2}, \sqrt{He_1{}^2 + He_2{}^2}\right)$;

(3) $M \times N = \left(Ex_1 Ex_2, \sqrt{(En_1 Ex_2)^2 + (En_2 Ex_1)^2}, \sqrt{(He_1 Ex_2)^2 + (He_2 Ex_1)^2}\right)$;

(4) $\lambda M = \left(\lambda Ex_1, \sqrt{\lambda} En_1, \sqrt{\lambda} He_1\right)$; *and*

(5) $M^\lambda = \left(Ex_1{}^\lambda, \sqrt{\lambda} Ex_1{}^{\lambda-1} En_1, \sqrt{\lambda} Ex_1{}^{\lambda-1} He_1\right)$.

2.4. Transformation Approach of Clouds

Definition 8 ([33]). *Let H_i be a linguistic term in $H = \{H_i | i = 1, 2, \ldots, 2t+1\}$, and f be a linguistic scale function. Then, the procedures for converting linguistic variables to clouds are presented below.*

(1) *Calculate θ_i: Map H_i to θ_i employing Equation (2) or (3) or (4).*

(2) *Calculate Ex_i: $Ex_i = X_{\min} + \theta_i (X_{\max} - X_{\min})$.*

(3) *Calculate En_i: Let (x, y) be a cloud droplet. Since $x \sim N(Ex_i, En_i'^2)$, we have $3En_i' = \max\{X_{\max} - Ex_i, Ex_i - X_{\min}\}$ in the light of 3σ principle of the normal distribution curve. Then, $En_i' = \begin{cases} \frac{(1-\theta_i)(X_{\max} - X_{\min})}{3} & 1 \leq i \leq t+1 \\ \frac{\theta_i (X_{\max} - X_{\min})}{3} & t+2 \leq i \leq 2t+1 \end{cases}$. Thus $En_i = \frac{En_{i-1}' + En_i' + En_{i+1}'}{3}$, $(1 < i < 2t+1)$, $En_i = \frac{En_i' + En_{i+1}'}{2}$, $(i = 1)$ and $En_i = \frac{En_{i-1}' + En_i'}{2}$, $(i = 2t+1)$ can be obtained.*

(4) *Calculate He_i: $He_i = \frac{(En'^{+} - En_i)}{3}$, where $En'^{+} = \max\{En_i'\}$.*

3. Simplified Neutrosophic Clouds and the Related Concepts

Based on SNLNs and the cloud transformation method, a novel concept of SNCs is proposed. Motivated by the existing studies, we provide the operations and comparison method for SNCs and investigate the distance measurement of SNCs.

3.1. SNCs and Their Operational Rules

Definition 9. *Let X be a space of points with a generic element x, $H = \{h_\tau | \tau = 1, 2, \cdots, 2t+1, t \in N^*\}$ be a linguistic term set, and $\langle h_C(x), (T_C(x), I_C(x), F_C(x))\rangle$ be a SNLN. In accordance with the cloud conversion method described in Section 2.4, the linguistic term $h_C(x) \in H$ can be converted into the cloud $\langle Ex, En, He\rangle$. Then, a simplified neutrosophic cloud (SNC) is defined as*

$$Y = (\langle Ex, En, He\rangle, \langle T, I, F\rangle)$$

Definition 10. *Let $a = \langle (Ex_1, En_1, He_1), (T_1, I_1, F_1)\rangle$ and $b = \langle (Ex_2, En_2, He_2), (T_2, I_2, F_2)\rangle$ be two SNCs, then the operations of SNC are defined as*

(1) $$a \oplus b = \left(\left\langle Ex_1 + Ex_2, \sqrt{En_1^2 + En_2^2}, \sqrt{He_1^2 + He_2^2}\right\rangle, \left\langle \frac{T_1(Ex_1+En_1^2+He_1^2)+T_2(Ex_2+En_2^2+He_2^2)}{Ex_1+Ex_2+En_1^2+He_1^2+En_2^2+He_2^2}, \frac{I_1(Ex_1+En_1^2+He_1^2)+I_2(Ex_2+En_2^2+He_2^2)}{Ex_1+Ex_2+En_1^2+He_1^2+En_2^2+He_2^2}, \frac{F_1(Ex_1+En_1^2+He_1^2)+F_2(Ex_2+En_2^2+He_2^2)}{Ex_1+Ex_2+En_1^2+He_1^2+En_2^2+He_2^2}\right\rangle\right);$$

(2) $a \otimes b = (\langle Ex_1 Ex_2, En_1 En_2, He_1 He_2\rangle, \langle\langle T_1 T_2, I_1 + I_2 - I_1 I_2, F_1 + F_2 - F_1 F_2\rangle)$;

(3) $\lambda a = \left(\left\langle \lambda Ex_1, \sqrt{\lambda} En_1, \sqrt{\lambda} He_1\right\rangle, \langle T_1, I_1, F_1\rangle\right)$; *and*

(4) $a^\lambda = \left(\langle Ex_1{}^\lambda, En_1{}^\lambda, He_1{}^\lambda\rangle, \left\langle T_1{}^\lambda, 1-(1-I_1)^\lambda, 1-(1-F_1)^\lambda\right\rangle\right)$.

Theorem 1. *Let $a = \langle (Ex_1, En_1, He_1), (T_1, I_1, F_1)\rangle$, $b = \langle (Ex_2, En_2, He_2), (T_2, I_2, F_2)\rangle$ and $c = \langle (Ex_3, En_3, He_3), (T_3, I_3, F_3)\rangle$ be three SNCs. Then, the following properties should be satisfied*

(1) $a + b = b + a$;
(2) $(a + b) + c = a + (b + c)$;
(3) $\lambda a + \lambda b = \lambda(a + b)$;
(4) $\lambda_1 a + \lambda_2 a = (\lambda_1 + \lambda_2)a$;
(5) $a \times b = b \times a$;
(6) $(a \times b) \times c = a \times (b \times c)$;
(7) $a^{\lambda_1} \times a^{\lambda_2} = a^{\lambda_1 + \lambda_2}$;
(8) $(a \times b)^\lambda = a^\lambda \times b^\lambda$.

3.2. Distance for SNCs

Definition 11. *Let $a = \langle (Ex_1, En_1, He_1), (T_1, I_1, F_1)\rangle$ and $b = \langle (Ex_2, En_2, He_2), (T_2, I_2, F_2)\rangle$ be two SNCs, then the generalized distance between a and b is defined as*

$$d(a,b) = |(1-\beta_1)Ex_1 - (1-\beta_2)Ex_2| + \left(\tfrac{1}{3}\left(|(1-\beta_1)Ex_1 T_1 - (1-\beta_2)Ex_2 T_2|^\lambda + |(1-\beta_1)Ex_1(1-I_1) - (1-\beta_2)Ex_2(1-I_2)|^\lambda + |(1-\beta_1)Ex_1(1-F_1) - (1-\beta_2)Ex_2(1-F_2)|^\lambda\right)\right)^{\frac{1}{\lambda}}, \tag{5}$$

where $\beta_1 = \frac{\sqrt{En_1^2+He_1^2}}{\sqrt{En_1^2+He_1^2}+\sqrt{En_2^2+He_2^2}}$ and $\beta_2 = \frac{\sqrt{En_2^2+He_2^2}}{\sqrt{En_1^2+He_1^2}+\sqrt{En_2^2+He_2^2}}$. When $\lambda = 1$ and 2, the generalized distance above becomes the Hamming distance and the Euclidean distance, respectively.

Theorem 2. *Let $a = \langle (Ex_1, En_1, He_1), (T_1, I_1, F_1)\rangle$, $b = \langle (Ex_2, En_2, He_2), (T_2, I_2, F_2)\rangle$, and $c = \langle (Ex_3, En_3, He_3), (T_3, I_3, F_3)\rangle$ be three SNCs. Then, the distance given in Definition 11 satisfies the following properties:*

(1) $d(a,b) \geq 0$;
(2) $d(a,b) = d(b,a)$; *and*

(3) *If* $Ex_1 \leq Ex_2 \leq Ex_3$, $En_1 \geq En_2 \geq En_3, He_1 \geq He_2 \geq He_3, T_1 \leq T_2 \leq T_3, I_1 \geq I_2 \geq I_3$, *and* $F_1 \geq F_2 \geq F_3$, *then* $d(a,b) \leq d(a,c)$, *and* $d(b,c) \leq d(a,c)$.

Proof. It is easy to prove that (1) and (2) in Theorem 2 are true. The proof of (3) in Theorem 2 is depicted in the following.

Let $\beta_{(a,b)1} = \frac{\sqrt{En_1{}^2+He_1{}^2}}{\sqrt{En_1{}^2+He_1{}^2}+\sqrt{En_2{}^2+He_2{}^2}}$, $\beta_{(a,b)2} = \frac{\sqrt{En_2{}^2+He_2{}^2}}{\sqrt{En_1{}^2+He_1{}^2}+\sqrt{En_2{}^2+He_2{}^2}}$, $\beta_{(a,c)1} = \frac{\sqrt{En_1{}^2+He_1{}^2}}{\sqrt{En_1{}^2+He_1{}^2}+\sqrt{En_3{}^2+He_3{}^2}}$, and $\beta_{(a,c)2} = \frac{\sqrt{En_3{}^2+He_3{}^2}}{\sqrt{En_1{}^2+He_1{}^2}+\sqrt{En_3{}^2+He_3{}^2}}$, then there are

$$\begin{aligned}
d(a,c) = &\left|\left(1-\beta_{(a,c)1}\right)Ex_1 - \left(1-\beta_{(a,c)2}\right)Ex_3\right| \\
&+\left(\tfrac{1}{3}\left(\left|\left(1-\beta_{(a,c)1}\right)Ex_1T_1 - \left(1-\beta_{(a,c)2}\right)Ex_3T_3\right|^\lambda\right.\right. \\
&+\left|\left(1-\beta_{(a,c)1}\right)Ex_1(1-I_1) - \left(1-\beta_{(a,c)2}\right)Ex_3(1-I_3)\right|^\lambda \\
&\left.\left.+\left|\left(1-\beta_{(a,c)1}\right)Ex_1(1-F_1) - \left(1-\beta_{(a,c)2}\right)Ex_3(1-F_3)\right|^\lambda\right)\right)^{\frac{1}{\lambda}},
\end{aligned}$$

$$\begin{aligned}
d(a,b) = &\left|\left(1-\beta_{(a,b)1}\right)Ex_1 - \left(1-\beta_{(a,b)2}\right)Ex_2\right| \\
&+\left(\tfrac{1}{3}\left(\left|\left(1-\beta_{(a,b)1}\right)Ex_1T_1 - \left(1-\beta_{(a,b)2}\right)Ex_2T_2\right|^\lambda + \left|\left(1-\beta_{(a,b)1}\right)\right.\right.\right. \\
&\left.+Ex_1(1-I_1) - \left(1-\beta_{(a,b)2}\right)Ex_2(1-I_2)\right|^\lambda \\
&\left.\left.+\left|\left(1-\beta_{(a,b)1}\right)Ex_1(1-F_1) - \left(1-\beta_{(a,b)2}\right)Ex_2(1-F_2)\right|^\lambda\right)\right)^{\frac{1}{\lambda}}.
\end{aligned}$$

Thus, we have

$$\begin{aligned}
d(a,c)-d(a,b) = &\left(1-\beta_{(a,b)1}\right)Ex_1 - \left(1-\beta_{(a,c)1}\right)Ex_1 \\
&+\left(1-\beta_{(a,c)2}\right)Ex_3 - \left(1-\beta_{(a,b)2}\right)Ex_2 \\
&+\left(\tfrac{1}{3}\left(\left|\left(1-\beta_{(a,c)1}\right)Ex_1T_1 - \left(1-\beta_{(a,c)2}\right)Ex_3T_3\right|^\lambda\right.\right. \\
&+\left|\left(1-\beta_{(a,c)1}\right)Ex_1(1-I_1) - \left(1-\beta_{(a,c)2}\right)Ex_3(1-I_3)\right|^\lambda \\
&\left.\left.+\left|\left(1-\beta_{(a,c)1}\right)Ex_1(1-F_1) - \left(1-\beta_{(a,c)2}\right)Ex_3(1-F_3)\right|^\lambda\right)\right)^{\frac{1}{\lambda}} \\
&-\left(\tfrac{1}{3}\left(\left|\left(1-\beta_{(a,b)1}\right)Ex_1T_1 - \left(1-\beta_{(a,b)2}\right)Ex_2T_2\right|^\lambda\right.\right. \\
&+\left|\left(1-\beta_{(a,b)1}\right)Ex_1(1-I_1) - \left(1-\beta_{(a,b)2}\right)Ex_2(1-I_2)\right|^\lambda \\
&\left.\left.+\left|\left(1-\beta_{(a,b)1}\right)Ex_1(1-F_1) - \left(1-\beta_{(a,b)2}\right)Ex_2(1-F_2)\right|^\lambda\right)\right)^{\frac{1}{\lambda}}.
\end{aligned}$$

Let

$$\begin{aligned}
p = &\left(1-\beta_{(a,b)1}\right)Ex_1 - \left(1-\beta_{(a,c)1}\right)Ex_1 + \left(1-\beta_{(a,c)2}\right)Ex_3 - \left(1-\beta_{(a,b)2}\right)Ex_2 \\
= &\left(1-\frac{\sqrt{En_1{}^2+He_1{}^2}}{\sqrt{En_1{}^2+He_1{}^2}+\sqrt{En_2{}^2+He_2{}^2}}\right)Ex_1 - \left(1-\frac{\sqrt{En_1{}^2+He_1{}^2}}{\sqrt{En_1{}^2+He_1{}^2}+\sqrt{En_3{}^2+He_3{}^2}}\right)Ex_1 \\
&+\left(1-\frac{\sqrt{En_3{}^2+He_3{}^2}}{\sqrt{En_1{}^2+He_1{}^2}+\sqrt{En_3{}^2+He_3{}^2}}\right)Ex_3 - \left(1-\frac{\sqrt{En_2{}^2+He_2{}^2}}{\sqrt{En_1{}^2+He_1{}^2}+\sqrt{En_2{}^2+He_2{}^2}}\right)Ex_2.
\end{aligned}$$

$$
\begin{aligned}
q = &\left(\tfrac{1}{3}\left(\left|\left(1-\beta_{(a,c)1}\right)Ex_1T_1-\left(1-\beta_{(a,c)2}\right)Ex_3T_3\right|^\lambda\right.\right.\\
&+\left|\left(1-\beta_{(a,c)1}\right)Ex_1(1-I_1)-\left(1-\beta_{(a,c)2}\right)Ex_3(1-I_3)\right|^\lambda\\
&\left.\left.+\left|\left(1-\beta_{(a,c)1}\right)Ex_1(1-F_1)-\left(1-\beta_{(a,c)2}\right)Ex_3(1-F_3)\right|^\lambda\right)\right)^{\frac{1}{\lambda}}\\
&-\left(\tfrac{1}{3}\left(\left|\left(1-\beta_{(a,b)1}\right)Ex_1T_1-\left(1-\beta_{(a,b)2}\right)Ex_2T_2\right|^\lambda\right.\right.\\
&+\left|\left(1-\beta_{(a,b)1}\right)Ex_1(1-I_1)-\left(1-\beta_{(a,b)2}\right)Ex_2(1-I_2)\right|^\lambda\\
&\left.\left.+\left|\left(1-\beta_{(a,b)1}\right)Ex_1(1-F_1)-\left(1-\beta_{(a,b)2}\right)Ex_2(1-F_2)\right|^\lambda\right)\right)^{\frac{1}{\lambda}},
\end{aligned}
$$

then $d(a,c)-d(a,b)=p+q$.

Simplifying the above equations, the following results can be obtained.

$$
\begin{aligned}
p = &\frac{\sqrt{En_2{}^2+He_2{}^2}}{\sqrt{En_1{}^2+He_1{}^2}+\sqrt{En_2{}^2+He_2{}^2}}Ex_1-\frac{\sqrt{En_3{}^2+He_3{}^2}}{\sqrt{En_1{}^2+He_1{}^2}+\sqrt{En_3{}^2+He_3{}^2}}Ex_1\\
&+\frac{\sqrt{En_1{}^2+He_1{}^2}}{\sqrt{En_1{}^2+He_1{}^2}+\sqrt{En_3{}^2+He_3{}^2}}Ex_3-\frac{\sqrt{En_1{}^2+He_1{}^2}}{\sqrt{En_1{}^2+He_1{}^2}+\sqrt{En_2{}^2+He_2{}^2}}Ex_2.
\end{aligned}
$$

Since $Ex_1 \le Ex_2 \le Ex_3$, $En_1 \ge En_2 \ge En_3$, and $He_1 \ge He_2 \ge He_3$, we have

$$
\frac{\sqrt{En_2{}^2+He_2{}^2}}{\sqrt{En_1{}^2+He_1{}^2}+\sqrt{En_2{}^2+He_2{}^2}}Ex_1-\frac{\sqrt{En_3{}^2+He_3{}^2}}{\sqrt{En_1{}^2+He_1{}^2}+\sqrt{En_3{}^2+He_3{}^2}}Ex_1 \ge 0,
$$

$$
\frac{\sqrt{En_1{}^2+He_1{}^2}}{\sqrt{En_1{}^2+He_1{}^2}+\sqrt{En_3{}^2+He_3{}^2}}Ex_3-\frac{\sqrt{En_1{}^2+He_1{}^2}}{\sqrt{En_1{}^2+He_1{}^2}+\sqrt{En_2{}^2+He_2{}^2}}Ex_2 \ge 0.
$$

Thus, $p \ge 0$ is determined.

According to $p = \left|\left(1-\beta_{(a,c)1}\right)Ex_1-\left(1-\beta_{(a,c)2}\right)Ex_3\right|-\left|\left(1-\beta_{(a,b)1}\right)Ex_1-\left(1-\beta_{(a,b)2}\right)Ex_2\right| \ge 0$, the following inequalities can be deduced.

$$
\left|\left(1-\beta_{(a,c)1}\right)Ex_1-\left(1-\beta_{(a,c)2}\right)Ex_3\right| \ge \left|\left(1-\beta_{(a,b)1}\right)Ex_1-\left(1-\beta_{(a,b)2}\right)Ex_2\right|,
$$

$$
\left|\left(1-\beta_{(a,c)1}\right)Ex_1-\left(1-\beta_{(a,c)2}\right)Ex_3\right|^\lambda \ge \left|\left(1-\beta_{(a,b)1}\right)Ex_1-\left(1-\beta_{(a,b)2}\right)Ex_2\right|^\lambda.
$$

Since $T_1 \le T_2 \le T_3$, the following inequality is true.

$$
\left|\left(1-\beta_{(a,c)1}\right)Ex_1T_1-\left(1-\beta_{(a,c)2}\right)Ex_3T_3\right|^\lambda \ge \left|\left(1-\beta_{(a,b)1}\right)Ex_1T_1-\left(1-\beta_{(a,b)2}\right)Ex_2T_2\right|^\lambda.
$$

In a similar manner, we can also obtain

$$
\left|\left(1-\beta_{(a,c)1}\right)Ex_1(1-I_1)-\left(1-\beta_{(a,c)2}\right)Ex_3(1-I_3)\right|^\lambda \ge \left|\left(1-\beta_{(a,b)1}\right)Ex_1(1-I_1)-\left(1-\beta_{(a,b)2}\right)Ex_2(1-I_2)\right|^\lambda,
$$

$$
\left|\left(1-\beta_{(a,c)1}\right)Ex_1(1-F_1)-\left(1-\beta_{(a,c)2}\right)Ex_3(1-F_3)\right|^\lambda \ge \left|\left(1-\beta_{(a,b)1}\right)Ex_1(1-F_1)-\left(1-\beta_{(a,b)2}\right)Ex_2(1-F_2)\right|^\lambda.
$$

Thus, there is

$$
\begin{aligned}
q = &\left(\tfrac{1}{3}\left(\left|\left(1-\beta_{(a,c)1}\right)Ex_1T_1-\left(1-\beta_{(a,c)2}\right)Ex_3T_3\right|^{\lambda}\right.\right.\\
&+\left|\left(1-\beta_{(a,c)1}\right)Ex_1(1-I_1)-\left(1-\beta_{(a,c)2}\right)Ex_3(1-I_3)\right|^{\lambda}\\
&\left.\left.+\left|\left(1-\beta_{(a,c)1}\right)Ex_1(1-F_1)-\left(1-\beta_{(a,c)2}\right)Ex_3(1-F_3)\right|^{\lambda}\right)\right)^{\frac{1}{\lambda}}\\
&-\left(\tfrac{1}{3}\left(\left|\left(1-\beta_{(a,b)1}\right)Ex_1T_1-\left(1-\beta_{(a,b)2}\right)Ex_2T_2\right|^{\lambda}\right.\right.\\
&+\left|\left(1-\beta_{(a,b)1}\right)Ex_1(1-I_1)-\left(1-\beta_{(a,b)2}\right)Ex_2(1-I_2)\right|^{\lambda}\\
&\left.\left.+\left|\left(1-\beta_{(a,b)1}\right)Ex_1(1-F_1)-\left(1-\beta_{(a,b)2}\right)Ex_2(1-F_2)\right|^{\lambda}\right)\right)^{\frac{1}{\lambda}}\\
&\geq 0.
\end{aligned}
$$

Thus, $d(a,c)-d(a,b)\geq 0 \Rightarrow d(a,c)\geq d(a,b)$. The inequality $d(a,c)\geq d(b,c)$ can be proved similarly. Hence, the proof of Theorem 2 is completed. □

Example 1. *Let* $a=\langle(0.5,0.2,0.1),(0.7,0.3,0.5)\rangle$, *and* $b=\langle(0.6,0.1,0.1),(0.8,0.2,0.4)\rangle$ *be two SNCs. Then, according to Definition 11, the Hamming distance* $d_{Hamming}(a,b)$ *and Euclidean distance* $d_{Euclidean}(a,b)$ *are calculated as*

$$d_{Hamming}(a,b)=0.4304, \text{ and } d_{Euclidean}(a,b)=0.3224.$$

4. SNCs Aggregation Operators

Maclaurin [38] introduced the MSM aggregation operator firstly. In this section, the MSM operator is expanded to process SNC information, and the SNCMSM operator and the weighted SNCMSM operator are then proposed.

Definition 12 ([38]). *Let* x_i $(i=1,2,\cdots,n)$ *be the set of nonnegative real numbers. A MSM aggregation operator of dimension n is mapping* $MSM^{(m)}:(R^+)^n\rightarrow R^+$, *and it can be defined as*

$$MSM^{(m)}(x_1,x_2,\cdots,x_n)=\left(\frac{\sum\limits_{1\leq i_1<\cdots<i_m\leq n}\prod\limits_{j=1}^{m}x_{i_j}}{C_n^m}\right)^{\frac{1}{m}}, \tag{6}$$

where $(i_1,i_2,\cdots,i_m)$ *traverses all the m-tuple combination of* $(i=1,2,\cdots,n)$, $C_n^m=\frac{n!}{m!(n-m)!}$ *is the binomial coefficient. In the subsequent analysis, assume that* $i_1<i_2<,\ldots,<i_m$. *In addition,* x_{i_j} *refers to the* i_j *th element in a particular arrangement.*

It is clear that $MSM^{(m)}$ *has the following properties:*

(1) *Idempotency. If* $x\geq 0$ *and* $x_i=x$ *for all i, then* $MSM^{(m)}(x,x,\ldots,x)=x$.
(2) *Monotonicity. If* $x_i\leq y_i$, *for all i,* $MSM^{(m)}(x_1,x_2,\ldots,x_n)\leq MSM^{(m)}(y_1,y_2,\ldots,y_n)$, *where* x_i *and* y_i *are nonnegative real numbers.*
(3) *Boundedness.* $MIN\{x_1,x_2,\ldots,x_n\}\leq MSM^{(m)}(x_1,x_2,\ldots,x_n)\leq MAX\{x_1,x_2,\ldots,x_n\}$.

4.1. SNCMSM Operator

In this subsection, the traditional $MSM^{(m)}$ operator is extended to accommodate the situations where the input variables are made up of SNCs. Then, the SNCMSM operator is developed.

Definition 13. *Let* $a_i = \langle (Ex_i, En_i, He_i), (T_i, I_i, F_i) \rangle (i = 1, 2, ..., n)$ *be a collection of SNCs. Then, the SNCMSM operator can be defined as*

$$SNCMSM^{(m)}(a_1, a_2, \cdots, a_n) = \left(\frac{\bigoplus_{1 \le i_1 < \cdots < i_m \le n} \left(\bigotimes_{j=1}^{m} a_{i_j} \right)}{C_n^m} \right)^{\frac{1}{m}}, \tag{7}$$

where $m = 1, 2, ..., n$ *and* $(i_1, i_2, \cdots, i_m)$ *traverses all the m-tuple combination of* $(i = 1, 2, \cdots, n)$, $C_n^m = \frac{n!}{m!(n-m)!}$ *is the binomial coefficient.*

In light of the operations of SNCs depicted in Definition 10, Theorem 3 can be acquired.

Theorem 3. *Let* $a_i = \langle (Ex_i, En_i, He_i), (T_i, I_i, F_i) \rangle (i = 1, 2, ..., n)$ *be a collection of SNCs, the aggregated value acquired by the SNCMSM operator is also a SNC and can be expressed as*

$$\begin{aligned}
&SNCMSM^{(m)}(a_1, a_2, \cdots, a_n) \\
&= \left(\left\langle \left(\frac{\sum_{k=1}^{C_n^m} \prod_{j=1}^{m} Ex_{i_j^{(k)}}}{C_n^m} \right)^{\frac{1}{m}}, \left(\frac{\sqrt{\sum_{k=1}^{C_n^m} \left(\prod_{j=1}^{m} En_{i_j^{(k)}} \right)^2}}{\sqrt{C_n^m}} \right)^{\frac{1}{m}}, \left(\frac{\sqrt{\sum_{k=1}^{C_n^m} \left(\prod_{j=1}^{m} He_{i_j^{(k)}} \right)^2}}{\sqrt{C_n^m}} \right)^{\frac{1}{m}} \right\rangle, \right. \\
&\left\langle \left(\frac{\sum_{k=1}^{C_n^m} \left(\prod_{j=1}^{m} T_{i_j^k} \left(\prod_{j=1}^{m} Ex_{i_j^{(k)}} + \left(\prod_{j=1}^{m} En_{i_j^{(k)}} \right)^2 + \left(\prod_{j=1}^{m} He_{i_j^{(k)}} \right)^2 \right) \right)}{\sum_{k=1}^{C_n^m} \left(\prod_{j=1}^{m} Ex_{i_j^{(k)}} + \left(\prod_{j=1}^{m} En_{i_j^{(k)}} \right)^2 + \left(\prod_{j=1}^{m} He_{i_j^{(k)}} \right)^2 \right)} \right)^{\frac{1}{m}}, \right. \\
&1 - \left(1 - \frac{\sum_{k=1}^{C_n^m} \left(\left(1 - \prod_{j=1}^{m} \left(1 - I_{i_j^k} \right) \right) \left(\prod_{j=1}^{m} Ex_{i_j^{(k)}} + \left(\prod_{j=1}^{m} En_{i_j^{(k)}} \right)^2 + \left(\prod_{j=1}^{m} He_{i_j^{(k)}} \right)^2 \right) \right)}{\sum_{k=1}^{C_n^m} \left(\prod_{j=1}^{m} Ex_{i_j^{(k)}} + \left(\prod_{j=1}^{m} En_{i_j^{(k)}} \right)^2 + \left(\prod_{j=1}^{m} He_{i_j^{(k)}} \right)^2 \right)} \right)^{\frac{1}{m}}, \\
&\left. \left. 1 - \left(1 - \frac{\sum_{k=1}^{C_n^m} \left(\left(1 - \prod_{j=1}^{m} \left(1 - F_{i_j^k} \right) \right) \left(\prod_{j=1}^{m} Ex_{i_j^{(k)}} + \left(\prod_{j=1}^{m} En_{i_j^{(k)}} \right)^2 + \left(\prod_{j=1}^{m} He_{i_j^{(k)}} \right)^2 \right) \right)}{\sum_{k=1}^{C_n^m} \left(\prod_{j=1}^{m} Ex_{i_j^{(k)}} + \left(\prod_{j=1}^{m} En_{i_j^{(k)}} \right)^2 + \left(\prod_{j=1}^{m} He_{i_j^{(k)}} \right)^2 \right)} \right)^{\frac{1}{m}} \right\rangle \right).
\end{aligned} \tag{8}$$

Proof.

$$a_{i_j^{(k)}} = \left(\left\langle Ex_{i_j^{(k)}}, En_{i_j^{(k)}}, He_{i_j^{(k)}} \right\rangle, \left\langle T_{i_j^{(k)}}, I_{i_j^{(k)}}, F_{i_j^{(k)}} \right\rangle \right), ((j = 1, 2, ..., m).$$

$$\begin{aligned}
\Rightarrow \bigotimes_{j=1}^{m} a_{i_j^{(k)}} = &\left(\left\langle \prod_{j=1}^{m} Ex_{i_j^{(k)}}, \prod_{j=1}^{m} En_{i_j^{(k)}}, \prod_{j=1}^{m} He_{i_j^{(k)}} \right\rangle, \right. \\
&\left. \left\langle \prod_{j=1}^{m} T_{i_j^{(k)}}, 1 - \prod_{j=1}^{m} \left(1 - I_{i_j^{(k)}} \right), 1 - \prod_{j=1}^{m} \left(1 - F_{i_j^{(k)}} \right) \right\rangle \right)
\end{aligned}$$

$$\begin{aligned}
\Rightarrow \bigoplus_{1 \le t_1 < \cdots < t_m \le n} \left(\bigotimes_{j=1}^{m} a_{i_j} \right) = &\left(\left\langle \sum_{k=1}^{C_n^m} \prod_{j=1}^{m} Ex_{i_j^{(k)}}, \sqrt{\sum_{k=1}^{C_n^m} \left(\prod_{j=1}^{m} En_{i_j^{(k)}} \right)^2}, \sqrt{\sum_{k=1}^{C_n^m} \left(\prod_{j=1}^{m} He_{i_j^{(k)}} \right)^2} \right\rangle, \right. \\
&\left\langle \frac{\sum_{k=1}^{C_n^m} \left(\prod_{j=1}^{m} T_{i_j^k} \left(\prod_{j=1}^{m} Ex_{i_j^{(k)}} + \left(\prod_{j=1}^{m} En_{i_j^{(k)}} \right)^2 + \left(\prod_{j=1}^{m} He_{i_j^{(k)}} \right)^2 \right) \right)}{\sum_{k=1}^{C_n^m} \left(\prod_{j=1}^{m} Ex_{i_j^{(k)}} + \left(\prod_{j=1}^{m} En_{i_j^{(k)}} \right)^2 + \left(\prod_{j=1}^{m} He_{i_j^{(k)}} \right)^2 \right)}, \right.
\end{aligned}$$

$$\frac{\sum_{k=1}^{C_n^m}\left(\left(1-\prod_{j=1}^{m}\left(1-I_{i_j^k}\right)\right)\left(\prod_{j=1}^{m}Ex_{i_j^{(k)}}+\left(\prod_{j=1}^{m}En_{i_j^{(k)}}\right)^2+\left(\prod_{j=1}^{m}He_{i_j^{(k)}}\right)^2\right)\right)}{\sum_{k=1}^{C_n^m}\left(\prod_{j=1}^{m}Ex_{i_j^{(k)}}+\left(\prod_{j=1}^{m}En_{i_j^{(k)}}\right)^2+\left(\prod_{j=1}^{m}He_{i_j^{(k)}}\right)^2\right)},$$

$$\left.\left.\frac{\sum_{k=1}^{C_n^m}\left(\left(1-\prod_{j=1}^{m}\left(1-F_{i_j^k}\right)\right)\left(\prod_{j=1}^{m}Ex_{i_j^{(k)}}+\left(\prod_{j=1}^{m}En_{i_j^{(k)}}\right)^2+\left(\prod_{j=1}^{m}He_{i_j^{(k)}}\right)^2\right)\right)}{\sum_{k=1}^{C_n^m}\left(\prod_{j=1}^{m}Ex_{i_j^{(k)}}+\left(\prod_{j=1}^{m}En_{i_j^{(k)}}\right)^2+\left(\prod_{j=1}^{m}He_{i_j^{(k)}}\right)^2\right)}\right\rangle\right)$$

$$\Rightarrow\left(\frac{\underset{1\le i_1<\cdots<i_m\le n}{\oplus}\left(\underset{j=1}{\overset{m}{\otimes}}\left(a_{i_j}\right)\right)}{C_n^m}\right)^{\frac{1}{m}}$$

$$=\left(\left\langle\left(\frac{\sum_{k=1}^{C_n^m}\prod_{j=1}^{m}Ex_{i_j^{(k)}}}{C_n^m}\right)^{\frac{1}{m}},\left(\frac{\sqrt{\sum_{k=1}^{C_n^m}\left(\prod_{j=1}^{m}En_{i_j^{(k)}}\right)^2}}{\sqrt{C_n^m}}\right)^{\frac{1}{m}},\left(\frac{\sqrt{\sum_{k=1}^{C_n^m}\left(\prod_{j=1}^{m}He_{i_j^{(k)}}\right)^2}}{\sqrt{C_n^m}}\right)^{\frac{1}{m}}\right\rangle,\right.$$

$$\left\langle\left(\frac{\sum_{k=1}^{C_n^m}\left(\prod_{j=1}^{m}T_{i_j^k}\left(\prod_{j=1}^{m}Ex_{i_j^{(k)}}+\left(\prod_{j=1}^{m}En_{i_j^{(k)}}\right)^2+\left(\prod_{j=1}^{m}He_{i_j^{(k)}}\right)^2\right)\right)}{\sum_{k=1}^{C_n^m}\left(\prod_{j=1}^{m}Ex_{i_j^{(k)}}+\left(\prod_{j=1}^{m}En_{i_j^{(k)}}\right)^2+\left(\prod_{j=1}^{m}He_{i_j^{(k)}}\right)^2\right)}\right)^{\frac{1}{m}},\right.$$

$$1-\left(1-\frac{\sum_{k=1}^{C_n^m}\left(\left(1-\prod_{j=1}^{m}\left(1-I_{i_j^k}\right)\right)\left(\prod_{j=1}^{m}Ex_{i_j^{(k)}}+\left(\prod_{j=1}^{m}En_{i_j^{(k)}}\right)^2+\left(\prod_{j=1}^{m}He_{i_j^{(k)}}\right)^2\right)\right)}{\sum_{k=1}^{C_n^m}\left(\prod_{j=1}^{m}Ex_{i_j^{(k)}}+\left(\prod_{j=1}^{m}En_{i_j^{(k)}}\right)^2+\left(\prod_{j=1}^{m}He_{i_j^{(k)}}\right)^2\right)}\right)^{\frac{1}{m}},$$

$$\left.\left.1-\left(1-\frac{\sum_{k=1}^{C_n^m}\left(\left(1-\prod_{j=1}^{m}\left(1-F_{i_j^k}\right)\right)\left(\prod_{j=1}^{m}Ex_{i_j^{(k)}}+\left(\prod_{j=1}^{m}En_{i_j^{(k)}}\right)^2+\left(\prod_{j=1}^{m}He_{i_j^{(k)}}\right)^2\right)\right)}{\sum_{k=1}^{C_n^m}\left(\prod_{j=1}^{m}Ex_{i_j^{(k)}}+\left(\prod_{j=1}^{m}En_{i_j^{(k)}}\right)^2+\left(\prod_{j=1}^{m}He_{i_j^{(k)}}\right)^2\right)}\right)^{\frac{1}{m}}\right\rangle\right).$$

The proof of Theorem 3 is completed. □

Theorem 4. *(Idempotency) If* $a_i = a = (\langle Ex_a, En_a, He_a\rangle, \langle T_a, I_a, F_a\rangle)$ *for all* $i = 1, 2, \ldots, n$, *then* $SNCMSM^{(m)}(a, a, \cdots, a) = a = (\langle Ex_a, En_a, He_a\rangle, \langle T_a, I_a, F_a\rangle)$.

Proof. Since $a_i = a$, there are

$$SNCMSM^{(m)}(a, a, \cdots, a)$$

$$=\left(\left\langle\left(\frac{\sum_{k=1}^{C_n^m}\prod_{j=1}^{m}Ex_a}{C_n^m}\right)^{\frac{1}{m}},\left(\frac{\sqrt{\sum_{k=1}^{C_n^m}\left(\prod_{j=1}^{m}En_a\right)^2}}{\sqrt{C_n^m}}\right)^{\frac{1}{m}},\left(\frac{\sqrt{\sum_{k=1}^{C_n^m}\left(\prod_{j=1}^{m}He_a\right)^2}}{\sqrt{C_n^m}}\right)^{\frac{1}{m}}\right\rangle,\right.$$

$$\left\langle\left(\frac{\sum_{k=1}^{C_n^m}\left(\prod_{j=1}^{m}T_a\left(\prod_{j=1}^{m}Ex_a+\left(\prod_{j=1}^{m}En_a\right)^2+\left(\prod_{j=1}^{m}He_a\right)^2\right)\right)}{\sum_{k=1}^{C_n^m}\left(\prod_{j=1}^{m}Ex_a+\left(\prod_{j=1}^{m}En_a\right)^2+\left(\prod_{j=1}^{m}He_a\right)^2\right)}\right)^{\frac{1}{m}},\right.$$

$$1-\left(1-\frac{\sum_{k=1}^{C_n^m}\left(\left(1-\prod_{j=1}^{m}\left(1-I_{i_j^k}\right)\right)\left(\prod_{j=1}^{m}Ex_a+\left(\prod_{j=1}^{m}En_a\right)^2+\left(\prod_{j=1}^{m}He_a\right)^2\right)\right)}{\sum_{k=1}^{C_n^m}\left(\prod_{j=1}^{m}Ex_a+\left(\prod_{j=1}^{m}En_a\right)^2+\left(\prod_{j=1}^{m}He_a\right)^2\right)}\right)^{\frac{1}{m}},$$

$$\left.1-\left(1-\frac{\sum_{k=1}^{C_n^m}\left(\left(1-\prod_{j=1}^{m}\left(1-I_{i_j^k}\right)\right)\left(\prod_{j=1}^{m}Ex_a+\left(\prod_{j=1}^{m}En_a\right)^2+\left(\prod_{j=1}^{m}He_a\right)^2\right)\right)}{\sum_{k=1}^{C_n^m}\left(\prod_{j=1}^{m}Ex_a+\left(\prod_{j=1}^{m}En_a\right)^2+\left(\prod_{j=1}^{m}He_a\right)^2\right)}\right)^{\frac{1}{m}}\right\rangle\Bigg)$$

$$=(\langle Ex_a, En_a, He_a\rangle, \langle T_a, I_a, F_a\rangle)=a.$$

□

Theorem 5. *(Commutativity). Let $(a'_1, a'_2, \cdots, a'_n)$ be any permutation of $(a_1, a_2, \cdots, a_n)$. Then, $SNCMSM^{(m)}(a'_1, a'_2, \cdots, a'_n) = SNCMSM^{(m)}(a_1, a_2, \cdots, a_n)$.*

Theorem 5 can be proved easily in accordance with Definition 13 and Theorem 3.

Three special cases of the SNCMSM operator are discussed below by selecting different values for the parameter m.

(1) If $m = 1$, then the SNCMSM operator becomes the simplest arithmetic average aggregation operator as follows:

$$\begin{aligned} SNCMSM^{(1)}(a_1, a_2, \cdots, a_n) &= \frac{\oplus_{i=1}^{n} a_i}{n} \\ = \Bigg(\left\langle \sum_{i=1}^{n} Ex_i, \sqrt{\sum_{i=1}^{n} En_i^2}, \sqrt{\sum_{i=1}^{n} He_i^2}\right\rangle, &\left\langle \frac{\sum_{i=1}^{n} T_i(Ex_i+En_i^2+He_i^2)}{\sum_{i=1}^{n}(Ex_i+En_i^2+He_i^2)}, \right. \\ \left. \frac{\sum_{i=1}^{n} I_i(Ex_i+En_i^2+He_i^2)}{\sum_{i=1}^{n}(Ex_i+En_i^2+He_i^2)}, \frac{\sum_{i=1}^{n} F_i(Ex_i+En_i^2+He_i^2)}{\sum_{i=1}^{n}(Ex_i+En_i^2+He_i^2)}\right\rangle\Bigg). \end{aligned} \tag{9}$$

(2) If $m = 2$, then the SNCMSM operator is degenerated to the following form:

$$SNCMSM^{(2)}(a_1, a_2, \cdots, a_n) = \left(\frac{\oplus_{i,j=1, i\neq j}^{n} a_i \otimes a_j}{n(n-1)}\right)^{\frac{1}{2}}$$

$$= \left(\left\langle \left(\frac{\sum_{i,j=1, i\neq j}^{n} Ex_iEx_j}{n(n-1)}\right)^{\frac{1}{2}}, \left(\sqrt{\frac{\sum_{i,j=1, i\neq j}^{n} (En_iEn_j)^2}{n(n-1)}}\right)^{\frac{1}{2}}, \left(\sqrt{\frac{\sum_{i,j=1, i\neq j}^{n} (He_iHe_j)^2}{n(n-1)}}\right)^{\frac{1}{2}}\right\rangle,\right.$$

$$\left\langle \left(\frac{\sum_{i,j=1, i\neq j}^{n} T_iT_j(Ex_iEx_j+En_i^2En_j^2+He_i^2He_j^2)}{\sum_{i,j=1, i\neq j}^{n} (Ex_iEx_j+En_i^2En_j^2+He_i^2He_j^2)}\right)^{\frac{1}{2}},\right.$$

$$1-\left(1-\frac{\sum\limits_{\substack{i,j=1\\ i\neq j}}^{n}\left[1-(1-I_i)(1-I_j)\right]\left(Ex_iEx_j+En_i^2En_j^2+He_i^2He_j^2\right)}{\sum\limits_{\substack{i,j=1\\ i\neq j}}^{n}\left(Ex_iEx_j+En_i^2En_j^2+He_i^2He_j^2\right)}\right)^{\frac{1}{2}},$$
$$\left.1-\left(1-\frac{\sum\limits_{\substack{i,j=1\\ i\neq j}}^{n}\left[1-(1-F_i)(1-F_j)\right]\left(Ex_iEx_j+En_i^2En_j^2+He_i^2He_j^2\right)}{\sum\limits_{\substack{i,j=1\\ i\neq j}}^{n}\left(Ex_iEx_j+En_i^2En_j^2+He_i^2He_j^2\right)}\right)^{\frac{1}{2}}\right\rangle\Bigg). \tag{10}$$

(3) If $m = n$, then the SNCMSM operator becomes the geometric average aggregation operator as follows:

$$\begin{aligned}&SNCMSM^{(n)}(a_1, a_2, \cdots, a_n) = (\otimes_{i=1}^{n} a_i)^{\frac{1}{n}}\\ &= \left(\left\langle\left(\prod_{i=1}^{n} Ex_i\right)^{\frac{1}{n}}, \left(\prod_{i=1}^{n} En_i\right)^{\frac{1}{n}}, \left(\prod_{i=1}^{n} He_i\right)^{\frac{1}{n}}\right\rangle,\right.\\ &\left.\left\langle\left(\prod_{i=1}^{n} T_i\right)^{\frac{1}{n}}, \left(1-\prod_{i=1}^{n}(1-I_i)\right)^{\frac{1}{n}}, \left(1-\prod_{i=1}^{n}(1-F_i)\right)^{\frac{1}{n}}\right\rangle\right).\end{aligned} \tag{11}$$

4.2. Weighted SNCMSM Operator

In this subsection, a weighted SNCMSM operator is investigated. Moreover, some desirable properties of this operator are analyzed.

Definition 14. *Let $a_i = \langle (Ex_i, En_i, He_i), (T_i, I_i, F_i)\rangle (i = 1, 2, ..., n)$ be a collection of SNCs, and $w = (w_1, w_2, ...w_n)^T$ be the weight vector, with $w_i \in [0, 1]$ and $\sum_{i=1}^{n} w_i = 1$. Then, the weighted simplified neutrosophic clouds Maclaurin symmetric mean (WSNCMSM) operator is defined as*

$$WSNCMSM_w{}^{(m)}(a_1, a_2, \cdots, a_n) = \left(\frac{\bigoplus\limits_{1\leq i_1<\cdots<i_m\leq n}\left(\bigotimes\limits_{j=1}^{m}\left(nw_{i_j}\cdot a_{i_j}\right)\right)}{C_n^m}\right)^{\frac{1}{m}}, \tag{12}$$

where $m = 1, 2, ..., n$ and $(i_1, i_2, \cdots, i_m)$ traverses all the m-tuple combination of $(i = 1, 2, \cdots, n)$, $C_n^m = \frac{n!}{m!(n-m)!}$ is the binomial coefficient.

The specific expression of the WSNCMSM operator can be obtained in accordance with the operations provided in Definition 10.

Theorem 6. *Let $a_i = \langle (Ex_i, En_i, He_i), (T_i, I_i, F_i)\rangle (i = 1, 2, ..., n)$ be a collection of SNCs, and $m = 1, 2, ..., n$. Then, the aggregated value acquired by the WSNCMSM operator can be expressed as*

$$WSNCMSM_w{}^{(m)}(a_1, a_2, \cdots, a_n)$$
$$= \left(\left\langle\left(\frac{\sum\limits_{k=1}^{C_n^m}\prod\limits_{j=1}^{m} nw_{i_j}Ex_{i_j^{(k)}}}{C_n^m}\right)^{\frac{1}{m}}, \left(\frac{\sqrt{\sum\limits_{k=1}^{C_n^m}\left(\prod\limits_{j=1}^{m}\sqrt{nw_{i_j}}En_{i_j^{(k)}}\right)^2}}{\sqrt{C_n^m}}\right)^{\frac{1}{m}}, \left(\frac{\sqrt{\sum\limits_{k=1}^{C_n^m}\left(\prod\limits_{j=1}^{m}\sqrt{nw_{i_j}}He_{i_j^{(k)}}\right)^2}}{\sqrt{C_n^m}}\right)^{\frac{1}{m}}\right\rangle,\right.$$

$$\left\langle \left(\frac{\sum_{k=1}^{C_n^m} \left(\prod_{j=1}^m T_{i_j^k} \left(\prod_{j=1}^m nw_{i_j} Ex_{i_j^{(k)}} + \left(\prod_{j=1}^m \sqrt{nw_{i_j}} En_{i_j^{(k)}} \right)^2 + \left(\prod_{j=1}^m \sqrt{nw_{i_j}} He_{i_j^{(k)}} \right)^2 \right) \right)}{\sum_{k=1}^{C_n^m} \left(\prod_{j=1}^m nw_{i_j} Ex_{i_j^{(k)}} + \left(\prod_{j=1}^m \sqrt{nw_{i_j}} En_{i_j^{(k)}} \right)^2 + \left(\prod_{j=1}^m \sqrt{nw_{i_j}} He_{i_j^{(k)}} \right)^2 \right)} \right)^{\frac{1}{m}}, \right.$$

$$1 - \left(1 - \frac{\sum_{k=1}^{C_n^m} \left(\left(1 - \prod_{j=1}^m \left(1 - I_{i_j^k} \right) \right) \left(\prod_{j=1}^m nw_{i_j} Ex_{i_j^{(k)}} + \left(\prod_{j=1}^m \sqrt{nw_{i_j}} En_{i_j^{(k)}} \right)^2 + \left(\prod_{j=1}^m \sqrt{nw_{i_j}} He_{i_j^{(k)}} \right)^2 \right) \right)}{\sum_{k=1}^{C_n^m} \left(\prod_{j=1}^m nw_{i_j} Ex_{i_j^{(k)}} + \left(\prod_{j=1}^m \sqrt{nw_{i_j}} En_{i_j^{(k)}} \right)^2 + \left(\prod_{j=1}^m \sqrt{nw_{i_j}} He_{i_j^{(k)}} \right)^2 \right)} \right)^{\frac{1}{m}}, \tag{13}$$

$$\left. 1 - \left(1 - \frac{\sum_{k=1}^{C_n^m} \left(\left(1 - \prod_{j=1}^m \left(1 - F_{i_j^k} \right) \right) \left(\prod_{j=1}^m nw_{i_j} Ex_{i_j^{(k)}} + \left(\prod_{j=1}^m \sqrt{nw_{i_j}} En_{i_j^{(k)}} \right)^2 + \left(\prod_{j=1}^m \sqrt{nw_{i_j}} He_{i_j^{(k)}} \right)^2 \right) \right)}{\sum_{k=1}^{C_n^m} \left(\prod_{j=1}^m nw_{i_j} Ex_{i_j^{(k)}} + \left(\prod_{j=1}^m \sqrt{nw_{i_j}} En_{i_j^{(k)}} \right)^2 + \left(\prod_{j=1}^m \sqrt{nw_{i_j}} He_{i_j^{(k)}} \right)^2 \right)} \right)^{\frac{1}{m}} \right\rangle \Bigg).$$

Theorem 6 can be proved similarly according to the proof procedures of Theorem 3.

Theorem 7. *(Reducibility) Let* $w = \left(\frac{1}{n}, \frac{1}{n}, \dots, \frac{1}{n}\right)^T$*, then,* $WSNCMSM_w{}^{(m)}(a_1, a_2, \dots, a_n) = SNCMSM^{(m)}(a_1, a_2, \dots, a_n)$.

Proof. When $w = \left(\frac{1}{n}, \frac{1}{n}, \dots, \frac{1}{n}\right)^T$,

$$WSNCMSM_w{}^{(m)}(a_1, a_2, \cdots, a_n)$$

$$= \left(\left\langle \left(\frac{\sum_{k=1}^{C_n^m} \prod_{j=1}^m n \cdot \frac{1}{n} Ex_{i_j^{(k)}}}{C_n^m} \right)^{\frac{1}{m}}, \left(\frac{\sqrt{\sum_{k=1}^{C_n^m} \left(\prod_{j=1}^m \sqrt{n \cdot \frac{1}{n}} En_{i_j^{(k)}} \right)^2}}{\sqrt{C_n^m}} \right)^{\frac{1}{m}}, \left(\frac{\sqrt{\sum_{k=1}^{C_n^m} \left(\prod_{j=1}^m \sqrt{n \cdot \frac{1}{n}} He_{i_j^{(k)}} \right)^2}}{\sqrt{C_n^m}} \right)^{\frac{1}{m}} \right\rangle, \right.$$

$$\left\langle \left(\frac{\sum_{k=1}^{C_n^m} \left(\prod_{j=1}^m T_{i_j^k} \left(\prod_{j=1}^m n \cdot \frac{1}{n} Ex_{i_j^{(k)}} + \left(\prod_{j=1}^m \sqrt{n \cdot \frac{1}{n}} En_{i_j^{(k)}} \right)^2 + \left(\prod_{j=1}^m \sqrt{n \cdot \frac{1}{n}} He_{i_j^{(k)}} \right)^2 \right) \right)}{\sum_{k=1}^{C_n^m} \left(\prod_{j=1}^m n \cdot \frac{1}{n} Ex_{i_j^{(k)}} + \left(\prod_{j=1}^m \sqrt{n \cdot \frac{1}{n}} En_{i_j^{(k)}} \right)^2 + \left(\prod_{j=1}^m \sqrt{n \cdot \frac{1}{n}} He_{i_j^{(k)}} \right)^2 \right)} \right)^{\frac{1}{m}}, \right.$$

$$1 - \left(1 - \frac{\sum_{k=1}^{C_n^m} \left(\left(1 - \prod_{j=1}^m \left(1 - I_{i_j^k} \right) \right) \left(\prod_{j=1}^m n \cdot \frac{1}{n} Ex_{i_j^{(k)}} + \left(\prod_{j=1}^m \sqrt{n \cdot \frac{1}{n}} En_{i_j^{(k)}} \right)^2 + \left(\prod_{j=1}^m \sqrt{n \cdot \frac{1}{n}} He_{i_j^{(k)}} \right)^2 \right) \right)}{\sum_{k=1}^{C_n^m} \left(\prod_{j=1}^m n \cdot \frac{1}{n} Ex_{i_j^{(k)}} + \left(\prod_{j=1}^m \sqrt{n \cdot \frac{1}{n}} En_{i_j^{(k)}} \right)^2 + \left(\prod_{j=1}^m \sqrt{n \cdot \frac{1}{n}} He_{i_j^{(k)}} \right)^2 \right)} \right)^{\frac{1}{m}},$$

$$\left. \left. 1 - \left(1 - \frac{\sum_{k=1}^{C_n^m} \left(\left(1 - \prod_{j=1}^m \left(1 - F_{i_j^k} \right) \right) \left(\prod_{j=1}^m n \cdot \frac{1}{n} Ex_{i_j^{(k)}} + \left(\prod_{j=1}^m \sqrt{n \cdot \frac{1}{n}} En_{i_j^{(k)}} \right)^2 + \left(\prod_{j=1}^m \sqrt{n \cdot \frac{1}{n}} He_{i_j^{(k)}} \right)^2 \right) \right)}{\sum_{k=1}^{C_n^m} \left(\prod_{j=1}^m n \cdot \frac{1}{n} Ex_{i_j^{(k)}} + \left(\prod_{j=1}^m \sqrt{n \cdot \frac{1}{n}} En_{i_j^{(k)}} \right)^2 + \left(\prod_{j=1}^m \sqrt{n \cdot \frac{1}{n}} He_{i_j^{(k)}} \right)^2 \right)} \right)^{\frac{1}{m}} \right\rangle \right)$$

$$= SNCMSM^{(m)}(a_1, a_2, \cdots, a_n).$$

The proof of Theorem 7 is completed. □

Definition 15. *Let* $a_i = \langle (Ex_i, En_i, He_i), (T_i, I_i, F_i) \rangle$ $(i = 1, 2, \dots, n)$ *be a collection of SNCs, and* $w = (w_1, w_2, \dots, w_n)^T$ *be the weight vector, which satisfies* $\sum_{i=1}^n w_i = 1$*, and* $w_i > 0$ $(i = 1, 2, \dots, n)$*. Then the* *generalized weighted simplified neutrosophic clouds Maclaurin symmetric mean (GWSNCMSM) operator is defined as*

$$GWSNCMSM^{(m,p_1,p_2,\ldots,p_m)}(a_1,\ldots,a_n) = \left(\frac{\oplus_{1\le i_1<\cdots<i_m\le n}\left(\otimes_{j=1}^{m}\left(nw_{i_j}\otimes a_{i_j}\right)^{p_j}\right)}{C_n^m} \right)^{\frac{1}{p_1+\cdots+p_m}}, \tag{14}$$

where $m = 1,2,\ldots,n$.

The specific expression of the GWSNCMSM operator can be obtained in accordance with the operations provided in Definition 10.

Theorem 8. *Let* $a_i = \langle (Ex_i, En_i, He_i), (T_i, I_i, F_i)\rangle$ $(i = 1,2,\ldots,n)$ *be a collection of SNCs, and* $m = 1,2,\ldots,n$. *Then, the aggregated value acquired by the GWSNCMSM operator can be expressed as*

$$\begin{aligned}
&GWSNCMSM^{(m,p_1,p_2,\ldots,p_m)}(a_1,\ldots,a_n) = \left\langle \left\langle \left(\frac{\sum_{k=1}^{C_n^m}\prod_{j=1}^{m}\left(nw_{i_j}Ex_{i_j^{(k)}}\right)^{p_j}}{C_n^m} \right)^{\frac{1}{p_1+\cdots+p_m}}, \right.\right.\\
&\left.\left(\frac{\sqrt{\sum_{k=1}^{C_n^m}\left(\prod_{j=1}^{m}\left(\sqrt{nw_{i_j}}En_{i_j^{(k)}}\right)^{p_j}\right)^2}}{\sqrt{C_n^m}}\right)^{\frac{1}{p_1+\cdots+p_m}}, \left(\frac{\sqrt{\sum_{k=1}^{C_n^m}\left(\prod_{j=1}^{m}\left(\sqrt{nw_{i_j}}He_{i_j^{(k)}}\right)^{p_j}\right)^2}}{\sqrt{C_n^m}}\right)^{\frac{1}{p_1+\cdots+p_m}} \right\rangle,\\
&\left\langle \left(\frac{\sum_{k=1}^{C_n^m}\left(\prod_{j=1}^{m}\left(T_{i_j^{(k)}}\right)^{p_j}\left(\prod_{j=1}^{m}\left(nw_{i_j}Ex_{i_j^{(k)}}\right)^{p_j}+\left(\prod_{j=1}^{m}\left(\sqrt{nw_{i_j}}En_{i_j^{(k)}}\right)^{p_j}\right)^2+\left(\prod_{j=1}^{m}\left(\sqrt{nw_{i_j}}He_{i_j^{(k)}}\right)^{p_j}\right)^2\right)\right)}{\sum_{k=1}^{C_n^m}\left(\prod_{j=1}^{m}\left(nw_{i_j}Ex_{i_j^{(k)}}\right)^{p_j}+\left(\prod_{j=1}^{m}\left(\sqrt{nw_{i_j}}En_{i_j^{(k)}}\right)^{p_j}\right)^2+\left(\prod_{j=1}^{m}\left(\sqrt{nw_{i_j}}He_{i_j^{(k)}}\right)^{p_j}\right)^2\right)} \right)^{\frac{1}{p_1+\cdots+p_m}}, \right.\\
&1-\left(1-\frac{\sum_{k=1}^{C_n^m}\left(\left(1-\prod_{j=1}^{m}\left(1-I_{i_j^{(k)}}\right)^{p_j}\right)\left(\prod_{j=1}^{m}\left(nw_{i_j}Ex_{i_j^{(k)}}\right)^{p_j}+\left(\prod_{j=1}^{m}\left(\sqrt{nw_{i_j}}En_{i_j^{(k)}}\right)^{p_j}\right)^2+\left(\prod_{j=1}^{m}\left(\sqrt{nw_{i_j}}He_{i_j^{(k)}}\right)^{p_j}\right)^2\right)\right)}{\sum_{k=1}^{C_n^m}\left(\prod_{j=1}^{m}\left(nw_{i_j}Ex_{i_j^{(k)}}\right)^{p_j}+\left(\prod_{j=1}^{m}\left(\sqrt{nw_{i_j}}En_{i_j^{(k)}}\right)^{p_j}\right)^2+\left(\prod_{j=1}^{m}\left(\sqrt{nw_{i_j}}He_{i_j^{(k)}}\right)^{p_j}\right)^2\right)}\right)^{\frac{1}{p_1+\cdots+p_m}},\\
&\left.\left. 1-\left(1-\frac{\sum_{k=1}^{C_n^m}\left(\left(1-\prod_{j=1}^{m}\left(1-F_{i_j^{(k)}}\right)^{p_j}\right)\left(\prod_{j=1}^{m}\left(nw_{i_j}Ex_{i_j^{(k)}}\right)^{p_j}+\left(\prod_{j=1}^{m}\left(\sqrt{nw_{i_j}}En_{i_j^{(k)}}\right)^{p_j}\right)^2+\left(\prod_{j=1}^{m}\left(\sqrt{nw_{i_j}}He_{i_j^{(k)}}\right)^{p_j}\right)^2\right)\right)}{\sum_{k=1}^{C_n^m}\left(\prod_{j=1}^{m}\left(nw_{i_j}Ex_{i_j^{(k)}}\right)^{p_j}+\left(\prod_{j=1}^{m}\left(\sqrt{nw_{i_j}}En_{i_j^{(k)}}\right)^{p_j}\right)^2+\left(\prod_{j=1}^{m}\left(\sqrt{nw_{i_j}}He_{i_j^{(k)}}\right)^{p_j}\right)^2\right)}\right)^{\frac{1}{p_1+\cdots+p_m}} \right\rangle \right\rangle.
\end{aligned} \tag{15}$$

Theorem 8 can be proved similarly according to the proof procedures of Theorem 3.

5. MCDM Approach under Simplified Neutrosophic Linguistic Circumstance

In this section, a MCDM approach is developed on the basis of the proposed simplified neutrosophic cloud aggregation operators to solve real-world problems. Consider a MCDM problem with simplified neutrosophic linguistic evaluation information, which can be converted to SNCs. Then, let $A = \{a_1, a_2, \ldots, a_m\}$ be a discrete set of alternatives, and $C = \{c_1, c_2, \ldots, c_n\}$ be the set of criteria. Suppose that the weight of the criteria is $w = (w_1, w_2, \ldots, w_s)^T$, where $w_k \ge 0$, and $\sum_{k=1}^{s} w_k = 1$. The original evaluation of alternative a_i under criterion c_j is expressed as SNLNs $\gamma_{ij} = \langle s_{ij}, (T_{ij}, I_{ij}, F_{ij})\rangle$ ($i = 1, 2, \ldots, m$; $j = 1, 2, \ldots, n$). The primary procedures of the developed method are presented in the following.

Step 1: Normalize the evaluation information.

Usually, two kinds of criteria—benefit criteria and cost criteria—exist in MCDM problems. Then, in accordance with the transformation principle of SNLNs [42], the normalization of original evaluation information can be shown as

$$\widetilde{\gamma}_{ij} = \begin{cases} \langle s_{ij}, (T_{ij}, I_{ij}, F_{ij}) \rangle, & \text{for benifit criterion,} \\ \left\langle h_{(2t+1-sub(s_{ij}))}, \left(T_{ij}, I_{ij}, F_{ij}\right) \right\rangle, & \text{for cos t criterion.} \end{cases} \tag{16}$$

Step 2: Convert SNLNs to SNCs.

Based on the transformation method described in Section 2.4 and Definition 9, we can convert SNLNs to SNCs. The SNC evaluation information can be obtained as $a_{ij} = \langle (Ex_{ij}, En_{ij}, He_{ij}), (T_{ij}, I_{ij}, F_{ij}) \rangle$ $(i = 1, 2, \ldots, m; j = 1, 2, \ldots, n)$.

Step 3: Acquire the comprehensive evaluation for each alternative.

The WSNCMSM operator or the GWSNCMSM operator can be employed to integrate the evaluation of $a_{ij}(j = 1, 2, ..., n)$ under all criteria and acquire the comprehensive evaluation $a_i = \langle (Ex_i, En_i, He_i), (T_i, I_i, F_i) \rangle$ for the alternative a_i.

Step 4: Compute the distance between the comprehensive evaluation of a_i and the PIS/NIS.

First, in accordance with the obtained overall evaluation values, the positive ideal solution (PIS) a^+ and negative ideal solution (NIS) a^- are determined as

$$a^+ = \langle (\max_i(Ex_i), \min_i(En_i), \min_i(He_i)), (\max_i(T_i), \min_i(I_i), \min_i(F_i)) \rangle,$$

$$a^- = \langle (\min_i(Ex_i), \max_i(En_i), \max_i(He_i)), (\min_i(T_i), \max_i(I_i), \max_i(F_i)) \rangle.$$

Second, in accordance with the proposed distance of SNCs, the distance $d(a_i, a^+)$ between a_i and a^+, and the distance $d(a_i, a^-)$ between a_i and a^- can be calculated.

Step 5: Compute the relative closeness of each alternative.

In the following, the relative closeness of each alternative can be calculated as

$$I_i = \frac{d(a_i, a^+)}{d(a_i, a^+) + d(a_i, a^-)} \tag{17}$$

where $d(a_i, a^+)$ and $d(a_i, a^-)$ are obtained in Step 4.

Step 6: Rank all the alternatives.

In accordance with the relative closeness I_i of each alternative, we can rank all the alternatives. The smaller the value of I_i, the better the alternative a_i is.

6. Illustrative Example

This section provides a real-world problem of hotel selection (adapted from Wang et al. [49]) to demonstrate the validity and feasibility of the developed approach.

6.1. Problem Description

Nowadays, consumers often book hotels online when traveling or on business trip. After they leave the hotel, they may evaluate the hotel and post the online reviews on the website. In this case, the online reviews are regard as the most important reference for the hotel selection decision of potential consumers. In order to enhance the accuracy of hotel recommendation in line with lots of online reviews, this study devotes to applying the proposed method to address hotel recommendation problems effectively. In practical hotel recommendation problems, many hotels (e.g., 10 hotels) need to be recommended for consumers. In order to save space, we select five hotels from a tourism website for recommendation here. The developed approach can be similarly applied to address hotel recommendation problems with many hotels. The five hotels are represented as a_1, a_2, a_3, a_4 and a_5. The employed linguistic term set is described as follows:

$S = \{s_1, s_2, s_3, s_4, s_5, s_6, s_7\}$ = {extremely poor, very poor, poor, fair good, very good, extremely good}

In this paper, we focus on the four hotel evaluation criteria including, c_1, location (such as near the downtown and is the traffic convenient or not); c_2, service (such as friendly staff and the breakfast); c_3, sleep quality (such as the soundproof effect of the room); and c_4, comfort degree (such as the softness of the bed and the shower). Wang et al. [49] introduced a text conversion technique to transform online reviews to neutrosophic linguistic information. Motivated by this idea, the online reviews of five hotels under four criteria can be described as SNLNs, as shown in Table 1. For simplicity, the weight information of the four criteria is assumed to be $w = (0.25, 0.22, 0.35, 0.18)^T$.

Table 1. Evaluation values in SNLNs.

a_i	c_1	c_2	c_3	c_4
a_1	$\langle s_4, (0.6, 0.6, 0.1)\rangle$	$\langle s_5, (0.6, 0.4, 0.3)\rangle$	$\langle s_4, (0.8, 0.5, 0.1)\rangle$	$\langle s_2, (0.8, 0.3, 0.1)\rangle$
a_2	$\langle s_2, (0.7, 0.5, 0.1)\rangle$	$\langle s_4, (0.6, 0.4, 0.2)\rangle$	$\langle s_3, (0.6, 0.2, 0.4)\rangle$	$\langle s_4, (0.7, 0.4, 0.3)\rangle$
a_3	$\langle s_3, (0.5, 0.1, 0.2)\rangle$	$\langle s_4, (0.6, 0.5, 0.3)\rangle$	$\langle s_6, (0.7, 0.6, 0.1)\rangle$	$\langle s_2, (0.5, 0.5, 0.2)\rangle$
a_4	$\langle s_2, (0.4, 0.5, 0.3)\rangle$	$\langle s_3, (0.5, 0.3, 0.4)\rangle$	$\langle s_4, (0.6, 0.8, 0.2)\rangle$	$\langle s_5, (0.9, 0.3, 0.1)\rangle$
a_5	$\langle s_5, (0.6, 0.4, 0.4)\rangle$	$\langle s_5, (0.8, 0.3, 0.1)\rangle$	$\langle s_3, (0.7, 0.5, 0.1)\rangle$	$\langle s_4, (0.6, 0.5, 0.2)\rangle$

6.2. Illustration of the Developed Methods

According to the steps of the developed method presented in Section 5, the optimal alternative from the five hotels can be determined.

6.2.1. Case 1—Approach based on the WSNCMSM Operator.

Let linguistic scale function be $f_1(h_x)$, and $m = 2$ in Equation (13) in the subsequent calculation. Then, the hotel selection problem can be addressed according to the following procedures.

Step 1: Normalize the evaluation information.

Obviously, the four criteria are the benefit type in the hotel selection problem above. Thus, the evaluation information does not need to be normalized.

Step 2: Convert SNLNs to SNCs.

Utilize the transformation method presented in Section 2.4, we transform the linguistic term s_i in SNLNs to the cloud model (Ex_i, En_i, He_i). The obtained results are shown as follows:

$$\begin{aligned}
s_1 &\to (Ex_1, En_1, He_1) = (0.833, 1.25, 0.231), \\
s_2 &\to (Ex_2, En_2, He_2) = (1.667, 1.11, 0.278), \\
s_3 &\to (Ex_3, En_3, He_3) = (2.5, 0.833, 0.37), \\
s_4 &\to (Ex_4, En_4, He_4) = (3.33, 0.556, 0.463), \\
s_5 &\to (Ex_5, En_5, He_5) = (4.167, 0.278, 0.556), \\
s_6 &\to (Ex_6, En_6, He_6) = (5, 0.741, 0.401), \\
s_7 &\to (Ex_7, En_7, He_7) = (5.833, 0.972, 0.324).
\end{aligned}$$

Then, according to Definition 9, SNLNs can be converted to SNCs, as presented in Table 2.

Table 2. Evaluation information in SNCs.

a_i	c_1	c_2	c_3	c_4
a_1	$\langle(3.33, 0.556, 0.463), (0.6, 0.6, 0.1)\rangle$	$\langle(4.167, 0.278, 0.556), (0.6, 0.4, 0.3)\rangle$	$\langle(3.33, 0.556, 0.463), (0.8, 0.5, 0.1)\rangle$	$\langle(1.667, 1.11, 0.278), (0.8, 0.3, 0.1)\rangle$
a_2	$\langle(1.667, 1.11, 0.278), (0.7, 0.5, 0.1)\rangle$	$\langle(3.33, 0.556, 0.463), (0.6, 0.4, 0.2)\rangle$	$\langle(2.5, 0.833, 0.37), (0.6, 0.2, 0.4)\rangle$	$\langle(3.33, 0.556, 0.463), (0.7, 0.4, 0.3)\rangle$
a_3	$\langle(2.5, 0.833, 0.37), (0.5, 0.1, 0.2)\rangle$	$\langle(3.33, 0.556, 0.463), (0.6, 0.5, 0.3)\rangle$	$\langle(5, 0.741, 0.401), (0.7, 0.6, 0.1)\rangle$	$\langle(1.667, 1.11, 0.278), (0.5, 0.5, 0.2)\rangle$
a_4	$\langle(1.667, 1.11, 0.278), (0.4, 0.5, 0.3)\rangle$	$\langle(2.5, 0.833, 0.37), (0.5, 0.3, 0.4)\rangle$	$\langle(3.33, 0.556, 0.463), (0.6, 0.8, 0.2)\rangle$	$\langle(4.167, 0.278, 0.556), (0.9, 0.3, 0.1)\rangle$
a_5	$\langle(4.167, 0.278, 0.556), (0.6, 0.4, 0.4)\rangle$	$\langle(4.167, 0.278, 0.556), (0.8, 0.3, 0.1)\rangle$	$\langle(2.5, 0.833, 0.37), (0.7, 0.5, 0.1)\rangle$	$\langle(3.33, 0.556, 0.463), (0.6, 0.5, 0.2)\rangle$

Step 3: Acquire the comprehensive evaluation for each alternative.

The WSNCMSM operator is employed to integrate the evaluations of alternative a_i under all the criteria. Then, the overall evaluation a_i^* for each alternative are obtained as

$$\begin{aligned}
a_1^* &= \langle(3.1311, 0.6228, 0.4509), (0.6866, 0.4765, 0.1589)\rangle, \\
a_2^* &= \langle(2.5946, 0.7909, 0.3881), (0.642, 0.3621, 0.2638)\rangle, \\
a_3^* &= \langle(3.1691, 0.801, 0.3835), (0.5986, 0.4584, 0.1895)\rangle, \\
a_4^* &= \langle(2.6569, 0.727, 0.4159), (0.6231, 0.5308, 0.2358)\rangle, \\
a_5^* &= \langle(3.4126, 0.5065, 0.4786), (0.6766, 0.4208, 0.2091)\rangle.
\end{aligned}$$

Step 4: Compute the distance between the comprehensive evaluation of a_i and the PIS/NIS.

First, the PIS a^+ and the NIS a^- are determined as $a^+ = \langle(3.4126, 0.5065, 0.3835), (0.6866, 0.3621, 0.1586)\rangle$, and $a^- = \langle(2.5946, 0.801, 0.4786), (0.5986, 0.5308, 0.2638)\rangle$, respectively. Then, based on Equation (5), the distance $d(a_i^*, a^+)$, and the distance $d(a_i^*, a^-)$ are computed as

$$\begin{aligned}
&d(a_1^*, a^+) = 0.8324,\ d(a_2^*, a^+) = 1.5966,\ d(a_3^*, a^+) = 1.2447,\ d(a_4^*, a^+) = 1.4864, \text{ and} \\
&d(a_5^*, a^+) = 0.3361;\ d(a_1^*, a^-) = 1.0135,\ d(a_2^*, a^-) = 0.2137,\ d(a_3^*, a^-) = 0.6535, \\
&d(a_4^*, a^-) = 0.3012, \text{ and } d(a_5^*, a^-) = 1.5101.
\end{aligned}$$

Step 5: Calculate the relative closeness of each alternative.

By using Equation (17), the relative closeness of each alternative is computed as

$$I_1 = 0.4509,\ I_2 = 0.882,\ I_3 = 0.6557,\ I_4 = 0.8315, \text{ and } I_5 = 0.1821.$$

Step 6: Rank all the alternatives.

On the basis of the comparison rule, the smaller the value of I_i, the better the alternative a_i is. We can rank the alternatives as $a_5 \succ a_1 \succ a_3 \succ a_4 \succ a_2$. The best one is a_5.

When $m = 3$ is used in Equation (13), the overall assessment value for each alternative a_i are derived as follows:

$$\begin{aligned}
a_1^* &= \langle(5.2615, 0.454, 0.2915), (0.5675, 0.6174, 0.229)\rangle, \\
a_2^* &= \langle(4.1045, 0.6629, 0.2384), (0.5177, 0.503, 0.3688)\rangle, \\
a_3^* &= \langle(5.1405, 0.6986, 0.2307), (0.4449, 0.5936, 0.2832)\rangle, \\
a_4^* &= \langle(4.0855, 0.5792, 0.2593), (0.468, 0.6791, 0.3475)\rangle, \\
a_5^* &= \langle(6.2421, 0.3334, 0.328), (0.5531, 0.5645, 0.2977)\rangle.
\end{aligned}$$

And the positive ideal point is determined as $a^+ = \langle(6.2421, 0.3334, 0.2307), (0.5675, 0.503, 0.229)\rangle$, the negative ideal point is determined as $a^- = \langle(4.0855, 0.6986, 0.328),$

$(0.4449, 0.6791, 0.3688)\rangle$. Then, the results of the distance between a_i^* and a^+, and the distance between a_i^* and a^- are obtained as

$$d(a_1^*, a^+) = 2.1919, d(a_2^*, a^+) = 4.064,\ d(a_3^*, a^+) = 3.7056,\ d(a_4^*, a^+) = 3.7812, \text{ and } d(a_5^*, a^+) = 0.8571;\ d(a_1^*, a^-) = 2.4095,\ d(a_2^*, a^-) = 0.4656,\ d(a_3^*, a^-) = 1.085, d(a_4^*, a^-) = 0.6172, \text{ and } d(a_5^*, a^-) = 3.8179.$$

Therefore, the relative closeness of each alternative is calculated as

$$I_1 = 0.4764,\ I_2 = 0.8972,\ I_3 = 0.7735,\ I_4 = 0.8597, \text{ and } I_5 = 0.1833$$

According to the results of I_i, we can rank the alternatives as $a_5 \succ a_1 \succ a_3 \succ a_4 \succ a_2$. The best one is a_5, which is the same as the obtained result in the situation $m = 2$.

6.2.2. Case 2—Approach Based on the GWSNCMSM Operator

Let the linguistic scale function be $f_1(h_x)$, and $m = 2$, $p_1 = 1$, $p_2 = 2$ in Equation (15) in the subsequent calculation. Then, the hotel selection problem can be addressed according to the following procedures.

Step 1: Normalize the evaluation information.

Obviously, the four criteria are the benefit type in the hotel selection problem above. Thus, the evaluation information does not need to normalize.

Step 2: Convert SNLNs to SNCs.

The obtained SNCs are the same as those in Case 1.

Step 3: Acquire the comprehensive evaluation for each alternative.

The GWSNCMSM operator is employed to integrate the evaluations of alternative a_i under all the criteria. Then, the overall evaluation a_i^* for each alternative are obtained as

$$\begin{aligned} a_1^* &= \langle (3.2899, 0.7006, 0.4668), (0.7068, 0.4812, 0.1544)\rangle, \\ a_2^* &= \langle (2.693, 0.805, 0.3968), (0.6395, 0.3374, 0.29)\rangle, \\ a_3^* &= \langle (3.7063, 0.8318, 0.3958), (0.6366, 0.5081, 0.1637)\rangle, \\ a_4^* &= \langle (2.9311, 0.7165, 0.4401), (0.6654, 0.5197, 0.2125)\rangle, \\ a_5^* &= \langle (3.3078, 0.5638, 0.4675), (0.6871, 0.4227, 0.1846)\rangle \end{aligned}$$

Step 4: Compute the distance between the comprehensive evaluation of a_i and the PIS/NIS.

First, the PIS a^+ and the NIS a^- are determined as $a^+ = \langle (3.7063, 0.5638, 0.3958),\ (0.7068, 0.3374, 0.1544)\rangle$, and $a^- = \langle (2.693, 0.8318, 0.4675),\ (0.6366, 0.5197, 0.29)\rangle$ respectively. Then, based on Equation (5), the distance $d(a_i^*, a^+)$, and the distance $d(a_i^*, a^-)$ are computed as

$$d(a_1^*, a^+) = 1.0407, d(a_2^*, a^+) = 1.6913,\ d(a_3^*, a^+) = 1.0619,\ d(a_4^*, a^+) = 1.371, \text{ and } d(a_5^*, a^+) = 0.6054;\ d(a_1^*, a^-) = 0.9235,\ d(a_2^*, a^-) = 0.2183,\ d(a_3^*, a^-) = 0.9925, d(a_4^*, a^-) = 0.5323, \text{ and } d(a_5^*, a^-) = 1.2871.$$

Step 5: Calculate the relative closeness of each alternative.

By using Equation (17), the relative closeness of each alternative is calculated as

$$I_1 = 0.5298,\ I_2 = 0.8857,\ I_3 = 0.5169,\ I_4 = 0.7203, \text{ and } I_4 = 0.7203$$

Step 6: Rank all the alternatives.

On the basis of the comparison rule, the smaller the value of I_i, the better the alternative a_i is. We can rank the alternatives as $a_5 \succ a_3 \succ a_1 \succ a_4 \succ a_2$, the best one is a_5.

Using the parameters $m = 2$, $p_1 = 1$, and $p_2 = 2$ in the aggregation operators, the ranking results acquired by the developed methods with the WSNCMSM operator and the GWSNCMSM operator are almost identical, and these rankings are described in Table 3. The basically identical ranking results indicate that the developed methods in this paper have a strong stability.

Table 3. Ranking results based on different operators.

Proposed Operators	m	p_1	p_2	Rankings
WSNCMSM	2	\	\	$a_5 \succ a_1 \succ a_3 \succ a_4 \succ a_2$
WSNCMSM	3	\	\	$a_5 \succ a_1 \succ a_3 \succ a_4 \succ a_2$
GWSNCMSM	2	1	2	$a_5 \succ a_3 \succ a_1 \succ a_4 \succ a_2$

6.3. Comparative Analysis and Sensitivity Analysis

This subsection implements a comparative study to verify the applicability and feasibility of the developed method. The developed method aims to improve the effectiveness of handling simplified neutrosophic linguistic information. Therefore, the proposed method can be demonstrated by comparing with the approaches in Wang et al. [21] and Tian et al. [20] that deal with SNLNs merely depend on the linguistic functions. The comparison between the developed method and two existed approaches is feasible because these three methods are based on the same information description tool and the aggregation operators developed in these methods have the same parameter characteristics. Two existing methods are employed to address the same hotel selection problem above, and the ranking results acquired by different approaches are described in Table 4.

Table 4. Ranking results obtained by different methods.

Methods	Rankings
Wang et al.'s method [21] (m = 2)	$a_5 \succ a_1 \succ a_3 \succ a_2 \succ a_4$
The proposed approach based on $WSNCMSM_w{}^{(m)}\,(m = 2)$	$a_5 \succ a_1 \succ a_3 \succ a_4 \succ a_2$
Wang et al.'s method [21] $(m = 2, p_1 = 1, p_2 = 1)$	$a_5 \succ a_3 \succ a_1 \succ a_2 \succ a_4$
Tian et al.'s method [20] $(m = 2, p_1 = 1, p_2 = 1)$	$a_5 \succ a_3 \succ a_1 \succ a_4 \succ a_2$
The proposed approach based on $GWSNCMSM^{(m,p_1,p_2,\ldots,p_m)}(m = 2, p_1 = 1, p_2 = 1)$	$a_5 \succ a_1 \succ a_3 \succ a_4 \succ a_2$

As described in Table 4, the rankings acquired by the developed approaches and that obtained by the existed approaches have obvious difference. However, the best alternative is always a_5, which demonstrates that the developed approach is reliable and effective for handling decision-making problems under simplified neutrosophic linguistic circumstance. There are still differences between the approaches developed in this paper and the methods presented by Wang et al. [21] and Tian et al. [20], which is that the proposed approaches use the cloud model instead of linguistic function to deal with linguistic information. The advantages of the proposed approaches in handling practical problems are summarized as follows:

First, comparing with the existing methods with SNLNs, the proposed approaches uses the cloud model to process qualitative evaluation information involved in SNLNs. The existing methods handle linguistic information merely depending on the relevant linguistic functions, which may result in loss and distortion of the original information. However, the cloud model depicts the randomness and fuzziness of a qualitative concept with three numerical characteristics perfectly, and it is more suitable to handle linguistic information than the linguistic function because it can reflect the vagueness and randomness of linguistic variables simultaneously.

Second, being compared with the simplified neutrosophic linguistic Bonferroni mean aggregation operator given in Tain et al. [20], the simplified neutrosophic clouds Maclaurin symmetric mean operator provided in this paper take more generalized forms and contain more flexible parameters that facilitate selecting the appropriate alternative.

In addition, being compared with SNLNs, SNCs not only provide the truth, indeterminacy, and falsity degrees for the evaluation object, but also utilize the cloud model to characterize linguistic information effectively.

The ranking results may vary with different values of parameters in the proposed aggregation operators. Thus, a sensitivity analysis will be implemented to analyze the influence of the parameter p_j on ranking results. The obtained results are presented in Table 5.

Table 5. Ranking results with different p_j under $m = 2$.

p_1	p_2	Rankings Based on GWSNCMSM
1	0	$a_5 \succ a_1 \succ a_3 \succ a_2 \succ a_4$
0	1	$a_4 \succ a_5 \succ a_3 \succ a_2 \succ a_1$
1	2	$a_5 \succ a_3 \succ a_1 \succ a_4 \succ a_2$
1	3	$a_3 \succ a_5 \succ a_1 \succ a_4 \succ a_2$
1	4	$a_3 \succ a_5 \succ a_1 \succ a_4 \succ a_2$
1	5	$a_3 \succ a_1 \succ a_5 \succ a_4 \succ a_2$
2	1	$a_5 \succ a_1 \succ a_3 \succ a_4 \succ a_2$
3	1	$a_5 \succ a_1 \succ a_3 \succ a_4 \succ a_2$
4	1	$a_1 \succ a_5 \succ a_3 \succ a_4 \succ a_2$
5	1	$a_1 \succ a_3 \succ a_5 \succ a_4 \succ a_2$
0.5	0.5	$a_5 \succ a_1 \succ a_3 \succ a_4 \succ a_2$
1	1	$a_5 \succ a_1 \succ a_3 \succ a_4 \succ a_2$
2	2	$a_5 \succ a_1 \succ a_3 \succ a_4 \succ a_2$
3	3	$a_5 \succ a_1 \succ a_3 \succ a_4 \succ a_2$
4	4	$a_5 \succ a_1 \succ a_3 \succ a_4 \succ a_2$
5	5	$a_5 \succ a_1 \succ a_3 \succ a_4 \succ a_2$

The data in Table 5 indicates that the best alternative is a_5 or a_1, and the worst one is a_2 when using the GWSNCMSM operator with different p_j under $m = 2$ to fuse evaluation information. When $p_1 = 0$, we can find the ranking result has obvious differences with other results. Therefore, $p_1 = 0$ is not used in practice. The data in Table 5 also suggests that the ranking vary obviously when the value of p_1 far exceeds the value of p_2. Thus, it can be concluded that the values of p_1 and p_2 should be selected as equally as possible in practical application. The difference of ranking results in Table 5 reveals that the values of p_1 and p_2 have great impact on the ranking results. As a result, selecting the appropriate parameters is a significant action when handling MCDM problems. In general, the values can be set as $p_1 = p_2 = 1$ or $p_1 = p_2 = 2$, which is not only simple and convenient but it also allows the interrelationship of criteria. It can be said that p_1 and p_2 are correlative with the thinking mode of the decision-maker; the bigger the values of p_1 and p_2, the more optimistic the decision-maker is; the smaller the values of p_1 and p_2, the more pessimistic the decision-maker is. Therefore, decision-makers can flexibly select the values of parameters based on the certain situations and their preferences and identify the most precise result.

7. Conclusions

SNLNs take linguistic terms into account on the basis of NSs, and they make the data description more complete and consistent with practical decision information than NSs. However, the cloud model, as an effective way to deal with linguistic information, has never been considered in combination with SNLNs. Motivated by the cloud model, we put forward a novel concept of SNCs based on SNLNs. Furthermore, the operation rules and distance of SNCs were defined. In addition, considering distinct importance of input variables, the WSNCMSM and GWSNCMSM operators were proposed and their properties and special cases were discussed. Finally, the developed approach was successfully applied to handle a practical hotel selection problem, and the validity of this approach was demonstrated.

The primary contributions of this paper can be summarized as follows. First, to process linguistic evaluation information involved in SNLNs, the cloud model is introduced and used. In this way, a new concept of SNCs is presented, and the operations and distance of SNCs are proposed. Being compared with other existing studies on SNLNs, the proposed method is more effective because the cloud model can comprehensively reflect the uncertainty of qualitative evaluation information. Second, based on the related studies, the MSM operator is extended to simplified neutrosophic cloud circumstances, and a series of SNCMSM aggregation operators are proposed. Third, a MCDM method is developed in light of the proposed aggregation operators, and its effectiveness and stability are demonstrated using the illustrative example, comparative analysis, and sensitivity analysis.

In some situations, asymmetrical and non-uniform linguistic information exists in practical problems. For example, customers pay more attention to negative comments when selecting hotels. In future study, we are going to introduce the unbalanced linguistic term sets to depict online linguistic comments and propose the hotel recommendation method.

Author Contributions: J.-Q.W., C.-Q.T., X.Z., H.-Y.Z., and T.-L.W. conceived and worked together to achieve this work; C.-Q.T. and X.Z. wrote the paper.

Conflicts of Interest: The authors declare no conflict of interest.

References

1. Smarandache, F. *Neutrosophy: Neutrosophic Probability, Set and Logic*; American Research Press: Rehoboth, DE, USA, 1998; pp. 1–105.
2. Wang, H.; Smarandache, F.; Zhang, Y.; Sunderraman, R. Single valued neutrosophic sets. *Multisp. Multistruct.* **2010**, *4*, 410–413.
3. Ye, J. Projection and bidirectional projection measures of single-valued neutrosophic sets and their decision-making method for mechanical design schemes. *J. Exp. Theor. Artif. Intell.* **2017**, *29*, 731–740. [CrossRef]
4. Ji, P.; Wang, J.; Zhang, H. Frank prioritized Bonferroni mean operator with single-valued neutrosophic sets and its application in selecting third-party logistics providers. *Neural Comput. Appl.* **2016**. [CrossRef]
5. Wang, F.H.; Smarandache, Y.; Zhang, Q.; Sunderraman, R. Interval neutrosophic sets and logic: Theory and applications in computing. *Comput. Sci.* **2005**, *65*, 86–87.
6. Ye, J. Trapezoidal neutrosophic set and its application to multiple attribute decision-making. *Neural Comput. Appl.* **2015**, *26*, 1157–1166. [CrossRef]
7. Liang, R.X.; Wang, J.Q.; Zhang, H.Y. A multi-criteria decision-making method based on single-valued trapezoidal neutrosophic preference relations with complete weight information. *Neural Comput. Appl.* **2017**. [CrossRef]
8. Peng, H.G.; Zhang, H.; Wang, J.Q. Probability multi-valued neutrosophic sets and its application in multi-criteria group decision-making problems. *Neural Comput. Appl.* **2016**. [CrossRef]
9. Wu, X.H.; Wang, J.Q.; Peng, J.J.; Qian, J. A novel group decision-making method with probability hesitant interval neutrosophic set and its application in middle level manager's selection. *Int. J. Uncertain. Quantif.* **2018**. [CrossRef]
10. Zhang, H.Y.; Ji, P.; Wang, J.Q.; Chen, X.H. A neutrosophic normal cloud and its application in decision-making. *Cogn. Comput.* **2016**, *8*, 649–669. [CrossRef]
11. Zhang, C.B.; Zhang, H.Y.; Wang, J.Q. Personalized restaurant recommendation combining group correlations and customer preferences. *Inf. Sci.* **2018**, *454–455*, 128–143. [CrossRef]
12. Ye, J. Hesitant interval neutrosophic linguistic set and its application in multiple attribute decision making. *Int. J. Mach. Learn. Cybern.* **2017**. [CrossRef]
13. Tan, R.; Zhang, W.; Chen, S. Some generalized single valued neutrosophic linguistic operators and their application to multiple attribute group decision making. *J. Syst. Sci. Inf.* **2017**, *5*, 148–162. [CrossRef]
14. Peng, H.G.; Wang, J.Q. Improved outranking decision-making method with Z-number cognitive information. *Cogn. Comput.* **2018**. [CrossRef]

15. Yu, S.M.; Wang, J.; Wang, J.Q.; Li, L. A multi-criteria decision-making model for hotel selection with linguistic distribution assessments. *Appl. Soft Comput.* **2018**, *67*, 739–753. [CrossRef]
16. Peng, H.G.; Wang, J.Q.; Cheng, P. A linguistic intuitionistic multi-criteria decision-making method based on the Frank Heronian mean operator and its application in evaluating coal mine safety. *Int. J. Mach. Learn. Cybern.* **2018**, *9*, 1053–1068. [CrossRef]
17. Ye, J. Some aggregation operators of interval neutrosophic linguistic numbers for multiple attribute decision making. *J. Intell. Fuzzy Syst.* **2014**, *27*, 2231–2241.
18. Broumi, S.; Ye, J.; Smarandache, F. An extended TOPSIS method for multiple attribute decision making based on interval neutrosophic uncertain linguistic variables. *Neutrosophic Sets Syst.* **2015**, *8*, 22–31.
19. Liu, P.; Khan, Q.; Ye, J. Group decision-making method under hesitant interval neutrosophic uncertain linguistic environment. *Int. J. Fuzzy Syst.* **2018**. [CrossRef]
20. Tian, Z.P.; Wang, J.; Wang, J.Q.; Zhang, H.Y. An improved multimoora approach for multi-criteria decision-making based on interdependent inputs of simplified neutrosophic linguistic information. *Neural Comput. Appl.* **2017**, *28*, 585–597. [CrossRef]
21. Wang, J.Q.; Yang, Y.; Li, L. Multi-criteria decision-making method based on single valued neutrosophic linguistic Maclaurin symmetric mean operators. *Neural Comput. Appl.* **2016**. [CrossRef]
22. Li, D.Y.; Liu, C.Y.; Gan, W.Y. A new cognitive model: Cloud model. *Int. J. Intell. Syst.* **2009**, *24*, 357–375. [CrossRef]
23. Wang, D.; Liu, D.; Ding, H. A cloud model-based approach for water quality assessment. *Environ. Res.* **2016**, *148*, 24–35. [CrossRef] [PubMed]
24. Li, L.; Fan, F.; Ma, L. Energy utilization evaluation of carbon performance in public projects by FAHP and cloud model. *Sustainability* **2016**, *8*, 630. [CrossRef]
25. Yan, F.; Xu, K.; Cui, Z. An improved layer of protection analysis based on a cloud model: Methodology and case study. *J. Loss Prev. Process Ind.* **2017**, *48*, 41–47. [CrossRef]
26. Peng, H.G.; Zhang, H.Y.; Wang, J.Q. Cloud decision support model for selecting hotels on TripAdvisor.com with probabilistic linguistic information. *Int. J. Hosp. Manag.* **2018**, *68*, 124–138. [CrossRef]
27. Wang, M.X.; Wang, J.Q. An evolving Takagi-Sugeno model based on aggregated trapezium clouds for anomaly detection in large datasets. *J. Intell. Fuzzy Syst.* **2017**, *32*, 2295–2308. [CrossRef]
28. Wang, M.X.; Wang, J.Q. New online recommendation approach based on unbalanced linguistic label with integrated cloud. *Kybernetes* **2018**. [CrossRef]
29. Chang, T.C.; Wang, H. A multi criteria group decision-making model for teacher evaluation in higher education based on cloud model and decision tree. *Eurasia J. Math. Sci. Technol. Educ.* **2016**, *12*, 1243–1262.
30. Wu, Y.; Xu, C.; Ke, Y. Multi-criteria decision-making on assessment of proposed tidal barrage schemes in terms of environmental impacts. *Mar. Pollut. Bull.* **2017**, *125*, 271–281. [CrossRef] [PubMed]
31. Ren, L.; He, L.; Chen, Y. A cloud model based multi-attribute decision making approach for selection and evaluation of groundwater management schemes. *J. Hydrol.* **2017**, *555*, 881–893.
32. Peng, H.G.; Wang, J.Q. Cloud decision model for selecting sustainable energy crop based on linguistic intuitionistic information. *Int. J. Syst. Sci.* **2017**, *48*, 3316–3333. [CrossRef]
33. Peng, H.G.; Wang, J.Q. A multi-criteria group decision-making method based on the normal cloud model with Zadeh's Z-numbers. *IEEE Trans. Fuzzy Syst.* **2018**. [CrossRef]
34. Liang, D.; Zhang, Y.; Xu, Z. Pythagorean fuzzy Bonferroni mean aggregation operator and its accelerative calculating algorithm with the multithreading. *Int. J. Intell. Syst.* **2018**, *33*, 615–633. [CrossRef]
35. Jiang, W.; Wei, B. Intuitionistic fuzzy evidential power aggregation operator and its application in multiple criteria decision-making. *Int. J. Syst. Sci.* **2018**, *49*, 582–594. [CrossRef]
36. Chen, S.M.; Kuo, L.W. Autocratic decision making using group recommendations based on interval type-2 fuzzy sets, enhanced Karnik–Mendel algorithms, and the ordered weighted aggregation operator. *Inf. Sci.* **2017**, *412*, 174–193. [CrossRef]
37. Lin, J.; Zhang, Q. Note on continuous interval-valued intuitionistic fuzzy aggregation operator. *Appl. Math. Model.* **2017**, *43*, 670–677. [CrossRef]
38. Maclaurin, C. A second letter to Martin Folkes, Esq.: Concerning the roots of equations, with the demonstration of other rules of algebra. *Philos. Trans. R. Soc. Lond. Ser. A* **1729**, *36*, 59–96.
39. Qin, J.; Liu, X.; Pedrycz, W. Hesitant Fuzzy Maclaurin Symmetric Mean Operators and Its Application to Multiple-Attribute Decision Making. *Int. J. Fuzzy Syst.* **2015**, *17*, 509–520. [CrossRef]

40. Qin, J.; Liu, X. Approaches to uncertain linguistic multiple attribute decision making based on dual Maclaurin symmetric mean. *J. Intell. Fuzzy Syst.* **2015**, *29*, 171–186. [CrossRef]
41. Liu, P.; Qin, X. Maclaurin symmetric mean operators of linguistic intuitionistic fuzzy numbers and their application to multiple-attribute decision-making. *J. Exp. Theor. Artif. Intell.* **2017**, *29*, 1173–1202. [CrossRef]
42. Wei, G.; Lu, M. Pythagorean fuzzy maclaurin symmetric mean operators in multiple attribute decision making. *Int. J. Intell. Syst.* **2018**, *33*, 1043–1070. [CrossRef]
43. Teng, F.; Liu, P.; Zhang, L. Multiple attribute decision-making methods with unbalanced linguistic variables based on maclaurin symmetric mean operators. *Int. J. Inf. Technol. Decis. Mak.* **2018**. [CrossRef]
44. Liu, P.; Zhang, X. Some Maclaurin symmetric mean operators for single-valued trapezoidal neutrosophic numbers and their applications to group decision making. *Int. J. Fuzzy Syst.* **2018**, *20*, 45–61. [CrossRef]
45. Ju, Y.; Liu, X.; Ju, D. Some new intuitionistic linguistic aggregation operators based on maclaurin symmetric mean and their applications to multiple attribute group decision making. *Soft Comput.* **2016**, *20*, 4521–4548. [CrossRef]
46. Yu, S.M.; Zhang, H.Y.; Wang, J.Q. Hesitant fuzzy linguistic maclaurin symmetric mean operators and their applications to multi-criteria decision-making problem. *Int. J. Intell. Syst.* **2018**, *33*, 953–982. [CrossRef]
47. Zadeh, L.A. The concept of a linguistic variable and its application to approximate reasoning. *Inf. Sci.* **1975**, *8*, 199–249. [CrossRef]
48. Wang, X.K.; Peng, H.G.; Wang, J.Q. Hesitant linguistic intuitionistic fuzzy sets and their application in multi-criteria decision-making problems. *Int. J. Uncertain. Quantif.* **2018**. [CrossRef]
49. Wang, J.Q.; Zhang, X.; Zhang, H.Y. Hotel recommendation approach based on the online consumer reviews using interval neutrosophic linguistic numbers. *J. Intell. Fuzzy Syst.* **2018**, *34*, 381–394. [CrossRef]

symmetry

MDPI

Article

DSmT Decision-Making Algorithms for Finding Grasping Configurations of Robot Dexterous Hands

Ionel-Alexandru Gal, Danut Bucur and Luige Vladareanu *

Institute of Solid Mechanics of the Romanian Academy, 010141 Bucharest S1, Romania; galexandru2003@yahoo.com (I.-A.G.); badbatx@yahoo.com (D.B.)

* Correspondence: luigiv2007@gmail.com; Tel.: +40-744-756-005

Received: 14 May 2018; Accepted: 29 May 2018; Published: 1 June 2018

Abstract: In this paper, we present a deciding technique for robotic dexterous hand configurations. This algorithm can be used to decide on how to configure a robotic hand so it can grasp objects in different scenarios. Receiving as input, several sensor signals that provide information on the object's shape, the DSmT decision-making algorithm passes the information through several steps before deciding what hand configuration should be used for a certain object and task. The proposed decision-making method for real time control will decrease the feedback time between the command and grasped object, and can be successfully applied on robot dexterous hands. For this, we have used the Dezert–Smarandache theory which can provide information even on contradictory or uncertain systems.

Keywords: neutrosophy; DSmT; decision-making algorithms; robotic dexterous hands; grasping configurations; grasp type

1. Introduction

The purpose of autonomous robotics is to build systems that can fulfill all kinds of tasks without human intervention, in different environments which were not specially build for robot interaction. A major challenge for this autonomous robotics field comes from high uncertainty within real environments. This is because the robot designer cannot know all the details regarding the environment. Most of the environment parameters are unknown, the position of humans and objects cannot be previously anticipated and the motion path might be blocked. Beside these, the accumulated sensor information can be uncertain and error prone. The quality of this information is influenced by noise, visual field limitations, observation conditions, and the complexity of interpretation technique.

The artificial intelligence and the heuristic techniques were used by many scientists in the field of robot control [1] and motion planning. Regarding grasping and object manipulations, the main research activities were to design a mechanism for hand [2–4] and dexterous finger motion [5], which are a high complexity research tasks in controlling robotic hands.

Currently, in the research area of robotics, there is a desire to develop robotic systems with applications in dynamic and unknown environments, in which human lives would be at risk, like natural or nuclear disaster areas, and also in different fields of work, ranging from house chores or agriculture to military applications. In any of these research areas, the robotic system must fulfill a series of tasks which implies object manipulation and transportation, or using equipment and tools. From here arises the necessity of development grasping systems [6] to reproduce, as well as possible human hand motion [7–9].

To achieve an accurate grasping system, a grasp taxonomy of the human hand was analyzed by Feix et al. [10] who found 33 different grasp types, sorted by opposition type, virtual finger assignments; type in terms of power, precision, or intermediate grasp; and the position of the thumb. While

Alvarez et al. [11] researched human grasp strategies within grasp types, Fermuller et al. [12] focused on manipulation action for human hand on different object types including hand pre-configuration. Tsai et al. [13] found that classifying objects into primitive shapes can provide a way to select the best grasping posture, but a general approach can also be used for hand–object geometry fitting [14]. This classification works well for grasping problems in constrained work space using visual data combined with force sensors [15] and also for under-actuated grasping which uses rotational stiffness [16]. However, for unknown objects, scientists found different approaches to solve the hand grasping problem. Choi et al. [17] used two different neural networks and data fusion to classify objects, Seredynski et al. [18] achieved fast grasp learning with probabilistic models, while Song et al. [19] used a tactile-based blind grasping along with a discrete-time controller. The same approach is used by Gu et al. [20] which proposed a blind haptic exploration of unknown objects for grasp planning of dexterous robotic hand. Using grasping methods, Yamakawa et al. [21] developed a robotic hand for knot manipulation, while Nacy et al. [22] used artificial neural network algorithms for slip prevention and Zaidi et al. [23] used a multi-fingered robot hand to grasp 3D deformable objects, applying the method on spheres and cubes.

While other scientists developed grasping strategies for different robotic hands [21–23], an anthropomorphic robotic hand has the potential to grasp regular objects of different shapes and sizes [24,25], but selecting the grasping method for a certain object is a difficult problem. A series of papers have approached this problem by developing algorithms for classifying the grasping by the contact points [26,27]. These algorithms are focused on finding a fix number of contact areas without taking into consideration the hand geometry. Other methods developed grasping systems for a certain robotic hand architecture, scaling down the problem to finding a grasping method with the tip of the fingers [27]. These methods are useful in certain object manipulation, but cannot be applied for a wide range of objects because it does not provide a stable grasping due to the face that it is not used, the finger's interior surface or the palm of the hand. A method for filtering the high number of hand configurations is to use predefined grasping hand configurations. Before grasping an object, humans, unconsciously simplify the grasping action, choosing one of the few hand positions which match the object's shape and the task to accomplish. In the scientific literature there are papers which have tried to log in the positioning for grasping and taxonomy, and one of the most known papers is [28]. Cutkosky and Weight [29] have extended Napier's [28] classification by adding the required taxonomy in the production environment, by studying the way in which the weight and geometry of the object affects choosing the grasping positioning. Iberall [30] has analyzed different grasping taxonomies and generalized them by using the virtual finger concept. Stransfield [31] has chosen a simpler classification and built a system based on rules which provided a grasping positioning set, starting from a simplified description of the object gained from a video system.

The developed algorithm presented in this paper has the purpose to determine the grasping position according to the object's shape. To prove the algorithm's efficiency we have chosen three types of objects for grasping: cylindrical, spherical, and prismatic. For this, we start from the hypothesis that the environment data are captured through a stereovision system [32] and a Kinect sensor [33]. On this data, which the two system observers provide, we apply a template matching algorithm [34]. This algorithm will provide a matching percentage of the object that needs to be grasped with a template object. Thus, each of the two sources will provide three matching values, for each of the three grasping types. These values represent the input for our detection algorithm, based of Dezert–Smarandache Theory (DSmT) [35] for data fusion. This algorithm has as input data from two or multiple observers and in the first phase they are processed through a process of neutrosofication which is similar with the fuzzification process. Then, the neutrosophic observers' data are passed through an algorithm which applies the classic DSm theory [35] in order to obtain a single data set on the system's states, by combining the observers' neutrosophic values. On this obtained data set, we apply the developed DSmT decision-making algorithm that decides on the category from which the target object is part of.

This decision facilitates the detection–recognition–grasping process which a robotic hand must follow, obtaining in the end a real-time decision that does not stop or delay the robot's task.

In recent years, using more sensors for a certain applications and then using data fusion is becoming more common in the military and nonmilitary research fields. The data fusion techniques combine the information received from different sensors with the purpose of eliminating disturbances and to improve precision compared to the situations when a single sensor is used [36,37]. This technique works on the same principle used by humans to feel the environment. For example, a human being cannot see over the corner or through vegetation, but with his hearing he can detect certain surrounding dangers. Beside the statistical advantage build from combining the details for a certain object (through redundant observations), using more types of sensors increases the precision with which an object can be observed and characterized. For example, an ultrasonic sensor can detect the distance to an object, but a video sensor can estimate its shape, and combining these two information sources will provide two distinct data on the same object.

The evolution on the new sensors, the hardware's processing techniques and capacity improvements facilitate more and more the real time data fusion. The latest progress were made in the area of computational and detection systems, and provide the ability to reproduce, in hardware and software, the data fusion capacity of humans and animals. The data fusion systems are used for target tracking [38], automatic target identification [39], and automated reasoning applications [40]. The data fusion applications are widespread, ranging from the military [41] applications (target recognition, autonomous moving vehicles, distance detection, battlefield surveillance, automatic danger detection) to civilian application (monitoring the production processes, complex tools maintenance based on certain conditions, robotics [42], and medical applications [43]). The data fusion techniques undertake classic elements like digital signal processing, statistical estimation, control theory, artificial intelligence, and numeric methods [44].

Combined data interpretation requires automated reasoning techniques taken from the area of artificial intelligence. The purpose of developing the recognition based systems, was to analyze issues like the data gathering context, the relationship between observed entities, hierarchical grouping of targets or objects and to predict future actions of these targets or entities. This kind of reasoning is encountered in humans, but the automated reasoning techniques can only closely reproduce it. Regardless of the used technique, for a knowledge based system, three elements are required: one or more reasoning diagrams, an automated evaluation process and a control diagram. The reasoning diagrams are techniques of facts representation, logical relations, procedural knowledge, and uncertainty. For these techniques, uncertainty from the observed data and from the logical relations can be represented using probabilities, fuzzy theory [45,46], Dempster–Shafer [47] evidence intervals or other methods. Dezert–Smarandache theory [35] comes to extend these methods, providing advanced techniques of uncertainty manipulation. The automated reasoning system's developing purpose is to reproduce the human capability of reasoning and decision making, by specifying rules and frames that define the studied situation. Having at hand an information database, an evaluation process is required so this information can be used. For this, there are formal diagrams developed on the formal logic, fuzzy logic, probabilistic reasoning, template based methods, case based reasoning, and many others. Each of these reasoning diagrams has a consistent internal formalism which describes how to use the knowledge database for obtaining the final conclusion. An automated reasoning system needs a control diagram to fulfill the thinking process. The used techniques include searching methods, systems for maintaining the truth based on assumptions and justifications, hierarchical decomposition, control theory, etc. Each of these methods has the purpose of controlling the reasoning evolution process.

The results presented in this paper, were obtained using the classic Dezert–Smarandache theory (DSmT) to combine inputs from two different observers that want to classify objects into three categories: sphere, parallelepiped, and cylinder. These categories were chosen to include most of the objects that a manipulator can grasp. The algorithm's inputs were transformed into belief values of certainty, falsity, uncertainty, and contradiction values. Using these four values and their combinations according to

DSmT, we applied Petri net diagram logic for taking decisions on the shape type of the analyzed objects. This type of algorithm has never been used before for real time decision on hand grasping taxonomy. Compared to other algorithms [13–15] and methods [16–18], ours has the advantage to detect high uncertainties and contradictions which in practice has a very low encounter rate but can have drastic effects on the decision type or robot, because if the object's shape is not detected properly, then the robot might not be able to grasp it, which can lead to serious consequences. In deciding how to grasp objects, researchers have used different methods to choose the grasping taxonomy using a blind haptic exploration [20] or in different applications for tying knots [21] or grasp deformable objects [23]. Because the proposed algorithm can detect anomalies of contradicting and uncertain input values, we can say that the proposed method transforms the deciding process into a less difficult problem of grasping method [24,25].

2. Objects Grasping and Its Classification

Mechanical hands have been developed to provide the robots with the ability of grasping objects with different geometrical and physical properties [48]. To make an anthropomorphic hand seem natural, its movement and the grasping type must match the human hand.

In this regard, grasping position taxonomy for human hands has been long studied and applied for robotic hands. Seventeen different categories of human hands grasping positions were studied. However, we must consider two important things: the first thing is that these categories are derived from human hand studies, which proves that they are more flexible and able to perform a multitude of movements than any other robotic hand, so that the grasping taxonomy for robot hands can be only a simple subset of the human hand. The second is that the human behavior studies of real object grasping have shown some differences between the real observations and the classified properties [49].

In conclusion, any proposed taxonomy is only a reference point which the robot hand must attain. Below the most used grasping positions are described (extracted from [50]), which should be considered when developing an able robotic hand:

1. Power grasping: The contact with the objects is made on large surfaces of the hand, including hand phalanges and the palm of the hand. For this kind of grasping, high forces can be exerted on the object.
    - Spherical grasping: used to grasp spherical objects;
    - Cylindrical grasping: used to grasp long objects which cannot be completely surrounded by the hand;
    - Lateral grasping: the thumb exerts a force towards the lateral side of the index finger.
2. Precision grasping: the contact is made only with the tip of the fingers.
    - Prismatic grasping (pinch): used to grasp long objects (with small diameter) or very small. Can be achieved with two to five fingers.
    - Circular grasping (tripod): used in grasping circular or round objects. Can be achieved with three, four, or five fingers.
3. No grasping:
    - Hook: the hand forms a hook on the object and the hand force is exerted against an external force, usually gravity.
    - Button pressing or pointing
    - Pushing with open hand.

In the Table 1, manipulation activities that the robotic hand can achieve are shown, correlated with the required activity grasping positions [51].

Table 1. Grasping position for certain tasks.

Object	Activity	Grasping Position
Bottles, cups, and mugs	Transport, pouring/filling	Force: Cylindrical grasping (from the side or the top)
Cups (using handles)	Pouring/filling	Force: Lateral grasping Precision: Prismatic grasping
Plates/trays	Transport Receiving from humans	Power: Lateral grasping Precision: Prismatic grasping No grasp: pushing (open hand)
Pens, cutlery	Transport	Precision: Prismatic grasping
Door handle	Open/Close	Force: Cylindrical grasping No grasp: Hook
Small objects	Transport	Power: Spherical grasping Precision: Circular grasping (tripod)
Switches, buttons	Pushing	No grasp: Button pressing
Round switches, bottle caps	Rotation	Force: Lateral grasping Precision: Circular grasping (tripod)

3. Object Detection Using Stereo-Vision and Kinect Sensor

Object recognition in artificial sight represents the task of searching a certain object in a picture or a video sequence. This problem can be approached as a learning problem. At first, the system is trained with sample images which belong to the target group, the system being taught to spot these among other pictures. Thus, when the system receives new images, it can 'feel' the presence of the searched object/sample/template.

Template matching is a techniques used to sort objects in an image. A model is an image region, and the goal is to find instances of this model in a larger picture. The template matching techniques represent a classic approach for localization problems and object recognition in a picture. These methods are used in applications like object tracking, image compression, stereograms, image segmentation [52], and other specific problems of artificial vision [53].

Object recognition is very important for a robot that must fulfill a certain task. To complete its task, the robot must avoid obstacles, obtain the size of the object, manipulate it, etc. For the case of detected object manipulation, the robot must detect the object's shape, size, and position in the environment. The main methods for achieving the depth information use stereoscopic cameras, laser scanners, and depth cameras.

To achieve the proposed decision-making algorithm, we assumed that the environment information is captured with a stereoscopic system and a Kinect sensor.

Stereovision systems [32] represents a passive technique of achieving a virtual 3D image of the environment in which the robot moves, by matching the common features of an image set of the same scene. Because this method works with images, it needs a high computational power. The depth information can be noisy in certain cases, because the method depends on the texture of the environment objects and on the ambient light.

Kinect [33] is a fairly easy to obtain platform, which makes it widespread. It uses a depth sensor based on structured light. By using an ASIC board, the Kinect sensor generates a depth map on 11 bits with a resolution of 640 × 480 pixels, at 30 Hz. Given the price of the device, the information quality is pretty good, but it has both advantages and disadvantages, meaning that the depth images contain areas where the depth reading could not be achieved. This problem appears from the fact that some materials do not reflect infrared light. When the device is moved really fast, like any other camera, it records blurry pictures, which also leads to missing information from the acquired picture.

4. Neutrosophic Logic and DSm Theory

4.1. Neutrosophic Logic

The neutrosophic triplet (truth, falsity, and uncertainty) idea appeared in 1764 when J.H. Lambert investigates a witness credibility which was affected by the testimony of another person. He generalized Hooper's rule of sample combination (1680), which was a non-Bayesian approach for finding a probabilistic model. Koopman in 1940 introduces the low and high probability, followed by Good and Dempster (1967) who gave a combination rule of two arguments. Shafer (1976) extended this rule to Dempster–Shafer Theory for trust functions by defining the trust and plausibility functions and using the inference rules of Dempster for combining two samples from two different sources. The trust function is a connection between the fuzzy reasoning and probability. Dempster–Shafer theory for trust functions is a generalization of Bayesian probability (Bayes 1760, Laplace 1780). It uses the mathematical probability in a more general way and it is based on the probabilistic combination of artificial intelligence samples.

Lambert one said that "there is a chance p that the witness can be trustworthy and fair, a chance q that he will be deceiving and a chance 1-p-q that he will be indifferent". This idea was taken by Shafer in 1986 and later, used by Smarandache to further develop the neutrosophic logic [54,55].

4.1.1. Neutrosophic Logic Definition

A logic in which each proposition has its percentage of truth in a subset T, its percentage of uncertainty in a subset I, and its percentage of falsity in a subset F is called neutrosophic logic [54,55].

This paper extends the general structure of the neutrophic robot control (RNC), known as the Vladareanu–Smarandache method [55–57] for the robot hybrid force-position control in a virtual platform [58,59], which applies neutrosophic science to robotics using the neutrosophic logic and set operators. Thus, using two observers, a stereovision system and a Kinect sensor, will provide three matching values for DSmT decision-making algorithms. A subset of truth, uncertainty and falsity is used instead of a single number because in many cases one cannot know with precision the percentage of truth or falsity, but these can be approximated. For example, a supposition can be 30% to 40% true and 60% to 70% false [60].

4.1.2. Neutrosophic Components Definition

Let T, I, F be three standard or non-standard subsets of $]^{-}0, 1^{+}[$ with

$$\begin{array}{ll} \sup T = t_{sup} & \inf T = t_{inf} \\ \sup I = i_{sup} & \inf I = i_{inf} \\ \sup F = f_{sup} & \inf F = f_{inf} \end{array}$$

and

$$\begin{aligned} n_{sup} &= t_{sup} + i_{sup} + f_{sup} \\ n_{inf} &= t_{inf} + i_{inf} + f_{inf} \end{aligned}$$

The $T, I,$ and F sets are not always intervals, but can be subsets: discrete or continuum; with a single element; finite or infinite (the elements are countable or uncountable); subsets union or intersection. Also, these subsets can overlap, and the real subsets represent the relative errors in finding the t, i, and f values (when the T, I, and F subsets are reduced to single points).

T, I, and F are called the neutrosophic components and represent the truth, uncertainty, and falsity values, when referring to neutrosophy, neutrosophic logic, neutrosophic sets, neutrosophic probability, or neutrosophic statistics.

This representation is closer to the human reasoning and defines knowledge imprecision or linguistic inaccuracy received from different observers (this is why T, I, and F are subsets and can be more that a set of points), the uncertainty given by incomplete knowledge or data acquisition errors (for this we have the set I) and the vagueness caused by missing edges or limits.

After defining the sets, we need to specify their superior (x_{sup}) and inferior (x_{inf}) limits because in most of the cases they will be needed [61,62].

4.2. Dezert–Smarandache Theory (DSmT)

To develop artificial cognitive systems a good management of sensor information is required. When the input data are gathered by different sensors, according to the environment certain situations may appear when one of the sensors cannot give correct information or the information is contradictory between sensors. To resolve this issue a strong mathematical model is required, especially when the information is inaccurate or uncertain.

The Dezert–Smarandache Theory (DSmT) [53,54,60] can be considered an extension of Dempster–Shafer theory (DST) [46]. DSmT allows information combining, gathered from different and independent sources as trust functions. DSmT can be used for solving information fusion on static or dynamic complex problems, especially when the information differences between the observers are very high.

DSmT starts by defining the notion of a DSm free model, denoted by $\mathcal{M}^f(\Theta)$ and says that Θ is a set of exhaustive elements θ_i, $i = 1, \ldots, n$ which cannot overlap. This model is *free* because there are no other suppositions over the hypothesis. As long as the DSm free model is fulfilled, we can apply the associative and commutative DSm rule of combination.

DSm theory [62] is based on defining the Dedekind lattice, known as the hyper power set of frame Θ. In DSmT, Θ is considered a set $\{\theta_1, \ldots, \theta_n\}$ of n exhaustive elements, without adding other constraints.

DSmT can tackle information samples, gathered from different information sources which do not allow the same interpretation of the set Θ elements. Let $\Theta = \theta_1, \theta_2$ be the simple case, made of two assumption, then [54]:

- the probability theory works (assuming exclusivity and completeness assumptions) with basic probability assignments (bpa) $m(.) \in [0,1]$ such that

$$m(\theta_1) + m(\theta_2) = 1$$

- the Dempster–Shafer theory works, (assuming exclusivity and completeness assumptions) with basic belief assignments (bba) $m(.) \in [0,1]$ such that

$$m(\theta_1) + m(\theta_2) + m(\theta_1 \cup \theta_2) = 1$$

- the DSm theory works (assuming exclusivity and completeness assumptions) with basic belief assignment (bba) $m(.) \in [0,1]$ such that

$$m(\theta_1) + m(\theta_2) + m(\theta_1 \cup \theta_2) + m(\theta_1 \cap \theta_2) = 1$$

4.2.1. The D^Θ Hyperpower Set Notion

One of the base elements of DSm theory is the notion of hyper power set. Let $\Theta = \{\theta_1, \ldots, \theta_n\}$ be a finite set (called frame) with n exhaustive elements. The Dedekind lattice, called hyper power set D^Θ within DSmT frame, is defined as the set of all built statements from the elements of set Θ with the $\cup$ and $\cap$ operators such that:

1. $\emptyset, \theta_1, \ldots, \theta_n \in D^\Theta$
2. If A, $B \in D^\Theta$, then $A \cap B \in D^\Theta$ and $A \cup B \in D^\Theta$.
3. No other element is included in D^Θ with the exception of those mentioned at 1 and 2.

D^{Θ} duals (obtained by changing within expressions the operator $\cap$ with the operator $\cup$) is D^{Θ}. In D^{Θ} there are elements that are dual with themselves. The cardinality of D^{Θ} increases with 2^n when the cardinality of Θ is n. Generating the D^{Θ} hyper power set is close connected with the Dedekind [54, 55] known problem of isotone Boolean function set. Because for any finite set Θ, $|D^{\Theta}| \geq |2^{\Theta}|$, we call D^{Θ} the hyper power set of Θ.

The $\theta_i, i = 1, \ldots, n$ elements from Θ form the finite set of suppositions/concepts that characterize the fusion problem. D^{Θ} represents the free model of DSm $\mathcal{M}^f(\Theta)$ and allows working with fuzzy concepts that describe the intrinsic and relative character. This kind of concept cannot be accurately distinguished within an absolute interpretation because of the unapproachable universal truth.

With all of this, there are certain particular fusion problems that imply discrete concepts, where the θ_i elements are exclusively true. In this case, all the exclusivity constraints of $\theta_i, i = 1, \ldots, n$ must be included in the previous model to properly characterize the truthiness character of the fusion problem and to match reality. For this, the hyper power set D^{Θ} is decreased to the classic power set 2^{Θ}, forming the smallest hybrid DSm model, noted with $\mathcal{M}^0(\Theta)$, and coincides with Shafer's model.

Besides the problem types that correspond with the Shaffer's model $\mathcal{M}^0(\Theta)$ and those that correspond with the DSm free model $\mathcal{M}^f(\Theta)$, there is an extensive class of fusion problems that include in Θ states, continuous fuzzy concepts and discrete hypothesis. In this case we must take into consideration certain exclusivity constraints and some non-existential constraints. Each fusion hybrid problem is described by a DSm hybrid model $\mathcal{M}(\Theta)$ with $\mathcal{M}(\Theta) \neq \mathcal{M}^f(\Theta)$ and $\mathcal{M}(\Theta) \neq \mathcal{M}^0(\Theta)$.

4.2.2. Generalized Belief Functions

Starting from a general frame Θ, we define a $D^{\Theta} \to [0,1]$ transformation associated with an information source $\mathcal{B}$ like [54]

$$m(\varnothing) = 0 \text{ and } \sum_{A \in D^{\Theta}} m(A) = 1. \tag{1}$$

The m(A) value is called generalized basic belief assignment of A.

The generalized trust and plausibility are defined in the same way as in Dempster–Shafer theory [47]

$$\text{Bel}(A) = \sum_{\substack{B \subseteq A \\ B \in D^{\Theta}}} m(B), \tag{2}$$

$$\text{Pl}(A) = \sum_{\substack{B \cap A \neq \varnothing \\ B \in D^{\Theta}}} m(B). \tag{3}$$

These definitions are compatible with the classic trust function definition from the Dempster–Shafer theory when D^{Θ} is reduced to 2^{Θ} for fusion problems where the Shafer model $\mathcal{M}^0(\Theta)$ can be applied. We still have $\forall\, A \in D^{\Theta}$, $\text{Bel}(A) \leq \text{Pl}(A)$. To notice that when we work with the free DSm $\mathcal{M}^f(\Theta)$ model, we will always have $\text{Pl}(A) = 1$, $\forall\, A \neq \varnothing \in D^{\Theta}$, which is normal [54].

4.2.3. DSm Classic Rule of Combination

When the DSm free model $\mathcal{M}^f(\Theta)$ can be applied, the combination rule $m_{\mathcal{M}^f(\Theta)} \equiv m(.) \triangleq [m_1 \otimes m_2](.)$ of two independent sources $\mathcal{B}_1$ and $\mathcal{B}_2$ that provide information on the same frame Θ with the belief functions $\text{Bel}_1(.)$ and $\text{Bel}_2(.)$ associated to gbba $m_1(.)$ and $m_2(.)$ correspond to the conjunctive consensus of sources. Data combinations are done by using the formula [54]

$$\forall C \in D^{\Theta},\ m_{\mathcal{M}^f(\Theta)}(C) \equiv m(C) = \sum_{\substack{A, B \in D^{\Theta} \\ A \cap B = C}} m_1(A) m_2(B). \tag{4}$$

Because D^{Θ} is closed under $\cap$ and $\cup$ operators, this new combination rule guarantees that $m(.)$ Is a generalized trust value, meaning that $m(.) : D^{\Theta} \to [0,1]$. This rule of combination is commutative and associative and can be used all the time for sources fusion which implies fuzzy concepts. This rule can be extended with ease for combining $k > 2$ independent information sources [55,56].

Because of the high number of elements in D^{Θ}, when the cardinality of Θ increases, the need of computational resources also increases for processing the DSm combination rule. This observation is true only if the core (the set of generalized basic belief assignment for the needed elements) $\mathcal{K}_1(m_1)$ and $\mathcal{K}_2(m_2)$ coincide with D^{Θ}, meaning that when $m_1(A) > 0$ and $m_2(A) > 0$ for any $A \neq \varnothing \in D^{\Theta}$. For most practical applications, the $\mathcal{K}_1(m_1)$ and $\mathcal{K}_2(m_2)$ dimensions are much smaller than $|D^{\Theta}|$ because the information sources provide most of the time the generalized basic belief assignment for only one subset of hyper power set. This facilitates the DSm classic rule implementation.

Figure 1 presents the DSm combination rule architecture. The first layer is formed by all the generalized basic belief assignment values of the needed elements A_i, $i = 1,\ldots,n$ of $m_1(.)$. The second layer is made out of all the generalized basic belief assignment values B_i, $i = 1,\ldots,k$ of $m_2(.)$. Each node from the first layer is connected with each node of the second layer. The output layer is created by combining the generalized basic belief assignment values of all the possible intersections $A_i \cap B_j$, $i = 1,\ldots,$ n and $j = 1,\ldots,$ k. If we would have a third source to provide generalized basic belief assignment values $m_3(.)$, this would have been combined by placing it between the output layer and the second one that provides the generalized basic belief assignment values $m_2(.)$. Due to the commutative and associative properties of DSm classic rule of combination, in developing the DSm network, a particular order of layers is not required [54].

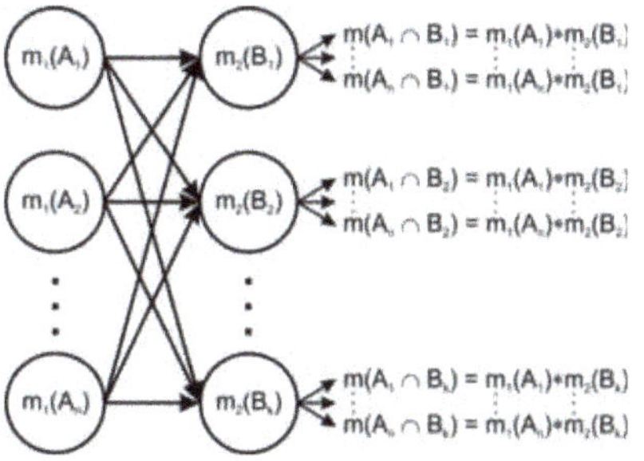

Figure 1. Graphical representation of DSm classic rule of combination for $\mathcal{M}^f(\Theta)$ [35].

5. Decision-Making Algorithm

As observed in this paper, according to the object shape and assigned task, grasping is divided into eight categories [63]: spherical grasping, cylindrical grasping, lateral grasp, prismatic grasp, circular grasping, hook grasping, button pressing, and pushing. From these grasping types, the most used ones are cylindrical and prismatic grasping (see Table 1). These can be used in almost any situation and we can say that spherical grasping is a particular grasping of these two. The spherical grasp is used for power grasping, when the contact with the object is achieved with all the fingers' phalanges and the hand's palm. This is why a requirement for classification by the shape of the object is needed. Due to the fact that these types are more often encountered, they were taken into consideration for the studied fusion problems.

The fusion problem aims to achieve a classification, by shape of objects to grasp, so that these can match with the other three types of grasping studied. The target objects are classified into three categories: sphere, parallelepiped, and cylinder. For each category, a grasping type is assigned [56].

Following the presented theory in Section 4, the information is provided by two independent sources (observers): a stereovision system and a Kinect sensor. The observers are presented in Section 3, and are used to scan the robot's work environment. By using the information provided by the two observers, a 3D virtual image of the environment is achieved, from which the human operator choses the object to be grasped, thus defining the grasping task that must be achieved by the robot. The 3D image of the object, isolated from the scene, is compared with three templates—formed by similar methods—which represent a sphere, a parallelepiped, and a cylinder. Afterwards, a template matching algorithm is applied to place the object in one of the three categories, with a certain matching percentage. This percentage can vary according to the conditions in which the images are obtained (weak light, object from which the light is reflected, etc.). The data taken from each sensor are then individual processed with a neutrosophication algorithm, with the purpose of obtaining the generalized basic belief assignment values for each hypothesis that can characterize the system. In the next step, having the basic belief assignment values, we combine the data provided by the two observers by using the classic DSm rule of combination. The next step is to apply a deneutrosophication algorithm on the obtained values, to achieve the decision on the shape of the object by placing it into the three categories mentioned above. The entire process is visually represented in Figure 2.

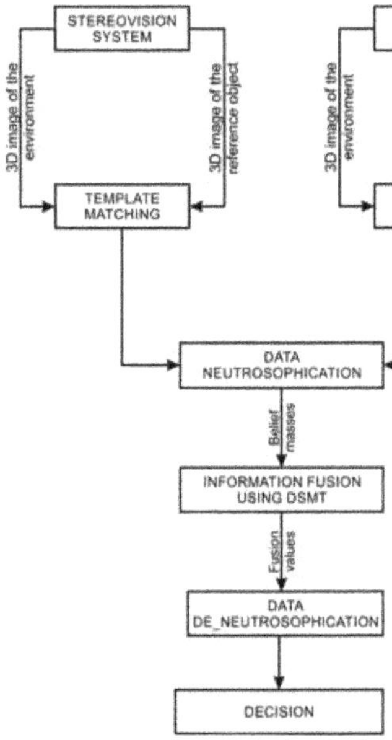

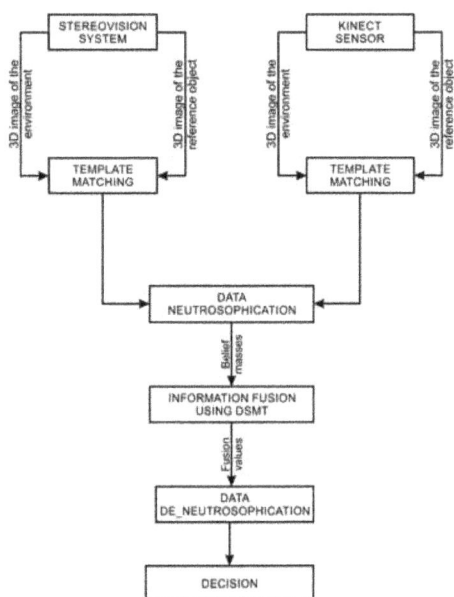

Figure 2. Diagram of the proposed algorithm.

5.1. Data Neutrosophication

Each observer provides a truth percentage for each system's state. The state set $\Theta = \{\theta_1, \theta_2, \theta_3\}$ that characterizes the fusion problem is

$$\Theta = \{Sp, Pa, Cy\}, \tag{5}$$

where Sp = sphere, Pa = parallelepiped, and Cy = cylinder.

To compute the belief values for the hyper power set D^Θ elements we developed an algorithm based on the neutrosophic logic. The hyper power set D^Θ is formed by using the method presented in Section 4.2.1 and has the form

$$D^\Theta = \{\varnothing, Sp,\ Pa,\ Cy, Sp \cup Pa, Sp \cup Cy, Cy \cup Pa, Sp \cap Pa, Sp \cap Cy, Cy \cap Pa, Sp \cap (Cy \cup Pa), Cy \cap (Sp \cup Pa), Pa \cap (Cy \cup Sp), Sp \cup Cy \cup Pa, Sp \cap Cy \cap Pa\}. \tag{6}$$

The statements of each observer are handled in ways of truth (T), uncertainty (I), and falsity (F), specific to the neutrosophic logic. Due to the fact that $F = 1 - T - I$, the statements of falsity are not taken into consideration.

The neutrosophic algorithm has as input the certainty probabilities (truth) provided by the observers on the system's states. These probabilities are then processed using the described rules in Figure 3. If the difference between the certainties probabilities used at a certain point by the processing algorithm is larger than a certain threshold found by trial and error, then we will consider that the uncertainty percentage between the compared states is null, and the probability that one of the states is true increases. In the case where this difference is not a set threshold, we compute the uncertainty probability by using the formula

$$m(A \cup B) = 1 - \frac{m(A) - m(B)}{const} \quad (7)$$

where $A, B \in \Theta$, and "*const*" depends of the chosen threshold. While the point determined by the two probabilities approaches the main diagonal, the uncertainty approaches the maximum probability value.

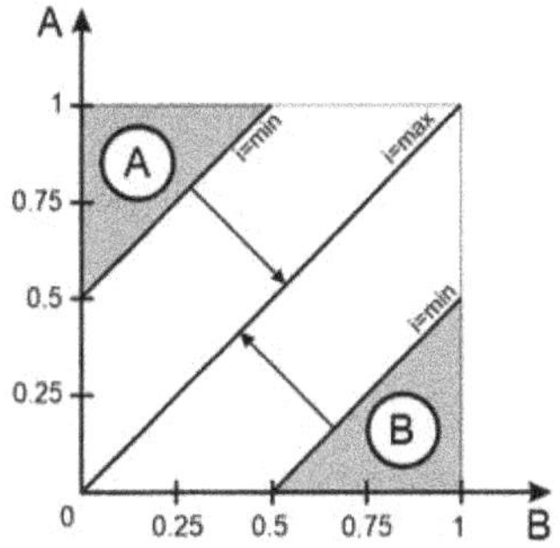

Figure 3. Data neutrosophication rule for the observer's data.

From the hyper power set D^{Θ}, we can determine the belief masses only for the values $Obs_i(D^{\Theta})$ (information obtained after observer's data interpretation) presented below, because the intersection operation $\cap$ represents contradiction in DSm theory and we cannot compute the contradiction values for a single observer using

$$Obs_i\left(D^{\Theta}\right) = \{Sp,\ Pa,\ Cy, Sp \cup Pa, Sp \cup Cy, Cy \cup Pa, Sp \cup Cy \cup Pa\}. \quad (8)$$

The neutrosophic probabilities are detailed in Table 2.

Table 2. Grasping position for certain tasks.

Mathematical Representation	Description
Sp	Certainty that the target object is a 'sphere'
Pa	Certainty that the target object is a 'parallelepiped'
Cy	Certainty that the target object is a 'cylinder'
$Sp \cup Pa$	Uncertainty that the target object is a 'sphere' or 'parallelepiped'
$Sp \cup Cy$	Uncertainty that the target object is a 'sphere' or 'cylinder'
$Cy \cup Pa$	Uncertainty that the target object is a 'cylinder' or 'parallelepiped'
$Sp \cup Cy \cup Pa$	Uncertainty that the target object is a 'sphere', 'cylinder', or 'parallelepiped'

5.2. Information Fusion

Having known the trust values of the hyper power set elements $Obs_i(D^{\Theta})$, presented in Table 2, we apply the fusion algorithm, using the classic DSm combination rule, detailed in Section 4.2.3.

Appling Equation (4), we get the following formulas for the combination values:

$$\begin{aligned} m(Sp) = \; & m_1(Sp)\cdot m_2(Sp) + m_1(Sp)\cdot m_2(Sp\cup Pa) + m_1(Sp\cup Pa)\cdot m_2(Sp) + m_1(Sp)\cdot m_2(Sp\cup Cy) \\ & + m_1(Sp\cup Cy)\cdot m_2(Sp) + m_1(Sp)\cdot m_2(Sp\cup Cy\cup Pa) + m_1(Sp\cup Cy\cup Pa)\cdot m_2(Sp) \\ & + m_1(Sp\cup Pa)\cdot m_2(Sp\cup Cy) + m_1(Sp\cup Cy)\cdot m_2(Sp\cup Pa) \end{aligned} \tag{9}$$

$$\begin{aligned} m(Pa) = \; & m_1(Pa)\cdot m_2(Pa) + m_1(Pa)\cdot m_2(Sp\cup Pa) + m_1(Sp\cup Pa)\cdot m_2(Pa) + m_1(Pa) \\ & \cdot m_2(Cy\cup Pa) + m_1(Cy\cup Pa)\cdot m_2(Sp) + m_1(Pa)\cdot m_2(Sp\cup Cy\cup Pa) \\ & + m_1(Sp\cup Cy\cup Pa)\cdot m_2(Pa) + m_1(Sp\cup Pa)\cdot m_2(Cy\cup Pa) \\ & + m_1(Cy\cup Pa)\cdot m_2(Sp\cup Pa) \end{aligned} \tag{10}$$

$$\begin{aligned} m(Cy) = \; & m_1(Cy)\cdot m_2(Cy) + m_1(Cy)\cdot m_2(Cy\cup Pa) + m_1(Cy\cup Pa)\cdot m_2(Cy) + m_1(Cy) \\ & \cdot m_2(Sp\cup Cy) + m_1(Sp\cup Cy)\cdot m_2(Cy) + m_1(Cy)\cdot m_2(Sp\cup Cy\cup Pa) \\ & + m_1(Sp\cup Cy\cup Pa)\cdot m_2(Cy) + m_1(Sp\cup Cy)\cdot m_2(Cy\cup Pa) \\ & + m_1(Cy\cup Pa)\cdot m_2(Sp\cup Cy) \end{aligned} \tag{11}$$

$$\begin{aligned} m(Sp\cup Pa) = \; & m_1(Sp\cup Pa)\cdot m_2(Sp\cup Pa) + m_1(Sp\cup Pa)\cdot m_2(Sp\cup Cy\cup Pa) \\ & + m_1(Sp\cup Cy\cup Pa)\cdot m_2(Sp\cup Pa) \end{aligned} \tag{12}$$

$$\begin{aligned} m(Sp\cup Cy) = \; & m_1(Sp\cup Cy)\cdot m_2(Sp\cup Cy) + m_1(Sp\cup Cy)\cdot m_2(Sp\cup Cy\cup Pa) \\ & + m_1(Sp\cup Cy\cup Pa)\cdot m_2(Sp\cup Cy) \end{aligned} \tag{13}$$

$$\begin{aligned} m(Cy\cup Pa) = \; & m_1(Cy\cup Pa)\cdot m_2(Cy\cup Pa) + m_1(Cy\cup Pa)\cdot m_2(Sp\cup Cy\cup Pa) \\ & + m_1(Sp\cup Cy\cup Pa)\cdot m_2(Cy\cup Pa) \end{aligned} \tag{14}$$

$$m(Sp\cup Cy\cup Pa) = m_1(Sp\cup Cy\cup Pa)\cdot m_2(Sp\cup Cy\cup Pa) \tag{15}$$

During the fusion process, between the information provided by the two observers contradiction situations may appear. These are included in the hyper power set D^{Θ} and are described in Table 3.

Table 3. Contradictions that may appear between the neutrosophic probabilities.

Mathematical Representation	Description
$Sp\cap Pa$	Contradiction between the certainties that the target object is a 'sphere' and 'parallelepiped'
$Sp\cap Cy$	Contradiction between the certainties that the target object is a 'sphere' and 'cylinder'
$Cy\cap Pa$	Contradiction between the certainties that the target object is a 'cylinder' and 'parallelepiped'
$Sp\cap (Cy\cup Pa)$	Contradiction between the certainty that the target object is a 'sphere' and the uncertainty that the target object is a 'cylinder' or 'parallelepiped'
$Pa\cap (Sp\cup Cy)$	Contradiction between the certainty that the target object is a 'parallelepiped' and the uncertainty that the target object is a 'sphere' or 'cylinder'
$Cy\cap (Pa\cup Sp)$	Contradiction between the certainty that the target object is 'cylinder' and the uncertainty that the target object is a 'parallelepiped' or 'sphere'
$Sp\cap Cy\cap Pa$	Contradiction between the certainties that the target object is a 'sphere', 'cylinder', and 'parallelepiped'

Fusion values for contradiction are determined as

$$m(Sp\cap Pa) = m_1(Sp)\cdot m_2(Pa) + m_1(Pa)\cdot m_2(Sp) \tag{16}$$

$$m(Sp \cap Cy) = m_1(Sp) \cdot m_2(Cy) + m_1(Cy) \cdot m_2(Sp) \tag{17}$$

$$m(Cy \cap Pa) = m_1(Cy) \cdot m_2(Pa) + m_1(Pa) \cdot m_2(Cy) \tag{18}$$

$$m(Sp \cap (Cy \cup Pa)) = m_1(Sp) \cdot m_2(Cy \cup Pa) + m_1(Cy \cup Pa) \cdot m_2(Sp) \tag{19}$$

$$m(Pa \cap (Sp \cup Cy)) = m_1(Pa) \cdot m_2(Sp \cup Cy) + m_1(Sp \cup Cy) \cdot m_2(Pa) \tag{20}$$

$$m(Cy \cap (Sp \cup Pa)) = m_1(Cy) \cdot m_2(Sp \cup Pa) + m_1(Sp \cup Pa) \cdot m_2(Cy) \tag{21}$$

5.3. Data Deneutrosophication and Decision-Making

The combination values found in the previous section are deneutrosophicated using the logic diagram presented in Figure 4. For the decision-making algorithm we opted to use Petri nets [64], for it is easier to notice the system's states transitions. The decision-making diagram proved to have a certain difficulty level, which required adding three sub diagrams:

1. sub_p1 (Figure 5)—this sub diagram deals with the contradiction between:
   - the certainty that the target object is a 'sphere' and the uncertainty that the target object is either a 'parallelepiped' or a 'cylinder'.
   - the certainty that the target object is a 'parallelepiped' and the uncertainty that the target object is either a 'sphere' or a 'cylinder'.
   - the certainty that the target object is a 'cylinder' and the uncertainty that the target object is either a 'parallelepiped' or a 'sphere'.
2. sub_p2 (Figure 6)—this sub diagram deals with the contradiction between:
   - The certainty that the target object is a 'sphere' and a 'parallelepiped'.
   - The certainty that the target object is a 'sphere' and a 'cylinder'.
   - The certainty that the target object is a 'cylinder' and a 'parallelepiped'.
3. sub_p3 (Figure 7)—this sub diagram deals with the uncertainty that the target object is:
   - a 'sphere' or a 'parallelepiped'
   - a 'sphere' or a 'cylinder'
   - a 'cylinder' or a 'parallelepiped'

To not overload Figures 4–7 we have the following notations:

$$\begin{aligned}
&A = \{m(Sp \cap (Cy \cup Pa)), m(Pa \cap (Sp \cup Cy)), m(Cy \cap (Pa \cup Sp))\},\\
&B = \{m(Sp \cap Pa), m(Sp \cap Cy), m(Cy \cap Pa)\},\\
&C = \{m(Sp \cup Pa), m(Sp \cup Cy), m(Cy \cup Pa)\},\\
&D = \{m(Sp), m(Pa), m(Cy)\},\\
&a = m(Sp \cup Cy),\\
&b = m(Pa \cup Sp),\\
&c = m(Cy \cup Pa).
\end{aligned}$$

With the help of the Petri diagram (Figure 4), we take the decision of sorting the target object in one of the three categories, as follows:

1. Determine $max(m(Sp \cap (Cy \cup Pa)), m(Pa \cap (Sp \cup Cy)), m(Cy \cap (Pa \cup Sp)))$.
   - If $max(m(Sp \cap (Cy \cup Pa)), m(Pa \cap (Sp \cup Cy)), m(Cy \cap (Pa \cup Sp))) = m(Sp \cap (Cy \cup Pa))$, the contradiction between the certainty value that the target object is a 'sphere' and the uncertainty value that the target object is a 'cylinder' or 'parallelepiped' is compared with

a threshold determined through an experimental trial-error process. If this is higher than or equal to the chosen threshold, the target object is a 'sphere'.

- If $max(m(Sp \cap (Cy \cup Pa)), m(Pa \cap (Sp \cup Cy)), m(Cy \cap (Pa \cup Sp))) = m(Pa \cap (Sp \cup Cy))$, the contradiction between the certainty value that the target object is 'parallelepiped' and the uncertainty value that the target object is a 'sphere' or 'cylinder' is compared with the threshold mentioned above. If this is higher than or equal to the chosen threshold, the target object is a 'parallelepiped'.
- If $max(m(Sp \cap (Cy \cup Pa)), m(Pa \cap (Sp \cup Cy)), m(Cy \cap (Pa \cup Sp))) = m(Cy \cap (Pa \cup Sp))$, the contradiction between the certainty value that the target object is a 'cylinder' and the uncertainty value that the target object is a 'parallelepiped' or 'sphere' is compared with the threshold mentioned above. If this is higher than or equal to the chosen threshold, the target object is a 'cylinder'.

If none of the three conditions are met, we proceed to the next step:

2. Determine $max(m(Sp \cap Pa), m(Sp \cap Cy), m(Cy \cap Pa))$

- If $max(m(Sp \cap Pa), m(Sp \cap Cy), m(Cy \cap Pa)) = m(Sp \cap Pa)$, the contradiction between the certainty values that the target object is a 'sphere' and 'parallelepiped' is compared with a threshold determined through an experimental trial-error process. If this is higher or equal with the chosen threshold, we check if $m(Sp) + m(Sp \cup Cy) > m(Pa) + m(Cy \cup Pa)$. If this condition if fulfilled, then the target objects is a 'sphere'. Otherwise, the target object is 'parallelepiped'.
- If $max(m(Sp \cap Pa), m(Sp \cap Cy), m(Cy \cap Pa)) = m(Sp \cap Cy)$, the contradiction between the certainty values that the target object is a 'sphere' and 'cylinder' is compared with the threshold mentioned above. If this is higher or equal with the chosen threshold, we check if $(Sp) + m(Sp \cup Pa) > m(Cy) + m(Cy \cup Pa)$. If this condition if fulfilled, then the target objects is a 'sphere'. Otherwise, the target object is a 'cylinder'.
- If $max(m(Sp \cap Pa), m(Sp \cap Cy), m(Cy \cap Pa)) = m(Cy \cap Pa)$, the contradiction between the certainty values that the target object is a 'cylinder' and a 'parallelepiped' is compared with the threshold mentioned above. If this is higher or equal with the chosen threshold, we check if $m(Cy) + m(Sp \cup Cy) > m(Pa) + m(Sp \cup Pa)$. If this condition if fulfilled, then the target objects is a 'cylinder'. Otherwise, the target object is a 'parallelepiped'.

If in none of the situations, the contradiction is not larger that the chosen threshold, we go to the next step:

3. Determine $max(m(Sp \cup Pa), m(Sp \cup Cy), m(Cy \cup Pa))$

- If $max(m(Sp \cup Pa), m(Sp \cup Cy), m(Cy \cup Pa)) = m(Sp \cup Pa)$, the uncertainty probability that the target object is a 'sphere' or ' parallelepiped' is larger than a threshold determined through an experimental trial-error process, we check if $m(Sp) > m(Pa)$. If the condition is fulfilled, the target object is a 'sphere'. Otherwise, the target object is a 'parallelepiped'.
- If $max(m(Sp \cup Pa), m(Sp \cup Cy), m(Cy \cup Pa)) = m(Sp \cup Cy)$, the uncertainty probability that the target object is a 'sphere' or 'cylinder' is larger than the threshold mentioned above, we check if $m(Sp) > m(Cy)$. If the condition is fulfilled, the target object is a 'sphere'. Otherwise, the target object is a 'cylinder'.
- If $max(m(Sp \cup Pa), m(Sp \cup Cy), m(Cy \cup Pa)) = m(Cy \cup Pa)$, the uncertainty probability that the target object is a 'cylinder' or ' parallelepiped' is larger than the threshold mentioned above, we check if $m(Cy) > m(Pa)$. If the condition is fulfilled, the target object is a 'cylinder'. Otherwise, the target object is a 'parallelepiped'.

If none of the hypotheses mentioned above are not fulfilled, we go to the next step:

4. Determine $max(m(Sp), m(Pa), m(Cy))$

   - If $max(m(Sp), m(Pa), m(Cy)) = m(Sp)$, the target object is a 'sphere'.
   - If $max(m(Sp), m(Pa), m(Cy)) = m(Pa)$, the target object is a 'parallelepiped'.
   - If $max(m(Sp), m(Pa), m(Cy)) = m(Cy)$, the target object is a 'cylinder'.

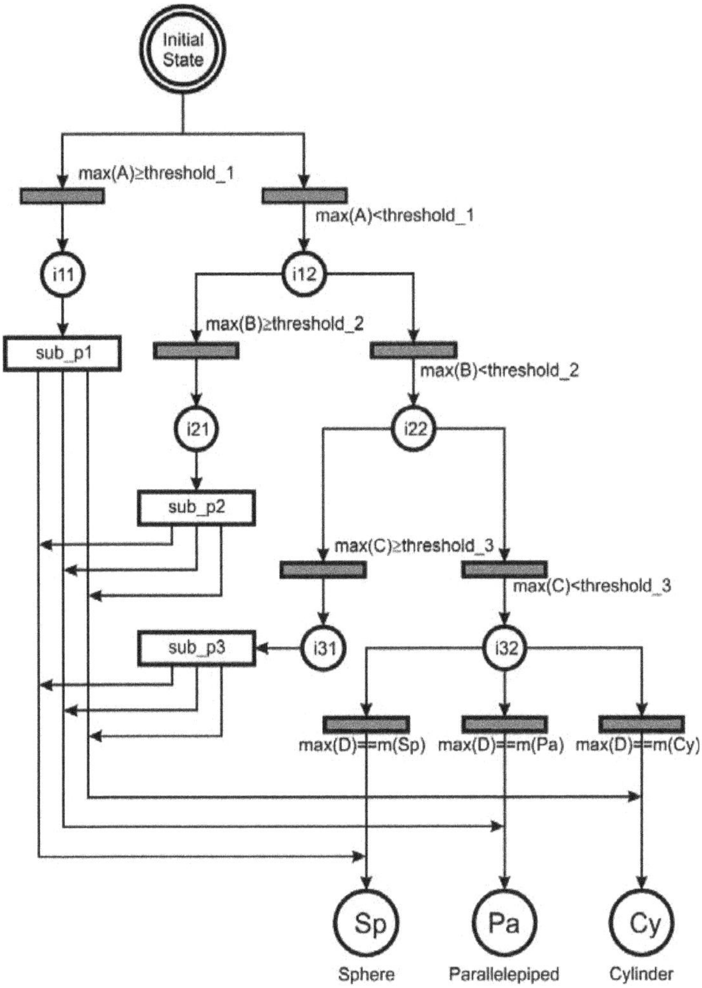

Figure 4. Petri diagram for decision-making algorithm.

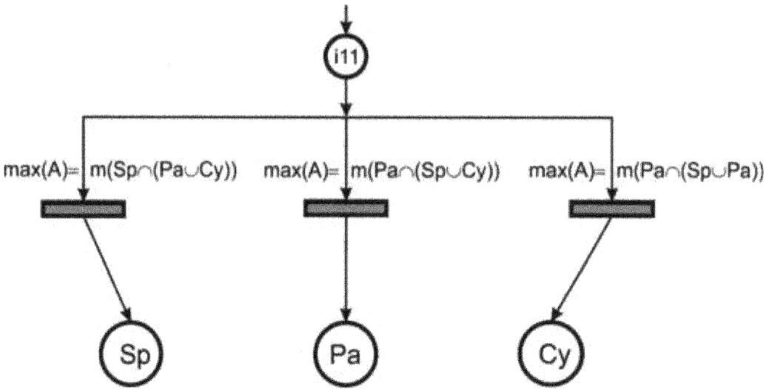

Figure 5. Petri net for sub_p1.

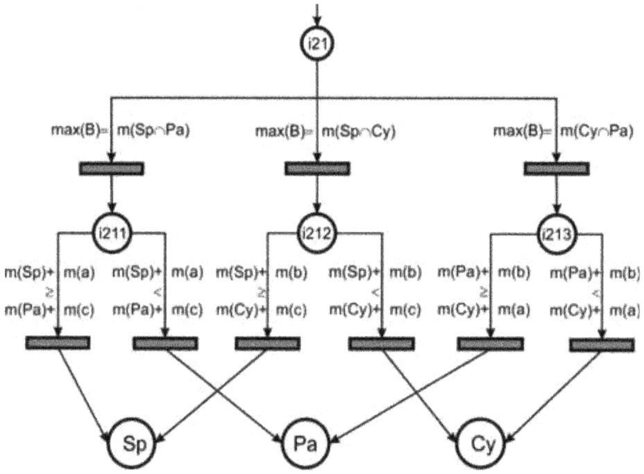

Figure 6. Petri net for sub_p2.

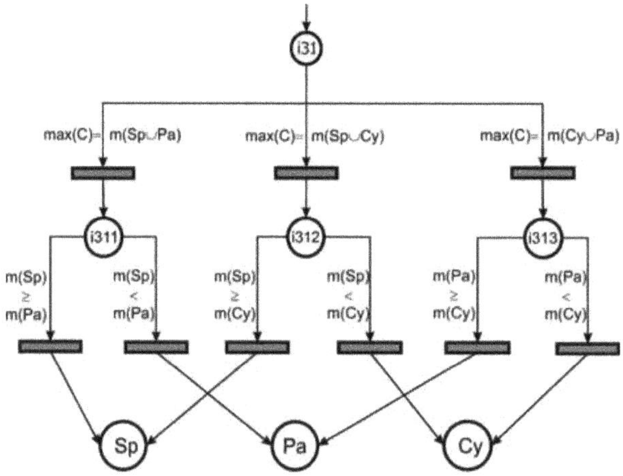

Figure 7. Petri net for sub_p3.

6. Discussion

As mentioned in the introduction chapter, the main goal of this paper is to find a way to grasp objects according to their shape. This is done by classifying the target objects into three main classes: sphere, parallelepiped, and cylinder.

To determine the shape of the target objects, the robot work environment was scanned with a stereovision system and a Kinect sensor, with the purpose of creating a 3D image of the surrounding space in which the robot must fulfill its task. From the two created images, the target object is selected and then it is compared with three templates, which represent a sphere, a cube, and a cylinder. With a template matching algorithm the matching percentage is determined for each of the templates. These percentages (Figure 8), represents the data gathered from the observers, for the fusion problem. Because we wanted to test and verify the decision-making algorithm for as many cases as possible, the observers' values were simulated using sine signals with different frequency and amplitude of 1 (Figure 8). This amplitude represents the maximum probability percentage that a certain type of object is found by the template matching algorithm.

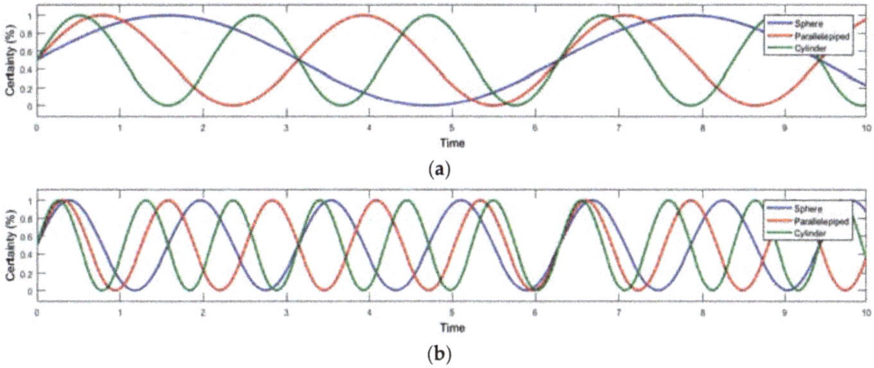

Figure 8. Simulation of the information provided by the two sensors/observers: (**a**) first observer detection; (**b**) second observer detection.

On these input data, we then apply a neutrosophication algorithm with the purpose of obtaining the generalized belief assignment values for each of the statements an observer is doing:

The certainty probability that the object is a 'sphere' (Figure 9a,h)
The certainty probability that the object is a 'parallelepiped' (Figure 9b,i)
The certainty probability that the object is a 'cylinder' (Figure 9c,j)
The uncertainty probability that the object is a 'sphere' or a 'parallelepiped' (Figure 9d,k)
The uncertainty probability that the object is a 'sphere' or a 'cylinder' (Figure 9e,l)
The uncertainty probability that the object is a 'cylinder' or a 'parallelepiped' (Figure 9f,m)
The uncertainty probability that the object is a 'sphere', a 'cylinder', or a 'parallelepiped' (Figure 9g,n).

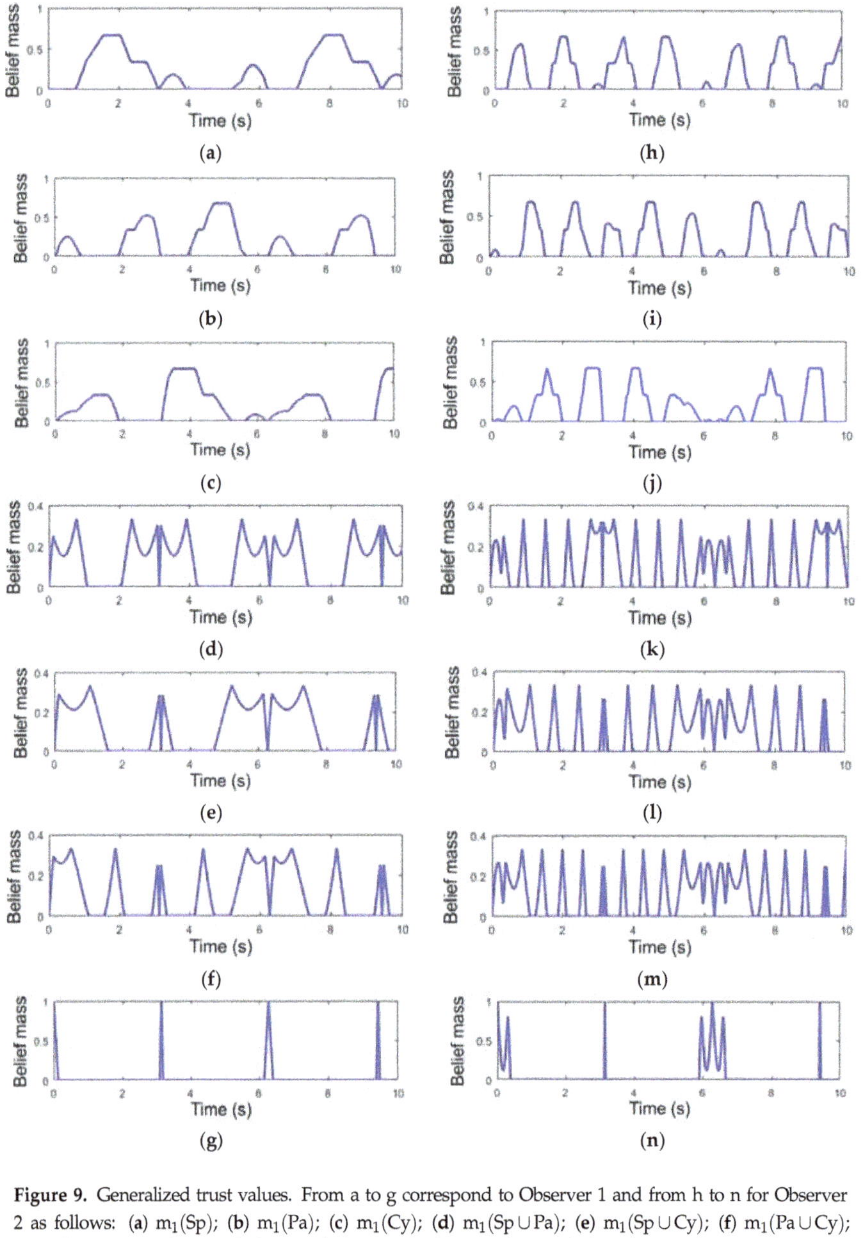

Figure 9. Generalized trust values. From a to g correspond to Observer 1 and from h to n for Observer 2 as follows: **(a)** $m_1(Sp)$; **(b)** $m_1(Pa)$; **(c)** $m_1(Cy)$; **(d)** $m_1(Sp \cup Pa)$; **(e)** $m_1(Sp \cup Cy)$; **(f)** $m_1(Pa \cup Cy)$; **(g)** $m_1(Sp \cup Pa \cup Cy)$; **(h)** $m_2(Sp)$; **(i)** $m_2(Pa)$; **(j)** $m_2(Cy)$; **(k)** $m_2(Sp \cup Pa)$; **(l)** $m_2(Sp \cup Cy)$; **(m)** $m_2(Pa \cup Cy)$; **(n)** $m_2(Sp \cup Pa \cup Cy)$.

After the belief values were computed for each statements of the observers, we go to the data fusion step (Figure 10).

With the help of belief values presented in Figure 10 and computed using the neutrosophication method presented in Sections 5.1 and 5.2, we find the fusion values, presented in Figure 11.

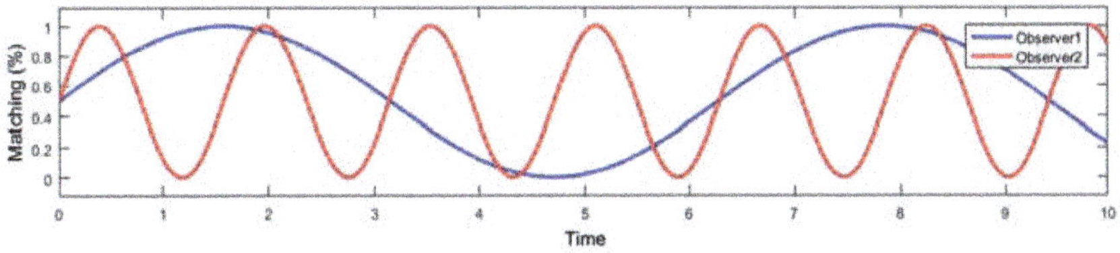

(**a**)

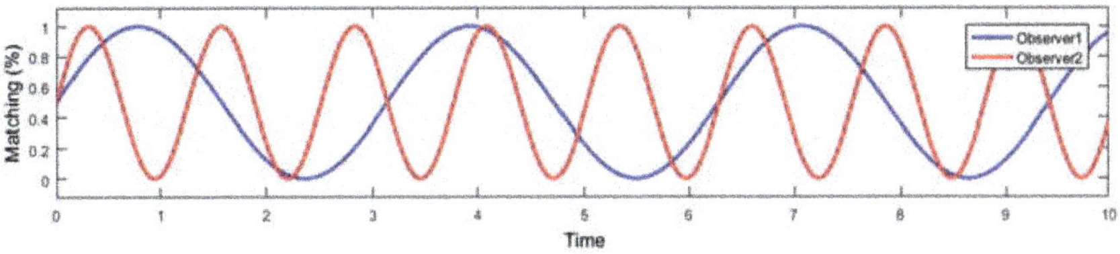

(**b**)

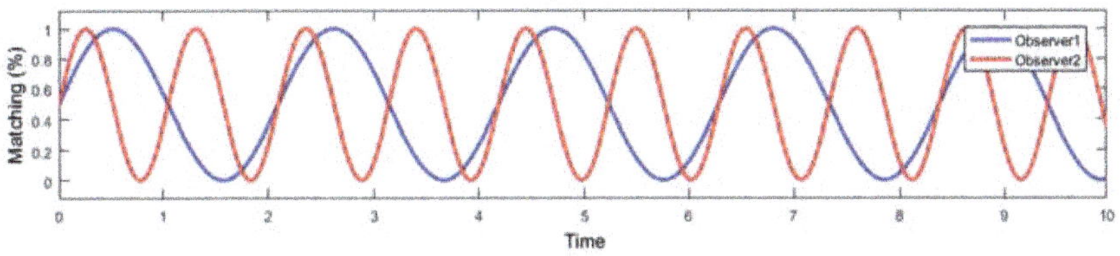

(**c**)

Figure 10. Data fusion: (**a**) Observer 1 vs. Observer 2 for sphere objects; (**b**) Observer 1 vs. Observer 2 for parallelepiped objects; (**c**) Observer 1 vs. Observer 2 for cylinder objects.

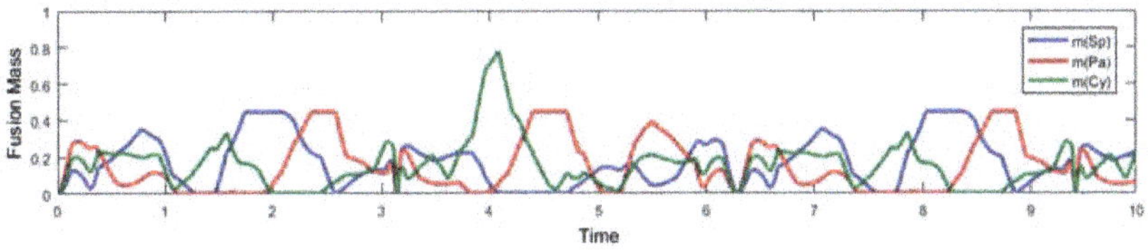

(**a**)

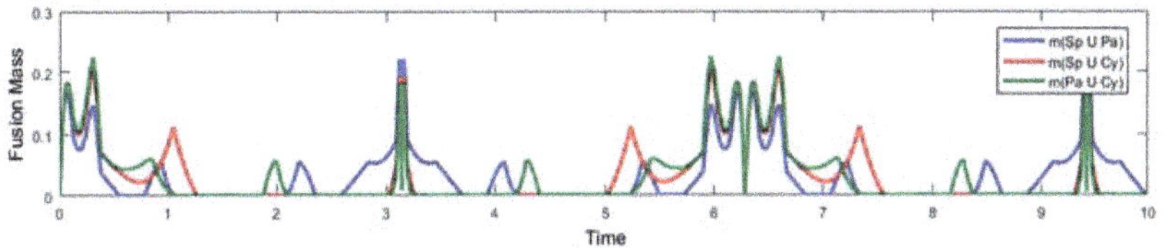

(**b**)

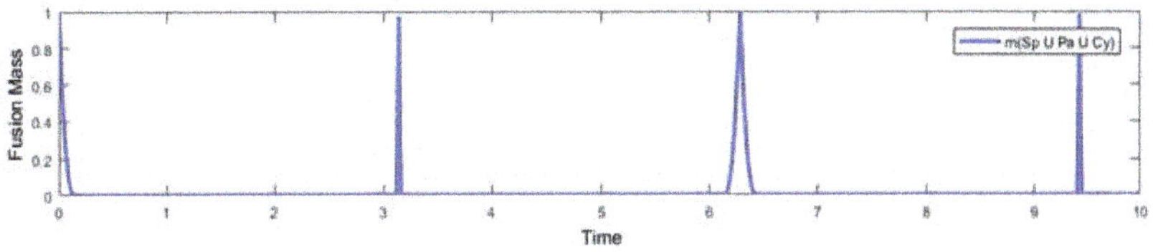

(**c**)

Figure 11. *Cont.*

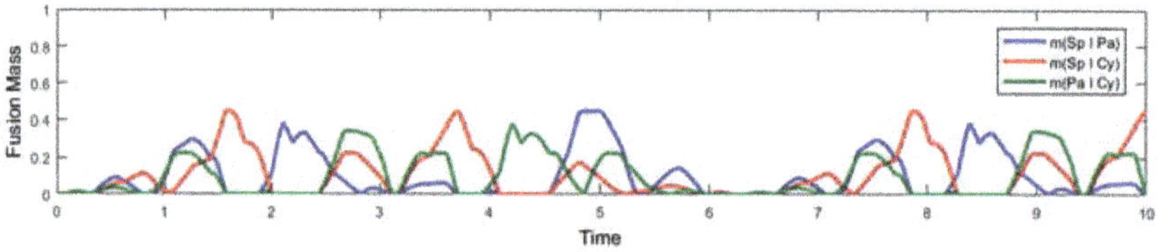

(**d**)

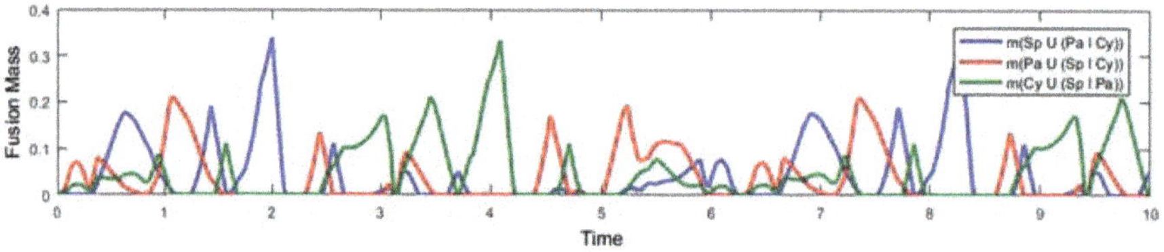

(**e**)

Figure 11. Fusion values: (**a**) Fusion values of m(Sp), m(Pa) and m(Cy); (**b**) Fusion values of m(Sp ∪ Pa), m(Sp ∪ Cy), m(Pa ∪ Cy); (**c**) Fusion value of m(Sp ∪ Pa ∪ Cy); (**d**) Fusion values of m(Sp ∪ Pa), m(Sp ∪ Cy), m(Pa ∪ Cy); (**e**) Fusion values of m(Sp ∪ (Pa ∪ Cy)), m(Pa ∪ (Sp ∪ Cy)), m(Cy ∪ (Sp ∪ Pa)).

Using the fusion values and the decision-making diagram (Figure 4), from Section 5.2, we can sort the desired object into the three categories: sphere, parallelepiped, and cylinder. The obtained results are presented in Figure 12.

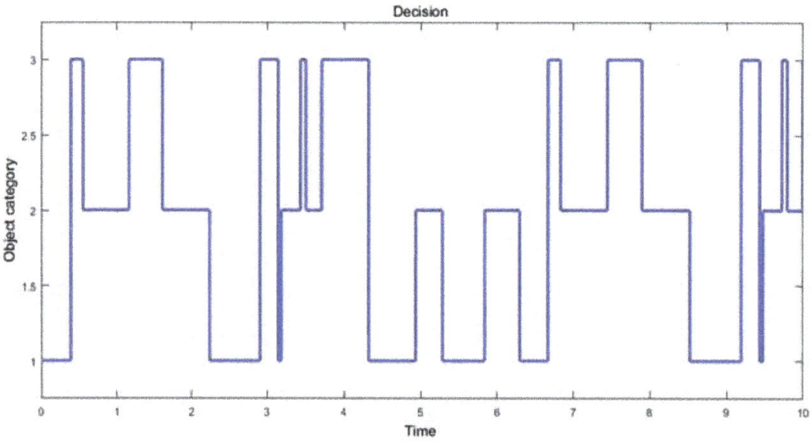

Figure 12. Object category decision, obtained from the proposed algorithm. Value 1 represents decision for sphere, Value 2 represents decision for parallelepiped, and Value 3 represents decision for cylinder.

As one can see in Figure 11, the fusion values for certainty, uncertainty and contradiction are minimum. The only exception is the fusion value for the uncertainty that the target object is *"sphere"* or *"parallelepiped"* or *"cylinder"*, $m(Sp \cup Cy \cup Pa)$, when the data received from the observers are identical and not contradicting, the uncertainty is maximum. This means

$$Obs_1 : 50\%\ sphere,\ 50\%\ parallelepiped,\ 50\%\ cylinder \text{ and}$$

$$Obs_2 : 50\%\ sphere,\ 50\%\ parallelepiped,\ 50\%\ cylinder.$$

Therefore, the system cannot decide on a single state. This is why the robotic hand will maintain its starting position until the system will decide the target object's category. This indecision period of time takes about 0.07 s. When the sensor values about the target object are changed from the equal values presented above, the algorithm is able to provide a solution.

The indecision also reaches high values at the time 3.14s, 6.28s, and 9.42s of the simulation, in the conditions that the observer's statements are close in value with the already presented case from above

$$Obs_1 : 50.08\%\ sphere,\ 50.24\%\ parallelepiped,\ 49.84\%\ cylinder$$

$$Obs_2 : 49.68\%\ sphere,\ 49.52\%\ parallelepiped,\ 50.4\%\ cylinder$$

for the moment 3.14 s

$$Obs_1 : 49.84\%\ sphere,\ 49.52\%\ parallelepiped,\ 49.68\%\ cylinder$$

$$Obs_2 : 49.36\%\ spehere,\ 49.04\%\ parallelepiped,\ 49.2\%\ cylinder$$

for the moment 6.28 s and

$$Obs_1 : 50.24\%\ spehere,\ 50.72\%\ parallelepiped,\ 49.52\%\ cylinder$$

$$Obs_2 : 49.04\%\ spehere,\ 48.57\%\ parallelepiped,\ 51.19\%\ cylinder$$

for the moment 9.42 s.

In Table 4, we present the general belief assignment values, the fusion values and the decision made by the algorithm for the situations previously mentioned.

Table 4. Generalized trust values, fusion values, and decisions for the analyzed situations.

Time		3.14 s		6.28 s		9.42 s	
Source / Hypothesis	State	Obs. 1	Obs. 2	Obs. 1	Obs. 2	Obs. 1	Obs. 2
	Sp	50.08%	49.68%	49.84%	49.36%	50.24%	49.04%
	Pa	50.24%	49.52%	49.52%	49.04%	50.72%	48.57%
	Cy	49.84%	50.4%	49.68%	49.2%	49.52%	51.19%
Generalized belief assignment values							
$m_i(Sp)$		0.0001	0.0001	0.0001	0.0001	0.0005	0.0007
$m_i(Pa)$		0.0001	0	0	0	0.0011	0.0067
$m_i(Cy)$		0	0.0008	0	0	0	0
$m_i(Sp \cup Pa)$		0.0106	0.0234	0.0085	0.0085	0.0317	0.0692
$m_i(Sp \cup Cy)$		0.0106	0.0231	0.0085	0.0085	0.0315	0.0669
$m_i(Cy \cup Pa)$		0.0106	0.0231	0.0085	0.0085	0.0312	0.0662
$m_i(Sp \cup Cy \cup Pa)$		0.9680	0.9296	0.9744	0.9744	0.9040	0.7904
Fusion values							
$m(Sp)$		0.0006		0.0001		0.0054	
$m(Pa)$		0.0006		0.0003		0.0053	
$m(Cy)$		0.0012		0.0002		0.0106	
$m(Sp \cup Pa)$		0.0328		0.0166		0.0898	
$m(Sp \cup Cy)$		0.0325		0.0166		0.0875	
$m(Cy \cup Pa)$		0.0324		0.0166		0.0866	
$m(Sp \cup Cy \cup Pa)$		0.8999		0.9495		0.7145	
$m(Sp \cap Pa)$		0		0		0	
$m(Sp \cap Cy)$		0		0		0	
$m(Cy \cap Pa)$		0		0		0	
$m(Sp \cap (Cy \cup Pa))$		0		0		0.0001	
$m(Pa \cap (Sp \cup Cy))$		0		0		0.0001	
$m(Cy \cap (Sp \cup Pa))$		0		0		0.0002	
Decision		Cylinder		Sphere		Cylinder	

In all three cases, the uncertainty is quite large, and the algorithm ask for restarting the decision process and keeps the decision taken in previous decision process. In our case the decision was that the object is a 'cylinder', 'sphere', and 'cylinder' for the three analyzed points.

Analyzing Figure 11a, at the time of 4.08 s the object is decided to be a 'cylinder' because the probability that the target object is a cylinder is very high, $m(Cy) = 0.7777$.

For the time interval of 4.3–4.9 s, where in Figure 11d the contradiction between the target object being a 'sphere' or a 'parallelepiped' is larger than that the target object is a 'sphere' or a 'cylinder' respectively a 'cylinder' or a 'parallelepiped', the object is decided to be a 'parallelepiped' at first because the probability for it being a 'parallelepiped' is larger than the probability of it being a 'sphere' or a 'cylinder'. This situation is changed starting with second 5 of the simulation, when the probability that the target object is a 'sphere' increase, the probability that the same object is a 'cylinder' remains low and the probability that the target object is a 'parallelepiped' decrease below the value of the 'sphere' probability.

7. Conclusions

Any robot, no mater of its purpose, has a task to fulfill. That task can be either of grasping and manipulation or just a transport task. To successfully complete its task, the robot must be equipped with a number of sensors that will provide enough information about the work environment in which the work is being done.

In this paper, we studied the situation in which the robot is equipped with a stereovision system and a Kinect sensor to detect the environment. The robot's job was to grab and manipulate certain objects. With the help of two different systems, two 3D images of the environment can be created, each one for the two sensor type. In these images, we isolated the target object and it is compared with three template images, obtained through similar methods as the environment images. The three template images represent the 3D virtual model of a sphere, a parallelepiped, and a cylinder. The comparison is achieved with a template matching method, and following that we obtain a matching percentage for each template tested against the desired image.

Because we wanted to develop the decision-making algorithm based on information received from certain template matching methods, we considered as known the information that these algorithms can provide. Moreover, to test different cases, we selected several sine signals that can provide all the different cases that can occur in practice as input for our decision-making algorithm and output for the template matching methods.

The goal of this paper is in part a data fusion problem with the purpose of classifying the objects in visual range of a humanoid robot, so it can fulfill his grasping and manipulation task. We also wanted to label the target object in one of three categories mentioned above, so that during the approach phase on the target object, the robotic hand can prepare for grasping the object, lowering the time needed to complete the task.

The stereovision system and the Kinect sensor presented in Section 3, represent the information sources, called in this paper, the observers, name taken from the neutrosophic logic. These observers specify the state in which the system is. One observer can specify seven states for the searched object.

With the help of neutrosophic logic, we determine the generalized belief values for each of the seven states. The neutrosophic algorithm is applied to information gathered from both of the sensors. We have chosen the neutrosophic logic, because it extends fuzzy logic, providing instruments for also approaching the uncertain situations besides the true and false ones.

Using these belief values, we compute the fusion values on which we apply the classic DSm combination rule, and build the decision-making algorithm presented in Section 5. To help develop this decision-making algorithm we used a Petri net which provided us a clear method of switching through system states under certain conditions.

The decision-making algorithm analyzes the probability of completing all the possible tasks that may appear in sensor data fusion and tackles these possibilities so that for every input the system will have an output.

The presented method can be used successfully in real time applications, because it provides a decision in all the cases in a very short time (Table 5). The algorithm can be extended so that it can use information received from multiple sources or provide a decision starting from a high number of system states. The number of observer/data sources is not limited nor is the system's states. However, while increasing the number of observers and system's states, the data to be processed is increased and the decision-making algorithm design is becoming a highly difficult task to achieve.

Table 5. Average execution time of the presented algorithm.

Method	Execution Time (s)
Data neutrosophication for Obs. 1	0.0026
Data neutrosophication for Obs. 2	0.0026
Data fusion using DSmT	0.0002
Data deneutrosophication/decision-making	0.0092
Total time	**0.0146**

In the case of autonomous robots, these must be taught what to do and how to complete their tasks. From this, the necessity of developing new intelligent and reasoning system arises. The developed algorithm in this paper can be used successfully for target identification applications, object sorting, image labeling, motion tracking, obstacle avoidance, edge detection, etc.

Author Contributions: Conceptualization, I.-A.G., D.B., and L.V.; Data curation, I.-A.G.; Formal analysis, I.-A.G.; Funding acquisition, L.V.; Investigation, I.-A.G., D.B., and L.V.; Methodology, L.V.; Resources, I.-A.G.; Software, I.-A.G.; Supervision, L.V.; Validation, I.-A.G., D.B., and L.V.; Visualization, I.-A.G., D.B., and L.V.; Writing—original draft, I.-A.G.; Writing—review and editing, L.V.

Funding: The paper was funded by the European Commission Marie Skłodowska-Curie SMOOTH project, Smart Robots for Fire-Fighting, H2020-MSCA-RISE-2016-734875, Yanshan University: "Joint Laboratory of Intelligent Rehabilitation Robot" project, KY201501009, Collaborative research agreement between Yanshan University, China and Romanian Academy by IMSAR, RO, and the UEFISCDI Multi-MonD2 Project, Multi-Agent Intelligent Systems Platform for Water Quality Monitoring on the Romanian Danube and Danube Delta, PCCDI 33/2018, PN-III-P1-1.2-PCCDI2017-2017-0637.

Acknowledgments: This work was developed with the support of the Institute of Solid Mechanics of the Romanian Academy, the European Commission Marie Skłodowska-Curie SMOOTH project, Smart Robots for Fire-Fighting, H2020-MSCA-RISE-2016-734875, Yanshan University: "Joint Laboratory of Intelligent Rehabilitation Robot" project, KY201501009, Collaborative research agreement between Yanshan University, China and Romanian Academy by IMSAR, RO, and the UEFISCDI Multi-MonD2 Project, Multi-Agent Intelligent Systems Platform for Water Quality Monitoring on the Romanian Danube and Danube Delta, PCCDI 33/2018, PN-III-P1-1.2-PCCDI2017-2017-0637.

Conflicts of Interest: The authors declare no conflict of interest.

References

1. Vladareanu, V.; Sandru, O.I.; Vladareanu, L.; Yu, H. Extension dynamical stability control strategy for the walking Robots. *Int. J. Technol. Manag.* **2013**, *12*, 1741–5276.
2. Xu, Z.; Kumar, V.; Todorov, E. A low-cost and modular, 20-DOF anthropomorphic robotic hand: Design, actuation and modeling. In Proceedings of the IEEE-RAS International Conference on Humanoid Robots (Humanoids), Atlanta, GA, USA, 15–17 October 2013.
3. Jaffar, A.; Bahari, M.S.; Low, C.Y.; Jaafar, R. Design and control of a Multifingered anthropomorphic Robotic hand international. *Int. J. Mech. Mechatron. Eng.* **2011**, *11*, 26–33.
4. Roa, M.A.; Argus, M.J.; Leidner, D.; Borst, C.; Hirzinger, G. Power grasp planning for anthropomorphic robot hands. In Proceedings of the 2012 IEEE International Conference on Robotics and Automation (ICRA), Saint Paul, MN, USA, 14–18 May 2012; pp. 563–569. [CrossRef]

5. Lippiello, V.; Siciliano, B.; Villani, L. Multi-fingered grasp synthesis based on the object Dynamic properties. *Robot. Auton. Syst.* **2013**, *61*, 626–636. [CrossRef]
6. Bullock, I.M.; Ma, R.R.; Dollar, A.M. A hand-centric classification of Human and Robot dexterous manipulation. *IEEE Trans. Haptics* **2013**, *6*, 129–144. [CrossRef] [PubMed]
7. Ormaechea, R.C. Robotic Hands. *Adv. Mech. Robot. Syst.* **2011**, *1*, 19–39. [CrossRef]
8. Enabling the Future. Available online: http://enablingthefuture.org (accessed on 26 March 2018).
9. Prattichizzo, D.; Trinkle, J.C. Grasping. In *Springer Handbook of Robotics*; Siciliano, B., Khatib, O., Eds.; Springer: Berlin, Germany, 2016; pp. 955–988. ISBN 978-3-319-32552-1.
10. Feix, T.; Romero, J.; Schmiedmayer, H.B.; Dollar, A.M.; Kragic, D. The GRASP taxonomy of human grasp types. *IEEE Trans. Hum.-Mach. Syst.* **2016**, *46*, 66–77. [CrossRef]
11. Alvarez, A.G.; Roby-Brami, A.; Robertson, J.; Roche, N. Functional classification of grasp strategies used by hemiplegic patients. *PLoS ONE* **2017**, *12*, e0187608. [CrossRef]
12. Fermuller, C.; Wang, F.; Yang, Y.Z.; Zampogiannis, K.; Zhang, Y.; Barranco, F.; Pfeiffer, M. Prediction of manipulation actions. *Int. J. Comput. Vis.* **2018**, *126*, 358–374. [CrossRef]
13. Tsai, J.R.; Lin, PC. A low-computation object grasping method by using primitive shapes and in-hand proximity sensing. In Proceedings of the IEEE ASME International Conference on Advanced Intelligent Mechatronics, Munich, Germany, 3–7 July 2017; pp. 497–502.
14. Song, P.; Fu, Z.Q.; Liu, LG. Grasp planning via hand-object geometric fitting. *Vis. Comput.* **2018**, *34*, 257–270. [CrossRef]
15. Ma, C.; Qiao, H.; Li, R.; Li, X.Q. Flexible robotic grasping strategy with constrained region in environment. *Int. J. Autom. Comput.* **2017**, *14*, 552–563. [CrossRef]
16. Stavenuiter, R.A.J.; Birglen, L.; Herder, J.L. A planar underactuated grasper with adjustable compliance. *Mech. Mach. Theory* **2017**, *112*, 295–306. [CrossRef]
17. Choi, C.; Yoon, S.H.; Chen, C.N.; Ramani, K. Robust hand pose estimation during the interaction with an unknown object. In Proceedings of the IEEE International Conference on Computer Vision, Venice, Italy, 22–29 October 2017; pp. 3142–3151.
18. Seredynski, D.; Szynkiewicz, W. Fast grasp learning for novel objects. In *Challenges in Automation, Robotics and Measurement Techniques, Proceedings of AUTOMATION-2016, Warsaw, Poland, 2–4 March 2016*; Advances in Intelligent Systems and Computing Book Series; Springer: Warsaw, Poland, 2016; Volume 440, pp. 681–692. ISBN 978-3-319-29357-8; 978-3-319-29356-1.
19. Shaw-Cortez, W.; Oetomo, D.; Manzie, C.; Choong, P. Tactile-based blind grasping: A discrete-time object manipulation controller for robotic hands. *IEEE Robot. Autom. Lett.* **2018**, *3*, 1064–1071. [CrossRef]
20. Gu, H.W.; Zhang, Y.F.; Fan, S.W.; Jin, M.H.; Liu, H. Grasp configurations optimization of dexterous robotic hand based on haptic exploration information. *Int. J. Hum. Robot.* **2017**, *14*. [CrossRef]
21. Yamakawa, Y.; Namiki, A.; Ishikawa, M.; Shimojo, M. Planning of knotting based on manipulation skills with consideration of robot mechanism/motion and its realization by a robot hand system. *Symmetry* **2017**, *9*, 194. [CrossRef]
22. Nacy, S.M.; Tawfik, M.A.; Baqer, I.A. A novel approach to control the robotic hand grasping process by using an artificial neural network algorithm. *J. Intell. Syst.* **2017**, *26*, 215–231. [CrossRef]
23. Zaidi, L.; Corrales, J.A.; Bouzgarrou, B.C.; Mezouar, Y.; Sabourin, L. Model-based strategy for grasping 3D deformable objects using a multi-fingered robotic hand. *Robot. Auton. Syst.* **2017**, *95*, 196–206. [CrossRef]
24. Liu, H.; Meusel, P.; Hirzinger, G.; Jin, M.; Liu, Y.; Xie, Z. The modular multisensory DLR-HIT-Hand: Hardware and software architecture. *IEEE/ASME Trans. Mechatron.* **2008**, *13*, 461–469. [CrossRef]
25. Zollo, L.; Roccella, S.; Guglielmelli, E.; Carrozza, M.C.; Dario, P. Biomechatronic design and control of an anthropomorphic artificial hand for prosthetic and robotic applications. *IEEE/ASME Trans. Mechatron.* **2007**, *12*, 418–429. [CrossRef]
26. Lopez-Damian, E.; Sidobre, D.; Alami, R. Grasp planning for non-convex objects. *Int. Symp. Robot.* **2009**, *36*. [CrossRef]
27. Miller, A.T.; Knoop, S.; Christensen, H.I.; Allen, P.K. Automatic grasp planning using shape primitives. In Proceedings of the 2003 IEEE International Conference on Robotics and Automation (Cat. No.03CH37422), Taipei, Taiwan, 14–19 September 2003.
28. Napier, J. The prehensile movements of the human hand. *J. Bone Jt. Surg.* **1956**, *38*, 902–913. [CrossRef]

29. Cutkosky, M.R.; Wright, P.K. Modeling manufacturing grips and correlation with the design of robotic hands. In Proceedings of the 1986 IEEE International Conference on Robotics and Automation, San Francisco, CA, USA, 7–10 April 1986; pp. 1533–1539.
30. Iberall, T. Human prehension and dexterous robot hands. *Int. J. Robot. Res.* **1997**, *16*, 285–299. [CrossRef]
31. Stansfield, S.A. Robotic grasping of unknown objects: A knowledge-based approach. *Int. J. Robot. Res.* **1991**, *10*, 314–326. [CrossRef]
32. Lai, X.; Wang, H.; Xu, Y. A real-time range finding system with binocular stereo vision. *Int. J. Adv. Robot Syst.* **2012**, *9*, 26. [CrossRef]
33. Oliver, A.; Wünsche, B.C.; Kang, S.; MacDonald, B. Using the Kinect as a navigation sensor for mobile robotics. In Proceedings of the 27th Conference on Image and Vision Computing New Zealand, Dunedin, New Zealand, 26–28 November 2012; pp. 509–514.
34. Aljarrah, I.A.; Ghorab, A.S.; Khater, I.M. Object recognition system using template matching based on signature and principal component analysis. *Int. J. Digit. Inf. Wirel. Commun.* **2012**, *2*, 156–163.
35. Smarandache, F.; Dezert, J. *Applications and Advances of DSmT for Information Fusion*; American Research Press: Rehoboth, DE, USA, 2009; Volume 3.
36. Khaleghi, B.; Khamis, A.; Karray, F.; Razavi, S.N. Multisensor data fusion: A review of the state-of-the-art. *Inf. Fusion* **2013**, *14*, 28–44. [CrossRef]
37. Jian, Z.; Hongbing, C.; Jie, S.; Haitao, L. Data fusion for magnetic sensor based on fuzzy logic theory. In Proceedings of the IEEE International Conference on Intelligent Computation Technology and Automation, Shenzhen, China, 28–29 March 2011.
38. Munz, M.; Dietmayer, K.; Mahlisch, M. Generalized fusion of heterogeneous sensor measurements for multi target tracking. In Proceedings of the 13th Conference on Information Fusion (FUSION), Edinburgh, UK, 26–29 July 2010; pp. 1–8.
39. Jiang, Y.; Wang, H.G.; Xi, N. Target object identification and location based on multi-sensor fusion. *Int. J. Autom. Smart Technol.* **2013**, *3*, 57–67. [CrossRef]
40. Hall, D.L.; Hellar, B.; McNeese, M.D. Rethinking the data overload problem: Closing the gap between situation assessment and decision-making. In Proceedings of the National Symposium on Sensor Data Fusion, McLean, VA, USA, 11–15 June 2007.
41. Esteban, J.; Starr, A.; Willetts, R.; Hannah, P.; Bryanston-Cross, P. A review of data fusion models and architectures: Towards engineering guidelines. *Neural Comput. Appl.* **2005**, *14*, 273–281. [CrossRef]
42. Chilian, A.; Hirschmuller, H.; Gorner, M. Multisensor data fusion for robust pose estimation of a six-legged walking robot. In Proceedings of the 2011 IEEE/RSJ International Conference on Intelligent Robots and Systems (IROS), San Francisco, CA, USA, 25–30 September 2011.
43. Dasarathy, B.V. Editorial: Information fusion in the realm of medical applications—A bibliographic glimpse at its growing appeal. *Inf. Fusion* **2012**, *13*, 1–9. [CrossRef]
44. Hall, D.L.; Linn, R.J. Survey of commercial software for multisensor data fusion. *Proc. SPIE* **1993**. [CrossRef]
45. Vladareanu, L.; Vladareanu, V.; Schiopu, P. Hybrid force-position dynamic control of the robots using fuzzy applications. *Appl. Mech. Mater.* **2013**, *245*, 15–23. [CrossRef]
46. Gaines, B.R. Fuzzy and probability uncertainty logics. *Inf. Control* **1978**, *38*, 154–169. [CrossRef]
47. Shafer, G. *A Mathematical Theory of Evidence*; Princeton University Press: Princeton, NJ, USA, 1976; Volume 73.
48. Sahbani, A.; El-Khoury, S.; Bidaud, P. An overview of 3D object grasp synthesis algorithms. *Robot. Auton. Syst.* **2012**, *60*, 326–336. [CrossRef]
49. De Souza, R.; El Khoury, S.; Billard, A. Towards comprehensive capture of human grasping and manipulation skills. In Proceedings of the Thirteenth International Symposium on the 3-D Analysis of Human Movement, Lausanne, Switzerland, 14–17 July 2014.
50. Bullock, I.M.; Dollar, A.M. Classifying human manipulation behavior. In Proceedings of the 2011 IEEE International Conference on Rehabilitation Robotics, Zurich, Switzerland, 29 June–1 July 2011; pp. 1–6.
51. Morales, A.; Azad, P.; Asfour, T.; Kraft, D.; Knoop, S.; Dillmann, R.; Kargov, A.; Pylatiuk, C.H.; Schulz, S. An anthropomorphic grasping approach for an assistant humanoid robot. *Int. Symp. Robot.* **2006**, *1956*, 149.
52. Etienne-Cwnmings, R.; Pouliquen, P.; Lewis, M. A Single chip for imaging, color segmentation, histogramming and template matching. *Electron. Lett.* **2002**, *2*, 172–174. [CrossRef]

53. Guskov, I. Kernel—based template alignment. In Proceedings of the 2006 IEEE Computer Society Conference on Computer Vision and Pattern Recognition (CVPR'06), New York, NY, USA, 17–22 June 2006; pp. 610–617.
54. Smarandache, F.; Dezert, J. *Applications and Advances of DSmT for Information Fusion*; American Research Press: Rehoboth, DE, USA, 2004.
55. Smarandache, F.; Vladareanu, L. Applications of neutrosophic logic to robotics: An introduction. In Proceedings of the 2011 IEEE International Conference on Granular Computing, Kaohsiung, Taiwan, 8–10 November 2011; pp. 607–612.
56. Gal, I.A.; Vladareanu, L.; Yu, H. Applications of neutrosophic logic approaches in 'RABOT' real time control. In Proceedings of the SISOM 2012 and Session of the Commission of Acoustics, Bucharest, Romania, 30–31 May 2013.
57. Gal, I.A.; Vladareanu, L.; Munteanu, R.I. Sliding motion control with bond graph modeling applied on a robot leg. *Rev. Roum. Sci. Tech.* **2010**, *60*, 215–224.
58. Vladareanu, V.; Dumitrache, I.; Vladareanu, L.; Sacala, I.S.; Tont, G.; Moisescu, M.A. Versatile intelligent portable robot control platform based on cyber physical systems principles. *Stud. Inform. Control* **2015**, *24*, 409–418. [CrossRef]
59. Melinte, D.O.; Vladareanu, L.; Munteanu, R.A.; Wang, H.; Smarandache, F.; Ali, M. Nao robot in virtual environment applied on VIPRO platform. In Proceedings of the Annual Symposium of the Institute of Solid Mechanics and Session of the Commission of Acoustics, Bucharest, Romania, 21–22 May 2016; Volume 57.
60. Smarandache, F. A unifying field in logics: Neutrosophic field. Multiple-valued logic. *Int. J.* **2002**, *8*, 385–438.
61. Wang, H.; Smarandache, F.; Zhang, Y.Q.; Sunderraman, R. *Interval Neutrosophic Sets and Logic: Theory and Application in Computing*; HEXIS Neutrosophic Book Series No.5; Georgia State University: Atlanta, GA, USA, 2005; ISBN 1-931233-94-2.
62. Smarandache, F. *Neutrosophy: Neutrosophic Probability, Set, and Logic: Analytic Synthesis & Synthetic Analysis*; American Research Press: Gallup, NM, USA, 1998; 105p, ISBN 1-87958-563-4.
63. Rosell, J.; Sierra, X.; Palomo, F.L. Finding grasping configurations of a dexterous hand and an industrial robot. In Proceedings of the 2005 IEEE International Conference on Robotics and Automation, Barcelona, Spain, 18–22 April 2005; pp. 1178–1183. [CrossRef]
64. Emadi, S.; Shams, F. Modeling of component diagrams using Petri Nets. *Indian J. Sci. Technol.* **2010**, *3*, 1151–1161. [CrossRef]

symmetry

MDPI

Article

Some Results on Neutrosophic Triplet Group and Their Applications

Tèmítópé Gbóláhàn Jaíyéolá [1,*] and Florentin Smarandache [2]

1 Department of Mathematics, Obafemi Awolowo University, Ile Ife 220005, Nigeria
2 Department of Mathematics and Science, University of New Mexico, 705 Gurley Ave., Gallup, NM 87301, USA; smarand@unm.edu
* Correspondence: tjayeola@oauife.edu.ng; Tel.: +234-813-961-1718

Received: 23 April 2018; Accepted: 15 May 2018; Published: 6 June 2018

Abstract: This article is based on new developments on a neutrosophic triplet group (NTG) and applications earlier introduced in 2016 by Smarandache and Ali. NTG sprang up from neutrosophic triplet set X: a collection of triplets $(b, neut(b), anti(b))$ for an $b \in X$ that obeys certain axioms (existence of neutral(s) and opposite(s)). Some results that are true in classical groups were investigated in NTG and were shown to be either universally true in NTG or true in some peculiar types of NTG. Distinguishing features between an NTG and some other algebraic structures such as: generalized group (GG), quasigroup, loop and group were investigated. Some neutrosophic triplet subgroups (NTSGs) of a neutrosophic triplet group were studied. In particular, for any arbitrarily fixed $a \in X$, the subsets $X_a = \{b \in X : neut(b) = neut(a)\}$ and $\ker f_a = \{b \in X | f(b) = neut(f(a))\}$ of X, where $f : X \to Y$ is a neutrosophic triplet group homomorphism, were shown to be NTSG and normal NTSG, respectively. Both X_a and $\ker f_a$ were shown to be a-normal NTSGs and found to partition X. Consequently, a Lagrange-like formula was found for a finite NTG X; $|X| = \sum_{a \in X} [X_a : \ker f_a] |\ker f_a|$ based on the fact that $|\ker f_a| \, | \, |X_a|$. The first isomorphism theorem $X/\ker f \cong \operatorname{Im} f$ was established for NTGs. Using an arbitrary non-abelian NTG X and its NTSG X_a, a Bol structure was constructed. Applications of the neutrosophic triplet set, and our results on NTG in relation to management and sports, are highlighted and discussed.

Keywords: generalized group; neutrosophic triplet set; neutrosophic triplet group; group

MSC: Primary 20N02; Secondary 20N05

1. Introduction

1.1. Generalized Group

Unified gauge theory has the algebraic structure of a generalized group abstrusely, in its physical background. It has been a challenge for physicists and mathematicians to find a desirable unified theory for twistor theory, isotopies theory, and so on. Generalized groups are instruments for constructions in unified geometric theory and electroweak theory. Completely simple semigroups are precisely generalized groups (Araujo et al. [1]). As recorded in Adeniran et al. [2], studies on the properties and structures of generalized groups have been carried out in the past, and these have been extended to smooth generalized groups and smooth generalized subgroups by Agboola [3,4], topological generalized groups by Molaei [5], Molaei and Tahmoresi [6], and quotient space of generalized groups by Maleki and Molaei [7].

Definition 1 (Generalized Group(GG)). *A generalized group X is a non-void set with a binary operation called multiplication obeying the set of rules given below.*

(i) $(ab)c = a(bc)$ *for all* $a, b, c \in X$.
(ii) *For each* $a \in X$ *there is a unique* $e(a) \in X$ *such that* $ae(a) = e(a)a = a$ *(existence and uniqueness of identity element).*
(iii) *For each* $a \in X$, *there is* $a^{-1} \in X$ *such that* $aa^{-1} = a^{-1}a = e(a)$ *(existence of inverse element).*

Definition 2. *Let X be a non-void set. Let* $(\cdot)$ *be a binary operation on X. Whenever* $a \cdot b \in X$ *for all* $a, b \in X$, *then* $(X, \cdot)$ *is called a groupoid.*

Whenever the equation $c \cdot x = d$ *(or* $y \cdot c = d$*) have unique solution with respect to x (or y) i.e., satisfies the left (or right) cancellation law, then* $(X, \cdot)$ *is called a left (or right) quasigroup. If a groupoid* $(X, \cdot)$ *is both a left quasigroup and right quasigroup, then it is called a quasigroup. If there is an element* $e \in X$ *called the identity element such that for all* $a \in X$, $a \cdot e = e \cdot a = a$, *then a quasigroup* $(X, \cdot)$ *is called a loop.*

Definition 3. *A loop is called a Bol loop whenever it satisfies the identity*

$$((ab)c)b = a((bc)b).$$

Remark 1. *One of the most studied classes of loops is the Bol loop.*

For more on quasigroups and loops, interested readers can check [8–15].

A generalized group X has the following properties:

(i) For each $a \in X$, there is a unique $a^{-1} \in X$.
(ii) $e(e(a)) = e(a)$ and $e(a^{-1}) = e(a)$ if $a \in X$.
(iii) If X is commutative, then X is a group.

1.2. Neutrosophic Triplet Group

Neutrosophy is a novel subdivision of philosophy that studies the nature, origination, and ambit of neutralities, including their interaction with ideational spectra. Florentin Smarandache [16] introduced the notion of neutrosophic logic and neutrosophic sets for the first time in 1995. As a matter of fact, the neutrosophic set is the generalization of classical sets [17], fuzzy sets [18], intuitionistic fuzzy sets [17,19], and interval valued fuzzy sets [17], to cite a few. The growth process of neutrosophic sets, fuzzy sets, and intuitionistic fuzzy sets are still evolving, with diverse applications. Some recent research findings in these directions are [20–27].

Smarandache and Ali [28] were the first to introduce the notion of the neutrosophic triplet, which they had earlier talked about at a conference. These neutrosophic triplets were used by them to introduce the neutrosophic triplet group, which differs from the classical group both in fundamental and structural properties. The distinction and comparison of the neutrosophic triplet group with the classical generalized group were given. They also drew a brief outline of the potential applications of the neutrosophic triplet group in other research fields. For discussions of results on neutrosophic triplet groups, neutrosophic quadruples, and neutrosophic duplets of algebraic structures, as well as new applications of neutrosophy, see Jaiyéọlá and Smarandache [29]. Jaiyéọlá and Smarandache [29] were the first to introduce and study inverse property neutrosophic triplet loops with applications to cryptography for the first time.

Definition 4 (Neutrosophic Triplet Set-NTS). *Let X be a non-void set together with a binary operation $\star$ defined on it. Then X is called a neutrosophic triplet set if, for any $a \in X$, there is a neutral of 'a' denoted by $neut(a)$ (not necessarily the identity element) and an opposite of 'a' denoted by $anti(a)$, with $neut(a), anti(a) \in X$ such that*

$$a \star neut(a) = neut(a) \star a = a \quad \text{and} \quad a \star anti(a) = anti(a) \star a = neut(a).$$

The elements $a, neut(a)$ and $anti(a)$ are together called neutrosophic triplet, and represented by $(a, neut(a), anti(a))$.

Remark 2. *For an $a \in X$, each of $neut(a)$ and $anti(a)$ may not be unique. In a neutrosophic triplet set $(X, \star)$, an element b (or c) is the second (or third) component of a neutrosophic triplet if $a, c \in X$ ($a, b \in X$) such that $a \star b = b \star a = a$ and $a \star c = c \star a = b$. Thus, (a, b, c) is a neutrosophic triplet.*

Example 1 (Smarandache and Ali [28]). *Consider $(\mathbb{Z}_6, \times_6)$ such that $\mathbb{Z}_6 = \{0, 1, 2, 3, 4, 5\}$ and $\times_6$ is multiplication in modulo 6. $(2, 4, 2)$, $(4, 4, 4)$, and $(0, 0, 0)$ are neutrosophic triplets, but 3 will not give rise to a neutrosophic triplet.*

Definition 5 (Neutrosophic Triplet Group—NTG). *Let $(X, \star)$ be a neutrosophic triplet set. Then $(X, \star)$ is referred to as a neutrosophic triplet group if $(X, \star)$ is a semigroup. Furthermore, if $(X, \star)$ obeys the commutativity law, then $(X, \star)$ is referred to as a commutative neutrosophic triplet group.*

Let $(X, \star)$ be a neutrosophic triplet group. Whenever $neut(ab) = neut(a)neut(b)$ for all $a, b \in X$, then X is referred to as a normal neutrosophic triplet group.

Let $(X, \star)$ be a neutrosophic triplet group and let $H \subseteq X$. H is referred to as a neutrosophic triplet subgroup (NTSG) of X if $(H, \star)$ is a neutrosophic triplet group. Whence, for any fixed $a \in X$, H is called a-normal NTSG of X, written $H \overset{a}{\triangleleft} X$ if $ay\, anti(a) \in H$ for all $y \in H$.

Remark 3. *An NTG is not necessarily a group. However, a group is an NTG where $neut(a) = e$, the general identity element for all $a \in X$, and $anti(a)$ is unique for each $a \in X$.*

Example 2 (Smarandache and Ali [28]). *Consider $(\mathbb{Z}_{10}, \otimes)$ such that $c \otimes d = 3cd \bmod 10$. $(\mathbb{Z}_{10}, \otimes)$ is a commutative NTG but neither a GG nor a classical group.*

Example 3 (Smarandache and Ali [28]). *Consider $(\mathbb{Z}_{10}, \star)$ such that $c \star d = 5c + d \bmod 10$. $(\mathbb{Z}_{10}, \star)$ is a non-commutative NTG but not a classical group.*

Definition 6 (Neutrosophic Triplet Group Homomorphism). *Let $f : X \to Y$ be a mapping such that X and Y are two neutrosophic triplet groups. Then f is referred to as a neutrosophic triplet group homomorphism if $f(cd) = f(c)f(d)$ for all $c, d \in X$. The kernel of f at $a \in X$ is defined by*

$$\ker f_a = \{x \in X : f(x) = neut(f(a))\}.$$

The Kernel of f is defined by

$$\ker f = \bigcup_{a \in X} \ker f_a$$

such that $f_a = f|_{X_a}$, where $X_a = \{x \in X : neut(x) = neut(a)\}$.

Remark 4. *The definition of neutrosophic triplet group homomorphism above is more general than that in Smarandache and Ali [28]. In Theorem 5, it is shown that, for an NTG homomorphism $f : X \to Y$, $f(neut(a)) = neut(f(a))$ and $f(anti(a)) = anti(f(a))$ for all $a \in X$.*

The present work is a continuation of the study of a neutrosophic triplet group (NTG) and its applications, which was introduced by Smarandache and Ali [28]. Some results that are true in classical groups were investigated in NTG and will be proved to be either generally true in NTG or true in some classes of NTG. Some applications of the neutrosophic triplet set, and our results on NTG in relation to management and sports will be discussed.

The first section introduces GG and NTG and highlights existing results that are relevant to the present study. Section 2 establishes new results on algebraic properties of NTGs and NTG homomorphisms, among which are Lagrange's Theorem and the first isomorphism theorem, and presents a method of the construction of Bol algebraic structures using an NTG. The third section describes applications of NTGs to human management and sports.

2. Main Results

We shall first establish the relationship among generalized groups, quasigroups, and loops with a neutrosophic triplet group assumed.

Lemma 1. *Let X be a neutrosophic triplet group.*

1. *X is a generalized group if it satisfies the left (or right) cancellation law or X is a left (or right) quasigroup.*
2. *X is a generalized group if and only if each element $x \in X$ has a unique $neut(x) \in X$.*
3. *Whenever X has the cancellation laws (or is a quasigroup), then X is a loop and group.*

Proof. 1. Let x have at least two neutral elements, say $neut(x), neut(x)' \in X$. Then $xx = xx \Rightarrow xx\,anti(x) = xx\,anti(x) \Rightarrow x\,neut(x) = x\,neut(x)' \overset{\text{left quasigroup}}{\underset{\text{left cancellation law}}{\Longrightarrow}} neut(x) = neut(x)'$. Therefore, X is a generalized group. Similarly, X is a generalized group if it is has the right cancellation law or if it is a right quasigroup.
2. This follows by definition.
3. This is straightforward because every associative quasigroup is a loop and group. □

2.1. Algebraic Properties of Neutrosophic Triplet Group

We now establish some new algebraic properties of NTGs.

Theorem 1. *Let X be a neutrosophic triplet group. For any $a \in X$, $anti(anti(a)) = a$.*

Proof. $anti(anti(a))anti(a) = neut(anti(a)) = neut(a)$ by Theorem 1 ([29]). After multiplying by a, we obtain

$$[anti(anti(a))anti(a)]a = neut(a)a = a. \quad (1)$$

$$\begin{aligned} LHS &= anti(anti(a))(anti(a)a) = anti(anti(a))neut(a) \\ &= anti(anti(a))neut(anti(a)) = anti(anti(a))neut\Big(anti(anti(a))\Big) = anti(anti(a)). \end{aligned} \quad (2)$$

Hence, based on Equations (1) and (2), $anti(anti(a)) = a$. □

Theorem 2. *Let X be a neutrosophic triplet group such that the left cancellation law is satisfied, and $neut(a) = neut(a\,anti(b))$ if and only if $a\,anti(b) = a$. Then X is an idempotent neutrosophic triplet group if and only if $neut(a)anti(b) = anti(b)neut(a)\ \forall\, a, b \in X$.*

Proof. $neut(a)anti(b) = anti(b)neut(a) \Leftrightarrow (a\,neut(a))anti(b) = a\,anti(b)neut(a) \Leftrightarrow a\,anti(b) = a\,anti(b)neut(a) \Leftrightarrow neut(a) = neut(a\,anti(b)) \Leftrightarrow a\,anti(b) = a \Leftrightarrow a\,anti(b)b = ab \Leftrightarrow a\,neut(b) = ab \Leftrightarrow anti(a)a\,neut(b) = anti(a)ab \Leftrightarrow neut(a)neut(b) = neut(a)b \Leftrightarrow neut(b) = b \Leftrightarrow b = bb.$ □

Theorem 3. *Let X be a normal neutrosophic triplet group in which* $neut(a)anti(b) = anti(b)neut(a)\ \forall\, a, b \in X$. *Then,* $anti(ab) = anti(b)anti(a)\ \forall\, a, b \in X$.

Proof. Since $anti(ab)(ab) = neut(ab)$, then by multiplying both sides of the equation on the right by $anti(b)anti(a)$, we obtain

$$\left[anti(ab)ab\right]anti(b)anti(a) = neut(ab)anti(b)anti(a). \tag{3}$$

Going by Theorem 1([29]),

$$\begin{aligned}\left[anti(ab)ab\right]anti(b)anti(a) &= anti(ab)a\big(b\, anti(b)\big)anti(a) = anti(ab)a\big(neut(b)anti(a)\big)\\ &= anti(ab)\big(a\, anti(a)\big)neut(b) = anti(ab)\big(neut(a)neut(b)\big)\\ &= anti(ab)neut(ab) = anti(ab)neut\big(anti(ab)\big) = anti(ab).\end{aligned} \tag{4}$$

Using Equations (3) and (4), we obtain

$$\begin{gathered}\left[anti(ab)ab\right]anti(b)anti(a) = anti(ab) \Rightarrow\\ neut(ab)\big(anti(b)anti(a)\big) = anti(ab) \Rightarrow anti(ab) = anti(b)anti(a).\end{gathered}$$

□

It is worth characterizing the neutrosophic triplet subgroup of a given neutrosophic triplet group to see how a new NTG can be obtained from existing NTGs.

Lemma 2. *Let H be a non-void subset of a neutrosophic triplet group X. The following are equivalent.*

(i) *H is a neutrosophic triplet subgroup of X.*
(ii) *For all* $a, b \in H$, $a\, anti(b) \in H$.
(iii) *For all* $a, b \in H$, $ab \in H$, *and* $anti(a) \in H$.

Proof. **(i)⇒ (ii)** If H is an NTSG of X and $a, b \in H$, then $anti(b) \in H$. Therefore, by closure property, $a\, anti(b) \in H\ \forall\, a, b \in H$.
(ii)⇒ (iii) If $H \neq \emptyset$, and $a, b \in H$, then we have $b\, anti(b) = neut(b) \in H$, $neut(b)anti(b) = anti(b) \in H$, and $ab = a\, anti(anti(b)) \in H$, i.e., $ab \in H$.
(iii)⇒ (i) $H \subseteq X$, so H is associative since X is associative. Obviously, for any $a \in H$, $anti(a) \in H$. Let $a \in H$, then $anti(a) \in H$. Therefore, $a\, anti(a) = anti(a)a = neut(a) \in H$. Thus, H is an NTSG of X.
□

Theorem 4. *Let G and H be neutrosophic triplet groups. The direct product of G and H defined by*

$$G \times H = \{(g, h) : g \in G \text{ and } h \in H\}$$

is a neutrosophic triplet group under the binary operation $\circ$ *defined by*

$$(g_1, h_1) \circ (g_2, h_2) = (g_1 g_2, h_1 h_2).$$

Proof. This is simply done by checking the axioms of neutrosophic triplet group for the pair $(G \times H, \circ)$, in which case $neut(g, h) = \big(neut(g), neut(h)\big)$ and $anti(g, h) = \big(anti(g), anti(h)\big)$. □

Lemma 3. *Let* $\mathcal{H} = \{H_i\}_{i \in \Omega}$ *be a family of neutrosophic triplet subgroups of a neutrosophic triplet group X such that* $\bigcap_{i \in \Omega} H_i \neq \emptyset$. *Then* $\bigcap_{i \in \Omega} H_i$ *is a neutrosophic triplet subgroup of X.*

Proof. This is a routine verification using Lemma 2. □

2.2. Neutrosophic Triplet Group Homomorphism

Let us now establish results on NTG homomorphisms, its kernels, and images, as well as a Lagrange-like formula and the First Isomorphism Theorem for NTGs.

Theorem 5. *Let $f : X \to Y$ be a homomorphism where X and Y are two neutrosophic triplet groups.*

1. $f(neut(a)) = neut(f(a))$ *for all* $a \in X$.
2. $f(anti(a)) = anti(f(a))$ *for all* $a \in X$.
3. *If H is a neutrosophic triplet subgroup of X, then $f(H)$ is a neutrosophic triplet subgroup of Y.*
4. *If K is a neutrosophic triplet subgroup of Y, then $\emptyset \neq f^{-1}(K)$ is a neutrosophic triplet subgroup of X.*
5. *If X is a normal neutrosophic triplet group and the set $X_f = \{(neut(a), f(a)) : a \in X\}$ with the product*
$$(neut(a), f(a))(neut(b), f(b)) := (neut(ab), f(ab)), \text{ then}$$
X_f *is a neutrosophic triplet group.*

Proof. Since f is an homomorphism, $f(ab) = f(a)f(b)$ for all $a, b \in X$.

1. Place $b = neut(a)$ in $f(ab) = f(a)f(b)$ to obtain $f(a\ neut(a)) = f(a)f(neut(a)) \Rightarrow f(a) = f(a)f(neut(a))$. Additionally, place $b = neut(a)$ in $f(ba) = f(b)f(a)$ to obtain $f(neut(a)a) = f(neut(a))f(a) \Rightarrow f(a) = f(neut(a))f(a)$. Thus, $f(neut(a)) = neut(f(a))$ for all $a \in X$.
2. Place $b = anti(a)$ in $f(ab) = f(a)f(b)$ to obtain $f(a\ anti(a)) = f(a)f(anti(a)) \Rightarrow f(neut(a)) = f(a)f(anti(a)) \Rightarrow neut(f(a)) = f(a)f(anti(a))$. Additionally, place $b = anti(a)$ in $f(ba) = f(b)f(a)$ to obtain $f(anti(a)a) = f(anti(a))f(a) \Rightarrow f(neut(a)) = f(a)f(anti(a)) \Rightarrow neut(f(a)) = f(anti(a))f(a)$. Thus, $f(anti(a)) = anti(f(a))$ for all $a \in X$.
3. If H is an NTSG of G, then $f(H) = \{f(h) \in Y : h \in H\}$. We shall prove that $f(H)$ is an NTSG of Y by Lemma 2.
Since $f(neut(a)) = neut(f(a)) \in f(H)$ for $a \in H$, $f(H) \neq \emptyset$. Let $a', b' \in f(H)$. Then $a' = f(a)$ and $b' = f(b)$. Thus, $a'\ anti(b') = f(a)anti(f(b)) = f(a)f(anti(b)) = f(a\ anti(b)) \in f(H)$. Therefore, $f(H)$ is an NTSG of Y.
4. If K is a neutrosophic triplet subgroup of Y, then $\emptyset \neq f^{-1}(K) = \{a \in X : f(a) \in K\}$. We shall prove that $f(H)$ is an NTSG of Y by Lemma 2.
Let $a, b \in f^{-1}(K)$. Then $a', b' \in K$ such that $a' = f(a)$ and $b' = f(b)$. Thus, $a'\ anti(b') = f(a)anti(f(b)) = f(a)f(anti(b)) = f(a\ anti(b)) \in K \Rightarrow a\ anti(b) \in f^{-1}(K)$. Therefore, $f^{-1}(K)$ is an NTSG of X.
5. Given the neutrosophic triplet group X and the set $X_f = \{(neut(a), f(a)) : a \in X\}$ with the product $(neut(a), f(a))(neut(b), f(b)) := (neut(ab), f(ab))$. X_f is a groupoid.
$(neut(a), f(a))(neut(b), f(b)) \cdot (neut(z), f(z)) = (neut(ab), f(ab))(neut(z), f(z)) = (neut(abz), f(abz))$
$= (neut(a), f(a))(neut(bz), f(bz)) = (neut(a), f(a)) \cdot (neut(b), f(b))(neut(z), f(z))$.
Therefore, X_f is a semigroup.
For $(neut(a), f(a)) \in X_f$, let $neut(neut(a), f(a)) = (neut(neut(a)), neut(f(a)))$. Then $neut(neut(a), f(a)) = (neut(a), (f(neut(a))) \in X_f$. Additionally, let $anti(neut(a), f(a)) = (anti(neut(a)), anti(f(a)))$. Then $anti(neut(a), f(a)) = (neut(a), f(anti(a))) \in X_f$.
Thus, $(neut(a), f(a))neut(neut(a), f(a)) = (neut(a), f(a))(neut(a), (f(neut(a))) = (neut(a), f(a))(neut(anti(a)), (f(neut(a))) = (neut(a\ \ anti(a)), f(a\ \ neut(a))) = (neut(neut(a)), f(a\ \ neut(a))) = (neut(a), f(a)) \Rightarrow (neut(a), f(a))neut(neut(a), f(a)) = (neut(a), f(a))$ and similarly, $neut(neut(a), f(a))(neut(a), f(a)) = (neut(a), f(a))$.

On the other hand, $(neut(a), f(a))anti(neut(a), f(a)) = (neut(a), f(a)) \cdot (neut(a), f(anti(a))) = (neut(a), f(a))(neut(anti(a)), (f(anti(a))) = (neut(a\ anti(a)), f(a\ anti(a))) = (neut(neut(a)), f(neut(a))) = (neut(a), (f(neut(a))) = neut(neut(a), f(a)) \Rightarrow (neut(a), f(a)) \cdot anti(neut(a), f(a)) = neut(neut(a), f(a))$ and similarly, $anti(neut(a), f(a)) \cdot (neut(a), f(a)) = neut(neut(a), f(a))$.

Therefore, X_f is a neutrosophic triplet group.

□

Theorem 6. *Let $f : X \to Y$ be a neutrosophic triplet group homomorphism.*

1. $\ker f_a \overset{a}{\triangleleft} X$.
2. $X_a \overset{a}{\triangleleft} X$.
3. *X_a is a normal neutrosophic triplet group.*
4. $anti(cd) = anti(d)anti(c)\ \forall\ c, d \in X_a$.
5. $X_a = \bigcup\limits_{c \in X_a} c \ker f_a$ *for all* $a \in X$.
6. *If X is finite,* $|X_a| = \sum_{c \in X_a} |c \ker f_a| = [X_a : \ker f_a]|\ker f_a|$ *for all* $a \in X$ *where* $[X_a : \ker f_a]$ *is the index of* $\ker f_a$ *in* X_a*, i.e., the number of distinct left cosets of* $\ker f_a$ *in* X_a.
7. $X = \bigcup\limits_{a \in X} X_a$.
8. *If X is finite,* $|X| = \sum\limits_{a \in X} [X_a : \ker f_a]|\ker f_a|$.

Proof. 1. $f(neut(a)) = neut(f(a)) = neut(neut(f(a))) = neut(f(neut(a))) \Rightarrow neut(a) \in \ker f_a \Rightarrow \ker f_a \neq \emptyset$. Let $c, d \in \ker f_a$, then $f(c) = f(d) = neut(f(a))$. We shall use Lemma 2.
$f(c\ anti(d)) = f(c)f(anti(d)) = f(c)anti(f(d)) = neut(f(a))anti(neut(f(a))) = neut(f(a))neut(f(a)) = neut(f(a)) \Rightarrow c\ anti(d) \in \ker f_a$.
Thus, $\ker f_a$ is a neutrosophic triplet subgroup of X. For the a-normality, let $d \in \ker f_a$, then $f(d) = neut(f(a))$. Therefore, $f(ad\ anti(a)) = f(a)f(d)f(anti(a)) = f(a)neut(f(a))anti(f(a)) = f(a)anti(f(a)) = neut(f(a)) \Rightarrow ad\ anti(a) \in \ker f_a$ for all $d \in \ker f_a$.
Therefore, $\ker f_a \overset{a}{\triangleleft} X$.

2. $X_a = \{c \in X : neut(c) = neut(a)\}$. $neut(neut(a)) = neut(a) \Rightarrow neut(a) \in X_a$. Therefore, $X_a \neq \emptyset$.
Let $c, d \in X_a$. Then $neut(c) = neut(a) = neut(d)$. $(cd)neut(a) = c(d\ neut(a)) = c(d\ neut(d)) = cd$, and $neut(a)(cd) = (neut(a)c)d = (neut(c)c)d = cd$. Therefore, $neut(cd) = neut(a)$.
$neut(anti(c)) = anti(neut(c)) = anti(neut(a)) = neut(a) \Rightarrow anti(c) \in X_a$. Thus, X_a is a neutrosophic triplet subgroup of X.
$neut(anti(a)) = neut(a) \Rightarrow anti(a) \in X_a$. Therefore, $(ac\ anti(a))neut(a) = (ac)(anti(a)neut(a)) = ac\ anti(a)$, and $neut(a)(ac\ anti(a)) = neut(a)a(c\ anti(a)) = ac\ anti(a)$.
Thus, $neut(ac\ anti(a)) = neut(a) \Rightarrow ac\ anti(a) \in X_a$. Therefore, $X_a \overset{a}{\triangleleft} X$.

3. Let $c, d \in X_a$. Then $neut(c) = neut(a) = neut(d)$. Therefore, $neut(cd) = neut(a) = neut(a)neut(a) = neut(c)neut(d)$. Thus, X_a is a normal NTG.

4. For all $c, d \in X_a$, $neut(c)anti(d) = neut(a)anti(d) = neut(d)anti(d) = anti(d) = anti(d)neut(d) = anti(d)neut(a) = anti(d)$. Therefore, based on Point 3 and Theorem 3, $anti(cd) = anti(d)anti(c)\ \forall\ c, d \in X_a$.

5. Define a relation $\asymp$ on X_a as follows: $c \asymp d$ if $anti(c)d \in \ker f_a$ for all $c, d \in X_a$. $anti(c)c = neut(c) = neut(a) \Rightarrow anti(c)c \in \ker f_a \Rightarrow c \asymp c$. Therefore, $\asymp$ is reflexive.
$c \asymp d \Rightarrow anti(c)d \in \ker f_a \overset{\text{by 4.}}{\Rightarrow} anti(anti(c)d) \in \ker f_a \Rightarrow anti(d)c \in \ker f_a \Rightarrow d \asymp c$. Therefore, $\asymp$ is symmetric.
$c \asymp d, d \asymp z \Rightarrow anti(c)d, anti(d)z \in \ker f_a \Rightarrow anti(c)d\ anti(d)z = anti(c)neut(d)z = anti(c)neut(a)z = anti(c)z \in \ker f_a \Rightarrow c \asymp z$. Therefore, $\asymp$ is transitive and $\asymp$ is an

equivalence relation.
The equivalence class $[c]_{f_a} = \{d : anti(c)d \in \ker f_a\} = \{d : c\ anti(c)d \in c \ker f_a\} = \{d : neut(c)d \in c \ker f_a\} = \{d : neut(a)d \in c \ker f_a\} = \{d : d \in c \ker f_a\} = c \ker f_a$. Therefore, $X_a / \asymp = \{[c]_{f_a}\}_{c \in X_a} = \{c \ker f_a\}_{c \in X_a}$.
Thus, $X_a = \bigcup_{c \in X_a} c \ker f_a$ for all $a \in X$.

6. If X is finite, then $|\ker f_a| = |c \ker f_a|$ for all $c \in X_a$. Thus, $|X_a| = \sum_{c \in X_a} |c \ker f_a| = [X_a : \ker f_a] |\ker f_a|$ for all $a \in X$ where $[X_a : \ker f_a]$ is the index of $\ker f_a$ in X_a, i.e., the number of distinct left cosets of $\ker f_a$ in X_a.
7. Define a relation $\sim$ on X: $c \sim d$ if $neut(c) = neut(d)$. $\sim$ is an equivalence relation on X, so $X / \sim = \{X_c\}_{c \in X}$ and, therefore, $X = \bigcup_{a \in X} X_a$.
8. Hence, based on Point 7, if X is finite, then $|X| = \sum_{a \in X} |X_a| = \sum_{a \in X} [X_a : \ker f_a] |\ker f_a|$.

□

Theorem 7. *Let $a \in X$ and $f : X \to Y$ be a neutrosophic triplet group homomorphism. Then*

1. *f is a monomorphism if and only if $\ker f_a = \{neut(a)\}$ for all $a \in X$;*
2. *the factor set $X / \ker f = \bigcup_{a \in X} X_a / \ker f_a$ is a neutrosophic triplet group (neutrosophic triplet factor group) under the operation defined by*

$$c \ker f_a \cdot d \ker f_b = (cd) \ker f_{ab}.$$

Proof. 1. Let $\ker f_a = \{neut(a)\}$ and let $c, d \in X$. If $f(c) = f(d)$, this implies that $f(c\ anti(d)) = f(d) anti(f(d)) = f(d\ anti(f(d))) \Rightarrow f(c\ anti(d)) = neut(f(d)) \Rightarrow c\ anti(d) \in \ker f_d \Rightarrow$

$$c\ anti(d) = neut(d) = neut(anti(d)). \tag{5}$$

Similarly, $f(anti(d)c) = neut(f(d)) \Rightarrow anti(d)c \in \ker f_d \Rightarrow$

$$anti(d)c = neut(anti(d)). \tag{6}$$

Using Equations (5) and (6), $c = anti(anti(d)) = d$. Therefore, f is a monomorphism.

Conversely, if f is mono, then $f(d) = f(c) \Rightarrow d = c$. Let $k \in \ker f_a$, $a \in X$. Then $f(k) = neut(f(a)) = f(neut(a)) \Rightarrow k = neut(a)$. Therefore, $\ker f_a = \{neut(a)\}$ for all $a \in X$.

2. Let $c \ker f_a, d \ker f_b, z \ker f_c \in X / \ker f = \bigcup_{a \in X} X_a / \ker f_a$.

Groupoid: Based on the multiplication $c \ker f_a \cdot d \ker f_b = (cd) \ker f_{ab}$, the factor set $X / \ker f$ is a groupoid.

Semigroup: $(c \ker f_a \cdot d \ker f_b) \cdot z \ker f_c = (cdz) \ker f_{abc} = c \ker f_a (d \ker f_b \cdot z \ker f_c)$.

Neutrality: Let $neut(c \ker f_a) = neut(c) \ker f_{neut(a)}$. Then $c \ker f_a \cdot neut(c \ker f_a) = c \ker f_a \cdot neut(c) \ker f_{neut(a)} = (c\ neut(c)) \ker f_{a\ neut(a)} = c \ker f_a$ and similarly, $neut(c \ker f_a) \cdot c \ker f_a = c \ker f_a$.

Opposite: Let $anti(c \ker f_a) = anti(c) \ker f_{anti(a)}$. Then $c \ker f_a \cdot anti(c \ker f_a) = c \ker f_a \cdot anti(c) \ker f_{anti(a)} = (c\ \ anti(c)) \ker f_{a\ anti(a)} = neut(c) \ker f_{neut(a)}$. Similarly, $anti(c \ker f_a)) \cdot c \ker f_a = neut(c) \ker f_{neut(a)}$.

$\therefore$ $(X / \ker f, \cdot)$ is an NTG.

□

Theorem 8. *Let $\phi : X \to Y$ be a neutrosophic triplet group homomorphism. Then $X/\ker\phi \cong Im\ \phi$.*

Proof. Based on Theorem 6(7), $X = \bigcup_{a\in X} X_a$. Similarly, define a relation $\approx$ on $\phi(X) = \mathrm{Im}\,\phi$: $\phi(c) \approx \phi(d)$ if $neut(\phi(c)) = neut(\phi(d))$. $\approx$ is an equivalence relation on $\phi(X)$, so $\phi(X)/\approx = \{\phi(X_c)\}_{c\in X}$ and $\mathrm{Im}\,\phi = \bigcup_{c\in X} \phi(X_c)$. It should be noted that $X_a \overset{a}{\triangleleft} X$ in Theorem 6(2).

Let $\bar{\phi}_a : X_a/\ker\phi_a \to \phi(X_a)$ given by $\bar{\phi}_a(c\ker\phi_a) = \phi(c)$. It should be noted that, by Theorem 6(1), $\ker\phi_a \overset{a}{\triangleleft} X$. Therefore, $c\ker\phi_a = d\ker\phi_a \Rightarrow anti(d)c\ker\phi_a = anti(d)d\ker\phi_a = neut(d)\ker\phi_a = \ker\phi_a \Rightarrow anti(d)c\ker\phi_a = \ker\phi_a \Rightarrow \phi(anti(d)c) = neut(\phi(a)) \Rightarrow anti(\phi(d))\phi(c) = neut(\phi(a)) \Rightarrow \phi(d)anti(\phi(d))\phi(c) = \phi(d)neut(\phi(a)) \Rightarrow neut(\phi(d))\phi(c) = \phi(d)neut(\phi(a)) \Rightarrow \phi(neut(d))\phi(c) = \phi(d)\phi(neut(a)) \Rightarrow \phi(neut(d)\ c)) = \phi(d\ neut(a)) \Rightarrow \phi(neut(a)\ c)) = \phi(d\ neut(a)) \Rightarrow \phi(neut(c)\ c)) = \phi(d\ neut(c)) \Rightarrow \phi(c) = \phi(d) \Rightarrow \bar{\phi}_a(c\ker\phi_a) = \bar{\phi}_a(d\ker\phi_a)$. Thus, $\bar{\phi}_a$ is well defined.

$\bar{\phi}_a(c\ker\phi_a) = \bar{\phi}_a(d\ker\phi_a) \Rightarrow \phi(c) = \phi(d) \Rightarrow anti(\phi(d))\phi(c) = anti(\phi(d))\phi(d) = neut(\phi(d)) \Rightarrow \phi(anti(d))\phi(c) = neut(\phi(d)) = \phi(neut(d)) = \phi(neut(a)) = neut(\phi(a)) \Rightarrow \phi(anti(d)\ c) = neut(\phi(a)) \Rightarrow anti(d)\ c \in \ker\phi_a \Rightarrow d\ anti(d)\ c \in d\ker\phi_a \Rightarrow neut(d)\ c \in d\ker\phi_a \Rightarrow neut(a)\ c \in d\ker\phi_a \Rightarrow c \in d\ker\phi_a \overset{\text{Theorem 6(1)}}{\Longrightarrow} c\ker\phi_a = d\ker\phi_a$. This means that $\bar{\phi}_a$ is 1-1. $\bar{\phi}_a$ is obviously onto. Thus, $\bar{\phi}_a$ is bijective.

Now, based on the above and Theorem 7(2), we have a bijection

$$\Phi = \bigcup_{a\in X} \bar{\phi}_a : X/\ker\phi = \bigcup_{a\in X} X_a/\ker\phi_a \to \mathrm{Im}\,\phi = \phi(X) = \bigcup_{a\in X} \phi(X_a)$$

defined by $\Phi(c\ker\phi_a) = \phi(c)$. Thus, if $c\ker\phi_a, d\ker\phi_b \in X/\ker\phi$, then

$$\Phi\Big(c\ker\phi_a \cdot d\ker\phi_b\Big) = \Phi(cd\ker\phi_a b) = \phi(cd) = \phi(c)\phi(d) = \Phi(c\ker\phi_a)\Phi(d\ker\phi_b).$$

$\therefore\ X/\ker\phi \cong \mathrm{Im}\,\phi$. □

2.3. Construction of Bol Algebraic Structures

We now present a method of constructing Bol algebraic structures using an NTG.

Theorem 9. *Let X be a non-abelian neutrosophic triplet group and let $A = X_a \times X$ for any fixed $a \in X$. For $(h_1, g_1), (h_2, g_2) \in A$, define $\circ$ on A as follows:*

$$(h_1, g_1) \circ (h_2, g_2) = (h_1 h_2, h_2 g_1\ anti(h_2) g_2).$$

Then $(A, \circ)$ is a Bol groupoid.

Proof. Let $a, b, c \in A$. By checking, it is true that $a \circ (b \circ c) \neq (a \circ b) \circ c$. Therefore, $(A, \circ)$ is non-associative. X_a is a normal neutrosophic triplet group by Theorem 6(3). A is a groupoid.

Let us now verify the Bol identity:

$$((a \circ b) \circ c) \circ b = a \circ ((b \circ c) \circ b)$$

$$\text{LHS} \quad = ((a \circ b) \circ c) \circ b = \Big(h_1 h_2 h_3 h_2, h_2 h_3 h_2 g_1\ anti(h_2) g_2\ anti(h_3) g_3\ anti(h_2) g_2\Big).$$

Following Theorem 6(4),

$$\text{RHS} \quad = a \circ ((b \circ c) \circ b) =$$
$$\Big(h_1h_2h_3h_2, h_2h_3h_2g_1\ anti\big(h_2h_3h_2\big)h_2h_3g_2\ anti(h_3)g_3\ anti(h_2)g_2\Big) =$$
$$\Big(h_1h_2h_3h_2, h_2h_3h_2g_1\ anti(h_2)\big(anti(h_3)\ anti(h_2)h_2h_3\big)g_2\ anti(h_3)g_3\ anti(h_2)g_2\Big) =$$
$$\Big(h_1h_2h_3h_2, h_2h_3h_2g_1\ anti(h_2)\big(anti(h_3)\ neut(h_2)h_3\big)g_2\ anti(h_3)g_3\ anti(h_2)g_2\Big) =$$
$$\Big(h_1h_2h_3h_2, h_2h_3h_2g_1\ anti(h_2)\big(anti(h_3)\ neut(a)h_3\big)g_2\ anti(h_3)g_3\ anti(h_2)g_2\Big) =$$
$$\Big(h_1h_2h_3h_2, h_2h_3h_2g_1\ anti(h_2)anti(h_3)h_3g_2\ anti(h_3)g_3\ anti(h_2)g_2\Big) =$$
$$\Big(h_1h_2h_3h_2, h_2h_3h_2g_1\ anti(h_2)neut(h_3)g_2\ anti(h_3)g_3\ anti(h_2)g_2\Big) =$$
$$\Big(h_1h_2h_3h_2, h_2h_3h_2g_1\ anti(h_2)neut(a)g_2\ anti(h_3)g_3\ anti(h_2)g_2\Big) =$$
$$\Big(h_1h_2h_3h_2, h_2h_3h_2g_1\ anti(h_2)g_2\ anti(h_3)g_3\ anti(h_2)g_2\Big).$$

Therefore, LHS = RHS. Hence, $(A, \circ)$ is a Bol groupoid. □

Corollary 1. *Let H be a subgroup of a non-abelian neutrosophic triplet group X, and let $A = H \times X$. For $(h_1, g_1), (h_2, g_2) \in A$, define $\circ$ on A as follows:*

$$(h_1, g_1) \circ (h_2, g_2) = (h_1h_2, h_2g_1\ anti(h_2)g_2).$$

Then $(A, \circ)$ is a Bol groupoid.

Proof. A subgroup H is a normal neutrosophic triplet group. The rest of the claim follows from Theorem 9. □

Corollary 2. *Let H be a neutrosophic triplet subgroup (which obeys the cancellation law) of a non-abelian neutrosophic triplet group X, and let $A = H \times X$. For $(h_1, g_1), (h_2, g_2) \in A$, define $\circ$ on A as follows:*

$$(h_1, g_1) \circ (h_2, g_2) = (h_1h_2, h_2g_1\ anti(h_2)g_2).$$

Then $(A, \circ)$ is a Bol groupoid.

Proof. By Theorem 1(3), H is a subgroup of X. Hence, following Corollary 1, $(A, \circ)$ is a Bol groupoid. □

Corollary 3. *Let H be a neutrosophic triplet subgroup of a non-abelian neutrosophic triplet group X that has the cancellation law and let $A = H \times X$. For $(h_1, g_1), (h_2, g_2) \in A$, define $\circ$ on A as follows:*

$$(h_1, g_1) \circ (h_2, g_2) = (h_1h_2, h_2g_1\ anti(h_2)g_2).$$

Then $(A, \circ)$ is a Bol loop.

Proof. By Theorem 1(3), X is a non-abelian group and H is a subgroup of X. Hence, $(A, \circ)$ is a loop and a Bol loop by Theorem 9. □

3. Applications in Management and Sports

3.1. One-Way Management and Division of Labor

Consider a company or work place consisting of a set of people X with $|X|$ number of people. A working unit or subgroup with a leader 'a' is denoted by X_a.

$neut(x)$ for any $x \in X$ represents a co-worker (or co-workers) who has (have) a good (non-critical) working relationship with x, while $anti(x)$ represents a co-worker (or co-workers) whom x considers as his/her personal critic(s) at work.

Hence, X_a can be said to include both critics and non-critics of each worker x. It should be noted that in X_a, $neut(a) = neut(x)$ for all $x \in X_a$. This means that every worker in X_a has a good relationship with the leader 'a'.

Thus, by Theorem 6(7)—$X = \bigcup_{a \in X} X_a$ and $|X| = \Sigma_{a \in X}|X_a|$—the company or work place X can be said to have a good division of labor for effective performance and maximum output based on the composition of its various units (X_a). See Figure 1.

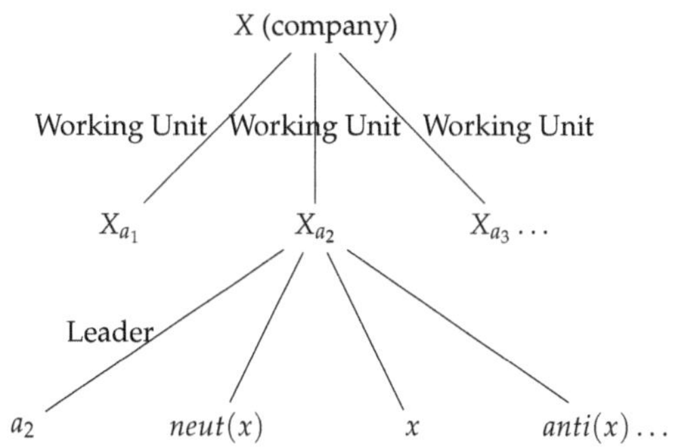

Figure 1. One-way management and division of labor.

3.2. Two-Way Management Division of Labor

Consider a company or work place consisting of a set of people X with $|X|$ number of people at a location A and another company or work place consisting of people Y with $|Y|$ number of people at another location B. Assume that both companies are owned by the same person f. Hence, $f : X \to Y$ can be considered as a movement (transfer) or working interaction between workers at A and at B. The fact that f is a neutrosophic triplet group homomorphism indicates that the working interaction between X and Y is preserved.

Let 'a' be a unit leader at A whose work correlates to another leader $f(a)$ at B. Then $Kerf_a$ represents the set of workers x in a unit at A under the leadership of 'a' such that there are other, corresponding workers $f(x)$ at B under the leadership of $f(a)$. Here, $f(x) = neut(f(a))$ means that workers $f(x)$ at B under the leadership of $f(a)$ are loyal and in a good working relationship. The mapping f_a shows that the operation of a subgroup leader (the operation is denoted by 'a') is subject to the modus operandi of the owner of the two companies, where the owner is denoted by f.

The final formula $|X| = \sum_{x \in X} [X_a : \ker f_a]|kerf_a|$ in Theorem 6(8) shows that the overall performance of the set of people X is determined by how the unit leaders 'a' at A properly harmonize with the unit leaders at B in the effective administration of $\ker f_a$ and X_a (Figure 2).

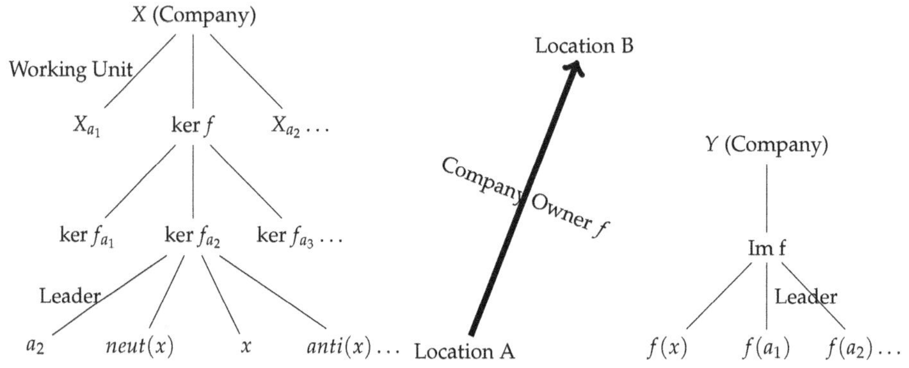

Figure 2. Two-way management division of labor.

3.3. *Sports*

In the composition of a team, a coach can take X_a as the set of players who play in a particular department (e.g., forward, middle field, or defence), where a is the leader of that department. Let $neut(x)$ represent player(s) whose performance is the same as that of player x, and let $anti(x)$ represent player(s) that can perform better than player x. It should be noted that the condition $neut(x) = neut(a)$ for all $x \in X_a$ means that the department X_a has player(s) who are equal in performance; i.e., those whose performance are equal to that of the departmental leader a. Hence, a neutrosophic triplet $(x, neut(x), anti(x))$ is a triple from which a coach can make a choice of his/her starting player and make a substitution. The neutrosophic triplet can also help a coach to make the best alternative choice when injuries arise. For instance, in the goal keeping department (for soccer/football), three goal keepers often make up the team for any international competition. Imagine an incomplete triplet $(x, neut(x), ?)$, i.e., no player is found to be better than x, which reduces to a duplet.

X_a can also be used for grouping teams in competitions in the preliminaries. If $x = team$, then $anti(x) = teams$ that can beat x and $neut(x) = teams$ that can play draw with x. Therefore, neutrosophic triplet $(x, neut(x), anti(x))$ is a triplet with which competition organizers can draw teams into groups for a balanced competition. The Fédération Internationale de Football Association (FIFA) often uses this template in drawing national teams into groups for its competitions. Club teams from various national leagues, to qualify for continental competitions (e.g., Union of European Football Associations (UEFA) Champions League and Confederation of African Football (CAF) Champions League), have to be among the five. This implies the application of duplets, triplets, quadruples, etc. (Figure 3).

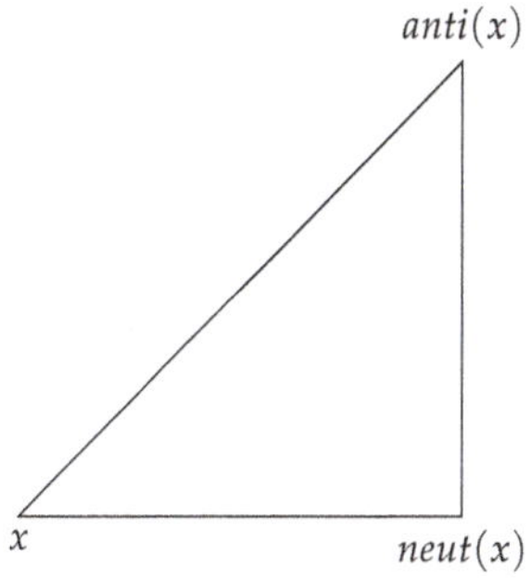

Figure 3. Sports.

Author Contributions: T.G.J. established new properties of neutrosophic triplet groups. He further presented applications of the neutrosophic triplet sets and groups to management and sports. F.S. cointroduced the neutrosophic triplet set and group, as well as their properties. He confirmed the relevance of the neutrosophic duplet and the quadruple in the applications of neutrosophic triplet set.

Conflicts of Interest: The authors declare no conflict of interest.

References

1. Araujo, J.; Konieczny, J. Molaei's Generalized Groups are Completely Simple Semigroups. *Bul. Inst. Polytech. Jassy Sect. I Mat. Mec. Teor. Fiz.* **2002**, *48*, 1–5.
2. Adeniran, J.O.; Akinmoyewa, J.T.; Solarin, A.R.T.; Jaiyeola, T.G. On some algebraic properties of generalized groups. *Acta Math. Acad. Paedagog. Nyhazi.* **2011**, *27*, 23–30.
3. Agboola, A.A.A. Smooth generalized groups. *J. Niger. Math. Soc.* **2004**, *23*, 6–76.
4. Agboola, A.A.A. Smooth generalized subgroups and homomorphisms. *Adv. Stud. Contemp. Math.* **2004**, *9*, 183–193.
5. Molaei, M.R. Topological generalized groups. *Int. J. Appl. Math.* **2000**, *2*, 1055–1060.
6. Molaei, M.R.; Tahmoresi, A. Connected topological generalized groups. *Gen. Math.* **2004**, *12*, 13–22.
7. Maleki, H.; Molaei, H. On the quotient space of a generalized action of a generalized group. *arXiv* **2014**, arXiv:1402.3408.
8. Jaiyéọlá, T.G. *A Study of New Concepts in Smarandache Quasigroups and Loops*; Books on Demand, ProQuest Information and Learning: Ann Arbor, MI, USA, 2009.
9. Shcherbacov, V. *Elements of Quasigroup Theory and Applications*; Monographs and Research Notes in Mathematics; CRC Press: Boca Raton, FL, USA, 2017; ISBN 978-1-4987-2155-4.
10. Chein, O.; Pflugfelder, H.O.; Smith, J.D.H. *Quasigroups and Loops: Theory and Applications*; Sigma Series in Pure Mathematics 8; Heldermann: Berlin, Germany, 1990; ISBN 3-88538-008-0.
11. Pflugfelder, H.O. *Quasigroups and Loops: Introduction*; Sigma Series in Pure Math. 7; Heldermann: Berlin, Germany, 1990.
12. Smith, J.D.H. *An introduction to Quasigroups and Their Representations*; Studies in Advanced Mathematics; Chapman & Hall/CRC: Boca Raton, FL, USA, 2007; ISBN 978-1-58488-537-5, 1-58488-537-8.
13. Smith, J.D.H. Poset Loops. *Order* **2017**, *34*, 265–285. [CrossRef]
14. Glukhov, M.M. On the multiplicative groups of free and free commutative quasigroups. *J. Math. Sci. (New York)* **2017**, *223*, 508–517. [CrossRef]
15. Greer, M. Semiautomorphic inverse property loops. *Comm. Algebra* **2017**, *45*, 2222–2237. [CrossRef]
16. Smarandache, F. *A Unifying Field in Logics. Neutrosophy: Neutrosophic Probability, Set and Logic*; American Research Press (ARP): Rehoboth, NM, USA, 2003; ISBN 1-879585-76-6.
17. Smarandache, F. Neutrosophic set, a generalization of the intuitionistic fuzzy set. In Proceedings of the 2006 IEEE International Conference on Granular Computing, Atlanta, GA, USA, 10–12 May 2006; pp. 38–42. GRC.2006.1635754. [CrossRef]
18. Zadeh, L.A. Information and control. *Fuzzy Sets* **1965**, *8*, 338–353.

19. Atanassov, A.K. Intuitionistic fuzzy sets. *Fuzzy Sets Syst.* **1986**, *20*, 87–96. [CrossRef]
20. Liu, P.; Liu, J.; Chen, S.M. Some intuitionistic fuzzy Dombi Bonferroni mean operators and their application to multi-attribute group decision making. *J. Oper. Res. Soc.* **2018**, *69*, 1–24. [CrossRef]
21. Liu, P.; Chen, S.M. Group decision making based on Heronian aggregation operators of intuitionistic fuzzy numbers. *IEEE Trans. Cybern.* **2017**, *47*, 2514–2530. [CrossRef] [PubMed]
22. Liu, P.; Chen, S.M. Multiattribute Group Decision Making Based on Intuitionistic 2-Tuple Linguistic Information. *Inf. Sci.* **2018**, *430–431*, 599–619. [CrossRef]
23. Liu, P. Multiple attribute group decision making method based on interval-valued intuitionistic fuzzy power Heronian aggregation operators. *Comput. Ind. Eng.* **2017**, *108*, 199–212. [CrossRef]
24. Liu, P.; Li, H. Interval-valued intuitionistic fuzzy power Bonferroni aggregation operators and their application to group decision making. *Cognit. Comput.* **2017**, *9*, 494–512. [CrossRef]
25. Liu, P.; Liu, J.; Merigo, J.M. Partitioned Heronian means based on linguistic intuitionistic fuzzy numbers for dealing with multi-attribute group decision making. *Appl. Soft Comput.* **2018**, *62*, 395-422. [CrossRef]
26. Liu, P.; Shi, L. Some Neutrosophic uncertain linguistic number Heronian mean operators and their application to multi-attribute group decision making. *Neural Comput. Appl.* **2017**, *28*, 1079–1093. [CrossRef]
27. Liu, P.; Zhang, L.; Liu, X.; Wang, P. Multi-valued Neutrosophic number Bonferroni mean operators and their application in multiple attribute group decision making. *Int. J. Inf. Technol. Decis. Mak.* **2016**, *15*, 1181–1210. [CrossRef]
28. Smarandache, F.; Ali, M. Neutrosophic triplet group. *Neural Comput. Appl.* **2016**. [CrossRef]
29. Jaiyéọlá, T. G.; Smarandache, F. Inverse Properties in Neutrosophic Triplet Loop and their Application to Cryptography. *Algorithms* **2018**, *11*, 32. [CrossRef]

symmetry

MDPI

Article

A Study on Neutrosophic Cubic Graphs with Real Life Applications in Industries

Muhammad Gulistan [1], Naveed Yaqoob [2], Zunaira Rashid [1,*], Florentin Smarandache [3] and Hafiz Abdul Wahab [1]

1. Department of Mathematics, Hazara University, Mansehra 21120, Pakistan; gulistanmath@hu.edu.pk (M.G.); wahab@hu.edu.pk (H.A.W.)
2. Department of Mathematics, College of Science, Majmaah University, Al-Zulfi 11952, Saudi Arabia; nayaqoob@ymail.com
3. Department of Mathematics, University of New Mexico, Albuquerque, NM 87301, USA; fsmarandache@gmail.com

* Correspondence: zunairarasheed95@gmail.com

Received: 27 April 2018; Accepted: 30 May 2018; Published: 5 June 2018

Abstract: Neutrosophic cubic sets are the more generalized tool by which one can handle imprecise information in a more effective way as compared to fuzzy sets and all other versions of fuzzy sets. Neutrosophic cubic sets have the more flexibility, precision and compatibility to the system as compared to previous existing fuzzy models. On the other hand the graphs represent a problem physically in the form of diagrams, matrices etc. which is very easy to understand and handle. So the authors applied the Neutrosophic cubic sets to graph theory in order to develop a more general approach where they can model imprecise information through graphs. We develop this model by introducing the idea of neutrosophic cubic graphs and introduce many fundamental binary operations like cartesian product, composition, union, join of neutrosophic cubic graphs, degree and order of neutrosophic cubic graphs and some results related with neutrosophic cubic graphs. One of very important futures of two neutrosophic cubic sets is the $R-$union that $R-$union of two neutrosophic cubic sets is again a neutrosophic cubic set, but here in our case we observe that $R-$union of two neutrosophic cubic graphs need not be a neutrosophic cubic graph. Since the purpose of this new model is to capture the uncertainty, so we provide applications in industries to test the applicability of our defined model based on present time and future prediction which is the main advantage of neutrosophic cubic sets.

Keywords: neutrosophic cubic set; neutrosophic cubic graphs; applications of neutrosophic cubic graphs

MSC: 68R10; 05C72; 03E72

1. Introduction

In 1965, Zadeh [1] published his seminal paper "Fuzzy Sets" which described fuzzy set theory and consequently fuzzy logic. The purpose of Zadeh's paper was to develop a theory which could deal with ambiguity and imprecision of certain classes or sets in human thinking, particularly in the domains of pattern recognition, communication of information and abstraction. This theory proposed making the grade of membership of an element in a subset of a universal set a value in the closed interval $[0,1]$ of real numbers. Zadeh's ideas have found applications in computer sciences, artificial intelligence, decision analysis, information sciences, system sciences, control engineering, expert systems, pattern recognition, management sciences, operations research and robotics. Theoretical mathematics has also been touched by fuzzy set theory. The ideas of fuzzy set theory have been introduced into topology, abstract algebra, geometry, graph theory and analysis. Further, he made the extension of fuzzy set to interval-valued fuzzy sets in 1975, where one is not bound to give a specific membership to a certain element. In 1975, Rosenfeld [2] discussed the concept of fuzzy graphs whose basic idea was introduced by Kauffmann [3] in 1973. The fuzzy relations between fuzzy sets were also considered by Rosenfeld and he developed the structure of fuzzy graphs obtaining analogs of several graph theoretical concepts [4]. Bhattacharya provided further studies on fuzzy graphs [5]. Akram and Dudek gave the idea of interval valued fuzzy graphs in 2011 where they used interval membership for an element in the vertex set [6]. Akram further extended the idea of interval valued fuzzy graphs to Interval-valued fuzzy line graphs in 2012. More detail of fuzzy graphs, we refer the reader to [7–12]. In 1986, Atanassov [13] use the notion of membership and non-membership of an element in a set X and gave the idea of intuitionistic fuzzy sets. He extended this idea to intuitionistic fuzzy graphs and for more detail in this direction, we refer the reader to [14–20]. Akram and Davvaz [21] introduced the notion of strong intuitionistic fuzzy graphs and investigated some of their properties. They discussed some propositions of self complementary and self weak complementary strong intuitionistic fuzzy graphs. In 1994, Zhang [22] started the theory of bipolar fuzzy sets as a generality of fuzzy sets. Bipolar fuzzy sets are postponement of fuzzy sets whose membership degree range is $[-1, 1]$. Akram [23,24] introduced the concepts of bipolar fuzzy graphs, where he introduced the notion of bipolar fuzzy graphs, described various methods of their construction, discussed the concept of isomorphisms of these graphs and investigated some of their important properties. He then introduced the notion of strong bipolar fuzzy graphs and studied some of their properties. He also discussed some propositions of self complementary and self weak complementary strong bipolar fuzzy graphs and applications, for example see [25]. Smarandache [26–28] extended the concept of Atanassov and gave the idea of neutrosophic sets. He proposed the term "neutrosophic" because "neutrosophic" etymologically comes from "neutrosophy" This comes from the French neutre $<$ Latin neuter, neutral, and Greek sophia, skill/wisdom, which means knowledge of neutral thought, and this third/neutral represents the main distinction between "fuzzy" and "intuitionistic fuzzy" logic/set, i.e., the included middle component (Lupasco-Nicolescu's logic in philosophy), i.e., the neutral/indeterminate/unknown part (besides the "truth"/"membership" and "falsehood"/"non-membership" components that both appear in fuzzy logic/set). See the Proceedings of the First International Conference on Neutrosophic Logic, The University of New Mexico, Gallup Campus, 1–3 December 2001, at http://www.gallup.unm.edu/~smarandache/FirstNeutConf.htm.

After that, many researchers used the idea of neutrosophic sets in different directions. The idea of neutrosophic graphs is provided by Kandasamy et al. in the book title as Neutrosophic graphs, where they introduce idea of neutrosophic graphs [29]. This study reveals that these neutrosophic graphs give a new dimension to graph theory. An important feature of this book is that it contains over 200 neutrosophic graphs to provide better understandings of these concepts. Akram and others discussed different aspects of neutrosophic graphs [30–33]. Further Jun et al. [34] gave the idea of cubic set and it was characterized by interval valued fuzzy set and fuzzy set, which is more general tool to capture uncertainty and vagueness, while fuzzy set deals with single value membership and interval valued fuzzy set ranges the membership in the form of interval. The hybrid platform provided by the cubic set is the main advantage, in that it contains more information then a fuzzy set and interval valued fuzzy set. By using this concept, we can solve different problems arising in several areas and can pick finest choice by means of cubic sets in various decision making problems. This hybrid nature of the cubic set attracted these researchers to work in this field. For more detail about cubic sets and their applications in different research areas, we refer the reader to [35–37]. Recently, Rashid et al. [38] introduced the notion of cubic graphs where they introduced many new types of graphs and provided their application. More recently Jun et al. [39,40] combined neutrosophic set with cubic sets and gave the idea of Neutrosophic cubic set and defined different operations.

Therefore, the need was felt to develop a model for neutrosophic cubic graphs which is a more generalized tool to handle uncertainty. In this paper, we introduce the idea of neutrosophic cubic graphs and introduce the fundamental binary operations, such as the cartesian product, composition, union, join of neutrosophic cubic graphs, degree, order of neutrosophic cubic graphs and some results related to neutrosophic cubic graphs. We observe that R-union of two neutrosophic cubic graphs need not to be a neutrosophic cubic graph. At the end, we provide applications of neutrosophic cubic graphs in industries to test the applicability of our presented model.

2. Preliminaries

We recall some basic definitions related to graphs, fuzzy graphs and neutrosophic cubic sets.

Definition 1. *A graph is an ordered pair $G^* = (V, E)$, where V is the set of vertices of G^* and E is the set of edges of G^*.*

Definition 2. *A fuzzy graph [2–4] with an underlying set V is defined to be a pair $G = (\mu, \nu)$ where μ is a fuzzy function in V and ν is a fuzzy function in $E \subseteq V \times V$ such that $\nu(\{x, y\}) \leq \min(\mu(x), \mu(y))$ for all $\{x, y\} \in E$.*

We call μ the fuzzy vertex function of V, ν the fuzzy edge function of E, respectively. Please note that ν is a symmetric fuzzy relation on μ. We use the notation xy for an element $\{x, y\}$ of E. Thus, $G = (\mu, \nu)$ is a fuzzy graph of $G^* = (V, E)$ if $\nu(xy) \leq \min(\mu(x), \mu(y))$ for all $xy \in E$.

Definition 3. *Let $G = (\mu, \nu)$ be a fuzzy graph. The order of a fuzzy graph [2–4] is defined by $O(G) = \sum_{x \in V} \mu(x)$. The degree of a vertex x in G is defined by $deg(x) = \sum_{xy \in E} \nu(xy)$.*

Definition 4. *Let μ_1 and μ_2 be two fuzzy functions of V_1 and V_2 and let ν_1 and ν_2 be fuzzy functions of E_1 and E_2, respectively. The Cartesian product of two fuzzy graphs G_1 and G_2 [2–4] of the graphs G_1^* and G_2^* is denoted by $G_1 \times G_2 = (\mu_1 \times \mu_2, \nu_1 \times \nu_2)$ and is defined as follows:*

(i) $(\mu_1 \times \mu_2)(x_1, x_2) = \min(\mu_1(x_1), \mu_2(x_2))$, *for all* $(x_1, x_2) \in V$.
(ii) $(\nu_1 \times \nu_2)((x, x_2)(x, y_2)) = \min(\mu_1(x), \nu_2(x_2y_2))$, *for all* $x \in V_1$, *for all* $x_2y_2 \in E_2$.
(iii) $(\nu_1 \times \nu_2)((x_1, z)(y_1, z)) = \min(\nu_1(x_1y_1), \mu_2(z))$, *for all* $z \in V_2$, *for all* $x_1y_1 \in E_1$.

Definition 5. *Let μ_1 and μ_2 be fuzzy functions of V_1 and V_2 and let ν_1 and ν_2 be fuzzy functions of E_1 and E_2, respectively. The composition of two fuzzy graphs G_1 and G_2 of the graphs G_1^* and G_2^* [2–4] is denoted by $G_1[G_2] = (\mu_1 \circ \mu_2, \nu_1 \circ \nu_2)$ and is defined as follows:*

(i) $(\mu_1 \circ \mu_2)(x_1, x_2) = \min(\mu_1(x_1), \mu_2(x_2))$, *for all* $(x_1, x_2) \in V$.
(ii) $(\nu_1 \circ \nu_2)((x, x_2)(x, y_2)) = \min(\mu_1(x), \nu_2(x_2y_2))$, *for all* $x \in V_1$, *for all* $x_2y_2 \in E_2$.
(iii) $(\nu_1 \circ \nu_2)((x_1, z)(y_1, z)) = \min(\nu_1(x_1y_1), \mu_2(z))$, *for all* $z \in V_2$, *for all* $x_1y_1 \in E_1$.
(iv) $(\nu_1 \circ \nu_2)((x_1, x_2)(y_1, y_2)) = \min(\mu_2(x_2), \mu_2(y_2), \nu_1(x_1y_1))$, *for all* $z \in V_2$, *for all* $(x_1, x_2)(y_1, y_2) \in E^0 - E$.

Definition 6. *Let μ_1 and μ_2 be fuzzy functions of V_1 and V_2 and let ν_1 and ν_2 be fuzzy functions of E_1 and E_2, respectively. Then union of two fuzzy graphs G_1 and G_2 of the graphs G_1^* and G_2^* [2–4] is denoted by $G_1 \cup G_2 = (\mu_1 \cup \mu_2, \nu_1 \cup \nu_2)$ and is defined as follows:*

(i) $(\mu_1 \cup \mu_2)(x) = \mu_1(x)$ *if* $x \in V_1 \cap V_2$,
(ii) $(\mu_1 \cup \mu_2)(x) = \mu_2(x)$ *if* $x \in V_2 \cap V_1$,
(iii) $(\mu_1 \cup \mu_2)(x) = \max(\mu_1(x), \mu_2(x))$ *if* $x \in V_1 \cap V_2$,
(iv) $(\nu_1 \cup \nu_2)(xy) = \nu_1(xy)$ *if* $xy \in E_1 \cap E_2$,
(v) $(\nu_1 \cup \nu_2)(xy) = \nu_2(xy)$ *if* $xy \in E_2 \cap E_1$,
(vi) $(\nu_1 \cup \nu_2)(xy) = \max(\nu_1(xy), \nu_2(xy))$ *if* $xy \in E_1 \cap E_2$.

Definition 7. *Let μ_1 and μ_2 be fuzzy functions of V_1 and V_2 and let ν_1 and ν_2 be fuzzy functions of E_1 and E_2, respectively. Then join of two fuzzy graphs G_1 and G_2 of the graphs G_1^* and G_2^* [2–4] is denoted by $G_1 + G_2 = (\mu_1 + \mu_2, \nu_1 + \nu_2)$ and is defined as follows:*

(i) $(\mu_1 + \mu_2)(x) = (\mu_1 \cup \mu_2)(x)$ *if* $x \in V_1 \cup V_2$,
(ii) $(\nu_1 + \nu_2)(xy) = (\nu_1 \cup \nu_2)(xy) = \nu_1(xy)$ *if* $xy \in E_1 \cup E_2$,
(iii) $(\nu_1 + \nu_2)(xy) = \min(\mu_1(x), \mu_2(y))$ *if* $xy \in E'$.

Definition 8. *Let X be a non-empty set. A neutrosophic cubic set (NCS) in X [39] is a pair* $A = (\mathbf{A}, \Lambda)$ *where* $\mathbf{A} = \{ \langle x, A_T(x), A_I(x), A_F(x)\rangle \mid x \in X\}$ *is an interval neutrosophic set in X and* $\Lambda = \{ \langle x, \lambda_T(x), \lambda_I(x), \lambda_F(x)\rangle \mid x \in X\}$ *is a neutrosophic set in X.*

3. Neutrosophic Cubic Graphs

The motivation behind this section is to combine the concept of neutrosophic cubic sets with graphs theory. We introduce the concept of neutrosophic cubic graphs, order and degree of neutrosophic cubic graph and different fundamental operations on neutrosophic cubic graphs with examples.

Definition 9. *Let* $G^* = (V, E)$ *be a graph. By neutrosophic cubic graph of* G^**, we mean a pair* $G = (M, N)$ *where* $M = (A, B) = ((\widetilde{T}_A, T_B), (\widetilde{I}_A, I_B), (\widetilde{F}_A, F_B))$ *is the neutrosophic cubic set representation of vertex set V and* $N = (C, D) = ((\widetilde{T}_C, T_D), (\widetilde{I}_C, I_D), (\widetilde{F}_C, F_D))$ *is the neutrosophic cubic set representation of edges set E such that;*

(i) $\left(\widetilde{T}_C(u_iv_i) \preceq rmin\{\widetilde{T}_A(u_i), \widetilde{T}_A(v_i)\}, T_D(u_iv_i) \leq \max\{T_B(u_i), T_B(v_i)\}\right)$,
(ii) $\left(\widetilde{I}_C(u_iv_i) \preceq rmin\{\widetilde{I}_A(u_i), \widetilde{I}_A(v_i)\}, I_D(u_iv_i) \leq \max\{I_B(u_i), I_B(v_i)\}\right)$,
(iii) $\left(\widetilde{F}_C(u_iv_i) \preceq rmax\{\widetilde{F}_A(u_i), \widetilde{F}_A(v_i)\}, F_D(u_iv_i) \leq \min\{F_B(u_i), F_B(v_i)\}\right)$.

Example 1. *Let* $G^* = (V, E)$ *be a graph where* $V = \{a, b, c, d\}$ *and* $E = \{ab, bc, ac, ad, cd\}$*, where*

$$M = \left\langle \begin{array}{c} \{a, ([0.2, 0.3], 0.5), ([0.1, 0.4], 0.6), ([0.5, 0.6], 0.3)\}, \\ \{b, ([0.1, 0.2], 0.4), ([0.4, 0.5], 0.6), ([0.7, 0.8], 0.4)\}, \\ \{c, ([0.4, 0.7], 0.1), ([0.7, 0.8], 0.9), ([0.3, 0.4]), 0.5)\}, \\ \{d, ([0.3, 0.5], 0.2), ([0.9, 1], 0.5), ([0.2, 0.4], 0.1)\} \end{array} \right\rangle$$

$$N = \left\langle \begin{array}{c} \{ab, ([0.1, 0.2], 0.5), ([0.1, 0.4], 0.6), ([0.7, 0.8], 0.3)\}, \\ \{ac, ([0.2, 0.3], 0.5), ([0.1, 0.4], 0.9), ([0.5, 0.6], 0.3)\}, \\ \{ad, ([0.2, 0.3], 0.5), ([0.1, 0.4], 0.6), ([0.5, 0.6]), 0.1)\}, \\ \{bc, ([0.1, 0.2], 0.4), ([0.4, 0.5], 0.9), ([0.7, 0.8], 0.4)\}, \\ \{bd, ([0.1, 0.2], 0.4), ([0.4, 0.5], 0.6), ([0.7, 0.8], 0.1)\}, \\ \{cd, ([0.3, 0.5], 0.2), ([0.7, 0.8], 0.9), ([0.3, 0.4], 0.1)\} \end{array} \right\rangle$$

Then clearly $G = (M, N)$ *is a neutrosophic cubic graph of* $G^* = (V, E)$ *as showin in Figure 1.*

Remark 1.

1. *If* $n \geq 3$ *in the vertex set and* $n \geq 3$ *in the set of edges then the graphs is a neutrosophic cubic polygon only when we join each vertex to the corresponding vertex through an edge.*
2. *If we have infinite elements in the vertex set and by joining the each and every edge with each other we get a neutrosophic cubic curve.*

Definition 10. *Let* $G = (M, N)$ *be a neutrosophic cubic graph. The order of neutrosophic cubic graph is defined by* $O(G) = \Sigma_{x \epsilon V}\{(\widetilde{T}_A, T_B)(x), (\widetilde{I}_A, I_B)(x), (\widetilde{F}_A, F_B)(x)\}$ *and degree of a vertex x in G is defined by* $\deg(x) = \Sigma_{xy \epsilon E}\{(\widetilde{T}_C, T_D)(xy), (\widetilde{I}_C, I_D)(xy), (\widetilde{F}_C, F_D)(xy))\}$.

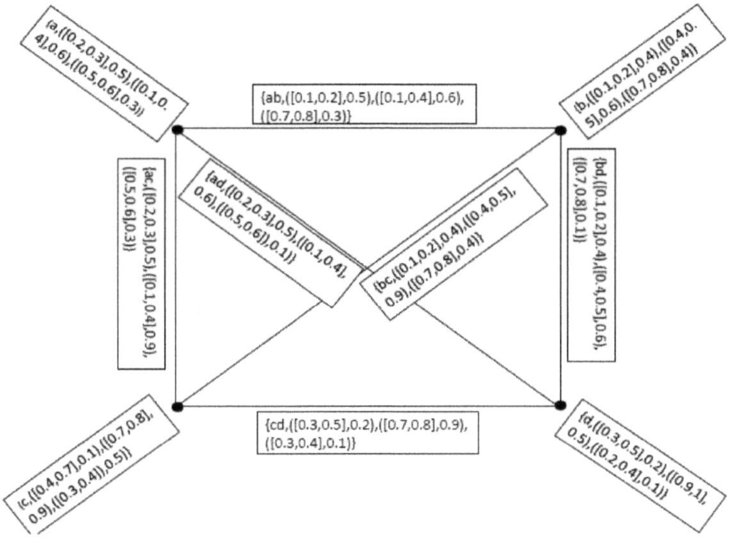

Figure 1. Neutrosophic Cubic Graph.

Example 2. *In Example 1, Order of a neutrosophic cubic graph is*

$$O(G) = \{([1.0, 1.7], 1.2), ([2.1, 1.8], 2.6), ([1.7, 2.2], 1.3)\}$$

and degree of each vertex in G is

$$\begin{aligned}
\deg(a) &= \{([0.5, 0.8], 1.5), ([0.3, 1.2], 2.1), ([1.7, 2.0], 0.7)\} \\
\deg(b) &= \{([0.3, 0.6], 1.3), ([0.9, 1.4], 2.1), ([2.1, 2.4], 0.8)\} \\
\deg(c) &= \{([0.6, 1.0], 1.1), ([1.2, 1.7], 2.7), ([1.5, 1.8], 0.8)\} \\
\deg(d) &= \{([0.6, 1.0], 1.1), ([1.2, 1.7], 2.1), ([1.5, 1.8], 0.3)\}
\end{aligned}$$

Definition 11. *Let $G_1 = (M_1, N_1)$ be a neutrosophic cubic graph of $G_1^* = (V_1, E_1)$, and $G_2 = (M_2, N_2)$ be a neutrosophic cubic graph of $G_2^* = (V_2, E_2)$. Then Cartesian product of G_1 and G_2 is denoted by*

$$\begin{aligned}
G_1 \times G_2 &= (M_1 \times M_2, N_1 \times N_2) = ((A_1, B_1) \times (A_2, B_2), (C_1, D_1) \times (C_2, D_2)) \\
&= ((A_1 \times A_2, B_1 \times B_2), (C_1 \times C_2, D_1 \times D_2)) \\
&= \left\langle \begin{array}{l} ((\widetilde{T}_{A_1 \times A_2}, T_{B_1 \times B_2}), (\widetilde{I}_{A_1 \times A_2}, I_{B_1 \times B_2}), (\widetilde{F}_{A_1 \times A_2}, F_{B_1 \times B_2})), \\ ((\widetilde{T}_{C_1 \times C_2}, T_{D_1 \times D_2}), (\widetilde{I}_{C_1 \times C_2}, I_{D_1 \times D_2}), (\widetilde{F}_{C_1 \times C_2}, F_{D_1 \times D_2})) \end{array} \right\rangle
\end{aligned}$$

and is defined as follow

(i) $\left(\widetilde{T}_{A_1 \times A_2}(x, y) = rmin(\widetilde{T}_{A_1}(x), \widetilde{T}_{A_2}(y)), T_{B_1 \times B_2}(x, y) = \max(T_{B_1}(x), T_{B_2}(y))\right)$,

(ii) $\left(\widetilde{I}_{A_1 \times A_2}(x, y) = rmin(\widetilde{I}_{A_1}(x), \widetilde{I}_{A_2}(y)), I_{B_1 \times B_2}(x, y) = \max(I_{B_1}(x), I_{B_2}(y))\right)$,

(iii) $\left(\widetilde{F}_{A_1 \times A_2}(x, y) = rmax(\widetilde{F}_{A_1}(x), \widetilde{F}_{A_2}(y)), F_{B_1 \times B_2}(x, y) = \min(F_{B_1}(x), F_{B_2}(y))\right)$,

(iv) $\left(\begin{array}{l} \widetilde{T}_{C_1 \times C_2}((x, y_1)(x, y_2)) = rmin(\widetilde{T}_{A_1}(x), \widetilde{T}_{C_2}(y_1 y_2)), \\ T_{D_1 \times D_2}((x, y_1)(x, y_2)) = \max(T_{B_1}(x), T_{D_2}(y_1 y_2)) \end{array}\right)$,

(v) $\left(\begin{array}{l} \widetilde{I}_{C_1 \times C_2}((x, y_1)(x, y_2)) = rmin(\widetilde{I}_{A_1}(x), \widetilde{I}_{C_2}(y_1 y_2)), \\ I_{D_1 \times D_2}((x, y_1)(x, y_2)) = \max(I_{B_1}(x), I_{D_2}(y_1 y_2)) \end{array}\right)$,

$$(vi)\left(\begin{array}{l}\widetilde{F}_{C_1\times C_2}((x,y_1)(x,y_2)) = rmax(\widetilde{F}_{A_1}(x),\widetilde{F}_{C_2}(y_1y_2)),\\ F_{D_1\times D_2}((x,y_1)(x,y_2)) = \min(F_{B_1}(x),F_{D_2}(y_1y_2))\end{array}\right),$$
$$(vii)\left(\begin{array}{l}\widetilde{T}_{C_1\times C_2}((x_1,y)(x_2,y)) = rmin(\widetilde{T}_{C_1}(x_1x_2),\widetilde{T}_{A_2}(y)),\\ T_{D_1\times D_2}((x_1,y)(x_2,y)) = \max(T_{D_1}(x_1x_2),T_{B_2}(y))\end{array}\right),$$
$$(viii)\left(\begin{array}{l}\widetilde{I}_{C_1\times C_2}((x_1,y)(x_2,y)) = rmin(\widetilde{I}_{C_1}(x_1x_2),\widetilde{I}_{A_2}(y)),\\ I_{D_1\times D_2}((x_1,y)(x_2,y)) = \max(I_{D_1}(x_1x_2),I_{B_2}(y))\end{array}\right),$$
$$(ix)\left(\begin{array}{l}\widetilde{F}_{C_1\times C_2}((x_1,y)(x_2,y)) = rmax(\widetilde{F}_{C_1}(x_1x_2),\widetilde{F}_{A_2}(y)),\\ F_{D_1\times D_2}((x_1,y)(x_2,y)) = \min(F_{D_1}(x_1x_2),F_{B_2}(y))\end{array}\right), \forall\ (x,y)\ \in\ (V_1,V_2)\ =\ V\ for\ (i)-$$
$(iii), \forall x \in V_1$ *and* $y_1y_2 \in E_2$ *for* $(iv)-(vi), \forall y \in V_2$ *and* $x_1x_2 \in E_1$ *for* $(vi)-(ix)$.

Example 3. *Let* $G_1 = (M_1, N_1)$ *be a neutrosophic cubic graph of* $G_1^* = (V_1, E_1)$ *as showin in Figure 2, where* $V_1 = \{a, b, c\}$, $E_1 = \{ab, bc, ac\}$

$$M_1 = \left\langle\begin{array}{l}\{a,([0.1,0.2],0.5),([0.4,0.5],0.3),([0.6,0.7],0.2)\},\\ \{b,([0.2,0.4],0.1),([0.5,0.6],0.4),([0.1,0.2],0.3)\},\\ \{c,([0.3,0.4],0.2),([0.1,0.3],0.7),([0.4,0.6],0.3)\}\end{array}\right\rangle$$
$$N_1 = \left\langle\begin{array}{l}\{ab,([0.1,0.2],0.5),([0.4,0.5],0.4),([0.6,0.7],0.2)\},\\ \{bc,([0.2,0.4],0.2),([0.1,0.3],0.7),([0.4,0.6]),0.3)\},\\ \{ac,([0.1,0.2],0.5),([0.1,0.3],0.7),([0.6,0.7],0.2)\}\end{array}\right\rangle$$

and $G_2 = (M_2, N_2)$ *be a neutrosophic cubic graph of* $G_2^* = (V_2, E_2)$ *as showin in Figure 3, where* $V_2 = \{x, y, z\}$ *and* $E_2 = \{xy, yz, xz\}$

$$M_2 = \left\langle\begin{array}{l}\{x,([0.7,0.8],0.6),([0.2,0.4],0.5),([0.3,0.4],0.7)\},\\ \{y,([0.2,0.3],0.4),([0.6,0.7],0.3),([0.9,1.0],0.5)\},\\ \{z,([0.4,0.5],0.2),([0.3,0.4],0.1),([0.6,0.7],0.4)\}\end{array}\right\rangle$$
$$N_2 = \left\langle\begin{array}{l}\{xy,([0.2,0.3],0.6),([0.2,0.4],0.5),([0,9,1.0],0.5)\},\\ \{yz,([0.2,0.3],0.4),([0.3,0.4],0.3),([0.9,1.0],0.4)\},\\ \{xz,([0.4,0.5],0.6),([0.2,0.4],0.5),([0.6,0.7],0.4)\}\end{array}\right\rangle$$

then $G_1 \times G_2$ *is a neutrosophic cubic graph of* $G_1^* \times G_2^*$*, as showin in Figure 4, where* $V_1 \times V_2 = \{(a,x),(a,y),(a,z),(b,x),(b,y),(b,z),(c,x),(c,y),(c,z)\}$ *and*

$$M_1\times M_2 = \left\langle\begin{array}{l}\{(a,x),([0.1,0.2],0.6),([0.2,0.4],0.5),([0.6,0.7],0.2)\},\\ \{(a,y),([0.1,0.2],0.5),([0.4,0.5],0.3),([0.9,1.0],0.2)\},\\ \{(a,z),([0.1,0.2],0.5),([0.3,0.4],0.3),([0.6,0.7],0.2)\},\\ \{(b,x),([0.2,0.4],0.6),([0.2,0.4],0.5),([0.3,0.4],0.3)\},\\ \{(b,y),([0.2,0.3],0.4),([0.5,0.6],0.4),([0.9,1.0],0.3)\},\\ \{(b,z),([0.2,0.4],0.2),([0.3,0.4],0.4),([0.6,0.7],0.3)\},\\ \{(c,x),([0.3,0.4],0.6),([0.1,0.3],0.7),([0.4,0.6]),0.3)\},\\ \{(c,y),([0.2,0.3],0.4),([0.1,0.3],0.7),([0.9,1.0],0.3)\},\\ \{(c,z),([0.3,0.4],0.2),([0.1,0.3],0.7),([0.6,0.7],0.3)\}\end{array}\right\rangle$$

$$N_1\times N_2 = \left\langle\begin{array}{l}\{((a,x)(a,y)),([0.1,0.2],0.6),([0.2,0.4],0.5),([0.9,1.0],0.2)\},\\ \{((a,y)(a,z)),([0.1,0.2],0.5),([0.3,0.4],0.3),([0.9,1.0],0.2)\},\\ \{((a,z)(b,z)),([0.1,0.2],0.5),([0.3,0.4],0.4),([0.6,0.7],0.2)\},\\ \{((b,x)(b,z)),([0.2,0.4],0.6),([0.2,0.4],0.5),([0.6,0.7],0.3)\},\\ \{((b,x)(b,y)),([0.2,0.3],0.6),([0.2,0.4],0.5),([0.9,1.0],0.3)\},\\ \{((b,y)(c,y)),([0.2,0.3],0.4),([0.1,0.3],0.7),([0.9,1.0],0.3)\},\\ \{((c,y)(c,z)),([0.2,0.3],0.4),([0.1,0.3],0.7),([0.9,1.0],0.3)\},\\ \{((c,x)(c,z)),([0.3,0.4],0.6),([0.1,0.3],0.7),([0.6,0.7],0.3)\},\\ \{((a,x)(c,x)),([0.1,0.2],0.6),([0.1,0.3],0.7),([0.6,0.7],0.2)\}\end{array}\right\rangle$$

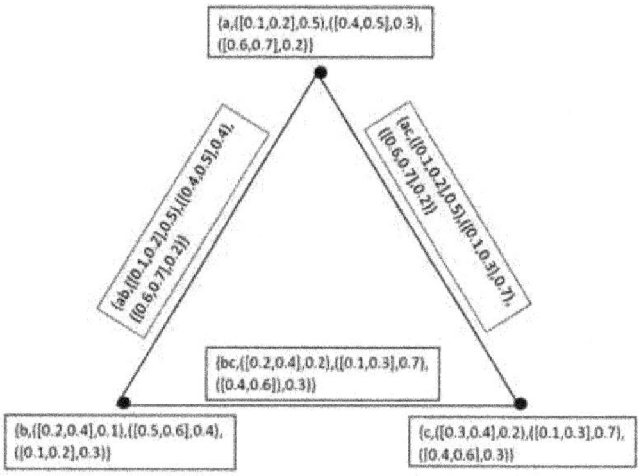

Figure 2. Neutrosophic Cubic Graph G_1.

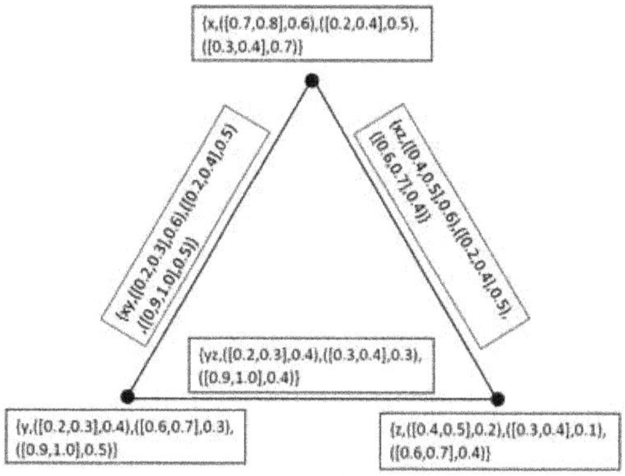

Figure 3. Neutrosophic Cubic Graph G_2.

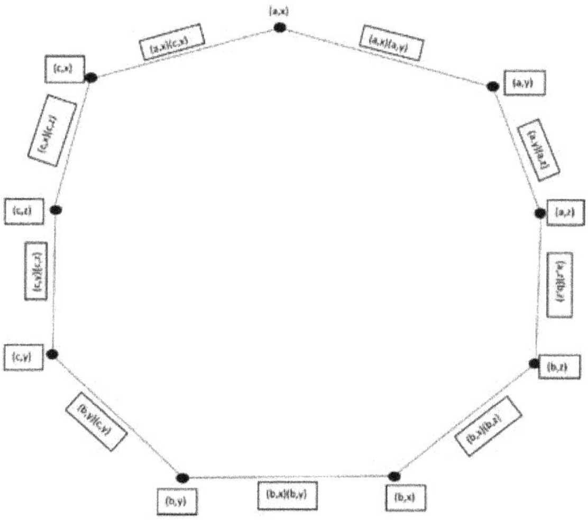

Figure 4. Cartesian Product of G_1 and G_2.

Proposition 1. *The cartesian product of two neutrosophic cubic graphs is again a neutrosophic cubic graph.*

Proof. Condition is obvious for $M_1 \times M_2$. Therefore we verify conditions only for $N_1 \times N_2$, where $N_1 \times N_2 = \{((\widetilde{T}_{C_1\times C_2}, T_{D_1\times D_2}), (\widetilde{I}_{C_1\times C_2}, I_{D_1\times D_2}), (\widetilde{F}_{C_1\times C_2}, F_{D_1\times D_2}))\}$. Let $x \in V_1$ and $x_2y_2 \in E_2$. Then

$$\begin{aligned}
\widetilde{T}_{C_1\times C_2}((x,x_2)(x,y_2)) &= rmin\{(\widetilde{T}_{A_1}(x), \widetilde{T}_{C_2}(x_2y_2))\}\\
&\preceq rmin\{(\widetilde{T}_{A_1}(x), rmin((\widetilde{T}_{A_2}(x_2), (\widetilde{T}_{A_2}(y_2))\}\\
&= rmin\{rmin((\widetilde{T}_{A_1}(x), (\widetilde{T}_{A_2}(x_2)), rmin((\widetilde{T}_{A_1}(x), (\widetilde{T}_{A_2}(y_2))\}\\
&= rmin\{(\widetilde{T}_{A_1} \times \widetilde{T}_{A_2})(x,x_2), ((\widetilde{T}_{A_1} \times \widetilde{T}_{A_2})(x,y_2)\}\\
T_{D_1\times D_2}((x,x_2)(x,y_2)) &= \max\{(T_{B_1}(x), T_{D_2}(x_2y_2))\}\\
&\leq \max\{(T_{B_1}(x), \max((T_{B_2}(x_2), (T_{B_2}(y_2))\}\\
&= \max\{\max((T_{B_1}(x), (T_{B_2}(x_2)), \max((T_{B_1}(x), (T_{B_2}(y_2))\}\\
&= \max\{(T_{B_1} \times T_{B_2})(x,x_2), ((T_{B_1} \times T_{B_2})(x,y_2)\}
\end{aligned}$$

$$\begin{aligned}
\tilde{I}_{C_1\times C_2}((x,x_2)(x,y_2)) &= rmin\{(\tilde{I}_{A_1}(x), \tilde{I}_{C_2}(x_2y_2))\}\\
&\preceq rmin\{(\tilde{I}_{A_1}(x), rmin((\tilde{I}_{A_2}(x_2), (\tilde{I}_{A_2}(y_2))\}\\
&= rmin\{rmin((\tilde{I}_{A_1}(x), (\tilde{I}_{A_2}(x_2)), rmin((\tilde{I}_{A_1}(x), (\tilde{I}_{A_2}(y_2))\}\\
&= rmin\{(\tilde{I}_{A_1} \times \tilde{I}_{A_2})(x,x_2), ((\tilde{I}_{A_1} \times \tilde{I}_{A_2})(x,y_2)\}\\
I_{D_1\times D_2}((x,x_2)(x,y_2)) &= \max\{(I_{B_1}(x), I_{D_2}(x_2y_2))\}\\
&\leq \max\{(I_{B_1}(x), \max((I_{B_2}(x_2), (I_{B_2}(y_2))\}\\
&= \max\{\max((I_{B_1}(x), (I_{B_2}(x_2)), \max((I_{B_1}(x), (I_{B_2}(y_2))\}\\
&= \max\{(I_{B_1} \times I_{B_2})(x,x_2), ((I_{B_1} \times I_{B_2})(x,y_2)\}
\end{aligned}$$

$$\begin{aligned}
\tilde{F}_{C_1\times C_2}((x,x_2)(x,y_2)) &= rmax\{(\tilde{F}_{A_1}(x), \tilde{F}_{C_2}(x_2y_2))\}\\
&\preceq rmax\{(\tilde{F}_{A_1}(x), rmax((\tilde{F}_{A_2}(x_2), (\tilde{F}_{A_2}(y_2))\}\\
&= rmax\{rmax((\tilde{F}_{A_1}(x), (\tilde{F}_{A_2}(x_2)), rmax((\tilde{F}_{A_1}(x), (\tilde{F}_{A_2}(y_2))\}\\
&= rmax\{(\tilde{F}_{A_1} \times \tilde{F}_{A_2})(x,x_2), ((\tilde{F}_{A_1} \times \tilde{F}_{A_2})(x,y_2)\}\\
F_{D_1\times D_2}((x,x_2)(x,y_2)) &= \min\{(F_{B_1}(x), F_{D_2}(x_2y_2))\}\\
&\leq \min\{(F_{B_1}(x), \min((F_{B_2}(x_2), (F_{B_2}(y_2))\}\\
&= \min\{\min((F_{B_1}(x), (F_{B_2}(x_2)), \min((F_{B_1}(x), (F_{B_2}(y_2))\}\\
&= \min\{(F_{B_1} \times F_{B_2})(x,x_2), (F_{B_1} \times F_{B_2})(x,y_2)\}
\end{aligned}$$

similarly we can prove it for $z \in V_2$ and $x_1y_1 \in E_1$. □

Definition 12. *Let $G_1 = (M_1, N_1)$ and $G_2 = (M_2, N_2)$ be two neutrosophic cubic graphs. The degree of a vertex in $G_1 \times G_2$ can be defined as follows, for any $(x_1, x_2) \in V_1 \times V_2$*

$$\begin{aligned}
\deg(\tilde{T}_{A_1} \times \tilde{T}_{A_2})(x_1,x_2) &= \Sigma_{(x_1,x_2)(y_1,y_2)\in E_2} rmax(\tilde{T}_{C_1} \times \tilde{T}_{C_2})((x_1,x_2)(y_1,y_2))\\
&= \Sigma_{x_1=y_1=x, x_2y_2\in E_2} rmax(\tilde{T}_{A_1}(x), \tilde{T}_{C_2}(x_2y_2))\\
&\quad +\Sigma_{x_2=y_2=z, x_1y_1\in E} rmax(\tilde{T}_{A_2}(z), \tilde{T}_{C_1}(x_1y_1))\\
&\quad +\Sigma_{x_1y_1\in E_1, x_2y_2\in E_2} rmax(\tilde{T}_{C_1}(x_1y_1), \tilde{T}_{C_2}(x_2y_2))
\end{aligned}$$

$$\begin{aligned}
\deg(T_{B_1} \times T_{B_2})(x_1, x_2) &= \Sigma_{(x_1,x_2)(y_1,y_2)\in E_2} \min(T_{D_1} \times T_{D_2})((x_1, x_2)(y_1, y_2)) \\
&= \Sigma_{x_1=y_1=x, x_2y_2\in E_2} \min(T_{B_1}(x), T_{D_2}(x_2y_2)) \\
&\quad + \Sigma_{x_2=y_2=z, x_1y_1\in E} \min(T_{B_2}(z), T_{D_1}(x_1y_1)) \\
&\quad + \Sigma_{x_1y_1\in E_1, x_2y_2\in E_2} \min(T_{D_1}(x_1y_1), T_{D_2}(x_2y_2))
\end{aligned}$$

$$\begin{aligned}
\deg(\tilde{I}_{A_1} \times \tilde{I}_{A_2})(x_1, x_2) &= \Sigma_{(x_1,x_2)(y_1,y_2)\in E_2} rmax(\tilde{I}_{C_1} \times \tilde{I}_{C_2})((x_1, x_2)(y_1, y_2)) \\
&= \Sigma_{x_1=y_1=x, x_2y_2\in E_2} rmax(\tilde{I}_{A_1}(x), \tilde{I}_{C_2}(x_2y_2)) \\
&\quad + \Sigma_{x_2=y_2=z, x_1y_1\in E} rmax(\tilde{I}_{A_2}(z), \tilde{I}_{C_1}(x_1y_1)) \\
&\quad + \Sigma_{x_1y_1\in E_1, x_2y_2\in E_2} rmax(\tilde{I}_{C_1}(x_1y_1), \tilde{I}_{C_2}(x_2y_2)) \\
\deg(I_{B_1} \times I_{B_2})(x_1, x_2) &= \Sigma_{(x_1,x_2)(y_1,y_2)\in E_2} \min(I_{D_1} \times I_{D_2})((x_1, x_2)(y_1, y_2)) \\
&= \Sigma_{x_1=y_1=x, x_2y_2\in E_2} \min(I_{B_1}(x), I_{D_2}(x_2y_2)) \\
&\quad + \Sigma_{x_2=y_2=z, x_1y_1\in E} \min(I_{B_2}(z), I_{D_1}(x_1y_1)) \\
&\quad + \Sigma_{x_1y_1\in E_1, x_2y_2\in E_2} \min(I_{D_1}(x_1y_1), I_{D_2}(x_2y_2))
\end{aligned}$$

$$\begin{aligned}
\deg(\tilde{F}_{A_1} \times \tilde{F}_{A_2})(x_1, x_2) &= \Sigma_{(x_1,x_2)(y_1,y_2)\in E_2} rmin(\tilde{F}_{C_1} \times \tilde{F}_{C_2})((x_1, x_2)(y_1, y_2)) \\
&= \Sigma_{x_1=y_1=x, x_2y_2\in E_2} rmin(F_{B_1}(x), F_{D_2}(x_2y_2)) \\
&\quad + \Sigma_{x_2=y_2=z, x_1y_1\in E} rmin(F_{B_2}(z), F_{D_1}(x_1y_1)) \\
&\quad + \Sigma_{x_1y_1\in E_1, x_2y_2\in E_2} rmin(F_{D_1}(x_1y_1), F_{D_2}(x_2y_2)) \\
\deg(F_{B_1} \times F_{B_2})(x_1, x_2) &= \Sigma_{(x_1,x_2)(y_1,y_2)\in E_2} \max(F_{D_1} \times F_{D_2})((x_1, x_2)(y_1, y_2)) \\
&= \Sigma_{x_1=y_1=x, x_2y_2\in E_2} \max(F_{B_1}(x), F_{D_2}(x_2y_2)) \\
&\quad + \Sigma_{x_2=y_2=z, x_1y_1\in E} \max(F_{B_2}(z), F_{D_1}(x_1y_1)) \\
&\quad + \Sigma_{x_1y_1\in E_1, x_2y_2\in E_2} \max(F_{D_1}(x_1y_1), F_{D_2}(x_2y_2))
\end{aligned}$$

Example 4. *In Example 3*

$$\begin{aligned}
d_{G_1\times G_2}(a, x) &= \{([0.9, 1.1], 1.0), ([0.6, 0.9], 0.8), ([0.9, 1.1], 1.2)\} \\
d_{G_1\times G_2}(a, y) &= \{([0.4, 0.6], 0.9), ([0.8, 1.0], 0.6), ([1.2, 1.4], 0.9)\} \\
d_{G_1\times G_2}(a, z) &= \{([0.6, 0.8], 0.6), ([0.8, 1.0], 0.4), ([1.2, 1.4], 0.8)\} \\
d_{G_1\times G_2}(b, z) &= \{([0.8, 1.0], 0.3), ([0.9, 1.1], 0.5), ([0.7, 0.9], 1.1)\} \\
d_{G_1\times G_2}(b, x) &= \{([0.6, 0.9], 0.6), ([1.0, 1.2], 0.7), ([0.2, 0.4], 1.2)\} \\
d_{G_1\times G_2}(b, y) &= \{([0.4, 0.8], 0.7), ([1.1, 1.3], 0.6), ([0.5, 0.8], 1.0)\} \\
d_{G_1\times G_2}(c, y) &= \{([0.5, 0.8], 0.4), ([0.9, 1.1], 0.6), ([0.8, 1.2], 0.9)\} \\
d_{G_1\times G_2}(c, z) &= \{([0.7, 0.9], 0.4), ([0.5, 0.8], 0.8), ([0.8, 1.2], 1.1)\} \\
d_{G_1\times G_2}(c, x) &= \{([1.1, 1.3], 0.7), ([0.4, 0.8], 1.0), ([0.7, 1.0], 1.4)\}
\end{aligned}$$

Definition 13. *Let $G_1 = (M_1, N_1)$ be a neutrosophic cubic graph of $G_1^* = (V_1, E_1)$ and $G_2 = (M_2, N_2)$ be a neutrosophic cubic graph of $G_2^* = (V_2, E_2)$. Then composition of G_1 and G_2 is denoted by $G_1[G_2]$ and defined as follow*

$$\begin{aligned}
G_1[G_2] &= (M_1, N_1)[(M_2, N_2)] = \{M_1[M_2], N_1[N_2]\} = \{(A_1, B_1)[(A_2, B_2)], (C_1, D_1)[(C_2, D_2)]\} \\
&= \{(A_1[A_2], B_1[B_2]), (C_1[C_2], D_1[D_2])\} \\
&= \left\{ \begin{array}{l} \langle((\tilde{T}_{A_1} \circ \tilde{T}_{A_2}), (T_{B_1} \circ T_{B_2})), ((\tilde{I}_{A_1} \circ \tilde{I}_{A_2}), (I_{B_1} \circ I_{B_2})), ((\tilde{F}_{A_1} \circ \tilde{F}_{A_2}), (F_{B_1} \circ F_{B_2}))\rangle, \\ \langle((\tilde{T}_{C_1} \circ \tilde{T}_{C_2}), (T_{D_1} \circ T_{D_2})), ((\tilde{I}_{C_1} \circ \tilde{I}_{C_2}), (I_{D_1} \circ I_{D_2})), (\tilde{F}_{C_1} \circ \tilde{F}_{C_2})), (F_{D_1} \circ F_{D_2}))\rangle \end{array} \right\}
\end{aligned}$$

where

(i) $\forall(x,y) \in (V_1, V_2) = V$,

$$(\tilde{T}_{A_1} \circ \tilde{T}_{A_2})(x,y) = rmin(\tilde{T}_{A_1}(x), \tilde{T}_{A_2}(y)), (T_{B_1} \circ T_{B_2})(x,y) = \max(T_{B_1}(x), T_{B_2}(y))$$

$$(\tilde{I}_{A_1} \circ \tilde{I}_{A_2})(x,y) = rmin(\tilde{I}_{A_1}(x), \tilde{I}_{A_2}(y)), (I_{B_1} \circ I_{B_2})(x,y) = \max(I_{B_1}(x), I_{B_2}(y))$$

$$(\tilde{F}_{A_1} \circ \tilde{F}_{A_2})(x,y) = rmax(\tilde{F}_{A_1}(x), \tilde{F}_{A_2}(y)), (F_{B_1} \circ F_{B_2})(x,y) = \min(F_{B_1}(x), F_{B_{F_2}}(y))$$

(ii) $\forall x \in V_1$ *and* $y_1y_2 \in E$

$$(\tilde{T}_{C_1} \circ \tilde{T}_{C_2})((x,y_1)(x,y_2)) = rmin(\tilde{T}_{A_1}(x), \tilde{T}_{C_2}(y_1y_2)), (T_{D_1} \circ T_{D_2})((x,y_1)(x,y_2)) = \max(T_{B_1}(x), T_{D_2}(y_1y_2))$$

$$(\tilde{I}_{C_1} \circ \tilde{I}_{C_2})((x,y_1)(x,y_2)) = rmin(\tilde{I}_{A_1}(x), \tilde{I}_{C_2}(y_1y_2)), (I_{D_1} \circ I_{D_2})((x,y_1)(x,y_2)) = \max(I_{B_1}(x), I_{D_2}(y_1y_2))$$

$$(\tilde{F}_{C_1} \circ \tilde{F}_{C_2})((x,y_1)(x,y_2)) = rmax(\tilde{F}_{A_1}(x), \tilde{F}_{C_2}(y_1y_2)), (F_{D_1} \circ F_{D_2})((x,y_1)(x,y_2)) = \min(F_{B_1}(x), F_{D_2}(y_1y_2))$$

(iii) $\forall y \in V_2$ *and* $x_1x_2 \in E_1$

$$(\tilde{T}_{C_1} \circ \tilde{T}_{C_2})((x_1,y)(x_2,y)) = rmin(\tilde{T}_{C_1}(x_1x_2), \tilde{T}_{A_2}(y)), (T_{D_1} \circ T_{D_2})((x_1,y)(x_2,y)) = \max(T_{D_1}(x_1x_2), T_{B_2}(y))$$

$$(\tilde{I}_{C_1} \circ \tilde{I}_{C_2})((x_1,y)(x_2,y)) = rmin(\tilde{I}_{C_1}(x_1x_2), \tilde{I}_{A_2}(y)), (I_{D_1} \circ I_{D_2})((x_1,y)(x_2,y)) = \max(I_{D_1}(x_1x_2), I_{B_2}(y))$$

$$(\tilde{F}_{C_1} \circ \tilde{F}_{C_2})((x_1,y)(x_2,y)) = rmax(\tilde{F}_{C_1}(x_1x_2), \tilde{F}_{A_2}(y)), (F_{D_1} \circ F_{D_2})((x_1,y)(x_2,y)) = \min(F_{D_1}(x_1x_2), F_{B_2}(y))$$

(iv) $\forall(x_1,y_1)(x_2,y_2) \in E^0 - E$

$$\begin{aligned}(\tilde{T}_{C_1} \circ \tilde{T}_{C_2})((x_1,y_1)(x_2,y_2)) &= rmin(\tilde{T}_{A_2}(y_1), \tilde{T}_{A_2}(y_2), \tilde{T}_{C_1}(x_1x_2)), (T_{D_1} \circ T_{D_2})((x_1,y_1)(x_2,y_2)) \\ &= \max(T_{B_2}(y_1), T_{B_2}(y_2), T_{D_1}(x_1x_2))\end{aligned}$$

$$\begin{aligned}(\tilde{I}_{C_1} \circ \tilde{I}_{C_2})((x_1,y_1)(x_2,y_2)) &= rmin(\tilde{I}_{A_2}(y_1), \tilde{I}_{A_2}(y_2), \tilde{I}_{C_1}(x_1x_2)), (I_{D_1} \circ I_{D_2})((x_1,y_1)(x_2,y_2)) \\ &= \max(I_{B_2}(y_1), I_{B_2}(y_2), I_{D_1}(x_1x_2))\end{aligned}$$

$$\begin{aligned}(\tilde{F}_{C_1} \circ \tilde{F}_{C_2})((x_1,y_1)(x_2,y_2)) &= rmax(\tilde{F}_{A_2}(y_1), \tilde{F}_{A_2}(y_2), \tilde{F}_{C_1}(x_1x_2)), (F_{D_1} \circ F_{D_2})((x_1,y_1)(x_2,y_2)) \\ &= \min(F_{B_2}(y_1), F_{B_2}(y_2), F_{D_1}(x_1x_2))\end{aligned}$$

Example 5. *Let* $G_1^* = (V_1, E_1)$ *and* $G_1^* = (V_2, E_2)$ *be two graphs as showin in Figure 5, where* $V_1 = (a,b)$ *and* $V_2 = (c,d)$. *Suppose* M_1 *and* M_2 *be the neutrosophic cubic set representations of* V_1 *and* V_2. *Also* N_1 *and* N_2 *be the neutrosophic cubic set representations of* E_1 *and* E_2 *defined as*

$$\begin{aligned}M_1 &= \left\langle \begin{array}{l}\{a, ([0.5, 0.6], 0.1), ([0.1, 0.2], 0.5), ([0.8, 0.9], 0.3)\}, \\ \{b, ([0.4, 0.5], 0.3), ([0.2, 0.3], 0.2), ([0.5, 0.6], 0.6)\}\end{array} \right\rangle \\ N_1 &= \langle \ \{ab, ([0.4, 0.5], 0.3), ([0.1, 0.2], 0.5), ([0.8, 0.9], 0.3)\} \ \rangle\end{aligned}$$

and

$$\begin{aligned}M_2 &= \left\langle \begin{array}{l}\{c, ([0.6, 0.7], 0.4), ([0.8, 0.9], 0.8), ([0.1, 0.2], 0.6)\}, \\ \{d, ([0.3, 0.4], 0.7), ([0.6, 0.7], 0.5), ([0.9, 1.0], 0.9)\}\end{array} \right\rangle \\ N_2 &= \langle \ \{cd, ([0.3, 0.4], 0.7), ([0.6, 0.7], 0.8), ([0.9, 1.0], 0.6)\} \ \rangle\end{aligned}$$

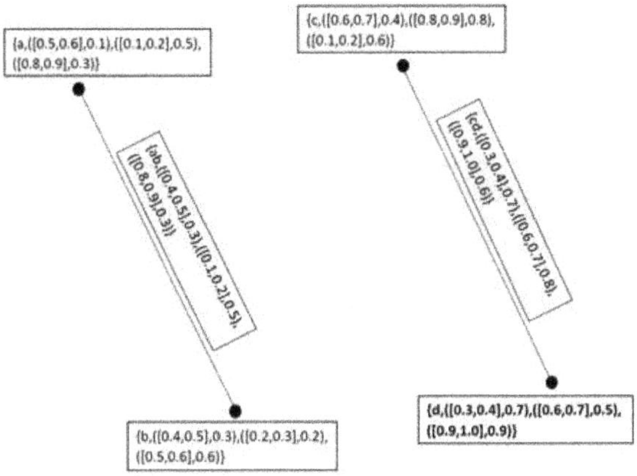

Figure 5. Neutrosophic Cubic Graph G_1 and G_2.

Clearly $G_1 = (M_1, N_1)$ and $G_2 = (M_2, N_2)$ are neutrosophic cubic graphs. So, the composition of two neutrosophic cubic graphs $G-1$ and $G-2$ is again a neutrosophic cubic graph as shown in Figure 6, where

$$M_1[M_2] = \left\langle \begin{array}{l} \{(a,c),([0.5,0.6],0.4),([0.1,0.2],0.8),([0.8,0.9],0.3)\}, \\ \{(a,d),([0.3,0.4],0.7),([0.1,0.2],0.5),([0.9,1.0],0.3)\}, \\ \{(b,c),([0.4,0.5],0.4),([0.2,0.3],0.8),[0.5,0.6],0.6)\}, \\ \{(b,d),([0.3,0.4],0.7),([0.2,0.3],0.5),([0.9,1.0],0.6)\} \end{array} \right\rangle$$

$$N_1[N_2] = \left\langle \begin{array}{l} \{((a,c)(a,d)),([0.3,0.4],0.7),([0.1,0.2],0.8),([0.9,1.0],0.3)\}, \\ \{((a,d)(b,d)),([0.3,0.4],0.7),([0.1,0.2],0.5),[0.9,1.0],0.3)\}, \\ \{((b,d)(b,c)),([0.3,0.4],0.7),([0.2,0.3],0.8),([0.9,1.0],0.6)\}, \\ \{((b,c)(a,c)),([0.4,0.5],0.4),([0.1,0.2],0.8),([0.8,0.9],0.3)\}, \\ \{((a,c)(b,d)),([0.3,0.4],0.7),([0.1,0.2],0.8),([0.9,1.0],0.3)\}, \\ \{((a,d)(b,c)),([0.3,0.4],0.7),([0.1,0.2],0.8),([0.9,1.0],0.3)\} \end{array} \right\rangle$$

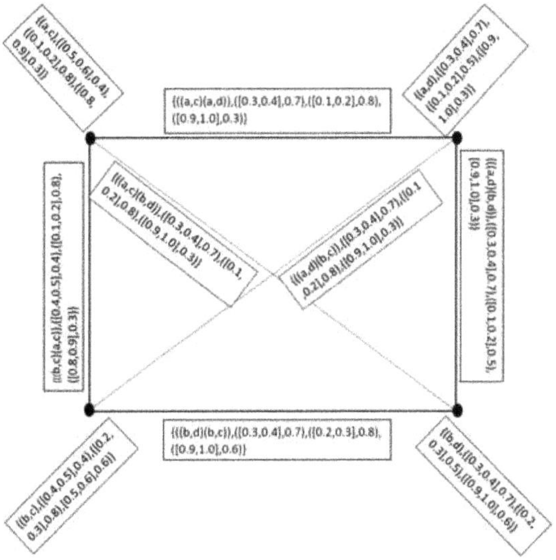

Figure 6. Composition of G_1 and G_2.

Proposition 2. *The composition of two neutrosophic cubic graphs is again a neutrosophic cubic graph.*

Definition 14. *Let $G_1 = (M_1, N_1)$ and $G_2 = (M_2, N_2)$ be two neutrosophic cubic graphs of the graphs G_1^* and G_2^* respectively. Then P-union is denoted by $G_1 \cup_P G_2$ and is defined as*

$$\begin{aligned} G_1 \cup_P G_2 &= \{(M_1, N_1) \cup_P (M_2, N_2)\} = \{M_1 \cup_P M_2, N_1 \cup_P N_2\} \\ &= \{\langle((\tilde{T}_{A_1} \cup_p \tilde{T}_{A_2}), (T_{B_1} \cup_p T_{B_2})), ((\tilde{I}_{A_1} \cup_p \tilde{I}_{A_2}), (I_{B_1} \cup_p I_{B_2})), ((\tilde{F}_{A_1} \cup_p \tilde{F}_{A_2}), (F_{B_1} \cup_p F_{B_2}))\rangle, \\ &\quad \langle((\tilde{T}_{C_1} \cup_p \tilde{T}_{C_2}), (T_{D_1} \cup_p T_{D_2})), ((\tilde{I}_{C_1} \cup_p \tilde{I}_{C_2}), (I_{D_1} \cup_p I_{D_2})), ((\tilde{F}_{C_1} \cup_p \tilde{F}_{C_2}), (F_{D_1} \cup_p F_{D_2}))\rangle\} \end{aligned}$$

where

$$(\tilde{T}_{A_1} \cup_p \tilde{T}_{A_2})(x) = \begin{cases} \tilde{T}_{A_1}(x) & \text{if } x \in V_1 - V_2 \\ \tilde{T}_{A_2}(x) & \text{if } x \in V_2 - V_1 \\ rmax\{\tilde{T}_{A_1}(x), \tilde{T}_{A_2}(x)\} & \text{if } x \in V_1 \cap V_2 \end{cases}$$

$$(T_{B_1} \cup_p T_{B_2})(x) = \begin{cases} T_{B_1}(x) & \text{if } x \in V_1 - V_2 \\ T_{B_2}(x) & \text{if } x \in V_2 - V_1 \\ \max\{T_{B_1}(x), T_{B_2}(x)\} & \text{if } x \in V_1 \cap V_2 \end{cases}$$

$$(\tilde{I}_{A_1} \cup_p \tilde{I}_{A_2})(x) = \begin{cases} \tilde{I}_{A_1}(x) & \text{if } x \in V_1 - V_2 \\ \tilde{I}_{A_2}(x) & \text{if } x \in V_2 - V_1 \\ rmax\{\tilde{I}_{A_1}(x), \tilde{I}_{A_2}(x)\} & \text{if } x \in V_1 \cap V_2 \end{cases}$$

$$(I_{B_1} \cup_p I_{B_2})(x) = \begin{cases} I_{B_1}(x) & \text{if } x \in V_1 - V_2 \\ I_{B_2}(x) & \text{if } x \in V_2 - V_1 \\ \max\{I_{B_1}(x), I_{B_2}(x)\} & \text{if } x \in V_1 \cap V_2 \end{cases}$$

$$(\tilde{F}_{A_1} \cup_p \tilde{F}_{A_2})(x) = \begin{cases} \tilde{F}_{A_1}(x) & \text{if } x \in V_1 - V_2 \\ \tilde{F}_{A_2}(x) & \text{if } x \in V_2 - V_1 \\ rmax\{\tilde{F}_{A_1}(x), \tilde{F}_{A_2}(x)\} & \text{if } x \in V_1 \cap V_2 \end{cases}$$

$$(F_{B_1} \cup_p F_{B_2})(x) = \begin{cases} F_{B_1}(x) & \text{if } x \in V_1 - V_2 \\ F_{B_2}(x) & \text{if } x \in V_2 - V_1 \\ \max\{F_{B_1}(x), F_{B_2}(x)\} & \text{if } x \in V_1 \cap V_2 \end{cases}$$

$$(\tilde{T}_{C_1} \cup_p \tilde{T}_{C_2})(x_2y_2) = \begin{cases} \tilde{T}_{C_1}(x_2y_2) & \text{if } x_2y_2 \in V_1 - V_2 \\ \tilde{T}_{C_2}(x_2y_2) & \text{if } x_2y_2 \in V_2 - V_1 \\ rmax\{\tilde{T}_{C_1}(x_2y_2), \tilde{T}_{C_2}(x_2y_2)\} & \text{if } x_2y_2 \in E_1 \cap E_2 \end{cases}$$

$$(T_{D_1} \cup_p T_{D_2})(x_2y_2) = \begin{cases} T_{D_1}(x_2y_2) & \text{if } x_2y_2 \in V_1 - V_2 \\ T_{D_2}(x_2y_2) & \text{if } x_2y_2 \in V_2 - V_1 \\ \max\{T_{D_1}(x_2y_2), T_{D_2}(x_2y_2)\} & \text{if } x_2y_2 \in E_1 \cap E_2 \end{cases}$$

$$(\tilde{I}_{C_1} \cup_p \tilde{I}_{C_2})(x_2y_2) = \begin{cases} \tilde{I}_{C_1}(x_2y_2) & \text{if } x_2y_2 \in V_1 - V_2 \\ \tilde{I}_{C_2}(x_2y_2) & \text{if } x_2y_2 \in V_2 - V_1 \\ rmax\{\tilde{I}_{C_1}(x_2y_2), \tilde{I}_{C_2}(x_2y_2)\} & \text{if } x_2y_2 \in E_1 \cap E_2 \end{cases}$$

$$(I_{D_1} \cup_p I_{D_2})(x_2y_2) = \begin{cases} I_{D_1}(x_2y_2) & \text{if } x_2y_2 \in V_1 - V_2 \\ I_{D_2}(x_2y_2) & \text{if } x_2y_2 \in V_2 - V_1 \\ \max\{I_{D_1}(x_2y_2), I_{D_2}(x_2y_2)\} & \text{if } x_2y_2 \in E_1 \cap E_2 \end{cases}$$

$$(\tilde{F}_{C_1} \cup_p \tilde{F}_{C_2})(x_2y_2) = \begin{cases} \tilde{F}_{C_1}(x_2y_2) & \text{if } x_2y_2 \in V_1 - V_2 \\ \tilde{F}_{C_2}(x_2y_2) & \text{if } x_2y_2 \in V_2 - V_1 \\ rmax\{\tilde{F}_{C_1}(x_2y_2), \tilde{F}_{C_2}(x_2y_2)\} & \text{if } x_2y_2 \in E_1 \cap E_2 \end{cases}$$

$$(F_{D_1} \cup_p F_{D_2})(x_2y_2) = \begin{cases} F_{D_1}(x_2y_2) & \text{if } x_2y_2 \in V_1 - V_2 \\ F_{D_2}(x_2y_2) & \text{if } x_2y_2 \in V_2 - V_1 \\ \max\{F_{D_1}(x_2y_2), F_{D_2}(x_2y_2)\} & \text{if } x_2y_2 \in E_1 \cap E_2 \end{cases}$$

and R-union is denoted by $G_1 \cup_R G_2$ and is defined by

$$\begin{aligned} G_1 \cup_R G_2 &= \{(M_1, N_1) \cup_R (M_2, N_2)\} = \{M_1 \cup_R M_2, N_1 \cup_R N_2\} \\ &= \{\langle((\tilde{T}_{A_1} \cup_R \tilde{T}_{A_2}), (T_{B_1} \cup_R T_{B_2})), ((\tilde{I}_{A_1} \cup_R \tilde{I}_{A_2}), (I_{B_1} \cup_R I_{B_2})), ((\tilde{F}_{A_1} \cup_R \tilde{F}_{A_2}), (F_{B_1} \cup_R F_{B_2}))\rangle, \\ &\quad \langle((\tilde{T}_{C_1} \cup_R \tilde{T}_{C_2}), (T_{D_1} \cup_R T_{D_2})), ((\tilde{I}_{C_1} \cup_R \tilde{I}_{C_2}), (I_{D_1} \cup_R I_{D_2})), ((\tilde{F}_{C_1} \cup_R \tilde{F}_{C_2}), (F_{D_1} \cup_R F_{D_2}))\rangle\} \end{aligned}$$

where

$$(\tilde{T}_{A_1} \cup_R \tilde{T}_{A_2})(x) = \begin{cases} \tilde{T}_{A_1}(x) & \text{if } x \in V_1 - V_2 \\ \tilde{T}_{A_2}(x) & \text{if } x \in V_2 - V_1 \\ rmax\{\tilde{T}_{A_1}(x), \tilde{T}_{A_2}(x)\} & \text{if } x \in V_1 \cap V_2 \end{cases}$$

$$(T_{B_1} \cup_R T_{B_2})(x) = \begin{cases} T_{B_1}(x) & \text{if } x \in V_1 - V_2 \\ T_{B_2}(x) & \text{if } x \in V_2 - V_1 \\ \min\{T_{B_1}(x), T_{B_2}(x)\} & \text{if } x \in V_1 \cap V_2 \end{cases}$$

$$(\tilde{I}_{A_1} \cup_R \tilde{I}_{A_2})(x) = \begin{cases} \tilde{I}_{A_1}(x) & \text{if } x \in V_1 - V_2 \\ \tilde{I}_{A_2}(x) & \text{if } x \in V_2 - V_1 \\ rmax\{\tilde{I}_{A_1}(x), \tilde{I}_{A_2}(x)\} & \text{if } x \in V_1 \cap V_2 \end{cases}$$

$$(I_{B_1} \cup_R I_{B_2})(x) = \begin{cases} I_{B_1}(x) & \text{if } x \in V_1 - V_2 \\ I_{B_2}(x) & \text{if } x \in V_2 - V_1 \\ \min\{I_{B_1}(x), I_{B_2}(x)\} & \text{if } x \in V_1 \cap V_2 \end{cases}$$

$$(\tilde{F}_{A_1} \cup_R M_{T_{F_2}})(x) = \begin{cases} \tilde{F}_{A_1}(x) & \text{if } x \in V_1 - V_2 \\ \tilde{F}_{A_2}(x) & \text{if } x \in V_2 - V_1 \\ rmax\{\tilde{F}_{A_1}(x), \tilde{F}_{A_2}(x)\} & \text{if } x \in V_1 \cap V_2 \end{cases}$$

$$(F_{B_1} \cup_R F_{B_2})(x) = \begin{cases} F_{B_1}(x) & \text{if } x \in V_1 - V_2 \\ F_{B_2}(x) & \text{if } x \in V_2 - V_1 \\ \min\{F_{B_1}(x), F_{B_2}(x)\} & \text{if } x \in V_1 \cap V_2 \end{cases}$$

$$(\tilde{T}_{C_1} \cup_R \tilde{T}_{C_2})(x_2y_2) = \begin{cases} \tilde{T}_{C_1}(x_2y_2) & \text{if } x_2y_2 \in V_1 - V_2 \\ \tilde{T}_{C_2}(x_2y_2) & \text{if } x_2y_2 \in V_2 - V_1 \\ rmax\{\tilde{T}_{C_1}(x_2y_2), \tilde{T}_{C_2}(x_2y_2)\} & \text{if } x_2y_2 \in E_1 \cap E_2 \end{cases}$$

$$(T_{D_1} \cup_R N_{D_2})(x_2y_2) = \begin{cases} T_{D_1}(x_2y_2) & \text{if } x_2y_2 \in V_1 - V_2 \\ T_{D_2}(x_2y_2) & \text{if } x_2y_2 \in V_2 - V_1 \\ \min\{T_{D_1}(x_2y_2), T_{D_2}(x_2y_2)\} & \text{if } x_2y_2 \in E_1 \cap E_2 \end{cases}$$

$$(\tilde{F}_{C_1} \cup_R \tilde{F}_{C_2})(x_2y_2) = \begin{cases} \tilde{F}_{C_1}(x_2y_2) & \text{if } x_2y_2 \in V_1 - V_2 \\ \tilde{F}_{C_2}(x_2y_2) & \text{if } x_2y_2 \in V_2 - V_1 \\ rmax\{\tilde{F}_{C_1}(x_2y_2), \tilde{F}_{C_2}(x_2y_2)\} & \text{if } x_2y_2 \in E_1 \cap E_2 \end{cases}$$

$$(F_{D_1} \cup_R F_{D_2})(x_2y_2) = \begin{cases} F_{D_1}(x_2y_2) & \text{if } x_2y_2 \in V_1 - V_2 \\ F_{D_2}(x_2y_2) & \text{if } x_2y_2 \in V_2 - V_1 \\ \min\{F_{D_1}(x_2y_2), F_{D_2}(x_2y_2)\} & \text{if } x_2y_2 \in E_1 \cap E_2 \end{cases}$$

Example 6. *Let G_1 and G_2 be two neutrosophic cubic graphs as represented by Figures 7 and 8, where*

$$M_1 = \left\langle \begin{array}{l} \{a,([0.2,0.3],0.5),([0.4,0.5],0.9),([0.1,0.3],0.2)\}, \\ \{b,([0.3,0.4],0.2),[0.1,0.2],0.1),[0.4,0.6],0.5)\}, \\ \{c,([0.2,0.4],0.6),([0.7,0.8],0.8),([0.3,0.5],0.7)\} \end{array} \right\rangle$$

$$N_1 = \left\langle \begin{array}{l} \{ab,([0.2,0.3],0.5),([0.1,0.2],0.9),([0.4,0.6],0.2)\}, \\ \{bc,([0.2,0.4],0.6),([0.1,0.2],0.8),([0.4,0.6],0.5)\}, \\ \{ac,([0.2,0.3],0.6),([0.4,0.5],0.9),([0.3,0.5],0.2)\} \end{array} \right\rangle$$

and

$$M_2 = \left\langle \begin{array}{l} \{a,([0.5,0.6],0.3),([0.1,0.2],0.6),([0.3,0.4],0.5)\}, \\ \{b,([0.6,0.7],0.6),([0.7,0.8],0.4),([0.1,0.2],0.5)\}, \\ \{c,([0.4,0.5],0.1),([0.2,0.5],0.5),([0.5,0.6],0.3)\} \end{array} \right\rangle$$

$$N_2 = \left\langle \begin{array}{l} \{ab,([0.5,0.6],0.6),([0.1,0.2],0.6),([0.3,0.4],0.5)\}, \\ \{bc,([0.4,0.5],0.6),([0.2,0.5],0.5),([0.5,0.6],0.3)\}, \\ \{ac,([0.4,0.5],0.3),([0.1,0.2],0.6),([0.5,0.6],0.3)\} \end{array} \right\rangle$$

then $G_1 \cup_p G_2$ will be a neutrosophic cubic graph as shown in Figure 9, where

$$M_1 \cup_p M_2 = \left\langle \begin{array}{l} \{a,([0.5,0.6],0.5),([0.4,0.5],0.9),([0.3,0.4],0.5)\}, \\ \{b,([0.6,0.7],0.6),([0.7,0.8],0.4),([0.4,0.6],0.5)\}, \\ \{c,([0.4,0.5],0.6),([0.7,0.8],0.8),([0.5,0.6],0.7)\} \end{array} \right\rangle$$

$$N_1 \cup_P N_2 = \left\langle \begin{array}{l} \{ab,([0.5,0.6],0.6),([0.1,0.2],0.9),([0.4,0.6],0.5)\}, \\ \{bc,([0.4,0.5],0.6),([0.2,0.5],0.8),[0.5,0.6],0.5)\}, \\ \{ac,([0.4,0.5],0.6),([0.4,0.5],0.9),([0.5,0.6],0.3)\} \end{array} \right\rangle$$

and $G_1 \cup_R G_2$ will be a neutrosophic cubic graph as shown in Figure 10, where

$$M_1 \cup_R M_2 = \begin{array}{l} \{a,([0.5,0.6],0.3),([0.4,0.5],0.6),([0.3,0.4],0.2)\}, \\ \{b,([0.6,0.7],0.2),([0.7,0.8],0.1),[0.4,0.6],0.5)\}, \\ \{c,([0.4,0.5],0.1),([0.7,0.8],0.5),([0.5,0.6],0.3)\} \end{array}$$

$$N_1 \cup_R N_2 = \begin{array}{l} \{ab,([0.5,0.6],0.5),([0.1,0.2],0.6),([0.4,0.6],0.2)\}, \\ \{bc,([0.4,0.5],0.6),([0.2,0.5],0.5),([0.5,0.6],0.3)\}, \\ \{ac,([0.4,0.5],0.3),([0.4,0.5],0.6),([0.5,0.6],0.2)\} \end{array}$$

Proposition 3. *The P-union of two neutrosophic cubic graphs is again a neutrosophic cubic graph.*

Remark 2. *The R-union of two neutrosophic cubic graphs may or may not be a neutrosophic cubic graph as in the Example 6 we see that*

$$T_{D_1 \cup_R D_2}(ab) = 0.5 \nleq \max\{0.3, 0.2\} = 0.3 = \max\{T_{D_1 \cup_R D_2}(a), T_{D_1 \cup_R D_2}(b)\}$$

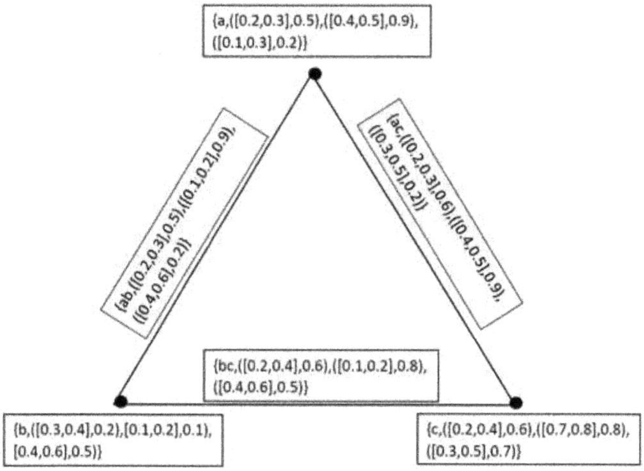

Figure 7. Neutrosophic Cubic Graph G_1.

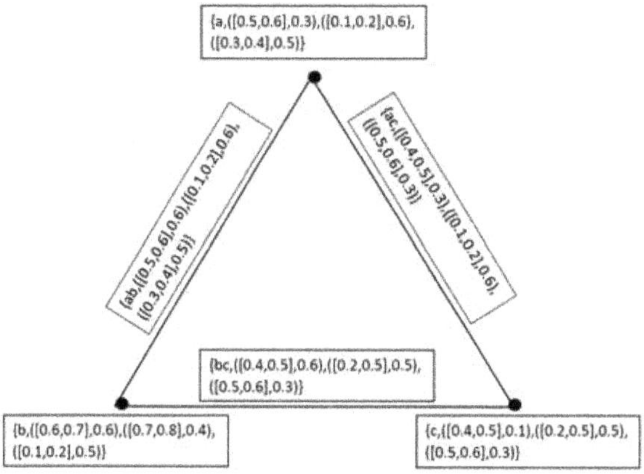

Figure 8. Neutrosophic Cubic Graph G_2.

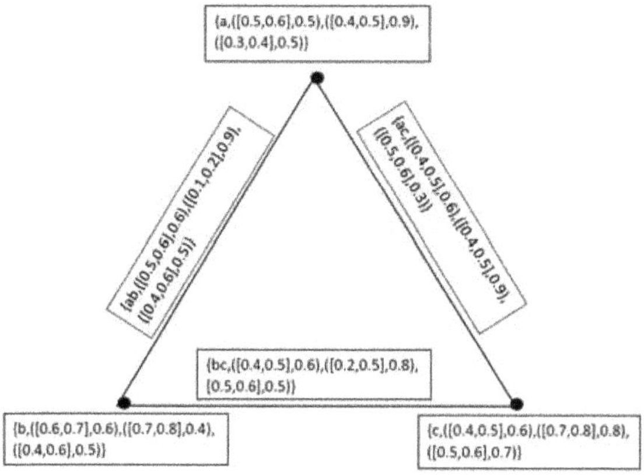

Figure 9. P-Union of G_1 and G_2.

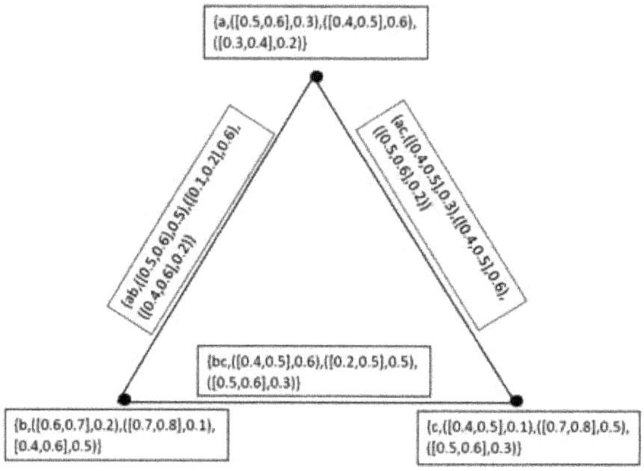

Figure 10. R-Union of G_1 and G_2.

Definition 15. *Let $G_1 = (M_1, N_1)$ and $G_2 = (M_2, N_2)$ be two neutrosophic cubic graphs of the graphs G_1^* and G_2^* respectively then P-join is denoted by $G_1 +_P G_2$ and is defined by*

$$\begin{aligned} G_1 +_P G_2 &= (M_1, N_1) +_P (M_2, N_2) = (M_1 +_P M_2, N_1 +_P N_2) \\ &= \{\langle((\tilde{T}_{A_1} +_P \tilde{T}_{A_2}), (T_{B_1} +_P T_{B_2})), ((\tilde{I}_{A_1} +_P \tilde{I}_{A_2}), (I_{B_1} +_P I_{B_2})), ((\tilde{F}_{A_1} +_P \tilde{F}_{A_2}), (F_{B_1} +_P F_{B_2}))\rangle, \\ &\quad \langle((\tilde{T}_{C_1} +_P \tilde{T}_{C_2}), (T_{D_1} +_P T_{D_2})), ((\tilde{I}_{C_1} +_P \tilde{I}_{C_2}), (I_{D_1} +_P I_{D_2})), ((\tilde{F}_{C_1} +_P \tilde{F}_{C_2}), (F_{D_1} +_P F_{D_2}))\rangle\} \end{aligned}$$

where

(i) if $x \in V_1 \cup V_2$

$$\begin{aligned} (\tilde{T}_{A_1} +_P \tilde{T}_{A_2})(x) &= (\tilde{T}_{A_1} \cup_P \tilde{T}_{A_2})(x), (T_{B_1} +_P T_{B_2})(x) = (T_{B_1} \cup_P T_{B_2})(x) \\ (\tilde{I}_{A_1} +_P \tilde{I}_{A_2})(x) &= (\tilde{I}_{A_1} \cup_P \tilde{I}_{A_2})(x), (I_{B_1} +_P I_{B_2})(x) = (I_{B_1} \cup_P I_{B_2})(x) \\ (\tilde{F}_{A_1} +_P \tilde{F}_{A_2})(x) &= (\tilde{F}_{A_1} \cup_P \tilde{F}_{A_2})(x), (F_{B_1} +_P F_{B_2})(x) = (F_{B_1} \cup_P F_{B_2})(x) \end{aligned}$$

(ii) if $xy \in E_1 \cup E_2$

$$\begin{aligned} (\tilde{T}_{C_1} +_P \tilde{T}_{C_2})(xy) &= (\tilde{T}_{C_1} \cup_P \tilde{T}_{C_2})(xy), (T_{D_1} +_P T_{D_2})(xy) = (T_{D_1} \cup_P T_{D_2})(xy) \\ (\tilde{I}_{C_1} +_P \tilde{I}_{C_2})(xy) &= (\tilde{I}_{C_1} \cup_P \tilde{I}_{C_2})(xy), (I_{D_1} +_P I_{D_2})(xy) = (I_{D_1} \cup_P I_{D_2})(xy) \\ (\tilde{F}_{C_1} +_P \tilde{F}_{C_2})(xy) &= (\tilde{F}_{C_1} \cup_P \tilde{F}_{C_2})(xy), (F_{D_1} +_P F_{D_2})(xy) = (F_{D_1} \cup_P F_{D_2})(xy) \end{aligned}$$

(iii) if $xy \in E^*$, where E^* is the set of all edges joining the vertices of V_1 and V_2

$$\begin{aligned} (\tilde{T}_{C_1} +_P \tilde{T}_{C_2})(xy) &= rmin\{\tilde{T}_{A_1}(x), \tilde{T}_{A_2}(y)\}, (T_{D_1} +_P T_{D_2})(xy) = \min\{T_{B_1}(x), T_{B_2}(y)\} \\ (\tilde{I}_{C_1} +_P \tilde{I}_{C_2})(xy) &= rmin\{\tilde{I}_{A_1}(x), \tilde{I}_{A_2}(y)\}, (I_{D_1} +_P I_{D_2})(xy) = \min\{I_{B_1}(x), I_{B_2}(y)\} \\ (\tilde{F}_{C_1} +_P \tilde{F}_{C_2})(xy) &= rmin\{\tilde{F}_{A_1}(x), \tilde{F}_{A_2}(y)\}, (F_{D_1} +_P F_{D_2})(xy) = \min\{F_{B_1}(x), F_{B_2}(y)\} \end{aligned}$$

Definition 16. *Let $G_1 = (M_1, N_1)$ and $G_2 = (M_2, N_2)$ be two neutrosophic cubic graphs of the graphs G_1^* and G_2^* respectively then R-join is denoted by $G_1 +_R G_2$ and is defined by*

$$\begin{aligned} G_1 +_R G_2 &= (M_1, N_1) +_R (M_2, N_2) = (M_1 +_R M_2, N_1 +_R N_2) \\ &= \{\langle((\tilde{T}_{A_1} +_R \tilde{T}_{A_2}), (T_{B_1} +_R T_{B_2})), ((\tilde{I}_{A_1} +_R \tilde{I}_{A_2}), (I_{B_1} +_R I_{B_2})), ((\tilde{F}_{A_1} +_R \tilde{F}_{A_2}), (F_{B_1} +_R F_{B_2}))\rangle, \\ &\quad \langle((\tilde{T}_{C_1} +_R \tilde{T}_{C_2}), (T_{D_1} +_R T_{D_2})), ((\tilde{I}_{C_1} +_R \tilde{I}_{C_2}), (I_{D_1} +_R I_{D_2})), ((\tilde{F}_{C_1} +_R \tilde{F}_{C_2}), (F_{D_1} +_R F_{D_2}))\rangle\} \end{aligned}$$

where

(i) if $x \in V_1 \cup V_2$

$$
\begin{aligned}
(\tilde{T}_{A_1} +_R \tilde{T}_{A_2})(x) &= (\tilde{T}_{A_1} \cup_R \tilde{T}_{A_2})(x), (T_{B_1} +_R T_{B_2})(x) = (T_{B_1} \cup_R T_{B_2})(x) \\
(\tilde{I}_{A_1} +_R \tilde{I}_{A_2})(x) &= (\tilde{I}_{A_1} \cup_R \tilde{I}_{A_2})(x), (I_{B_1} +_R I_{B_2})(x) = (I_{B_1} \cup_R I_{B_2})(x) \\
(\tilde{F}_{A_1} +_R \tilde{F}_{A_2})(x) &= (\tilde{F}_{A_1} \cup_R \tilde{F}_{A_2})(x), (F_{B_1} +_R F_{B_2})(x) = (F_{B_1} \cup_R F_{B_2})(x)
\end{aligned}
$$

(ii) if $xy \in E_1 \cup E_2$

$$
\begin{aligned}
&2a.\ \{(\tilde{T}_{C_1} +_R \tilde{T}_{C_2})(xy) = (\tilde{T}_{C_1} \cup_R \tilde{T}_{C_2})(xy), (T_{D_1} +_R T_{D_2})(xy) = (T_{D_1} \cup_R T_{D_2})(xy) \\
&2b.\ \{(\tilde{I}_{C_1} +_R \tilde{I}_{C_2})(xy) = (\tilde{I}_{C_1} \cup_R \tilde{I}_{C_2})(xy), (I_{D_1} +_R I_{D_2})(xy) = (I_{D_1} \cup_R I_{D_2})(xy) \\
&2c.\ \{(\tilde{F}_{C_1} +_R \tilde{F}_{C_2})(xy) = (\tilde{F}_{C_1} \cup_R \tilde{F}_{C_2})(xy), (F_{D_1} +_R F_{D_2})(xy) = (F_{D_1} \cup_R F_{D_2})(xy)
\end{aligned}
$$

(iii) if $xy \in E^*$, where E^* is the set of all edges joining the vertices of V_1 and V_2

$$
3a. \begin{cases} (\tilde{T}_{C_1} +_R \tilde{T}_{C_2})(xy) = rmin\{\tilde{T}_{A_1}(x), \tilde{T}_{A_2}(y)\}, \\ (T_{D_1} +_R T_{D_2})(xy) = \max\{T_{B_1}(x), T_{B_2}(y)\} \end{cases}
$$

$$
3b. \begin{cases} (\tilde{I}_{C_1} +_R \tilde{I}_{C_2})(xy) = rmin\{\tilde{I}_{A_1}(x), \tilde{I}_{A_2}(y)\}, \\ (I_{D_1} +_R I_{D_2})(xy) = \max\{I_{B_1}(x), I_{B_2}(y)\} \end{cases}
$$

$$
3c. \begin{cases} (\tilde{F}_{C_1} +_R \tilde{F}_{C_2})(xy) = rmin\{\tilde{F}_{A_1}(x), \tilde{F}_{A_2}(y)\}, \\ (F_{D_1} +_R F_{D_2})(xy) = \max\{F_{B_1}(x), F_{B_2}(y)\} \end{cases}
$$

Proposition 4. *The P-join and R-join of two neutrosophic cubic graphs is again a neutrosophic cubic graph.*

4. Applications

Fuzzy graph theory is an effective field having a vast range of applications in Mathematics. Neutrosophic cubic graphs are more general and effective approach used in daily life very effectively.

Here in this section we test the applicability of our proposed model by providing applications in industries.

Example 7. *Let us suppose a set of three industries representing a vertex set $V = \{A, B, C\}$ and let the truth-membership of each vertex in V denotes "win win" situation of industry, where they do not harm each other and do not capture other's customers. Indetermined-membership of members of vertex set represents the situation in which industry works in a diplomatic and social way, that is, they are ally being social and competitive being industry. Falsity-membership shows a brutal competition where price war starts among industries. We want to observe the effect of one industry on other industry with respect to their business power and strategies. Let we have a neutrosophic cubic graph for industries having the following data with respect to business strategies*

$$
M = \left\langle \begin{array}{l} \{A, ([0.3, 0.4], 0.3), ([0.5, 0.7], 0.6), ([0.4, 0.5], 0.2)\}, \\ \{B, ([0.4, 0.5], 0.4), ([0.7, 0.8], 0.5), ([0.2, 0.3], 0.3)\}, \\ \{C, ([0.6, 0.8], 0.8), ([0.4, 0.5], 0.3), ([0.1, 0.2], 0.1)\} \end{array} \right\rangle
$$

where interval memberships indicate the business strength and strategies of industries for the present time while fixed membership indicates the business strength and strategies of industries for future based on given information. So on the basis of M we get a set of edges defined as

$$
N = \left\langle \begin{array}{l} \{AB, ([0.3, 0.4], 0.4), ([0.5, 0.7], 0.6), ([0.4, 0.5], 0.2)\}, \\ \{BC, ([0.4, 0.5], 0.8), ([0.4, 0.5], 0.5), ([0.2, 0.3], 0.1)\}, \\ \{AC, ([0.3, 0.4], 0.8), ([0.4, 0.5], 0.6), ([0.4, 0.5], 0.1)\} \end{array} \right\rangle
$$

where interval memberships indicate the business strength and strategies of industries for the present time while fixed membership indicate the business strength and strategies of industries for future when it will be the time of more competition. It is represented in Figure 11.

Finally we see that the business strategies of one industry strongly affect its business with other industries. Here

$$order(G) = \{([1.3, 1.7], 1.5), ([1.6, 2.0], 1.4), ([0.7, 1.0], 0.6)\}$$

and

$$\begin{aligned} \deg(A) &= \{([0.6, 0.8], 1.2), ([0.9, 1.2], 1.2), ([0.8, 1.0], 0.3)\} \\ \deg(B) &= \{([0.7, 0.9], 1.2), ([0.9, 1.2], 1.1), ([0.6, 0.8], 0.3)\} \\ \deg(C) &= \{([0.7, 0.9], 1.6), ([0.8, 1.0], 1.1), ([0.6, 0.8], 0.2)\} \end{aligned}$$

Order of G represents the overall effect on market of above given industries A, B and C. Degree of A represents the effect of other industries on A link through an edge with the industry A. The minimum degree of A is 0 when it has no link with any other.

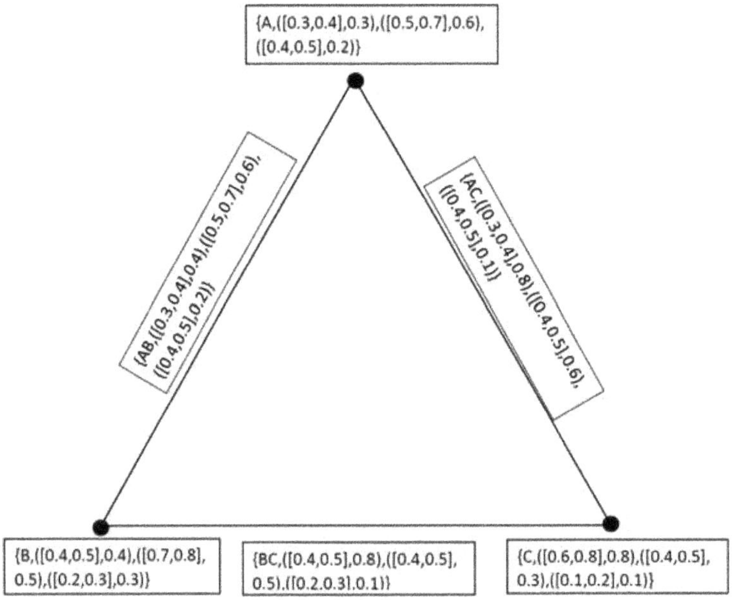

Figure 11. Neutrosophic Cubic Graph.

Example 8. *Let us take an industry and we want to evaluate its overall performance. There are a lot of factors affecting it. However, some of the important factors influencing industrial productivity are with neutrosophic cubic sets as under, where the data is provided in the form of interval based on future prediction and data given in the form of a number from the unit interval* $[0, 1]$ *is dependent on the present time after a careful testing of different models as a sample in each case,*

1. *Technological Development* $A = ((\widetilde{T}_A, T_A), (\widetilde{I}_A, I_A), (\widetilde{F}_A, F_A)) =$ *((degree of mechanization), (technical know-how), (product design))* $= \{A, ([0.3, 0.4], 0.3), ([0.5, 0.7], 0.6), ([0.4, 0.5], 0.2)\}$,
2. *Quality of Human Resources* $B = ((\widetilde{T}_B, T_B), (\widetilde{I}_B, I_B), (\widetilde{F}_B, F_B)) =$ *((ability of the worker), (willingness of the worker), (the environment under which he has to work))* $= \{B, ([0.4, 0.5], 0.4), ([0.7, 0.8], 0.5), ([0.2, 0.3], 0.3)\}$,

3. *Availability of Finance* $C = ((\widetilde{T}_C, T_C), (\widetilde{I}_C, I_C), (\widetilde{F}_C, F_C))$ = ((*advertisement campaign*), (*better working conditions to the workers*), (*up-keep of plant and machinery*)) $= \{C, ([0.6, 0.8], 0.8), ([0.4, 0.5], 0.3), ([0.1, 0.2], 0.1)\}$,
4. *Managerial Talent* $D = ((\widetilde{T}_D, T_D), (\widetilde{I}_D, I_D), (\widetilde{F}_D, F_D))$ = ((*devoted towards their profession*), (*Links with workers, customers and suppliers*), (*conceptual, human relations and technical skills*)) $= \{D, ([0.3, 0.6], 0.4), ([0.2, 0.7], 0.9), ([0.3, 0.5], 0.6)\}$,
5. *Government Policy*= $E = ((\widetilde{T}_E, T_E), (\widetilde{I}_E, I_E), (\widetilde{F}_E, F_E))$ =*Government Policy*= ((*favorable conditions for saving*), (*investment*), (*flow of capital from one industrial sector to another*)) $= \{E, ([0.2, 0.4], 0.5), ([0.5, 0.6], 0.1), ([0.4, 0.5], 0.2)\}$,
6. *Natural Factors*= $F = ((\widetilde{T}_F, T_F), (\widetilde{I}_F, I_F), (\widetilde{F}_F, F_F))$ = ((*physical*), (*geographical*), (*climatic exercise*)) $= \{F, ([0.1, 0.4], 0.8), ([0.5, 0.7], 0.2), ([0.4, 0.5], 0.2)\}$. *As these factors affecting industrial productivity are inter-related and inter-dependent, it is a difficult task to evaluate the influence of each individual factor on the overall productivity of industrial units. The use of neutrosophic cubic graphs give us a more reliable information as under. Let* $X = \{A, B, C, D, F, E\}$ *we have a neutrosophic cubic set for the vertex set as under*

$$M = \left\langle \begin{array}{c} \{A, ([0.3, 0.4], 0.3), ([0.5, 0.7], 0.6), ([0.4, 0.5], 0.2)\}, \\ \{B, ([0.4, 0.5], 0.4), ([0.7, 0.8], 0.5), ([0.2, 0.3], 0.3)\}, \\ \{C, ([0.6, 0.8], 0.8), ([0.4, 0.5], 0.3), ([0.1, 0.2], 0.1)\}, \\ \{D, ([0.3, 0.6], 0.4), ([0.2, 0.7], 0.9), ([0.3, 0.5], 0.6)\}, \\ \{E, ([0.2, 0.4], 0.5), ([0.5, 0.6], 0.1), ([0.4, 0.5], 0.2)\}, \\ \{F, ([0.1, 0.4], 0.8), ([0.5, 0.7], 0.2), ([0.4, 0.5], 0.2)\} \end{array} \right\rangle$$

Now, in order to find the combined effect of all these factors we need to use neutrosophic cubic sets for edges as under

$$N = \left\langle \begin{array}{c} \{AB, ([0.3, 0.4], 0.4), ([0.5, 0.7], 0.6), ([0.4, 0.5], 0.2)\}, \\ \{AC, ([0.3, 0.4], 0.8), ([0.4, 0.5], 0.6), ([0.4, 0.5], 0.1)\}, \\ \{AD, ([0.3, 0.4], 0.4), ([0.2, 0.7], 0.9), ([0.4, 0.5], 0.2)\}, \\ \{AE, ([0.2, 0.4], 0.5), ([0.5, 0.6], 0.6), ([0.4, 0.5], 0.2)\}, \\ \{AF, ([0.1, 0.4], 0.8), ([0.5, 0.7], 0.6), ([0.4, 0.5], 0.2)\}, \\ \{BC, ([0.4, 0.5], 0.8), ([0.4, 0.5], 0.5), ([0.2, 0.3], 0.1)\}, \\ \{BD, ([0.3, 0.5], 0.4), ([0.2, 0.7], 0.9), ([0.3, 0.5], 0.3)\}, \\ \{BE, ([0.2, 0.4], 0.5), ([0.5, 0.6], 0.5), ([0.4, 0.5], 0.2)\}, \\ \{BF, ([0.1, 0.4], 0.8), ([0.5, 0.7], 0.5), ([0.4, 0.5], 0.2)\}, \\ \{CD, ([0.3, 0.6], 0.8), ([0.2, 0.5], 0.9), ([0.3, 0.5], 0.1)\}, \\ \{CE, ([0.2, 0.4], 0.8), ([0.4, 0.5], 0.3), ([0.4, 0.5], 0.1)\}, \\ \{CF, ([0.1, 0.4], 0.8), ([0.4, 0.5], 0.3), ([0.4, 0.5], 0.1)\}, \\ \{DE, ([0.2, 0.4], 0.5), ([0.2, 0.6], 0.9), ([0.4, 0.5], 0.2)\}, \\ \{DF, ([0.1, 0.4], 0.8), ([0.2, 0.7], 0.9), ([0.4, 0.5], 0.2)\}, \\ \{EF, ([0.1, 0.4], 0.8), ([0.5, 0.6], 0.2), ([0.4, 0.5], 0.2)\} \end{array} \right\rangle$$

where the edge $\{AB, ([0.3, 0.4], 0.4), ([0.5, 0.7], 0.6), ([0.4, 0.5], 0.2)\}$ *denotes the combined effect of technological development and quality of human resources on the productivity of the industry. Now, if we are interested to find which factors are more effective to the productivity of the industry, we may use the score and accuracy of the neutrosophic cubic sets, which will give us a closer view of the factors. It is represented in Figure 12.*

Remark. We used degree and order of the neutrosophic cubic graphs in an application see Example 7 and if we have two different sets of industries having finite number of elements, we can easily find the applications of cartesian product, composition, union, join, order and degree of neutrosophic cubic graphs.

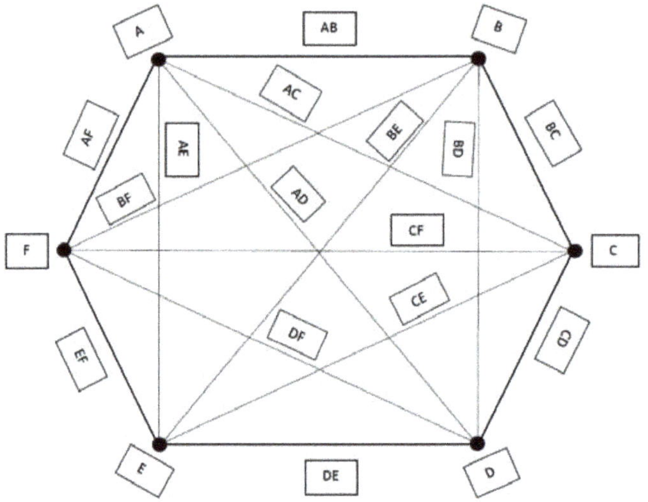

Figure 12. Neutrosophic Cubic Graph.

5. Comparison Analysis

In 1975, Rosenfeld discussed the concept of fuzzy graphs whose basic idea was introduced by [3] Kauffmann in 1973. Atanassov extended this idea to intuitionistic fuzzy graphs [14] in 1995. The idea of neutrosophic graphs provided by Kandasamy et al. in the book [29]. [38] Recently Rashid et al., introduced the notion of cubic graphs. In this paper, we introduced the study of neutrosophic cubic graphs. We claim that our model is more generalized from the previous models, as if we both indeterminacy and falsity part of neutrosophic cubic graphs $G = (M, N)$ where $M = (A, B) = ((\widetilde{T}_A, T_B), (\widetilde{I}_A, I_B), (\widetilde{F}_A, F_B))$ is the neutrosophic cubic set representation of vertex set V and $N = (C, D) = ((\widetilde{T}_C, T_D), (\widetilde{I}_C, I_D), (\widetilde{F}_C, F_D))$ is the neutrosophic cubic set representation of edges set E vanishes we get a cubic graph provided by Rashid et al., in [38]. Similarly, by imposing certain conditions on cubic graphs, we may obtain intuitionistic fuzzy graphs provided by Atanassov in 1995 and after that fuzzy graphs provided by Rosenfeld in 1975. So our proposed model is a generalized model and it has the ability to capture the uncertainty in a better way.

6. Conclusions

A generalization of the old concepts is the main motive of research. So in this paper, we proposed a generalized model of neutrosophic cubic graphs with different binary operations. We also provided applications of neutrosophic cubic graphs in industries. We also discussed conditions under which our model reduces to the previous models. In future, we will try to discuss different types of neutrosophic cubic graphs such as internal neutrosophic cubic graphs, external neutrosophic cubic graphs and many more with applications.

Author Contributions: Write up Z.R., Methodology, N.Y.; Project administration, F.S.; Supervision, M.G.; English Editing, H.A.W.

Conflicts of Interest: The authors declare no conflict of interest.

References

1. Zadeh, L.A. Fuzzy sets. *Inf. Control* **1965**, *8*, 338–353.
2. Rosenfeld, A. *Fuzzy Graphs, Fuzzy Sets and Their Applications*; Academic Press: New York, NY, USA, 1975; pp. 77–95.

3. Kauffman, A. *Introduction a la Theorie des Sous-Emsembles Flous*; Masson: Issy-les-Moulineaux, French, 1973; Volume 1.
4. Bhattacharya, P. Some remarks on fuzzy graphs. *Pattern Recognit. Lett.* **1987**, *6*, 297–302.
5. Akram, M.; Dudek, W.A. Interval-valued fuzzy graphs. *Comput. Math. Appl.* **2011**, *61*, 289–299.
6. Akram, M. Interval-valued fuzzy line graphs. *Neural Comput. Appl.* **2012**, *21*, 145–150.
7. Mordeson, J.N.; Nair, P.S. *Fuzzy Graphs and Fuzzy Hypergraphs*; Springer: Berlin/Heidelberg, Germany, 2001.
8. Sunitha, M.S.; Sameena, K. Characterization of g-self centered fuzzy graphs. *J. Fuzzy Math.* **2008**, *16*, 787–791.
9. Borzooei, R.A.; Rashmanlou, H. Cayley interval-valued fuzzy threshold graphs, U.P.B. *Sci. Bull. Ser. A* **2016**, *78*, 83–94.
10. Pal, M.; Samanta, S.; Rashmanlou, H. Some results on interval-valued fuzzy graphs. *Int. J. Comput. Sci. Electron. Eng.* **2015**, *3*, 2320–4028.
11. Pramanik, T.; Pal, M.; Mondal, S. Inteval-valued fuzzy threshold graph. *Pac. Sci. Rev. A Nat. Sci. Eng.* **2016**, *18*, 66–71.
12. Pramanik, T.; Samanta, S.; Pal, M. Interval-valued fuzzy planar graphs. *Int. J. Mach. Learn. Cybern.* **2016**, *7*, 653–664.
13. Atanassov, K.T. Intuitionistic fuzzy sets. *Fuzzy Sets Syst.* **1986**, *20*, 87–96.
14. Atanassov, K.T. On Intuitionistic Fuzzy Graphs and Intuitionistic Fuzzy Relations. In Proceedings of the VI IFSA World Congress, Sao Paulo, Brazil, 22–28 July 1995; Volume 1, pp. 551–554.
15. Atanassov, K.T.; Shannon, A. On a generalization of intuitionistic fuzzy graphs. *Notes Intuit. Fuzzy Sets* **2006**, *12*, 24–29.
16. Karunambigai, M.G.; Parvathi, R. Intuitionistic Fuzzy Graphs. In *Computational Intelligence, Theory and Applications*; Springer: Berlin/Heidelberg, Germany, 2006; Volume 20, pp. 139–150.
17. Shannon, A.; Atanassov, K.T. A first step to a theory of the intutionistic fuzzy graphs. In Proceedings of the 1st Workshop on Fuzzy Based Expert Systems, Sofia, Bulgaria, 26–29 June 1994; pp. 59–61.
18. Mishra, S.N.; Rashmanlou, H.; Pal, A. Coherent category of interval-valued intuitionistic fuzzy graphs. *J. Mult. Val. Log. Soft Comput.* **2017**, *29*, 355–372.
19. Parvathi, R.; Karunambigai, M.G.; Atanassov, K. Operations on Intuitionistic Fuzzy Graphs. In Proceedings of the IEEE International Conference on Fuzzy Systems, Jeju Island, Korea, 20–24 August 2009; pp. 1396–1401.
20. Sahoo, S.; Pal, M. Product of intiutionistic fuzzy graphs and degree. *J. Intell. Fuzzy Syst.* **2017**, *32*, 1059–1067.
21. Akram, M.; Davvaz, B. Strong intuitionistic fuzzy graphs. *Filomat* **2012**, *26*, 177–196.
22. Zhang, W.R. Bipolar Fuzzy Sets and Relations: A Computational Framework for Coginitive Modeling and Multiagent Decision Analysis. In Proceedings of the IEEE Industrial Fuzzy Control and Intelligent Systems Conference, and the NASA Joint Technology Workshop on Neural Networks and Fuzzy Logic, Fuzzy Information Processing Society Biannual Conference, San Antonio, TX, USA, 18–21 December 1994; pp. 305–309.
23. Akram, M. Bipolar fuzzy graphs. *Inf. Sci.* **2011**, *181*, 5548–5564.
24. Akram, M. Bipolar fuzzy graphs with applications. *Knowl. Based Syst.* **2013**, *39*, 1–8.
25. Akram, M.; Karunambigai, M.G. Metric in bipolar fuzzy graphs. *World Appl. Sci. J.* **2012**, *14*, 1920–1927.
26. Smarandache, F. *A Unifying Field in Logics: Neutrosophic Logic. Neutrosophy, Neutrosophic Set, Neutrosophic Probability*; American Research Press: Rehoboth, NM, USA, 1999.
27. Smarandache, F. Neutrosophic set-a generalization of the intuitionistic fuzzy set. *Int. J. Pure Appl. Math.* **2005**, *24*, 287–297.
28. Wang, H.; Smarandache, F.; Zhang, Y.Q.; Sunderraman, R. *Interval Neutrosophic Sets and Logic: Theory nd Applications in Computing*; Hexis: phoenix, AZ, USA, 2005.
29. Kandasamy, W.B.V.; Ilanthenral, K.; Smarandache, F. *Neutrosophic Graphs: A New Dimension to Graph Theory*; EuropaNova ASBL: Bruxelles, Belgium, 2015.
30. Akram, M.; Rafique, S.; Davvaz, B. New concepts in neutrosophic graphs with application. *J. Appl. Math. Comput.* **2018**, *57*, 279–302.
31. Akram, M.; Nasir, M. Concepts of Interval-Valued Neutrosophic Graphs. *Int. J. Algebra Stat.* **2017**, *6*, 22–41.
32. Akram, M. Single-valued neutrosophic planar graphs. *Int. J. Algebra Stat.* **2016**, *5*, 157–167.
33. Akram, M.; Shahzadi, S. Neutrosophic soft graphs with application. *J. Intell. Fuzzy Syst.* **2017**, *32*, 841–858.
34. Jun, Y.B.; Kim, C.S.; Yang, K.O. Cubic Sets. *Ann. Fuzzy Math. Inf.* **2012**, *4*, 83–98.

35. Jun, Y.B.; Kim, C.S.; Kang, M.S. Cubic subalgebras and ideals of BCK/BCI-algebras. *Far East J. Math. Sci.* **2010**, *44*, 239–250.
36. Jun, Y.B.; Lee, K.J.; Kang, M.S. Cubic structures applied to ideals of BCI-algebras. *Comput. Math. Appl.* **2011**, *62*, 3334–3342.
37. Kang, J.G.; Kim, C.S. Mappings of cubic sets. *Commun. Korean Math. Soc.* **2016**, *31*, 423–431.
38. Rashid, S.; Yaqoob, N.; Akram, M.; Gulistan, M. Cubic Graphs with Application. *Int. J. Anal. Appl.* **2018**, in press.
39. Jun, Y.B.; Smarandache, F.; Kim, C.S. Neutrosophic cubic sets. *New Math. Nat. Comput.* **2017**, *13*, 41–54.
40. Jun, Y.B.; Smarandache, F.; Kim, C.S. P-union and P-intersection of neutrosophic cubic sets. *Anal. Univ. Ovid. Constant. Seria Mat.* **2017**, *25*, 99–115.

symmetry

MDPI

Article

Medical Diagnosis Based on Single-Valued Neutrosophic Probabilistic Rough Multisets over Two Universes

Chao Zhang [1], Deyu Li [1,*], Said Broumi [2] and Arun Kumar Sangaiah [3]

1. Key Laboratory of Computational Intelligence and Chinese Information Processing of Ministry of Education, Research Institute of Big Data Science and Industry, School of Computer and Information Technology, Shanxi University, Taiyuan 030006, China; zhch3276152@163.com
2. Laboratory of Information Processing, Faculty of Science Ben M'Sik, University Hassan II, B.P 7955, Sidi Othman, Casablanca 20000, Morocco; broumisaid78@gmail.com
3. School of Computing Science and Engineering, VIT University, Vellore 632014, India; arunkumarsangaiah@gmail.com

* Correspondence: lidysxu@163.com; Tel.: +86-351-701-8775

Received: 17 May 2018; Accepted: 11 June 2018; Published: 12 June 2018

Abstract: In real-world diagnostic procedures, due to the limitation of human cognitive competence, a medical expert may not conveniently use some crisp numbers to express the diagnostic information, and plenty of research has indicated that generalized fuzzy numbers play a significant role in describing complex diagnostic information. To deal with medical diagnosis problems based on generalized fuzzy sets (FSs), the notion of single-valued neutrosophic multisets (SVNMs) is firstly used to express the diagnostic information in this article. Then the model of probabilistic rough sets (PRSs) over two universes is applied to analyze SVNMs, and the concepts of single-valued neutrosophic rough multisets (SVNRMs) over two universes and probabilistic rough single-valued neutrosophic multisets (PRSVNMs) over two universes are introduced. Based on SVNRMs over two universes and PRSVNMs over two universes, single-valued neutrosophic probabilistic rough multisets (SVNPRMs) over two universes are further established. Next, a three-way decisions model by virtue of SVNPRMs over two universes in the context of medical diagnosis is constructed. Finally, a practical case study along with a comparative study are carried out to reveal the accuracy and reliability of the constructed three-way decisions model.

Keywords: single-valued neutrosophic multisets; medical diagnosis; probabilistic rough sets over two universes; three-way decisions

1. Introduction

In medical science and technology, it is acknowledged that disease diagnosis is a rather complicated activity for medical experts who are faced with tasks in handling varieties of uncertain diagnostic information. In order to seek the accurate diagnosis for the considered patients, it is essential for medical experts to take into account a number of related symptoms simultaneously, and this procedure might take a long time to reach a final diagnostic outcome. Considering it is meaningful to cope with the above complex decision making situation within the background of medical diagnosis, plenty of practitioners are likely to focus on the relationship between the diagnosis set and the symptom set, and varieties of achievements have been made on the basis of fuzzy approaches [1–3]. According to the FS theory established by Zadeh [4], fuzzy approaches have been extensively used in lots of medical diagnosis situations. However, the modeling tools of classical FSs are confined when multiple kinds of uncertainties emerge at the same time. Thus, several new notions of generalized FSs were put forward one after another during the past decades [5].

Among various generalized FSs, in view of intuitionistic fuzzy sets (IFSs) [6] lack a reasonable scheme to effectively process inconsistent and indeterminate information embedded in realistic scenarios, Smarandache [7] initiated the framework of neutrosophic sets (NSs) and neutrosophic logic, which could be seen as a generalized form of FSs, IFSs, fuzzy logic and intuitionistic fuzzy logic [8–10]. Compared with IFSs, through adding an indeterminacy membership function which is focused on separately, NSs are able to express incomplete, inconsistent and indeterminate information efficiently. Further, in order to utilize the idea of NSs to solve a broader range of practical issues, Wang et al. [11] presented a novel branch of NSs called single-valued neutrosophic sets (SVNSs), whose values of three membership functions belong to $[0,1]$. Ever since the establishment of SVNSs, many enlightening research results have been made in many real-world areas [12–21]. Recently, inspired by one typical solution to obtain an accurate diagnosis for a patient is to arrange the medical examination at different parts on a day to day basis (e.g., morning, noon and evening), Ye [22] introduced the notion of SVNMs by taking advantage of fuzzy multisets (FMs) [23] and n-valued refined neutrosophic logic [24]. With the support of SVNMs, the SVN membership values occur one or multiple times, which is favourable to the expression of the above SVN diagnostic information at different time intervals, hence SVNMs could process the uncertain information well, and offer medical experts a rather powerful tool to record a complicated medical diagnosis knowledge base. Until now, the studies of SVNMs are mainly concentrated on algebraic properties, similarity measures, neutrosophic multiple relations, cosine measures, and so on [25–28].

In general, medical experts are often confronted with the following two challenges in practical medical diagnosis, one is to make an accurate diagnostic conclusion for the considered patients, another one is to provide a reasonable explanation on how to obtain the result under uncertain scenarios [29–31]. Through designing possible and deterministic decision rules, rough set theory has illustrated its powerful performances for solving various decision making situations in the above challenges [32–42]. In addition, among varieties of specific rough set models, it is worth noticing that lots of them are often too strict and might require additional information when constructing approximations originated from the classical rough set theory. In order to effectively handle the above issue, by combining the rough sets with probability theory, the concept of PRSs is initiated by Wong and Ziarko [43] to let rough set models possess the error tolerance capability when processing the noisy data. Further, PRSs are developed by virtue of more powerful soft computing tools such as Bayesian decision theory, graded set inclusions, Bayesian confirmation measures, etc [44–48]. Compared with other types of rough sets, through introducing the probability theory to estimate the rough membership, PRSs permit the existence of the error tolerance by means of the introduction of thresholds. In recent years, motivated by the notion of three-way decisions [49,50], several decision making methods by means of PRSs-based three-way decisions have been put forward to promote the solving efficiency of real-world problems [51–54].

In this paper, inspired by the idea of PRSs-based three-way decisions, we systematically study various probabilistic rough approximations in the background of SVNMs information by integrating SVNMs with PRSs over two universes, and propose the model of SVNPRMs over two universes. Then, we aim to investigate a three-way decisions method by utilizing the proposed SVNPRMs over two universes under the context of medical diagnosis. In light of the above discussion, we arrange the structure of the article below. In Section 2, we revisit several fundamental notions about SVNMs and PRSs. In the next section, we present the notion of SVNRMs over two universes and PRSVNMs over two universes at first, then present the notion of SVNPRMs over two universes. Section 4 constructs a medical diagnosis algorithm based on SVNPRMs over two universes, then a case study and its corresponding comparative study are carried out to show the validity of the constructed medical diagnosis algorithm. The last section concludes the contributions of the work.

2. Preliminaries

In the following, we revisit several fundamental concepts which will be utilized in the latter part of the paper.

2.1. SVNMs

As a generalization of many concepts such as FSs, IFSs, SVNSs, etc., the definition of SVNMs and their related operations are presented below.

Definition 1. [22] *Suppose that U is a finite and nonempty set, a SVNM A is featured by count truth-membership of CT_A, count indeterminacy-membership of CI_A, count falsity-membership of CF_A, where $CT_A, CI_A, CF_A : U \to R$ for all $x \in U$. Then a SVNM A is given by*

$$A = \left\{ \left\langle \begin{array}{c} x, \left(T_A^1(x), T_A^2(x), \ldots, T_A^q(x)\right), \\ \left(I_A^1(x), I_A^2(x), \ldots, I_A^q(x)\right), \\ \left(F_A^1(x), F_A^2(x), \ldots, F_A^q(x)\right) \end{array} \right\rangle | x \in U \right\},$$

where the truth-membership sequence, the indeterminacy-membership sequence and the falsity-membership sequence $\left(T_A^1(x), T_A^2(x), \ldots, T_A^q(x)\right)$, $\left(I_A^1(x), I_A^2(x), \ldots, I_A^q(x)\right)$ and $\left(F_A^1(x), F_A^2(x), \ldots, F_A^q(x)\right)$ are arranged in an increasing or decreasing order. Additionally, for each $i = 1, 2, \ldots, q$, the sum of $T_A^i(x), I_A^i(x), F_A^i(x) \in [0,1]$ fulfills the requirement $0 \leq T_A^i(x) + I_A^i(x) + F_A^i(x) \leq 3$. For the sake of convenience, a simplified form of SVNM could be expressed as $A = \left\{\langle x, T_A^i(x), I_A^i(x), F_A^i(x)\rangle | x \in U, i = 1, 2, \ldots, q\right\}$. Furthermore, we represent the set of all SVNMs on U as $SVNM(U)$. For all $x \in U$, A is called a full SVNM if and only if $x = \langle 1, 0, 0\rangle$, while A is called an empty SVNM if and only if $x = \langle 0, 1, 1\rangle$.

Definition 2. [22] *The length of x in a SVNM A is represented by $L(x : A)$, where $L(x : A) = |CT_A(x)| = |CI_A(x)| = |CF_A(x)|$ ($|CT_A(x)|$, $|CI_A(x)|$ and $|CF_A(x)|$ represent the cardinality of $CT_A(x)$, $CI_A(x)$ and $CF_A(x)$). In addition, for any $A, B \in SVNM(U)$, $L(x : A, B) = \max\{L(x : A), L(x : B)\}$.*

For any two SVNMs A and B, it is noted that $L(x : A)$ might be different from $L(x : B)$ in many situations. Through adding the maximum number for the indeterminacy-membership value and the falsity-membership value, and further adding the minimum number for the truth-membership value, we could make $L(x : A) = L(x : B)$.

Definition 3. [22] *For any two SVNMs A and B in the universe U, we have*

1. $A \oplus B = \left\{\langle x, T_A^i(x) + T_B^i(x) - T_A^i(x) T_B^i(x), I_A^i(x) I_B^i(x), F_A^i(x) F_B^i(x)\rangle | x \in U, i = 1, 2, \ldots, q\right\}$;
2. $A \otimes B = \{\langle x, T_A^i(x) T_B^i(x), I_A^i(x) + I_B^i(x) - I_A^i(x) I_B^i(x), F_A^i(x) + F_B^i(x) - F_A^i(x) F_B^i(x)\rangle | x \in U, i = 1, 2, \ldots, q\}$;
3. *the complement of A is represented by A^c such that $\forall x \in U$,* $A^c = \left\{\langle x, F_A^i(x), 1 - I_A^i(x), T_A^i(x)\rangle | x \in U, i = 1, 2, \ldots, q\right\}$;
4. *the union of A and B is represented by $A \cup B$ such that $\forall x \in U$,* $A \cup B = \left\{\langle x, T_A^i(x) \vee T_B^i(x), I_A^i(x) \wedge I_B^i(x), F_A^i(x) \wedge F_B^i(x)\rangle | x \in U, i = 1, 2, \ldots, q\right\}$;
5. *the intersection of A and B is represented by $A \cap B$ such that $\forall x \in U$,* $A \cap B = \left\{\langle x, T_A^i(x) \wedge T_B^i(x), I_A^i(x) \vee I_B^i(x), F_A^i(x) \vee F_B^i(x)\rangle | x \in U, i = 1, 2, \ldots, q\right\}$.

2.2. PRSs

In view of classical rough sets being rather rigorous when constructing lower/upper approximations and often requiring some additional information, Wong and Ziarko [43] took advantage of the probabilistic measure theory, and presented a probabilistic model of rough sets to update the method for obtaining a related rough set region.

Definition 4. *[43] Suppose that U is the universe of discourse, and P is the probabilistic measure based on the σ algebra. Then (U, R, P) is named a probabilistic approximation space. For any $0 \leq \beta \leq \alpha \leq 1$, $X \subseteq U$, the lower and upper approximations of X are given by*

$$\underline{P}_{\alpha}(X) = \{P(X|[x]) \geq \alpha \,|x \in U\},$$
$$\overline{P}_{\beta}(X) = \{P(X|[x]) > \beta \,|x \in U\},$$

where the pair $\left(\underline{P}_{\alpha}(X), \overline{P}_{\beta}(X)\right)$ is named a PRS. Moreover, by virtue of the above approximations, the positive region, negative region and boundary region of X are further given by

$$POS(X, \alpha, \beta) = \underline{P}_{\alpha}(X) = \{P(X|[x]) \geq \alpha \,|x \in U\},$$
$$NEG(X, \alpha, \beta) = U - \overline{P}_{\beta}(X) = \{P(X|[x]) \leq \beta \,|x \in U\},$$
$$BND(X, \alpha, \beta) = \overline{P}_{\beta}(X) - \underline{P}_{\alpha}(X) = \{\beta < P(X|[x]) < \alpha \,|x \in U\},$$

it is noted that the above approximations reduce to classical rough sets when $\alpha = 1$ and $\beta = 0$, thus classical rough sets act as a special case of PRSs.

3. Probabilistic Rough Approximations of a SVNM under the Background of Two Universes

In this section, we aim to put forward the probabilistic rough approximations of a SVNM under the background of two universes, and eventually develop the model of SVNPRMs over two universes. To facilitate our discussion, based on the proposed relation on SVNMs from the universe U to the universe V and some operations, we first discuss the general rough approximations of a SVNM under the background of two universes and present the model of SVNRMs over two universes. Then, we investigate rough single-valued neutrosophic multisets on a probabilistic approximation space over two universes and propose the model of PRSVNMs over two universes. Lastly, the definition of SVNPRMs over two universes is put forward and several significant propositions of the presented model are explored.

3.1. Relations on SVNMs Based on Two Universes and Some Operations

In what follows, we introduce the arbitrary relation on SVNMs based on two related universes.

Definition 5. *Suppose that U, V are two universes of discourse, and R is a relation on SVNMs. Then R is given by*

$$R = \left\{\left\langle (x, y), T_R^i(x, y), I_R^i(x, y), F_R^i(x, y)\right\rangle |(x, y) \in U \times V, i = 1, 2, \ldots, q\right\},$$

Moreover, we denote the family of all relations on SVNMs from U to V as $SVNM(U \times V)$.

In order to facilitate the latter discussions of the paper, we present subtraction and division operations, and the corresponding score functions for SVNMs as follows.

Definition 6. *For any two SVNMs A and B in the universe U, we have*

1. $A \ominus B = \left\{\left\langle x, \frac{T_A^i(x) - T_B^i(x)}{1 - T_B^i(x)}, \frac{I_A^i(x)}{I_B^i(x)}, \frac{F_A^i(x)}{F_B^i(x)}\right\rangle |x \in U, i = 1, 2, \ldots, q\right\}$, *which is valid under the requirements $A \geq B$, $T_B^i(x) \neq 1$, $I_B^i(x) \neq 0$ and $F_B^i(x) \neq 0$;*

2. $A \oslash B = \left\{ \left\langle x, \frac{T_A^i(x)}{T_B^i(x)}, \frac{I_A^i(x)-I_B^i(x)}{1-I_B^i(x)}, \frac{F_A^i(x)-F_B^i(x)}{1-F_B^i(x)} \right\rangle | x \in U, i = 1,2,\ldots,q \right\}$, *which is valid under the requirements* $A \leq B$, $T_B^i(x) \neq 0$, $I_B^i(x) \neq 1$ *and* $F_B^i(x) \neq 1$.

In the following, the corresponding score function is proposed to rank different elements in SVNMs.

Definition 7. *Suppose that* $x = \langle T_A^i(x), I_A^i(x), F_A^i(x) \rangle$ *is an element in a SVNM, the corresponding score function of x is defined below.*

$$s(x) = \left[\sum_{i=1}^{q} T_A^i(x) + \sum_{i=1}^{q} \left(1 - I_A^i(x)\right) + \sum_{i=1}^{q} \left(1 - F_A^i(x)\right) \right] / 3q.$$

For two elements in a SVNM x_1 *and* x_2, *we have*

1. *If* $s(x_1) < s(x_2)$, *then* $x_1 < x_2$;
2. *If* $s(x_1) = s(x_2)$, *then* $x_1 = x_2$;
3. *If* $s(x_1) > s(x_2)$, *then* $x_1 > x_2$.

3.2. SVNRMs over Two Universes

By virtue of the above presented relations on SVNMs based on two universes, the definition of SVNRMs over two universes is put forward below.

Definition 8. *Suppose that* U, V *are two universes of discourse, and* $R \in SVNM(U \times V)$ *is a relation on SVNMs. Then* (U, V, R) *is named a general approximation space over two universes based on SVNMs. For any* $A \in SVNM(V)$, $x \in U$, $y \in V$, *the lower and upper approximations of A with respect to* (U, V, R) *are given by*

$$\underline{R}(A) = \left\{ \left\langle x, T_{\underline{R}(A)}^i(x), I_{\underline{R}(A)}^i(x), F_{\underline{R}(A)}^i(x) \right\rangle | x \in U, i = 1,2,\ldots,q \right\},$$
$$\overline{R}(A) = \left\{ \left\langle x, T_{\overline{R}(A)}^i(x), I_{\overline{R}(A)}^i(x), F_{\overline{R}(A)}^i(x) \right\rangle | x \in U, i = 1,2,\ldots,q \right\},$$

where $T_{\underline{R}(A)}^i(x) = \wedge_{y \in V} \left[F_R^i(x,y) \vee T_A^i(y) \right]$, $I_{\underline{R}(A)}^i(x) = \vee_{y \in V} \left[\left(1 - I_R^i(x,y)\right) \wedge I_A^i(y) \right]$, $F_{\underline{R}(A)}^i(x) = \vee_{y \in V} \left[T_R^i(x,y) \wedge F_A^i(y) \right]$, $T_{\overline{R}(A)}^i(x) = \vee_{y \in V} \left[T_R^i(x,y) \wedge T_A^i(y) \right]$, $I_{\overline{R}(A)}^i(x) = \wedge_{y \in V} \left[I_R^i(x,y) \vee I_A^i(y) \right]$ *and* $F_{\overline{R}(A)}^i(x) = \wedge_{y \in V} \left[F_R^i(x,y) \vee F_A^i(y) \right]$. *Based on the above statements, the pair* $\left(\underline{R}(A), \overline{R}(A)\right)$ *is named a SVNRM over two universes.*

3.3. PRSVNMs over Two Universes

Considering the various advantages of PRSs, we then extend the presented SVNRMs over two universes to the context of PRSs. In what follows, we first investigate rough single-valued neutrosophic multisets on a probabilistic approximation space over two universes.

Definition 9. *Suppose that* U, V *are two universes of discourse,* $R \subseteq U \times V$ *is a binary relation on two universes, and P is a probabilistic measure based on the* σ *algebra. Then* (U, V, R, P) *is named a probabilistic approximation space over two universes.*

Next, we develop the approach to obtain the conditional probability $P(A | R(x))$ of an event expressed by SVNMs given the description $R(x) \in 2^V$.

Definition 10. *Suppose that* (U, V, R, P) *is a probabilistic approximation space. For any* $A \in SVNM(V)$, $x \in U, y \in V$, *the conditional probability* $P(A|R(x))$ *is given by*

$$P(A|R(x)) = \frac{\sum_{y\in R(x)} A(y)}{|R(x)|}.$$

By virtue of the proposed conditional probability $P(A|R(x))$, the definition of PRSVNMs over two universes is put forward as follows.

Definition 11. *Suppose that* (U, V, R, P) *is a probabilistic approximation space over two universes. For any* $0 \leq \beta \leq \alpha \leq 1$, $A \in SVNM(V)$, $x \in U$, $y \in V$, *the lower and upper approximations of A with respect to* (U, V, R, P) *are given by*

$$\underline{SVNM}_P^{\alpha}(A) = \{P(A|R(x)) \geq \alpha | x \in U, y \in V\} = \left\{\frac{\sum_{y\in R(x)} A(y)}{|R(x)|} \geq \alpha | x \in U, y \in V\right\},$$

$$\overline{SVNM}_P^{\beta}(A) = \{P(A|R(x)) > \beta | x \in U, y \in V\} = \left\{\frac{\sum_{y\in R(x)} A(y)}{|R(x)|} > \beta | x \in U, y \in V\right\},$$

where the pair $\left(\underline{SVNM}_P^{\alpha}(A), \overline{SVNM}_P^{\beta}(A)\right)$ *is named a PRSVNM over two universes. Moreover, by virtue of the above approximations, the positive region, negative region and boundary region of A are further given by*

$$POS_{SVNM}(A, \alpha, \beta) = \underline{SVNM}_P^{\alpha}(A) = \left\{\frac{\sum_{y\in R(x)} A(y)}{|R(x)|} \geq \alpha | x \in U, y \in V\right\},$$

$$NEG_{SVNM}(A, \alpha, \beta) = U - \overline{SVNM}_P^{\beta}(A) = \left\{\frac{\sum_{y\in R(x)} A(y)}{|R(x)|} < \beta | x \in U, y \in V\right\},$$

$$BND_{SVNM}(A, \alpha, \beta) = \overline{SVNM}_P^{\beta}(A) - \underline{SVNM}_P^{\alpha}(A) = \left\{\beta < \frac{\sum_{y\in R(x)} A(y)}{|R(x)|} < \alpha | x \in U, y \in V\right\}.$$

3.4. SVNPRMs over Two Universes

In the previous descriptions, we explore rough single-valued neutrosophic multisets on a probabilistic approximation space over two universes. However, the probabilistic approximation space over two universes (U, V, R, P) could only express the crisp relation of the elements from the universe U to the universe V. Since there exist lots of relations based on SVNMs, it is necessary to study SVNPRMs over two universes. In what follows, we extend the general probabilistic approximation space over two universes to the probabilistic approximation space over two universes on SVNMs,

Definition 12. *Suppose that U, V are two universes of discourse,* $R \in SVNM(U \times V)$ *is a relation on SVNMs, and P is a probabilistic measure based on the* σ *algebra. Then* (U, V, R, P) *is named a probabilistic approximation space over two universes based on SVNMs.*

Definition 13. *Suppose that* (U, V, R, P) *is a probabilistic approximation space over two universes based on SVNMs. For any* $A \in SVNM(V)$, $x \in U$, $y \in V$, *the conditional probability* $P(A|R(x,y))$ *is given by*

$$P(A|R(x,y)) = \frac{\sum_{y\in V} A(y) R(x,y)}{\sum_{y\in V} R(x,y)}.$$

By virtue of the proposed conditional probability $P(A|R(x,y))$, the definition of SVNPRMs over two universes is put forward as follows.

Definition 14. *Suppose that (U, V, R, P) is a probabilistic approximation space over two universes based on SVNMs. For any $0 \leq \beta \leq \alpha \leq 1$, $A \in SVNM(V)$, $x \in U$, $y \in V$, the lower and upper approximations of A with respect to (U, V, R, P) are given by*

$$\underline{SVNMR_p^\alpha}(A) = \{P(A|R(x,y)) \geq \alpha | x \in U, y \in V\} = \left\{\frac{\sum_{y\in V} A(y) R(x,y)}{\sum_{y\in V} R(x,y)} \geq \alpha | x \in U, y \in V\right\},$$

$$\overline{SVNMR_p^\beta}(A) = \{P(A|R(x,y)) > \beta | x \in U, y \in V\} = \left\{\frac{\sum_{y\in V} A(y) R(x,y)}{\sum_{y\in V} R(x,y)} > \beta | x \in U, y \in V\right\},$$

where the pair $\left(\underline{SVNMR_p^\alpha}(A), \overline{SVNMR_p^\beta}(A)\right)$ is named a SVNPRM over two universes. Moreover, by virtue of the above approximations, the positive region, negative region and boundary region of A are further given by

$$POS_{SVNMR}(A, \alpha, \beta) = \underline{SVNMR_p^\alpha}(A) = \left\{\frac{\sum_{y\in V} A(y)R(x,y)}{\sum_{y\in V} R(x,y)} \geq \alpha | x \in U, y \in V\right\},$$
$$NEG_{SVNMR}(A, \alpha, \beta) = U - \overline{SVNMR_p^\beta}(A) = \left\{\frac{\sum_{y\in V} A(y)R(x,y)}{\sum_{y\in V} R(x,y)} \leq \beta | x \in U, y \in V\right\},$$
$$BND_{SVNMR}(A, \alpha, \beta) = \overline{SVNMR_p^\beta}(A) - \underline{SVNMR_p^\alpha}(A) = \left\{\beta < \frac{\sum_{y\in V} A(y)R(x,y)}{\sum_{y\in V} R(x,y)} < \alpha | x \in U, y \in V\right\}.$$

It is noted that the parameters α and β in the above definitions can be determined in advance by experts based on their experience and knowledge in realistic decision making situations.

According to the above definitions, a simple example is offered as follows.

Example 1. *Suppose that $U = \{x_1.x_2, x_3\}$ and $V = \{x_1.x_2, x_3\}$ are two universes, $R \in SVNM(U \times V)$ is a relation on SVNMs based on two related universes, where*

$$R = \left\{\begin{array}{cccc} & y_1 & y_2 & y_3 \\ x_1 & \left\langle\begin{array}{l}(0.6,0.5,0.4),\\(0.3,0.2,0.2),\\((0.2,0.1,0.1))\end{array}\right\rangle & \left\langle\begin{array}{l}(0.4,0.3,0.2),\\(0.7,0.6,0.6),\\((0.8,0.7,0.7))\end{array}\right\rangle & \left\langle\begin{array}{l}(0.8,0.8,0.7),\\(0.2,0.2,0.1),\\((0.1,0.1,0.0))\end{array}\right\rangle \\ x_2 & \left\langle\begin{array}{l}(0.3,0.3,0.2),\\(0.4,0.4,0.3),\\((0.6,0.5,0.4))\end{array}\right\rangle & \left\langle\begin{array}{l}(0.8,0.7,0.6),\\(0.3,0.2,0.1),\\((0.2,0.2,0.1))\end{array}\right\rangle & \left\langle\begin{array}{l}(0.5,0.5,0.4),\\(0.3,0.3,0.2),\\((0.6,0.6,0.5))\end{array}\right\rangle \\ x_3 & \left\langle\begin{array}{l}(0.7,0.7,0.6),\\(0.3,0.2,0.1),\\((0.1,0.1,0.0))\end{array}\right\rangle & \left\langle\begin{array}{l}(0.4,0.4,0.3),\\(0.5,0.5,0.4),\\((0.6,0.6,0.5))\end{array}\right\rangle & \left\langle\begin{array}{l}(0.3,0.2,0.2),\\(0.6,0.6,0.5),\\((0.9,0.9,0.7))\end{array}\right\rangle \end{array}\right\}.$$

A SVNM A in the universe V is provided below.

$$A = \{\langle y_1, (0.5), (0.5), (0.2)\rangle, \langle y_2, (0.8), (0.3), (0.1)\rangle, \langle y_3, (0.3), (0.6), (0.7)\rangle\}.$$

By virtue of Definition 13, we obtain

$$P(A|R(x_1,y)) = \frac{\sum_{y\in V} A(y) R(x_1,y)}{\sum_{y\in V} R(x_1,y)} = \langle (0.67,0.61,0.55), (0.32,0.28,0.27), (0.20,0.14,0.14)\rangle,$$

$$P(A|R(x_2,y)) = \frac{\sum_{y\in V} A(y) R(x_2,y)}{\sum_{y\in V} R(x_2,y)} = \langle (0.80,0.76,0.73), (0.23,0.20,0.16), (0.10,0.09,0.07)\rangle,$$

$$P(A|R(x_3,y)) = \frac{\sum_{y\in V} A(y) R(x_3,y)}{\sum_{y\in V} R(x_3,y)} = \langle (0.69,0.68,0.64), (0.29,0.29,0.24), (0.13,0.13,0.10)\rangle.$$

If we assume $\alpha = \langle(0.76, 0.74, 0.72), (0.28, 0.27, 0.26), (0.13, 0.12, 0.12)\rangle$ *and* $\beta = \langle(0.68, 0.65, 0.6), (0.32, 0.31, 0.26), (0.18, 0.14, 0.12)\rangle$, *we obtain* $\underline{SVNMR}_P^{\alpha}(A) = \{x_2\}$ *and* $\overline{SVNMR}_P^{\beta}(A) = \{x_2, x_3\}$. *Then it is not difficult to obtain* $POS_{SVNMR}(A, \alpha, \beta) = \{x_2\}$, $NEG_{SVNMR}(A, \alpha, \beta) = \{x_1\}$ *and* $BND_{SVNMR}(A, \alpha, \beta) = \{x_3\}$.

In what follows, we show some common properties that are owned by the presented SVNPRMs over two universes.

Proposition 1. *Suppose that* (U, V, R, P) *is a probabilistic approximation space over two universes based on SVNMs. For any* $0 \leq \beta \leq \alpha \leq 1$, $A \in SVNM(V)$, $x \in U$, $y \in V$, *we have the following propositions:*

1. $A \subseteq B \Rightarrow \underline{SVNMR}_P^{\alpha}(A) \subseteq \underline{SVNMR}_P^{\alpha}(B)$, $\overline{SVNMR}_P^{\beta}(A) \subseteq \overline{SVNMR}_P^{\beta}(B)$;
2. $\underline{SVNMR}_P^{\alpha}(\emptyset) = \emptyset$, $\overline{SVNMR}_P^{\beta}(V) = U$;
3. $\underline{SVNMR}_P^{\alpha}(A) \subseteq \overline{SVNMR}_P^{\beta}(A)$;
4. $\underline{SVNMR}_P^{\alpha}(A \cap B) \subseteq \underline{SVNMR}_P^{\alpha}(A) \cap \underline{SVNMR}_P^{\alpha}(B)$, $\overline{SVNMR}_P^{\beta}(A \cup B) \supseteq \overline{SVNMR}_P^{\beta}(A) \cup \overline{SVNMR}_P^{\beta}(B)$;
5. $\alpha_1 \leq \alpha_2 \Rightarrow \underline{SVNMR}_P^{\alpha_2}(A) \subseteq \underline{SVNMR}_P^{\alpha_1}(A)$, $\beta_1 \leq \beta_2 \Rightarrow \overline{SVNMR}_P^{\beta_2}(A) \subseteq \overline{SVNMR}_P^{\beta_1}(A)$.

The detailed proofs of Proposition 1 are included in the Appendix A at the end of the paper.

4. Medical Diagnosis Based on SVNPRMs over Two Universes

4.1. Medical Diagnosis Model

In the following, we explore a reasonable and effective medical diagnosis approach by means of SVNPRMs over two universes. As pointed out in the earlier statements, SVNPRMs over two universes take advantage of SVNMs and PRSs at the same time. For one thing, SVNMs are able to provide medical experts with a more powerful tool to describe a complicated medical diagnosis knowledge base, i.e., the SVNMs information could not only handle the uncertain situation well, but also record the diagnostic information at different time intervals reasonably. For another, PRSs-based three-way decisions could further overcome the drawbacks of classical rough sets, and provide a robust decision result by considering the decision risk, hence PRSs-based three-way decisions act as an effectual way to analyze the above SVNMs information.

In the medical diagnosis procedures, suppose that $U = \{x_1, x_2, \ldots, x_m\}$ is a diagnosis set, and $V = \{y_1, y_2, \ldots, y_n\}$ is a symptom set. Then based on the universe U and the universe V, medical experts are likely to construct the relationship between the diagnosis set and the symptom set by means of the SVNMs information, which is represented by $R \in SVNM(U \times V)$. Moreover, suppose that P is a probabilistic measure based on the σ algebra. Hence we establish a medical diagnosis probabilistic approximation space over two universes based on SVNMs (U, V, R, P).

For a given patient, the symptoms of the patient are expressed by a SVNM A in the universe V. Next suppose that α and β are the thresholds provided in advance by medical experts according to their experience and knowledge in realistic medical diagnosis situations. In light of the above statements, it is not difficult to calculate the two approximations of A in terms of (U, V, R, P), which are denoted by $\underline{SVNMR}_P^{\alpha}(A)$ and $\overline{SVNMR}_P^{\beta}(A)$. Further, we obtain the positive region, negative region and boundary region of A according to Definition 14, which are expressed as $POS_{SVNMR}(A, \alpha, \beta)$, $NEG_{SVNMR}(A, \alpha, \beta)$ and $BND_{SVNMR}(A, \alpha, \beta)$. Lastly, the following medical diagnosis rule could be concluded by virtue of the three-way decisions theory originated by Yao [49,50].

(*P*) If $x_i \in POS_{SVNMR}(A, \alpha, \beta)$, $i = 1, 2, \ldots, m$, then x_i is the determined diagnostic conclusion;
(*N*) If $x_i \in NEG_{SVNMR}(A, \alpha, \beta)$, $i = 1, 2, \ldots, m$, then x_i is the excluded diagnostic conclusion;

(*B*) If $x_i \in BND_{SVNMR}(A,\alpha,\beta), i = 1,2,\ldots,m$, then medical experts are unable to confirm whether x_i is the determined or excluded diagnostic conclusion, they need more additional medical examinations to confirm the final diagnostic conclusion

4.2. Algorithm for Medical Diagnosis Model

To summarize, we conclude the medical diagnosis approach for a given patient based on SVNPRMs over two universes (Algorithm 1).

Algorithm 1 Medical diagnosis based on SVNPRMs over two universes.

Require: (U,V,R,P) and A.
Ensure: The determined diagnostic conclusion.
Step 1. Calculating the conditional probability $P(A\,|R(x,y))$;
Step 2. Presenting the thresholds α and β;
Step 3. Calculating the lower and upper approximations of A in terms of (U,V,R,P). i.e., $\underline{SVNMR}_p^{\alpha}(A)$ and $\overline{SVNMR}_p^{\beta}(A)$;
Step 4. Calculating the positive region, negative region and boundary region of A, i.e., $POS_{SVNMR}(A,\alpha,\beta)$, $NEG_{SVNMR}(A,\alpha,\beta)$ and $BND_{SVNMR}(A,\alpha,\beta)$;
Step 5. Confirming the determined diagnostic conclusion on the basis of the proposed medical diagnosis rule (P), (N) and (B).

4.3. An Illustrative Example

In the following, a case study within the context of medical diagnosis is illustrated as the demonstration of the presented medical diagnosis approach based on SVNPRMs over two universes. The content of the illustrative example (adapted from [26]) is shown as follows.

Let $U = \{x_1,x_2,x_3,x_4\}$ be a diagnosis set (where $x_i\,(i = 1,2,3,4)$ denotes "viral fever", "tuberculosis", "typhoid", and "throat disease", respectively), and $V = \{y_1,y_2,y_3,y_4,y_5\}$ be a symptom set (where $y_i\,(i = 1,2,3,4,5)$ denotes "temperature", "cough", "throat pain", "headache", and "body pain", respectively). Let $R \in SVNM(U \times V)$ be a relation on SVNMs based on two related universes, which is recorded in the Table 1.

Table 1. Relationship between the considered diseases and symptoms represented by SVNMs.

R	y_1	y_2	y_3	y_4	y_5
x_1	$\langle(0.8),(0.1),(0.1)\rangle$	$\langle(0.2),(0.7),(0.1)\rangle$	$\langle(0.3),(0.5),(0.2)\rangle$	$\langle(0.5),(0.3),(0.2)\rangle$	$\langle(0.5),(0.4),(0.1)\rangle$
x_2	$\langle(0.2),(0.7),(0.1)\rangle$	$\langle(0.9),(0.0),(0.1)\rangle$	$\langle(0.7),(0.2),(0.1)\rangle$	$\langle(0.6),(0.3),(0.1)\rangle$	$\langle(0.7),(0.2),(0.1)\rangle$
x_3	$\langle(0.5),(0.3),(0.2)\rangle$	$\langle(0.3),(0.5),(0.2)\rangle$	$\langle(0.2),(0.7),(0.1)\rangle$	$\langle(0.2),(0.6),(0.2)\rangle$	$\langle(0.4),(0.4),(0.2)\rangle$
x_4	$\langle(0.1),(0.7),(0.2)\rangle$	$\langle(0.3),(0.6),(0.1)\rangle$	$\langle(0.8),(0.1),(0.1)\rangle$	$\langle(0.1),(0.8),(0.1)\rangle$	$\langle(0.1),(0.8),(0.1)\rangle$

Suppose that the symptoms of the patient is denoted by a SVNM A in the universe V below.

$$A = \left\{ \begin{array}{l} \left\langle \begin{array}{l} y_1,(0.8,0.6,0.5), \\ (0.3,0.2,0.1), \\ (0.4,0.2,0.1) \end{array} \right\rangle, \left\langle \begin{array}{l} y_2,(0.5,0.4,0.3), \\ (0.4,0.4,0.3), \\ (0.6,0.3,0.4) \end{array} \right\rangle, \left\langle \begin{array}{l} y_3,(0.2,0.1,0.0), \\ (0.3,0.2,0.2), \\ (0.8,0.7,0.7) \end{array} \right\rangle, \\ \left\langle \begin{array}{l} y_4,(0.7,0.6,0.5), \\ (0.3,0.2,0.1), \\ (0.4,0.3,0.2) \end{array} \right\rangle, \left\langle \begin{array}{l} y_5,(0.4,0.3,0.2), \\ (0.6,0.5,0.5), \\ (0.6,0.4,0.4) \end{array} \right\rangle \end{array} \right\}.$$

According to the steps of the proposed medical diagnosis approach (Algorithm 1), we first calculate the conditional probability as follows.

$$P(A\,|R(x_1,y)) = \frac{\sum_{y\in V} A(y)\,R(x_1,y)}{\sum_{y\in V} R(x_1,y)} = \langle (0.87,0.74,0.64),(0.07,0.04,0.02),(0.08,0.02,0.01)\rangle,$$
$$P(A\,|R(x_2,y)) = \frac{\sum_{y\in V} A(y)\,R(x_2,y)}{\sum_{y\in V} R(x_2,y)} = \langle (0.84,0.74,0.61),(0.05,0.03,0.02),(0.07,0.01,0.01)\rangle,$$
$$P(A\,|R(x_3,y)) = \frac{\sum_{y\in V} A(y)\,R(x_3,y)}{\sum_{y\in V} R(x_3,y)} = \langle (0.75,0.62,0.5),(0.13,0.09,0.06),(0.1,0.03,0.02)\rangle,$$
$$P(A\,|R(x_4,y)) = \frac{\sum_{y\in V} A(y)\,R(x_4,y)}{\sum_{y\in V} R(x_4,y)} = \langle (0.46,0.34,0.22),(0.15,0.1,0.08),(0.08,0.02,0.01)\rangle.$$

Then we present the thresholds α and β below.

$$\alpha = \langle (0.85,0.74,0.63),(0.06,0.03,0.02),(0.075,0.02,0.01)\rangle,$$
$$\beta = \langle (0.5,0.4,0.3),(0.13,0.1,0.07),(0.09,0.02,0.01)\rangle.$$

Next the two approximations of A in terms of (U,V,R,P) could be obtained.

$$\underline{SVNMR}_P^{\alpha}(A) = \{P(A\,|R(x,y)) \geq \alpha\,|x\in U, y\in V\} = \{x_1\},$$
$$\overline{SVNMR}_P^{\beta}(A) = \{P(A\,|R(x,y)) > \beta\,|x\in U, y\in V\} = \{x_1,x_2,x_3\}.$$

Based on the above calculated data, we further obtain

$$POS_{SVNMR}(A,\alpha,\beta) = \underline{SVNMR}_P^{\alpha}(A) = \{x_1\},$$
$$NEG_{SVNMR}(A,\alpha,\beta) = U - \overline{SVNMR}_P^{\beta}(A) = \{x_4\},$$
$$BND_{SVNMR}(A,\alpha,\beta) = \overline{SVNMR}_P^{\beta}(A) - \underline{SVNMR}_P^{\alpha}(A) = \{x_2,x_3\}.$$

Finally, we could obtain the diagnostic conclusion by means of the proposed medical diagnosis rule (P), (N) and (B).

- (P) The patient is suffering from viral fever, medical experts need to pay close attention to the diagnosis;
- (N) The same patient shows no signs of having throat disease, which does not need close attention at the current stage;
- (B) Medical experts are unable to confirm whether the considered patient is suffering from tuberculosis and typhoid or not due to insufficient diagnostic information, they might organize an expert consultation to confirm the determined diagnostic conclusion at a later stage.

4.4. *Comparative Analysis*

In what follows, in order to show the applicability and validity of the constructed medical diagnosis approach, we compare the approach based on SVNPRMs over two universes with two significant and common approaches (similarity measures and cosine measures) presented in literature [26,28] respectively.

4.4.1. Comparison with Other Approaches in Literature [26]

As presented by Ye et al. [26], suppose that any two SVNMs in the universe $U = \{x_1,x_2,\ldots,x_m\}$ could be expressed by $A = \{\langle x_j, T_A^i(x_j), I_A^i(x_j), F_A^i(x_j)\rangle\,|x_j\in U, i=1,2,\ldots,q\}$ and $B =$

$\{\langle x_j, T_B^i(x_j), I_B^i(x_j), F_B^i(x_j)\rangle \,|x_j \in U, i = 1,2,\ldots,q\}$, then the generalized distance measure between A and B is defined as

$$d_p(A,B) = \left[\frac{1}{m}\sum_{j=1}^{m}\frac{1}{3l_j}\sum_{i=1}^{l_j}\left(\left|T_A^i(x_j) - T_B^i(x_j)\right|^p + \left|I_A^i(x_j) - I_B^i(x_j)\right|^p + \left|F_A^i(x_j) - F_B^i(x_j)\right|^p\right)\right]^{1/p},$$

where $l_j = L(x : A,B) = \max\{L(x : A), L(x : B)\}, j = 1,2,\ldots,m$. Then based on the generalized distance measure between A and B, two similarity measures are defined as

$$\begin{aligned} s_1(A,B) &= 1 - d_p(A,B) \\ &= 1 - \left[\frac{1}{m}\sum_{j=1}^{m}\frac{1}{3l_j}\sum_{i=1}^{l_j}\left(\left|T_A^i(x_j) - T_B^i(x_j)\right|^p + \left|I_A^i(x_j) - I_B^i(x_j)\right|^p + \left|F_A^i(x_j) - F_B^i(x_j)\right|^p\right)\right]^{1/p}, \\ s_2(A,B) &= \frac{1 - d_p(A,B)}{1 + d_p(A,B)} \\ &= \frac{1 - \left[\frac{1}{m}\sum_{j=1}^{m}\frac{1}{3l_j}\sum_{i=1}^{l_j}\left(|T_A^i(x_j) - T_B^i(x_j)|^p + |I_A^i(x_j) - I_B^i(x_j)|^p + |F_A^i(x_j) - F_B^i(x_j)|^p\right)\right]^{1/p}}{1 + \left[\frac{1}{m}\sum_{j=1}^{m}\frac{1}{3l_j}\sum_{i=1}^{l_j}\left(|T_A^i(x_j) - T_B^i(x_j)|^p + |I_A^i(x_j) - I_B^i(x_j)|^p + |F_A^i(x_j) - F_B^i(x_j)|^p\right)\right]^{1/p}}. \end{aligned}$$

According to the above stated similarity measures, if we take $p = 2$, the largest value of similarity measures between the symptoms of the patient and each potential diagnosis could be regarded as the determined diagnostic conclusion. Since the overall ranking result of the similarity measure shows $x_1 \succ x_3 \succ x_2 \succ x_4$, the results of two similarity measures indicate the patient is suffering from viral fever, which is identical with the determined diagnostic conclusion obtained from our proposed approach.

4.4.2. Comparison with Other Approaches in Literature [28]

In literature [28], the authors mainly proposed a novel decision making method based on cosine measures of SVNMs. We also suppose that any two SVNMs in the universe $U = \{x_1, x_2, \ldots, x_m\}$ are described as $A = \{x_j, (p_{A1}, \langle T_{A1}(x_j), I_{A1}(x_j), F_{A1}(x_j)\rangle), (p_{A2}, \langle T_{A2}(x_j), I_{A2}(x_j), F_{A2}(x_j)\rangle), \ldots,$ $(p_{A1}, \langle T_{Ai}(x_j), I_{Ai}(x_j), F_{Ai}(x_j)\rangle)\,|x_j \in U\}$ and $B = \{x_j, (p_{B1}, \langle T_{B1}(x_j), I_{B1}(x_j), F_{B1}(x_j)\rangle),$ $(p_{B2}, \langle T_{B2}(x_j), I_{B2}(x_j), F_{B2}(x_j)\rangle), \ldots, (p_{B1}, \langle T_{Bi}(x_j), I_{Bi}(x_j), F_{Bi}(x_j)\rangle)\,|x_j \in U\}$, where p denotes the repeated times with the same neutrosophic components. Based on that, the cosine measure between two SVNMs A and B is defined as

$$\rho(A,B) = \frac{1}{m}\sum_{i=1}^{m}\cos\left\{\frac{\pi}{6}\left(\begin{array}{l}\left|\prod_{k=1}^{i}(1 - T_{Ak}(x_j))^{p_{Ak}} - \prod_{k=1}^{i}(1 - T_{Bk}(x_j))^{p_{Bk}}\right| + \\ \left|\prod_{k=1}^{i}(I_{Ak}(x_j))^{p_{Ak}} - \prod_{k=1}^{i}(I_{Bk}(x_j))^{p_{Bk}}\right| + \\ \left|\prod_{k=1}^{i}(F_{Ak}(x_j))^{p_{Ak}} - \prod_{k=1}^{i}(F_{Bk}(x_j))^{p_{Bk}}\right|\end{array}\right)\right\},$$

As described in the case study section in literature [28], one customer intends to purchase a car, let $U = \{x_1, x_2, x_3, x_4\}$ be a car set with four possible alternatives. Then we let $V = \{y_1, y_2, y_3, y_4\}$ be an attribute set (where y_i $(i = 1,2,3,4)$ denotes "fuel economy", "price", "comfort", and "safety", respectively). Let $R \in SVNM(U \times V)$ be a relation on SVNMs based on two related universes, which is recorded in the Table 2.

Table 2. Relationship between alternatives and attributes represented by SVNMs for purchasing a car.

R	y_1	y_2	y_3	y_4
x_1	$\langle(0.7,0.5),(0.7,0.3),(0.6,0.2)\rangle$	$\langle(0.5),(0.4),(0.4)\rangle$	$\langle(0.8,0.7),(0.7,0.7),(0.6,0.5)\rangle$	$\langle(0.5,0.1),(0.5,0.2),(0.8,0.7)\rangle$
x_2	$\langle(0.9,0.7),(0.7,0.7),(0.5,0.1)\rangle$	$\langle(0.8),(0.7),(0.6)\rangle$	$\langle(0.9,0.9),(0.6,0.6),(0.4,0.4)\rangle$	$\langle(0.5,0.5),(0.2,0.1),(0.9,0.7)\rangle$
x_3	$\langle(0.6,0.3),(0.4,0.3),(0.7,0.2)\rangle$	$\langle(0.2),(0.2),(0.2)\rangle$	$\langle(0.9,0.6),(0.5,0.5),(0.5,0.2)\rangle$	$\langle(0.7,0.4),(0.5,0.2),(0.3,0.2)\rangle$
x_4	$\langle(0.9,0.8),(0.7,0.6),(0.2,0.1)\rangle$	$\langle(0.5),(0.3),(0.2)\rangle$	$\langle(0.5,0.1),(0.7,0.4),(0.5,0.2)\rangle$	$\langle(0.8,0.8),(0.4,0.4),(0.2,0.2)\rangle$

Suppose that the ideal attribute set is denoted by a SVNM A in the universe V below.

$$A = \left\{ \begin{array}{l} \langle y_1,(0.98),(0.12),(0.02)\rangle, \langle y_2,(0.7),(0.2),(0.2)\rangle, \\ \langle y_3,(0.99),(0.16),(0.1)\rangle, \langle y_4,(0.82),(0.02),(0.06)\rangle \end{array} \right\}.$$

Based on the steps of our proposed approach, we first calculate the conditional probability as follows.

$$\begin{aligned}
P(A\,|R(x_1,y)) &= \frac{\sum_{y\in V} A(y)\,R(x_1,y)}{\sum_{y\in V} R(x_1,y)} = \langle(1,0.97),(0.05,0.02),(0.06,0.02)\rangle, \\
P(A\,|R(x_2,y)) &= \frac{\sum_{y\in V} A(y)\,R(x_2,y)}{\sum_{y\in V} R(x_2,y)} = \langle(1,0.99),(0.02,0.02),(0.04,0.01)\rangle, \\
P(A\,|R(x_3,y)) &= \frac{\sum_{y\in V} A(y)\,R(x_3,y)}{\sum_{y\in V} R(x_3,y)} = \langle(1,0.98),(0.03,0.01),(0.03,0)\rangle, \\
P(A\,|R(x_4,y)) &= \frac{\sum_{y\in V} A(y)\,R(x_4,y)}{\sum_{y\in V} R(x_4,y)} = \langle(0.99,0.97),(0.04,0.03),(0,0)\rangle.
\end{aligned}$$

Then we present the thresholds α and β below.

$$\begin{aligned}
\alpha &= \langle(1,0.99),(0.02,0.02),(0.025,0.02)\rangle, \\
\beta &= \langle(0.98,0.97),(0.03,0.02),(0.02,0.01)\rangle.
\end{aligned}$$

Next we further calculate the two approximations of A in terms of (U,V,R,P) and the three divided regions.

$$\begin{aligned}
&\underline{SVNMR_p^{\alpha}}(A) = \{P(A\,|R(x,y)) \geq \alpha\,|x \in U, y \in V\} = \{x_3\}, \\
&\overline{SVNMR_p^{\beta}}(A) = \{P(A\,|R(x,y)) > \beta\,|x \in U, y \in V\} = \{x_2, x_3, x_4\}, \\
&POS_{SVNMR}(A,\alpha,\beta) = \underline{SVNMR_p^{\alpha}}(A) = \{x_3\}, \\
&NEG_{SVNMR}(A,\alpha,\beta) = U - \overline{SVNMR_p^{\beta}}(A) = \{x_1\}, \\
&BND_{SVNMR}(A,\alpha,\beta) = \overline{SVNMR_p^{\beta}}(A) - \underline{SVNMR_p^{\alpha}}(A) = \{x_2, x_4\}.
\end{aligned}$$

Lastly, we could obtain the decision making recommendation by means of the proposed three-way decisions rule (P), (N) and (B).

- (P) The customer is suggested to purchase the third car;
- (N) The same customer is not suggested to by the first car at present;
- (B) The same customer is not sure whether the second car and the forth car are the ideal selections, he or she might collect some additional information to make a final conclusion at a later stage.

According to the above stated cosine measures of SVNMs, the largest value of cosine measures between the alternatives and attributes for purchasing a car could be regarded as the final decision making conclusion. It is noted that the overall ranking result of the cosine measure shows $x_3 \succ x_4 \succ$

$x_1 \succ x_2$, which indicates the customer should buy the third car, the decision making result is also the same as the determined decision making conclusion obtained from our proposed approach.

In conclusion, it is noted that in the above two comparative analyses, though the final decision making result by virtue of our proposed method is the same as the approaches of similarity measures and cosine measures, the overall ranking result by using our proposed method is slightly different from the approaches of similarity measures and cosine measures based on SVNMs. To be specific, if we rank alternatives according to the corresponding values of conditional probability in our presented approach, all the results are shown in the following Tables 3 and 4.

Table 3. The ranking orders by utilizing two different methods in the first comparison analysis.

Different Methods	Ranking Results of Alternatives	The Best Alternative
Method 1 based on similarity measures in [26]	$x_1 \succ x_3 \succ x_2 \succ x_4$	x_1
The proposed method	$x_1 \succ x_2 \succ x_3 \succ x_4$	x_1

Table 4. The ranking orders by utilizing two different methods in the second comparison analysis.

Different Methods	Ranking Results of Alternatives	The Best Alternative
Method 2 based on cosine measures in [28]	$x_3 \succ x_4 \succ x_1 \succ x_2$	x_3
The proposed method	$x_3 \succ x_2 \succ x_4 \succ x_1$	x_3

Compared with the approaches of similarity measures and cosine measures based on SVNMs, the approaches of similarity measures and cosine measures lack the ability of processing decision risks and noisy decision making data. In addition, the proposed approach based on SVNPRMs over two universes offers a reasonable and efficient tool for analyzing the SVNMs information, which not only considers the decision risks by introducing a three-way decision tactic, but also enhances the performance of handling various noisy SVNMs data by introducing the thresholds. Moreover, our presented medical diagnosis approach could be seen as another similarity measures approach, i.e., the conditional probability expresses the similarity of the symptoms of the patient A with the relationship between the considered diseases and symptoms, by further adding the thresholds α and β, the ability of processing risk preferences of medical experts could be improved. Thus, the constructed approach based on SVNPRMs over two universes is able to enhance the reliability and accuracy of medical diagnosis efficiently.

5. Conclusions

In this article, we mainly investigate a PRSs-based method to analyze the SVNMs information within the medical diagnosis context. Specifically, after revisiting several fundamental concepts about SVNMs and PRSs, we first put forward the notion of SVNRMs over two universes and PRSVNMs over two universes. Based on that, the notion of SVNPRMs over two universes is further established. Then some common propositions of the presented SVNPRMs over two universes are further explored. Next, based on the proposed SVNPRMs over two universes, we construct a medical diagnosis approach by means of the three-way decisions strategy. At last, an illustrative case analysis along with a comparative study is carried out to reveal the practicability and effectiveness of the constructed medical diagnosis approach. In future work, it is necessary to establish some more PRSs-based theoretical models based on neutrosophic triplet structures and neutrosophic duplet structures, and it is also meaningful to apply other valid decision making tools to handle various complicated decision making situations.

Author Contributions: C.Z. conducted the modelings and wrote the paper. D.L. provided innovation points of the work. S.B. and A.K. Sangaiah offered plenty of advice for enhancing the readability of the work. All authors approved the publication of the work.

Funding: This work was supported in part by the National Natural Science Foundation of China (Nos. 61672331, 61432011, 61573231 and U1435212), and the Natural Science Foundation of Shanxi Province (Nos. 201601D021076, 2015091001-0102 and 201601D021072).

Acknowledgments: The authors would like to thank anonymous reviewers for their valuable comments and suggestions which have significantly improved the quality and presentation of this paper. This work was supported in part by the National Natural Science Foundation of China (Nos. 61672331, 61432011, 61573231 and U1435212), and the Natural Science Foundation of Shanxi Province (Nos. 201601D021076, 2015091001-0102 and 201601D021072).

Conflicts of Interest: The authors declare no conflict of interest.

Appendix A

The proofs of Proposition 1 are listed as follows.

Proof.

1. Since $A \subseteq B$, according to Definition 14, we have

   $\underline{SVNMR}_p^{\alpha}(A) = \left\{\frac{\sum_{y\in V} A(y)R(x,y)}{\sum_{y\in V} R(x,y)} \geq \alpha \,|x \in U, y \in V\right\} \subseteq \left\{\frac{\sum_{y\in V} B(y)R(x,y)}{\sum_{y\in V} R(x,y)} \geq \alpha \,|x \in U, y \in V\right\} = \underline{SVNMR}_p^{\alpha}(B)$.

   Thus we obtain $A \subseteq B \Rightarrow \underline{SVNMR}_p^{\alpha}(A) \subseteq \underline{SVNMR}_p^{\alpha}(B)$. Similarly, we could also obtain $\overline{SVNMR}_p^{\beta}(A) \subseteq \overline{SVNMR}_p^{\beta}(B)$.

2. $\underline{SVNMR}_p^{\alpha}(\emptyset) = \left\{\frac{\sum_{y\in V} \emptyset R(x,y)}{\sum_{y\in V} R(x,y)} \geq \alpha \,|x \in U, y \in V\right\} = \emptyset$,

   $\overline{SVNMR}_p^{\beta}(V) = \left\{\frac{\sum_{y\in V} VR(x,y)}{\sum_{y\in V} R(x,y)} > \beta \,|x \in U, y \in V\right\} = U$.

   Hence $\underline{SVNMR}_p^{\alpha}(\emptyset) = \emptyset$ and $\overline{SVNMR}_p^{\beta}(V) = U$ could be obtained.

3. Since $0 \leq \beta \leq \alpha \leq 1$, it is not difficult to obtain

   $\underline{SVNMR}_p^{\alpha}(A) = \left\{\frac{\sum_{y\in V} A(y)R(x,y)}{\sum_{y\in V} R(x,y)} \geq \alpha \,|x \in U, y \in V\right\} \subseteq \left\{\frac{\sum_{y\in V} B(y)R(x,y)}{\sum_{y\in V} R(x,y)} > \beta \,|x \in U, y \in V\right\} = \overline{SVNMR}_p^{\beta}(B)$.

   Therefore, $\underline{SVNMR}_p^{\alpha}(A) \subseteq \overline{SVNMR}_p^{\beta}(A)$ could be obtained.

4. If $\underline{SVNMR}_p^{\alpha}(A \cap B)$ holds, we have $P\left((A \cap B)\,|R(x,y)\right) \geq \alpha$, then it is not difficult to obtain

   $\alpha \leq \frac{\sum_{y\in V} (A\cap B)(y)R(x,y)}{\sum_{y\in V} R(x,y)} \leq \frac{\sum_{y\in V} A(y)R(x,y)}{\sum_{y\in V} R(x,y)}$ and $\alpha \leq \frac{\sum_{y\in V} (A\cap B)(y)R(x,y)}{\sum_{y\in V} R(x,y)} \leq \frac{\sum_{y\in V} B(y)R(x,y)}{\sum_{y\in V} R(x,y)}$.

   Hence we obtain $\underline{SVNMR}_p^{\alpha}(A \cap B) \subseteq \underline{SVNMR}_p^{\alpha}(A) \cap \underline{SVNMR}_p^{\alpha}(B)$. In an identical fashion, $\overline{SVNMR}_p^{\beta}(A \cup B) \supseteq \overline{SVNMR}_p^{\beta}(A) \cup \overline{SVNMR}_p^{\beta}(B)$ could also be obtained.

5. Since $\alpha_1 \leq \alpha_2$, we have

   $\underline{SVNMR}_p^{\alpha_2}(A) = \left\{\frac{\sum_{y\in V} A(y)R(x,y)}{\sum_{y\in V} R(x,y)} \geq \alpha_2 \,|x \in U, y \in V\right\} \subseteq \left\{\frac{\sum_{y\in V} A(y)R(x,y)}{\sum_{y\in V} R(x,y)} \geq \alpha_1 \,|x \in U, y \in V\right\} = \underline{SVNMR}_p^{\alpha_1}(A)$.

   Hence we have $\alpha_1 \leq \alpha_2 \Rightarrow \underline{SVNMR}_p^{\alpha_2}(A) \subseteq \underline{SVNMR}_p^{\alpha_1}(A)$, and $\beta_1 \leq \beta_2 \Rightarrow \overline{SVNMR}_p^{\beta_2}(A) \subseteq \overline{SVNMR}_p^{\beta_1}(A)$ could be proved in a similar way.

□

References

1. Mak, D.K. A fuzzy probabilistic method for medical diagnosis. *J. Med. Syst.* **2015**, *39*, 26. [CrossRef] [PubMed]
2. Le, H.S.; Thong, N.T. Intuitionistic fuzzy recommender systems: An effective tool for medical diagnosis. *Knowl. Based Syst.* **2015**, *74*, 133–150.

3. Choi, H.; Han, K.; Choi, K.; Ahn, N. A fuzzy medical diagnosis based on quantiles of diagnostic measures. *J. Intell. Fuzzy Syst.* **2016**, *31*, 3197–3202. [CrossRef]
4. Zadeh, L.A. Fuzzy sets. *Inform. Control* **1965**, *8*, 338–353. [CrossRef]
5. Bustince, H.; Barrenechea, E.; Pagola, M.; Fernandez, J.; Xu, Z.; Bedregal, B.; Montero, J.; Hagras, H.; Herrera, F.; De Baets, B. A historical account of types of fuzzy sets and their relationships. *IEEE. Trans. Fuzzy Syst.* **2016**, *24*, 179–194. [CrossRef]
6. Atanassov, K.T. Intuitionistic fuzzy sets. *Fuzzy Set. Syst.* **1986**, *20*, 87–96. [CrossRef]
7. Smarandache, F. *A Unifying Field in Logics. Neutrosophy: Neutrosophic Probability, Set and Logic*; American Research Press: Rehoboth, DE, USA, 1998.
8. Smarandache, F. Neutrosophic logic-generalization of the intuitionistic fuzzy logic. In *Extractive Metallurgy of Nickel Cobalt and Platinum Group Metals*; Elsevier: New York, NY, USA, 2012; Volume 369, pp. 49–53.
9. Rogatko, A.; Smarandache, F.; Sunderraman, R. A neutrosophic description logic. *New Math. Natl. Comput.* **2012**, *4*, 273–290.
10. Kavitha, B.; Karthikeyan, S.; Maybell, S. An ensemble design of intrusion detection system for handling uncertainty using neutrosophic logic classifier. *Knowl. Based Syst.* **2012**, *28*, 88–96. [CrossRef]
11. Wang, H.B.; Smarandache, F.; Zhang, Y.Q.; Sunderraman, R. *Interval Neutrosophic Sets and Logic: Theory and Applications in Computing*; Hexis: Phoenix, AZ, USA, 2005; pp. 1–87.
12. Broumi, S.; Smarandache, F.; Talea, M.; Bakali, A. An introduction to bipolar single valued neutrosophic graph theory. *Appl. Mech. Mater.* **2016**, *841*, 184–191. [CrossRef]
13. Broumi, S.; Smarandache, F.; Talea, M.; Bakali, A. Single valued neutrosophic graphs: Degree, order and size. In Proceedings of the IEEE International Conference on Fuzzy Systems, Vancouver, BC, Canada, 24–29 July 2016.
14. Zhang, C.; Zhai, Y.H.; Li, D.Y.; Mu, Y. Steam turbine fault diagnosis based on single-valued neutrosophic multigranulation rough sets over two universes. *J. Intell. Fuzzy Syst.* **2016**, *31*, 2829–2837. [CrossRef]
15. Chen, J.Q.; Ye, J. Some Single-valued neutrosophic dombi weighted aggregation operators for multiple attribute decision-making. *Symmetry* **2017**, *9*, 82. [CrossRef]
16. Thanh, N.D.; Ali, M.; Le, H.S. A novel clustering algorithm in a neutrosophic recommender system for medical diagnosis. *Cogn. Comput.* **2017**, *9*, 526–544. [CrossRef]
17. Ali, M.; Smarandache, F. Complex neutrosophic set. *Neural Comput. Appl.* **2017**, *28*, 1817–1834. [CrossRef]
18. Li, X.; Zhang, X.H. Single-valued neutrosophic hesitant fuzzy choquet aggregation operators for multi-attribute decision making. *Symmetry* **2018**, *10*, 50. [CrossRef]
19. Abdel-Basset, M.; Mohamed, M.; Smarandache, F. An extension of neutrosophic AHP-SWOT analysis for strategic planning and decision-making. *Symmetry* **2018**, *10*, 116. [CrossRef]
20. Wang, Y.; Liu, P. Linguistic neutrosophic generalized partitioned bonferroni mean operators and their application to multi-attribute group decision making. *Symmetry* **2018**, *10*, 160. [CrossRef]
21. Zhang, X.; Bo, C.; Smarandache, F.; Park, C. New operations of totally dependent-neutrosophic sets and totally dependent-neutrosophic soft sets. *Symmetry* **2018**, *10*, 187. [CrossRef]
22. Ye, S.; Ye, J. Dice similarity measure between single valued neutrosophic multisets and its application in medical diagnosis. *Neutrosophic Sets Syst.* **2014**, *6*, 48–53.
23. Yager, R.R. On the theory of bags. *Int. J. Gen. Syst.* **1986**, *13*, 23–37. [CrossRef]
24. Smarandache, F. *n*-Valued Refined Neutrosophic Logic and Its Applications in Physics. *Prog. Phys.* **2013**, *4*, 143–146.
25. Chatterjee, R.; Majumdar, P.; Samanta, S.K. Single valued neutrosophic multisets. *Ann. Fuzzy Math. Inform.* **2015**, *10*, 499–514.
26. Ye, S.; Fu, J.; Ye, J. Medical diagnosis using distance-based similarity measures of single valued neutrosophic multisets. *Neutrosophic Sets Syst.* **2015**, *7*, 47–52.
27. Broumi, S.; Deli, I.; Smarandache, F. Relations on neutrosophic multi sets with properties. *arXiv* **2015**, arxiv:1506.04025. [CrossRef]
28. Fan, C.X.; Fan, E.; Ye, J. The cosine measure of single-valued neutrosophic multisets for multiple attribute decision-making. *Symmetry* **2018**, *10*, 154. [CrossRef]
29. Zhang, C.; Li, D.Y.; Yan, Y. A dual hesitant fuzzy multigranulation rough set over two-universe model for medical diagnoses. *Comput. Math. Method Med.* **2015**, *2015*; doi:10.1155/2015/292710. [CrossRef] [PubMed]

30. Guo, Z.L.; Liu, Y.L.; Yang, H.L. A novel rough set model in generalized single valued neutrosophic approximation spaces and its application. *Symmetry* **2017**, *9*, 119. [CrossRef]
31. Lu, J.; Li, D.Y.; Zhai, Y.H.; Bai, H.-J. Granular structure of type-2 fuzzy rough sets over two universes. *Symmetry* **2017**, *9*, 284. [CrossRef]
32. Sun, B.Z.; Ma, W.M.; Zhao, H.Y. A fuzzy rough set approach to emergency material demand prediction over two universes. *Appl. Math. Model.* **2013**, *37*, 7062–7070. [CrossRef]
33. Sun, B.Z.; Ma, W.M. Multigranulation rough set theory over two universes. *J. Intell. Fuzzy Syst.* **2015**, *28*, 1251–1269.
34. Zhang, C.; Li, D.Y.; Ren, R. Pythagorean fuzzy multigranulation rough set over two universes and its applications in merger and acquisition. *Int. J. Intell. Syst.* **2016**, *31*, 921–943. [CrossRef]
35. Sun, B.Z.; Ma, W.M.; Qian, Y.H. Multigranulation fuzzy rough set over two universes and its application to decision making. *Knowl. Based Syst.* **2017**, *123*, 61–74. [CrossRef]
36. Zhang, C.; Li, D.Y.; Mu, Y.M.; Song, D. An interval-valued hesitant fuzzy multigranulation rough set over two universes model for steam turbine fault diagnosis. *Appl. Math. Model.* **2017**, *42*, 693–704. [CrossRef]
37. Zhang, C.; Li, D.Y.; Sangaiah, A.; Broumi, S. Merger and acquisition target selection based on interval neutrosophic multigranulation rough sets over two universes. *Symmetry* **2017**, *9*, 126. [CrossRef]
38. Zhang, C.; Li, D.Y.; Zhai, Y.H.; Yang, Y. Multigranulation rough set model in hesitant fuzzy information systems and its application in person-job fit. *Int. J. Mach. Learn. Cyber.* **2017**. [CrossRef]
39. Zhang, F.W.; Chen, J.H.; Zhu, Y.H.; Li, J.; Li, Q.; Zhuang, Z. A dual hesitant fuzzy rough pattern recognition approach based on deviation theories and its application in urban traffic modes recognition. *Symmetry* **2017**, *9*, 262. [CrossRef]
40. Zeljko, S.; Pamucar, D.; Zavadskas, E.K.; Ćirović, G.; Prentkovskis, O. The selection of wagons for the internal transport of a logistics company: A novel approach based on rough BWM and rough SAW methods. *Symmetry* **2017**, *9*, 264.
41. Akram, M.; Ali, G.; Alshehri, N.O. A new multi-attribute decision-making method based on m-polar fuzzy soft rough sets. *Symmetry* **2017**, *9*, 271. [CrossRef]
42. Zhang, C.; Li, D.Y.; Liang, J.Y. Hesitant fuzzy linguistic rough set over two universes model and its applications. *Int. J. Mach. Learn. Cyber.* **2018**, *9*, 577–588. [CrossRef]
43. Wong, S.K.M.; Ziarko, W. Comparison of the probabilistic approximate classification and the fuzzy set model. *Fuzzy Sets Syst.* **1987**, *21*, 357–362. [CrossRef]
44. Yao, Y.Y.; Wong, S.K.M. A decision theoretic framework for approximating concepts. *Int. J. Man Mach. Stud.* **1992**, *37*, 793–809. [CrossRef]
45. Ziarko, W. Variable precision rough sets model. *J. Comput. Syst. Sci.* **1993**, *46*, 39–59. [CrossRef]
46. Slezak, D.; Ziarko, W. The investigation of the bayesian rough set model. *Int. J. Approx. Reason.* **2005**, *40*, 81–91. [CrossRef]
47. Greco, S.; Pawlak, Z.; Slowinski, R. Can bayesian confirmation measures be useful for rough set decision rules? *Eng. Appl. Artif. Intel.* **2004**, *17*, 345–361. [CrossRef]
48. Yao, Y.Y.; Zhou, B. Two Bayesian approaches to rough sets. *Eur. J. Oper. Res.* **2016**, *251*, 904–917. [CrossRef]
49. Yao, Y.Y. Three-way decisions with probabilistic rough sets. *Inform. Sci.* **2010**, *180*, 314–353. [CrossRef]
50. Yao, Y.Y. Three-way decisions and cognitive computing. *Cogn. Comput.* **2016**, *8*, 543–554. [CrossRef]
51. Yang, H.L.; Liao, X.W.; Wang, S.Y.; Wang, J. Fuzzy probabilistic rough set model on two universes and its applications. *Int. J. Approx. Reason.* **2013**, *54*, 1410–1420. [CrossRef]
52. Sun, B.Z.; Ma, W.M.; Chen, X.T. Fuzzy rough set on probabilistic approximation space over two universes and its application to emergency decision-making. *Expert Syst.* **2015**, *32*, 507–521. [CrossRef]
53. Sun, B.Z.; Ma, W.M.; Zhao, H.Y. An approach to emergency decision making based on decision-theoretic rough set over two universes. *Soft Comput.* **2016**, *20*, 3617–3628. [CrossRef]
54. Sun, B.Z.; Ma, W.M.; Xiao, X. Three-way group decision making based on multigranulation fuzzy decision-theoretic rough set over two universes. *Int. J. Approx. Reason.* **2017**, *81*, 87–102. [CrossRef]

symmetry

MDPI

Article

Multiple Attribute Decision-Making Method Using Similarity Measures of Neutrosophic Cubic Sets

Angyan Tu [1,2], Jun Ye [2] and Bing Wang [1,*]

1 School of Mechatronic Engineering and Automation, Shanghai University, 149 Yanchang Road, Shanghai 200072, China; lucytu@shu.edu.cn
2 Department of Electrical and Information Engineering, Shaoxing University, 508 Huancheng West Road, Shaoxing 312000, China; yejun@usx.edu.cn
* Correspondence: susanbwang@shu.edu.cn

Received: 18 May 2018; Accepted: 11 June 2018; Published: 12 June 2018

Abstract: In inconsistent and indeterminate settings, as a usual tool, the neutrosophic cubic set (NCS) containing single-valued neutrosophic numbers and interval neutrosophic numbers can be applied in decision-making to present its partial indeterminate and partial determinate information. However, a few researchers have studied neutrosophic cubic decision-making problems, where the similarity measure of NCSs is one of the useful measure methods. For this work, we propose the Dice, cotangent, and Jaccard measures between NCSs, and indicate their properties. Then, under an NCS environment, the similarity measures-based decision-making method of multiple attributes is developed. In the decision-making process, all the alternatives are ranked by the similarity measure of each alternative and the ideal solution to obtain the best one. Finally, two practical examples are applied to indicate the feasibility and effectiveness of the developed method.

Keywords: similarity measures; neutrosophic cubic set; decision-making

1. Introduction

The classic fuzzy set [1] is expressed by its membership degree in the unit interval [0,1]. But in many complicated cases of the real world, the data often are vague and uncertain, and are difficult to express as classic fuzzy sets. Thus, the neutrosophic set (NS) concept was presented by Smarandache [2], which is an extension of the fuzzy set and (interval-valued) intuitionistic fuzzy sets. He defined the indeterminacy, falsity, and truth degrees of NS in the nonstandard interval $]^{-}0,1^{+}[$ and standard interval [0,1]. However, the nonstandard interval is difficult to apply in real situations, so a simplified neutrosophic set (SNS), including single-valued and interval neutrosophic sets, was presented by Ye [3], which is depicted by the truth, indeterminacy, and falsity degrees in the interval [0,1], to conveniently apply it in science and engineering fields, such as decision-making [4–8], medical diagnoses [9,10], image processing [11,12], and clustering analyses [13]. Meanwhile, different measures were constantly proposed, such as similarity measures, cross entropy measures, correlation coefficients, and various aggregation operators for multiple attribute decision-making (MADM) problems [14–21]. Then, various simplified neutrosophic decision-making methods were presented, such as the technique for order preference by similarity to an ideal solution (TOPSIS) method [22], the projection and bidirectional projection measures [23], and the VIKOR method [24].

In recent years, (fuzzy) cubic sets (CSs) presented by Jun et al. [25] have received much attention due to the vague properties of human hesitant judgments. Since CS implies its partial certain and partial uncertain information, it is depicted by the hybrid form composed of an exact value and an interval value. Hence, CSs are very well suited for the representation of its partial indeterminate and partial determinate information in fuzzy environments. But many scientific problems in the real world are very complex. To handle more complicated problems with incomplete, inconsistent, and indeterminate information, Jun et al. [26] and Ali et al. [27] have introduced neutrosophic cubic sets (NCSs) which contain both single-valued neutrosophic information and interval neutrosophic information, as introduced in References [2,28,29]. Lu and Ye [30] used cosine measures for NCSs for the first time to handle decision-making problems in an NCS setting. Banerjee et al. [31] presented MADM problems regarding grey relational analysis in an NCS setting. Pramanik et al. [32] introduced a multiple attribute group decision-making method regarding the distance-based similarity measure of NCSs. Ye [33] put forward the operational laws and weighted aggregation operators of NCSs and their MADM method in an NCS setting. Then, Shi and Ye [34] further proposed the Dombi aggregation operators of NCSs and their MADM method. However, few researchers have studied neutrosophic cubic MADM problems, where the similarity measure of NCSs is one of the useful measure methods. On the other hand, Ye proposed the cosine, Dice, and Jaccard measures of single-valued and interval neutrosophic sets [35], the generalized Dice measure of SNSs [36], and the single-valued neutrosophic cotangent measures [37]. Since NCS is combined with an interval neutrosophic set (INS) and a single-valued neutrosophic set (SVNS), we can extend them to NCSs. Motivated by the similarity measures of INSs and SVNSs in the literature [35,37], we propose the Dice, cotangent, and Jaccard measures between NCSs to enrich the existing similarity measures of NCSs. Then, a MADM method is developed based on the proposed similarity measures in an NCS setting. Their difference is that the similarity measures in the literature [30] only use three cosine measures for MADM problems, but this work proposes the Dice, cotangent, and Jaccard measures for MADM problems in an NCS setting. By comparison with existing decision-making methods [30], the decision results show that our similarity measures have better decision-making robustness and discrimination than existing cosine measures [30].

The contents of this paper are organized as follows: Section 2 introduces basic definitions of CSs and NCSs. The similarity measures of NCSs and their properties are presented in Section 3. A MADM method is developed by using the three measures of the Dice, cotangent, and Jaccard measures in Section 4. In Section 5, a practical example is given in an NCS setting to present the applications and the effectiveness of the developed method. Finally, Section 6 indicates conclusions and future work.

2. Basic Definitions of CSs and NCSs

Based on the combination of both a fuzzy value and an interval-valued fuzzy number (IVFN), a CS was defined by Jun et al. [25].

The CS Z in a universe of discourse Y is defined by the following form [25]:

$$Z = \{y, T(y), \mu(y) | y \in Y\},$$

where $\mu(y)$ is a fuzzy value and $T(y) = [T^-(y), T^+(y)]$ is an IVFN for $y \in Y$. Then, we define

(i) $Z = \{y, T(y), \mu(y) | y \in Y\}$ as an internal CS if $T^-(y) \leq \mu(y) \leq T^+(y)$ for $y \in Y$;
(ii) $Z = \{y, T(y), \mu(y) | y \in Y\}$ as an external CS if $\mu(y) \notin [T^-(y), T^+(y)]$ for $y \in Y$.

When combining a single-valued neutrosophic number (SVNN) with an interval neutrosophic number (INN), CS was extended to NCS by Jun et al. [26] and Ali et al. [27], which is constructed as an NCS Z in Y by the following form [26,27]:

$$R = \{y, < T(y), U(y), F(y) > t(y), u(y), f(y) > y \in Y\},$$

where $< T(y), U(y), F(y) >$ is an INN for the truth-interval $T(y) = [T^-(y), T^+(y)] \subseteq [0,1]$, the falsity-interval $F(y) = [F^-(y), F^+(y)] \subseteq [0,1]$, the indeterminacy-interval $U(y) = [U^-(y), U^+(y)] \subseteq [0,1]$, $y \in Y$ and $< t(y), u(y), f(y) >$ is an SVNN for the truth, falsity, and indeterminacy degrees $t(y), f(y), u(y) \in [0,1]$ and $y \in Y$.

An NCS $R = \{y, < T(y), U(y), F(y) > t(y), u(y), f(y) > y \in Y\}$ is called [26,27]:

(i) An internal NCS $R = \{y, < T(y), U(y), F(y) > t(y), u(y), f(y) > y \in Y\}$ if $T^-(y) \le t(y) \le T^+(y), U^-(y) \le u(y) \le U^+(y)$, and $F^-(y) \le f(y) \le F^+(y)$ for $y \in Y$;

(ii) An external NCS $R = \{y, < T(y), U(y), F(y) > t(y), u(y), f(y) > y \in Y\}$ if $t(y) \notin [T^-(y), T^+(y)], u(y) \notin [U^-(y), U^+(y)]$, and $f(y) \notin [F^-(y), F^+(y)]$ for $y \in Y$.

For the simplified expression, a basic element $(y, < T(y), U(y), F(y) > t(y), u(y), f(y) >$ in an NCS R is denoted as $r = (< T, U, F > t, u, f >$, which is called a neutrosophic cubic number (NCN), where $T, U, F \subseteq [0,1]$ and $t, u, f \in [0,1]$, satisfying $0 \le T^+(y) + U^+(y) + F^+(y) \le 3$ and $0 \le t + u + f \le 3$.

Let $r_1 = (< T_1, U_1, F_1 > t_1, u_1, f_1 >$ and $r_2 = (< T_2, U_2, F_2 > t_2, u_2, f_2 >$ be two NCNs. We can indicate the following relations [26,27]:

(1) $r_1^c = (< F_1^-, F_1^+], [1 - U_1^+, 1 - U_1^-], [T_1^-, T_1^+] > f_1, 1 - u_1, t_1)$ (the complement of r_1);
(2) $r_1 \subseteq r_2$ if and only if $T_1 \subseteq T_2, U_1 \supseteq U_2, F_1 \supseteq F_2, t_1 \le t_2, u_1 \ge u_2$, and $f_1 \ge f_2$ (P-order);
(3) $r_1 = r_2$ if and only if $r_1 \subseteq r_2$ and $r_2 \subseteq r_1$, i.e., $< T_1, U_1, F_1 > T_2, U_2, F_2 >$ and $< t_1, u_1, f_1 > t_2, u_2, f_2 >$.

3. Similarity Measures of NCSs

Based on the Dice and Jaccard measures of SVNSs and INSs (SNSs) [35], and the single-valued neutrosophic cotangent measures [37] proposed by Ye, we can extend them to NCSs to present the Dice, Jaccard, and cotangent measures between NCSs in this section.

Definition 1. *Let two NCSs be* $R = \{r1, r2, r3, \cdots, rn\}$ *and* $H = \{h1, h2, h3, \cdots, hn\}$ *in the universe of discourse* $Y = \{y1, y2, y3, \cdots, yn\}$, *where* $r_i = (< T_{ri}, U_{ri}, F_{ri} > t_{ri}, u_{ri}, f_{ri} >$ *and* $h_i = (< T_{hi}, U_{hi}, F_{hi} > t_{hi}, u_{hi}, f_{hi} >$ *are two NCNs for i = 1, 2, ... , n. Thus, the similarity measures of the NCSs R and H are presented as follows:*

(1) *Dice Measure between the NCSs R and H*

$$Z_1(R,H) = \frac{1}{2n}\left\{\sum_{i=1}^{n} \frac{2(T_{ri}^- T_{hi}^- + T_{ri}^+ T_{hi}^+ + U_{ri}^- U_{hi}^- + U_{ri}^+ U_{hi}^+ + F_{ri}^- F_{hi}^- + F_{ri}^+ F_{hi}^+)}{\left\{\begin{array}{l}(T_{ri}^-)^2 + (T_{ri}^+)^2 + (U_{ri}^-)^2 + (U_{ri}^+)^2 + (F_{ri}^-)^2 + (F_{ri}^+)^2 \\ + (T_{hi}^-)^2 + (T_{hi}^+)^2 + (U_{hi}^-)^2 + (U_{hi}^+)^2 + (F_{hi}^-)^2 + (F_{hi}^+)^2\end{array}\right\}} + \sum_{i=1}^{n} \frac{2(t_{ri}t_{hi} + u_{ri}u_{hi} + f_{ri}f_{hi})}{t_{ri}^2 + u_{ri}^2 + f_{ri}^2 + t_{hi}^2 + u_{hi}^2 + f_{hi}^2}\right\} \quad (1)$$

(2) *Cotangent Measure between the NCSs R and H*

$$Z_2(R,H) = \frac{1}{2n}\sum_{i=1}^{n}\left\{\begin{array}{l}\cot\left[\frac{\pi}{4} + \frac{\pi}{24}\left(|T_{ri}^- - T_{hi}^-| + |T_{ri}^+ - T_{hi}^+| + |U_{ri}^- - U_{hi}^-| + |U_{ri}^+ - U_{hi}^+| + |F_{ri}^- - F_{hi}^-| + |F_{ri}^+ - F_{hi}^+|\right)\right] \\ + \cot\left[\frac{\pi}{4} + \frac{\pi}{12}\left(|t_{ri} - t_{hi}| + |u_{ri} - u_{hi}| + |f_{ri} - f_{hi}|\right)\right]\end{array}\right\} \quad (2)$$

(3) *Jaccard Measure between the NCSs R and H*

$$Z_3(R,H)=\frac{1}{2n}\left\{\sum_{i=1}^{n}\frac{T_{ri}^-T_{hi}^- + T_{ri}^+T_{hi}^+ + U_{ri}^-U_{hi}^- + U_{ri}^+U_{hi}^+ + F_{ri}^-F_{hi}^- + F_{ri}^+F_{hi}^+}{\left\{\begin{array}{l}(T_{ri}^-)^2+(T_{ri}^+)^2+(U_{ri}^-)^2+(U_{ri}^+)^2+(F_{ri}^-)^2+(F_{ri}^+)^2\\+(T_{hi}^-)^2+(T_{hi}^+)^2+(U_{hi}^-)^2+(U_{hi}^+)^2+(F_{hi}^-)^2+(F_{hi}^+)^2\\-T_{ri}^-T_{hi}^- - T_{ri}^+T_{hi}^+ - U_{ri}^-U_{hi}^- - U_{ri}^+U_{hi}^+ - F_{ri}^-F_{hi}^- - F_{ri}^+F_{hi}^+\end{array}\right\}}+\sum_{i=1}^{n}\frac{t_{ri}t_{hi}+u_{ri}u_{hi}+f_{ri}f_{hi}}{\left\{\begin{array}{l}t_{ri}^2+u_{ri}^2+f_{ri}^2+t_{hi}^2+u_{hi}^2+f_{hi}^2\\-t_{ri}t_{hi}-u_{ri}u_{hi}-f_{ri}f_{hi}\end{array}\right\}}\right\} \quad (3)$$

Theorem 1. *The three measures* $Z_m(R,H)$ *(m = 1, 2, 3) satisfy the three properties (I)–(III):*

(I) $0 \le Z_m(R,H) \le 1$;

(II) $Z_m(R,H) = Z_m(H,R)$;

(III) $Z_m(R,H) = 1$ *if R = H, i.e.,* $< T_{ri}, U_{ri}, F_{ri} > = < T_{hi}, U_{hi}, F_{hi} >$ *and* $< t_{ri}, u_{ri}, f_{ri} > < t_{hi}, u_{hi}, f_{hi} >$.

Proof.

Firstly, we prove the properties (I)–(III) of $Z_1(R,H)$).

(I) The inequality $Z_1(R,H) \ge 0$ is obvious. Then, we only prove $Z_1(R,H) \le 1$.

Based on the basic inequality $2x_iy_i \le x_i^2 + y_i^2$ for $i = 1, 2, \ldots, n$, where $(x_1, x_2, x_3, \ldots, x_n) \in R^n$ and $(y_1, y_2, y_3, \ldots, y_n) \in R^n$, it is extended to the NCNs, and then the following inequality is obtained:

$$2(T_{ri}^- T_{hi}^-) \le (T_{ri}^-)^2 + (T_{hi}^-)^2$$

When T_{ri}^- and T_{hi}^- are not equal to zero, we obtain the following inequality:

$$\frac{2(T_{ri}^- T_{hi}^-)}{(T_{ri}^-)^2 + (T_{hi}^-)^2} \le 1$$

Similarly, we have these inequalities $2(T_{ri}^+T_{hi}^+) \le (T_{ri}^+)^2 + (T_{hi}^+)^2$, $2(U_{ri}^-U_{hi}^-) \le (U_{ri}^-)^2 + (U_{hi}^-)^2$, $2(U_{ri}^+U_{hi}^+) \le (U_{ri}^+)^2 + (U_{hi}^+)^2$, $2(F_{ri}^-F_{hi}^-) \le (F_{ri}^-)^2 + (F_{hi}^-)^2$, and $2(F_{ri}^+F_{hi}^+) \le (F_{ri}^+)^2 + (F_{hi}^+)^2$.

Then, we get the following sum of the six inequalities with both sides.

$$\begin{array}{c}2(T_{ri}^-T_{hi}^-)+2(T_{ri}^+T_{hi}^+)+2(U_{ri}^-U_{hi}^-)+2(U_{ri}^+U_{hi}^+)+2(F_{ri}^-F_{hi}^-)+2(F_{ri}^+F_{hi}^+)\le\\(T_{ri}^-)^2+(T_{hi}^-)^2+(T_{ri}^+)^2+(T_{hi}^+)^2+(U_{ri}^-)^2+(U_{hi}^-)^2+(U_{ri}^+)^2+(U_{hi}^+)^2+(F_{ri}^-)^2+(F_{hi}^-)^2+(F_{ri}^+)^2+(F_{hi}^+)^2\end{array}.$$

Thus, we have the following result:

$$\frac{2(T_{ri}^-T_{hi}^-)+2(T_{ri}^+T_{hi}^+)+2(U_{ri}^-U_{hi}^-)+2(U_{ri}^+U_{hi}^+)+2(F_{ri}^-F_{hi}^-)+2(F_{ri}^+F_{hi}^+)}{\left\{\begin{array}{c}(T_{ri}^-)^2+(T_{hi}^-)^2+(T_{ri}^+)^2+(T_{hi}^+)^2+(U_{ri}^-)^2+(U_{hi}^-)^2\\+(U_{ri}^+)^2+(U_{hi}^+)^2+(F_{ri}^-)^2+(F_{hi}^-)^2+(F_{ri}^+)^2+(F_{hi}^+)^2\end{array}\right\}}\le 1.$$

So, we can further get the result:

$$\frac{1}{n}\sum_{i=1}^{n}\frac{2(T_{ri}^-T_{hi}^-)+2(T_{ri}^+T_{hi}^+)+2(U_{ri}^-U_{hi}^-)+2(U_{ri}^+U_{hi}^+)+2(F_{ri}^-F_{hi}^-)+2(F_{ri}^+F_{hi}^+)}{\left\{\begin{array}{c}(T_{ri}^-)^2+(T_{hi}^-)^2+(T_{ri}^+)^2+(T_{hi}^+)^2+(U_{ri}^-)^2+(U_{hi}^-)^2\\+(U_{ri}^+)^2+(U_{hi}^+)^2+(F_{ri}^-)^2+(F_{hi}^-)^2+(F_{ri}^+)^2+(F_{hi}^+)^2\end{array}\right\}}\le 1.$$

Similarly, we have the following inequalities:

$$\frac{1}{n}\sum_{i=1}^{n}\frac{2(t_{ri}t_{hi}+u_{ri}u_{hi}+f_{ri}f_{hi})}{t_{ri}^2+u_{ri}^2+f_{ri}^2+t_{hi}^2+u_{hi}^2+f_{hi}^2}\leq 1.$$

Thus, we have $Z_1(R,H)\leq 1$, and then $0\leq Z_1(R,H)\leq 1$ holds.

(II) The equality is obvious.

(III) When $R = H$, we have $\langle T_{ri}, U_{ri}, F_{ri}\rangle = \langle T_{hi}, U_{hi}, F_{hi}\rangle$ and $< t_{ri}, u_{ri}, f_{ri} >< t_{hi}, u_{hi}, f_{hi} >$. Thus $T_{ri} = T_{hi}$, $U_{ri} = U_{hi}$, $F_{ri} = F_{hi}$, $t_{ri} = t_{hi}$, $u_{ri} = u_{hi}$, and $f_{ri} = f_{hi}$ for $i = 1, 2, \ldots, n$. Hence $Z_1(R,H) = 1$ holds.

Secondly, the properties (I)–(III) of $Z_2(R,H)$ can be proved as follows:

(I) The inequality $0\leq|T_{ri}^- - T_{hi}^-|\leq 1$ is obvious. Similarly, we obtain other inequalities $0\leq|T_{ri}^+ - T_{hi}^+|\leq 1$, $0\leq|U_{ri}^- - U_{hi}^-|\leq 1$, $0\leq|U_{ri}^+ - U_{hi}^+|\leq 1$, $0\leq|F_{ri}^- - F_{hi}^-|\leq 1$, and $0\leq|F_{ri}^+ - F_{hi}^+|\leq 1$.

Based on these inequalities, we get the inequality:
$0\leq|T_{ri}^- - T_{hi}^-|+|T_{ri}^+ - T_{hi}^+|+|U_{ri}^- - U_{hi}^-|+|U_{ri}^+ - U_{hi}^+|+|F_{ri}^- - F_{hi}^-|+|F_{ri}^+ - F_{hi}^+|\leq 6$,
and then obtain the inequality:
$0\leq\frac{1}{24}\left(\left|T_{ri}^- - T_{hi}^-\right|+\left|T_{ri}^+ - T_{hi}^+\right|+\left|U_{ri}^- - U_{hi}^-\right|+\left|U_{ri}^+ - U_{hi}^+\right|+\left|F_{ri}^- - F_{hi}^-\right|+\left|F_{ri}^+ - F_{hi}^+\right|\right)\leq\frac{1}{4}$
and the following inequality:
$0\leq\frac{\pi}{24}(|T_{ri}^- - T_{hi}^-|+|T_{ri}^+ - T_{hi}^+|+|U_{ri}^- - U_{hi}^-|+|U_{ri}^+ - U_{hi}^+|+|F_{ri}^- - F_{hi}^-|+|F_{ri}^+ - F_{hi}^+|)\leq\frac{\pi}{4}$.

Hence, the result is obtained as follows:

$\cot(\frac{\pi}{2})\leq\cot[\frac{\pi}{4}+\frac{\pi}{24}(|T_{ri}^- - T_{hi}^-|+|T_{ri}^+ - T_{hi}^+|+|U_{ri}^- - U_{hi}^-|+|U_{ri}^+ - U_{hi}^+|+|F_{ri}^- - F_{hi}^-|+|F_{ri}^+ - F_{hi}^+|)]\leq\cot(\frac{\pi}{4})$.

Simplifying the above inequality, we get the simplified inequality:

$0\leq\cot[\frac{\pi}{4}+\frac{\pi}{24}(|T_{ri}^- - T_{hi}^-|+|T_{ri}^+ - T_{hi}^+|+|U_{ri}^- - U_{hi}^-|+|U_{ri}^+ - U_{hi}^+|+|F_{ri}^- - F_{hi}^-|+|F_{ri}^+ - F_{hi}^+|)]\leq 1$.

Let us prove the other inequality $0\leq\cot[\frac{\pi}{4}+\frac{\pi}{12}(|t_{ri}-t_{hi}|+|u_{ri}-u_{hi}|+|f_{ri}-f_{hi}|)]\leq 1$. Because there are the inequalities $0\leq|t_{ri}-t_{hi}|\leq 1$, $0\leq|u_{ri}-u_{hi}|\leq 1$, and $0\leq|f_{ri}-f_{hi}|\leq 1$, we get the inequality $0\leq|t_{ri}-t_{hi}|+|u_{ri}-u_{hi}|+|f_{ri}-f_{hi}|\leq 1$ and $0\leq\frac{\pi}{12}(|t_{ri}-t_{hi}|+|u_{ri}-u_{hi}|+|f_{ri}-f_{hi}|)\leq\frac{3\pi}{12}$, and then $\cot(\frac{\pi}{2})\leq\cot[\frac{\pi}{4}+\frac{\pi}{12}(|t_{ri}-t_{hi}|+|u_{ri}-u_{hi}|+|f_{ri}-f_{hi}|)]\leq\cot(\frac{\pi}{4})$. Thus, the other form is $0\leq\cot[\frac{\pi}{4}+\frac{\pi}{12}(|t_{ri}-t_{hi}|+|u_{ri}-u_{hi}|+|f_{ri}-f_{hi}|)]\leq 1$. Hence $0\leq Z_2(R,H)\leq 1$ holds.

Thirdly, the properties (I)–(III) of $Z_3(R,H)$ can be proved below.

Based on the inequality $xy\leq x^2+y^2-xy$, we get such an inequality $T_{ri}^-T_{hi}^-\leq(T_{ri}^-)^2+(T_{hi}^-)^2-T_{ri}^-T_{hi}^-$. When T_{ri}^- and T_{hi}^- are not equal to zero, we obtain the inequality:

$$\frac{T_{ri}^-T_{hi}^-}{(T_{ri}^-)^2+(T_{hi}^-)^2-T_{ri}^-T_{hi}^-}\leq 1.$$

Thus, we can get the following inequality:

$$\frac{T_{ri}^-T_{hi}^-+T_{ri}^+T_{hi}^++U_{ri}^-U_{hi}^-+U_{ri}^+U_{hi}^++F_{ri}^-F_{hi}^-+F_{ri}^+F_{hi}^+}{\left\{\begin{array}{c}(T_{ri}^-)^2+(T_{ri}^+)^2+(U_{ri}^-)^2+(U_{ri}^+)^2+(F_{ri}^-)^2+(F_{ri}^+)^2\\+(T_{hi}^-)^2+(T_{hi}^+)^2+(U_{hi}^-)^2+(U_{hi}^+)^2+(F_{hi}^-)^2+(F_{hi}^+)^2\\-T_{ri}^-T_{hi}^--T_{ri}^+T_{hi}^+-U_{ri}^-U_{hi}^--U_{ri}^+U_{hi}^+-F_{ri}^-F_{hi}^--F_{ri}^+F_{hi}^+\end{array}\right\}}\leq 1.$$

Similarly, because the inequality $\frac{t_{ri}t_{hi}}{t_{ri}^2+t_{hi}^2-t_{ri}t_{hi}} \leq 1$ holds, the inequality $\frac{t_{ri}t_{hi}+u_{ri}u_{hi}+f_{ri}f_{hi}}{t_{ri}^2+u_{ri}^2+f_{ri}^2+t_{hi}^2+u_{hi}^2+f_{hi}^2-t_{ri}t_{hi}-u_{ri}u_{hi}-f_{ri}f_{hi}} \leq 1$ also holds. Hence, there is the following inequality:

$$\left\{ \sum_{i=1}^{n} \frac{T_{ri}^{-}T_{hi}^{-} + T_{ri}^{+}T_{hi}^{+} + U_{ri}^{-}U_{hi}^{-} + U_{ri}^{+}U_{hi}^{+} + F_{ri}^{-}F_{hi}^{-} + F_{ri}^{+}F_{hi}^{+}}{\left\{ \begin{array}{l} (T_{ri}^{-})^2 + (T_{ri}^{+})^2 + (U_{ri}^{-})^2 + (U_{ri}^{+})^2 + (F_{ri}^{-})^2 + (F_{ri}^{+})^2 \\ + (T_{hi}^{-})^2 + (T_{hi}^{+})^2 + (U_{hi}^{-})^2 + (U_{hi}^{+})^2 + (F_{hi}^{-})^2 + (F_{hi}^{+})^2 \\ - T_{ri}^{-}T_{hi}^{-} - T_{ri}^{+}T_{hi}^{+} - U_{ri}^{-}U_{hi}^{-} - U_{ri}^{+}U_{hi}^{+} - F_{ri}^{-}F_{hi}^{-} - F_{ri}^{+}F_{hi}^{+} \end{array} \right\}} + \sum_{i=1}^{n} \frac{t_{ri}t_{hi} + u_{ri}u_{hi} + f_{ri}f_{hi}}{t_{ri}^2 + u_{ri}^2 + f_{ri}^2 + t_{hi}^2 + u_{hi}^2 + f_{hi}^2 - t_{ri}t_{hi} - u_{ri}u_{hi} - f_{ri}f_{hi}} \right\} \leq 2n.$$

Thus, we have $Z_3(R, H) \leq 1$. Then, $0 \leq Z_3(R, H) \leq 1$ holds.

If we consider $\theta = \{\theta_1, \theta_2, \cdots, \theta_n\}$ as the weights of the elements r_i and h_i with $\theta_i \in [0,1]$ and $\sum_{i=1}^{n} \theta_i = 1$, the corresponding three measures $Z_{\theta m}(R, H)$ (m = 1, 2, 3) are given as follows:

$$Z_{\theta 1}(R,H) = \frac{1}{2}\left\{ \sum_{i=1}^{n} \theta_i \frac{2(T_{ri}^{-}T_{hi}^{-} + T_{ri}^{+}T_{hi}^{+} + U_{ri}^{-}U_{hi}^{-} + U_{ri}^{+}U_{hi}^{+} + F_{ri}^{-}F_{hi}^{-} + F_{ri}^{+}F_{hi}^{+})}{\left\{ \begin{array}{l} (T_{ri}^{-})^2 + (T_{ri}^{+})^2 + (U_{ri}^{-})^2 + (U_{ri}^{+})^2 + (F_{ri}^{-})^2 + (F_{ri}^{+})^2 \\ + (T_{hi}^{-})^2 + (T_{hi}^{+})^2 + (U_{hi}^{-})^2 + (U_{hi}^{+})^2 + (F_{hi}^{-})^2 + (F_{hi}^{+})^2 \end{array} \right\}} + \sum_{i=1}^{n} \theta_i \frac{2(t_{ri}t_{hi} + u_{ri}u_{hi} + f_{ri}f_{hi})}{t_{ri}^2 + u_{ri}^2 + f_{ri}^2 + t_{hi}^2 + u_{hi}^2 + f_{hi}^2} \right\} \quad (4)$$

$$Z_{\theta 2}(R,H) = \frac{1}{2}\sum_{i=1}^{n} \theta_i \left\{ \begin{array}{l} \cot\left[\frac{\pi}{4} + \frac{\pi}{24}\left(|T_{ri}^{-} - T_{hi}^{-}| + |T_{ri}^{+} - T_{hi}^{+}| + |U_{ri}^{-} - U_{hi}^{-}| + |U_{ri}^{+} - U_{hi}^{+}| + |F_{ri}^{-} - F_{hi}^{-}| + |F_{ri}^{+} - F_{hi}^{+}|\right)\right] \\ + \cot\left[\frac{\pi}{4} + \frac{\pi}{12}\left(|t_{ri} - t_{hi}| + |u_{ri} - u_{hi}| + |f_{ri} - f_{hi}|\right)\right] \end{array} \right\} \quad (5)$$

$$Z_{\theta 3}(R,H) = \frac{1}{2}\left\{ \sum_{i=1}^{n} \theta_i \frac{T_{ri}^{-}T_{hi}^{-} + T_{ri}^{+}T_{hi}^{+} + U_{ri}^{-}U_{hi}^{-} + U_{ri}^{+}U_{hi}^{+} + F_{ri}^{-}F_{hi}^{-} + F_{ri}^{+}F_{hi}^{+}}{\left\{ \begin{array}{l} (T_{ri}^{-})^2 + (T_{ri}^{+})^2 + (U_{ri}^{-})^2 + (U_{ri}^{+})^2 + (F_{ri}^{-})^2 + (F_{ri}^{+})^2 \\ + (T_{hi}^{-})^2 + (T_{hi}^{+})^2 + (U_{hi}^{-})^2 + (U_{hi}^{+})^2 + (F_{hi}^{-})^2 + (F_{hi}^{+})^2 \\ - T_{ri}^{-}T_{hi}^{-} - T_{ri}^{+}T_{hi}^{+} - U_{ri}^{-}U_{hi}^{-} - U_{ri}^{+}U_{hi}^{+} - F_{ri}^{-}F_{hi}^{-} - F_{ri}^{+}F_{hi}^{+} \end{array} \right\}} + \sum_{i=1}^{n} \theta_i \frac{t_{ri}t_{hi} + u_{ri}u_{hi} + f_{ri}f_{hi}}{\left\{ \begin{array}{l} t_{ri}^2 + u_{ri}^2 + f_{ri}^2 + t_{hi}^2 + u_{hi}^2 + f_{hi}^2 \\ - t_{ri}t_{hi} - u_{ri}u_{hi} - f_{ri}f_{hi} \end{array} \right\}} \right\} \quad (6)$$

Obviously, the three measures $Z_{\theta m}(R, H)$ (m = 1, 2, 3) also conform to the following properties (I)–(III):

(I) $0 \leq Z_{\theta m}(R, H) \leq 1$;
(II) $Z_{\theta m}(R, H) = Z_{\theta m}(H, R)$;
(III) $Z_{\theta m}(R, H) = 1$ if $R = H$, i.e., $\langle t_{ri}, u_{ri}, f_{ri} \rangle = \langle t_{hi}, u_{hi}, f_{hi} \rangle$ and $\langle T_{ri}, U_{ri}, F_{ri} \rangle = \langle T_{hi}, U_{hi}, F_{hi} \rangle$. □

The proofs of the three properties are similar, so we omitted them here.

4. MADM Method Using the Proposed Measures of NCSs

The proposed weighted measures of NCSs are applied in MADM problems with NCSs in this section.

In a MADM problem, there are the set of m alternatives $R = \{R_1, R_2, \ldots, R_m\}$ and the set of n attributes $B = \{B_1, B_2, \ldots, B_n\}$. Then, the weight of the attributes θ_t with $\theta_t \in [0,1]$ and $\sum_{t=1}^{n} \theta_t = 1$ is considered. The evaluation information of each alternative on each attribute in the MADM problem can be represented by a NCN $r_{st} = (< T_{st}, U_{st}, F_{st} >, < t_{st}, u_{st}, f_{st} >)$ ($t = 1, 2, \ldots, n$; $s = 1, 2, \ldots, m$) with $T_{st}, U_{st}, F_{st} \subseteq [0,1]$ and $t_{st}, u_{st}, f_{st} \subseteq [0,1]$. So, the decision matrix with neutrosophic cubic information can be expressed as $R = (r_{st})_{m \times n}$. Thus the decision procedures are listed in the following:

Step 1: By considering the benefit and cost types of attributes, setup an ideal solution (ideal alternative) $r^* = \{r_1^*, r_2^*, \cdots, r_n^*\}$, where the desired NCNs $r_t^*(t = 1, 2, \ldots, n)$ are expressed by

$$r_t^* = \begin{pmatrix} < [\max_s(T_{st}^-), \max_s(T_{st}^+)], [\min_s(U_{st}^-), \min_s(U_{st}^+)], \\ [\min_s(F_{st}^-), \min_s(F_{st}^+)] >, < \max_s(t_{st}), \min_s(u_{st}), \min_s(f_{st}) > \end{pmatrix} \text{ for the benefit attributes}$$

or

$$r_t^* = \begin{pmatrix} < [\min_s(T_{st}^-), \min_s(T_{st}^+)], [\max_s(U_{st}^-), \max_s(U_{st}^+)], [\max_s(F_{st}^-), \max_s(F_{st}^+)] >, \\ < \min_s(t_{st}), \max_s(u_{st}), \max_s(f_{st}) > \end{pmatrix} \text{ for the cost attributes.}$$

Step 2: Compute the measure value between an alternative R_s ($s = 1, 2, \ldots, m$) and the ideal solution R^* by using Equation (4) or Equation (5) or Equation (6), and then obtain the values of $Z_{\theta 1}(R_s, R^*)$ or $Z_{\theta 2}(R_s, R^*)$ or $Z_{\theta 3}(R_s, R^*)$ ($s = 1, 2, \ldots, m$).
Step 3: Corresponding to the measure values of $Z_{\theta 1}(R_s, R^*)$ or $Z_{\theta 2}(R_s, R^*)$ or $Z_{\theta 3}(R_s, R^*)$, rank the alternatives in descending order and choose the best one regarding the bigger measure value.
Step 4: End.

5. Decision-Making Example

Two practical decision-making examples in real environments are given in this section to illustrate the applications of the developed MADM method in an NCS setting.

5.1. Practical Example 1

We consider the practical decision-making example adapted from Reference [30] for convenient comparison. Suppose that a sum of money is invested by an investment company for one of four potential alternatives: R_1 (a food company), R_2 (a transportation company), R_3 (a software company), and R_4 (a manufacturing company). Then the four alternatives are evaluated over the set of the three attributes: H_1 (the potential risk as the benefit type), H_2 (the growth as the benefit type), and H_3 (the environmental impact as the cost type). Then the importance of the three attributes is indicated by the weight vector $\theta = (0.32, 0.38, 0.3)$. The evaluation values of the four alternatives over the three attributes are given by NCSs $r_{st} = (< T_{st}, U_{st}, F_{st} >, < t_{st}, u_{st}, f_{st} >)$ ($t = 1, 2, 3$; $s = 1, 2, 3, 4$). Thus, the neutrosophic cubic decision matrix can be constructed as follows:

$$R = (r_{st})_{4\times 3} = \begin{bmatrix} (< [0.5,0.6], [0.1,0.3], [0.2,0.4] >, < 0.6,0.2,0.3 >) & (< [0.5,0.6], [0.1,0.3], [0.2,0.4] >, < 0.6,0.2,0.3 >) & (< [0.6,0.8], [0.2,0.3], [0.1,0.2] >, < 0.7,0.2,0.1 >) \\ (< [0.6,0.8], [0.1,0.2], [0.2,0.3] >, < 0.7,0.1,0.2 >) & (< [0.6,0.7], [0.1,0.2], [0.2,0.3] >, < 0.6,0.1,0.2 >) & (< [0.6,0.7], [0.3,0.4], [0.1,0.2] >, < 0.7,0.4,0.1 >) \\ (< [0.4,0.6], [0.2,0.3], [0.1,0.3] >, < 0.6,0.2,0.2 >) & (< [0.5,0.6], [0.2,0.3], [0.3,0.4] >, < 0.6,0.3,0.4 >) & (< [0.5,0.7], [0.2,0.3], [0.3,0.4] >, < 0.6,0.2,0.3 >) \\ (< [0.7,0.8], [0.1,0.2], [0.1,0.2] >, < 0.8,0.1,0.2 >) & (< [0.6,0.7], [0.1,0.2], [0.1,0.3] >, < 0.7,0.1,0.2 >) & (< [0.6,0.7], [0.3,0.4], [0.2,0.3] >, < 0.7,0.3,0.2 >) \end{bmatrix}.$$

By the following steps, we use the proposed MADM method to judge which one is the best investment under an NCS environment.

First, when the ideal NCNs $r_t^*(t = 1, 2, 3)$ of three attributes H_1, H_2, H_3 are obtained by

$$r_t^* = \begin{pmatrix} <[\max_s(T_{st}^-), \max_s(T_{st}^+)], [\min_s(U_{st}^-), \min_s(U_{st}^+)], [\min_s(F_{st}^-), \min_s(F_{st}^+)]>, \\ <\max_s(t_{st}), \min_s(u_{st}), \min_s(f_{st})> \end{pmatrix}$$ for the benefit attributes H_1, H_2

or

$$r_t^* = \begin{pmatrix} <[\min_s(T_{st}^-), \min_s(T_{st}^+)], [\max_s(U_{st}^-), \max_s(U_{st}^+)], [\max_s(F_{st}^-), \max_s(F_{st}^+)]>, \\ <\min_s(t_{st}), \max_s(u_{st}), \max_s(f_{st})> \end{pmatrix}$$ for the cost attribute H_3.

We can obtain an ideal solution (an ideal alternative) as follows:

$$R^* = \{r_1^*, r_2^*, r_3^*\} = \left\{ \begin{array}{c} (<[0.7, 0.8], [0.1, 0.2], [0.1, 0.2]>, <0.8, 0.1, 0.2>), \\ (<[0.6, 0.7], [0.1, 0.2], [0.1, 0.3]>, <0.7, 0.1, 0.2>), \\ (<[0.5, 0.7], [0.3, 0.4], [0.3, 0.4]>, <0.6, 0.4, 0.3>) \end{array} \right\}.$$

Second, by Equation (4) or Equation (5) or Equation (6), we compute the measure value between an alternative R_s (s = 1, 2, 3, 4) and the ideal solution R^*. Then the measure values of $Z_{\theta 1}(R_s,R^*)$ or $Z_{\theta 2}(R_s,R^*)$ or $Z_{\theta 3}(R_s,R^*)$ (s = 1, 2, 3, 4) and the ranking of the alternatives are indicated in Table 1.

Table 1. Measure results between the two NCSs R_s and R^* and ranking.

$Z_{\theta m}(R_s,R^*)$	Measure Result	Ranking	The Best One
$Z_{\theta 1}(R_s,R^*)$	0.9517,0.9822,0.9498,0.9945	$Z_4 > Z_2 > Z_1 > Z_3$	Z_4
$Z_{\theta 2}(R_s,R^*)$	0.8246,0.9248,0.8474,0.9668	$Z_4 > Z_2 > Z_3 > Z_1$	Z_4
$Z_{\theta 3}(R_s,R^*)$	0.9085,0.9654,0.9054,0.9893	$Z_4 > Z_2 > Z_1 > Z_3$	Z_4

According to the results of Table 1, the two alternatives Z_4 and Z_2 have the same ranking orders in all the measures, and Z_4 is the best choice.

5.2. Related Comparison

For convenient comparison, we select the MADM method introduced in the literature [30] as the related comparison. Then, we can get the measure values between R_s and R^* by the cosine measure $S_{ws}(R_s,R^*)$ (s = 1, 2, 3, 4) in [30], the standard deviation (SD), and the best choice, which are given in Table 2. Obviously, the SD values of our measures are bigger than the SD values of existing cosine measures. Therefore, our measures not only have good discrimination, but also get the same as the best choice (Z_4), while existing cosine measures [30] indicate the different best choices (Z_4 or Z_2). Thus, our measures have better decision-making robustness and discrimination than existing cosine measures [30].

Table 2. Related comparison of our measure results with existing cosine measure results.

Measure	Measure Value	Ranking Order	SD	The Best One
$Z_{\theta 1}(R_s,R^*)$	0.9945,0.9822,0.9517,0.9498	$Z_4 > Z_2 > Z_1 > Z_3$	0.0193	Z_4
$Z_{\theta 2}(R_s,R^*)$	0.9668,0.9248,0.8474,0.8246	$Z_4 > Z_2 > Z_3 > Z_1$	0.0574	Z_4
$Z_{\theta 3}(R_s,R^*)$	0.9085,0.9654,0.9054,0.9893	$Z_4 > Z_2 > Z_1 > Z_3$	0.0362	Z_4
$S_{w1}(R_1,R^*)$ [30]	0.9451, 0.9794, 0.9524, 0.9846	$Z_4 > Z_2 > Z_3 > Z_1$	0.0169	Z_4
$S_{w2}(R_2,R^*)$ [30]	0.9700, 0.9906, 0.9732, 0.9877	$Z_2 > Z_4 > Z_3 > Z_1$	0.0089	Z_2
$S_{w2}(R_2,R^*)$ [30]	0.9867, 0.9942, 0.9877, 0.9968	$Z_4 > Z_2 > Z_3 > Z_1$	0.0043	Z_4

5.3. Practical Example 2

Further, we give a real case about a punching machine to clearly demonstrate the usefulness of the proposed measures. There are four alternatives (design schemes), R_1, R_2, R_3, and R_4 in Table 3. Then the four alternatives are evaluated over the set of three attributes: H_1 (manufacturing cost),

H_2 (structure complexity), and H_3 (reliability). Then, the importance of the three attributes is indicated by the weight vector θ = (0.36, 0.3, 0.34). By the suitable evaluation of the four alternatives over the three attributes regarding NCNs $r_{st} = (< T_{st}, U_{st}, F_{st} >, < t_{st}, u_{st}, f_{st} >)$ (t = 1, 2, 3; s = 1, 2, 3, 4), the neutrosophic cubic decision matrix which is adapted from the literature [23] can be constructed as follows:

$$R = (r_{st})_{4\times 3} = \begin{bmatrix} (<[0.7,0.8],[0.0,0.2],[0.3,0.5]>,<0.75,0.1,0.4>) & (<[0.7,0.9],[0.0,0.3],[0.2,0.4]>,<0.80,0.1,0.3>) & (<[0.8,0.9],[0.0,0.2],[0.2,0.4]>,<0.85,0.1,0.3>) \\ (<[0.6,0.8],[0.0,0.2],[0.4,0.6]>,<0.70,0.1,0.5>) & (<[0.7,0.8],[0.0,0.3],[0.0,0.2]>,<0.75,0.1,0.1>) & (<[0.7,0.8],[0.0,0.2],[0.0,0.2]>,<0.80,0.1,0.1>) \\ (<[0.7,0.9],[0.1,0.3],[0.2,0.4]>,<0.80,0.2,0.3>) & (<[0.7,0.8],[0.0,0.2],[0.1,0.3]>,<0.78,0.1,0.2>) & (<[0.7,0.9],[0.1,0.3],[0.1,0.3]>,<0.80,0.2,0.2>) \\ (<[0.8,1.0],[0.0,0.2],[0.1,0.3]>,<0.90,0.1,0.2>) & (<[0.8,0.9],[0.0,0.2],[0.0,0.2]>,<0.85,0.1,0.1>) & (<[0.8,0.9],[0.0,0.2],[0.2,0.4]>,<0.85,0.1,0.3>) \end{bmatrix}.$$

Table 3. Four alternatives (design schemes) of a punching machine [23].

Alternative	R_1	R_2	R_3	R_4
Reducing mechanism	Gear reducer	Gear head motor	Gear reducer	Gear head motor
Punching mechanism	Crank-slider mechanism	Six bar punching mechanism	Six bar punching mechanism	Crank-slider mechanism
Dial feed intermittent mechanism	Sheave mechanism		Ratchet feed mechanism	

By the following steps, we use the proposed MADM method to judge which one is the best design scheme under an NCS environment.

First, because we use a suitable evaluation of the four alternatives over the three attributes, all the benefit attributes are given in this decision problem. Thus, when the ideal NCNs $r_t^*(t = 1, 2, 3)$ of the three attributes H_1, H_2, H_3 are obtained by $r_t^* = \left(\begin{array}{l} < [\max_s(T_{st}^-), \max_s(T_{st}^+)], [\min_s(U_{st}^-), \min_s(U_{st}^+)], [\min_s(F_{st}^-), \min_s(F_{st}^+)] >, \\ < \max_s(t_{st}), \min_s(u_{st}), \min_s(f_{st}) > \end{array} \right)$, we can obtain an ideal solution (an ideal alternative) as follows:

$$R^* = \{r_1^*, r_2^*, r_3^*\} = \left\{ \begin{array}{l} (<[0.8,1.0],[0.0,0.2],[0.1,0.3]>,<0.90,0.1,0.2>), \\ (<[0.8,0.9],[0.0,0.2],[0.0,0.2]>,<0.85,0.1,0.1>), \\ (<[0.8,0.9],[0.0,0.2],[0.0,0.2]>,<0.85,0.1,0.1>) \end{array} \right\}.$$

According to Equation (4) or Equation (5) or Equation (6), we can obtain the measure values of $Z_{\theta 1}(R_s, R^*)$ or $Z_{\theta 2}(R_s, R^*)$ or $Z_{\theta 3}(R_s, R^*)$ (s = 1, 2, 3, 4) and the ranking of all the alternatives, which are indicated in Table 4.

Table 4. Measure values between the two NCSs R_s and R^* and ranking.

$Z_{\theta m}(R_s, R^*)$	Measure Value	Ranking	The Best One
$Z_{\theta 1}(R_s, R^*)$	0.9683,0.9704,0.9847,0.9924	$Z_4 > Z_3 > Z_2 > Z_1$	Z_4
$Z_{\theta 2}(R_s, R^*)$	0.8652,0.8937,0.8813,0.9701	$Z_4 > Z_2 > Z_3 > Z_1$	Z_4
$Z_{\theta 3}(R_s, R^*)$	0.9386,0.9445,0.9699,0.9853	$Z_4 > Z_3 > Z_2 > Z_1$	Z_4

According to the decision results in Table 4, they show that the two alternatives Z_4 and Z_1 have the same ranking orders in all the measures, with the best choice Z_4 and the worst choice Z_1.

If we set the same importance (θ_t = 1/3 for t = 1, 2, 3) of three attributes without considering the three attribute weights, we also obtained the same ranking with the attribute weights and without considering the three attribute weights in Table 5. It is obvious that the decision results of the proposed measures imply better robustness and lower sensitivity regarding attribute weights.

Table 5. Measure values based on the different weights of the three attributes and ranking.

$Z_{\theta m}(R_s,R^*)$	Measure Value Based on θ = (0.36, 0.3, 0.34)	Measure Value Based on θ = (1/3, 1/3, 1/3)	Ranking	The Best One
$Z_{\theta 1}(R_s,R^*)$	0.9683,0.9704,0.9847,0.9924	0.9684,0.9697,0.9845,0.991	$Z_4 > Z_3 > Z_2 > Z_1$	Z_4
$Z_{\theta 2}(R_s,R^*)$	0.8652,0.8937,0.8813,0.9701	0.8659,0.8927,0.8795,0.966	$Z_4 > Z_2 > Z_3 > Z_1$	Z_4
$Z_{\theta 3}(R_s,R^*)$	0.9386,0.9445,0.9699,0.9853	0.9387,0.9432,0.9695,0.983	$Z_4 > Z_3 > Z_2 > Z_1$	Z_4

6. Conclusions

This work proposed the Dice measure, cotangent measure, and Jaccard measure between two NCSs and discussed their properties. Then, we developed a MADM method based on one of three measures and applied it in real cases with neutrosophic cubic information. By comparison with an existing related MADM method, the proposed measures imply better robustness and lower sensitivity regarding attribute weights.

In this work, our main contributions are to enrich the neutrosophic cubic similarity measures and their decision-making method under NCS environments. In future work, the developed measures will be extended to medical/fault diagnosis and image processing.

Author Contributions: A.T. proposed the three measures of SNSs and their MADM method, and finished the initial draft; J.Y. provided the practical decision-making examples and related comparison; B.W. supervised the research activity planning and execution; then we wrote this paper together.

Funding: This research received no external funding.

Conflicts of Interest: The authors declare no conflict of interest.

References

1. Zadeh, L.A. Fuzzy sets. *Inf. Control* **1965**, *8*, 338–353. [CrossRef]
2. Smarandache, F. *Neutrosophy: Neutrosophic Probability, Set, and Logic*; American Research Press: Rehoboth, DE, USA, 1998.
3. Ye, J. A multicriteria decision-making method using aggregation operators for simplified neutrosophic sets. *J. Intell. Fuzzy Syst.* **2014**, *26*, 2459–2466.
4. Ye, J. Multicriteria decision-making method using the correlation coefficient under single-valued neutrosophic environment. *Int. J. Gen. Syst.* **2013**, *42*, 386–394. [CrossRef]
5. Liu, P.-D.; Wang, Y.-M. Multiple attribute decision-making method based on single valued neutrosophic normalized weighted Bonferroni mean. *Neural Comput. Appl.* **2014**, *25*, 2001–2010. [CrossRef]
6. Liu, P.D.; Wang, Y.M. Interval neutrosophic prioritized OWA operator and its application to multiple attribute decision making. *J. Sci. Complex.* **2016**, *29*, 681–697. [CrossRef]
7. Sahin, R. Cross-entropy measure on interval neutrosophic sets and its applications in multicriteria decision making. *Neural Comput. Appl.* **2017**, *28*, 1177–1187. [CrossRef]
8. Stanujkic, D.; Zavadskas, E.K.; Smarandache, F.; Brauers, W.K.M.; Karabasevic, D. A neutrosophic extension of the MULTIMOORA method. *Informatica* **2017**, *28*, 181–192. [CrossRef]
9. Ye, J. Improved cosine similarity measures of simplified neutrosophic sets for medical diagnoses. *Artif. Intell. Med.* **2015**, *63*, 171–179. [CrossRef] [PubMed]
10. Ye, J.; Fu, J. Multi-period medical diagnosis method using a single valued neutrosophic similarity measure based on tangent function. *Comput. Methods Progr. Biomed.* **2016**, *123*, 142–149. [CrossRef] [PubMed]
11. Cheng, H.D.; Guo, Y. A new neutrosophic approach to image thresholding. *New Math. Nat. Comput.* **2008**, *4*, 291–308. [CrossRef]
12. Guo, Y.; Sengur, A.; Ye, J. A novel image thresholding algorithm based on neutrosophic similarity score. *Measurement* **2014**, *58*, 175–186. [CrossRef]
13. Ye, J. Clustering methods using distance-based similarity measures of single-valued neutrosophic sets. *J. Intell. Syst.* **2014**, *23*, 379–389. [CrossRef]
14. Ye, J. Similarity measures between interval neutrosophic sets and their applications in multicriteria decision-making. *J. Intell. Fuzzy Syst.* **2014**, *26*, 165–172.

15. Ye, J. Single valued neutrosophic cross-entropy for multicriteria decision making problems. *Appl. Math. Model.* **2014**, *38*, 1170–1175. [CrossRef]
16. Ye, J. Improved correlation coefficients of single valued neutrosophic sets and interval neutrosophic sets for multiple attribute decision making. *J. Intell. Fuzzy Syst.* **2014**, *27*, 2453–2462.
17. Zhang, H.-Y.; Ji, P.; Wang, J.-Q.; Chen, X.-H. An improved weighted correlation coefficient based on integrated weight for interval neutrosophic sets and its application in multi-criteria decision making problems. *Int. J. Comput. Intell. Syst.* **2015**, *8*, 1027–1043. [CrossRef]
18. Chen, J.; Ye, J. Some single-valued neutrosophic Dombi weighted aggregation operators for multiple attribute decision-making. *Symmetry* **2017**, *9*, 82. [CrossRef]
19. Zhang, H.-Y.; Wang, J.-Q.; Chen, X.-H. Interval neutrosophic sets and their application in multicriteria decision making problems. *Sci. World J.* **2014**, *2014*. [CrossRef] [PubMed]
20. Zhao, A.-W.; Du, J.-G.; Guan, H.-J. Interval valued neutrosophic sets and multi-attribute decision-making based on generalized weighted aggregation operator. *J. Intell. Fuzzy Syst.* **2015**, *29*, 2697–2706.
21. Sun, H.-X.; Yang, H.-X.; Wu, J.-Z.; Yao, O.-Y. Interval neutrosophic numbers Choquet integral operator for multi-criteria decision making. *J. Intell. Fuzzy Syst.* **2015**, *28*, 2443–2455. [CrossRef]
22. Biswas, P.; Pramanik, S.; Giri, B.C. TOPSIS method for multi-attribute group decision-making under single-valued neutrosophic environment. *Neural Comput. Appl.* **2016**, *27*, 727–737. [CrossRef]
23. Ye, J. Projection and bidirectional projection measures of single valued neutrosophic sets and their decision-making method for mechanical design schemes. *J. Exp. Theor. Artif. Intell.* **2017**, *29*, 731–740. [CrossRef]
24. Pouresmaeil, H.; Shivanian, E.; Khorram, E.; Fathabadi, H.S. An extended method using TOPSIS and VIKOR for multiple attribute decision making with multiple decision makers and single valued neutrosophic numbers. *Adv. Appl. Stat.* **2017**, *50*, 261–292. [CrossRef]
25. Jun, Y.B.; Kim, C.S.; Yang, K.O. Cubic sets. *Ann. Fuzzy Math. Inform.* **2012**, *4*, 83–98.
26. Jun, Y.B.; Smarandache, F.; Kim, C.S. Neutrosophic cubic sets. *New Math. Nat. Comput.* **2017**, *13*, 41–45. [CrossRef]
27. Ali, M.; Deli, I.; Smarandache, F. The theory of neutrosophic cubic sets and their applications in pattern recognition. *J. Intell. Fuzzy Syst.* **2016**, *30*, 1957–1963. [CrossRef]
28. Wang, H.; Smarandache, F.; Zhang, Y.-Q.; Sunderraman, R. *Interval Neutrosophic Sets and Logic: Theory and Applications in Computing*; Hexis: Phoenix, AZ, USA, 2005.
29. Wang, H.; Smarandache, F.; Zhang, Y.-Q.; Sunderraman, R. Single valued neutrosophic sets. *Multispace Multistruct.* **2010**, *4*, 410–413.
30. Lu, Z.K.; Ye, J. Cosine measures of neutrosophic cubic Sets for multiple attribute decision-making. *Symmetry* **2017**, *9*, 121. [CrossRef]
31. Banerjee, D.; Giri, B.C.; Pramanik, S.; Smarandache, F. GRA for multi attribute decision making in neutrosophic cubic set environment. *Neutrosophic Sets Syst.* **2017**, *15*, 60–69.
32. Pramanik, S.; Dalapati, S.; Alam, S.; Roy, T.K.; Smarandache, F. Neutrosophic cubic MCGDM method based on similarity measure. *Neutrosophic Sets Syst.* **2017**, *16*, 44–56.
33. Ye, J. Operations and aggregation method of neutrosophic cubic numbers for multiple attribute decision-making. *Soft Comput.* **2018**. [CrossRef]
34. Shi, L.L.; Ye, J. Dombi aggregation operators of neutrosophic cubic sets for multiple attribute decision-making. *Algorithms* **2018**, *11*, 29. [CrossRef]
35. Ye, J. Vector similarity measures of simplified neutrosophic sets and their application in multicriteria decision making. *Int. J. Fuzzy Syst.* **2014**, *16*, 204–211.
36. Ye, J. The generalized Dice measures for multiple attribute decision making under simplified neutrosophic environments. *J. Intell. Fuzzy Syst.* **2016**, *31*, 663–671. [CrossRef]
37. Ye, J. Single valued neutrosophic similarity measures based on cotangent function and their application in the fault diagnosis of steam turbine. *Soft Comput.* **2017**, *21*, 817–825. [CrossRef]

symmetry

MDPI

Article

Decision-Making via Neutrosophic Support Soft Topological Spaces

Parimala Mani [1,*], Karthika Muthusamy [1], Saeid Jafari [2], Florentin Smarandache [3] and Udhayakumar Ramalingam [4]

1 Department of Mathematics, Bannari Amman Institute of Technology, Sathyamangalam 638401, Tamil Nadu, India; karthikamuthusamy1991@gmail.com
2 College of Vestsjaelland South, Herrestraede 11, 4200 Slagelse, Denmark; jafaripersia@gmail.com
3 Mathematics & Science Department, University of New Maxico, 705 Gurley Ave, Gallup, NM 87301, USA; fsmarandache@gmail.com
4 Department of Mathematics, School of Advanced Sciences, Vellore Institute of Technology, Vellore 632012, Tamil Nadu and India; udhayaram_v@yahoo.co.in
* Correspondence: rishwanthpari@gmail.com

Received: 29 April 2018; Accepted: 7 June 2018; Published: 13 June 2018

Abstract: The concept of interval neutrosophic sets has been studied and the introduction of a new kind of set in topological spaces called the interval valued neutrosophic support soft set has been suggested. We study some of its basic properties. The main purpose of this paper is to give the optimum solution to decision-making in real life problems the using interval valued neutrosophic support soft set.

Keywords: soft sets; support soft sets; interval valued neutrosophic support soft sets

2010 AMS Classification: 06D72; 54A05; 54A40; 54C10

1. Introduction

To deal with uncertainties, many theories have been recently developed, including the theory of probability, the theory of fuzzy sets, the theory of rough sets, and so on. However, difficulties are still arising due to the inadequacy of parameters. The concept of fuzzy sets, which deals with the nonprobabilistic uncertainty, was introduced by Zadeh [1] in 1965. Since then, many researchers have defined the concept of fuzzy topology that has been widely used in the fields of neural networks, artificial intelligence, transportation, etc.The intuitionistic fuzzy set (IFS for short) on a universe X was introduced by K. Atanaasov [2] in 1983 as a generalization of the fuzzy set in addition to the degree of membership and the degree of nonmembership of each element.

In 1999, Molodtsov [3] successfully proposed a completely new theory called soft set theory using classical sets. This theory is a relatively new mathematical model for dealing with uncertainty from a parametrization point of view. After Molodtsov, many researchers have shown interest in soft sets and their applications. Maji [4,5] introduced neutrosophic soft sets with operators, which are free from difficulties since neutrosophic sets [6–9] can handle indeterminate information. However, the neutrosophic sets and operators are hard to apply in real life applications. Therefore, Smarandache [10] proposed the concept of interval valued neutrosophic sets which can represent uncertain, imprecise, incomplete, and inconsistent information.

Nguyen [11] introduced the new concept in a type of soft computing, called the support-neutrosophic set. Deli [12] defined a generalized concept of the interval-valued neutrosophic soft set. In this paper, we combine interval-valued neutrosophic soft sets and support sets to yield the interval-valued neutrosophic support soft set, and we study some of its basic operations. Our main aim of this paper is to make decisions using interval-valued neutrosophic support soft topological spaces.

2. Preliminaries

In this paper, we provide the basic definitions of neutrosophic and soft sets. These are very useful for what follows.

Definition 1. *([13]) Let X be a non-empty set. A neutrosophic set, A, in X is of the form* $A = \{\langle x, \mu_A(x), \sigma_A(x), \omega_A(x), \gamma_A(x); x \in X\rangle\}$, *where* $\mu_A : X \to [0,1]$, $\sigma_A : X \to [0,1]$ *and* $\gamma_A : X \to [0,1]$ *represent the degree of membership function, degree of indeterminacy, and degree of non-membership function, respectively and* $0 \leq \sup \mu_A(x) + \sup \sigma_A(x) + \sup \gamma_A(x) \leq 3, \forall x \in X$.

Definition 2. *([5]) Let X be a non-empty set , let P(X) be the power set of X, and let E be a set of parameters, and* $A \subseteq E$. *The soft set function,* f_X, *is defined by*

$$f_X : A \to P(X) \; such \; that \; f_X(x) = \emptyset \; if x \notin X.$$

The function f_X *may be arbitrary. Some of them may be empty and may have non-empty intersections. A soft set over X can be represented as the set of order pairs* $F_X = \{(x, f_X(x)) : x \in X, f_X(x) \in P(X)\}$.

Example 1. *Consider the soft set* $\langle F, A\rangle$, *where X is a set of six mobile phone models under consideration to be purchased by decision makers, which is denoted by* $X = \{x_1, x_2, x_3, x_4, x_5, x_6\}$, *and A is the parameter set, where* $A = \{y_1, y_2, y_3, y_4, y_5\} = \{price, look, \; camera, efficiency, processsor\}$. *A soft set,* F_X, *can be constructed such that* $f_X(y_1) = \{x_1, x_2\}$,$f_X(y_2) = \{x_1, x_4, x_5, x_6\}$, $f_X(y_3) = \emptyset$, $f_X(y_4) = X$, *and* $f_X(y_5) = \{x_1, x_2, x_3, x_4, x_5\}$. *Then,*

$$F_X = \{(y_1, x_1, x_2), (y_2, x_1, x_4, x_5, x_6), (y_3, \emptyset), (y_4, X), (y_5, x_1, x_2, x_3, x_4, x_5)\}.$$

X	x_1	x_2	x_3	x_4	x_5	x_6
y_1	1	1	0	0	0	0
y_2	1	0	0	1	1	1
y_3	0	0	0	0	0	0
y_4	1	1	1	1	1	1
y_5	1	1	1	1	1	0

Definition 3. *([4]) Let X be a non-empty set, and* $A = \{y_1, y_2, y_3, ..., y_n\}$, *the subset of X and* F_X *is a soft set over X. For any* $y_i \in A$, $f_X(y_i)$ *is a subset of X. Then, the choice value of an object,* $x_i \in X$, *is* $C_{V_i} = \sum_j x_{ij}$, *where* x_{ij} *are the entries in the table of* F_X:

$$x_{ij} = \begin{cases} 1, & if \;\; x_i \in f_X(y_j) \\ 0, & if \;\; x_i \notin f_X(y_j). \end{cases}$$

Example 2. *Consider Example 2. Clearly,* $C_{V_1} = \sum_{j=1}^{5} x_{1j} = 4$, $C_{V_3} = C_{V_6} = \sum_{j=1}^{5} x_{3j} = \sum_{j=1}^{5} x_{6j} = 2$, $C_{V_2} = C_{V_4} = C_{V_5} = \sum_{j=1}^{5} x_{2j} = \sum_{j=1}^{5} x_{4j} = \sum_{j=1}^{5} x_{5j} = 3$.

Definition 4. *([13]) Let* F_X *and* F_Y *be two soft sets over X and Y. Then,*

(1) *The complement of* F_X *is defined by* $F_{X^c}(x) = X \setminus f_X(x)$ *for allx* $\in A$;
(2) *The union of two soft sets is defined by* $f_{X\cup Y}(x) = f_X(x) \cup f_Y(x)$ *for all* $x \in A$;
(3) *The intersection of two soft sets is defined by* $f_{X\cap Y}(x) = f_X(x) \cap f_Y(x)$ *for all* $x \in A$.

3. Interval Valued Neutrosophic Support Soft Set

In this paper, we provide the definition of a interval-valued neutrosophic support soft set and perform some operations along with an example.

Definition 5. *Let X be a non-empty fixed set with a generic element in X denoted by a. An interval-valued neutrosophic support set, A, in X is of the form*

$$A = \{\langle x, \mu_A(x), \sigma_A(x), \omega_A(x), \gamma_A(x)\rangle / a; a \in X\}.$$

For each point, $a \in X, x, \mu_A(x), \sigma_A(x), \omega_A(x),$ *and* $\gamma_A(x) \in [0,1]$.

Example 3. *Let* $X = \{a, b\}$ *be a non-empty set, where* $a, b \subseteq [0,1]$. *An interval valued neutrosophic support set,* $A \subseteq X$, *constructed according to the degree of membership function,* $(\mu_A(x))$, *indeterminacy* $(\sigma_A(x))$, *support function* $(\omega_A(x))$, *and non-membership function* $(\gamma_A(x))$ *is as follows:*
$A = \{\langle(0.2, 1.0), (0.2, 0.4), (0.1, 0.7), (0.5, 0.7)\rangle / a, \langle(0.6, 0.8), (0.8, 1.0), (0.4, 0.6), (0.4, 0.6)\rangle / b\}$.

Definition 6. *Let X be a non-empty set; the interval-valued neutrosophic support set A in X is of the form* $A = \{\langle x, \mu_A(x), \sigma_A(x), \omega_A(x), \gamma_A(x); x \in X\rangle\}$.

(i) *An empty set A, denoted by* $A = \emptyset$, *is defined by*
$\emptyset = \{\langle(0,0), (1,1), (0,0), (1,1)\rangle / x : x \in X\}$.
(ii) *The universal set is defined by*
$U = \{\langle(1,1), (0,0), (1,1), (0,0)\rangle / x : x \in X\}$.
(iii) *The complement of A is defined by*
$A^c = \{\langle(\inf \gamma_A(x), \sup \gamma_A(x)), (1 - \sup \sigma_A(x), 1 - \inf \sigma_A(x)), (1 - \sup \omega_A(x), 1 - \inf \omega_A(x)),$
$(\inf \mu_A(x), \sup \mu_A(x))\rangle / x : x \in X\}$.
(iv) *A and B are two interval-valued neutrosophic support sets of X. A is a subset of B if*
$\mu_A(x) \leq \mu_B(x), \sigma_A(x) \geq \sigma_B(x), \omega_A(x) \leq \omega_B(x), \gamma_A(x) \geq \gamma_B(x)$.
(v) *Two interval-valued neutrosophic support sets A and B in X are said to be equal if* $A \subseteq B$ *and* $B \subseteq A$.

Definition 7. *Let A and B be two interval-valued neutrosophic support sets. Then, for every* $x \in X$

(i) *The intersection of A and B is defined by*
$A \cap B = \{\langle(\min[\inf \mu_A(x), \inf \mu_B(x)], \min[\sup \mu_A, \sup \mu_B(x)]), (\max[\inf \sigma_A(x), \inf \sigma_B(x)],$
$\max[\sup \sigma_A(x), \sup \sigma_B(x)]), (\min[\inf \omega_A(x), \inf \omega_B(x)], \min[\sup \omega_A(x), \sup \omega_B(x)]),$
$(\max[\inf \gamma_A(x), \inf \gamma_B(x)], \max[\sup$
$\gamma_A(x), \sup \gamma_B(x)])\rangle / x : x \in X\}$.
(ii) *The union of A and B is defined by*
$A \cup B = \{\langle(\max[\inf \mu_A(x), \inf \mu_B(x)], \max[\sup \mu_A(x), \sup \mu_B(x)]), (\min[\inf \sigma_A(x),$
$\inf \sigma_B(x)], \min[\sup \sigma_A(x), \sigma_B(x)]), (\max[\inf \omega_A(x), \inf \omega_B(x)], \max[\sup \omega_A(x),$
$\sup \omega_B(x)]), (\min[\inf \gamma_A(x), \inf \gamma_B(x)], \min[\sup \gamma_A(x), \sup \gamma_B(x)])\rangle / x : x \in X\}$.
(iii) *A difference, B, is defined by*
$A \setminus B = \{\langle(\min[\inf \mu_A(x), \inf \gamma_B(x)], \min[\sup \mu_A(x), \sup \gamma_B(x)]), (\max[\inf \sigma_A(x), 1 - \sup \sigma_B(x)],$
$\max[\sup \sigma_A(x), 1 - \inf \sigma_B(x)]), (\min[\inf \omega_A(x), 1 - \sup \omega_B(x)], \min[\sup \omega_A(x), 1 - \inf \omega_B(x)]),$
$(\max[\inf \gamma_A(x), \inf \mu_B(x)], \max[\sup \gamma_B(x), \sup \mu_B(x)])\rangle / x : x \in X\}$.
(iv) *Scalar multiplication of A is defined by*
$A.a = \{\langle(\min[\inf \mu_A(x).a, 1], \min[\sup \mu_A(x).a, 1]), (\min[\inf \sigma_A(x).a, 1], \min[\sup \sigma_A(x).a, 1]),$
$(\min[\inf \omega_A(x).a, 1], \min[\sup \omega_A(x).a, 1]), (\min[\inf \gamma_A(x).a, 1], \min[\sup \gamma_A(x).a, 1])\rangle / x : x \in X\}$.
(v) *Scalar division of A is defined by*
$A/a = \{\langle(\min[\inf \mu_A(x)/a, 1], \min[\sup \mu_A(x)/a, 1]), (\min[\inf \sigma_A(x)/a, 1], \min[\sup \sigma_A(x)/a, 1]),$
$(\min[\inf \omega_A(x)/a, 1], \min[\sup \omega_A(x)/a, 1]), (\min[\inf \gamma_A(x)/a, 1], \min[\sup \gamma_A(x)/a, 1])\rangle / x : x \in X\}$.

Definition 8. *Let X be a non-empty set; IVNSS(X) denotes the set of all interval-valued neutrosophic support soft sets of X and a subset, A, of X . The soft set function is*

$$g_i : A \to IVNSS(x).$$

The interval valued neutrosophic support soft setover X can be represented by

$$G_i = \{(y, g_i(y)) : y \in A\},\ such\ that\ g_i(y) = \emptyset\ if x \notin X.$$

Example 4. *Consider the interval-valued neutrosophic support soft set, $\langle G_i, A\rangle$, where X is a set of two brands of mobile phones being considered by a decision maker to purchase, which is denoted by $X = \{a, b\}$, and A is a parameter set, where $A = \{y_1 = price,\ y_2 = camera\ specification,\ y_3 = Efficency,$ and $y_4 = size,$ $y_5 = processsor\}$. In this case, we define a set G_i over X as follows:*

G_i	a	b
y_1	[0.6,0.8],[0.8,0.9][0.5,0.6][0.1,0.5]	[0.6,0.8][0.1,0.8][0.3,0.7][0.1,0.7]
y_2	[0.2,0.4][0.5,0.8][0.4,0.3][0.3,0.8]	[0.2,0.8][0.6,0.9][0.5,0.8][0.2,0.3]
y_3	[0.1,0.9][0.2,0.5][0.5,0.7][0.6,0.8]	[0.4,0.9][0.2,0.6][0.5,0.6][0.5,0.7]
y_4	[0.6,0.8][0.8,0.9][0.1,0.9][0.8,0.9]	[0.5,0.7][0.6,0.8][0.7,0.9][0.1,0.8]
y_5	[0.0,0.9][1.0,0.1][1.0,0.9][1.0,1.0]	[0.0,0.9][0.8,1.0][0.3,0.5][0.2,0.5]

Clearly, we can see that the exact evaluation of each object on each parameter is unknown, while the lower limit and upper limit of such an evaluation are given. For instance, we cannot give the exact membership degree, support, indeterminacy and nonmembership degree of price 'a'; however, the price of model 'a' is at least on the membership degree of 0.6 and at most on the membership degree of 0.8.

Definition 9. *Let G_i be a interval valued neutrosophic support soft set of X. Then, G_i is known as an empty interval valued neutrosophic support soft set, if $g_i(y) = \emptyset$.*

Definition 10. *Let G_i be a interval valued neutrosophic support soft set of X. Then, G_i is known as the universal interval valued neutrosophic support soft set, if $g_i(y) = X$.*

Definition 11. *Let G_i ,G_j be two interval valued neutrosophic support soft set of X. Then, G_i is said to be subset of G_j, if $g_i(y) \subseteq g_j(y)$.*

Example 5. *Two interval-valued neutrosophic support soft sets, G_i and G_j, are constructed as follows:*

G_i	a	b
y_1	[0.6,0.8],[0.8,0.9][0.5,0.6][0.1,0.5]	[0.6,0.8][0.1,0.8][0.3,0.7][0.1,0.7]
y_2	[0.2,0.4][0.5,0.8][0.4,0.3][0.3,0.8]	[0.2,0.8][0.6,0.9][0.5,0.8][0.2,0.3]
y_3	[0.1,0.9][0.2,0.5][0.5,0.7][0.6,0.8]	[0.4,0.9][0.2,0.6][0.5,0.6][0.5,0.7]
y_4	[0.6,0.8][0.8,0.9][0.1,0.9][0.8,0.9]	[0.5,0.7][0.6,0.8][0.7,0.9][0.1,0.8]
y_5	[0.0,0.9][1.0,0.1][1.0,0.9][1.0,1.0]	[0.0,0.9][0.8,1.0][0.3,0.5][0.2,0.5]

G_j	a	b
y_1	[0.7,0.8],[0.7,0.9][0.6,0.6][0.1,0.5]	[0.7,0.9][0.0,0.8][0.4,0.8][0.1,0.6]
y_2	[0.3,0.6][0.5,0.5][0.5,0.3][0.2,0.6]	[0.4,0.8][0.6,0.9][0.5,0.8][0.1,0.2]
y_3	[0.2,1.0][0.2,0.5][0.5,0.7][0.5,0.7]	[0.5,0.9][0.2,0.6][0.6,0.6][0.5,0.5]
y_4	[0.6,0.8][0.8,0.9][0.1,0.7][0.8,0.9]	[0.6,0.8][0.6,0.8][0.9,0.9][0.1,0.4]
y_5	[0.1,1.0][0.9,0.1][1.0,1.0][0.9,0.8]	[0.2,0.9][0.7,0.9][0.3,0.5][0.2,0.5]

Following Definition 11, G_i is a subset of G_j.

Definition 12. *The two interval valued neutrosophic support soft sets, G_i, G_j, such that $G_i \subseteq G_j$, is said to be classical subset of X where every element of G_i does not need to be an element of G_j*

Proposition 1. *Let G_i, G_j, G_k be an interval valued neutrosophic support soft set of X. Then,*

(1) Each G_n is a subset of G_X, where n= i,j,k;

(2) *Each G_n is a superset of $G_{\emptyset}$, where n= i,j,k;*
(3) *If G_i is a subset of G_j and G_j is a subset of G_k, then, G_i is a subset of G_k.*

Proof. The proof of this proposition is obvious. □

Definition 13. *The two interval valued neutrosophic support soft sets of X are said to be equal, if and only if $g_i = g_j$, for all $i, j \in X$*

Proposition 2. *Let X be a non-empty set and G_i, G_j be an interval valued neutrosophic support soft set of X. G_i is a subset of G_j, and G_j is a subset of G_i, if and only if G_i is equal to G_j*

Definition 14. *The complement of the interval valued neutrosophic support soft set, G_i, of X is denoted by G_{i^c}, for all $i \in A$*

(i) *The complement of the empty interval valued neutrosophic support soft set of X is the universal interval valued neutrosophic support soft setof X.*
(ii) *The complement of the universal interval valued neutrosophic support soft set of X is the empty interval valued neutrosophic support soft set of X.*

Theorem 1. *Let G_i, G_j be an interval valued neutrosophic support soft set of X. Then, G_i is a subset of G_j and the complement of G_j is a subset of the complement of G_i.*

Proof. Let G_i, and G_j be an interval valued neutrosophic support soft set of X. By definition, 3.7 G_i is a subset of G_j if $g_i(y) \subseteq g_j(y)$. Then, the complement of $g_i(y) \subseteq g_j(y)$ is $g_i^c(y) \supseteq g_j^c(y)$. Hence, the complement of G_j is a subset of the complement of G_i. □

Example 6. *From Example 4, the complement of G_i is constructed as follows:*

G_{i^c}	a	b
y_1	[0.1,0.5],[0.1,0.2][0.4,0.5][0.6,0.8]	[0.1,0.7][0.2,0.9][0.3,0.7][0.6,0.8]
y_2	[0.3,0.8][0.2,0.5][0.6,0.7][0.2,0.4]	[0.2,0.3][0.1,0.4][0.2,0.5][0.2,0.8]
y_3	[0.6,0.8][0.5,0.8][0.3,0.5][0.1,0.9]	[0.5,0.7][0.4,0.8][0.4,0.5][0.4,0.9]
y_4	[0.8,0.9][0.1,0.2][0.3,0.9][0.6,0.8]	[0.1,0.8][0.2,0.4][0.1,0.3][0.5,0.7]
y_5	[1.0,1.0]0.0,0.9][0.0,0.1][0.0,0.9]	[0.2,0.5][0.0,0.2][0.5,0.7][0.0,0.8]

Definition 15. *The union of the interval valued neutrosophic support soft set of X is denoted by $G_i \cup G_j$ and is defined by $g_i(y) \cup g_j(y) = g_j(y) \cup g_i(y)$ for all $y \in A$.*

Proposition 3. *Let G_i,G_j,G_k be an interval valued neutrosophic support soft set of X. Then,*

(i) $G_i \cup G_{\emptyset} = G_i$.
(ii) $G_i \cup G_X = G_X$.
(iii) $G_i \cup G_j = G_j \cup G_i$.
(iv) $(G_i \cup G_j) \cup G_k = G_i \cup (G_j \cup G_k)$.

Example 7. *From Example 4, the union of two sets is represented as follows:*

$G_i \cup G_j$	a	b
y_1	[0.7,0.8],[0.7,0.9][0.6,0.6][0.1,0.5]	[0.7,0.9][0.0,0.8][0.4,0.8][0.1,0.6]
y_2	[0.3,0.6][0.5,0.5][0.5,0.3][0.2,0.6]	[0.4,0.8][0.6,0.9][0.5,0.8][0.1,0.2]
y_3	[0.2,1.0][0.2,0.5][0.5,0.7][0.5,0.7]	[0.5,0.9][0.2,0.6][0.6,0.6][0.5,0.5]
y_4	[0.6,0.8][0.8,0.9][0.1,0.7][0.8,0.9]	[0.6,0.8][0.6,0.8][0.9,0.9][0.1,0.4]
y_5	[0.1,1.0]0.1,0.9][1.0,1.0][0.8,0.9]	[0.2,0.9][0.7,0.9][0.3,0.5][0.2,0.5]

Definition 16. *Let G_i,G_j be an interval valued neutrosophic support soft set of X. Then, the intersection of two sets denoted by $G_i \cap G_j$ is defined as $g_i(y) \cap g_j(y) = g_j(y) \cap g_i(y)$ for all $y \in A$.*

Proposition 4. *Let G_i,G_j,G_k be an interval valued neutrosophic support soft set of X. Then,*

(i) $G_i \cap G_\emptyset = G_\emptyset$.
(ii) $G_i \cap G_X = G_i$.
(iii) $G_i \cap G_j = G_j \cap G_i$.
(iv) $(G_i \cap G_j) \cap G_k = G_i \cap (G_j \cap G_k)$.

Proof. The proof is obvious. □

Example 8. *In accordance with Example 4, the intersection operation is performed as follows:*

$G_i \cap G_j$	a	b
y_1	[0.6,0.8],[0.8,0.9][0.5,0.6][0.1,0.5]	[0.6,0.8][0.1,0.8][0.3,0.7][0.1,0.7]
y_2	[0.2,0.4][0.5,0.8][0.3,0.4][0.3,0.8]	[0.2,0.8][0.6,0.9][0.5,0.8][0.2,0.3]
y_3	[0.1,0.9][0.2,0.5][0.5,0.7][0.6,0.8]	[0.4,0.9][0.2,0.6][0.5,0.6][0.5,0.7]
y_4	[0.6,0.8][0.8,0.9][0.1,0.7][0.8,0.9]	[0.5,0.7][0.6,0.8][0.7,0.9][0.1,0.8]
y_5	[0.0,0.9]0.1,0.9][0.9,1.0][1.0,1.0]	[0.0,0.8][0.8,1.0][0.3,0.5][0.2,0.5]

Definition 17. *Let G_i be an interval valued neutrosophic support soft set of X. Then, the union of interval valued neutrosophic support soft setand its complement is not a universal set and it is not mutually disjoint.*

Proposition 5. *Let G_i, G_j be an interval valued neutrosophic support soft set of X. Then, the D'Margan Laws hold.*

(i) $(G_i \cup G_j)^c = G_i^c \cap G_j^c$.
(ii) $(G_i \cap G_j)^c = G_i^c \cup G_j^c$.

Proposition 6. *Let G_i,G_j, G_k be an interval valued neutrosophic support soft set of X. Then, the following hold.*

(i) $G_i \cup (G_j \cap G_k) = (G_i \cup G_j) \cap (G_i \cap G_k)$.
(ii) $G_i \cap (G_i \cup G_j) = (G_i \cap G_j) \cup (G_i \cap G_k)$

Definition 18. *Let G_i, G_j be an interval valued neutrosophic support soft set of X. Then, the difference between two sets is denoted by G_i/G_j and is defined by*

$$g_{i/j}(y) = g_i(y)/g_j(y)$$

for all $y \in A$.

Definition 19. *Let G_i, G_j be an interval valued neutrosophic support soft set of X. Then the addition of two sets are denoted by $G_i + G_j$ and is defined by*

$$g_{i+j}(y) = g_i(y) + g_j(y)$$

for all $y \in A$.

Definition 20. *Let G_i be an interval valued neutrosophic support soft set of X. Then, the scalar division of G_I is denoted by G_i/a and is defined by*

$$g_{i/a}(y) = g_i(y)/a$$

for all $y \in A$.

4. Decision-Making

In this paper, we provide the definition of relationship between the interval valued neutrosophic support soft set, the average interval valued neutrosophic support soft setand the algorithm to get the optimum decision.

Definition 21. *Let G_i be an interval valued neutrosophic support soft set of X. Then, the relationship, R, for G_i is defined by*

$$R_{G_i} = \{r_{G_i}(y,a) : r_{G_i}(y,a) \in \ interval\ valued\ neutrosophic\ support\ set.\ y \in A, a \in X\}$$

where $r_{G_i} : A \setminus X \Rightarrow$ interval valued neutrosophic support soft set (X) and $r_{G_i}(y,a) = g_{i(y)}(a)$ for all $y \in A$ and $a \in X$

Example 9. *From Example 4, the relationship for the interval valued neutrosophic support soft set of X is given below.*
$g_{i(y_1)}(a) = \langle [0.6,0.8], [0.8,0.9], [0.5,0.6], [0.1,0.5] \rangle$,
$g_{i(y_1)}(b) = \langle [0.6,0.8], [0.1,0.8], [0.3,0.7], [0.1,0.7] \rangle$,
$g_{i(y_2)}(a) = \langle [0.2,0.4], [0.5,0.8], [0.4,0.3], [0.3,0.8] \rangle$,
$g_{i(y_2)}(b) = \langle [0.2,0.8], [0.6,0.9], [0.5,0.8], [0.4,0.3] \rangle$,
$g_{i(y_3)}(a) = \langle [0.1,0.9], [0.2,0.5], [0.5,0.7], [0.6,0.8] \rangle$,
$g_{i(y_3)}(b) = \langle [0.4,0.9], [0.2,0.6], [0.5,0.6], [0.5,0.7] \rangle$,
$g_{i(y_4)}(a) = \langle [0.6,0.8], [0.8,0.9], [0.1,0.7], [0.8,0.9] \rangle$,
$g_{i(y_4)}(b) = \langle [0.5,0.7], [0.6,0.8], [0.7,0.9], [0.1,0.8] \rangle$,
$g_{i(y_5)}(a) = \langle [0.0,0.9], [1.0,0.1], [1.0,0.9], [1.0,1.0] \rangle$,
$g_{i(y_5)}(b) = \langle [0.0,0.8], [0.8,1.0], [0.3,0.5], [0.2,0.5] \rangle$.

Definition 22. *Let G_i be an interval valued neutrosophic support soft set of X. For $\mu, \sigma, \omega, \gamma \subseteq [0,1]$, the $(\mu, \sigma, \omega, \gamma)$-level support soft set of G_i defined by $\langle G_i; (\mu, \sigma, \omega, \gamma) \rangle = \{(y_i, \{a_{ij}:a_{ij} \in X, \mu(a_{ij}) = 1\}):y \in A\}$, where*

$$\mu(a_{ij}) = \begin{cases} 1, & if\ (\mu, \sigma, \omega, \gamma) \leq g_i(y_i)(a_j) \\ 0, & if\ otherwise \end{cases}. \text{ For all } a_j \in X.$$

Definition 23. *Let G_i be an interval valued neutrosophic support soft set of X. The average interval valued neutrosophic support soft set is defined by $\langle \mu, \sigma, \omega, \gamma \rangle Avg_{G_i}(y_i) = \sum_{a \in X} g_{i(y_i)}(a)/|X|$ for all $y \in A$*

Example 10. *Considering Example 4, the average interval valued neutrosophic support soft set is calculated as follows:*

$$\langle \mu, \sigma, \omega, \gamma \rangle Avg_{G_i}(y_1) = \sum_{i=1}^{2} g_{i(y_1)}(a)/|X| = \langle [0.6,0.8], [0.45,0.85], [0.4,0.65], [0.1,0.6] \rangle$$

$$\langle \mu, \sigma, \omega, \gamma \rangle Avg_{G_i}(y_2) = \sum_{i=1}^{2} g_{i(y_2)}(a)/|X| = \langle [0.2,0.6], [0.55,0.85], [0.45,0.55], [0.25,0.55] \rangle$$

$$\langle \mu, \sigma, \omega, \gamma \rangle Avg_{G_i}(y_3) = \sum_{i=1}^{2} g_{i(y_3)}(a)/|X| = \langle [0.25,0.9], [0.2,0.55], [0.5,0.65], [0.55,0.75] \rangle$$

$$\langle \mu, \sigma, \omega, \gamma \rangle Avg_{G_i}(y_4) = \sum_{i=1}^{2} g_{i(y_4)}(a)/|X| = \langle [0.55,0.75], [0.7,0.85], [0.4,0.8], [0.45,0.85] \rangle$$

$$\langle \mu, \sigma, \omega, \gamma \rangle Avg_{G_i}(y_5) = \sum_{i=1}^{2} g_{i(y_5)}(a)/|X| = \langle [0.0, 0.85], [0.9, 0.55], [0.65, 0.7], [0.6, 0.75] \rangle$$

Theorem 2. *Let X be a non-empty set and G_i, G_j be an interval valued neutrosophic support soft set of X. $\{G_i; \langle \mu_1, \sigma_1, \omega_1, \gamma_1 \rangle\}$ and $\{G_i; \langle \mu_2, \sigma_2, \omega_2, \gamma_2 \rangle\}$ are level support soft sets if $\langle \mu_1, \sigma_1, \omega_1, \gamma_1 \rangle \leq \langle \mu_2, \sigma_2, \omega_2, \gamma_2 \rangle$. Then, $\{G_i; \langle \mu_1, \sigma_1, \omega_1, \gamma_1 \rangle\} \leq \{G_i; \langle \mu_2, \sigma_2, \omega_2, \gamma_2 \rangle\}$.*

Proof. Let G_i and G_j be an interval valued neutrosophic support soft set of X. In accordance with Definition 3.2 (iv), each function is $\mu_1 \leq \mu_2$, $\sigma_1 \leq \sigma_2$, $\omega_1 \leq \omega_2$, $\gamma_1 \geq \gamma_2$. Thus, the corresponding interval valued neutrosophic support soft set is $\{G_i; \langle \mu_1, \sigma_1, \omega_1, \gamma_1 \rangle\} \leq \{G_j; \langle \mu_2, \sigma_2, \omega_2, \gamma_2 \rangle\}$. Hence, the proof. □

The following algorithm is used to make decisions in an interval-valued neutrosophic support soft set.

Algorithm 1:

(1) Enter the interval valued neutrosophic support soft set, G_i;
(2) Enter the average interval valued neutrosophic support soft set, $\langle \mu, \sigma, \omega, \gamma \rangle Avg_{G_i}$, using average-level decision rules to make decisions;
(3) Determine the average-level support soft set, $G_i; \langle \mu, \sigma, \omega, \gamma \rangle Avg_{G_i}$;
(4) Present the level support soft set in tabular form;
(5) Determine the choice value, C_{v_i}, of a_i for any $a \in X$;
(6) Select the optimum value for the optimum decision, $C_{v_i} = \max_{a_i \in X} C_{v_i}$.

Example 11. *People who are affected by cancer, have a combination of treatments, such as surgery with chemotherapy and/or radiation therapy, hormone therapy, and immunotherapy. Our main objective is to find the best treatment from the above mentioned therapies. However, all the treatments can cause side effects. Our goal is to find the best treatment which cause the least side effects, reduce the cost of the treatment, extend the patient's life, cure the cancer and control its growth using an interval-valued neutrosophic support soft set.*

G_i	a	b
y_1	[0.4,0.7][0.8,0.8][0.4,0.8][0.3,0.5]	[0.3,0.6][0.3,0.8][0.3,0.7][0.3,0.8]
y_2	[0.1,0.3][0.6,0.7][0.2,0.3][0.3,0.8]	[0.2,0.7][0.7,0.9][0.3,0.6][0.3,0.4]
y_3	[0.2,0.6][0.4,0.5][0.1,0.5][0.7,0.8]	[0.4,0.9][0.1,0.6][0.3,0.8][0.5,0.7]
y_4	[0.6,0.9][0.6,0.9][0.6,0.9][0.6,0.9]	[0.5,0.9][0.6,0.8][0.2,0.8][0.1,0.7]
y_5	[0.0,0.9]1.0,1.0][1.0,1.0][1.0,1.0]	[0.0,0.9][0.8,1.0][0.1,0.4][0.2,0.5]

G_i	c	d
y_1	[0.5,0.7][0.8,0.9][0.4,0.8][0.2,0.5]	[0.3,0.6][0.3,0.9][0.2,0.8][0.2,0.8]
y_2	[0.0,0.3][0.6,0.8][0.1,0.4][0.3,0.9]	[0.1,0.8][0.8,0.9][0.2,0.9][0.3,0.5]
y_3	[0.1,0.7][0.4,0.5][0.2,0.8][0.8,0.9]	[0.2,0.5][0.5,07][0.3,0.6][0.6,0.8]
y_4	[0.2,0.4][0.7,0.9][0.6,0.8][0.6,0.9]	[0.3,0.9][0.6,0.9][0.2,0.8][0.3,0.9]
y_5	[0.0,0.2][1.0,1.0][1.0,1.0][1.0,1.0]	[0.0,0.1][0.9,1.0][0.2,0.2][0.2,0.9]

1. *The average interval valued neutrosophic support soft set is determined as follows:*

$$\langle \mu, \sigma, \omega, \gamma \rangle Avg_{G_i} = \{\langle (0.375, 0.65), (0.55, 0.85), (0.325, 0.775), (0.25, 0.6), \rangle / y_1, \langle (0.125, 0.575),$$

$$(0.675, 0.825), (0.2, 0.5), (0.3, 0.65) \rangle / y_2, \langle (0.225, 0.675), (0.35, 0.575), (0.225, 0.675), (0.65, 0.8) \rangle$$

$$/ y_3, \langle (0.4, 0.775), (0.625, 0.875), (0.4, 0.825), (0.4, 0.85) \rangle / y_4, \langle (0.0, 0.525), (0.825, 1.0),$$

$$(0.575, 0.625), (0.6, 0.85)\rangle / y_5\};$$

2. $\{G_i; \langle \mu, \sigma, \omega, \gamma \rangle Avg_{G_i}\} = \{(y_2, b), (y_3, b), (y_4, a), (y_5, b)\};$
3. *The average-level support soft set,* $\{G_i; \langle \mu, \sigma, \omega, \gamma \rangle Avg_{G_i}\}$ *is represented in tabular form.*

X	a	b	c	d
y_1	0	0	0	0
y_2	0	1	0	0
y_3	0	1	0	0
y_4	1	0	0	0
y_5	0	1	0	0

4. *Compute the choice value,* C_{v_i}, *of* a_i *for all* $a_i \in X$ *as*

$$C_{v_3} = C_{v_4} = \sum_{j=1}^{4} a_{3j} = \sum_{j=1}^{4} a_{4j} = 0,\ C_{v_1} = \sum_{j=1}^{4} a_{1j} = 1,\ C_{v_2} = \sum_{j=1}^{4} a_{2j} = 3;$$

5. C_{v_2} *gives the maximum value. Therefore b is the optimum choice.*

   Now, we conclude that there are a few ways to get rid of cancer, but surgery chemotherapy is preferred by most of the physicians with respect to the cost of treatment and extending the life of the patient with the least side effects. Moreover, side effects will be reduced or vanish completely after finished chemotherapy, and the cancer and its growth will be controlled.

5. Conclusions and Future Work

Fuzzy sets are inadequate for representing some parameters. Therefore, intuitionistic fuzzy sets were introduced to overcome this inadequacy. Further, neutrosophic sets were introduced to represent the indeterminacy. In order to make decisions efficiently, we offer this new research work which does not violate the basic definitions of neutrosophic sets and their properties. In this paper, we add one more function called the support function in interval-valued neutrosophic soft set, and we also provide the basic definition of interval valued neutrosophic support soft set and some of its properties. Further, we framed an algorithm for making decisions in medical science with a real-life problem. Here, we found the best treatment for cancer under some constraints using interval valued neutrosophic support soft set. In the future, motivated by the interval valued neutrosophic support soft set, we aim to develop interval valued neutrosophic support soft set in ideal topological spaces. In addition, weaker forms of open sets, different types of functions and theorems can be developed using interval valued neutrosophic support soft set to allow continuous function. This concept may be applied in operations research, data analytics, medical sciences, etc. Industry may adopt this technique to minimize the cost of investment and maximize the profit.

Author Contributions: All authors have contributed equally to this paper. The individual responsibilities and contributions of all authors can be described as follows: the idea of this whole paper was put forward by M.P. and M.K. R.U. and M.K. completed the preparatory work of the paper. F.S. and S.J. analyzed the existing work. The revision and submission of this paper was completed by M.P. and F.S.

Conflicts of Interest: The authors declare no conflict of interest.

References

1. Zadeh, L.A. Fuzzy sets. *Inf. Control* **1965**, *8*, 338–353. [CrossRef]
2. Atanassov, K.T. Intuitionstic fuzzy sets. *Fuzzy Sets Syst.* **1986**, *20*, 87–96. [CrossRef]
3. Molodtsov, D. Soft set theory—First results. *Comput. Math. Appl.* **1999**, *37*, 19–31. [CrossRef]
4. Maji, P.K. Neutrosophic soft sets. *Ann. Fuzzy Math. Inf.* **2013**, *5*, 157–168.
5. Maji, P. K.; Roy, A.R.; Biswass, R. An application of soft sets in a decision making problem. *Comput. Math. Appl.* **2002**, *44*, 1077–1083. [CrossRef]

6. Parimala, M.; Smarandache, F.; Jafari, S.; Udhayakumar, R. On Neutrosophic $\alpha\psi$ -Closed Sets. *Information* **2018**, *9*, 103. [CrossRef]
7. Parimala, M.; Karthika, M.; Dhavaseelan, R.; Jafari, S. On neutrosophic supra pre-continuous functions in neutrosophic topological spaces. In *New Trends in Neutrosophic Theory and Applications*; European Union: Brussels, Belgium, 2018; Volume 2, pp. 371–383.
8. Smarandache, F. *Neutrosophy. Neutrosophic Probability, Set, and Logic*; ProQuest Information & Learning: Ann Arbor, MI, USA, 1998; 105p.
9. Broumi, S.; Smarandache, F. Intuitionistic neutrosophic soft set. *J. Inf. Comput. Sc.* **2013**, *8*, 130–140.
10. Wang, H.; Smarandache, F.; Zhang, Y.Q.; Sunderraman, R. *Interval Neutrosophic Sets and Logic: Theory and Applications in Computing*; Neutrosophic Book Series, No. 5; Hexis: Staffordshire, UK, 2005.
11. Thao, N.X.; Smarandache, F.; Dinh, N.V. Support-Neutrosophic Set: A New Concept in Soft Computing. *Neutrosophic Sets Syst.* **2017**, *16*, 93–98.
12. Deli, I. Interval-valued neutrosophic soft sets and its decision making. *Int. J. Mach. Learn. Cyber.* **2017**, *8*, 665–676. [CrossRef]
13. Cagman, N.; Citak, F.; Enginoglu, S. FP-soft set theory and its applications. *Ann. Fuzzy Math. Inform.* **2011**, *2*, 219–226.

symmetry

MDPI

Article

A Hybrid Neutrosophic Group ANP-TOPSIS Framework for Supplier Selection Problems

Mohamed Abdel-Basset [1,*], **Mai Mohamed** [1] **and Florentin Smarandache** [2]

[1] Department of Operations Research, Faculty of Computers and Informatics, Zagazig University, Sharqiyah 44519, Egypt; mmgaafar@zu.edu.eg
[2] Math & Science Department, University of New Mexico, Gallup, NM 87301, USA; smarand@unm.edu
* Correspondence: analyst_mohamed@yahoo.com

Received: 9 May 2018; Accepted: 12 June 2018; Published: 15 June 2018

Abstract: One of the most significant competitive strategies for organizations is sustainable supply chain management (SSCM). The vital part in the administration of a sustainable supply chain is the sustainable supplier selection, which is a multi-criteria decision-making issue, including many conflicting criteria. The valuation and selection of sustainable suppliers are difficult problems due to vague, inconsistent and imprecise knowledge of decision makers. In the literature on supply chain management for measuring green performance, the requirement for methodological analysis of how sustainable variables affect each other, and how to consider vague, imprecise and inconsistent knowledge, is still unresolved. This research provides an incorporated multi-criteria decision-making procedure for sustainable supplier selection problems (SSSPs). An integrated framework is presented via interval-valued neutrosophic sets to deal with vague, imprecise and inconsistent information that exists usually in real world. The analytic network process (ANP) is employed to calculate weights of selected criteria by considering their interdependencies. For ranking alternatives and avoiding additional comparisons of analytic network processes, the technique for order preference by similarity to ideal solution (TOPSIS) is used. The proposed framework is turned to account for analyzing and selecting the optimal supplier. An actual case study of a dairy company in Egypt is examined within the proposed framework. Comparison with other existing methods is implemented to confirm the effectiveness and efficiency of the proposed approach.

Keywords: sustainable supplier selection problems (SSSPs); analytic network process; interdependency of criteria; TOPSIS; neutrosophic set

1. Introduction

The major priority for decision makers and managers in many fields such as agriculture, tourism, business development or manufacturing is the management of environmental and social issues, and the emergency to address them with the economic factors [1]. The sustainability is the synthesis of social, environmental and economic development [2]. The sustainability applies to all pertinent supply chain sides in supply chain management [3]. In sustainable supply chain management, managers seek to enhance the economic realization of their organization not only to survive, but also to succeed in close and distant future. The social and environmental activities that can enhance economic goals of organizations should be undertaken by managers in sustainable supply chain management [4]. Selecting the sustainable suppliers is very significant when designing new strategies and models in the case of lack of available knowledge and resources. Thus, the most important part in sustainable supply chain management is to construct and implement an effective and efficient supplier section process [5]. The supplier selection problems, combining social and environmental factors for estimating and ranking suppliers to select the best, can be regarded as a sustainable supplier selection problems (SSSPs). The selection process of sustainable suppliers involves several conflicting criteria. The evaluation and selection of suppliers is very difficult due to vague, inconsistent and imprecise knowledge of decision makers. In order to deal with vague information, Zadeh introduced the theory of fuzzy sets in 1965 [6]. It is difficult to identify the truth-membership degree of a fuzzy set to a specific value. Therefore, Turksen introduced interval-valued fuzzy sets in 1986 [7]. Because fuzzy set only considers the truth-membership (membership) degree and fails to consider falsity-membership (non-membership) degree, Atanassov introduced intuitionistic fuzzy sets [8]. Moreover, intuitionistic fuzzy sets were expanded to interval-valued intuitionistic fuzzy sets [9]. The intuitionistic fuzzy sets have been exercised to disband multi-criteria decision-making problems [10–12]. The fuzzy and intuitionistic fuzzy sets fail to treat all types of uncertainties such as indeterminacy and inconsistency that exist usually in natural decision-making processes. For instance, when a decision maker gives his/her judgment toward anything, he/she may say that: this statement is 50% correct, 60% false and 20% I am not sure [13]. From this concept, Smarandache suggested the neutrosophic logic, probability and sets [14–16]. In neutrosophy, the indeterminacy degree is independent of truth and falsity degrees [17]. To facilitate the practical side of neutrosophic sets, a single-valued neutrosophic set (SVNS) was presented [13,18]. In real life problems, the statement could not be accurately defined by a certain degree of truth, indeterminacy and falsity, but indicated by various interval values. Therefore, interval neutrosophic set (INS) was conceptualized. The interval neutrosophic set (INS) was introduced by Wang et al. [19]. The authors in [17] used interval-valued neutrosophic set to present multi-criteria decision-making (MCDM) problems using aggregation operators. The neutrosophic linguistic environment was used by Broumi and Smarandache [20] to deal with multi-criteria decision-making problems. Zhang et al. [21] introduced an outranking technique to solve MCDM problems by using an interval-valued neutrosophic set. However, the current literature did not advance the integration of ANP and TOPSIS using INS for solving sustainable supplier selection problems. Consequently, we are the first to use an interval-valued neutrosophic set for representing a group ANP-TOPSIS framework for sustainable supplier selection.

Research Contribution

Our contribution can be summed up as follows:

- The sustainable supplier selection is a multi-criteria decision-making issue including many conflicting criteria. The valuation and selection of sustainable suppliers is a difficult problem due to vague, inconsistent and imprecise knowledge of decision makers. The literature on supply chain management for measuring green performance, the requirement for methodological analysis of how sustainable variables affect each other and of how to consider vague, imprecise and inconsistent knowledge is somehow inconclusive, but these drawbacks have been treated in our research.
- In most cases, the truth, falsity and indeterminacy degrees cannot be defined precisely in the real selection of sustainable suppliers, but denoted by several possible interval values. Therefore, we presented ANP TOPSIS, and combined them with interval-valued neutrosophic sets to select sustainable suppliers for the first time.
- The integrated framework leads to accurate decisions due to the way it treats uncertainty. The sustainable criteria for selecting suppliers are determined from the cited literature and the features of organizations under analysis. Then, the decision makers gather data and information.
- We select ANP and TOPSIS for solving sustainable supplier selection problems for the following reasons:
    - Since the independent concept of criteria is not constantly right and in actual life, there exist criteria dependent on each other, and we used ANP for precise weighting of criteria.

- The ANP needs many pairwise comparison matrices based on numerals and interdependence of criteria and alternatives, and, to escape this drawback, the TOPSIS was used to rank alternatives.
- The main problem of sustainable supplier selection problems is how to design and implement a flexible model for evaluating all available suppliers; since it considers the uncertainty that usually exists in real life, our model is the best.
- The proposed framework is used to study the case of a dairy and foodstuff company in Egypt, and can be employed to solve any sustainable supplier selection problem of any other company.
- Comparison with other existing methods, which are popular and attractive, was presented to validate our model.

The plan of this research is as follows: a literature review on the multi-criteria decision-making techniques to disband sustainable supplier selection problems is presented in Section 2. The basic concepts and definitions of interval-valued neutrosophic sets and its operations are discussed in Section 3. The ANP and TOPSIS methods are described in Section 4. The proposed framework for selecting optimal suppliers is presented in Section 5. An actual case study of a dairy and foodstuff company in Egypt is examined in Section 6. The conclusion and future directions are presented in Section 7.

2. Literature Review

Many research works intensify a supplier selection problem using various MCDM methods. For listing the optimal supplier under environmental factors, Govindan et al. [22] proposed a fuzzy TOPSIS framework. For evaluating sustainable suppliers' performance in a supply chain, Erol et al. [23] validated a multi-criteria setting based on fuzzy multi-attribute utility. The fuzzy inference system, the fuzzy logic and ranking method are used to address the subjectivity of DM estimation.

To handle sustainable supplier selection in a group decision environment, Wen et al. [24] proposed a fuzzy intuitionistic TOPSIS model. To analyze sustainability criteria and select the optimal sustainable supplier, Orji and Wei [25] used fuzzy logic, decision-making trial and evaluation laboratory (DEMATEL) and TOPSIS.

To bridge the gap between numerous existing research works on supplier selection and others who depend on environmental issues, Shaw et al. [26] were the first to employ AHP in fuzzy environment for green supplier selection. The fuzzy ANP and multi-person decision-making schema through imperfect preference relations are used by Buyukozkan and Cifci [27].

The requirements of company stakeholders are translated into multiple criteria for supplier selection by Ho et al. [28] by using a QFD approach. A family group decision-making model was developed by Dursun and Karsak [29] by using a QFD method to determine the characteristics that a product must hold to achieve customer needs and construct the assessment criteria for suppliers. A two-stage structure including data envelopment analysis (DEA) and rough set theory was proposed by Bai and Sarkis [30] to determine and evaluate relative performance of suppliers.

To rank sustainable suppliers, Kumar et al. [31] proposed a unified green DEA model. A fuzzy DEA model was used by Azadi et al. [32] to measure the efficiency, effectiveness and productivity of sustainable suppliers. To optimize supplier selection processes, numerous models have been integrated. The integrated analytic frameworks were combined through the recent research: ANP and/or AHP integrated with QFD by many researchers [33–38]. The DEMATEL was integrated with fuzzy ANP and TOPSIS as in [39]. Kumaraswamy et al. [40] integrated QFD with TOPSIS.

The integration of a fuzzy Delphi approach, ANP and TOPSIS were proposed by Chung et al. [41] for supplier selection. A review of multi-attribute decision-making techniques for evaluating and selecting suppliers in fuzzy environment is presented in [42]. In addition, the ANP was integrated with intuitionistic fuzzy TOPSIS by Rouyendegh [43] for selecting an optimal supplier. Tavana et al. [44] integrated ANP with QFD for sustainable supplier selection.

A neutrosophic group decision-making technique based on TOPSIS was proposed by Şahin and Yiğider for a supplier selection problem [45]. A hybrid multi-criteria group decision-making technique based on interval-valued neutrosophic sets was proposed by Reddy et al. [46] for lean supplier selection. An extended version of EDAS using an interval valued neutrosophic set for a supplier selection problem is presented in [47]. A quality function deployment technique for supplier selection and evaluation based on an interval neutrosophic set is presented in [48]. To develop supplier selection criteria, the DEMATEL technique is presented in neutrosophic environment, as in [49].

The main criteria for supplier selection problems have been identified in many studies. The economic factors, which were considered in traditional supplier selection methods, are as follows:

- Cost,
- Quality,
- Flexibility,
- Technology capability.

There exist environmental factors for sustainable supplier selection as follows:

- Defilement production,
- Resource exhaustion,
- Eco-design and environmental administration.

The critical aspects of selecting green sustainable factors of supply chain design were provided by Dey and Ho [38] in a review of the recent research development.

3. Preliminaries

The significant definitions of interval-valued neutrosophic sets and its operations are presented in this section.

3.1. *Interval-Valued Neutrosophic Sets (INS)*

The interval-valued neutrosophic set V in X is described by truth $T_V(x)$, indeterminacy $I_V(x)$ and falsity $F_V(x)$ membership degrees for each $x \in X$. Here, $T_V(x)=[T_V^L(x), T_V^U(x) \subseteq [0,\ 1]]$, $I_V(x)=[I_V^L(x), I_V^U(x) \subseteq [0,\ 1]]$ and $F_V(x)=[F_V^L(x), F_V^U(x) \subseteq [0,\ 1]]$. Then, we can write interval-valued neutrosophic set as $V = < [T_V^L(x), T_V^U(x)], [I_V^L(x), I_V^U(x)], [F_V^L(x), F_V^U(x)] >$.

The INS is a neutrosophic set.

3.2. *The Related Operations of Interval-Valued Neutrosophic Sets*

- Addition

  Let A_1, A_2 be two INSs, where

  $A_1 = < \left[T_{A_1}^L, T_{A_1}^U\right], \left[I_{A_1}^L, I_{A_1}^U\right], \left[F_{A_1}^L, F_{A_1}^U\right] >$, $A_2 = < \left[T_{A_2}^L, T_{A_2}^U\right], \left[I_{A_2}^L, I_{A_1}^U\right], \left[F_{A_2}^L, F_{A_2}^U\right] >$ then $A_1 + A_2 = < \left[T_{A_1}^L + T_{A_2}^L - T_{A_1}^L T_{A_2}^L, T_{A_1}^U + T_{A_2}^U - T_{A_1}^U T_{A_2}^U\right], \left[I_{A_1}^L I_{A_2}^L, I_{A_1}^U I_{A_2}^U\right], \left[F_{A_1}^L F_{A_2}^L, F_{A_1}^U F_{A_2}^U\right] >$.
- Subset

  $A_1 \subseteq A_2$ if and only if $T_{A_1}^L \le T_{A_2}^L, T_{A_1}^U \le T_{A_2}^U$; $I_{A_1}^L \ge I_{A_2}^L, I_{A_1}^U \ge I_{A_2}^U$; $F_{A_1}^L \ge F_{A_2}^L, F_{A_1}^U \ge F_{A_2}^U$.
- Equality

  $A_1 = A_2$ if and only if $A_1 \subseteq A_2$ and $A_2 \subseteq A_1$.
- Complement

  Let $V = < [T_V^L(x), T_V^U(x)], [I_V^L(x), I_V^U(x)], [F_V^L(x), F_V^U(x)] >$, then

$V^c = < [F_V^L(x), F_V^U(x)], [1 - I_V^U(x), 1 - I_V^L(x)], [T_V^L(x), T_V^U(x)] >.$

- Multiplication

$A_1 \times A_2 =< \left[T_{A_1}^L T_{A_2}^L, T_{A_1}^U T_{A_2}^U\right], \left[I_{A_1}^L + I_{A_2}^L - I_{A_1}^L I_{A_2}^L, I_{A_1}^U + I_{A_2}^U - I_{A_1}^U I_{A_2}^U\right],$

$\left[F_{A_1}^L + F_{A_2}^L - F_{A_1}^L F_{A_2}^L, F_{A_1}^U + F_{A_2}^U - F_{A_1}^U F_{A_2}^U\right] > .$

- Subtraction

$A_1 - A_2 = < \left[T_{A_1}^L - F_{A_2}^U, T_{A_1}^U - F_{A_2}^L\right], \left[\max\left(I_{A_1}^L, I_{A_2}^l\right), \max\left(I_{A_1}^U, I_{A_2}^U\right)\right], \left[F_{A_1}^L - T_{A_2}^U, F_{A_1}^U - T_{A_2}^L\right] >.$

- Multiplication by a constant value

$\lambda A_1 = < \left[1 - \left(1 - T_{A_1}^L\right)^\lambda, 1 - \left(1 - T_{A_1}^U\right)^\lambda\right], \left[\left(I_{A_1}^L\right)^\lambda, \left(I_{A_1}^U\right)^\lambda\right], \left[\left(F_{A_1}^L\right)^\lambda, \left(F_{A_1}^U\right)^\lambda\right] >,$

where $\lambda > 0$.

- Addition

Let A_1, A_2 two INSs where

$A_1 = < \left[T_{A_1}^L, T_{A_1}^U\right], \left[I_{A_1}^L, I_{A_1}^U\right], \left[F_{A_1}^L, F_{A_1}^U\right] >$, $A_2 = < \left[T_{A_2}^L, T_{A_2}^U\right], \left[I_{A_2}^L, I_{A_1}^U\right], \left[F_{A_2}^L, F_{A_2}^U\right] >$ then
$A_1 + A_2 = < \left[T_{A_1}^L + T_{A_2}^L - T_{A_1}^L T_{A_2}^L, T_{A_1}^U + T_{A_2}^U - T_{A_1}^U T_{A_2}^U\right], \left[I_{A_1}^L I_{A_2}^L, I_{A_1}^U I_{A_2}^U\right], \left[F_{A_1}^L F_{A_2}^L, F_{A_1}^U F_{A_2}^U\right] >.$

- Subset

$A_1 \subseteq A_2$ if and only if $T_{A_1}^L \leq T_{A_2}^L, T_{A_1}^U \leq T_{A_2}^U; I_{A_1}^L \geq I_{A_2}^L, I_{A_1}^U \geq I_{A_2}^U; F_{A_1}^L \geq F_{A_2}^L, F_{A_1}^U \geq F_{A_2}^U$.

- Equality

$A_1 = A_2$ if and only if $A_1 \subseteq A_2$ and $A_2 \subseteq A_1$.

- Complement

Let $V = < [T_V^L(x), T_V^U(x)], [I_V^L(x), I_V^U(x)], [F_V^L(x), F_V^U(x)] >$,

then $V^c = < [F_V^L(x), F_V^U(x)], [1 - I_V^U(x), 1 - I_V^L(x)], [T_V^L(x), T_V^U(x)] >$.

- Multiplication

$A_1 \times A_2 =< \left[T_{A_1}^L T_{A_2}^L, T_{A_1}^U T_{A_2}^U\right], \left[I_{A_1}^L + I_{A_2}^L - I_{A_1}^L I_{A_2}^L, I_{A_1}^U + I_{A_2}^U - I_{A_1}^U I_{A_2}^U\right], \left[F_{A_1}^L + F_{A_2}^L - F_{A_1}^L F_{A_2}^L, F_{A_1}^U + F_{A_2}^U - F_{A_1}^U F_{A_2}^U\right] >.$

- Subtraction

$A_1 - A_2 = < \left[T_{A_1}^L - F_{A_2}^U, T_{A_1}^U - F_{A_2}^L\right], \left[\max\left(I_{A_1}^L, I_{A_2}^l\right), \max\left(I_{A_1}^U, I_{A_2}^U\right)\right], \left[F_{A_1}^L - T_{A_2}^U, F_{A_1}^U - T_{A_2}^L\right] >.$

- Multiplication by a constant value

$\lambda A_1 = < \left[1 - \left(1 - T_{A_1}^L\right)^\lambda, 1 - \left(1 - T_{A_1}^U\right)^\lambda\right], \left[\left(I_{A_1}^L\right)^\lambda, \left(I_{A_1}^U\right)^\lambda\right], \left[\left(F_{A_1}^L\right)^\lambda, \left(F_{A_1}^U\right)^\lambda\right] >$, where $\lambda > 0$.

3.3. Weighted Average for Interval-Valued Neutrosophic Numbers (INN)

Let $y_j = < \left[T_j^L, T_j^U\right], \left[I_j^L, I_j^U\right], \left[F_j^L, F_j^U\right] >$ be a group of interval-valued neutrosophic numbers, $j = 1, 2 \ldots, n$ is the number of decision makers. The weighted arithmetic average of interval-valued neutrosophic number

$$\text{INNWAA}(y_1, y_2, \ldots, y_n) = \sum_{k=1}^{n} w_k y_j = < \left[1 - \prod_{k=1}^{n}\left(1 - T_j^L\right)^{w_k}, 1 - \prod_{k=1}^{n}(1 - T_j^U)^{w_k}, \right] \left[\prod_{k=1}^{n}\left(I_j^L\right)^{w_k}, \prod_{k=1}^{n}\left(I_j^U\right)^{w_k}\right], \left[\prod_{k=1}^{n}\left(F_j^L\right)^{w_k}, \prod_{k=1}^{n}\left(F_j^U\right)^{w_k}\right] >, \quad (1)$$

where w_k is the decision maker's weight vector.

3.4. INS Deneutrosophication Function

The deneutrosophication function converts each interval-valued neutrosophic number into crisp number. Let $A = < [T_A^L, T_A^U], [I_A^L, I_A^U], [F_A^L, F_A^U] >$ be an interval-valued neutrosophic number, then the deneutrosophication function $D(A)$ will be defined by

$$D(A) = 10^{\left(\frac{2+(T_A^L+T_A^U)-2(I_A^L+I_A^U)-(F_A^L,F_A^U)}{4}\right)}. \tag{2}$$

3.5. Ranking Method for Interval-Valued Neutrosophic Numbers

Let A_1, A_2 be interval-valued neutrosophic numbers, then,

- if $D(A_1)$ greater than $D(A_2)$, then $A_1 > A_2$;
- if $D(A_1)$ less than $D(A_2)$, then $A_1 < A_2$;
- if $D(A_1)$ equals $D(A_2)$, then $A_1 = A_2$.

4. The ANP and TOPSIS Methods

In this section, we present an overview of the two techniques used in our proposed research.

4.1. The Analytic Network Process (ANP)

The ANP is a development of analytic hierarchy process (AHP), and it was advanced by Saaty in 1996 for considering dependency and feedback among decision-making problem's elements. The ANP structures the problem as a network, not as hierarchies as with the AHP. In the analytic hierarchy process, it is assumed that the alternatives depend on criteria and criteria depend on goal. Therefore, in AHP, the criteria do not depend on alternatives, criteria do not affect (depend on) each other, and alternatives do not depend on each other. Nevertheless, in the analytic network process, the dependencies between decision-making elements are allowed. The differences between ANP and AHP are presented with the structural graph in Figure 1. The upper side of Figure 1 shows the hierarchy of AHP in which elements from the lower level have an influence on the higher level or, in other words, the upper level depends on the lower level. However, in the lower side of Figure 1, which shows the network model of ANP, we have a cluster network, and there exists some dependencies between them. The dependencies may be inner-dependencies when the cluster influence itself or may be outer-dependencies when cluster depends on another one. The complex decision-making problem in real life may contain dependencies between problem's elements, but AHP does not consider them, so it may lead to less optimal decisions, and ANP is more appropriate.

The general steps of ANP [50]:

1. The decision-making problem should be structured as a network that consists of a main objective, criteria for achieving this objective and can be divided to sub-criteria, and finally all available alternatives. The feedback among network elements should be considered here.
2. To calculate criteria's and alternatives' weights, the comparisons matrices should be constructed utilizing the 1–9 scale of Saaty. After then, we should check the consistency ratio of these matrices, and it must be ≤ 0.1 for each comparison matrix. The comparison matrix's eigenvector should be calculated after that by summing up the columns of comparison matrix. A new matrix is constructed by dividing each value in a column by the summation of that column, and then taking the average of new matrix rows. For more information, see [51]. The ANP comparison matrices may be constructed for comparing:

    - Criteria with respect to goal,
    - Sub-criteria with respect to criterion from the same cluster,

- Alternatives with respect to each criterion,
- Criteria that belong to the same cluster with respect to each alternative.

3. Use the eigenvectors calculated in the previous step for constructing the super-matrix columns. For obtaining a weighted super-matrix, a normalization process must be established. Then, raise the weighted matrix to a larger power until the raw values will be equal to each column values of super-matrix for obtaining the limiting matrix.
4. Finally, choose the best alternative by depending on weight values.

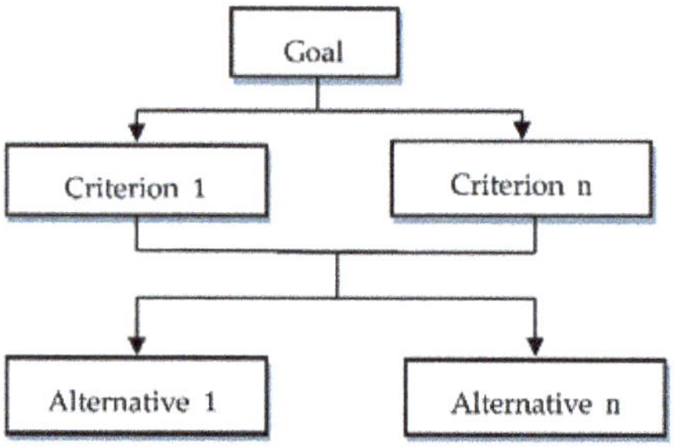

(a) The AHP hierarchy.

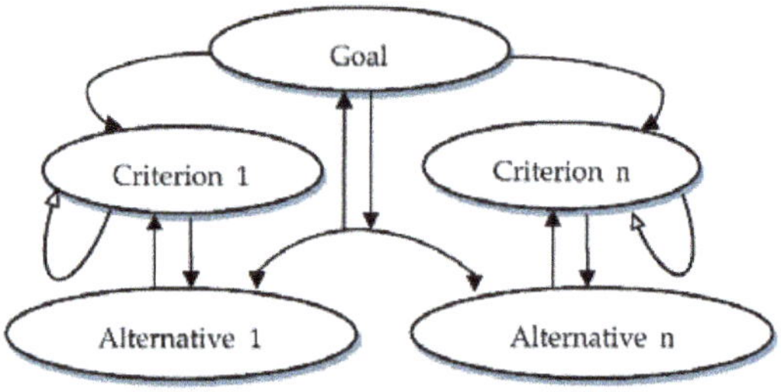

(b) The ANP network.

Figure 1. The structural difference between hierarchy and network model.

4.2. The TOPSIS Technique

The technique for order preference by similarity to ideal solution (TOPSIS) is proposed by Hwang and Yoon for aiding decision makers in determining positive (A^+) and negative (A^-) ideal solutions [52]. The chosen alternative is the one with the least distance from the positive ideal solution and the greatest distance from the negative ideal solution. The TOPSIS steps summarized as follows:

1. The decision makers should construct the evaluation matrix that consists of m alternatives and n criteria. The intersection of each alternative and criterion is denoted as x_{ij}, and then we have $(x_{ij})_{m*n}$ matrix.
2. Use the following equation for obtaining the normalized evaluation matrix:

$$r_{ij} = \frac{x_{ij}}{\sqrt{\sum_{i=1}^{m} x_{ij}^2}};\ i = 1, 2, \ldots, m;\ j = 1, 2, \ldots, n. \quad (3)$$

3. Structure the weighted matrix through multiplying criteria's weights w_j, by the normalized decision matrix r_{ij} as follows:
$$v_{ij} = w_j \times r_{ij}. \tag{4}$$
4. Calculate the positive A^+ and negative ideal solution A^- using the following:
$$A^+ = \{ < \max(v_{ij} | i = 1, 2, \ldots, m) | j \in J^+ >, < \min(v_{ij} | i = 1, 2, \ldots, m) | j \in J^- > \}, \tag{5}$$
$$A^- = \{ < \min(v_{ij} | i = 1, 2, \ldots, m) | j \in J^+ >, < \max(v_{ij} | i = 1, 2, \ldots, m) | j \in J^- > \}, \tag{6}$$
where J^+ associated with the criteria that have a beneficial influence and J^- associated with the criteria that have a non-beneficial influence.
5. Calculate the Euclidean distance among positive (d_i^+) and negative ideal solution (d_i^-) as follows:
$$d_i^+ = \sqrt{\sum_{j=1}^{n} \left(v_{ij} - v_j^+\right)^2} \; i = 1, 2, \ldots, m, \tag{7}$$
$$d_i^- = \sqrt{\sum_{j=1}^{n} \left(v_{ij} - v_j^-\right)^2} \; i = 1, 2, \ldots, m. \tag{8}$$
6. Calculate the relative closeness to the ideal solution and make the final ranking of alternatives

   $c_i = \frac{d_i^-}{d_i^+ + d_i^-}$ for $i = 1, 2, \ldots, m$, and based on the largest c_i value, begin to rank alternatives. (9)
7. According to your rank of alternatives, take your final decision.

5. The Proposed Framework

The steps of the proposed interval-valued neutrosophic ANP-TOPSIS framework are presented with details in this section.

The proposed framework consists of four phases, which contains a number of steps as follows:

Phase 1: For better understanding of a complex problem, we must firstly breakdown it.

Step 1.1. Select a group of experts to share in making decisions. If we select n experts, then we have the panel = $[e_1, e_2, \ldots, e_n]$.

Step 1.2. Use the literature review to determine problem's criteria and ask experts for confirming these criteria.

Step 1.3. Determine the alternatives of the problem.

Step 1.4. Begin to structure the hierarchy of the problem.

In an analytic hierarchy process, it is assumed that the alternatives depend on criteria, criteria affects goal, and in real complex problems, there likely is a dependency between a problem's elements. In order to overcome this drawback of AHP, we utilized ANP for solving the problem. Figure 2 presents a sample of an ANP network.

Phase 2: Calculate the weight of problem's elements as follows:

Step 2.1. The interval-valued comparison matrices should be constructed according to each expert and then aggregate experts' matrices by using Equation (1).

In this step, we compare criteria according to overall goals, sub-criteria according to criteria, and alternatives according to criteria. In addition, the interdependencies among problem's elements must be pair-wisely compared. The 9-point scale of Saaty [53] was used to represent comparisons in traditional ANP.

In our research, we used the interval-valued neutrosophic numbers for clarifying pair-wise comparisons as presented in Table 1, and these values returned to authors' opinions. When comparing alternative 1 with alternative 2, and the first alternative was "Very strongly significant" than second one, then the truth degree is high and indeterminacy degree is very small because the term "Very strongly important" means that the decision makers are very confident of comparison results in a large percentage. Therefore, we represented this linguistic term using interval-neutrosophic number equals ([0.8, 0.9], [0.0, 0.1], [0.0, 0.1]), as it appears in Table 1. All other values in Table 1 were scaled with the same approach.

Step 2.2. Use the de-neutrosophication function for transforming the interval-valued neutrosophic numbers to crisp numbers as in Equation (2).

Step 2.3. Use super decision software, which is available here (http://www.superdecisions.com/downloads/) to check the consistency of comparison matrices.

Step 2.4. Calculate the eigenvectors for determining weight that will be used in building a super-matrix.

Step 2.5. The super-matrix of interdependencies should be constructed after then.

Step 2.6. Multiply the local weight, which was obtained from experts' comparison matrices of criteria according to goal, by the weight of interdependence matrix of criteria for calculating global weight of criteria. In addition, calculate the global weights of sub-criteria by multiplying its local weight by the inner interdependent weight of the criterion to which it belongs.

Table 1. The interval-valued neutrosophic scale for comparison matrix.

Linguistic Variables	Interval-Valued Neutrosophic Numbers for Relative Importance <T,I,F>
Evenly significant	([0.5,0.5], [0.5,0.5], [0.5,0.5])
Low significant	([0.4,0.5], [0.1,0.2], [0.2,0.3])
Basically important	([0.6,0.7], [0.0,0.1], [0.0,0.1])
Very strongly significant	([0.8,0.9], [0.0,0.1], [0.0,0.1])
Absolutely significant	([1,1], [0.0,0.1], [0.0,0.0])
Intermediate values	([0.3,0.4], [0.1,0.2], [0.6,0.7]), ([0.6,0.7], [0.1,0.2], [0.0,0.1]), ([0.7,0.8], [0.0,0.1], [0.0,0.1]), ([0.9,1], [0.0,0.1], [0.0,0.1]).

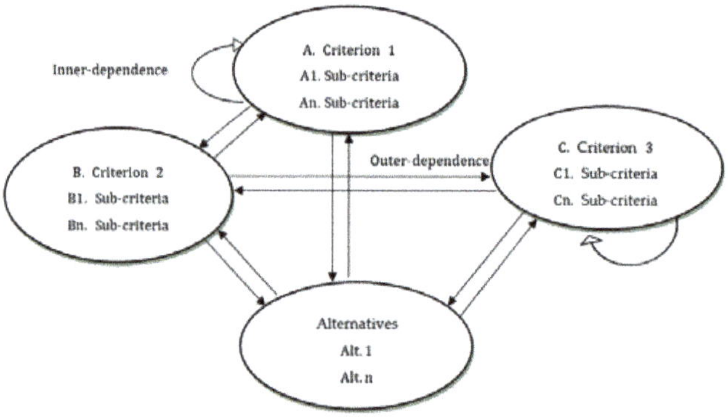

Figure 2. An example of ANP interdependencies.

Phase 3: Rank alternatives of problems.

Step 3.1. Make the evaluation matrix, and then a normalization process must be performed for obtaining the normalized evaluation matrix using Equation (3).

Step 3.2. Multiply criteria's weights, which was obtained from ANP by the normalized evaluation matrix as in Equation (4) to construct the weighted matrix.
Step 3.3. Determine positive and negative ideal solutions using Equations (5) and (6).
Step 3.4. Calculate the Euclidean distance between positive solution (d_i^+) and negative ideal solution (d_i^-) using Equations (7) and (8).
Step 3.5.Make the final ranking of alternatives based on closeness coefficient.

Phase 4: Compare the proposed method with other existing methods for validating it. The framework of the suggested method is presented in Figure 3.

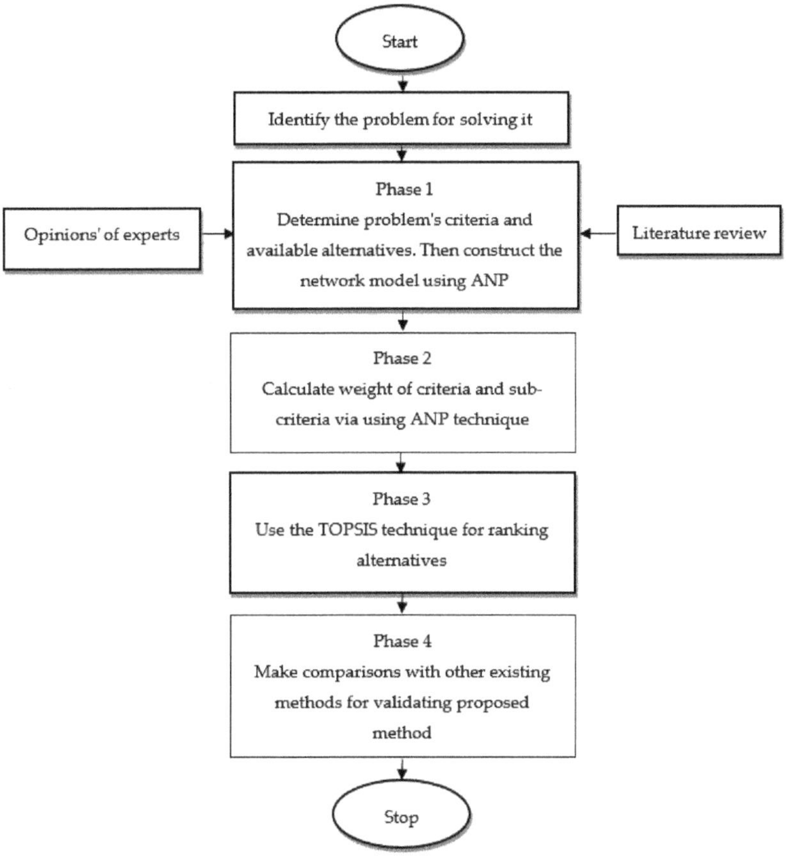

Figure 3. The framework's proposed phases.

6. The Case Study: Results and Analysis

The proposed framework has been applied to a real sustainable supplier selection problem, and the results are analyzed in this section.

An Egyptian dairy and foodstuff corporation was founded in 1999 and is based in 10th of Ramadan City, Egypt. The corporation products include cream and skimmed milk, flavored milk, juice nectars, junior milk and juices, and tomato paste. The procurement department of the corporation is responsible for providing the required raw materials with the lowest possible cost, and purchasing corporation's required equipment. The types of equipment are material-handling, laboratory, technical parts and machinery. The procurement department supplies packaging pure materials, pure materials and manufacturing technology. The dairy and foodstuff corporation must evaluate available suppliers and their sustainability to improve their productivity and be more competitive. Therefore, improving a system to assess and identify the superior suppliers is a significant component of this corporation's objectives. The corporation consulted the executive manager and asked three experts to help in gathering required information for this study. The experts are in marketing, manufacturing and strategy with more than five years of experience. There are four suppliers, denoted in this study by $A_1 \ldots A_4$.

Phase 1: Breakdown the complex problem for understanding it better.

The criteria and available suppliers which are relevant to our case study are identified from the literature review. The experts vote to confirm the information. The criteria, sub-criteria and available suppliers are presented in Figure 4. In order to determine how criteria and sub-criteria influence each other and correlate, for being able to apply the ANP and weighting them, we interviewed the experts.

Phase 2: Calculate the weights of problem elements.

The verdicts of experts were applied through using the interval-valued neutrosophic numbers in Table 1. We used interval-valued neutrosophic numbers because they are more realistic and accurate than crisp values, and can deal efficiently and effectively with vague and inconsistent information.

Let experts express their judgments by constructing the pairwise comparison matrices using the presented scale in Table 1—after that, aggregate comparison matrices using Equation (1). The aggregated comparison matrices of experts are presented in Tables 2–11.

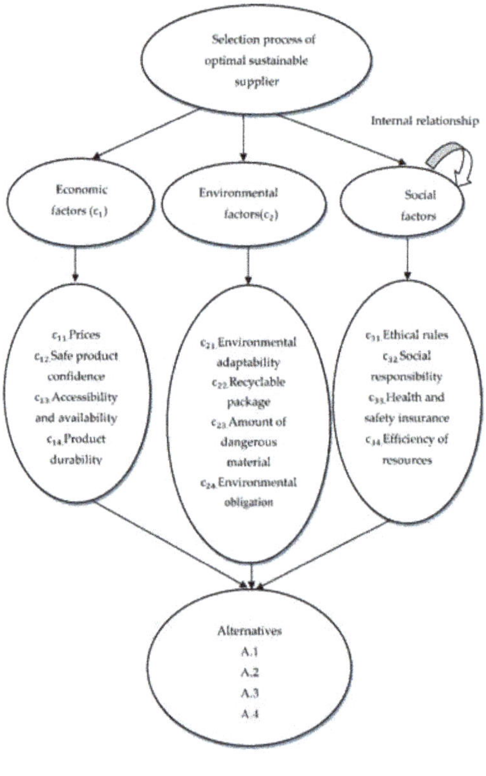

Figure 4. Hierarchy for dairy and foodstuff corporation to select the optimal supplier.

Table 2. The pairwise comparison matrix of criteria with respect to goal.

Goal	C_1	C_2	C_3
C_1	[0.5,0.5], [0.5,0.5], [0.5,0.5]	[0.3,0.4], [0.1,0.2], [0.6,0.7]	[0.7,0.8], [0.0,0.1], [0.0,0.1]
C_2		[0.5,0.5], [0.5,0.5], [0.5,0.5]	[0.6,0.7], [0.1,0.2], [0.0,0.1]
C_3			[0.5,0.5], [0.5,0.5], [0.5,0.5]

By using the deneutrosophication function through Equation (2), we will obtain the crisp matrix of comparison as in Table 3.

Table 3. The equivalent crisp matrix of criteria with respect to goal.

Goal	C_1	C_2	C_3	Weights
C_1	1	2	6	0.59
C_2	0.5	1	4	0.32
C_3	0.17	0.25	1	0.09

By checking consistency of the previous matrix using super decision software, we noted that the matrix is consistent with consistency ratio (CR) = 1%.

The inner interdependency of main criteria according to C_1 is presented in Table 4.

Table 4. Internal interdependencies of criteria with respect to C_1 .

C_1	C_2	C_3
C_2	[0.5,0.5], [0.5,0.5], [0.5,0.5]	[0.7,0.8], [0.0,0.1], [0.0,0.1]
C_3		[0.5,0.5], [0.5,0.5], [0.5,0.5]

Table 5. The crisp interdependencies values of factors with respect to C_1 .

C_1	C_2	C_3	Weights
C_2	1	6	0.86
C_3	0.17	1	0.14

Table 6. Internal interdependencies of criteria with respect to C_2 .

C_2	C_1	C_3
C_1	[0.5,0.5], [0.5,0.5], [0.5,0.5]	[0.6,0.7], [0.1,0.2], [0.0,0.1]
C_3		[0.5,0.5], [0.5,0.5], [0.5,0.5]

Table 7. The crisp interdependencies values of factors with respect to C_2.

C_2	C_1	C_3	Weights
C_1	1	4	0.8
C_3	0.25	1	0.2

Table 8. Internal interdependencies of criteria with respect to C_3 .

C_3	C_1	C_2
C_1	[0.5,0.5], [0.5,0.5], [0.5,0.5]	[1,1], [0.0,0.1], [0.0,0.0]
C_2		[0.5,0.5], [0.5,0.5], [0.5,0.5]

Table 9. The crisp interdependencies values of factors with respect to C_3 .

C_3	C_1	C_2	**Weights**
C_1	1	9	0.9
C_2	0.11	1	0.1

Table 10. The relative impact of decision criteria.

	C_1	C_2	C_3
C_1	1	0.8	0.9
C_2	0.86	1	0.1
C_3	0.14	0.2	1

Table 11. The normalized relative impact of decision criteria.

	C_1	C_2	C_3
C_1	0.5	0.4	0.45
C_2	0.43	0.5	0.05
C_3	0.07	0.1	0.5

Then, the weights of decision criteria based on their inner interdependencies are as follows:

$$w_{criteria} = \begin{bmatrix} economical \\ environmental \\ social \end{bmatrix} = \begin{bmatrix} 0.5 & 0.4 & 0.45 \\ 0.43 & 0.5 & 0.05 \\ 0.07 & 0.1 & 0.5 \end{bmatrix} \times \begin{bmatrix} 0.59 \\ 0.32 \\ 0.09 \end{bmatrix} = \begin{bmatrix} 0.46 \\ 0.42 \\ 0.12 \end{bmatrix}.$$

It is obvious that the economic factors are the most significant factors when evaluating suppliers, followed by environmental and social factors, according to experts' opinions.

We should also note the influence of inner interdependencies of criteria on its weights. It changed the weights of main criteria from (0.59, 0.32, 0.09) to (0.46, 0.42, 0.12).

The comparison matrices and local weights of sub-criteria relevant to their clusters are expressed in Tables 12–17.

Table 12. The comparison matrix and local weight of C_1 indicators.

C_1	C_{11}	C_{12}	C_{13}	C_{14}
C_{11}	[0.5,0.5], [0.5,0.5], [0.5,0.5]	[0.4,0.5], [0.1,0.2], [0.2,0.3]	[0.6,0.7], [0.1,0.2], [0.0,0.1]	[0.6,0.7], [0.0,0.1], [0.0,0.1]
C_{12}		[0.5,0.5], [0.5,0.5], [0.5,0.5]	[0.3,0.4], [0.1,0.2], [0.6,0.7]	[0.6,0.7], [0.1,0.2], [0.0,0.1]
C_{13}			[0.5,0.5], [0.5,0.5], [0.5,0.5]	[0.3,0.4], [0.1,0.2], [0.6,0.7]
C_{14}				[0.5,0.5], [0.5,0.5], [0.5,0.5]

Table 13. The crisp comparison matrix and local weight of C_1 indicators.

C_1	C_{11}	C_{12}	C_{13}	C_{14}	**Weights**
C_{11}	1	3	4	5	0.54
C_{12}	0.33	1	2	4	0.23
C_{13}	0.25	0.50	1	2	0.13
C_{14}	0.20	0.25	0.5	1	0.08

The consistency ratio (CR) of previous matrix = 0.03.

Table 14. The comparison matrix and local weight of C_2 indicators.

C_2	C_{21}	C_{22}	C_{23}	C_{24}
C_{21}	[0.5,0.5], [0.5,0.5], [0.5,0.5]	[0.4,0.5], [0.1,0.2], [0.2,0.3]	[0.8,0.9], [0.0,0.1], [0.0,0.1]	[1,1], [0.0,0.1], [0.0,0.0]
C_{22}		[0.5,0.5], [0.5,0.5], [0.5,0.5]	[0.6,0.7], [0.0,0.1], [0.0,0.1]	[0.8,0.9], [0.0,0.1], [0.0,0.1]
C_{23}			[0.5,0.5], [0.5,0.5], [0.5,0.5]	[0.3,0.4], [0.1,0.2], [0.6,0.7]
C_{24}				[0.5,0.5], [0.5,0.5], [0.5,0.5]

Table 15. The crisp comparison matrix and local weight of C_2 indicators.

C_2	C_{21}	C_{22}	C_{23}	C_{24}	**Weights**
C_{21}	1	3	7	9	0.59
C_{22}	0.33	1	5	7	0.29
C_{23}	0.14	0.20	1	2	0.08
C_{24}	0.11	0.14	0.50	1	0.05

The consistency ratio (CR) of previous matrix = 0.04.

Table 16. The comparison matrix and local weight of C_3 indicators.

C_3	C_{31}	C_{32}	C_{33}	C_{34}
C_{31}	[0.5,0.5], [0.5,0.5], [0.5,0.5]	[0.3,0.4], [0.1,0.2], [0.6,0.7]	[0.4,0.5], [0.1,0.2], [0.2,0.3]	[1,1], [0.0,0.1], [0.0,0.0]
C_{32}		[0.5,0.5], [0.5,0.5], [0.5,0.5]	[0.3,0.4], [0.1,0.2], [0.6,0.7]	[0.7,0.8], [0.0,0.1], [0.0,0.1]
C_{33}			[0.5,0.5], [0.5,0.5], [0.5,0.5]	[0.4,0.5], [0.1,0.2], [0.2,0.3]
C_{34}				[0.5,0.5], [0.5,0.5], [0.5,0.5]

Table 17. The crisp comparison matrix and local weight of C_3 indicators.

C_3	C_{31}	C_{32}	C_{33}	C_{34}	**Weights**
C_{31}	1	2	3	9	0.50
C_{32}	0.50	1	2	6	0.29
C_{33}	0.33	0.50	1	3	0.15
C_{34}	0.11	0.17	0.33	1	0.05

The consistency ratio (CR) of previous matrix = 0.004.

Each sub-criteria global weight is calculated via multiplying its local weight by the inner interdependent weight of the criterion to which it belongs as in Table 18.

Table 18. The sub-criteria global weights.

Criteria Local Weight	Sub-Criteria	Local Weight	Global Weight
Economic factors (0.46)	C_{11}	0.54	0.25
	C_{12}	0.23	0.11
	C_{13}	0.13	0.06
	C_{14}	0.08	0.04
Environmental factors (0.42)	C_{21}	0.59	0.25
	C_{22}	0.29	0.12
	C_{23}	0.08	0.03
	C_{24}	0.05	0.02
Social factors (0.12)	C_{31}	0.50	0.06
	C_{32}	0.29	0.03
	C_{33}	0.15	0.02
	C_{34}	0.05	0.006

Phase 3: Rank alternatives of problems.

Let each expert build the evaluation matrix via comparing the four alternatives relative to each criterion, by utilizing the interval-valued scale, which is presented in Table 1. After that, use Equation (1) to aggregate the evaluation matrices and obtain the final evaluation matrix relevant to experts' committee. Proceed to deneutrosophication function to convert the interval-valued neutrosophic evaluation matrix to its crisp form using Equation (2). Then, make a normalization process to obtain the normalized evaluation matrix using Equation (3), as observed in Table 19.

Table 19. The normalized evaluation matrix.

	C_{11}	C_{12}	C_{13}	C_{14}	C_{21}	C_{22}	C_{23}	C_{24}	C_{31}	C_{32}	C_{33}	C_{34}
A_1	0.53	0.46	0.46	0.43	0.52	0.54	0.45	0.58	0.48	0.59	0.59	0.51
A_2	0.46	0.58	0.53	0.48	0.43	0.58	0.59	0.52	0.54	0.54	0.46	0.64
A_3	0.44	0.43	0.56	0.53	0.49	0.45	0.36	0.46	0.49	0.38	0.47	0.47
A_4	0.56	0.52	0.43	0.55	0.54	0.41	0.56	0.43	0.48	0.45	0.46	0.32

Then, build the weighted matrix by multiplying the weights of criteria, obtained from ANP by the normalized evaluation matrix using Equation (4), as in Table 20.

Table 20. The weighted evaluation matrix.

	C_{11}	C_{12}	C_{13}	C_{14}	C_{21}	C_{22}	C_{23}	C_{24}	C_{31}	C_{32}	C_{33}	C_{34}
A_1	0.13	0.05	0.03	0.02	0.13	0.06	0.01	0.01	0.03	0.02	0.01	0.003
A_2	0.11	0.06	0.03	0.02	0.11	0.07	0.02	0.01	0.03	0.02	0.01	0.004
A_3	0.11	0.05	0.03	0.02	0.12	0.05	0.01	0.01	0.03	0.01	0.01	0.003
A_4	0.14	0.06	0.03	0.02	0.13	0.05	0.02	0.01	0.03	0.01	0.01	0.002

Determine the ideal solutions using Equations (5) and (6) as follows:

$$A^{+} = \{0.14, 0.06, 003, 0.02, 0.13, 0.07, 0.02, 0.01, 0.03, 0.02, 0.01, 0.004\},$$

$$A^{-} = \{0.11, 0.05, 0.03, 0.02, 0.11, 0.05, 0.01, 0.01, 0.03, 0.01, 0.01, 0.002\}.$$

After that, measure the Euclidean distance between positive solution (d_i^+) and negative ideal solution (d_i^-) using Equations (7) and (8) as follows:

$$d_1^+ = \{0.020\},\ d_2^+ = \{0.036\},\ d_3^+ = \{0.041\},\ d_4^+ = \{0.022\},$$

$$d_1^- = \{0.032\},\ d_2^- = \{0.026\},\ d_3^- = \{0.010\},\ d_4^- = \{0.040\}.$$

Step 3.6. Calculate the closeness coefficient using Equation (9), and make the final ranking of alternatives as in Table 21.

Table 21. TOPSIS results and ranking of alternatives.

	d_i^+	d_i^-	c_i	Rank
A_1	0.020	0.032	0.615	2
A_2	0.036	0.026	0.419	3
A_3	0.041	0.010	0.196	4
A_4	0.022	0.040	0.645	1

The ranking for the optimal sustainable suppliers of dairy and foodstuff corporation is Alternative 4, Alternative 1, Alternative 2 and Alternative 3, as shown in Figure 5.

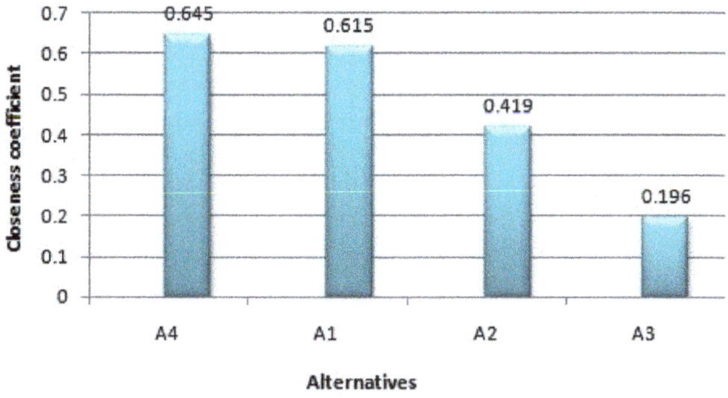

Figure 5. The ranking for the optimal alternatives of dairy and foodstuff corporation.

Phase 4: Validate the model and make comparisons with other existing methods.

In this phase, the obtained ranking of optimal suppliers by the proposed framework is compared with the obtained results by the analytic hierarchy process, the analytic network process, MOORA and MOOSRA techniques.

The obtained ranking of suppliers by using an AHP technique is as follows:

Since AHP does not consider inner interdependency between problem's elements, then weights of sub-criteria are as follows:

$$\begin{bmatrix} 0.32 \\ 0.14 \\ 0.08 \\ 0.47 \\ 0.19 \\ 0.09 \\ 0.03 \\ 0.02 \\ 0.04 \\ 0.03 \\ 0.01 \\ 0.00 \end{bmatrix}.$$

The comparison matrix of alternatives relevant to each sub-criterion is as follows:

$$\begin{bmatrix} 0.53 & 0.46 & 0.46 & 0.43 & 0.52 & 0.54 & 0.45 & 0.58 & 0.48 & 0.59 & 0.59 & 0.51 \\ 0.46 & 0.58 & 0.53 & 0.48 & 0.43 & 0.58 & 0.59 & 0.52 & 0.54 & 0.54 & 0.46 & 0.64 \\ 0.44 & 0.43 & 0.56 & 0.53 & 0.49 & 0.45 & 0.36 & 0.46 & 0.49 & 0.38 & 0.47 & 0.47 \\ 0.56 & 0.52 & 0.43 & 0.55 & 0.54 & 0.41 & 0.56 & 0.43 & 0.48 & 0.45 & 0.46 & 0.32 \end{bmatrix}.$$

The final weights of alternatives after multiplying two previous matrices and making normalization of results are as in Table 22.

Table 22. Ranking alternatives relevant to AHP.

Alternatives	Weights	Rank
A_1	0.245	3
A_2	0.250	2
A_3	0.244	4
A_4	0.267	1

Our proposed framework and the analytic hierarchy process agreed that the Alternative 3 is the worst alternative for the company. The two methods are different in ranking the optimal alternative due to the inner interdependencies between the problem's criteria effect on the global weight of alternatives, and, in our case study, it reduced weights of main criteria from (0.59, 0.32, 0.09) to (0.46, 0.42, 0.12), and this surely regarded the global weight of sub-criteria and also ranking of alternatives.

The weights of sub-criteria when we applied the analytic network process are as follows (see also Table 18):

$$\begin{bmatrix} 0.25 \\ 0.11 \\ 0.06 \\ 0.04 \\ 0.25 \\ 0.12 \\ 0.03 \\ 0.02 \\ 0.06 \\ 0.03 \\ 0.02 \\ 0.006 \end{bmatrix}.$$

In addition, the comparison matrix of alternatives relevant to each sub-criterion is as follows:

$$\begin{bmatrix} 0.53 & 0.46 & 0.46 & 0.43 & 0.52 & 0.54 & 0.45 & 0.58 & 0.48 & 0.59 & 0.59 & 0.51 \\ 0.46 & 0.58 & 0.53 & 0.48 & 0.43 & 0.58 & 0.59 & 0.52 & 0.54 & 0.54 & 0.46 & 0.64 \\ 0.44 & 0.43 & 0.56 & 0.53 & 0.49 & 0.45 & 0.36 & 0.46 & 0.49 & 0.38 & 0.47 & 0.47 \\ 0.56 & 0.52 & 0.43 & 0.55 & 0.54 & 0.41 & 0.56 & 0.43 & 0.48 & 0.45 & 0.46 & 0.32 \end{bmatrix}.$$

After proceeding to the normalization process, the ranking of alternatives relevant to the ANP technique is presented in Table 23.

Table 23. Ranking alternatives relevant to ANP.

Alternatives	Weights	Rank
A_1	0.26	1
A_2	0.25	2
A_3	0.23	3
A_4	0.26	1

By using the ANP technique for solving the same case study, we noted that Alternative 1 and Alternative 4 have the same rank and are the best alternatives, followed by Alternative 2 and finally Alternative 3. The proposed framework and the ANP agreed that Alternative 3 is the worst alternative.

We not only used the AHP and ANP techniques for solving the case study of a dairy and foodstuff corporation, but also two other multi-objective decision-making techniques.

The first technique is the multi-objective optimization based on simple ratio analysis (MOORA), proposed by Brauers and Zavadskas [54]. There are two approaches under the MOORA: the ratio system and the reference point approaches [53]. Here, we used the ratio system method of the MOORA to validate our proposed framework.

The normalized weighted matrix and ranking of alternatives using the MOORA technique are presented in Tables 24 and 25. The equations that we used in our calculation of MOORA normalized weighted matrix, and the equations that we employed in the ranking process are available with details in [53].

Table 24. The weighted normalized matrix under the MOORA technique.

	C_{11}	C_{12}	C_{13}	C_{14}	C_{21}	C_{22}	C_{23}	C_{24}	C_{31}	C_{32}	C_{33}	C_{34}
A_1	0.13	0.05	0.03	0.02	0.13	0.06	0.01	0.01	0.03	0.02	0.01	0.003
A_2	0.11	0.06	0.03	0.02	0.11	0.07	0.02	0.01	0.03	0.02	0.01	0.004
A_3	0.11	0.05	0.03	0.02	0.12	0.05	0.01	0.01	0.03	0.01	0.01	0.003
A_4	0.14	0.06	0.03	0.02	0.13	0.05	0.02	0.01	0.03	0.01	0.01	0.002

Table 25. The ranking of alternatives using the MOORA technique.

	$\sum_{j=1}^{g} xij^*$	$\sum_{j=g+1}^{n} xij^*$	p_i^*	**Ranking**
A_1	0.43	0.073	0.357	2
A_2	0.41	0.084	0.326	4
A_3	0.39	0.063	0.327	3
A_4	0.44	0.072	0.368	1

The fourth column in Table 25 is the index of the total performance p_i^* and equals the difference between beneficial criteria summation and non-beneficial criteria summation. The beneficial and non-beneficial criteria were determined according to experts' weights of criteria. In other words, the total performance p_i^* is the difference between the second column and third column values in Table 25.

The other technique we applied to the same case study for validating our proposed framework is MOOSRA. The MOOSRA technique determines the simple ratio of beneficial and non-beneficial criteria. The MOOSRA is a multi-objective optimization technique. The steps of the MOOSRA technique are similar to the MOORA technique, except in calculating total performance index p_i^*. For more details, see [53]. The ranking of alternatives using MOOSRA technique is presented in Table 26.

Table 26. The ranking of alternatives using the MOOSRA technique.

	$\sum_{j=1}^{g} xij^*$	$\sum_{j=g+1}^{n} xij^*$	p_i^*	**Ranking**
A_1	0.43	0.073	5.89	3
A_2	0.41	0.084	4.88	4
A_3	0.39	0.063	6.19	1
A_4	0.44	0.072	6.11	2

The ranking of suppliers using the proposed framework and the other four techniques are aggregated in Table 27. The correlation coefficient between the proposed framework and other techniques is presented in Table 28; we calculated it using Microsoft Excel (version, Manufacturer, City, US State abbrev. if applicable, Country) by using the CORREL() function.

Table 27. The ranking of alternatives relevant to various applied techniques.

Suppliers	**Proposed Technique (1)**	**AHP (2)**	**ANP (3)**	**MOORA (4)**	**MOOSRA (5)**
A_1	2	3	1	2	3
A_2	3	2	2	4	4
A_3	4	4	3	3	1
A_4	1	1	1	1	2

Table 28. The correlation coefficients between the proposed model and other applied techniques.

Correlation (1, 2)	**Correlation (1, 3)**	**Correlation (1, 4)**	**Correlation (1, 5)**
0.8	0.9	0.8	0.2

The proposed framework and the first three applied techniques (i.e., AHP, ANP, MOORA) agreed that Alternative 4 is the best alternative. The correlation coefficients help to measure the efficiency of various MCDM techniques. The correlation coefficients between our proposed framework and AHP, ANP, MOORA are very high, as shown in Table 28. The high value of Spearman correlation coefficients reflects the high consistency and validity of the proposed framework. However, the correlation coefficient between our proposed model and MOOSRA is low. Our framework is valid and consistent because the proposed framework and the first three applied techniques agreed that Alternative 4 is the optimal supplier for the dairy and foodstuff corporation.

7. Conclusions and Future Directions

For solving the sustainable supplier selection problem, many steps must be performed: the sustainability criteria must be determined; the interdependencies between these criteria must be identified—ranking and evaluating supplier performance. For more accuracy, we have suggested a framework consisting of four phases, by integrating ANP with TOPSIS using the interval-valued neutrosophic numbers. The ANP is used to weight problem criteria and sub-criteria because of its capability to consider interdependencies between problem's elements. The TOPSIS is used to rank available suppliers for avoiding additional comparisons of analytic network process. The suggested method provides a reliable and easy to implement procedure, which is suitable for a broad range of real life applications. A case study of a dairy and foodstuff corporation has been solved employing the proposed framework. The dairy corporation trying to earn an important market share and competitive benefits faces competition from other corporations. The objectives of food corporation are to improve the green food process, to get the standard certificate. Many customers consider the ISO standard as a priority for them. Suppliers are a great part of the production process; consequently, they must be sorted and analyzed carefully using efficient framework. The selection process of experts is not an easy matter. Therefore, the provided data and information from experts must be more accurate; otherwise, it will affect the selection process of optimal suppliers. Because real life has a great amount of vague and inconsistent information and surely affects experts' judgment, we presented our suggested framework using interval-valued neutrosophic numbers. Neutrosophic sets make a simulation of natural decision-making process, since it considers all aspects of making a decision (i.e., agree, not sure and falsity). In the future, we plan to solve the sustainable supplier selection problem with more difficult and complex dependencies between criteria using different multi-criteria decision-making techniques and presenting them in a neutrosophic environment using the alpha cut method.

Author Contributions: All authors contributed equally to this paper. The individual responsibilities and contributions of each author are described as follows: the idea of the paper was put forward by M.A.-B.; F.S. completed the preparatory work of the paper; M.M. analyzed the existing work; and the revision and submission of the paper was completed by M.A.-B.

Acknowledgments: The authors would like to thank the anonymous referees, the Chief-Editor, and support Editors for their constructive suggestions and propositions that have helped to improve the quality of this research. Thanks to the Symmetry international journal open source, we had the chance to read many papers in a neutrosophic environment, which inspired us to write this paper.

Conflicts of Interest: The authors declare no conflict of interest.

References

1. Amindoust, A.; Ahmed, S.; Saghafinia, A.; Bahreininejad, A. Sustainable supplier selection: A ranking model based on fuzzy inference system. *Appl. Soft Comput.* **2012**, *12*, 1668–1677. [CrossRef]
2. Abdel-Basset, M.; Manogaran, G.; Mohamed, M. Internet of Things (IoT) and its impact on supply chain: A framework for building smart, secure and efficient systems. *Future Gener. Comput. Syst.* **2018**, in press. [CrossRef]
3. Lin, C.-Y.; Ho, Y.-H. Determinants of green practice adoption for logistics companies in China. *J. Bus. Ethics* **2011**, *98*, 67–83. [CrossRef]

4. Carter, C.R.; Easton, P.L. Sustainable supply chain management: Evolution and future directions. *Int. J. Phys. Distrib. Logist. Manag.* **2011**, *41*, 46–62. [CrossRef]
5. El-Hefenawy, N.; Metwally, M.A.; Ahmed, Z.M.; El-Henawy, I.M. A review on the applications of neutrosophic sets. *J. Comput. Theor. Nanosci.* **2016**, *13*, 936–944. [CrossRef]
6. Zadeh, L.A. Information and control. *Fuzzy Sets* **1965**, *8*, 338–353.
7. Turksen, I.B. Interval valued fuzzy sets based on normal forms. *Fuzzy Sets Syst.* **1986**, *20*, 191–210. [CrossRef]
8. Atanassov, K.T. Intuitionistic fuzzy sets. *Fuzzy Sets Syst.* **1986**, *20*, 87–96. [CrossRef]
9. Atanassov, K.; Gargov, G. Interval valued intuitionistic fuzzy sets. *Fuzzy Sets Syst.* **1989**, *31*, 343–349. [CrossRef]
10. Liu, H.-W.; Wang, G.-J. Multi-criteria decision-making methods based on intuitionistic fuzzy sets. *Eur. J. Oper. Res.* **2007**, *179*, 220–233. [CrossRef]
11. Pei, Z.; Zheng, L. A novel approach to multi-attribute decision-making based on intuitionistic fuzzy sets. *Expert Syst. Appl.* **2012**, *39*, 2560–2566. [CrossRef]
12. Chen, T.-Y. An outcome-oriented approach to multicriteria decision analysis with intuitionistic fuzzy optimistic/pessimistic operators. *Expert Syst. Appl.* **2010**, *37*, 7762–7774. [CrossRef]
13. Wang, H.; Smarandache, F.; Zhang, Y.; Sunderraman, R. *Single Valued Neutrosophic Sets*; Review of the Air Force Academy: Colorado Springs, CO, USA, 2010; p. 10.
14. Smarandache, F. *Neutrosophy: A Unifying Field in Logics: Neutrosophic Logic. Neutrosophy, Neutrosophic Set, Neutrosophic Probability*; American Research Press: Santa Fe, NM, USA, 1999.
15. Abdel-Basset, M.; Mohamed, M.; Smarandache, F.; Chang, V. Neutrosophic Association Rule Mining Algorithm for Big Data Analysis. *Symmetry* **2018**, *10*, 106. [CrossRef]
16. Abdel-Basset, M.; Mohamed, M.; Smarandache, F. An Extension of Neutrosophic AHP–SWOT Analysis for Strategic Planning and Decision-Making. *Symmetry* **2018**, *10*, 116. [CrossRef]
17. Zhang, H.-Y.; Wang, J.-Q.; Chen, X.-H. Interval neutrosophic sets and their application in multicriteria decision-making problems. *Sci. World J.* **2014**, *2014*. [CrossRef] [PubMed]
18. Majumdar, P.; Samanta, S.K. On similarity and entropy of neutrosophic sets. *J. Intell. Fuzzy Syst.* **2014**, *26*, 1245–1252.
19. Wang, H.; Smarandache, F.; Zhang, Y.; Sunderraman, R. *Interval Neutrosophic Sets and Logic: Theory and Applications in Computing*; Hexis: Phoenix, AZ, USA, 2005.
20. Broumi, S.; Smarandache, F. Single Valued Neutrosophic Trapezoid Linguistic Aggregation Operators Based Multi-Attribute Decision Making. *Bull. Pure Appl. Sci.-Math.* **2014**, *33*, 135–155. [CrossRef]
21. Zhang, H.; Wang, J.; Chen, X. An outranking approach for multi-criteria decision-making problems with interval-valued neutrosophic sets. *Neural Comput. Appl.* **2016**, *27*, 615–627. [CrossRef]
22. Govindan, K.; Khodaverdi, R.; Jafarian, A. A fuzzy multi criteria approach for measuring sustainability performance of a supplier based on triple bottom line approach. *J. Clean. Prod.* **2013**, *47*, 345–354. [CrossRef]
23. Erol, I.; Sencer, S.; Sari, R. A new fuzzy multi-criteria framework for measuring sustainability performance of a supply chain. *Ecol. Econ.* **2011**, *70*, 1088–1100. [CrossRef]
24. Wen, L.; Xu, L.; Wang, R. Sustainable supplier evaluation based on intuitionistic fuzzy sets group decision methods. *J. Inf. Comput. Sci.* **2013**, *10*, 3209–3220. [CrossRef]
25. Orji, I.; Wei, S. A decision support tool for sustainable supplier selection in manufacturing firms. *J. Ind. Eng. Manag.* **2014**, *7*, 1293. [CrossRef]
26. Shaw, K.; Shankar, R.; Yadav, S.S.; Thakur, L.S. Supplier selection using fuzzy AHP and fuzzy multi-objective linear programming for developing low carbon supply chain. *Expert Syst. Appl.* **2012**, *39*, 8182–8192. [CrossRef]
27. Büyüközkan, G.; Çifçi, G. A novel fuzzy multi-criteria decision framework for sustainable supplier selection with incomplete information. *Comput. Ind.* **2011**, *62*, 164–174. [CrossRef]
28. Ho, W.; He, T.; Lee, C.K.M.; Emrouznejad, A. Strategic logistics outsourcing: An integrated QFD and fuzzy AHP approach. *Expert Syst. Appl.* **2012**, *39*, 10841–10850. [CrossRef]
29. Dursun, M.; Karsak, E.E. A QFD-based fuzzy MCDM approach for supplier selection. *Appl. Math. Model.* **2013**, *37*, 5864–5875. [CrossRef]
30. Dai, J.; Blackhurst, J. A four-phase AHP-QFD approach for supplier assessment: A sustainability perspective. *Int. J. Prod. Res.* **2012**, *50*, 5474–5490. [CrossRef]

31. Kumar, A.; Jain, V.; Kumar, S. A comprehensive environment friendly approach for supplier selection. *Omega* **2014**, *42*, 109–123. [CrossRef]
32. Azadi, M.; Jafarian, M.; Saen, R.F.; Mirhedayatian, S.M. A new fuzzy DEA model for evaluation of efficiency and effectiveness of suppliers in sustainable supply chain management context. *Comput. Oper. Res.* **2015**, *54*, 274–285. [CrossRef]
33. Choy, K.L.; Lee, W.; Lo, V. An enterprise collaborative management system—A case study of supplier relationship management. *J. Enterp. Inf. Manag.* **2004**, *17*, 191–207. [CrossRef]
34. Bhattacharya, A.; Geraghty, J.; Young, P. Supplier selection paradigm: An integrated hierarchical QFD methodology under multiple-criteria environment. *Appl. Soft Comput.* **2010**, *10*, 1013–1027. [CrossRef]
35. Amin, S.H.; Razmi, J.; Zhang, G. Supplier selection and order allocation based on fuzzy SWOT analysis and fuzzy linear programming. *Expert Syst. Appl.* **2011**, *38*, 334–342. [CrossRef]
36. Rajesh, G.; Malliga, P. Supplier selection based on AHP QFD methodology. *Procedia Eng.* **2013**, *64*, 1283–1292. [CrossRef]
37. Taghizadeh, H.; Ershadi, M.S.S. Selection in Supply Chain with Combined QFD and ANP Approaches (Case study). *Res. J. Recent Sci.* **2013**, *2*, 66–76.
38. Dey, P.K.; Bhattacharya, A.; Ho, W. Strategic supplier performance evaluation: A case-based action research of a UK manufacturing organisation. *Int. J. Prod. Econ.* **2015**, *166*, 192–214. [CrossRef]
39. Bayrak, M.; Celebi, N.; Taşkin, H. A fuzzy approach method for supplier selection. *Prod. Plan. Control* **2007**, *18*, 54–63. [CrossRef]
40. Kumaraswamy, A.H.; Bhattacharya, A.; Kumar, V.; Brady, M. An integrated QFD-TOPSIS methodology for supplier selection in SMEs. In Proceedings of the 2011 Third International Conference on Computational Intelligence, Modelling and Simulation (CIMSiM), Langkawi, Malaysia, 20–22 September 2011; pp. 271–276.
41. Wu, C.-M.; Hsieh, C.-L.; Chang, K.-L. A hybrid multiple criteria decision-making model for supplier selection. *Math. Probl. Eng.* **2013**, *2013*. [CrossRef]
42. Ghorabaee, M.K.; Amiri, M.; Zavadskas, E.K.; Antucheviciene, J. Supplier evaluation and selection in fuzzy environments: A review of MADM approaches. *Econ. Res.-Ekon. Istraž.* **2017**, *30*, 1073–1118.
43. Rouyendegh, B.D. Developing an integrated ANP and intuitionistic fuzzy TOPSIS model for supplier selection. *J. Test. Eval.* **2014**, *43*, 664–672. [CrossRef]
44. Tavana, M.; Yazdani, M.; di Caprio, D. An application of an integrated ANP-QFD framework for sustainable supplier selection. *Int. J. Logist. Res. Appl.* **2017**, *20*, 254–275. [CrossRef]
45. Şahin, R.; Yiğider, M. A Multi-criteria neutrosophic group decision-making method based TOPSIS for supplier selection. *arXiv*, 2014.
46. Reddy, R.; Reddy, D.; Krishnaiah, G. Lean supplier selection based on hybrid MCGDM approach using interval valued neutrosophic sets: A case study. *Int. J. Innov. Res. Dev.* **2016**, *5*, 291–296.
47. Karaşan, A.; Kahraman, C. Interval-Valued Neutrosophic Extension of EDAS Method. In *Advances in Fuzzy Logic and Technology 2017*; Springer: Berlin, Germany, 2017; pp. 343–357.
48. Van, L.H.; Yu, V.F.; Dat, L.Q.; Dung, C.C.; Chou, S.-Y.; Loc, N.V. New Integrated Quality Function Deployment Approach Based on Interval Neutrosophic Set for Green Supplier Evaluation and Selection. *Sustainability* **2018**, *10*, 838. [CrossRef]
49. Abdel-Basset, M.; Manogaran, G.; Gamal, A.; Smarandache, F. A hybrid approach of neutrosophic sets and DEMATEL method for developing supplier selection criteria. In *Design Automation for Embedded Systems*; Springer: Berlin, Germany, 2018; pp. 1–22.
50. Saaty, T.L. Decision making—The analytic hierarchy and network processes (AHP/ANP). *J. Syst. Sci. Syst. Eng.* **2004**, *13*, 1–35. [CrossRef]
51. Saaty, T.L. What is the analytic hierarchy process. In *Mathematical Models for Decision Support*; Springer: Berlin, Germany, 1988; pp. 109–121.
52. Yoon, K.; Hwang, C.-L. *Multiple Attribute Decision Making: Methods and Applications*; Springer: Berlin, Germany, 1981.

53. Brauers, W.K.; Zavadskas, E.K. The MOORA method and its application to privatization in a transition economy. *Control Cybern.* **2006**, *35*, 445–469.
54. Adalı, E.A.; Işık, A.T. The multi-objective decision-making methods based on MULTIMOORA and MOOSRA for the laptop selection problem. *J. Ind. Eng. Int.* **2017**, *13*, 229–237. [CrossRef]

symmetry

MDPI

Article

An Extended Technique for Order Preference by Similarity to an Ideal Solution (TOPSIS) with Maximizing Deviation Method Based on Integrated Weight Measure for Single-Valued Neutrosophic Sets

Ganeshsree Selvachandran [1,*], Shio Gai Quek [2], Florentin Smarandache [3] and Said Broumi [4]

1 Department of Actuarial Science and Applied Statistics, Faculty of Business & Information Science, UCSI University, Jalan Menara Gading, 56000 Kuala Lumpur, Malaysia

2 A-Level Academy, UCSI College KL Campus, Lot 12734, Jalan Choo Lip Kung, Taman Taynton View, 56000 Kuala Lumpur, Malaysia; queksg@ucsicollege.edu.my

3 Department of Mathematics, University of New Mexico, 705 Gurley Avenue, Gallup, NM 87301, USA; fsmarandache@gmail.com or smarand@unm.edu

4 Laboratory of Information Processing, Faculty of Science Ben M'Sik, University Hassan II, B.P 7955, Sidi Othman, Casablanca 20000, Morocco; broumisaid78@gmail.com

* Correspondence: Ganeshsree@ucsiuniversity.edu.my; Tel.: +60-391-018-880

Received: 7 June 2018; Accepted: 20 June 2018; Published: 22 June 2018

Abstract: A single-valued neutrosophic set (SVNS) is a special case of a neutrosophic set which is characterized by a truth, indeterminacy, and falsity membership function, each of which lies in the standard interval of [0, 1]. This paper presents a modified Technique for Order Preference by Similarity to an Ideal Solution (TOPSIS) with maximizing deviation method based on the single-valued neutrosophic set (SVNS) model. An integrated weight measure approach that takes into consideration both the objective and subjective weights of the attributes is used. The maximizing deviation method is used to compute the objective weight of the attributes, and the non-linear weighted comprehensive method is used to determine the combined weights for each attributes. The use of the maximizing deviation method allows our proposed method to handle situations in which information pertaining to the weight coefficients of the attributes are completely unknown or only partially known. The proposed method is then applied to a multi-attribute decision-making (MADM) problem. Lastly, a comprehensive comparative studies is presented, in which the performance of our proposed algorithm is compared and contrasted with other recent approaches involving SVNSs in literature.

Keywords: 2ingle-valued neutrosophic set; Technique for Order Preference by Similarity to an Ideal Solution (TOPSIS); integrated weight; maximizing deviation; multi-attribute decision-making (MADM)

1. Introduction

The study of fuzzy set theory proposed by Zadeh [1] was an important milestone in the study of uncertainty and vagueness. The widespread success of this theory has led to the introduction of many extensions of fuzzy sets such as the intuitionistic fuzzy set (IFS) [2], interval-valued fuzzy set (IV-FS) [3], vague set [4], and hesitant fuzzy set [5]. The most widely used among these models is the IFS model which has also spawned other extensions such as the interval-valued intuitionistic fuzzy set [6] and bipolar intuitionistic fuzzy set [7]. Smarandache [8] then introduced an improvement to IFS theory called neutrosophic set theory which loosely refers to neutral knowledge. The study of the neutrality aspect of knowledge is the main distinguishing criteria between the theory of fuzzy sets,

IFSs, and neutrosophic sets. The classical neutrosophic set (NS) is characterized by three membership functions which describe the degree of truth (T), the degree of indeterminacy (I), and the degree of falsity (F), whereby all of these functions assume values in the non-standard interval of $]0^-$, $1^+[$. The truth and falsity membership functions in a NS are analogous to the membership and non-membership functions in an IFS, and expresses the degree of belongingness and non-belongingness of the elements, whereas the indeterminacy membership function expresses the degree of neutrality in the information. This additional indeterminacy membership function gives NSs the ability to handle the neutrality aspects of the information, which fuzzy sets and its extensions are unable to handle. Another distinguishing factor between NSs and other fuzzy-based models is the fact that all the three membership functions in a NS are entirely independent of one another, unlike the membership and non-membership functions in an IFS or other fuzzy-based models in which values of the membership and non-membership functions are dependent on one another. This gives NSs the ability to handle uncertain, imprecise, inconsistent, and indeterminate information, particularly in situations whereby the factors affecting these aspects of the information are independent of one another. This also makes the NS more versatile compared to IFSs and other fuzzy- or IF-based models in literature.

Smarandache [8] and Wang et al. [9] pointed out that the non-standard interval of $]0^-, 1^+[$ in which the NS is defined in, makes it impractical to be used in real-life problems. Furthermore, values in this non-standard interval are less intuitive and the significance of values in this interval can be difficult to be interpreted. This led to the conceptualization of the single-valued neutrosophic set (SVNS). The SVNS is a straightforward extension of NS which is defined in the standard unit interval of [0, 1]. As values in [0, 1] are compatible with the range of acceptable values in conventional fuzzy set theory and IFS theory, it is better able to capture the intuitiveness of the process of assigning membership values. This makes the SVNS model easier to be applied in modelling real-life problems as the results obtained are a lot easier to be interpreted compared to values in the interval $]0^-, 1^+[$.

The SVNS model has garnered a lot of attention since its introduction in [9], and has been actively applied in various multi-attribute decision-making (MADM) problems using a myriad of different approaches. Wang et al. [9] introduced some set theoretic operators for SVNSs, and studied some additional properties of the SVNS model. Ye [10,11] introduced a decision-making algorithm based on the correlation coefficients for SVNSs, and applied this algorithm in solving some MADM problems. Ye [12,13] introduced a clustering method and also some decision-making methods that are based on the similarity measures of SVNSs, whereas Huang [14] introduced a new decision-making method for SVNSs and applied this method in clustering analysis and MADM problems. Peng and Liu [15] on the other hand proposed three decision-making methods based on a new similarity measure, the EDAS method and level soft sets for neutrosophic soft sets, and applied this new measure to MADM problems set in a neutrosophic environment. The relations between SVNSs and its properties were first studied by Yang et al. [16], whereas the graph theory of SVNSs and bipolar SVNSs were introduced by Broumi et al. in [17–19] and [20–22], respectively. The aggregation operators of simplified neutrosophic sets (SNSs) were studied by Tian et al. [23] and Wu et al. [24]. Tian et al. [23] introduced a generalized prioritized aggregation operator for SNSs and applied this operator in a MADM problem set in an uncertain linguistic environment, whereas Wu et al. [24] introduced a cross-entropy measure and a prioritized aggregation operator for SNSs and applied these in a MADM problem. Sahin and Kucuk [25] proposed a subsethood measure for SVNSs and applied these to MADM problems.

The fuzzy Technique for Order Preference by Similarity to an Ideal Solution (TOPSIS) method for SVNSs were studied by Ye [26] and Biswas et al. [27]. Ye [26] introduced the TOPSIS method for group decision-making (MAGDM) that is based on single-valued neutrosophic linguistic numbers, to deal with linguistic decision-making. This TOPSIS method uses subjective weighting method whereby attribute weights are randomly assigned by the users. Maximizing deviation method or any other objective weighting methods are not used. Biswas et al. [27] proposed a TOPSIS method for group decision-making (MAGDM) based on the SVNS model. This TOPSIS method is based on the original fuzzy TOPSIS method and does not use the maximizing deviation method to calculate the objective weights for each attribute. The subjective weight of each attribute is determined by using the single-valued neutrosophic weighted averaging aggregation operator to calculate the aggregated weights of the attributes using the subjective weights that are assigned by each decision maker.

The process of assigning weights to the attributes is an important phase of decision making. Most research in this area usually use either objective or subjective weights. However, considering the fact that different values for the weights of the attributes has a significant influence on the ranking of the alternatives, it is imperative that both the objective and subjective weights of the attributes are taken into account in the decision-making process. In view of this, we consider the attributes' subjective weights which are assigned by the decision makers, and the objective weights which are computed using the maximizing deviation method. These weights are then combined using the non-linear weighted comprehensive method to obtain the integrated weight of the attributes.

The advantages and drawbacks of the methods that were introduced in the works described above served as the main motivation for the work proposed in this paper, as we seek to introduce an effective SVNS-based decision-making method that is free of all the problems that are inherent in the other existing methods in literature. In addition to these advantages and drawbacks, the works described above have the added disadvantage of not being able to function (i.e., provide reasonable solutions) under all circumstances. In view of this, the objective of this paper is to introduce a novel TOPSIS with maximizing deviation method for SVNSs that is able to provide effective solutions under any circumstances. Our proposed TOPSIS method is designed to handle MADM problems, and uses the maximizing deviation method to calculate the objective weights of attributes, utilizing an integrated weight measure that takes into consideration both the subjective and objective weights of the attributes. The robustness of our TOPSIS method is verified through a comprehensive series of tests which proves that our proposed method is the only method that shows compliance to all the tests, and is able to provide effective solutions under all different types of situations, thus out-performing all of the other considered methods.

The remainder of this paper is organized as follows. In Section 2, we recapitulate some of the fundamental concepts related to neutrosophic sets and SVNSs. In Section 3, we define an SVNS-based TOPSIS and maximizing deviation methods and an accompanying decision-making algorithm. The proposed decision-making method is applied to a supplier selection problem in Section 4. In Section 5, a comprehensive comparative analysis of the results obtained via our proposed method and other recent approaches is presented. The similarities and differences in the performance of the existing algorithms and our algorithm is discussed, and it is proved that our algorithm is effective and provides reliable results in every type of situation. Concluding remarks are given in Section 6, followed by the acknowledgements and list of references.

2. Preliminaries

In this section, we recapitulate some important concepts pertaining to the theory of neutrosophic sets and SVNSs. We refer the readers to [8,9] for further details pertaining to these models.

The neutrosophic set model [8] is a relatively new tool for representing and measuring uncertainty and vagueness of information. It is fast becoming a preferred general framework for the analysis of uncertainty in data sets due to its capability in the handling big data sets, as well as its ability in representing all the different types of uncertainties that exists in data, in an effective and concise manner via a triple membership structure. This triple membership structure captures not only the degree of belongingness and non-belongingness of the objects in a data set, but also the degree of neutrality and indeterminacy that exists in the data set, thereby making it superior to ordinary fuzzy sets [1] and its extensions such as IFSs [2], vague sets [4], and interval-valued fuzzy sets [3]. The formal definition of a neutrosophic set is as given below.

Let U be a universe of discourse, with a class of elements in U denoted by x.

Definition 1. [8] *A neutrosophic set A is an object having the form* $A = \{x,\ T_A(x),\ I_A(x),\ F_A(x) : x \in U\}$, *where the functions* $T,\ I,\ F\ : U \rightarrow]^-0,\ 1^+[$ *denote the truth, indeterminacy, and falsity membership functions, respectively, of the element* $x \in U$ *with respect to A. The membership functions must satisfy the condition* $^-0 \leq T_A(x) + I_A(x) + F_A(x) \leq 3^+$.

Definition 2. [8] *A neutrosophic set A is contained in another neutrosophic set B, if* $T_A(x) \leq T_B(x),\ I_A(x) \geq I_B(x)$, *and* $F_A(x) \geq F_B(x)$, *for all* $x \in U$. *This relationship is denoted as* $A \subseteq B$.

Wang et al. [9] then introduced a special case of the NS model called the single-valued neutrosophic set (SVNS) model, which is as defined below. This SVNS model is better suited to applied in real-life problems compared to NSs due to the structure of its membership functions which are defined in the standard unit interval of [0, 1].

Definition 3. [9] *A SVNS A is a neutrosophic set that is characterized by a truth-membership function* $T_A(x)$, *an indeterminacy-membership function* $I_A(x)$, *and a falsity-membership function* $F_A(x)$, *where* $T_A(x),\ I_A(x),\ F_A(x) \in [0,\ 1]$. *This set A can thus be written as*

$$A = \{\langle x,\ T_A(x),\ I_A(x),\ F_A(x)\rangle : x \in U\}. \tag{1}$$

The sum of $T_A(x),\ I_A(x)$ and $F_A(x)$ must fulfill the condition $0 \leq T_A(x) + I_A(x) + F_A(x) \leq 3$. For a SVNS A in U, the triplet $(T_A(x),\ I_A(x),\ F_A(x))$ is called a single-valued neutrosophic number (SVNN). For the sake of convenience, we simply let $x = (T_x,\ I_x,\ F_x)$ to represent a SVNN as an element in the SVNS A.

Next, we present some important results pertaining to the concepts and operations of SVNSs. The subset, equality, complement, union, and intersection of SVNSs, and some additional operations between SVNSs were all defined by Wang et al. [9], and these are presented in Definitions 4 and 5, respectively.

Definition 4. [9] *Let A and B be two SVNSs over a universe U.*

(i) *A is contained in B, if* $T_A(x) \leq T_B(x),\ I_A(x) \geq I_B(x)$, *and* $F_A(x) \geq F_B(x)$, *for all* $x \in U$. *This relationship is denoted as* $A \subseteq B$.
(ii) *A and B are said to be equal if* $A \subseteq B$ *and* $B \subseteq A$.
(iii) $A^c = (x,\ (F_A(x),\ 1 - I_A(x),\ T_A(x)))$, *for all* $x \in U$.
(iv) $A \cup B = (x,\ (\max(T_A,\ T_B), \min(I_A,\ I_B), \min(F_A,\ F_B)))$, *for all* $x \in U$.
(v) $A \cap B = (x,\ (\min(T_A,\ T_B), \max(I_A,\ I_B), \max(F_A,\ F_B)))$, *for all* $x \in U$.

Definition 5. [9] *Let* $x = (T_x,\ I_x,\ F_x)$ *and* $y = (T_y,\ I_y,\ F_y)$ *be two SVNNs. The operations for SVNNs can be defined as follows:*

(i) $x \bigoplus y = (T_x + T_y - T_x * T_y,\ I_x * I_y,\ F_x * F_y)$
(ii) $x \bigotimes y = (T_x * T_y,\ I_x + I_y - I_x * I_y,\ F_x + F_y - F_x * F_y)$
(iii) $\lambda x = \left(1 - (1 - T_x)^\lambda,\ (I_x)^\lambda,\ (F_x)^\lambda\right)$, *where* $\lambda > 0$
(iv) $x^\lambda = \left((T_x)^\lambda,\ 1 - (1 - I_x)^\lambda,\ 1 - (1 - F_x)^\lambda\right)$, *where* $\lambda > 0$.

Majumdar and Samanta [28] introduced the information measures of distance, similarity, and entropy for SVNSs. Here we only present the definition of the distance measures between SVNSs as it is the only component that is relevant to this paper.

Definition 6. [28] *Let A and B be two SVNSs over a finite universe* $U = \{x_1,\ x_2, \ldots,\ x_n\}$. *Then the various distance measures between A and B are defined as follows:*

(i) *The Hamming distance between A and B are defined as:*

$$d_H(A, B) = \sum_{i=1}^{n}\{|T_A(x_i) - T_B(x_i)| + |I_A(x_i) - I_B(x_i)| + |F_A(x_i) - F_B(x_i)|\} \tag{2}$$

(ii) *The normalized Hamming distance between A and B are defined as:*

$$d_H^N(A, B) = \frac{1}{3n}\sum_{i=1}^{n}\{|T_A(x_i) - T_B(x_i)| + |I_A(x_i) - I_B(x_i)| + |F_A(x_i) - F_B(x_i)|\} \tag{3}$$

(ii) *The Euclidean distance between A and B are defined as:*

$$d_E(A, B) = \sqrt{\sum_{i=1}^{n}\left\{(T_A(x_i) - T_B(x_i))^2 + (I_A(x_i) - I_B(x_i))^2 + (F_A(x_i) - F_B(x_i))^2\right\}} \tag{4}$$

(iv) *The normalized Euclidean distance between A and B are defined as:*

$$d_E^N(A, B) = \sqrt{\frac{1}{3n}\sum_{i=1}^{n}\left\{(T_A(x_i) - T_B(x_i))^2 + (I_A(x_i) - I_B(x_i))^2 + (F_A(x_i) - F_B(x_i))^2\right\}} \tag{5}$$

3. A TOPSIS Method for Single-Valued Neutrosophic Sets

In this section, we present the description of the problem that is being studied followed by our proposed TOPSIS method for SVNSs. The accompanying decision-making algorithm which is based on the proposed TOPSIS method is presented. This algorithm uses the maximizing deviation method to systematically determine the objective weight coefficients for the attributes.

3.1. Description of Problem

Let $U = \{u_1, u_2, \ldots, u_m\}$ denote a finite set of m alternatives, $A = \{e_1, e_2, \ldots, e_n\}$ be a set of n parameters, with the weight parameter w_j of each e_j completely unknown or only partially known, $w_j \in [0, 1]$, and $\sum_{j=1}^{n} w_j = 1$.

Let A be an SVNS in which $x_{ij} = (T_{ij}, I_{ij}, F_{ij})$ represents the SVNN that represents the information pertaining to the ith alternative x_i that satisfies the corresponding jth parameter e_j. The tabular representation of A is as given in Table 1.

Table 1. Tabular representation of the Single Valued Neutrosophic Set (SVNS) A.

U	e_1	e_2	$\ldots$	e_n
x_1	(T_{11}, I_{11}, F_{11})	(T_{12}, I_{12}, F_{12})	$\ldots$	(T_{1n}, I_{1n}, F_{1n})
x_2	(T_{21}, I_{21}, F_{21})	(T_{22}, I_{22}, F_{22})	$\ldots$	(T_{2n}, I_{2n}, F_{2n})
$\vdots$	$\vdots$	$\vdots$	$\ddots$	$\vdots$
x_m.	(T_{m1}, I_{m1}, F_{m1})	(T_{m2}, I_{m2}, F_{m2})	$\ldots$	(T_{mn}, I_{mn}, F_{mn})

3.2. The Maximizing Deviation Method for Computing Incomplete or Completely Unknown Attribute Weights

The maximizing deviation method was proposed by Wang [29] with the aim of applying it in MADM problems in which the weights of the attributes are completely unknown or only partially known. This method uses the law of input arguments i.e., it takes into account the magnitude of the membership functions of each alternative for each attribute, and uses this information to obtain exact and reliable evaluation results pertaining to the weight coefficients for each attribute. As such, this method is able to compute the weight coefficients of the attributes without any subjectivity, in a fair and objective manner.

The maximizing deviation method used in this paper is a modification of the original version introduced in Wang [29] that has been made compatible with the structure of the SVNS model. The definitions of the important concepts involved in this method are as given below.

Definition 7. *For the parameter $e_j \in A$, the deviation of the alternative x_i to all the other alternatives is defined as:*

$$D_{ij}(w_j) = \sum_{k=1}^{m} w_j\, d\left(x_{ij},\, x_{kj}\right), \tag{6}$$

where x_{ij}, x_{kj} are the elements of the SVNS A, $i = 1, 2, \ldots, m$, $j = 1, 2, \ldots, n$ and $d\left(x_{ij},\, x_{kj}\right)$ denotes the distance between elements x_{ij} and x_{kj}.

The other deviation values include the deviation value of all alternatives to other alternatives, and the total deviation value of all parameters to all alternatives, both of which are as defined below:

(i) The deviation value of all alternatives to other alternatives for the parameter $e_j \in A$, denoted by $D_j(w_j)$, is defined as:

$$D_j(w_j) = \sum_{i=1}^{m} D_{ij}(w_j) = \sum_{i=1}^{m}\sum_{k=1}^{m} w_j\, d\left(x_{ij},\, x_{kj}\right), \tag{7}$$

where $j = 1, 2, \ldots, n$.

(ii) The total deviation value of all parameters to all alternatives, denoted by $D(w_j)$, is defined as:

$$(w_j) = \sum_{j=1}^{n} D_j(w_j) = \sum_{j=1}^{n}\sum_{i=1}^{m}\sum_{k=1}^{m} w_j\, d\left(x_{ij},\, x_{kj}\right), \tag{8}$$

where w_j represents the weight of the parameter $e_j \in A$.

(iii) The individual objective weight of each parameter $e_j \in A$, denoted by θ_j, is defined as:

$$\theta_j = \frac{\sum_{i=1}^{m}\sum_{k=1}^{m} d\left(x_{ij},\, x_{kj}\right)}{\sum_{j=1}^{n}\sum_{i=1}^{m}\sum_{k=1}^{m} d\left(x_{ij},\, x_{kj}\right)} \tag{9}$$

It should be noted that any valid distance measure between SVNSs can be used in Equations (6)–(9). However, to improve the effective resolution of the decision-making process, in this paper, we use the normalized Euclidean distance measure given in Equation (5) in the computation of Equations (6)–(9).

3.3. TOPSIS Method for MADM Problems with Incomplete Weight Information

The Technique for Order of Preference by Similarity to Ideal Solution (TOPSIS) was originally introduced by Hwang and Yoon [30], and has since been extended to fuzzy sets, IFSs, and other fuzzy-based models. The TOPSIS method works by ranking the alternatives based on their distance from the positive ideal solution and the negative ideal solution. The basic guiding principle is that the most preferred alternative should have the shortest distance from the positive ideal solution and the farthest distance from the negative ideal solution (Hwang and Yoon [30], Chen and Tzeng [31]). In this section, we present a decision-making algorithm for solving MADM problems in single-valued neutrosophic environments, with incomplete or completely unknown weight information.

3.3.1. The Proposed TOPSIS Method for SVNSs

After obtaining information pertaining to the weight values for each parameter based on the maximizing deviation method, we develop a modified TOPSIS method for the SVNS model. To achieve our goal, we introduce several definitions that are the important components of our proposed TOPSIS method.

Let the relative neutrosophic positive ideal solution (RNPIS) and relative neutrosophic negative ideal solution (RNNIS) be denoted by b^+ and b^-, respectively, where these solutions are as defined below:

$$b^+ = \left\{ \left(\max_i T_{ij},\ \min_i I_{ij},\ \min_i F_{ij} \right) \middle| \, j = 1, 2, \ldots, n \right\}, \tag{10}$$

and

$$b^- = \left\{ \left(\min_i T_{ij},\ \max_i I_{ij},\ \max_i F_{ij} \right) \middle| \, j = 1, 2, \ldots, n \right\} \tag{11}$$

The difference between each object and the RNPIS, denoted by D_i^+, and the difference between each object and the RNNIS, denoted by D_i^-, can then be calculated using the normalized Euclidean distance given in Equation (5) and by the formula given in Equations (12) and (13).

$$D_i^+ = \sum_{j=1}^{n} w_j\, d_{NE}\left(b_{ij},\, b_j^+\right), \qquad i = 1, 2, \ldots, m \tag{12}$$

and

$$D_i^- = \sum_{j=1}^{n} w_j\, d_{NE}\left(b_{ij},\, b_j^-\right), \qquad i = 1, 2, \ldots, m \tag{13}$$

Here, w_j denotes the integrated weight for each of the attributes.

The optimal alternative can then be found using the measure of the relative closeness coefficient of each alternative, denoted by C_i, which is as defined below:

$$C_i = \frac{D_i^-}{\max_j D_j^-} - \frac{D_i^+}{\min_j D_j^+}, \quad i, j = 1, 2, \ldots, m \tag{14}$$

From the structure of the closeness coefficient in Equation (14), it is obvious that the larger the difference between an alternative and the fuzzy negative ideal object, the larger the value of the closeness coefficient of the said alternative. Therefore, by the principal of maximum similarity between an alternative and the fuzzy positive ideal object, the objective of the algorithm is to determine the alternative with the maximum closeness coefficient. This alternative would then be chosen as the optimal alternative.

3.3.2. Attribute Weight Determination Method: An Integrated WEIGHT MEASure

In any decision-making process, there are two main types of weight coefficients, namely the subjective and objective weights that need to be taken into consideration. Subjective weight refers to the values assigned to each attribute by the decision makers based on their individual preferences and experience, and is very much dependent on the risk attitude of the decision makers. Objective weight refers to the weights of the attributes that are computed mathematically using any appropriate computation method. Objective weighting methods uses the law of input arguments (i.e., the input values of the data) as it determines the attribute weights based on the magnitude of the membership functions that are assigned to each alternative for each attribute.

Therefore, using only subjective weighting in the decision-making process would be inaccurate as it only reflects the opinions of the decision makers while ignoring the importance of each attribute that are reflected by the input values. Using only objective weighting would also be inaccurate as it only reflects the relative importance of the attributes based on the law of input arguments, but fails to take into consideration the preferences and risk attitude of the decision makers.

To overcome this drawback and improve the accuracy and reliability of the decision-making process, we use an integrated weight measure which combines the subjective and objective weights of the attributes. This factor makes our decision-making algorithm more accurate compared to most of the other existing methods in literature that only take into consideration either the objective or subjective weights.

Based on the formula and weighting method given above, we develop a practical and effective decision-making algorithm based on the TOPSIS approach for the SVNS model with incomplete weight information. The proposed Algorithm 1 is as given below.

Algorithm 1. (based on a modified TOPSIS approach).

Step 1. Input the SVNS A which represents the information pertaining to the problem.
Step 2. Input the subjective weight h_j for each of the attributes $e_j \in A$ as given by the decision makers.
Step 3. Compute the objective weight θ_j for each of the attributes $e_j \in A$, using Equation (9).
Step 4. The integrated weight coefficient w_j for each of the attributes $e_j \in A$, is computed using Equation as follow:

$$w_j = \frac{h_j\,\theta_j}{\sum_{j=1}^{n} h_j\,\theta_j}$$

Step 5. The values of RNPIS b^+ and RNNIS b^- are computed using Equations (10) and (11).
Step 6. The difference between each alternative and the RNPIS, D^+ and the RNNIS D^- are computed using Equations (12) and (13), respectively.
Step 7. The relative closeness coefficient C_i for each alternative is calculated using Equation (14).
Step 8. Choose the optimal alternative based on the principal of maximum closeness coefficient.

4. Application of the Topsis Method in a Made Problem

The implementation process and utility of our proposed decision-making algorithm is illustrated via an example related to a supplier selection problem.

4.1. Illustrative Example

In today's extremely competitive business environment, firms must be able to produce good quality products at reasonable prices in order to be successful. Since the quality of the products is directly dependent on the effectiveness and performance of its suppliers, the importance of supplier selection has become increasingly recognized. In recent years, this problem has been handled using various mathematical tools. Some of the recent research in this area can be found in [32–38].

Example 1. *A manufacturing company is looking to select a supplier for one of the products manufactured by the company. The company has shortlisted ten suppliers from an initial list of suppliers. These ten suppliers form the set of alternatives U that are under consideration,*

$$U = \{x_1, x_2, x_3, x_4, x_5, x_6, x_7, x_8, x_9, x_{10}\}.$$

The procurement manager and his team of buyers evaluate the suppliers based on a set of evaluation attributes E which is defined as:

$$E = \{e_1 = \text{service quality},\ e_2 = \text{pricing and cos t structure},\ e_3 = \text{financial stability},$$
$$e_4 = \text{environmental regulation compliance},\ e_5 = \text{reliability},$$
$$e_6 = \text{relevant experience}\}.$$

The firm then evaluates each of the alternatives x_i $(i = 1, 2, \ldots, 10)$, with respect to the attributes e_j $(j = 1, 2, \ldots, 6)$. The evaluation done by the procurement team is expressed in the form of SVNNs in a SVNS A.

Now suppose that the company would like to select one of the five shortlisted suppliers to be their supplier. We apply the proposed Algorithm 1 outlined in Section 3.3 to this problem with the aim of selecting a supplier that best satisfies the specific needs and requirements of the company. The steps involved in the implementation process of this algorithm are outlined below (Algorithm 2).

Algorithm 2. (based on the modified TOPSIS approach).

Step 1. The SVNS A constructed for this problem is given in tabular form in Table 2

Step 2. The subjective weight h_j for each attribute $e_j \in A$ as given by the procurement team (the decision makers) are $h = \{h_1 = 0.15,\ h_2 = 0.15,\ h_3 = 0.22,\ h_4 = 0.25,\ h_5 = 0.14,\ h_6 = 0.09\}$.

Step 3. The objective weight θ_j for each attribute $e_j \in A$ is computed using Equation (9) are as given below:

$$\theta = \{\theta_1 = 0.139072,\ \theta_2 = 0.170256,\ \theta_3 = 0.198570,\ \theta_4 = 0.169934, \theta_5 = 0.142685,\ \theta_6 = 0.179484\}.$$

Step 4. The integrated weight w_j for each attribute $e_j \in A$ is computed using Equation (15). The integrated weight coefficent obtained for each attribute is:

$$w = \{w_1 = 0.123658,\ w_2 = 0.151386,\ w_3 = 0.258957,\ w_4 = 0.251833,\ w_5 = 0.118412,\ w_6 = 0.0957547\}.$$

Step 5. Use Equations (10) and (11) to compute the values of b^+ and b^- from the neutrosophic numbers given in Table 2. The values are as given below:

$$b^+ = \{b_1^+ = [0.7,\ 0.2,\ 0.1],\ b_2^+ = [0.9,\ 0,\ 0.1],\ b_3^+ = [0.8,\ 0,\ 0],\ b_4^+ = [0.9,\ 0.3,\ 0],\ b_5^+ = [0.7,\ 0.2,\ 0.2],\ b_6^+ = [0.8,\ 0.2\ 0.1\}$$

and

$$b^- = \{b_1^- = [0.5,\ 0.8,\ 0.5],\ b_2^- = [0.6,\ 0.8,\ 0.5],\ b_3^- = [0.1,\ 0.7,\ 0.5],\ b_4^- = [0.3,\ 0.8,\ 0.7],\ b_5^- = [0.5,\ 0.8,\ 0.7],\ b_6^- = [0.5,\ 0.8,\ 0.9]\}.$$

Step 6. Use Equations (12) and (13) to compute the difference between each alternative and the RNPIS and the RNNIS, respectively. The values of D^+ and D^- are as given below:

$$D^+ = \{D_1^+ = 0.262072,\ D_2^+ = 0.306496,\ D_3^+ = 0.340921,\ D_4^+ = 0.276215,\ D_5^+ = 0.292443,\ D_6^+ = 0.345226,\ D_7^+ = 0.303001,\ D_8^+ = 0.346428,\ D_9^+ = 0.271012,\ D_{10}^+ = 0.339093\}.$$

and

$$D^- = \{D_1^- = 0.374468,\ D_2^- = 0.307641,\ D_3^- = 0.294889,\ D_4^- = 0.355857,\ D_5^- = 0.323740\ D_6^- = 0.348903,\ D_7^- = 0.360103,\ D_8^- = 0.338725,\ D_9^- = 0.379516,\ D_{10}^- = 0.349703\}.$$

Step 7. Using Equation (14), the closeness coefficient C_i for each alternative is:

$C_1 = -0.0133,\ C_2 = -0.3589,\ C_3 = -0.5239,\ C_4 = -0.1163,\ C_5 = -0.2629,$
$C_6 = -0.3980,\ C_7 = -0.2073,\ C_8 = -0.4294,\ C_9 = -0.0341,\ C_{10} = -0.3725.$

Step 8. The ranking of the alternatives obtained from the closeness coefficient is as given below:

$$x_1 > x_9 > x_4 > x_7 > x_5 > x_2 > x_{10} > x_6 > x_8 > x_3.$$

Therefore the optimal decision is to select supplier x_1.

Table 2. Tabular representation of SVNS A.

U	e1	e2	e3
x_1	(0.7, 0.5, 0.1)	(0.7, 0.5, 0.3)	(0.8, 0.6, 0.2)
x_2	(0.6, 0.5, 0.2)	(0.7, 0.5, 0.1)	(0.6, 0.3, 0.5)
x_3	(0.6, 0.2, 0.3)	(0.6, 0.6, 0.4)	(0.7, 0.7, 0.2)
x_4	(0.5, 0.5, 0.4)	(0.6, 0.4, 0.4)	(0.7, 0.7, 0.3)
x_5	(0.7, 0.5, 0.5)	(0.8, 0.3, 0.1)	(0.7, 0.6, 0.2)
U	e1	e2	e3
x_6	(0.5, 0.5, 0.5)	(0.7, 0.8, 0.1)	(0.7, 0.3, 0.5)
x_7	(0.6, 0.8, 0.1)	(0.7, 0.2, 0.1)	(0.6, 0.3, 0.4)
x_8	(0.7, 0.8, 0.3)	(0.6, 0.6, 0.5)	(0.8, 0, 0.5)
x_9	(0.6, 0.7, 0.1)	(0.7, 0, 0.1)	(0.6, 0.7, 0)
x_{10}	(0.5, 0.7, 0.4)	(0.9, 0, 0.3)	(1, 0, 0)

Table 2. *Cont.*

U	e_4	e_5	e_6
x_1	(0.9, 0.4, 0.2)	(0.6, 0.4, 0.7)	(0.6, 0.5, 0.4)
x_2	(0.6, 0.4, 0.3)	(0.7, 0.5, 0.4)	(0.7, 0.8, 0.9)
x_3	(0.5, 0.5, 0.3)	(0.6, 0.8, 0.6)	(0.7, 0.2, 0.5)
x_4	(0.9, 0.4, 0.2)	(0.7, 0.3, 0.5)	(0.6, 0.4, 0.4)
x_5	(0.7, 0.5, 0.2)	(0.7, 0.5, 0.6)	(0.6, 0.7, 0.8)
U	e_4	e_5	e_6
x_6	(0.4, 0.8, 0)	(0.7, 0.4, 0.2)	(0.5, 0.6, 0.3)
x_7	(0.3, 0.5, 0.1)	(0.6, 0.3, 0.6)	(0.5, 0.2, 0.6)
x_8	(0.7, 0.3, 0.6)	(0.6, 0.8, 0.5)	(0.6, 0.2, 0.4)
x_9	(0.7, 0.4, 0.3)	(0.6, 0.6, 0.7)	(0.7, 0.3, 0.2)
x_{10}	(0.5, 0.6, 0.7)	(0.5, 0.2, 0.7)	(0.8, 0.4, 0.1)

4.2. Adaptation of the Algorithm to Non-Integrated Weight Measure

In this section, we present an adaptation of our algorithm introduced in Section 4.1 to cases where only the objective weights or subjective weights of the attributes are taken into consideration. The results obtained via these two new variants are then compared to the results obtained via the original algorithm in Section 4.1. Further, we also compare the results obtained via these two new variants of the algorithm to the results obtained via the other methods in literature that are compared in Section 5.

To adapt our proposed algorithm in Section 3 for these special cases, we hereby represent the objective-only and subjective-only adaptations of the algorithm. This is done by taking only the objective (subjective) weight is to be used, then simply take $w_j = \theta_j$ ($w_j = h_j$). The two adaptations of the algorithm are once again applied to the dataset for SVNS A given in Table 2.

4.2.1. Objective-Only Adaptation of Our Algorithm

All the steps remain the same as the original algorithm; however, only the objective weights of the attributes are used, i.e., we take $w_j = \theta_j$.

The results of applying this variant of the algorithm produces the ranking given below:

$$x_9 > x_1 > x_4 > x_{10} > x_7 > x_6 > x_5 > x_8 > x_3 > x_2.$$

Therefore, if only the objective weight is to be considered, then the optimal decision is to select supplier x_9.

4.2.2. Subjective-Only Adaptation of Our Algorithm

All the steps remain the same as the original algorithm; however, only the subjective weights of the attributes are used, i.e., we take $w_j = h_j$.

The results of applying this variant of the algorithm produces the ranking given below:

$$x_1 > x_9 > x_4 > x_7 > x_5 > x_2 > x_6 > x_{10} > x_8 > x_3$$

Therefore, if only the objective weight is to be considered, then the optimal decision is to select supplier x_1.

From the results obtained above, it can be observed that the ranking of the alternatives are clearly affected by the decision of the decision maker to use only the objective weights, only the subjective weights of the attributes, or an integrated weight measure that takes into consideration both the objective and subjective weights of the attributes.

5. Comparatives Studies

In this section, we present a brief comparative analysis of some of the recent works in this area and our proposed method. These recent approaches are applied to our Example 1, and the limitations that exist in these methods are elaborated, and the advantages of our proposed method are discussed and analyzed. The results obtained are summarized in Table 3.

5.1. Comparison of Results Obtained Through Different Methods

Table 3. The results obtained using different methods for Example 1.

Method	The Final Ranking	The Best Alternative
Ye [39] (i) WAAO * (ii) WGAO **	$x_1 > x_4 > x_9 > x_5 > x_7 > x_2 > x_{10} > x_8 > x_3 > x_6$ $x_{10} > x_9 > x_8 > x_1 > x_5 > x_7 > x_4 > x_2 > x_6 > x_3$	x_1 x_{10}
Ye [10] (i) Weighted correlation coefficient (ii) Weighted cosine similarity measure	$x_1 > x_4 > x_5 > x_9 > x_2 > x_8 > x_7 > x_3 > x_6 > x_{10}$ $x_1 > x_9 > x_4 > x_5 > x_2 > x_{10} > x_8 > x_3 > x_7 > x_6$	x_1 x_1
Ye [11]	$x_1 > x_9 > x_4 > x_7 > x_5 > x_2 > x_8 > x_6 > x_3 > x_{10}$	x_1
Huang [14]	$x_1 > x_9 > x_4 > x_5 > x_2 > x_7 > x_8 > x_6 > x_3 > x_{10}$	x_1
Peng et al. [40] (i) GSNNWA *** (ii) GSNNWG ****	$x_9 > x_{10} > x_8 > x_6 > x_1 > x_7 > x_4 > x_5 > x_2 > x_3$ $x_1 > x_9 > x_4 > x_5 > x_7 > x_2 > x_8 > x_3 > x_6 > x_{10}$	x_9 x_1
Peng & Liu [15] (i) EDAS (ii) Similarity measure	$x_1 > x_4 > x_6 > x_9 > x_{10} > x_3 > x_2 > x_7 > x_5 > x_8$ $x_{10} > x_8 > x_7 > x_4 > x_1 > x_2 > x_5 > x_9 > x_3 > x_6$	x_1 x_{10}
Maji [41]	$x_5 > x_1 > x_9 > x_6 > x_2 > x_4 > x_3 > x_8 > x_7 > x_{10}$	x_5
Karaaslan [42]	$x_1 > x_9 > x_4 > x_5 > x_7 > x_2 > x_8 > x_3 > x_6 > x_{10}$	x_1
Ye [43]	$x_1 > x_9 > x_4 > x_5 > x_7 > x_2 > x_8 > x_3 > x_6 > x_{10}$	x_1
Biswas et al. [44]	$x_{10} > x_9 > x_7 > x_1 > x_4 > x_6 > x_5 > x_8 > x_2 > x_3$	x_{10}
Ye [45]	$x_9 > x_7 > x_1 > x_4 > x_2 > x_{10} > x_5 > x_8 > x_3 > x_6$	x_9
Adaptation of our algorithm (objective weights only)	$x_9 > x_1 > x_4 > x_{10} > x_7 > x_6 > x_5 > x_8 > x_3 > x_2$	x_9
Adaptation of our algorithm (subjective weights only)	$x_1 > x_9 > x_4 > x_7 > x_5 > x_2 > x_6 > x_{10} > x_8 > x_3$	x_1
Our proposed method (using integrated weight measure)	$x_1 > x_9 > x_4 > x_7 > x_5 > x_2 > x_{10} > x_6 > x_8 > x_3$	x_1

* WAAO = weighted arithmetic average operator; ** WGAO = weighted geometric average operator; *** GSNNWA = generalized simplified neutrosophic number weighted averaging operator; **** GSNNWG = generalized simplified neutrosophic number weighted geometric operator.

5.2. Discussion of Results

From the results obtained in Table 3, it can be observed that different rankings and optimal alternatives were obtained from the different methods that were compared. This difference is due to a number of reasons. These are summarized briefly below:

(i) The method proposed in this paper uses an integrated weight measure which considers both the subjective and objective weights of the attributes, as opposed to some of the methods that only consider the subjective weights or objective weights.

(ii) Different operators emphasizes different aspects of the information which ultimately leads to different rankings. For example, in [40], the GSNNWA operator used is based on an arithmetic average which emphasizes the characteristics of the group (i.e., the whole information), whereas the GSNNWG operator is based on a geometric operator which emphasizes the characteristics of each individual alternative and attribute. As our method places more importance on the characteristics of the individual alternatives and attributes, instead of the entire information as a whole, our method produces the same ranking as the GSNNWG operator but different results from the GSNNWA operator.

5.3. Analysis of the Performance and Reliability of Different Methods

The performance of these methods and the reliability of the results obtained via these methods are further investigated in this section.

Analysis

In all of the 11 papers that were compared in this section, the different authors used different types of measurements and parameters to determine the performance of their respective algorithms. However, all of these inputs *always* contain a tensor with at least three degrees. This tensor can refer to different types of neutrosophic sets depending on the context discussed in the respective papers, e.g., simplified neutrosophic sets, single-valued neutrosophic sets, neutrosophic sets, or INSs. For the sake of simplicity, we shall denote them simply as S.

Furthermore, all of these methods consider a weighted approach i.e., the weight of each attribute is taken into account in the decision-making process. The decision-making algorithms proposed in [10,11,14,39,40,43,45] use the subjective weighting method, the algorithms proposed in [42,44] use the objective weighting method, whereas only the decision-making methods proposed in [15] use an integrated weighting method which considers both the subjective and objective weights of the attributes. The method proposed by Maji [41] did not take the attribute weights into consideration in the decision-making process.

In this section, we first apply the inputs of those papers into our own algorithm. We then compare the results obtained via our proposed algorithm with their results, with the aim of justifying the effectiveness of our algorithm. The different methods and their algorithms are analyzed below:

(i) The algorithms in [10,11,39] all use the data given below as inputs

$$S = \left\{ \begin{array}{l} [0.4,\ 0.2,\ 0.3],\ [0.4,\ 0.2,\ 0.3],\ [0.2,\ 0.2,\ 05] \\ [0.6,\ 0.1,\ 0.2],\ [0.6,\ 0.1,\ 0.2],\ [0.5,\ 0.2,\ 0.2] \\ [0.3,\ 0.2,\ 0.3],\ [0.5,\ 0.2,\ 0.3],\ [0.5,\ 0.3,\ 0.2] \\ [0.7,\ 0.0,\ 0.1],\ [0.6,\ 0.1,\ 0.2],\ [0.4,\ 0.3,\ 0.2] \end{array} \right\}$$

The subjective weights w_j of the attributes are given by $w_1 = 0.35,\ w_2 = 0.25,\ w_3 = 0.40$. All the five algorithms from papers [10,11,39] yields either one of the following rankings:

$$A_4 > A_2 > A_3 > A_1 \quad \text{or} \quad A_2 > A_4 > A_3 > A_1$$

Our algorithm yields the ranking $A_4 > A_2 > A_3 > A_1$ which is consistent with the results obtained through the methods given above.

(ii) The method proposed in [44] also uses the data given in S above as inputs but ignores the opinions of the decision makers as it does not take into account the subjective weights of the attributes. The algorithm from this paper yields the ranking of $A_4 > A_2 > A_3 > A_1$. To fit this data into our algorithm, we randomly assigned the subjective weights of the attributes as $w_j = \frac{1}{3}$ for $j = 1, 2, 3$. A ranking of $A_4 > A_2 > A_3 > A_1$ was nonetheless obtained from our algorithm.

(iii) The methods introduced in [14,43,45] all use the data given below as input values:

$$S = \left\{ \begin{array}{l} [0.5,\ 0.1,\ 0.3],\ [0.5,\ 0.1,\ 0.4],\ [0.7,\ 0.1,\ 02],\ [0.3,\ 0.2,\ 0.1] \\ [0.4,\ 0.2,\ 0.3],\ [0.3,\ 0.2,\ 0.4],\ [0.9,\ 0.0,\ 0.1],\ [0.5,\ 0.3,\ 0.2] \\ [0.4,\ 0.3,\ 0.1],\ [0.5,\ 0.1,\ 0.3],\ [0.5,\ 0.0,\ 0.4],\ [0.6,\ 0.2,\ 0.2] \\ [0.6,\ 0.1,\ 0.2],\ [0.2,\ 0.2,\ 0.5],\ [0.4,\ 0.3,\ 0.2],\ [0.7,\ 0.2,\ 0.1] \end{array} \right\}$$

The subjective weights w_j of the attributes are given by $w_1 = 0.30$, $w_2 = 0.25$, $w_3 = 0.25$ and $w_4 = 0.20$.

In this case, all of the three algorithms produces a ranking of $A_1 > A_3 > A_2 > A_4$.

This result is however not very reliable as all of these methods only considered the subjective weights of the attributes and ignored the objective weight which is a vital measurement of the relative importance of an attribute e_j relative to the other attributes in an objective manner i.e., without "prejudice".

When we calculated the objective weights using our own algorithm we have the following objective weights:

$$a_j = [0.203909,\ 0.213627,\ 0.357796,\ 0.224667]$$

In fact, it is indeed <0.9, 0.0, 0.1> that mainly contributes to the largeness of the objective weight of attribute e_3 compared to the other values of e_j. Hence, when we calculate the integrated weight, the weight of attribute e_3 is still the largest.

Since $[0.9,\ 0.0,\ 0.1]$ is in the second row, our algorithm yields a ranking of $A_2 > A_1 > A_3 > A_4$ as a result.

We therefore conclude that our algorithm is more effective and the results obtained via our algorithm is more reliable than the ones obtained in [14,43,45], as we consider both the objective and subjective weights.

(iv) It can be observed that for the methods introduced in [10,11,39,44], we have $0.8 \leq T_{ij} + I_{ij} + F_{ij} \leq 1$ for all the entries. A similar trend can be observed in [14,43,45], where $0.6 \leq T_{ij} + I_{ij} + F_{ij} \leq 1$ for all the entries. Therefore, we are not certain about the results obtained through the decision making algorithms in these papers when the value of $T_{ij} + I_{ij} + F_{ij}$ deviates very far from 1.

Another aspect to be considered is the weighting method that is used in the decision making process. As mentioned above, most of the current decision making methods involving SVNSs use subjective weighting, a few use objective weighting and only two methods introduced in [15] uses an integrated weighting method to arrive at the final decision. In view of this, we proceeded to investigate if all of the algorithms that were compared in this section are able to produce reliable results when both the subjective and objective weights are taken into consideration. Specifically, we investigate if these algorithms are able to perform effectively in situations where the subjective weights clearly prioritize over the objective weights, and vice-versa. To achieve this, we tested all of the algorithms with three sets of inputs as given below:

Test 1: A scenario containing a very small value of $T_{ij} + I_{ij} + F_{ij}$.

$$S_1 = \left\{ \begin{array}{c} A_1 = ([0.5,\ 0.5,\ 0.5],\ [\mathbf{0.9999},\ 0.0001,\ 0.000]) \\ A_2 = ([0.5,\ 0.5,\ 0.5],\ [\mathbf{0.9999},\ 0.0001,\ \mathbf{0.0001}]) \\ A_3 = ([0.5,\ 0.5,\ 0.5],\ [\mathbf{0.9999},\ 0.0000,\ \mathbf{0.0001}]) \\ A_4 = ([0.5,\ 0.5,\ 0.5],\ [0.0001,\ 0.0000,\ 0.000]) \end{array} \right\}$$

The subjective weight in this case is assigned as: $a_j = [0.5,\ 0.5]$.

By observation alone, it is possible to tell that an effective algorithm should produce A_4 as the least favoured alternative, and A_2 should be second least-favoured alternative.

Test 2: A scenario where subjective weights *prioritize* over objective weight.

$$S_2 = \left\{ \begin{array}{c} A_1 = ([0.80,\ 0.10,\ 0.10],\ [0.19,\ 0.50,\ 0.50]) \\ A_2 = ([0.20,\ 0.50,\ 0.50],\ [0.81,\ 0.10,\ 0.10]) \end{array} \right\}$$

The subjective weight in this case is assigned as: $a_j = [0.99,\ 0.01]$.

By observation alone, we can tell that an effective algorithm should produce a ranking of $A_1 > A_2$.

Test 3: This test is based on a real-life situation.

Suppose a procurement committee is looking to select the best supplier to supply two raw materials e_1 and e_2. In this context, the triplet $[T, I, F]$ represents the following:

T : the track record of the suppliers that is approved by the committee
I : the track record of the suppliers that the committee feels is questionable
F : the track record of the suppliers that is rejected by the committee

Based on their experience, the committee is of the opinion that raw material e_1 is slightly more important than raw material e_2, and assigned subjective weights of $w_1^{sub} = 0.5001$ and $w_2^{sub} = 0.4999$.

After an intensive search around the country, the committee shortlisted 20 candidates (A_1 to A_{20}). After checking all of the candidates' track records and analyzing their past performances, the committee assigned the following values for each of the suppliers.

$$S_3 = \left\{ \begin{array}{c} A_1 = ([0.90,\ 0.00,\ 0.10],\ [0.80,\ 0.00,\ 0.10]),\ A_2 = ([0.80,\ 0.00,\ 0.10],\ [0.90,\ 0.00,\ 0.10]) \\ A_3 = ([0.50,\ 0.50,\ 0.50],\ [0.00,\ 0.90,\ 0.90]), A_4 = ([0.50,\ 0.50,\ 0.50],\ [0.10,\ 0.90,\ 0.80]) \\ A_5 = ([0.50,\ 0.50,\ 0.50],\ [0.20,\ 0.90,\ 0.70]), A_6 = ([0.50,\ 0.50,\ 0.50],\ [0.30,\ 0.90,\ 0.60]) \\ A_7 = ([0.50,\ 0.50,\ 0.50],\ [0.40,\ 0.90,\ 0.50]), A_8 = ([0.50,\ 0.50,\ 0.50],\ [0.50,\ 0.90,\ 0.40]) \\ A_9 = ([0.50,\ 0.50,\ 0.50],\ [0.60,\ 0.90,\ 0.30]), A_{10} = ([0.50,\ 0.50,\ 0.50],\ [0.70,\ 0.30,\ 0.90]) \\ A_{11} = ([0.50,\ 0.50,\ 0.50],\ [0.70,\ 0.90,\ 0.30]),\ A_{12} = ([0.50,\ 0.50,\ 0.50],\ [0.00,\ 0.30,\ 0.30]) \\ A_{13} = ([0.50,\ 0.50,\ 0.50],\ [0.70,\ 0.90,\ 0.90]),\ A_{14} = ([0.50,\ 0.50,\ 0.50],\ [0.70,\ 0.30,\ 0.30]) \\ A_{15} = ([0.50,\ 0.50,\ 0.50],\ [0.60,\ 0.40,\ 0.30]),\ A_{16} = ([0.50,\ 0.50,\ 0.50],\ [0.50,\ 0.50,\ 0.30]) \\ A_{17} = ([0.50,\ 0.50,\ 0.50],\ [0.40,\ 0.60,\ 0.30]),\ A_{18} = ([0.50,\ 0.50,\ 0.50],\ [0.30,\ 0.70,\ 0.30]) \\ A_{19} = ([0.50,\ 0.50,\ 0.50],\ [0.20,\ 0.80,\ 0.30]),\ A_{20} = ([0.50,\ 0.50,\ 0.50],\ [0.10,\ 0.90,\ 0.30]) \end{array} \right\}$$

The objective weights for this scenario was calculated based on our algorithm and the values are $w_1^{obj} = 0.1793$ and $w_2^{obj} = 0.8207$.

Now it can be observed that suppliers A_1 and A_2 are the ones that received the best evaluation scores from the committee. Supplier A_1 received better evaluation scores from the committee compared to supplier A_2 for attribute e_1. Attribute e_1 was deemed to be more important than attribute e_2 by the committee, and hence had a higher subjective weight. However, the objective weight of attribute e_2 is much higher than e_1. This resulted in supplier A_2 ultimately being chosen as the best supplier. This is an example of a scenario where the objective weights are prioritized over the subjective weights, and has a greater influence on the decision-making process.

Therefore, in the scenario described above, an effective algorithm should select A_2 as the optimal supplier, followed by A_1. All of the remaining choices have values of $T < 0.8$, $I > 0.0$ and $F > 0.1$. As such, an effective algorithm should rank all of these remaining 18 choices behind A_1.

We applied the three tests mentioned above and the data set for S_3 given above to the decision-making methods introduced in the 11 papers that were compared in the previous section. The results obtained are given in Table 4.

Thus it can be concluded that our proposed algorithm is the most effective algorithm and the one that yields the most reliable results in all the different types of scenario. Hence, our proposed algorithm provides a robust framework that can be used to handle any type of situation and data, and produce accurate and reliable results for any type of situation and data.

Finally, we look at the context of the scenario described in Example 1. The structure of our data (given in Table 2) is more generalized, by theory, having $0 \leq T_{ij} + I_{ij} + F_{ij} \leq 1$ and $0 \leq T_{ij} + I_{ij} + F_{ij} \leq 3$, and is similar to the structure of the data used in [15,40–42]. Hence, our choice of input data serves as a more faithful indicator of how each algorithm works under all sorts of possible conditions.

Table 4. Compliance to Tests 1, 2, and 3.

Paper		Test 1 Compliance	Test 2 Compliance	Test 3 Compliance
Ye [39]	WAAO *	Y	Y	N
	WGAO *	N	Y	N
Ye [10]	Weighted correlation coefficient	Y	Y	N
	Weighted cosine similarity measure	N	Y	N
Ye [11]		Y	Y	N
Huang [14]		Y	Y	N
Peng et al. [40]	GSNNWA **	Y	Y	N
	GSNNWG **	Y	Y	N
Peng & Liu [15]	EDAS	Y	Y	N
	Similarity measure	N	Y	Y
Maji [41]		N	N	N
Karaaslan [42]		Y	Y	N
Ye [43]		Y	Y	N
Biswas et al. [44]		Y	N	Y
Ye [45]		Y	Y	N
Adaptation of our proposed algorithm (objective weights only)		Y	N	Y
Adaptation of our proposed algorithm (subjective weights only)		Y	Y	N
Our proposed algorithm		Y	Y	Y

Remarks: Y = Yes (which indicates compliance to Test); N = No (which indicates non-compliance to Test); * WAAO = weighted arithmetic average operator; * WGAO = weighted geometric average operator; ** GSNNWA = generalized simplified neutrosophic number weighted averaging operator; ** GSNNWG = generalized simplified neutrosophic number weighted geometric operator.

6. Conclusions

The concluding remarks and the significant contributions that were made in this paper are expounded below.

(i) A novel TOPSIS method for the SVNS model is introduced, with the maximizing deviation method used to determine the objective weight of the attributes. Through thorough analysis, we have proven that our algorithm is compliant with all of the three tests that were discussed in Section 5.3. This clearly indicates that our proposed decision-making algorithm is not only an effective algorithm but one that produces the most reliable and accurate results in all the different types of situation and data inputs.

(ii) Unlike other methods in the existing literature which reduces the elements from single-valued neutrosophic numbers (SVNNs) to fuzzy numbers, or interval neutrosophic numbers (INNs) to neutrosophic numbers or fuzzy numbers, in our version of the TOPSIS method the input data is in the form of SVNNs and this form is maintained throughout the decision-making process. This prevents information loss and enables the original information to be retained, thereby ensuring a higher level of accuracy for the results that are obtained.

(iii) The objective weighting method (e.g., the ones used in [10,11,14,39,40,43,45]) only takes into consideration the values of the membership functions while ignoring the preferences of the decision makers. Through the subjective weighting method (e.g., the ones used in [42,44]), the attribute weights are given by the decision makers based on their individual preferences and experiences. Very few approaches in the existing literature (e.g., [15]) consider both the objective and subjective weighting methods. Our proposed method uses an integrated weighting model that considers both the objective and subjective weights of the attributes, and this accurately reflects the input values of the alternatives as well as the preferences and risk attitude of the decision makers.

Author Contributions: Conceptualization, Methodology, Writing-Original Draft Preparation: G.S.; Investigation, Validation, and Visualization: S.G.Q.; Writing-Review: S.B.; Editing: F.S.; Funding Acquisition: G.S. and S.G.Q.

Funding: This research was funded by the Ministry of Education, Malaysia under grant no. FRGS/1/2017/STG06/UCSI/03/1.

Acknowledgments: The authors would like to thank the Editor-in-Chief and the anonymous reviewers for their valuable comments and suggestions.

Conflicts of Interest: The authors declare that there is no conflict of interest.

References

1. Zadeh, L.A. Fuzzy sets. *Inf. Control* **1965**, *8*, 338–353. [CrossRef]
2. Atanassov, K.T. Intuitionistic fuzzy sets. *Fuzzy Sets Syst.* **1986**, *20*, 87–96. [CrossRef]
3. Gorzalczany, M.B. A method of inference in approximate reasoning based on interval-valued fuzzy sets. *Fuzzy Sets Syst.* **1987**, *21*, 1–17. [CrossRef]
4. Gau, W.L.; Buehrer, D.J. Vague sets. *IEEE Trans. Syst. Man Cybern.* **1993**, *23*, 610–614. [CrossRef]
5. Torra, V. Hesitant fuzzy sets. *Int. J. Intell. Syst.* **2010**, *25*, 529–539. [CrossRef]
6. Atanassov, K.; Gargov, G. Interval valued intuitionistic fuzzy sets. *Fuzzy Sets Syst.* **1989**, *31*, 343–349. [CrossRef]
7. Ezhilmaran, D.; Sankar, K. Morphism of bipolar intuitionistic fuzzy graphs. *J. Discret. Math. Sci. Cryptogr.* **2015**, *18*, 605–621. [CrossRef]
8. Smarandache, F. *Neutrosophy. Neutrosophic Probability, Set, and Logic*; ProQuest Information & Learning: Ann Arbor, MI, USA, 1998; 105p. Available online: http://fs.gallup.unm.edu/eBook-neutrosophics6.pdf (accessed on 7 June 2018).
9. Wang, H.; Smarandache, F.; Zhang, Y.Q.; Sunderraman, R. Single valued neutrosophic sets. *Multisp. Multistruct.* **2010**, *4*, 410–413.
10. Ye, J. Multicriteria decision-making method using the correlation coefficient under single-valued neutrosophic environment. *Int. J. Gen. Syst.* **2013**, *42*, 386–394. [CrossRef]
11. Ye, J. Improved correlation coefficients of single valued neutrosophic sets and interval neutrosophic sets for multiple attribute decision making. *J. Intell. Fuzzy Syst.* **2014**, *27*, 2453–2462.
12. Ye, J. Clustering methods using distance-based similarity measures of single-valued neutrosophic sets. *J. Intell. Fuzzy Syst.* **2014**, *23*, 379–389. [CrossRef]
13. Ye, J. Multiple attribute group decision-making method with completely unknown weights based on similarity measures under single valued neutrosophic environment. *J. Intell. Fuzzy Syst.* **2014**, *27*, 2927–2935.
14. Huang, H.L. New distance measure of single-valued neutrosophic sets and its application. *Int. J. Gen. Syst.* **2016**, *31*, 1021–1032. [CrossRef]
15. Peng, X.; Liu, C. Algorithms for neutrosophic soft decision making based on EDAS, new similarity measure and level soft set. *J. Intell. Fuzzy Syst.* **2017**, *32*, 955–968. [CrossRef]
16. Yang, H.L.; Guo, Z.L.; She, Y.H.; Liao, X.W. On single valued neutrosophic relations. *J. Intell. Fuzzy Syst.* **2016**, *30*, 1045–1056. [CrossRef]
17. Broumi, S.; Smarandache, F.; Talea, M.; Bakali, A. Single valued neutrosophic graph: Degree, order and size. In Proceedings of the IEEE International Conference on Fuzzy Systems, Vancouver, BC, Canada, 24–29 July 2016; pp. 2444–2451.
18. Broumi, S.; Bakali, A.; Talea, M.; Smarandache, F. Isolated single valued neutrosophic graphs. *Neutrosophic Sets Syst.* **2016**, *11*, 74–78.
19. Broumi, S.; Talea, M.; Bakali, A.; Smarandache, F. Single valued neutrosophic graphs. *J. New Theory* **2016**, *10*, 86–101.
20. Broumi, S.; Smarandache, F.; Talea, M.; Bakali, A. An introduction to bipolar single valued neutrosophic graph theory. *Appl. Mech. Mater.* **2016**, *841*, 184–191. [CrossRef]
21. Broumi, S.; Talea, M.; Bakali, A.; Smarandache, F. On bipolar single valued neutrosophic graphs. *J. New Theory* **2016**, *11*, 84–102.
22. Hassan, A.; Malik, M.A.; Broumi, S.; Bakali, A.; Talea, M.; Smarandache, F. Special types of bipolar single valued neutrosophic graphs. *Ann. Fuzzy Math. Inform.* **2017**, *14*, 55–73.
23. Tian, Z.P.; Wang, J.; Zhang, H.Y.; Wang, J.Q. Multi-criteria decision-making based on generalized prioritized aggregation operators under simplified neutrosophic uncertain linguistic environment. *Int. J. Mach. Learn. Cybern.* **2016**. [CrossRef]

24. Wu, X.H.; Wang, J.; Peng, J.J.; Chen, X.H. Cross-entropy and prioritized aggregation operator with simplified neutrosophic sets and their application in multi-criteria decision-making problems. *Int. J. Fuzzy Syst.* **2016**, *18*, 1104–1116. [CrossRef]
25. Sahin, R.; Kucuk, A. Subsethood measure for single valued neutrosophic sets. *J. Intell. Fuzzy Syst.* **2015**, *29*, 525–530. [CrossRef]
26. Ye, J. An extended TOPSIS method for multiple attribute group decision making based on single valued neutrosophic linguistic numbers. *J. Intell. Fuzzy Syst.* **2015**, *28*, 247–255.
27. Biswas, P.; Pramanik, S.; Giri, B.C. TOPSIS method for multi-attribute group decision-making under single-valued neutrosophic environment. *Neural Comput. Appl.* **2016**, *27*, 727–737. [CrossRef]
28. Majumdar, P.; Samanta, S.K. On similarity and entropy of neutrosophic sets. *J. Intell. Fuzzy Syst.* **2014**, *26*, 1245–1252.
29. Wang, Y.M. Using the method of maximizing deviations to make decision for multiindices. *Syst. Eng. Electron.* **1997**, *8*, 21–26.
30. Hwang, C.L.; Yoon, K. *Multiple Attribute Decision Making: Methods and Applications*; Springer-Verlag: New York, NY, USA, 1981.
31. Chen, M.F.; Tzeng, G.H. Combining grey relation and TOPSIS concepts for selecting an expatriate host country. *Math. Comput. Model.* **2004**, *40*, 1473–1490. [CrossRef]
32. Shaw, K.; Shankar, R.; Yadav, S.S.; Thakur, L.S. Supplier selection using fuzzy AHP and fuzzy multi-objective linear programming for developing low carbon supply chain. *Expert Syst. Appl.* **2012**, *39*, 8182–8192. [CrossRef]
33. Rouyendegh, B.D.; Saputro, T.E. Supplier selection using fuzzy TOPSIS and MCGP: A case study. *Procedia Soc. Behav. Sci.* **2014**, *116*, 3957–3970. [CrossRef]
34. Dargi, A.; Anjomshoae, A.; Galankashi, M.R.; Memari, A.; Tap, M.B.M. Supplier selection: A fuzzy-ANP approach. *Procedia Comput. Sci.* **2014**, *31*, 691–700. [CrossRef]
35. Kaur, P. Selection of vendor based on intuitionistic fuzzy analytical hierarchy process. *Adv. Oper. Res.* **2014**, *2014*. [CrossRef]
36. Kaur, P.; Rachana, K.N.L. An intuitionistic fuzzy optimization approach to vendor selection problem. *Perspect. Sci.* **2016**, *8*, 348–350. [CrossRef]
37. Dweiri, F.; Kumar, S.; Khan, S.A.; Jain, V. Designing an integrated AHP based decision support system for supplier selection in automotive industry. *Expert Syst. Appl.* **2016**, *62*, 273–283. [CrossRef]
38. Junior, F.R.L.; Osiro, L.; Carpinetti, L.C.R. A comparison between fuzzy AHP and fuzzy TOPSIS methods to supplier selection. *Appl. Soft Comput.* **2014**, *21*, 194–209. [CrossRef]
39. Ye, J. A multicriteria decision-making method using aggregation operators for simplified neutrosophic sets. *J. Intell. Fuzzy Syst.* **2014**, *26*, 2459–2466.
40. Peng, J.J.; Wang, J.; Wang, J.; Zhang, H.; Chen, X.H. Simplified neutrosophic sets and their applications in multi-criteria group decision-making problems. *Int. J. Syst. Sci.* **2016**, *47*, 2342–2358. [CrossRef]
41. Maji, P.K. A neutrosophic soft set approach to a decision making problem. *Ann. Fuzzy Math. Inform.* **2012**, *3*, 313–319.
42. Karaaslan, F. Neutrosophic soft sets with applications in decision making. *Int. J. Inf. Sci. Intell. Syst.* **2015**, *4*, 1–20.
43. Ye, J. Single valued neutrosophic cross-entropy for multicriteria decision making problems. *Appl. Math. Model.* **2014**, *38*, 1170–1175. [CrossRef]
44. Biswas, P.; Pramanik, S.; Giri, B.C. Entropy based grey relational analysis method for multi-attribute decision making under single valued neutrosophic assessments. *Neutrosophic Sets Syst.* **2014**, *2*, 102–110.
45. Ye, J. Improved cross entropy measures of single valued neutrosophic sets and interval neutrosophic sets and their multicriteria decision making methods. *Cybern. Inf. Technol.* **2015**, *15*, 13–26. [CrossRef]

symmetry

MDPI

Article

Fixed Point Theorem for Neutrosophic Triplet Partial Metric Space

Memet Şahin [1,*], Abdullah Kargın [1] and Mehmet Ali Çoban [2]

[1] Department of Mathematics, Gaziantep University, Gaziantep 27310, Turkey; abdullahkargin27@gmail.com
[2] Department of Computer Programming, Gaziantep University, Gaziantep 27310, Turkey; coban@gantep.edu.tr
* Correspondence: mesahin@gantep.edu.tr

Received: 4 June 2018; Accepted: 19 June 2018; Published: 25 June 2018

Abstract: Neutrosphic triplet is a new theory in neutrosophy. In a neutrosophic triplet set, there is a neutral element and antielement for each element. In this study, the concept of neutrosophic triplet partial metric space (NTPMS) is given and the properties of NTPMS are studied. We show that both classical metric and neutrosophic triplet metric (NTM) are different from NTPM. Also, we show that NTPMS can be defined with each NTMS. Furthermore, we define a contraction for NTPMS and we give a fixed point theory (FPT) for NTPMS. The FPT has been revealed as a very powerful tool in the study of nonlinear phenomena. This study is also part of the "Algebraic Structures of Neutrosophic Triplets, Neutrosophic Duplets, or Neutrosophic Multisets" which is a special issue.

Keywords: neutrosophic triplet set (NTS); partial metric spaces (PMS); fixed point theory (FPT)

1. Introduction

Neutrosophy was first studied by Smarandache in [1]. Neutrosophy consists of neutrosophic logic, probability, and sets. Actually, neutrosophy is generalization of fuzzy set in [2] and intuitionistic fuzzy set in [3]. Also, researchers have introduced neutrosophic theory in [4–6]. Recently, Olgun and Bal introduced the neutrosophic module in [7], Şahin, Uluçay, Olgun, and Kılıçman introduced neutrosophic soft lattices in [8], and Uluçay, Şahin, and Olgun studied soft normed rings in [9]. Furthermore, Smarandache and Ali studied NT theory in [10] and NT groups (NTG) in [11,12]. The greatest difference between NTG and classical groups is that there can be more than one unit element. That is, each element in a neutrosophic triplet group can be a separate unit element. In addition, the unit elements in the NTG must be different from the unit elements in the classical group. Also, a lot of researchers have introduced NT theory in [13–16]. Recently, Smarandache, Şahin, and Kargın studied neutrosophic triplet G-module in [17], and Bal, Shalla, and Olgun introduced neutrosophic triplet cosets and quotient groups in [18].

Matthew introduced the concept of partial metric spaces (PMS) in [19]. It is a generalization of usual metric space since self-distance cannot be zero in PMS. The most important use of PMS is to transfer mathematical techniques to computer science. Also, Matthew introduced Banach contraction theorem for PMS and a lot of researchers introduced PMS and its topological properties and FPT for PMS in [20–23]. If f is a mapping from a set E into itself, any element x of E, such that $f(x) = x$, is called a fixed point of f. Many problems, including nonlinear partial differential equations problems, may be recast as problems of finding a fixed point of a mapping in a space. Recently, Shukla introduced FPT for ordered contractions in partial b-metric space in [24]. Kim, Okeke, and Lim introduced common coupled FPT for w-compatible mappings in PMS in [25]. Pant, Shukla, and Panicker introduced new FPT in PMS in [26].

In this paper, we first introduced PMS and contraction in NT theory. So, we obtained a new structure for developing NT theory. Thus, researchers can arrive at nonlinear partial differential equations problem solutions in NT theory. In Section 2, we give some basic results and definitions for NTPM and NTM. In Section 3, NTPMS is defined and some properties of a NTPMS are given. It was shown that both the classical metric and NTM are different from the NTPM, and NTPMS can be defined with each NTMS. Furthermore, the convergent sequence and Cauchy sequence in NTPMS are defined. Also, complete NTPMS are defined. Later, we define contractions for NTPM and we give some properties of these contractions. Furthermore, we give a FPT for NTPMS. In Section 4, we give conclusions.

2. Preliminaries

We give some basic results and definitions for NTPM and NTM in this section.

Definition 1 ([19]). *Let A be nonempty set. If the function p_m:AxA $\rightarrow \mathbb{R}^+$ satisfies the conditions given below; p is called a PM.* $\forall a, b, c \in A$;

(i) $p_m(a, a) = p_m(b, b) = p_m(a, b) = p_m(b, a) \Longleftrightarrow a = b$;
(ii) $p_m(a, a) \leq p_m(a, b)$;
(iii) $p_m(a, b) = p_m(b, a)$;
(iv) $p_m(a, c) \leq p_m(a, b) + p_m(b, c) - p_m(b, b)$;

Also, (A, p_m) is called a PMS.

Definition 2 ([12]). *Let N be a nonempty and # be a binary operation. Then, N is called a NT if the given below conditions are satisfied.*

(i) *There is neutral element (neut(x)) for $x \in N$ such that x*neut(x) = neut(x)* x = x.*
(ii) *There is anti element (anti(x)) for $x \in N$ such that x*anti(x) = anti(x)* x = neut(x).*

NT is shown by (x, neut(x), anti(x)).

Definition 3 ([15]). *Let (M, #) be a NTS and a#b $\in$ N, $\forall a, b \in M$. NTM is a map d_T:MxM $\rightarrow \mathbb{R}^+ \cup \{0\}$ such that $\forall a, b, c \in M$,*

(a) $d_T(a, b) \geq 0$
(b) *If a = b, then* $d_T(a, b) = 0$
(c) $d_T(a, b) = d_T(a, b)$
(d) *If there exists any element c in M such that* $d_T(a, c) \leq d_T(a, c*neut(b))$, *then* $d_T(a, c*neut(b)) \leq d_T(a, b) + d_T(b, c)$.

Also, ((M,), d_T) space is called NTMS.*

3. Neutrosophic Triplet Partial Metric Space

Partial metric is the generalization of usual metric space, since self-distance cannot be zero in partial metric space. The most important use of PMS is to transfer mathematical techniques to computer science. Also, If *f* is a mapping from a set *E* into itself, any element *x* of *E* such that *f*(*x*) = *x* is called a fixed point of *f*. Many problems, including nonlinear partial differential equations problems, may be recast as problems of finding a fixed point of a mapping in a space. In this section, we introduced firstly PMS and FPT in NT theory. So, we obtained a new structure for developing NT theory. Thus, researchers can arrive at nonlinear partial differential equations problem solutions in NT theory.

Definition 4. *Let (A, #) be a NTS and a#b in A, ∀a, b in A. NTPM is a map* p_N*: AxA* $\rightarrow \mathbb{R}^+ \cup \{0\}$ *such that ∀a, b, c ∈ A*

(i) $0 \leq p_N(a, a) \leq p_N(a, b)$
(ii) *If* $p_N(a, a) = p_N(a, b) = p_N(b, b) = 0$*, then there exits any a, b such that a = b.*
(iii) $p_N(a, b) = p_N(a, b)$
(iv) *If there exists any element b in A such that* $p_N(a, c) \leq p_N(a, c\#neut(b))$*, then* $p_N(a, c\#neut(b)) \leq p_N(a, b) + p_N(b, c) - p_N(b, b)$

Additionally, ((A, #), p_N*) is called NTPMS.*

Example 1. *Let A be a nonempty set and P(A) be power set of A and m(X) be cardinal of X ∈ P(A). Where, it is clear that X ∪ X = X. Thus; we give that neut(X) = X and anti(X) = X for X ∈ P(A). So, (P(A), ∪) is a NTS. We give the function* p_N*: P(A)x P(A)* $\rightarrow \mathbb{R}^+ \cup \{0\}$ *such that* $p_N(X,Y) = max\{m(X), m(Y)\}$*. From Definition 4,*

(i), (ii) and (iii) are apparent.
(iv) Let ∅ be empty element of P(X). Then, $p_N(X, Y) = p_N(X, Y \cup \emptyset)$ *since for* $p_N(X, Y \cup \emptyset) = p_N(X, Y) = max\{m(X), m(Y)\}$*. Also, it is clear that*
$max\{m(X), m(Y)\} \leq max\{m(X), m(Z)\} + max\{m(Z), m(Y)\} - max\{m(\emptyset), m(\emptyset)\}$.

Therefore, $p_N(X, Y \cup \emptyset) \leq p_N(X, \emptyset) + p_N(\emptyset, Y) - p_N(\emptyset, \emptyset)$*. Thus, ((P(A), ∪),* p_N*) is a NTPMS.*

Corollary 1. *NTPM is different from the partial metric. Because there isn't a "#"binary operation and neutral of x in PMS.*

Corollary 2. *Generally the NTPM is different from NT metric, since for* $p_N(x, x) \geq 0$*.*

Theorem 1. *Let A be a nonempty set and P(A) be power set of A and m(X) be cardinal of X ∈ P (A) and (P(A), #), d) be a NT metric space (NTMS). If there exists any Z ∈ P(A) such that* $m(Y\#neut(Z) = m(Y)$*; then ((P(A), #),* p_N*) is a NTPMS such that*

$$p_N(X, Y) = \frac{d(X, Y) + m(X) + m(Y)}{2}$$

Proof.

(i) $p_N(X, X) = \frac{d(X, X)+m(X)+m(X)}{2} = m(X) \leq \frac{d(X, Y)+m(X)+m(Y)}{2} = p_N(X, Y)$, since for $d(X,X) = 0$. Thus; $0 \leq p_N(X, X) \leq p_N(X, Y)$ for X, Y ∈ P(A).
(ii) If $p_N(X, X) = p_N(X, Y) = p_N(Y, Y) = 0$, then
(iii) $\frac{d(X, X)+m(X)+m(X)}{2} = \frac{d(X, Y)+m(X)+m(Y)}{2} = \frac{d(Y, Y)+m(Y)+m(Y)}{2} = 0$ and $d(X, Y) + m(X) + m(Y) = 0$. Where, $m(X) = 0$, $m(Y) = 0$ and $d(X, Y) = 0$. Thus, X = Y = ∅ (empty set).
(iv) $p_N(X, Y) = \frac{d(X, Y)+m(X)+m(Y)}{2} = \frac{d(Y, X)+m(Y)+m(X)}{2} = p_N(Y, X)$, since for $d(X, Y) = d(Y, X)$.
(v) We suppose that there exists any Z ∈ P(A) such that $m(Y\#neut(Z)) = m(Y)$ and $p_N(X, Y) \leq p_N(X, Y\#neut(Z))$. Thus,

$$\frac{d(X, Y) + m(X) + m(Y)}{2} \leq \frac{d(X, Y\#neut(Z)) + m(X) + m(Y\#neut(Z))}{2} \tag{1}$$

From (1), $d(X, Y) \leq d(X, Y\#neut(Z))$. Since (P(A), #), d) is a NTMS,

$$d(X, Y\#neut(Z)) \leq d(X, Z) + d(X, Z) \tag{2}$$

From (1), (2)

$$\frac{d(X,\,Y)+m(X)+m(Y)}{2}\le\frac{d(X,\,Y\#\text{neut}(Z))+m(X)+m(Y\#\text{neut}(Z))}{2}\le\frac{d(X,\,Z)+d(Z,\,Y)+m(X)+m(Y)+m(Z)}{2}=\frac{d(X,\,Z)+m(X)+m(Z)}{2}+\frac{d(Z,\,Y)+m(Z)+m(Y)}{2}-m(Z).\text{ Where, } p_N(Z,\,Z)=m(Z).$$

Thus, $p_N(X, Y^*\text{neut}(Z)) \le p_N(X, Z) + p_N(Z, Y) - p_N(Z, Z)$. Hence, $((P(A), \#), p_N)$ is a NTPMS. □

Theorem 2. *Let $(A, \#)$ be a NT set, $k \in \mathbb{R}^+$ and $((A, \#), d_T)$ be a NTMS. Then; $((A, \#), p_N)$ is a NTPMS such that*

$p_N(a, b) = d_T(a, b) + k, \forall\, a, b \in A.$

Proof.

(i) Since for $d_T(a, a) = 0$, $0 \le p_N(a, a) = d_T(a, a) + k = k \le p_N(a, b) = d_T(a, b) + k$. Thus;

(ii) $0 \le p_N(a, a) \le p_N(a, b)$.

(iii) There do not exists $a, b \in A$ such that $p_N(a, a) = p_N(a, b) = p_N(b, b) = 0$ since for $k \in \mathbb{R}^+$ and $d_T(a, a) = 0$.

(iv) $p_N(a, b) = d_T(a, b) + k = d_T(b, a) + k$, since for $d_T(a, b) = d_T(b, a)$.

(v) Suppose that there exists any element c *in* A such that $p_N(a, b) \le p_N(a, b\#\text{neut}(c))$. Then $d_T(a, b) + k \le d_T(a, b\#\text{neut}(c)) + k$. Thus,

$$d_T(a,\,b)\le d_T(a,\,b\#\text{neut}(c)) \tag{3}$$

Also,

$$d_T(a,\,b\#\text{neut}(c))\le d_T(a,\,c)\,+\,d_T(c,\,b) \tag{4}$$

since for $((A, \#), d_T)$ is a NTMS.

From (3) and (4),

$$p_N(a,\,b)\le p_N(a,\,b\#\text{neut}(c)) = d_T(a,\,b\#\text{neut}(c))+k\le d_T(a,\,c)\,+\,d_T(c,\,b)= p_N(a,\,c)\,+\,p_N(c,\,b)-k$$

where, $p_N(c, c) = k$. Thus;

$p_N(a, b\#\text{neut}(c)) \le p_N(a, c) + p_N(c, b) - p_N(c, c)$. Hence, $((A, \#), p_N)$ is a NTPMS. □

Corollary 3. *From Theorem 2, we can define NTPMS with each NTMS.*

Definition 5. *Let $((A, \#), p_N)$ be a NTPMS, $\{x_n\}$ be a sequence in NTPMS and a in A. If for $\forall\varepsilon > 0$ and $\forall n \ge M$, there exist a M in $\mathbb{N}$ such that $p_N(a, \{x_n\}) < \varepsilon + p_N(a, a)$, then $\{x_n\}$* converges to a *in* A. It is shown by

$$\lim_{n\to\infty} x_n = a \text{ or } x_n \to a.$$

Definition 6. *Let $((A, \#), p_N)$ be a NTPMS, $\{x_n\}$ be a sequence in NTPMS and a in A. If for $\forall\varepsilon > 0$ and $\forall n, m \ge M$, there exist a M in $\mathbb{N}$ such that $p_N(\{x_m\}, \{x_n\}) < \varepsilon + p_N(a, a)$; then $\{x_n\}$ is a Cauchy sequence in $((A, \#), p_N)$.*

Theorem 3. *Let $((A, \#), p_N)$ be a NTPMS, $\{x_n\}$ be a convergent sequence in NTPMS and $p_N(\{x_m\}, \{x_n\}) \le p_N(\{x_m\}, \{x_n\})\ {}^*\text{neut}(a))$ for any a in A. Then $\{x_n\}$ is a Cauchy sequence in NTPMS.*

Proof. It is clear that

$$p_N(a,\,\{x_n\})<\varepsilon/2+p_N(a,\,a) \tag{5}$$

for each $n \geq M$ or

$$p_N(a,\ \{x_m\}) < \varepsilon/2 + p_N(a,\ a) \tag{6}$$

for each $m \geq M$

Because $\{x_n\}$ is a convergent. Then, we suppose that $p_N(\{x_m\}, \{x_n\}) \leq p_N(\{x_m\}, \{x_n\})$ *neut(a)) for any a *in* A. It is clear that for $n, m \geq M$;

$$p_N(\{x_m\},\ \{x_n\}) \leq p_N(\{x_m\},\ \{x_n\}) * \text{neut}(a)) \leq p_N(a,\ \{x_n\}) + p_N(a,\ \{x_m\}) - p_N(a,\ a) \tag{7}$$

Because $((A, \#), p_N)$ is a NTPMS. From (5)–(7),

$p_N(\{x_m\}, \{x_n\}) < \varepsilon/2 + p_N(a, a) + \varepsilon/2 + p_N(a, a) - p_N(a, a) = \varepsilon + p_N(a, a)$. Thus; $\{x_n\}$ is a Cauchy sequence in $((A, \#), p_N)$. □

Definition 7. *Let $((A, \#), p_N)$ be a NTPMS and $\{x_n\}$ be a Cauchy sequence in NTPMS. If every $\{x_n\}$ is convergent in $((A, \#), p_N)$, then $((A, \#), p_N)$ is called a complete NTPMS.*

Definition 8. *Let $((A, \#), p_N)$ be a NTPMS and m: $A \to A$ be a map. If the map m and the NTPM p_N satisfy the conditions given below, then m is called a contraction for $((A, \#), p_N)$.*

(i) *There exists any element c in A such that $p_N(a, b) \leq p_N(a, b{*}neut(c))$; $\forall a$, b in A.*
(ii) *There exists k in [0, 1) such that $p_N(m(a), m(b)) \leq k.\ p_N(a, b)$; $\forall a$, b in A.*

Example 2. *Let $A = \{\varnothing, \{x\}, \{x, y\}\}$ be a set and m(X) be cardinal of X in A. Where, it is clear that $X \cap X = X$. Thus, we give that neut(X) = X and anti(X) = X. So, $(A, \cap)$ is a NTS. We give the function p_N: $AxA \to \mathbb{R}^+ \cup \{0\}$ such that $p_N(X, Y) = max\{2^{2-m(X)} - 1, 2^{2-m(Y)} - 1\}$. From Definition 4,*

(i), (ii) and (iii) are apparent.

(iv) $p_N(X, \{x, y\}) = p_N(X, Y \cap \{x, y\})$ since for X, YinA. Furthermore, it is clear that

$max\{2^{2-m(X)} - 1, 2^{2-m(Y)} - 1\} \leq max\{2^{2-m(X)} - 1, 2^{2-m(\{x,\ y\})} - 1\} + max\{2^{2-m(Z)} - 1, 2^{2-m(\{x,y\})} - 1\} - max\{2^{2-m(\{x,y\})} - 1, 2^{2-m(\{x,y\})} - 1\}$. Thus,

$p_N(X, Y \cap \{x, y\}) \leq p_N(X, \{x, y\}) + p_N(\{x, y\}, B) - p_N(\{x, y\}, \{x, y\})$. Furthermore, $((A, \cap), p_N)$ is a NTPMS.

Let m: $A \to A$ be a map such that $m(X) = \begin{cases} \{x,y\},\ X = \{x,y\} \\ \{x\},\ X = \varnothing \\ \{x,y\},\ X = \{x\} \end{cases}$

For k = 0, 2

$p_N(m(\varnothing), m(\varnothing)) = p_N(\{x\}, \{x\}) = 1 \leq 0, 2.\ p_N(\varnothing, \varnothing) = 1, 5$
$p_N(m(\varnothing), m(\{x\})) = p_N(\{x\}, \{x, y\}) = 1 \leq 0, 2.\ p_N(\varnothing, \{x\}) = 1, 5$
$p_N(m(\varnothing), m(\{x, y\})) = p_N(\{x\}, \{x, y\}) = 1 \leq 0, 2.\ p_N(\varnothing, \{x, y\}) = 1, 5$
$p_N(m(\{x\}), m(\{x\})) = p_N(\{x, y\}, \{x, y\}) = 0 \leq 0, 2.\ p_N(\{x\}, \{x\}) = 0, 5$
$p_N(m(\{x\}), m(\{x, y\})) = p_N(\{x, y\}, \{x, y\}) = 0 \leq 0, 2.\ p_N(\{x\}, \{x,y\}) = 0, 5$
$p_N(m(\{x, y\}), m(\{x, y\})) = p_N(\{x, y\}, \{x, y\}) = 0 \leq 0, 2.\ p_N(\{x, y\}, \{x, y\}) = 0, 5$
Thus, m is a contraction for $((A, \cap), p_N)$

Theorem 4. *For each contraction m over a complete NTPMS $((A, \#), p_N)$, there exists a unique x in A such that x = m(x). Also, $p_N(x, x) = 0$.*

Proof. Let m be a contraction for $((A, \#), p_N)$ complete NTPMS and $x_n = m(x_{n-1})$ and $x_0 \in A$ be a unique element. Also, we can take

$$p_N(x_n,\ x_k) \leq p_N\ (x_n,\ x_k{*}\text{neut}(x_{n-1})) \tag{8}$$

since for m is a contraction over $((A, \#), p_N)$ complete NTPMS. Then,

$p_N(x_2, x_1) = p_N(m(x_1), m(x_0)) \leq$ c. $p_N(x_1, x_0)$ and

$p_N(x_3, x_2) = p_N(m(x_2), m(x_1)) \leq$ c. $p_N(x_2, x_1) \leq c^2$. $p_N(x_1, x_0)$. From mathematical induction, $n \geq m$; $p_N(x_{m+1}, x_m) = p_N(m(x_m), m(x_{m-1})) \leq$ c. $p_N(x_m, x_{m-1}) \leq c^m$. $p_N(x_1, x_0)$. Thus; from (8) and definition of NTPMS,

$$\begin{aligned}
p_N(x_n, x_m) \leq p_N(x_n x_m * \text{neut}(x_{n-1})) &\leq p_N(x_n, x_{n-1}) + p_N(x_{n-1}, x_m) - p_N(x_{n-1}, x_{n-1}) \\
&\leq c^{n-1}.p_N(x_1, x_0) + p_N(x_{n-1}, x_m) - p_N(x_{n-1}, x_{n-1}) \\
&\leq c^{n-1}.\ p_N(x_1, x_0) + p_N(x_{n-1}, x_{n-2}) + \ldots + p_N(x_m, x_{m-1}) \\
&\leq (c^{n-1} + c^{n-2} + \ldots + c^{m-1} + c^m).\ p_N(x_1, x_0) - \textstyle\sum_{i=m}^{n-1} p_N(x_i, x_i) \\
&\leq \textstyle\sum_{i=m}^{n-1} c^i.p_N(x_1, x_0) - \sum_{i=m}^{n-1} p_N(x_i, x_i) \\
&\leq \textstyle\sum_{i=m}^{n-1} c^i.p_N(x_1, x_0) + p_N(x_0, x_0) \\
&= \textstyle\sum_{i=m}^{n-1} c^i.p_N(x_1, x_0) + p_N(x_0, x_0)\,(\text{For } n,\ m \to \infty) \\
&= \tfrac{c^m}{1-c}\, p_N(x_1, x_0) + p_N(x_0, x_0) \to p_N(x_0, x_0)
\end{aligned}$$

Thus $\{x_n\}$ is a cauchy sequence. Also $\{x_n\}$ is convergent such that $x_n \to x$. Because $((A, \#), p_N)$ is complete NTPMS. Thus; $m(x_n) \to m(x)$ since for $x_n = m(x_{n-1})$; $m(x_n) = x_{n+1} \to x$. Thus; $m(x) = x$. Suppose that $m(x) = x$ or $m(y) = y$ for $x, y \in x_n$. Where;

$p_N(x, y) = p_N(m(x), m(y)) \leq c.\ p_N(x, y)$. $p_N(x, y) > 0$, $c \geq 1$ and it is a contradiction. Thus; $p_N(x, y) = p_N(x, x) = p_N(y, y) = 0$ and $x = y$. Therefore, $p_N(x, x) = 0$. □

4. Conclusions

In this paper, we introduced NTPMS. We also show that both the classical metric and NTM are different from the NT partial metric. This NT notion has more features than the classical notion. We also introduced contraction for PMS and we give a fixed point theory for PMS in NT theory. So, we obtained a new structure for developing NT theory. Thus, researchers can arrive at nonlinear partial differential equations problem solutions in NT theory thanks to NTPMS and FPT for NTPMS.

Author Contributions: In this paper, each Author contributed equally. M.S. introduced NTPMS and provided examples. A.K. introduced contraction for NTPMS and provided examples. M.A.C. gave fixed point theory for NTPMS and organized the paper.

Funding: This research received no external funding.

Conflicts of Interest: The authors are not report a conflict of interest.

References

1. Smarandache, F. *Neutrosophy: Neutrosophic Probability, Set and Logic;* ProQuest Information & Learning: Ann Arbor, MI, USA, 1998; p. 105.
2. Zadeh, L.A. Fuzzy sets. *Inf. Control* **1965**, *8*, 338–353. [CrossRef]
3. Atanassov, T.K. Intuitionistic fuzzy sets. *Fuzzy Sets Syst.* **1986**, *20*, 87–96. [CrossRef]
4. Kandasamy, W.B.V.; Smarandache, F. *Some Neutrosophic Algebraic Structures and Neutrosophic N-Algebraic Structures;* Hexis: Phoenix, AZ, USA, 2006; p. 209.
5. Şahin, M.; Olgun, N.; Uluçay, V.; Kargın, A.; Smarandache, F. A new similarity measure based on falsity value between single valued neutrosophic sets based on the centroid points of transformed single valued neutrosophic numbers with applications to pattern recognition. *Neutrosophic Sets Syst.* **2017**, *15*, 31–48. [CrossRef]
6. Sahin, M.; Deli, I.; Ulucay, V. Similarity measure of bipolar neutrosophic sets and their application to multiple criteria decision making. *Neural Comput. Appl.* **2016**, *29*, 739–748. [CrossRef]
7. Olgun, N.; Bal, M. Neutrosophic modules. *Neutrosophic Oper. Res.* **2017**, *2*, 181–192.
8. Şahin, M.; Uluçay, V.; Olgun, N.; Kilicman, A. On neutrosophic soft lattices. *Afr. Matematika* **2017**, *28*, 379–388.
9. Uluçay, V.; Şahin, M.; Olgun, N. Soft normed ring. *SpringerPlus* **2016**, *5*, 1950. [CrossRef] [PubMed]
10. Smarandache, F.; Ali, M. Neutrosophic triplet as extension of matter plasma, unmatter plasma and antimatter plasma. In Proceedings of the APS Gaseous Electronics Conference, Bochum, Germany, 10–14 October 2016.
11. Smarandache, F.; Ali, M. *The Neutrosophic Triplet Group and its Application to Physics;* Universidad National de Quilmes, Department of Science and Technology: Buenos Aires, Argentina, 2014.
12. Smarandache, F.; Ali, M. Neutrosophic triplet group. *Neural Comput. Appl.* **2016**, *29*, 595–601. [CrossRef]

13. Smarandache, F.; Ali, M. Neutrosophic Triplet Field Used in Physical Applications, (Log Number: NWS17-2017-000061). In Proceedings of the 18th Annual Meeting of the APS Northwest Section, Pacific University, Forest Grove, OR, USA, 1–3 June 2017; Available online: http://meetings.aps.org/Meeting/NWS17/Session/D1.1 (accessed on 25 June 2018).
14. Smarandache, F.; Ali, M. Neutrosophic Triplet Ring and Its Applications, (Log Number: NWS17-2017-000062). In Proceedings of the 18th Annual Meeting of the APS Northwest Section, Pacific University, Forest Grove, OR, USA, 1–3 June 2017; Available online: http://meetings.aps.org/Meeting/NWS17/Session/D1.2 (accessed on 25 June 2018).
15. Şahin, M.; Kargın, A. Neutrosophic triplet normed space. *Open Phys.* **2017**, *15*, 697–704.
16. Şahin, M.; Kargın, A. Neutrosophic triplet inner product space. *Neutrosophic Oper. Res.* **2017**, *2*, 193–215.
17. Smarandache, F.; Şahin, M.; Kargın, A. Neutrosophic Triplet G-Module. *Mathematics* **2018**, *6*, 53. [CrossRef]
18. Bal, M.; Shalla, M.M.; Olgun, N. Neutrosophic triplet cosets and quotient groups. *Symmetry* **2018**, *10*, 126. [CrossRef]
19. Matthews, S.G. Partial metric topology. *Ann. N. Y. Acad. Sci.* **1994**, *728*, 183–197. [CrossRef]
20. Kopperman, H.D.; Matthews, S.G.; Pajoohesh, K. Partial metrizability in value quantales. *Appl. Gen. Topol.* **2004**, *5*, 115–127. [CrossRef]
21. Altun, I.; Sola, F.; Simsek, H. Generalized contractions on partial metric space. *Topol. Appl.* **2010**, *157*, 2778–2785. [CrossRef]
22. Romeguera, S. A Kirk type characterization of completeness for partial metric space. *Fixed Point Theory Appl.* **2010**, *2010*, 493298. [CrossRef]
23. Romeguera, S. Fixed point theorems for generalized contractions on partial metric space. *Appl. Gen. Topol.* **2012**, *3*, 91–112. [CrossRef]
24. Shukla, S. Some fixed point theorems for ordered contractions in partial b-metric space. *Gazi Univ. J. Sci.* **2017**, *30*, 345–354.
25. Kim, J.K.; Okeke, G.A.; Lim, W.H. Common couplet fixed point theorems for w-compatible mapping in partial metric spaces. *Glob. J. Pure Appl. Math.* **2017**, *13*, 519–536.
26. Pant, R.; Shukla, R.; Nashine, H.K.; Panicker, R. Some new fixed point theorems in partial metric space with applications. *J. Funct. Spaces* **2017**, *2017*, 1072750. [CrossRef]

symmetry

MDPI

Article

Left (Right)-Quasi Neutrosophic Triplet Loops (Groups) and Generalized BE-Algebras

Xiaohong Zhang [1,2,*], Xiaoying Wu [1], Florentin Smarandache [3] and Minghao Hu [1]

1 Department of Mathematics, Shaanxi University of Science & Technology, Xi'an 710021, China; 46018@sust.edu.cn(X.W.); huminghao@sust.edu.cn (M.H.)
2 Department of Mathematics, Shanghai Maritime University, Shanghai 201306, China
3 Department of Mathematics, University of New Mexico, Gallup, NM 87301, USA; smarand@unm.edu
* Correspondence: zhangxiaohong@sust.edu.cn or zhangxh@shmtu.edu.cn; Tel.: +86-029-8616-8320

Received: 29 May 2018; Accepted: 19 June 2018; Published: 26 June 2018

Abstract: The new notion of a neutrosophic triplet group (NTG) is proposed by Florentin Smarandache; it is a new algebraic structure different from the classical group. The aim of this paper is to further expand this new concept and to study its application in related logic algebra systems. Some new notions of left (right)-quasi neutrosophic triplet loops and left (right)-quasi neutrosophic triplet groups are introduced, and some properties are presented. As a corollary of these properties, the following important result are proved: for any commutative neutrosophic triplet group, its every element has a unique neutral element. Moreover, some left (right)-quasi neutrosophic triplet structures in BE-algebras and generalized BE-algebras (including CI-algebras and pseudo CI-algebras) are established, and the adjoint semigroups of the BE-algebras and generalized BE-algebras are investigated for the first time.

Keywords: neutrosophic triplet; quasi neutrosophic triplet loop; quasi neutrosophic triplet group; BE-algebra; CI-algebra

---

1. Introduction

The symmetry exists in the real world, and group theory is a mathematical tool for describing symmetry. At the same time, in order to describe the generalized symmetry, the concept of group is popularized in different ways, for example, the notion of a generalized group is introduced (see [1–4]). Recently, F. Smarandache [5,6] introduced another new algebraic structure, namely: neutrosophic triplet group, which comes from the theory of the neutrosophic set (see [7–11]). As a new extension of the concept of group, the neutrosophic triplet group has attracted the attention of many scholars, and a series of related papers have been published [12–15].

On the other hand, in the last twenty years, the non-classical logics, such as various fuzzy logics, have made great progress. At the same time, the research on non-classical logic algebras that are related to it have also made great achievements [16–26]. As a generalization of BCK-algebra, H.S. Kim and Y.H. Kim [27] introduced the notion of BE-algebra. Since then, some scholars have studied ideals (filters), congruence relations of BE-algebras, and various special BE-algebras have been proposed, these research results are included in the literature [28–31] and monograph [32]. In 2013 and 2016, the new notions of pseudo BE-algebra and commutative pseudo BE-algebra were introduced, and some new properties were obtained [33,34]. Similar to BCI-algebra as a generalization of BCK-algebra, B.L. Meng introduced the concept of CI-algebra, which is as a generalization of BE-algebra, and studied the structures and closed filters of CI-algebras [35–37]. After that, the CI-algebras and their related algebraic structures (such as Q-algebras, pseudo Q-algebras, pseudo CI-algebras, and pseudo BCH-algebras) have been extensively studied [38–46].

This paper will combine the above two directions to study general neutrosophic triplet structures and the relationships between these structures and generalized BE-algebras. On the one hand, we introduce various general neutrosophic triplet structures, such as (*l-l*)-type, (*l-r*)-type, (*r-l*)-type, (*r-r*)-type, (*l-lr*)-type, (*r-lr*)-type, (*lr-l*)-type, and (*lr-r*)-type quasi neutrosophic triplet loops (groups), and investigate their basic properties. Moreover, we get an important corollary, namely: that for any commutative neutrosophic triplet group, its every element has a unique neutral element. On the other hand, we further study the properties of (pseudo) BE-algebras and (pseudo) CI-algebras, and the general neutrosophic triplet structures that are contained in a BE-algebra (CI-algebra) and pseudo BE-algebra (pseudo CI-algebra). Moreover, for the first time, we introduce the concepts of adjoint semigroups of BE-algebras and generalized BE-algebras (including CI-algebras, pseudo BE-algebras, and pseudo CI-algebras) and discuss some interesting topics.

2. Basic Concepts

Definition 1. *([5,6]) Let N be a set together with a binary operation *. Then, N is called a neutrosophic triplet set if, for any a∈N, there exists a neutral of 'a', called neut(a), and an opposite of 'a', called anti(a), with neut(a) and anti(a), belonging to N, such that:*

$$a * neut(a) = neut(a) * a = a;$$

$$a * anti(a) = anti(a) * a = neut(a).$$

It should be noted that *neut*(*a*) and *anti*(*a*) may not be unique here for some *a*∈*N*. We call (*a*, *neut*(*a*), and *anti*(*a*)) a neutrosophic triplet for the determined *neut*(*a*) and *anti*(*a*).

Remark 1. *In the original definition, the neutral element is different from the unit element in the traditional group theory. The above definition of this paper takes away such restriction, please see the Remark 3 in Ref. [12].*

Definition 2. *([5,6,13]) Let (N, *) be a neutrosophic triplet set.*

(1) *If * is well-defined, that is, for any a, b ∈ N, one has a * b ∈ N. Then, N is called a neutrosophic triplet loop.*
(2) *If N is a neutrosophic triplet loop, and * is associative, that is, (a * b) * c= a * (b * c) for all a, b, c ∈ N. Then, N is called a neutrosophic triplet group.*
(3) *If N is a neutrosophic triplet group, and * is commutative, that is, a * b = b * a for all a, b ∈ N. Then, N is called a commutative neutrosophic triplet group.*

Definition 3. *([27,35,41,42]) A CI-algebra (dual Q-algebra) is an algebra (X; →, 1) of type (2, 0), satisfying the following conditions:*

(i) $x \to x = 1$,
(ii) $1 \to x = x$,
(iii) $x \to (y \to z) = y \to (x \to z)$, *for all* $x, y, z \in X$.

A CI-algebra (X; →, 1)is called a BE-algebra, if it satisfies the following axiom:

(iv) $x \to 1 = 1$, *for all* $x \in X$.

A CI-algebra (X; →, 1)is called a dual BCH-algebra, if it satisfies the following axiom:

(v) $x \to y = y \to x = 1 \Rightarrow x = y$.

A binary relation ≤ on CI-algebra (BE-algebra) X, is defined by $x \leq y$ if, and only if, $x \to y = 1$.

Definition 4. *([33,43,45]) An algebra (X; $\rightarrow$, $\rightsquigarrow$, 1) of type (2, 2, 0) is called a dual pseudo Q-algebra if, for all x, y, z $\in$ X, it satisfies the following axioms:*

(dpsQ1) $x \rightarrow x = x \rightsquigarrow x = 1$,
(dpsQ2) $1 \rightarrow x = 1 \rightsquigarrow x = x$,
(dpsQ3) $x \rightarrow (y \rightsquigarrow z) = y \rightsquigarrow (x \rightarrow z)$.

A dual pseudo Q-algebra X is called a pseudo CI-algebra, if it satisfies the following condition:

(psCI) $x \rightarrow y = 1 \Leftrightarrow x \rightsquigarrow y = 1$.

A pseudo CI-algebra X is called a pseudo BE-algebra, if it satisfies the following condition:

(psBE) $x \rightarrow 1 = x \rightsquigarrow 1 = 1$, *for all* $x \in X$.

A pseudo CI-algebra X is called a pseudo BCH-algebra, if it satisfies the following condition:

(psBCH) $x \rightarrow y = y \rightsquigarrow x = 1 \Rightarrow x = y$.

In a dual pseudo-Q algebra, one can define the following binary relations:

$$x \leq_{\rightarrow} y \Leftrightarrow x \rightarrow y = 1. \ x \leq_{\rightsquigarrow} y \Leftrightarrow x \rightsquigarrow y = 1.$$

Obviously, a dual pseudo-Q algebra X is a pseudo CI-algebra if, and only if, $\leq_{\rightarrow} = \leq_{\rightsquigarrow}$.

3. Various Quasi Neutrosophic Triplet Loops (Groups)

Definition 5. *Let N be a set together with a binary operation * (that is, (N, *) be a loop) and a $\in$ N.*

(1) *If exist b, c $\in$ N, such that a * b = a and a * c = b, then a is called an NT-element with (r-r)- property;*
(2) *If exist b, c $\in$ N, such that a * b = a and c * a = b, then a is called an NT-element with (r-l)- property;*
(3) *If exist b, c $\in$ N, such that b * a = a and c * a = b, then a is called an NT-element with (l-l)- property;*
(4) *If exist b, c $\in$ N, such that b * a = a and a * c = b, then a is called an NT-element with (l-r)- property;*
(5) *If exist b, c $\in$ N, such that a * b = b * a = a and c * a = b, then a is called an NT-element with (lr-l)-property;*
(6) *If exist b, c $\in$ N, such that a * b = b * a = a and a * c = b, then a is called an NT-element with (lr-r)-property;*
(7) *If exist b, c $\in$ N, such that b * a = a and a * c = c * a = b, then a is called an NT-element with (l-lr)-property;*
(8) *If exist b, c $\in$ N, such that a * b = a and a * c = c * a = b, then a is called an NT-element with (r-lr)-property;*
(9) *If exist b, c $\in$ N, such that a * b = b * a = a and a * c = c * a = b, then a is called an NT-element with (lr-lr)-property.*

It is easy to verify that, (i) if *a* is an NT-element with (*l-lr*)-property, then *a* is an NT-element with (*l-l*)-property and (*l-r*)-property; if *a* is an NT-element with (*lr-l*)-property, then *a* is an NT-element with (*l-l*)-property and (*r-l*)-property; and so on; (ii) a neutrosophic triplet loop (*N*, *) is a neutrosophic triplet group if, and only if, every element in *N* is an NT-element with (*lr-lr*)-property; (iii) if * is commutative, then the above properties coincide. Moreover, the following example shows that (*r-l*)-property and (*r-r*)-property cannot infer to (*r-lr*)-property, and (*r-r*)-property and (*l-lr*)-property cannot infer to (*lr-lr*)-property.

Example 1. *Let N = {a, b, c, d}. The operation * on N is defined as Table 1. Then, (N, *) is a loop, and a is an NT-element with (lr-lr)-property; b is an NT-element with (lr-r)-property; c is an NT-element with (r-l)-property and (r-r)-property, but c is not an NT-element with (r-lr)-property; and d is an NT-element with (r-r)-property and (l-lr)-property, but d is not an NT-element with (lr-lr)-property.*

Table 1. Neutrosophic triplet (NT)-elements in a loop.

*	a	b	c	d
a	a	a	a	d
b	c	a	b	c
c	c	b	d	a
d	a	d	b	a

Definition 6. *Let (N, *) be a loop (semi-group). If for every element a in N, a is an NT-element with (r-r)-property, then (N, *) is called (r-r)-quasi neutrosophic triplet loop (group). Similarly, if for every element a in N, a is an NT-element with (r-l)-, (l-l)-, (l-r)-, (lr-l)-, (lr-r)-, (l-lr)-, (r-lr)-property, then (N, *) is called (r-l)-, (l-l)-, (l-r)-, (lr-l)-, (lr-r)-, (l-lr)-, (r-lr)-quasi neutrosophic triplet loop (group), respectively. All of these generalized neutrosophic triplet loops (groups) are collectively known as quasi neutrosophic triplet loops (groups).*

Remark 2. *For quasi neutrosophic triplet loops (groups), we will use the notations like neutrosophic triplet loops (groups), for example, to denote a (r-r)-neutral of 'a' by $neut_{(r\text{-}r)}(a)$, denote a (r-r)-opposite of 'a' by $anti_{(r\text{-}r)}(a)$, where 'a' is an NT-element with (r-r)-property. If $neut_{(r\text{-}r)}(a)$ and $anti_{(r\text{-}r)}(a)$ are not unique, then denote the set of all (r-r)-neutral of 'a' by $\{neut_{(r\text{-}r)}(a)\}$, denote the set of all (r-r)-opposite of 'a' by $\{anti_{(r\text{-}r)}(a)\}$.*

For the loop (N, *) in Example 1, we can verify that (N, *) is a (*r*-*r*)-quasi neutrosophic triplet loop, and we have the following:

$$neut_{(r\text{-}r)}(a) = a,\ anti_{(r\text{-}r)}(a) = a;\ neut_{(r\text{-}r)}(b) = c,\ \{anti_{(r\text{-}r)}(b)\} = \{a, d\};$$

$$neut_{(r\text{-}r)}(c) = a,\ anti_{(r\text{-}r)}(c) = d;\ neut_{(r\text{-}r)}(d) = b,\ anti_{(r\text{-}r)}(d) = c.$$

Theorem 1. *If (N, *) is a (l-lr)-quasi neutrosophic triplet group, then (N, *) is a neutrosophic triplet group. Moreover, if (N, *) is a (r-lr)-quasi neutrosophic triplet group, then (N, *) is a neutrosophic triplet group.*

Proof. Suppose that (N, *) is a (*l*-*lr*)-quasi neutrosophic triplet group. For any $a \in N$, by Definitions 5 and 6, we have the following:

$$neut_{(l\text{-}lr)}(a) * a = a,\ anti_{(l\text{-}lr)}(a) * a = a * anti_{(l\text{-}lr)}(a) = neut_{(l\text{-}lr)}(a).$$

Here, $neut_{(l\text{-}lr)}(a) \in \{neut_{(l\text{-}lr)}(a)\}$, $anti_{(l\text{-}lr)}(a) \in \{anti_{(l\text{-}lr)}(a)\}$. Applying associative law we get the following:

$$a * neut_{(l\text{-}lr)}(a) = a * (anti_{(l\text{-}lr)}(a) * a) = (a * anti_{(l\text{-}lr)}(a)) * a = neut_{(l\text{-}lr)}(a) * a = a.$$

This means that $neut_{(l\text{-}lr)}(a)$ is a right neutral of '*a*'. From the arbitrariness of *a*, it is known that (*N*, *) is a neutrosophic triplet group.

Another result can be proved similarly. □

Theorem 2. *Let (N, *) be a (r-lr)-quasi neutrosophic triplet group such that:*

$$(s * p) * a = a * (s * p),\ \forall\, s \in \{neut_{(r\text{-}lr)}(a)\},\ \forall\, p \in \{anti_{(r\text{-}lr)}(a)\}.$$

Then,

(1) *for any $a \in N$, $s \in \{neut_{(r\text{-}lr)}(a)\} \Rightarrow s * s = s$.*
(2) *for any $a \in N$, $s, t \in \{neut_{(r\text{-}lr)}(a)\} \Rightarrow s * t = t$.*
(3) *when * is commutative, for any $a \in N$, $neut_{(r\text{-}lr)}(a)$ is unique.*

Proof. (1) Assume $s \in \{neut_{(r\text{-}lr)}(a)\}$, then $a * s = a$, and exist $p \in N$, such that $p * a = a * p = s$. Thus,

$$(s * p) * a = s * (p * a) = s * s,$$

$$a * (s * p) = (a * s) * p = a * p = s.$$

According to the hypothesis, $(s * p) * a = a * (s * p)$, it follows that $s * s = s$.

(2) Assume $s, t \in \{neut_{(r\text{-}lr)}(a)\}$, then $a * s = a$, $a * t = a$, and exist $p, q \in N$, such that $p * a = a * p = s$, $q * a = a * q = t$. Thus,

$$(s * q) * a = s * (q * a) = s * t,$$

$$a * (s * q) = (a * s) * q = a * q = t.$$

According to the hypothesis, $(s * p) * a = a * (s * p)$, it follows that $s * t = t$.

(3) Suppose $a \in N$, $s, t \in \{neut_{(r\text{-}lr)}(a)\}$. Applying Theorem (2) to s and t we have $s * t = t$. Moreover, applying Therorem (2) to t and s we have $t * s = s$. Hence, when $*$ is commutative, $s * t = t * s$. Therefore, $s = t$, that is, $neut_{(r\text{-}lr)}(a)$ is unique. □

Corollary 1. *Let (N, *) be a commutative neutrosophic triplet group. Then neut(a) is unique for any* $a \in N$.

Proof. Since all neutrosophic triplet groups are *(r-lr)*-quasi neutrosophic triplet groups, and * is commutative, then the assumption conditions in Theorem 2 are valid for *N*, so applying Theorem 2 (3), we get that *neut(a)* is unique for any $a \in N$. □

The following examples show that the neutral element may be not unique in the neutrosophic triplet loop.

Example 2. *Let N = {1, 2, 3}. Define binary operation * on N as following Table 2. Then, (N, *) is a commutative neutrosophic triplet loop, and {neut(1)} = {1, 2}. Since* $(1 * 3) * 3 \neq 1 * (3 * 3)$, *so (N, *) is not a neutrosophic triplet group.*

Table 2. Commutative neutrosophic triplet loop.

*	1	2	3
1	1	1	2
2	1	2	3
3	2	3	3

Example 3. *Let N = {1, 2, 3, 4}. Define binary operation * on N as following Table 3. Then, (N, *) is a neutrosophic triplet loop, and {neut(4)} = {2, 3}. Since* $(4 * 1) * 1 \neq 4 * (1 * 1)$, *so (N, *) is not a neutrosophic triplet group.*

Table 3. Non-commutative neutrosophic triplet loop.

*	1	2	3	4
1	3	1	1	3
2	4	2	2	4
3	1	3	3	4
4	3	4	4	2

4. Quasi Neutrosophic Triplet Structures in BE-Algebras and CI-Algebras

From the definition of BE-algebra and CI-algebra (see Definition 3), we can see that '1' is a left neutral element of every element, that is, BE-algebras and CI-algebras are directly related to quasi neutrosophic triplet structures. This section will reveal the various internal connections among them.

4.1. BE-Algebras (CI-Algebras) and (l-l)-Quasi Neutrosophic Triplet Loops

Theorem 3. *Let (X; →, 1) be a BE-algebra. Then (X, →) is a (l-l)-quasi neutrosophic triplet loop. And, when* $|X|>1$*, (X, →) is not a (lr-l)-quasi neutrosophic triplet loop with neutral element 1.*

Proof. By Definition 3, for all $x \in X$, $1 \to x = x$ and $x \to x = 1$. According Definition 6, we know that (X, →) is a (*l-l*)-quasi neutrosophic triplet loop, such that:

$$1 \in \{neut_{(l\text{-}l)}(x)\},\ x \in \{anti_{(l\text{-}l)}(x)\}, \text{ for any } x \in X.$$

If $|X| > 1$, then exist $x \in X$, such that $x \neq 1$. Using Definition 3 (iv), $x \to 1 = 1 \neq x$, this means that 1 is not a right neutral element of x. Hence, (X, →) is not a (*lr-l*)-quasi neutrosophic triplet loop with neutral element 1. □

Example 4. *Let X = {a, b, c, 1}. Define binary operation * on N as following Table 4. Then, (X; →, 1) is a BE-algebra, and (X, →) is a (l-l)-quasi neutrosophic triplet loop, such that:*

$$\{neut_{(l\text{-}l)}(a)\} = \{1\}, \{anti_{(l\text{-}l)}(a)\} = \{a, c\}; \{neut_{(l\text{-}l)}(b)\} = \{1\}, \{anti_{(l\text{-}l)}(b)\} = \{b, c\};$$

$$\{neut_{(l\text{-}l)}(c)\} = \{1\}, \{anti_{(l\text{-}l)}(c)\} = \{c\}; \{neut_{(l\text{-}l)}(1)\} = \{1\}, \{anti_{(l\text{-}l)}(1)\} = \{1\}.$$

Table 4. BE-algebra and (*l-l*)-quasi neutrosophic triplet loop (1).

→	*a*	*b*	*c*	*1*
a	1	*b*	*b*	1
b	*a*	1	*a*	1
c	1	1	1	1
1	*a*	*b*	*c*	1

Example 5. *Let X = {a, b, c, 1}. Define binary operation * on N as following Table 5. Then, (X; →, 1) is a BE-algebra, and (X, →) is a (l-l)-quasi neutrosophic triplet loop such that:*

$$\{neut_{(l\text{-}l)}(a)\} = \{1\}, \{anti_{(l\text{-}l)}(a)\} = \{a\}; \{neut_{(l\text{-}l)}(b)\} = \{1\}, \{anti_{(l\text{-}l)}(b)\} = \{b\};$$

$$\{neut_{(l\text{-}l)}(c)\} = \{1\}, \{anti_{(l\text{-}l)}(c)\} = \{c\}; \{neut_{(l\text{-}l)}(1)\} = \{1\}, \{anti_{(l\text{-}l)}(1)\} = \{1\}.$$

Table 5. BE-algebra and (*l-l*)-quasi neutrosophic triplet loop (2).

→	*a*	*b*	*c*	*1*
a	1	*b*	*c*	1
b	*a*	1	*c*	1
c	*a*	*b*	1	1
1	*a*	*b*	*c*	1

Definition 7. *([36]) Let $(X; \rightarrow, 1)$ be a CI-algebra and $a \in X$. If for any $x \in X$, $a \rightarrow x = 1$ implies $a = x$, then a is called an atom in X. Denote $A(X) = \{a \in X \mid a \text{ is an atom in } X\}$, it is called the singular part of X. A CI-algebra $(X; \rightarrow, 1)$ is said to be singular if every element of X is an atom.*

Lemma 1. *([35–37]) If $(X; \rightarrow, 1)$ is a CI-algebra, then for all $x, y \in X$:*

(1) $x \rightarrow ((x \rightarrow y) \rightarrow y) = 1$,
(2) $1 \rightarrow x = 1$ *(or equivalently,* $1 \leq x$*) implies* $x = 1$,
(3) $(x \rightarrow y) \rightarrow 1 = (x \rightarrow 1) \rightarrow (y \rightarrow 1)$.

Lemma 2. *([36]) Let $(X; \rightarrow, 1)$ be a CI-algebra. If $a, b \in X$ are atoms in X, then the following are true:*

(1) $a = (a \rightarrow 1) \rightarrow 1$,
(2) $(a \rightarrow b) \rightarrow 1 = b \rightarrow a$,
(3) $((a \rightarrow b) \rightarrow 1) \rightarrow 1 = a \rightarrow b$,
(4) *for any* $x \in X$, $(a \rightarrow x) \rightarrow (b \rightarrow x) = b \rightarrow a$,
(5) *for any* $x \in X$, $(a \rightarrow x) \rightarrow b = (b \rightarrow x) \rightarrow a$,
(6) *for any* $x \in X$, $(a \rightarrow x) \rightarrow (y \rightarrow b) = (b \rightarrow x) \rightarrow (y \rightarrow a)$.

Definition 8. *Let $(X; \rightarrow, 1)$ be a CI-algebra. If for any $x \in X$, $x \rightarrow 1 = x$, then $(X; \rightarrow, 1)$ is said to be a strong singular.*

Proposition 1. *If $(X; \rightarrow, 1)$ is a strong singular CI-algebra. Then $(X; \rightarrow, 1)$ is a singular CI-algebra.*

Proof. For any $x \in X$, assume that $a \rightarrow x = 1$, where $a \in X$. By Definition 8, we have $x \rightarrow 1 = x$, $a \rightarrow 1 = a$. Hence, applying Definition 3,

$$a = a \rightarrow 1 = a \rightarrow (x \rightarrow x) = x \rightarrow (a \rightarrow x) = x \rightarrow 1 = x.$$

By Definition 7, x is an atom. Therefore, $(X; \rightarrow, 1)$ is singular CI-algebra. □

Proposition 2. *Let $(X; \rightarrow, 1)$ be a CI-algebra. Then $(X; \rightarrow, 1)$ is a strong singular CI-algebra if, and only if, $(X; \rightarrow, 1)$ is an associative BCI-algebra.*

Proof. Obviously, every associative BCI-algebra is a strong singular CI-algebra (see [36] and Proposition 1 in Ref. [12]).

Assume that $(X; \rightarrow, 1)$ is a strong singular CI-algebra.

(1) For any $x, y \in X$, if $x \rightarrow y = y \rightarrow x = 1$, then, by Definitions 8 and 3, we have the following:

$$x = x \rightarrow 1 = x \rightarrow (y \rightarrow x) = y \rightarrow (x \rightarrow x) = y \rightarrow 1 = y.$$

(2) For any $x, y, z \in X$, by Proposition 1 and Lemma 2 (4), we can get the following:

$$(y \rightarrow z) \rightarrow ((z \rightarrow x) \rightarrow (y \rightarrow x)) = (y \rightarrow z) \rightarrow (y \rightarrow z) = 1.$$

Combining Proof (1) and (2), we know that $(X; \rightarrow, 1)$ is a BCI-algebra. From this, applying Definition 8 and Proposition 1 in Ref. [12], $(X; \rightarrow, 1)$ is an associative BCI-algebra. □

Theorem 4. *Let $(X; \rightarrow, 1)$ be a CI-algebra. Then, $(X, \rightarrow)$ is a (l-l)-quasi neutrosophic triplet loop. Moreover, $(X, \rightarrow)$ is a neutrosophic triplet group if, and only if, $(X; \rightarrow, 1)$ is a strong singular CI-algebra (associative BCI-algebra).*

Proof. It is similar to the proof of Theorem 3, and we know that $(X, \rightarrow)$ is a (*l*-*l*)-quasi neutrosophic triplet loop.

If $(X; \rightarrow, 1)$ is a strong singular CI-algebra, using Proposition 2, $(X; \rightarrow, 1)$ is an associative BCI-algebra. Hence, $\rightarrow$ is associative and commutative, it follows that $(X, \rightarrow)$ is a neutrosophic triplet group.

Conversely, if $(X, \rightarrow)$ is a neutrosophic triplet group, then $\rightarrow$ is associative, thus

$$x \rightarrow 1 = x \rightarrow (x \rightarrow x) = (x \rightarrow x) \rightarrow x = 1 \rightarrow x = x.$$

By Definition 8 we know that $(X; \rightarrow, 1)$ is a strong singular CI-algebra. □

Example 6. *Let X = {a, b, c, d, e,1}. Define operation → on X, as following Table 6. Then, (X; →, 1) is a CI-algebra, and (X, →) is a (l-l)-quasi neutrosophic triplet loop, such that*

$$\{neut_{(l\text{-}l)}(a)\} = \{1\}, \{anti_{(l\text{-}l)}(a)\} = \{a,b\}; \{neut_{(l\text{-}l)}(b)\} = \{1\}, \{anti_{(l\text{-}l)}(b)\} = \{a,b,c\};$$

$$\{neut_{(l\text{-}l)}(c)\} = \{1\}, \{anti_{(l\text{-}l)}(c)\} = \{c,d,e\}; \{neut_{(l\text{-}l)}(d)\} = \{1\}, \{anti_{(l\text{-}l)}(d)\} = \{d,e\};$$

$$\{neut_{(l\text{-}l)}(e)\}=\{1\}, \{anti_{(l\text{-}l)}(e)\}=\{d,e\}; \{neut_{(l\text{-}l)}(1)\}=\{1\}, \{anti_{(l\text{-}l)}(1)\}=\{1\}.$$

Table 6. CI-algebra and (*l*-*l*)-quasi neutrosophic triplet loop.

→	*a*	*b*	*c*	*d*	*e*	**1**
a	1	1	c	c	c	1
b	1	1	c	c	c	1
c	d	1	1	a	b	c
d	c	c	1	1	1	c
e	c	c	1	1	1	c
1	a	b	c	d	e	1

4.2. BE-Algebras (CI-Algebras) and Their Adjoint Semi-Groups

I. Fleischer [16] studied the relationship between BCK-algebras and semigroups, and W. Huang [17] studied the close connection between the BCI-algebras and semigroups. In this section, we have studied the adjoint semigroups of the BE-algebras and CI-algebras, and will give some interesting examples.

For any BE-algebra or CI-algebra $(X; \rightarrow, 1)$, and any element a in X, we use p_a to denote the self-map of X defined by the following:

$$p_a\text{: } X \rightarrow X; \mapsto a \rightarrow x, \text{ for all } x \in X.$$

Theorem 5. *Let (X; →, 1) be a BE-algebra (or CI-algebra), and M(X) be the set of finite products* $p_a * \ldots * p_b$ *of self-map of X with a, ..., b ∈X, where * represents the composition operation of mappings. Then, (M(X), *) is a commutative semigroup with identity* p_1.

Proof. Since the composition operation of mappings satisfies the associative law, (M(X), *) is a semigroup. Moreover, since

$$p_1\text{: } X \rightarrow X \mapsto 1 \rightarrow x, \text{ for all } x \in X.$$

Applying Definition 3 (ii), we get that $p_1(x)=x$ for any $x \in X$. Hence, $p_1*m= p_1*m=m$ for any $m \in M(X)$.

For any $a, b \in X$, using Definition 3 (iii) we have ($\forall x \in X$) the following:

$$(p_a{}^* p_b)(x) = p_a(b \to x) = a \to (b \to x) = b \to (a \to x) = p_b(a \to x) = (p_b{}^* p_a)(x).$$

Therefore, ($M(X)$, *) is a commutative semigroup with identity p_1. □

Now, we call ($M(X)$, *) the adjoint semigroup of X.

Example 7. *Let X = {a, b, c, 1}. Define operation → on X, as following Table 7. Then, (X; →, 1) is a BE-algebra, and*

p_a: $X \to X; a \mapsto 1, b \mapsto 1, c \mapsto 1, 1 \mapsto 1$. It is abbreviated to $p_a = (1, 1, 1, 1)$.
p_b: $X \to X; a \mapsto c, b \mapsto 1, c \mapsto a, 1 \mapsto 1$. It is abbreviated to $p_b = (c, 1, a, 1)$.
p_c: $X \to X; a \mapsto 1, b \mapsto 1, c \mapsto 1, 1 \mapsto 1$. It is abbreviated to $p_c = (1, 1, 1, 1)$.
p_1: $X \to X; a \mapsto a, b \mapsto b, c \mapsto c, 1 \mapsto 1$. It is abbreviated to $p_1 = (a, b, c, 1)$.

We can verify that $p_a * p_a = p_a$, $p_a * p_b = p_a$, $p_a * p_c = p_a$; $p_b * p_b = (a, 1, c, 1)$, $p_b * p_c = p_c = p_a$; $p_a * (p_b * p_b) = p_a$, $p_b * (p_b * p_b) = p_b$, $p_c * (p_b * p_b) = p_c = p_a$. Denote $p_{bb} = p_b{}^* p_b = (a, 1, c, 1)$, then $M(X) = \{p_a, p_b, p_{bb}, p_1\}$, and its Cayley table is Table 8. Obviously, ($M(X)$, *) is a commutative neutrosophic triplet group and

$neut(p_a) = p_a$, $anti(p_a) = p_a$; $neut(p_b) = p_{bb}$, $anti(p_b) = p_b$; $neut(p_{bb}) = p_{bb}$, $anti(p_{bb}) = p_{bb}$; $neut(p_1) = p_1$, $anti(p_1) = p_1$.

Table 7. BE-algebra.

→	*a*	*b*	*c*	*1*
a	1	1	1	1
b	*c*	1	*a*	1
c	1	1	1	1
1	*a*	*b*	*c*	1

Table 8. Adjoint semigroup of the above BE-algebra.

*	p_a	p_b	p_{bb}	p_1
p_a	p_a	p_a	p_a	p_a
p_b	p_a	p_{bb}	p_b	p_b
p_{bb}	p_a	p_b	p_{bb}	p_{bb}
p_1	p_a	p_b	p_{bb}	p_1

Example 8. *Let X = {a, b, 1}. Define operation → on X, as following Table 9. Then, (X; →, 1) is a CI-algebra, and*

p_a: $X \to X; a \mapsto 1, b \mapsto a, 1 \mapsto b$. It is abbreviated to $p_a = (1, a, b)$.
p_b: $X \to X; a \mapsto b, b \mapsto 1, 1 \mapsto a$. It is abbreviated to $p_b = (b, 1, a)$.
p_1: $X \to X; a \mapsto a, b \mapsto b, 1 \mapsto 1$. It is abbreviated to $p_1 = (a, b, 1)$.

We can verify that $p_a * p_a = p_b$, $p_a * p_b = p_1$; $p_b * p_b = p_a$. Then $M(X) = \{p_a, p_b, p_1\}$ and its Cayley table is Table 10. Obviously, ($M(X)$, *) is a commutative group with identity p_1 and $(p_a)^{-1} = p_b$, $(p_b)^{-1} = p_a$.

Table 9. CI-algebra.

$\rightarrow$	a	b	1
a	1	a	b
b	b	1	a
1	a	b	1

Table 10. Adjoint semigroup of the above CI-algebra.

*	p_a	p_b	p_1
p_a	p_b	p_1	p_a
p_b	p_1	p_a	p_b
p_1	p_a	p_b	p_1

Theorem 6. *Let $(X; \rightarrow, 1)$ be a singular CI-algebra, and $M(X)$ be the adjoint semigroup. Then $(M(X), *)$ is a commutative group with identity p_1, where $M(X) = \{p_a \mid a \in X\}$ and $|M(X)| = |X|$.*

Proof. (1) First, we prove that for any singular CI-algebra, $a \rightarrow (b \rightarrow x) = ((a \rightarrow 1) \rightarrow b) \rightarrow x$, $\forall\, a, b, x \in X$.

In fact, by Definition 7 and Lemma 2, we have the following:

$$\begin{aligned}((a \rightarrow 1) \rightarrow b) \rightarrow x &= ((a \rightarrow 1) \rightarrow b) \rightarrow ((x \rightarrow 1) \rightarrow 1) \\ &= (x \rightarrow 1) \rightarrow (((a \rightarrow 1) \rightarrow b) \rightarrow 1) \\ &= (x \rightarrow 1) \rightarrow (((a \rightarrow 1) \rightarrow 1) \rightarrow (b \rightarrow 1)) \\ &= (x \rightarrow 1) \rightarrow (a \rightarrow (b \rightarrow 1)) \\ &= a \rightarrow ((x \rightarrow 1) \rightarrow (b \rightarrow 1)) \\ &= a \rightarrow (b \rightarrow x).\end{aligned}$$

(2) Second, we prove that for any singular CI-algebra, $a \neq b \Rightarrow p_a \neq p_b$, $\forall\, a, b \in X$.
Assume $p_a = p_b$, $a, b \in X$. Then, for all x in X, $p_a(x) = p_b(x)$. Hence,

$$a \rightarrow b = p_a(b) = p_b(b) = b \rightarrow b = 1.$$

From this, applying Lemma 2 (1) and (6) we get

$$a = (a \rightarrow 1) \rightarrow 1 = (a \rightarrow 1) \rightarrow (a \rightarrow b) = (b \rightarrow 1) \rightarrow (a \rightarrow a) = (b \rightarrow 1) \rightarrow 1 = b.$$

(3) Using Lemma 2 (1), we know that for any $a, b \in X$, there exist $c \in X$, such that $p_a * p_b = p_c$, where $c = (a \rightarrow 1) \rightarrow b$. This means that $M(X) \subseteq \{p_a \mid a \in X\}$. By the definition of $M(X)$, $\{p_a\ a \in X\} \subseteq M(X)$. Hence, $M(X) = \{p_a \mid a \in X\}$.

(4) Using Lemma 2 (2) and (3), we know that $|M(X)| = |X|$. □

5. Quasi Neutrosophic Triplet Structures in Pseudo BE-Algebras and Pseudo CI-Algebras

Like the above Section 4, we can discuss the relationships between pseudo BE-algebras (pseudo CI-algebras) and quasi neutrosophic triplet structures. This section will give some related results and examples, but part of the simple proofs will be omitted.

5.1. Pseudo BE-Algebras (Pseudo CI-Algebras) and (l-l)-Quasi Neutrosophic Triplet Loops

Theorem 7. *Let (X; $\rightarrow$, $\rightsquigarrow$, 1) be pseudo BE-algebra. Then (X, $\rightarrow$) and (X, $\rightsquigarrow$) are (l-l)-quasi neutrosophic triplet loops. And, when $|X| > 1$, (X, $\rightarrow$) and (X, $\rightsquigarrow$) are not (lr-l)-quasi neutrosophic triplet loops with neutral element 1.*

Example 9. *Let X = {a, b, c, 1}. Define operations $\rightarrow$ and $\rightsquigarrow$ on X as following* Tables 11 and 12. *Then, (X; $\rightarrow$, $\rightsquigarrow$, 1) is a pseudo BE-algebra, and (X, $\rightarrow$) and (X, $\rightsquigarrow$) are (l-l)-quasi neutrosophic triplet loops.*

Table 11. Pseudo BE-algebra (1).

$\rightarrow$	a	b	c	1
a	1	1	b	1
b	a	1	c	1
c	1	1	1	1
1	a	b	c	1

Table 12. Pseudo BE-algebra (2).

$\rightsquigarrow$	a	b	c	1
a	1	1	a	1
b	a	1	a	1
c	1	1	1	1
1	a	b	c	1

Definition 9. *([44,46]) Let a be an element of a pseudo CI-algebra (X; $\rightarrow$, $\rightsquigarrow$, 1). a is said to be an atom in X if for any $x \in X$, $a \rightarrow x = 1$ implies $a = x$.*

Applying the results in Ref. [44–46] we have the following propositions (the proofs are omitted).

Proposition 3. *If (X; $\rightarrow$,$\rightsquigarrow$, 1) is a pseudo CI-algebra, then for all x, y $\in$X*

(1) $x \leq (x \rightarrow y) \rightsquigarrow y$, $x \leq (x \rightsquigarrow y) \rightarrow y$,
(2) $x \leq y \rightarrow z \Leftrightarrow y \leq x \rightsquigarrow z$,
(3) $(x \rightarrow y) \rightarrow 1 = (x \rightarrow 1) \rightsquigarrow (y \rightsquigarrow 1)$, $(x \rightsquigarrow y) \rightsquigarrow 1 = (x \rightsquigarrow 1) \rightarrow (y \rightarrow 1)$,
(4) $x \rightarrow 1 = x \rightsquigarrow 1$,
(5) $x \leq y$ *implies* $x \rightarrow 1 = y \rightarrow 1$.

Proposition 4. *Let (X; $\rightarrow$,$\rightsquigarrow$, 1) be a pseudo CI-algebra. If a, b $\in$X are atoms in X, then the following are true:*

(1) $a = (a \rightarrow 1) \rightarrow 1$,
(2) *for any* $x \in X$, $(a \rightarrow x) \rightsquigarrow x = a$, $(a \rightsquigarrow x) \rightarrow x = a$,
(3) *for any* $x \in X$, $(a \rightarrow x) \rightsquigarrow 1 = x \rightarrow a$, $(a \rightsquigarrow x) \rightarrow 1 = x \rightsquigarrow a$,
(4) *for any* $x \in X$, $x \rightarrow a = (a \rightarrow 1) \rightsquigarrow (x \rightarrow 1)$, $x \rightsquigarrow a = (a \rightsquigarrow 1) \rightarrow (x \rightsquigarrow 1)$.

Definition 10. *A pseudo CI-algebra (X; $\rightarrow$,$\rightsquigarrow$, 1) is said to be singular if every element of X is an atom. A pseudo CI-algebra (X; $\rightarrow$,$\rightsquigarrow$, 1) is said to be strong singular if for any x $\in$X, $x \rightarrow 1 = x = x \rightsquigarrow 1$.*

Proposition 5. *If (X; $\rightarrow$,$\rightsquigarrow$, 1) is a strong singular pseudo CI-algebra. Then (X; $\rightarrow$,$\rightsquigarrow$, 1) is singular.*

Proof. For any $x \in X$, assume that $a \rightarrow x = 1$, where $a \in X$. It follows from Definition 10,

$$x \rightarrow 1 = x = x \rightsquigarrow 1, a \rightarrow 1 = a = a \rightsquigarrow 1.$$

Hence, applying Definition 4 and Proposition 3,

$$a = a \rightarrow 1 = a \rightarrow (x \rightsquigarrow x) = x \rightsquigarrow (a \rightarrow x) = x \rightsquigarrow 1 = x.$$

By Definition 9, x is an atom. Therefore, $(X; \rightarrow, \rightsquigarrow, 1)$ is singular pseudo CI-algebra. □

Applying Theorem 3.11 in Ref. [46], we can get the following:

Lemma 3. *Let $(X; \rightarrow, \rightsquigarrow, 1)$ be a pseudo CI-algebra. Then the following statements are equivalent:*

(1) $x \rightarrow (y \rightarrow z) = (x \rightarrow y) \rightarrow z$, *for all* x, y, z *in* X;
(2) $x \rightarrow 1 = x = x \rightsquigarrow 1$, *for every* x *in* X;
(3) $x \rightarrow y = x \rightsquigarrow y = y \rightarrow x$, *for all* x, y *in* X;
(4) $x \rightsquigarrow (y \rightsquigarrow z) = (x \rightsquigarrow y) \rightsquigarrow z$, *for all* x, y, z *in* X.

Proposition 6. *Let $(X; \rightarrow, \rightsquigarrow, 1)$ be a pseudo CI-algebra. Then $(X; \rightarrow, \rightsquigarrow, 1)$ is a strong singular pseudo CI-algebra if, and only if, $\rightarrow = \rightsquigarrow$ and $(X; \rightarrow, 1)$ is an associative BCI-algebra.*

Proof. We know that every associative BCI-algebra is a strong singular pseudo CI-algebra. □

Now, suppose that $(X; \rightarrow, 1)$ is a strong singular pseudo CI-algebra. By Definition 10 and Lemma 3 (3), $x \rightarrow y = x \rightsquigarrow y$, $\forall x, y \in X$. That is, $\rightarrow = \rightsquigarrow$. Hence, $(X; \rightarrow, 1)$ is a strong singular CI-algebra. It follows that $(X; \rightarrow, 1)$ is an associative BCI-algebra (using Proposition 2).

Theorem 8. *Let $(X; \rightarrow, \rightsquigarrow, 1)$ be a pseudo CI-algebra. Then $(X, \rightarrow)$ and $(X, \rightsquigarrow)$ are(l-l)-quasi neutrosophic triplet loops. Moreover, $(X, \rightarrow)$ and $(X, \rightsquigarrow)$ are neutrosophic triplet groups if, and only if, $(X; \rightarrow, \rightsquigarrow, 1)$ is a strong singular pseudo CI-algebra (associative BCI-algebra).*

Proof. Applying Lemma 3, and the proof is omitted. □

5.2. Pseudo BE-Algebras (Pseudo CI-Algebras) and Their Adjoint Semi-Groups

For any pseudo BE-algebra or pseudo CI-algebra $(X; \rightarrow, \rightsquigarrow, 1)$ as well as any element a in X, we use $p_a^{\rightarrow}$ and $p_a^{\rightsquigarrow}$ to denote the self-map of X, which is defined by the following:

$$p_a^{\rightarrow}: X \rightarrow X; \mapsto a \rightarrow x, \text{ for all } x \in X.$$

$$p_a^{\rightsquigarrow}: X \rightarrow X; \mapsto a \rightsquigarrow x, \text{ for all } x \in X.$$

Theorem 9. *Let $(X; \rightarrow, \rightsquigarrow, 1)$ be a pseudo BE-algebra (or pseudo CI-algebra), and*

$$M^{\rightarrow}(X) = \{\text{finite products } p_a^{\rightarrow} * \ldots * p_b^{\rightarrow} \text{ of self-map of } X \mid a, \ldots, b \in X\},$$

$$M^{\rightsquigarrow}(X) = \{\text{finite products } p_a^{\rightsquigarrow} * \ldots * p_b^{\rightsquigarrow} \text{ of self-map of } X \mid a, \ldots, b \in X\},$$

$$M(X) = \{\text{finite products } p_a^{\rightarrow} \text{ (or } p_a^{\rightsquigarrow}) * \ldots * p_b^{\rightarrow} \text{ (or } p_b^{\rightsquigarrow}) \text{ of self-map of } X \mid a, \ldots, b \in X\},$$

where * represents the composition operation of mappings. Then $(M^{\rightarrow}(X), *)$, $(M^{\rightsquigarrow}(X), *)$, and $(M(X), *)$ are all semigroups with the identity $p_1 = p_1^{\rightarrow} = p_1^{\rightsquigarrow}$.

Proof. It is similar to Theorem 5. □

Now, we call $(M^{\rightarrow}(X), *)$, $(M^{\rightsquigarrow}(X), *)$, *and* $(M(X), *)$ the adjoint semigroups of X.

Example 10. *Let X = {a, b, c, 1}. Define operations → and ⇝ on X as following* Tables 13 and 14. *Then, (X; →, ⇝, 1) is a pseudo BE-algebra, and*

$$p_a{}^{\rightarrow} = (1, b, b, 1), p_b{}^{\rightarrow} = (a, 1, c, 1), p_c{}^{\rightarrow} = (1, 1, 1, 1), p_1{}^{\rightarrow} = (a, b, c, 1).$$

We can verify the following:

$$p_a{}^{\rightarrow} * p_a{}^{\rightarrow} = p_a{}^{\rightarrow}, p_a{}^{\rightarrow} * p_b{}^{\rightarrow} = (1, 1, b, 1), p_a{}^{\rightarrow} * p_c{}^{\rightarrow} = p_c{}^{\rightarrow}, p_a{}^{\rightarrow} * p_1{}^{\rightarrow} = p_a{}^{\rightarrow};$$

$$p_b{}^{\rightarrow} * p_a{}^{\rightarrow} = p_c{}^{\rightarrow}, p_b{}^{\rightarrow} * p_b{}^{\rightarrow} = p_b{}^{\rightarrow}, p_b{}^{\rightarrow} * p_c{}^{\rightarrow} = p_c{}^{\rightarrow}, p_b{}^{\rightarrow} * p_1{}^{\rightarrow} = p_b{}^{\rightarrow};$$

$$p_c{}^{\rightarrow} * p_a{}^{\rightarrow} = p_c{}^{\rightarrow}, p_c{}^{\rightarrow} * p_b{}^{\rightarrow} = p_c{}^{\rightarrow}, p_c{}^{\rightarrow} * p_c{}^{\rightarrow} = p_c{}^{\rightarrow}, p_c{}^{\rightarrow} * p_1{}^{\rightarrow} = p_c{}^{\rightarrow};$$

$$p_1{}^{\rightarrow} * p_a{}^{\rightarrow} = p_a{}^{\rightarrow}, p_1{}^{\rightarrow} * p_b{}^{\rightarrow} = p_b{}^{\rightarrow}, p_1{}^{\rightarrow} * p_c{}^{\rightarrow} = p_c{}^{\rightarrow}, p_1{}^{\rightarrow} * p_1{}^{\rightarrow} = p_1{}^{\rightarrow}.$$

Denote $p_{ab}{}^{\rightarrow} = p_a{}^{\rightarrow} * p_b{}^{\rightarrow} = (1, 1, b, 1)$, then $p_{ab}{}^{\rightarrow} * p_a{}^{\rightarrow} = p_c{}^{\rightarrow}$, $p_{ab}{}^{\rightarrow} * p_b{}^{\rightarrow} = p_{ab}{}^{\rightarrow}$, $p_{ab}{}^{\rightarrow} * p_{ab}{}^{\rightarrow} = p^{\rightarrow}$, $p_{ab}{}^{\rightarrow} * p_c{}^{\rightarrow} = p_c{}^{\rightarrow}$. Hence, $M^{\rightarrow}(X) = \{p_a{}^{\rightarrow}, p_b{}^{\rightarrow}, p_{ab}{}^{\rightarrow}, p_c{}^{\rightarrow}, p_1{}^{\rightarrow}\}$ and its Cayley table is Table 15. Obviously, $(M^{\rightarrow}(X), *)$ is a non-commutative semigroup, but it is not a neutrosophic triplet group.

Table 13. Pseudo BE-algebra and adjoint semigroups (1).

→	a	b	c	1
a	1	b	b	1
b	a	1	c	1
c	1	1	1	1
1	a	b	c	1

Table 14. Pseudo BE-algebra and adjoint semigroups (2).

⇝	a	b	c	1
a	1	b	c	1
b	a	1	a	1
c	1	1	1	1
1	a	b	c	1

Table 15. Pseudo BE-algebra and adjoint semigroups (3).

*	$p_a{}^{\rightarrow}$	$p_b{}^{\rightarrow}$	$p_{ab}{}^{\rightarrow}$	$p_c{}^{\rightarrow}$	$p_1{}^{\rightarrow}$
$p_a{}^{\rightarrow}$	$p_a{}^{\rightarrow}$	$p_{ab}{}^{\rightarrow}$	$p_{ab}{}^{\rightarrow}$	$p_c{}^{\rightarrow}$	$p_a{}^{\rightarrow}$
$p_b{}^{\rightarrow}$	$p_c{}^{\rightarrow}$	$p_b{}^{\rightarrow}$	$p_c{}^{\rightarrow}$	$p_c{}^{\rightarrow}$	$p_b{}^{\rightarrow}$
$p_{ab}{}^{\rightarrow}$	$p_c{}^{\rightarrow}$	$p_{ab}{}^{\rightarrow}$	$p_c{}^{\rightarrow}$	$p_c{}^{\rightarrow}$	$p_{ab}{}^{\rightarrow}$
$p_c{}^{\rightarrow}$	$p_c{}^{\rightarrow}$	$p_c{}^{\rightarrow}$	$p_c{}^{\rightarrow}$	$p_c{}^{\rightarrow}$	$p_c{}^{\rightarrow}$
$p_1{}^{\rightarrow}$	$p_a{}^{\rightarrow}$	$p_b{}^{\rightarrow}$	$p_{ab}{}^{\rightarrow}$	$p_c{}^{\rightarrow}$	$p_1{}^{\rightarrow}$

Similarly, we can verify that

$$p_a{}^{\rightsquigarrow} = (1, b, c, 1), p_b{}^{\rightsquigarrow} = (a, 1, a, 1), p_c{}^{\rightsquigarrow} = (1, 1, 1, 1), p_1{}^{\rightsquigarrow} = (a, b, c, 1).$$

$$p_a{}^{\rightsquigarrow} * p_a{}^{\rightsquigarrow} = p_a{}^{\rightsquigarrow}, p_a{}^{\rightsquigarrow} * p_b{}^{\rightsquigarrow} = p_a{}^{\rightsquigarrow} * p_c{}^{\rightsquigarrow} = (1, 1, 1, 1), p_a{}^{\rightsquigarrow} * p_1{}^{\rightsquigarrow} = p_a{}^{\rightsquigarrow};$$

$$p_b{}^{\rightsquigarrow} * p_a{}^{\rightsquigarrow} = (1, 1, a, 1), p_b{}^{\rightsquigarrow} * p_b{}^{\rightsquigarrow} = p_b{}^{\rightsquigarrow}, p_b{}^{\rightsquigarrow} * p_c{}^{\rightsquigarrow} = p_c{}^{\rightsquigarrow}, p_b{}^{\rightsquigarrow} * p_1{}^{\rightsquigarrow} = p_b{}^{\rightsquigarrow};$$

$$p_c^{\rightsquigarrow} * p_a^{\rightsquigarrow} = p_c^{\rightsquigarrow}, p_c^{\rightsquigarrow} * p_b^{\rightsquigarrow} = p_c^{\rightsquigarrow}, p_c^{\rightsquigarrow} * p_c^{\rightsquigarrow} = p_c^{\rightsquigarrow}, p_c^{\rightsquigarrow} * p_1^{\rightsquigarrow} = p_c^{\rightsquigarrow}.$$

Denote $p_{ba}^{\rightsquigarrow} = p_b^{\rightsquigarrow} * p_a^{\rightsquigarrow} = (1, 1, a, 1)$, then $p_{ba}^{\rightsquigarrow} * p_a^{\rightsquigarrow} = p_{ba}^{\rightsquigarrow}$, $p_a^{\rightsquigarrow} * p_{ba}^{\rightsquigarrow} = p_c^{\rightsquigarrow}$; $p_{ba}^{\rightsquigarrow} * p_b^{\rightsquigarrow} = p_c^{\rightsquigarrow}$, $p_b^{\rightsquigarrow} * p_{ba}^{\rightsquigarrow} = p_{ba}^{\rightsquigarrow}$; $p_{ba}^{\rightsquigarrow} * p_{ba}^{\rightsquigarrow} = p_c^{\rightsquigarrow}$; $p_{ba}^{\rightsquigarrow} * p_c^{\rightsquigarrow} = p_c^{\rightsquigarrow}$, $p_c^{\rightsquigarrow} * p_{ba}^{\rightsquigarrow} = p_c^{\rightsquigarrow}$. Hence, $M^{\rightsquigarrow}(X) = \{p_a^{\rightsquigarrow}, p_b^{\rightsquigarrow}, p_{ba}^{\rightsquigarrow}, p_c^{\rightsquigarrow}, p_1^{\rightsquigarrow}\}$ and its Cayley table is Table 16. Obviously, $(M^{\rightsquigarrow}(X), *)$ is a non-commutative semigroup, but it is not a neutrosophic triplet group.

Table 16. Pseudo BE-algebra and adjoint semigroups (4).

*	$p_a^{\rightsquigarrow}$	$p_b^{\rightsquigarrow}$	$p_{ba}^{\rightsquigarrow}$	$p_c^{\rightsquigarrow}$	$p_1^{\rightsquigarrow}$
$p_a^{\rightsquigarrow}$	$p_a^{\rightsquigarrow}$	$p_c^{\rightsquigarrow}$	$p_c^{\rightsquigarrow}$	$p_c^{\rightsquigarrow}$	$p_a^{\rightsquigarrow}$
$p_b^{\rightsquigarrow}$	$p_{ba}^{\rightsquigarrow}$	$p_b^{\rightsquigarrow}$	$p_{ba}^{\rightsquigarrow}$	$p_c^{\rightsquigarrow}$	$p_b^{\rightsquigarrow}$
$p_{ba}^{\rightsquigarrow}$	$p_{ba}^{\rightsquigarrow}$	$p_c^{\rightsquigarrow}$	$p_c^{\rightsquigarrow}$	$p_c^{\rightsquigarrow}$	$p_{ba}^{\rightsquigarrow}$
$p_c^{\rightsquigarrow}$	$p_c^{\rightsquigarrow}$	$p_c^{\rightsquigarrow}$	$p_c^{\rightsquigarrow}$	$p_c^{\rightsquigarrow}$	$p_c^{\rightsquigarrow}$
$p_1^{\rightsquigarrow}$	$p_a^{\rightsquigarrow}$	$p_b^{\rightsquigarrow}$	$p_{ba}^{\rightsquigarrow}$	$p_c^{\rightsquigarrow}$	$p_1^{\rightsquigarrow}$

Now, we consider $M(X)$. Since

$$p_c^{\rightarrow} = (1, 1, 1, 1) = p_c^{\rightsquigarrow}, p_1^{\rightarrow} = (a, b, c, 1) = p_1^{\rightsquigarrow};$$

$$p_a^{\rightarrow} * p_a^{\rightsquigarrow} = p_a^{\rightarrow}, p_a^{\rightsquigarrow} * p_a^{\rightarrow} = p_a^{\rightarrow};$$

$$p_a^{\rightarrow} * p_b^{\rightsquigarrow} = (1, 1, 1, 1) = p_c^{\rightarrow}, p_b^{\rightsquigarrow} * p_a^{\rightarrow} = (1, 1, 1, 1) = p_c^{\rightarrow};$$

$$p_a^{\rightsquigarrow} * p_b^{\rightarrow} = p_b^{\rightarrow} * p_a^{\rightsquigarrow} = (1, 1, c, 1);$$

$$p_a^{\rightsquigarrow} * p_{ab}^{\rightarrow} = p_{ab}^{\rightarrow}, p_{ab}^{\rightarrow} * p_a^{\rightsquigarrow} = p_{ab}^{\rightarrow}; p_b^{\rightarrow} * p_b^{\rightsquigarrow} = p_b^{\rightsquigarrow}, p_b^{\rightsquigarrow} * p_b^{\rightarrow} = p_b^{\rightsquigarrow};$$

$$p_{ab}^{\rightarrow} * p_b^{\rightsquigarrow} = (1, 1, 1, 1) = p_c^{\rightarrow}, p_b^{\rightsquigarrow} * p_{ab}^{\rightarrow} = (1, 1, 1, 1) = p_c^{\rightarrow};$$

$$p_a^{\rightarrow} * p_{ba}^{\rightsquigarrow} = (1, 1, 1, 1) = p_c^{\rightarrow}, p_{ba}^{\rightsquigarrow} * p_a^{\rightarrow} = (1, 1, 1, 1) = p_c^{\rightarrow};$$

$$p_b^{\rightarrow} * p_{ba}^{\rightsquigarrow} = p_{ba}^{\rightsquigarrow}, p_{ba}^{\rightsquigarrow} * p_b^{\rightarrow} = p_{ba}^{\rightsquigarrow};$$

$$p_{ab}^{\rightarrow} * p_{ba}^{\rightsquigarrow} = (1, 1, 1, 1) = p_c^{\rightarrow}, p_{ba}^{\rightsquigarrow} * p_{ab}^{\rightarrow} = (1, 1, 1, 1) = p_c^{\rightarrow}.$$

Denote $p = (1, 1, c, 1)$, then $M(X) = \{p_a^{\rightarrow}, p_a^{\rightsquigarrow}, p_b^{\rightarrow}, p_b^{\rightsquigarrow}, p_{ab}^{\rightarrow}, p_{ba}^{\rightsquigarrow}, p, p_c^{\rightarrow}, p_1^{\rightarrow}\}$, and Table 17 is its Cayley table (it is a non-commutative semigroup, but it is not a neutrosophic triplet group).

Table 17. Pseudo BE-algebra and adjoint semigroups (5).

*	$p_a^{\rightarrow}$	$p_a^{\rightsquigarrow}$	$p_b^{\rightarrow}$	$p_b^{\rightsquigarrow}$	$p_{ab}^{\rightarrow}$	$p_{ba}^{\rightsquigarrow}$	p	$p_c^{\rightarrow}$	$p_1^{\rightarrow}$
$p_a^{\rightarrow}$	$p_a^{\rightarrow}$	$p_a^{\rightarrow}$	$p_{ab}^{\rightarrow}$	$p_c^{\rightarrow}$	$p_{ab}^{\rightarrow}$	$p_c^{\rightarrow}$	$p_{ab}^{\rightarrow}$	$p_c^{\rightarrow}$	$p_a^{\rightarrow}$
$p_a^{\rightsquigarrow}$	$p_a^{\rightarrow}$	$p_a^{\rightsquigarrow}$	p	$p_c^{\rightarrow}$	$p_{ab}^{\rightarrow}$	$p_{ba}^{\rightsquigarrow}$	p	$p_c^{\rightarrow}$	$p_a^{\rightsquigarrow}$
$p_b^{\rightarrow}$	$p_c^{\rightarrow}$	p	$p_b^{\rightarrow}$	$p_b^{\rightsquigarrow}$	$p_c^{\rightarrow}$	$p_{ba}^{\rightsquigarrow}$	p	$p_c^{\rightarrow}$	$p_b^{\rightarrow}$
$p_b^{\rightsquigarrow}$	$p_c^{\rightarrow}$	$p_{ba}^{\rightsquigarrow}$	$p_b^{\rightsquigarrow}$	$p_b^{\rightsquigarrow}$	$p_c^{\rightarrow}$	$p_{ba}^{\rightsquigarrow}$	$p_{ba}^{\rightsquigarrow}$	$p_c^{\rightarrow}$	$p_b^{\rightsquigarrow}$
$p_{ab}^{\rightarrow}$	$p_c^{\rightarrow}$	$p_{ab}^{\rightarrow}$	$p_{ab}^{\rightarrow}$	$p_c^{\rightarrow}$	$p_c^{\rightarrow}$	$p_c^{\rightarrow}$	$p_{ab}^{\rightarrow}$	$p_c^{\rightarrow}$	$p_{ab}^{\rightarrow}$
$p_{ba}^{\rightsquigarrow}$	$p_c^{\rightarrow}$	$p_{ba}^{\rightsquigarrow}$	$p_{ba}^{\rightsquigarrow}$	$p_c^{\rightarrow}$	$p_c^{\rightarrow}$	$p_c^{\rightarrow}$	$p_{ba}^{\rightsquigarrow}$	$p_c^{\rightarrow}$	$p_{ba}^{\rightsquigarrow}$
p	$p_c^{\rightarrow}$	p	p	$p_c^{\rightarrow}$	$p_c^{\rightarrow}$	$p_c^{\rightarrow}$	p	$p_c^{\rightarrow}$	p
$p_c^{\rightarrow}$	$p_c^{\rightarrow}$	$p_c^{\rightarrow}$	$p_c^{\rightarrow}$	$p_c^{\rightarrow}$	$p_c^{\rightarrow}$	$p_c^{\rightarrow}$	$p_c^{\rightarrow}$	$p_c^{\rightarrow}$	$p_c^{\rightarrow}$
$p_1^{\rightarrow}$	$p_a^{\rightarrow}$	$p_a^{\rightsquigarrow}$	$p_b^{\rightarrow}$	$p_b^{\rightsquigarrow}$	$p_{ab}^{\rightarrow}$	$p_{ba}^{\rightsquigarrow}$	p	$p_c^{\rightarrow}$	$p_1^{\rightarrow}$

The following example shows that the adjoint semigroups of a pseudo BE-algebra may be a commutative neutrosophic triplet group.

Example 11. *Let X= {a, b, c, d, 1}. Define operations* $\rightarrow$ *and* $\rightsquigarrow$ *on X as* Tables 18 and 19. *Then, (X;* $\rightarrow$, $\rightsquigarrow$*, 1) is a pseudo BE-algebra, as well as the following:*

$$p_a^{\rightarrow} = (1, c, c, 1, 1), p_b^{\rightarrow} = (d, 1, 1, d, 1), p_c^{\rightarrow} = (d, 1, 1, d, 1), p_d^{\rightarrow} = (1, c, c, 1, 1), p_1^{\rightarrow} = (a, b, c, d, 1).$$

We can verify the following:

$$p_a^{\rightarrow} * p_a^{\rightarrow} = p_a^{\rightarrow}, p_a^{\rightarrow} * p_b^{\rightarrow} = p_a^{\rightarrow} * p_c^{\rightarrow} = (1, 1, 1, 1, 1), p_a^{\rightarrow} * p_d^{\rightarrow} = p_a^{\rightarrow}, p_a^{\rightarrow} * p_1^{\rightarrow} = p_a^{\rightarrow};$$

$$p_b^{\rightarrow} * p_a^{\rightarrow} = (1, 1, 1, 1, 1), p_b^{\rightarrow} * p_b^{\rightarrow} = p_b^{\rightarrow} * p_c^{\rightarrow} = p_b^{\rightarrow}, p_b^{\rightarrow} * p_d^{\rightarrow} = (1, 1, 1, 1, 1), p_b^{\rightarrow} * p_1^{\rightarrow} = p_b^{\rightarrow};$$

$$p_c^{\rightarrow} * p_a^{\rightarrow} = (1, 1, 1, 1, 1), p_c^{\rightarrow} * p_b^{\rightarrow} = p_c^{\rightarrow} * p_c^{\rightarrow} = p_c^{\rightarrow}, p_c^{\rightarrow} * p_d^{\rightarrow} = (1, 1, 1, 1, 1), p_c^{\rightarrow} * p_1^{\rightarrow} = p_b^{\rightarrow};$$

$$p_d^{\rightarrow} * p_a^{\rightarrow} = p_d^{\rightarrow}, p_d^{\rightarrow} * p_b^{\rightarrow} = p_d^{\rightarrow} * p_c^{\rightarrow} = (1, 1, 1, 1, 1), p_d^{\rightarrow} * p_d^{\rightarrow} = p_d^{\rightarrow}, p_d^{\rightarrow} * p_1^{\rightarrow} = p_d^{\rightarrow}.$$

Denote $p_{ab}^{\rightarrow} = p_a^{\rightarrow} * p_b^{\rightarrow} = (1, 1, 1, 1, 1)$, then $p_{ab}^{\rightarrow} * p_a^{\rightarrow} = p_{ab}^{\rightarrow} * p_b^{\rightarrow} = p_{ab}^{\rightarrow} * p_c^{\rightarrow} = p_{ab}^{\rightarrow} * p_d^{\rightarrow} = p_{ab}^{\rightarrow} * p_{ab}^{\rightarrow} = p_{ab}^{\rightarrow} * p_1^{\rightarrow} = p_{ab}^{\rightarrow}$. Hence, $M^{\rightarrow}(X) = \{p_a^{\rightarrow}, p_b^{\rightarrow}, p_{ab}^{\rightarrow}, p_1^{\rightarrow}\}$ and its Cayley table is Table 20. Obviously, $(M^{\rightarrow}(X), *)$ is a commutative neutrosophic triplet group.

Table 18. Pseudo BE-algebra and commutative neutrosophic triplet groups (1).

$\rightarrow$	a	b	c	d	1
a	1	c	c	1	1
b	d	1	1	d	1
c	d	1	1	d	1
d	1	c	c	1	1
1	a	b	c	d	1

Table 19. Pseudo BE-algebra and commutative neutrosophic triplet groups (2).

$\rightsquigarrow$	a	b	c	d	1
a	1	b	c	1	1
b	d	1	1	d	1
c	d	1	1	d	1
d	1	b	c	1	1
1	a	b	c	d	1

Table 20. Pseudo BE-algebra and commutative neutrosophic triplet groups (3).

*	$p_a^{\rightarrow}$	$p_b^{\rightarrow}$	$p_{ab}^{\rightarrow}$	$p_1^{\rightarrow}$
$p_a^{\rightarrow}$	$p_a^{\rightarrow}$	$p_{ab}^{\rightarrow}$	$p_{ab}^{\rightarrow}$	$p_a^{\rightarrow}$
$p_b^{\rightarrow}$	$p_{ab}^{\rightarrow}$	$p_b^{\rightarrow}$	$p_{ab}^{\rightarrow}$	$p_b^{\rightarrow}$
$p_{ab}^{\rightarrow}$	$p_{ab}^{\rightarrow}$	$p_{ab}^{\rightarrow}$	$p_{ab}^{\rightarrow}$	$p_{ab}^{\rightarrow}$
$p_1^{\rightarrow}$	$p_a^{\rightarrow}$	$p_b^{\rightarrow}$	$p_{ab}^{\rightarrow}$	$p_1^{\rightarrow}$

Similarly, we can verify the following:

$$p_a^{\rightsquigarrow} = (1, b, c, 1, 1), p_b^{\rightsquigarrow} = (d, 1, 1, d, 1), p_c^{\rightsquigarrow} = (d, 1, 1, d, 1), p_d^{\rightsquigarrow} = (1, b, c, 1, 1), p_1^{\rightsquigarrow} = (a, b, c, d, 1).$$

$$p_a^{\rightsquigarrow} * p_a^{\rightsquigarrow} = p_a^{\rightsquigarrow}, p_a^{\rightsquigarrow} * p_b^{\rightsquigarrow} = p_a^{\rightsquigarrow} * p_c^{\rightsquigarrow} = (1, 1, 1, 1, 1), p_a^{\rightsquigarrow} * p_d^{\rightsquigarrow} = p_a^{\rightsquigarrow};$$

$$p_b^{\rightsquigarrow} * p_a^{\rightsquigarrow} = (1, 1, 1, 1, 1), p_b^{\rightsquigarrow} * p_b^{\rightsquigarrow} = p_b^{\rightsquigarrow} * p_c^{\rightsquigarrow} = p_b^{\rightsquigarrow}, p_b^{\rightsquigarrow} * p_d^{\rightsquigarrow} = (1, 1, 1, 1, 1).$$

Denote $p_{ab}^{\rightsquigarrow} = p_a^{\rightsquigarrow} * p_b^{\rightsquigarrow} = (1, 1, 1, 1, 1)$, then $M^{\rightsquigarrow}(X) = \{p_a^{\rightsquigarrow}, p_b^{\rightsquigarrow}, p_{ab}^{\rightsquigarrow}, p_1^{\rightsquigarrow}\}$ and its Cayley table is Table 21. Obviously, $(M^{\rightsquigarrow}(X), *)$ is a commutative neutrosophic triplet group.

Table 21. Pseudo BE-algebra and commutative neutrosophic triplet groups (4).

*	$p_a^{\rightsquigarrow}$	$p_b^{\rightsquigarrow}$	$p_{ab}^{\rightsquigarrow}$	$p_1^{\rightsquigarrow}$
$p_a^{\rightsquigarrow}$	$p_a^{\rightsquigarrow}$	$p_{ab}^{\rightsquigarrow}$	$p_{ab}^{\rightsquigarrow}$	$p_a^{\rightsquigarrow}$
$p_b^{\rightsquigarrow}$	$p_{ab}^{\rightsquigarrow}$	$p_b^{\rightsquigarrow}$	$p_{ab}^{\rightsquigarrow}$	$p_b^{\rightsquigarrow}$
$p_{ab}^{\rightsquigarrow}$	$p_{ab}^{\rightsquigarrow}$	$p_{ab}^{\rightsquigarrow}$	$p_{ab}^{\rightsquigarrow}$	$p_{ab}^{\rightsquigarrow}$
$p_1^{\rightsquigarrow}$	$p_a^{\rightsquigarrow}$	$p_b^{\rightsquigarrow}$	$p_{ab}^{\rightsquigarrow}$	$p_1^{\rightsquigarrow}$

Now, we consider $M(X)$. Since the following:

$$p_b^{\rightarrow} = p_c^{\rightarrow} = (d, 1, 1, d, 1) = p_b^{\rightsquigarrow} = p_c^{\rightsquigarrow}, p_a^{\rightarrow} = p_d^{\rightarrow} = (1, c, c, 1, 1), p_a^{\rightsquigarrow} = p_d^{\rightsquigarrow} = (1, b, c, 1, 1);$$

$$p_a^{\rightarrow} * p_a^{\rightsquigarrow} = p_a^{\rightarrow}, p_a^{\rightsquigarrow} * p_a^{\rightarrow} = p_a^{\rightarrow}; p_a^{\rightarrow} * p_b^{\rightsquigarrow} = (1, 1, 1, 1, 1) = p_{ab}^{\rightarrow} = p_{ab}^{\rightsquigarrow}, p_b^{\rightsquigarrow} * p_a^{\rightarrow} = (1, 1, 1, 1, 1).$$

Hence, $M(X) = \{p_a^{\rightarrow}, p_a^{\rightsquigarrow}, p_b^{\rightarrow}, p_{ab}^{\rightarrow}, p_1^{\rightarrow}\}$, and Table 22 is its Cayley table (it is a commutative neutrosophic triplet group).

Table 22. Pseudo BE-algebra and commutative neutrosophic triplet groups (5).

*	$p_a^{\rightarrow}$	$p_a^{\rightsquigarrow}$	$p_b^{\rightarrow}$	$p_{ab}^{\rightarrow}$	$p_1^{\rightarrow}$
$p_a^{\rightarrow}$	$p_a^{\rightarrow}$	$p_a^{\rightarrow}$	$p_{ab}^{\rightarrow}$	$p_{ab}^{\rightarrow}$	$p_a^{\rightarrow}$
$p_a^{\rightsquigarrow}$	$p_a^{\rightarrow}$	$p_a^{\rightsquigarrow}$	$p_{ab}^{\rightarrow}$	$p_{ab}^{\rightarrow}$	$p_a^{\rightsquigarrow}$
$p_b^{\rightarrow}$	$p_{ab}^{\rightarrow}$	$p_{ab}^{\rightarrow}$	$p_b^{\rightarrow}$	$p_{ab}^{\rightarrow}$	$p_b^{\rightarrow}$
$p_{ab}^{\rightarrow}$	$p_{ab}^{\rightarrow}$	$p_{ab}^{\rightarrow}$	$p_{ab}^{\rightarrow}$	$p_{ab}^{\rightarrow}$	$p_{ab}^{\rightarrow}$
$p_1^{\rightarrow}$	$p_a^{\rightarrow}$	$p_a^{\rightsquigarrow}$	$p_b^{\rightarrow}$	$p_{ab}^{\rightarrow}$	$p_1^{\rightarrow}$

Remark 3. *Through the discussions of Examples 10 and 11 above, we get the following important revelations: (1) ($M^{\rightarrow}(X)$, *), ($M^{\rightsquigarrow}(X)$, *), and ($M(X)$, *) are usually three different semi-groups; (2) ($M^{\rightarrow}(X)$, *) and ($M^{\rightsquigarrow}(X)$, *) are all sub-semi-groups of ($M(X)$, *), which can also be proved from their definitions; (3) ($M^{\rightarrow}(X)$, *), ($M^{\rightsquigarrow}(X)$, *), and ($M(X)$, *) may be neutrosophic triplet groups. Under what circumstances they will become neutrosophic triplet groups, will be examined in the next study.*

6. Conclusions

In this paper, the concepts of neutrosophic triplet loops (groups) are further generalized, and some new concepts of generalized neutrosophic triplet structures are proposed, including (*l-l*)-type, (*l-r*)-type, (*r-l*)-type, (*r-r*)-type, (*l-lr*)-type, (*r-lr*)-type, (*lr-l*)-type, and (*lr-r*)-type quasi neutrosophic triplet loops (groups), and their basic properties are discussed. In particular, as a corollary of these new properties, an important result is proved. For any commutative neutrosophic triplet group, its every element has only one neutral element. At the same time, the BE-algebras and its various extensions (including CI-algebras, pseudo BE-algebras, and pseudo CI-algebras) have been studied, and some related generalized neutrosophic triplet structures that are contained in these algebras are presented. Moreover, the concept of adjoint semigroups of (generalized) BE-algebras are proposed for the first time, abundant examples are given, and some new results are obtained.

Author Contributions: X.Z. and X.W. initiated the research and wrote the paper; F.S. supervised the research work and provided helpful suggestions; and M.H. participated in some of the research work.

Funding: This research received no external funding.

Acknowledgments: This work was supported by the National Natural Science Foundation of China (Grant Nos. 61573240, 61473239).

Conflicts of Interest: The authors declare no conflict of interest.

References

1. Molaei, M.R. Generalized groups. *Bull. Inst. Polit. Di. Iase Fasc.* **1999**, *3*, 21–24.
2. Molaei, M.R. Generalized actions. In Proceedings of the First International Conference on Geometry, Integrability and Quantization, Varna, Bulgaria, 1–10 September 1999; pp. 175–180.
3. Araujo, J.; Konieczny, J. Molaei's Generalized groups are completely simple semigroups. *Bull. Polytech. Inst. Iassy* **2002**, *48*, 1–5.
4. Adeniran, J.O.; Akinmoyewa, J.T.; Solarin, A.R.T.; Jaiyeola, T.G. On some algebraic properties of generalized groups. *Acta Math. Acad.* **2011**, *27*, 23–30.
5. Smarandache, F. *Neutrosophic Perspectives: Triplets, Duplets, Multisets, Hybrid Operators, Modal Logic, Hedge Algebras. and Applications*; Pons Publishing House: Brussels, Belgium, 2017.
6. Smarandache, F.; Ali, M. Neutrosophic triplet group. *Neural Comput. Appl.* **2018**, *29*, 595–601. [CrossRef]
7. Smarandache, F. Neutrosophic set—A generialization of the intuituionistics fuzzy sets. *Int. J. Pure Appl. Math.* **2005**, *3*, 287–297.
8. Liu, P.D.; Shi, L.L. Some Neutrosophic uncertain linguistic number Heronian mean operators and their application to multi-attribute group decision making. *Neural Comput. Appl.* **2017**, *28*, 1079–1093. [CrossRef]
9. Ye, J.; Du, S. Some distances, similarity and entropy measures for interval-valued neutrosophic sets and their relationship. *Int. J. Mach. Learn Cybern.* **2018**. [CrossRef]
10. Zhang, X.H.; Ma, Y.C.; Smarandache, F. Neutrosophic regular filters and fuzzy regular filters in pseudo-BCI algebras. *Neutrosophic Sets Syst.* **2017**, *17*, 10–15.
11. Zhang, X.H.; Bo, C.X.; Smarandache, F.; Dai, J.H. New inclusion relation of neutrosophic sets with applications and related lattice structure. *Int. J. Mach. Learn. Cyber.* **2018**. [CrossRef]
12. Zhang, X.H.; Smarandache, F.; Liang, X.L. Neutrosophic duplet semi-group and cancellable neutrosophic triplet groups. *Symmetry* **2017**, *9*, 275. [CrossRef]
13. Jaiyeola, T.G.; Smarandache, F. Inverse properties in neutrosophic triplet loop and their application to cryptography. *Algorithms* **2018**, *11*, 32. [CrossRef]
14. Bal, M.; Shalla, M.M.; Olgun, N. Neutrosophic triplet cosets and quotient groups. *Symmetry* **2018**, *10*, 126. [CrossRef]
15. Zhang, X.H.; Smarandache, F.; Ali, M.; Liang, X.L. Commutative neutrosophic triplet group and neutro-homomorphism basic theorem. *Ital. J. Pure Appl. Math.* **2018**, in press.
16. Fleischer, I. Every BCK-algebra is a set of residuables in an integral pomonoid. *J. Algebra* **1980**, *119*, 360–365. [CrossRef]
17. Huang, W. On BCI-algebras and semigroups. *Math. Jpn.* **1995**, *42*, 59–64.
18. Zhang, X.H.; Ye, R.F. BZ-algebra and group. *J. Math. Phys. Sci.* **1995**, *29*, 223–233.
19. Dudek, W.A.; Zhang, X.H. On atoms in BCC-algebras. *Discuss. Math. Algebra Stoch. Methods* **1995**, *15*, 81–85.
20. Huang, W.; Liu, F. On the adjoint semigroups of p-separable BCI-algebras. *Semigroup Forum* **1999**, *58*, 317–322. [CrossRef]
21. Zhang, X.H.; Wang, Y.Q.; Dudek, W.A. T-ideals in BZ-algebras and T-type BZ-algebras. *Indian J. Pure Appl. Math.* **2003**, *34*, 1559–1570.
22. Zhang, X.H. *Fuzzy Logics and Algebraic Analysis*; Science Press: Beijing, China, 2008.
23. Zhang, X.H.; Dudek, W.A. BIK+-logic and non-commutative fuzzy logics. *Fuzzy Syst. Math.* **2009**, *23*, 8–20.
24. Zhang, X.H.; Jun, Y.B. Anti-grouped pseudo-BCI algebras and anti-grouped pseudo-BCI filters. *Fuzzy Syst. Math.* **2014**, *28*, 21–33.
25. Zhang, X.H. Fuzzy anti-grouped filters and fuzzy normal filters in pseudo-BCI algebras. *J. Intell. Fuzzy Syst.* **2017**, *33*, 1767–1774. [CrossRef]
26. Zhang, X.H.; Park, C.; Wu, S.P. Soft set theoretical approach to pseudo-BCI algebras. *J. Intell. Fuzzy Syst.* **2018**, *34*, 559–568. [CrossRef]
27. Kim, H.S.; Kim, Y.H. On BE-algebras. *Sci. Math. Jpn.* **2007**, *66*, 113–116.
28. Ahn, S.S.; So, Y.H. On ideals and upper sets in BE-algebras. *Sci. Math. Jpn.* **2008**, *68*, 279–285.
29. Walendziak, A. On commutative BE-algebras. *Sci. Math. Jpn.* **2009**, *69*, 281–284.
30. Meng, B.L. On filters in BE-algebras. *Sci. Math. Jpn.* **2010**, *71*, 201–207.
31. Walendziak, A. On normal filters and congruence relations in BE-algebras. *Comment. Math.* **2012**, *52*, 199–205.
32. Sambasiva Rao, M. *A Course in BE-Algebras*; Springer: Berlin, Germany, 2018.

33. Borzooei, R.A.; Saeid, A.B.; Rezaei, A.; Radfar, A.; Ameri, R. On pseudo BE-algebras. *Discuss. Math. Gen. Algebra Appl.* **2013**, *33*, 95–108. [CrossRef]
34. Ciungu, L.C. Commutative pseudo BE-algebras. *Iran. J. Fuzzy Syst.* **2016**, *13*, 131–144.
35. Meng, B.L. CI-algebras. *Sci. Math. Jpn.* **2010**, *71*, 11–17.
36. Meng, B.L. Atoms in CI-algebras and singular CI-algebras. *Sci. Math. Jpn.* **2010**, *72*, 67–72.
37. Meng, B.L. Closed filters in CI-algebras. *Sci. Math. Jpn.* **2010**, *71*, 265–270.
38. Kim, K.H. A note on CI-algebras. *Int. Math. Forum* **2011**, *6*, 1–5.
39. Jun, Y.B.; Lee, K.J.; Roh, E.H. Ideals and filters in CI-algebras based on bipolar-valued fuzzy sets. *Ann. Fuzz. Math. Inf.* **2012**, *4*, 109–121.
40. Sabhapandit, P.; Pathak, K. On homomorphisms in CI-algebras. *Int. J. Math. Arch.* **2018**, *9*, 33–36.
41. Neggers, J.; Ahn, S.S.; Kim, H.S. On Q-algebras. *Int. J. Math. Math. Sci.* **2001**, *27*, 749–757. [CrossRef]
42. Saeid, A.B. CI-algebra is equivalent to dual Q-algebra. *J. Egypt. Math. Soc.* **2013**, *21*, 1–2. [CrossRef]
43. Walendziak, A. Pseudo-BCH-algebras. *Discuss. Math. Gen. Algebra Appl.* **2015**, *35*, 5–19. [CrossRef]
44. Jun, Y.B.; Kim, H.S.; Ahn, S.S. Structures of pseudo ideal and pseudo atom in a pseudo Q-algebra. *Kyungpook Math. J.* **2016**, *56*, 95–106. [CrossRef]
45. Rezaei, A.; Saeid, A.B.; Walendziak, A. Some results on pseudo-Q algebras. *Ann. Univ. Paedagog. Crac. Stud. Math.* **2017**, *16*, 61–72. [CrossRef]
46. Bajalan, S.A.; Ozbal, S.A. Some properties and homomorphisms of pseudo-Q algebras. *J. Cont. Appl. Math.* **2016**, *6*, 3–17.

MDPI
St. Alban-Anlage 66
4052 Basel
Switzerland
Tel. +41 61 683 77 34
Fax +41 61 302 89 18
www.mdpi.com

Symmetry Editorial Office
E-mail: symmetry@mdpi.com
www.mdpi.com/journal/symmetry

www.ingramcontent.com/pod-product-compliance
Lightning Source LLC
Chambersburg PA
CBHW051702210326
41597CB00032B/5343